Asheville-Buncombe
Technical Community College
Learning Resources Center
340 Victoria Rd.
Asheville, NC 28801

Steel Design Handbook

Other McGraw-Hill Books of Interest

Breyer • DESIGN OF WOOD STRUCTURES
Brockenbrough & Merritt • STRUCTURAL STEEL DESIGNER'S HANDBOOK
Brown • PRACTICAL FOUNDATION ENGINEERING HANDBOOK
Gaylord & Gaylord • STRUCTURAL ENGINEERING HANDBOOK
Faherty & Williamson • WOOD ENGINEERING AND CONSTRUCTION HANDBOOK
Newman • STANDARD HANDBOOK OF STRUCTURAL DETAILS FOR BUILDING CONSTRUCTION
Newman • DESIGN AND CONSTRUCTION OF WOOD FRAMED BUILDINGS
Merritt • STANDARD HANDBOOK FOR CIVIL ENGINEERS
Sharp • BEHAVIOR AND DESIGN OF ALUMINUM STRUCTURES
Tonias • BRIDGE ENGINEERING

Steel Design Handbook

LRFD Method

Editor
Akbar R. Tamboli P.E. FASCE
Consulting Engineer
Princeton, New Jersey

McGraw-Hill

New York • San Francisco • Washington, D.C • Auckland • Bogotá
Caracas • Lisbon • London • Madrid • Mexico City • Milan
Montreal • New Delhi • San Juan • Singapore
Sydney • Tokyo • Toronto

Library of Congress Cataloging in Publication Data

Tamboli, Akbar R.
    Steel Design Handbook : LRFD Method / Akbar R. Tamboli, editor,
        p. cm.
    ISBN 0-07-061400-8 (hardcover)
    1. Building, Iron and steel.   2. Load factor design.   3. Steel
Structural.   I. Title
TA684.T345   1997
624.1'821—dc20   96-34616
                                                                     CIP

McGraw-Hill

A Division of The McGraw-Hill Companies

Copyright © 1997 by The McGraw-Hill Companies, Inc. All rights reserved. Printed in the United States of America. Except as permitted under the United States Copyright Act of 1976, no part of this publication may be reproduced or distributed in any form or by any means, or stored in a data base or retrieval system, without the prior written permission of the publisher.

1 2 3 4 5 6 7 8 9 0   DOC/DOC   9 0 1 0 9 8 7 6

P/N 061409-1
PART OF
ISBN 0-07-061400-8

The sponsoring editor for this book was Larry S. Hager, the editing supervisor was Paul R. Sobel, and the production supervisor was Suzanne W. B. Rapcavage. It was set in Palatino by HMC Group Ltd.

Printed and bound by R. R. Donnelley & Sons Company.

McGraw-Hill books are available at special quantity discounts to use as premiums and sales proportions, or for use in corporate training programs. For more information, please write to the Director of Special Sales, McGraw-Hill, 11 West 19th Street, New York, NY 10011. Or contact your local bookstore.

Information contained in this work has been obtained by The McGraw-Hill Companies, Inc. ("McGraw-Hill"), from sources believed to be reliable. However, neither McGraw-Hill nor its authors guarantees the accuracy or completeness of any information published herein, and neither McGraw-Hill nor its authors shall be responsible for any errors, omissions, or damages arising out of use of this information. This work is published with the understanding that McGraw-Hill and its authors are supplying information, but are not attempting to render engineering or other professional services. If such services are required, the assistance of an appropriate professional should be sought.

 This book is printed on recycled, acid-free paper containing a minimum of 50% recycled de-inked fiber.

Contents

Contributors ix

Preface xi

Acknowledgments xiii

Chapter 1. History—Design Concepts and Material Behavior
by Akbar R. Tamboli 1-1

    1.1 History / 1-1
    1.2 Design Concepts / 1-17
    1.3 Material Behavior / 1-23

Chapter 2. Design of Flexural Members by Jack C. McCormac 2-1

    2.1 Beam Behavior / 2-1

Chapter 3. Design of Tension and Compression Members
by Jack C. McCormac 3-1

    3.1 Tension Members / 3-1
    3.2 Design Topics / 3-14
    3.3 Compression Members / 3-28
    3.4 Tension and Compression Members with Flexure / 3-64

Chapter 4. Torsion by Louis F. Geschwinder 4-1

    4.1 Introduction / 4-1
    4.2 Pure Torsion / 4-7
    4.3 Warping Torsion / 4-11
    4.4 Combined Stresses / 4-17
    4.5 Bending Analogy / 4-23
    4.6 Serviceability / 4-25
    4.7 LRFD Design / 4-25
    4.8 Simplified Equations / 4-26

Chapter 5. Building Design Loading Criteria and General Considerations by Stephen J. Y. Tang, Ian R. Chin, Jerome W. Rasgus, and Richard F. Rowe 5-1

    5.1 Planning Building Structures / 5-1
    5.2 Loads / 5-10
    5.3 Miscellaneous Considerations / 5-29

Chapter 6. LRFD—Limit Design of Frames *by J.Y. Richard Liew and W.F. Chen* **6-1**

6.1 General Principles / 6-1
6.2 Single-Story Frames / 6-16
6.3 Multistory Frames / 6-41

Chapter 7. LRFD—Limit Design of Connections *by William A. Thornton and Thomas Kane* **7-1**

7.1 Introduction / 7-1
7.2 Bolted Connections / 7-2
7.3 Welded Connections / 7-12
7.4 Connection Design / 7-29

Chapter 8. Fatigue and Fracture *by Robert Dexter and John Fisher* **8-1**

8.1 Introduction / 8-1
8.2 Evaluation of Structural Details for Fatigue / 8-8
8.3 Evaluation of Structural Details for Fracture / 8-25
8.4 Summary / 8-48

Chapter 9. Serviceability Considerations for Floor and Roof Systems *by Linda M. Hanagan and Thomas M. Murray* **9-1**

9.1 Introduction / 9-1
9.2 Serviceability Requirements for Static Deflections / 9-2
9.3 Dynamic Behavior of Floor Systems / 9-3
9.4 Reiher-Meister Scale for Steady-State Vibrations / 9-12
9.5 Modified Reiher-Meister Scale for Transient Vibrations / 9-13
9.6 Wiss and Parmelee Rating Factor for Transient Vibrations / 9-14
9.7 ISO Scale for Human Response to Building Vibrations / 9-15
9.8 Vibration Criteria for Sensitive Equipment and Facilities / 9-17
9.9 CSA Criterion for Walking Vibrations / 9-17
9.10 Murray Criterion for Walking Vibrations / 9-20
9.11 Allen and Murray Criterion for Walking Vibrations / 9-23
9.12 Vibration Considerations for Rhythmic Excitations / 9-29
9.13 Design Examples / 9-31

Chapter 10. Composite Design *by Ali A. K. Haris* **10-1**

10.1 Introduction / 10-1
10.2 Composite Beams / 10-3
10.3 Reinforcing Existing Composite Beams / 10-26
10.4 Computer Program (HCOMPL) / 10-37
10.5 Composite Columns / 10-38

Chapter 11. Limit Design of Steel Highway Bridges *by Demetrios E. Tonias* **11-1**

11.1 Overview of Limit States Design of Highway Bridges / 11-1
11.2 Steel Highway Bridges / 11-8
11.3 AASHTO LRFD Code / 11-13

11.4 Load Rating / *11-48*
11.5 LRFD and Working Stress in Bridge Engineering / *11-54*

Chapter 12. Fabrication, Erection, and Quality Control
by Robert E. Shaw, Jr.   **12-1**

12.1 Shop Detailing Processes / *12-1*
12.2 Shop Fabrication / *12-4*
12.3 Bolted Connections / *12-8*
12.4 Welded Connections / *12-15*
12.5 Materials, Fabrication, and Erection Tolerances / *12-27*
12.6 Coating Systems / *12-30*
12.7 Erection Procedures / *12-36*
12.8 Certification, Qualification, and Inspection Programs / *12-37*

Appendix A. Steel Deck Design   by Richard B. Heagler   **A-1**

A.1 General / *A-1*
A.2 Roof Deck Design / *A-2*
A.3 Form Deck Design / *A-11*
A.4 Composite Deck Design / *A-16*

Appendix B. Special Welding Issues for Seismically Resistant Structures   by Omer W. Blodgett and Duane K. Miller   **B-1**

B.1 Introduction and Background / *B-1*
B.2 General Review of Welding Engineering Principles / *B-3*
B.3 Unique Aspects of Seismically Loaded Welded Structures / *B-7*
B.4 Design of Seismically Resistant Welded Structures / *B-8*
B.5 Materials / *B-21*
B.6 Workmanship / *B-26*
B.7 Inspection / *B-28*
B.8 Post-Northridge Details / *B-31*

Appendix C. Load Factor Design Selection Table   **C-1**

Appendix D. SI Metric Conversion Table   **D-1**

Appendix E. Seismic Design of Steel Buildings Using LRFD
by Roy Becker   **E-1**

E.1 General Design Information / *E-1*
E.2 Special Moment Frames (SMF) / *E-4*
E.3 Ordinary Moment Frames / *E-33*
E.4 Braced Frames (CBF) / *E-35*
E.5 Conclusion / *E-52*

Appendix F. Nomenclature   **F-1**

Index follows Appendix F   **I-1**

Contributors

Roy Becker, P.E. *Principal, Becker and Pritchett, Structural Engineers, Inc., Lake Forest, California* (APPENDIX E)

Omer W. Blodgett, P.E. *The Lincoln Electric Company, Cleveland, Ohio* (APPENDIX B)

W.F. Chen, Ph.D., P.E. *George E. Goodwin Distinguished Professor of Civil Engineering, Department of Civil Engineering, Purdue University, West Lafayette, Indiana* (CHAPTER 6)

Ian R. Chin, P.E. *Vice President and Principal, Wiss, Janney, Elstner Associates, Chicago, Illinois* (CHAPTER 5)

Robert Dexter, Ph.D., P.E. *ATLSS Engineering Research Center, Lehigh University, Bethlehem, Pennsylvania* (CHAPTER 8)

John Fisher, Ph.D., P.E. *ATLSS Engineering Research Center, Lehigh University, Bethlehem, Pennsylvania* (CHAPTER 8)

Louis F. Geschwindner, Ph.D., P.E. *Professor, Department of Architecture-Engineering, The Pennsylvania State University, University Park, Pennsylvania* (CHAPTER 4)

Linda M. Hanagan, Ph.D., P.E. *Assistant Professor, Department of Civil and Architectural Engineering, University of Miami, Coral Gables, Florida* (CHAPTER 9)

Ali A. K. Haris, Ph.D., P.E. *Principal, Haris Engineering Services Company, Prarie Village, Kansas* (CHAPTER 10)

Richard B. Heagler, P.E. *Director of Engineering, Nichols J. Bouras Inc., Summit, New Jersey* (APPENDIX A)

Thomas Kane *Technical Manager, Cives Steel Company, Roswell, Georgia* (CHAPTER 7)

J. Y. Richard Liew, Ph.D. *Lecturer, Department of Civil Engineering, National University of Singapore, Singapore* (CHAPTER 6)

Jack C. McCormac *Department of Civil Engineering, Clemson University, Clemson, South Carolina* (CHAPTERS 3, 4)

Duane K. Miller, P.E. *The Lincoln Electric Company, Cleveland, Ohio* (APPENDIX B)

Thomas M. Murray, Ph.D., P.E. *Montague Betts Professor of Structural Steel, Department of Civil Engineering, Virginia Polytechnic Institute, Blacksburg, Virginia* (CHAPTER 9)

Jerome W. Rasgus, P.E. *Structural Department Manager, URS, Inc., Washington, D.C.* (CHAPTER 5)

Research Engineers, Inc. *Yorba Linda, California, Software Development Group* (CD-ROM Disk)

Contributors

Richard F. Rowe, P.E. *Principal, Chief Structural Engineer, The Kling Lindquist Partnership, Inc., Philadelphia, Pennsylvania* (CHAPTER 5)

Robert E. Shaw, Jr., P.E. *President, Steel Structures Technology Centers, Inc., Novi, Michigan* (CHAPTER 12)

Akbar R. Tamboli, P.E., FASCE *Consulting Engineer, CUH2A, Inc., Princeton, New Jersey* (CHAPTER 1)

William A. Thornton, Ph.D., P.E. *Chief Engineer, Cives Steel Company, Roswell, Georgia* (CHAPTER 7)

Demetrios E. Tonias, P.E. *President, HMC Group Ltd., Brookline, Massachusetts* (CHAPTER 11)

Preface

Load and Resistance Factor Design (LRFD) method has been used in the United States for the design of steel structures since its use was allowed in 1986 by the American Institute of Steel Construction (AISC). This handbook is developed to serve as a comprehensive reference source for LRFD design of steel structures. Each topic is written by leading experts in the field. Emphasis is given to provide practical design examples for cost effective approach. The theory and criteria are explained as well as cross-references to equations to AISC are given where applicable

The book starts with history, design concepts and material behavior. It then goes into design of flexural members and design of tension compression members. Detail design aspects are covered in two sections for these topics.

Torsion design is treated as a separate topic. State-of-the-art design examples are given for use in daily practice. Building design loading criteria and general considerations of design are covered in detail, with State-of-the-art aspects are covered for the use in general consulting practice.

Limit states design of frames is explained including drift design and P-Δ analysis in a separate section.

LRFD—Design of Connections is covered in detail including practical examples. Emphasis is given on state-of-the-art and the current practice in connection design.

Brittle fracture and fatigue is becoming critical in design. Therefore, a separate section is devoted to these topics with many practical examples on how to design building and bridge connections for fatigue and fracture applications.

The use of high strength steel and long spans lead to the importance of checking floor vibration and deflections. This has been dealt with in several examples in the section on deflection and serviceability considerations.

Steel encased with or acting composite has many economic applications. The composite design section covers the design of composite steel sections and composite columns and includes practical examples

AASHTO has adopted the LRFD-limit states design approach. The design of steel beam and girder bridges has been covered with references to the AASHTO Code.

Fabrication, erection and quality control during the construction process has been covered in the final chapter of this handbook. This chapter deals with the most important aspect of controlling the quality of the design during actual erection in the field.

One of the reference sections deals with practical aspects of metal deck. At present, steel deck is the most- commonly used for floor system of steel buildings. State-of-the-art design of deck is covered in this section.

Seismic aspects of welded connections are covered in a separate appendix section. Critical design concepts are explained in detail to prevent connection problems in actual earthquake situations.

The editor gratefully acknowledges the efforts of contributors in preparing excellent manuscripts. Thanks are due to the management and staff at CUH2A, Inc.

The editor and authors are indebted to several sources for the information presented. Space considerations preclude listing all, but credit is given wherever feasible, especially in references throughout the book.

Users of this handbook are urged to communicate with the editor regarding all aspects of this book, particularly any error and suggestion for improvement.

Akbar R. Tamboli

Acknowledgments

The editor would like to acknowledge the input and help received from the many people, specifically those listed below for the time and encouragement they provided:

- Theodore Galmbos, University of Minnesota
- Lynn Beedle, Lehigh University

Appreciation in expressed to Al Perry of CUH2A, Inc., Princeton, New Jersey for his encouragement during the handbook preparation, and Irwin Cantor, Ysrael Seinuk of Cantor-Seinuk Group of New York City for encouraging the use of LRFD approach in major projects like Seven World Trade Center, New York City; Newport Office Tower, Jersey City, New Jersey; and Chase Metrotech Complex, Brooklyn, New York.

The editor would also like to acknowledge the help and assistance provided by Larry S. Hager, sponsoring editor of this handbook, who had put forth invaluable support during process of preparing the manuscript. Also thanks to the many other individuals at McGraw-Hill responsible for bringing this book to press, including Sybil P. Parker, Publisher; Margaret Webster-Shapiro, in preparing cover artwork; Paul Sobel, editing supervisor; Suzanne Rapcavage, production supervisor; and production manager, Thomas Kowalczyk.

Finally, the editor wishes to extend his thanks and appreciation to his wife, Rounkbi; his children Tahira, Ajim, and Alamgir for their patience and understanding during preparation of this handbook.

Steel
Design Handbook

1

Akbar R. Tamboli

History—Design Concepts and Material Behavior

1.1 History

Evolution of Structural Design ■ The first development of load factor (limit) design started 70 years back in Europe. Beams with fixed ends were tested, and it was found that they had much more capacity than elastic analysis indicated. The theoretical method of analysis was developed and used for design of apartment buildings.

Maier Leibnitz tests in Germany were used to show that ultimate strength of continuous beams is not affected by settlement of supports. This was the beginning of plastic design and analysis. However, to reach limited design, several hundred years of development in the structural analysis and design field had taken place. The brief account of this historical progress is given in the following pages. From prehistoric times, people have found it necessary to have systems for determining dimensions of different members of structures.

1.1.1 Early Empirical Period

Egyptians used empirical rules to build pyramids and temples. Many of these structures still exist today. The Greeks had developed rules of statics which were used to further advance the art of building. Archimedes (287–212 B.C.) discovered the principle of the lever and explained the methods for finding the center of gravity of bodies. Methods based on these principles were used in transporting the stone columns in this era.

1-2 ■ Chapter One

Romans built the great monuments, roads, bridges, and aqueducts; some of them even remain today. Some glimpse of their building methods could be seen in the book of the famous Roman architect engineer Vitruvius in the period of Emperor Augustus. Arches were commonly used and the semicircular shape was prevalent. Sizes were much heavier than indicated by today's theory of structures. Other shapes for effective stress line were not yet evolved.

Most of the structural engineering knowledge of Greek and Roman times remained dormant until it was rediscovered in the Renaissance Era. During the Renaissance, Leonardo da Vinci (1452–1519) made an outstanding contribution to the field of architecture, engineering, and art. Most of this contribution he wrote in his notebooks. He used the principle of virtual displacement to analyze systems of pulleys and levers in hoisting devices.

Engineers of the fifteenth and sixteenth centuries did not use da Vinci's discoveries but continued using empirical methods. The first influential book, *Two New Sciences* by Galileo, may be considered the effective early beginning of the present practice of analysis and design.

1.1.2 Historical Period between 1600 and 1699

Galileo (1564–1642) investigated the cantilever beam. He assumed that at fracture of the beam resistance is uniformly distributed over the entire cross section as shown in Fig. 1.1. According to Hooke's law, if material follows linear distribution up to fracture, stress distribution will be linear as shown in Fig. 1.2. The resisting moment couple will be one-third of what Galileo assumed. However, stress distribution at fracture is different from that shown in Fig. 1.2, and the difference between Galileo's theory and the true buckling load is not much different.

Robert Hooke (1635–1703) established that there is a linear relationship between force and deformation which became the so-called Hooke's law. Further development of elastic bodies was built upon this hypothesis. *Mariotte* (1620–1684) confirmed Hooke's law and experimented with different types of beams. He found that the fixed-end beam has two times the ultimate load-carrying capacity compared to the simply supported beam of the same size and material.

The *Bernoulli* family contributed to the knowledge of calculus and pointed out that the deflection curve at each point is proportional to the bending moment at that point. This was used as the basis for later development of the subject by other researchers.

1.1.3 Historical Period between 1700 and 1799

Euler (1707–1783) contributed greatly to the method of analysis as he applied calculus to solving general problems. His solution to analysis of slender columns is still accepted universally. He gave the equation of elastic curves at which buckling occurs:

Fig. 1.1 Galileo's experiment on cantilever beam.

$$P = C \frac{\Pi^2}{4L^2}$$

The above equation indicates the load P the column can carry is inversely proportional to the square of the height of the column. This formula has wide applications in analysis of elastic stability of engineering structures.

Lagrange (1736–1813) introduced the concepts of generalized coordinates based on virtual displacements and generalized forces, after the use of d'Alembert's principle developed a theory of mechanics which contains general formulas from which the necessary equations in any problem could be developed. The most important contribution made by Lagrange is his explanation of elastic curves. He described how, with prismatic bars having hinges at the ends and assuming there is small deflection under the action of the axial compressive force, it is possible to produce an infinite number of buckling curves (Fig. 1.3a, b, c).

To produce the curve with one half wave, as in Fig. 1.3a, it is necessary to apply a load four times larger than that calculated by Euler for the case where one end is built in. To get the curve in Fig. 1.3b, the required load is 16 times Euler's load, and so on. Lagrange did not limit himself to calculation of the critical values of the load P but went on to investigate deflections which will exist if load P exceeds the critical value.

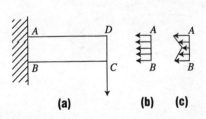

Fig. 1.2 Stress distribution at fracture.

Coulomb (1736–1806) contributed substantially in the eighteenth century to the science of elastic bodies. His experiments found that the effect of shearing forces on the strength of a beam can be neglected if its depth is small in comparison with its length. In his book of memoirs on torsion, he explained his observation on experiments carried out on a metal cylinder suspended by wire.

He concluded for each kind of wire the torsional limit of elasticity beyond which some permanent set occurs. Further, he showed if the wire is twisted initially far beyond the elastic limit, the material becomes harder and its elastic limit is raised whereas elastic modules remains unchanged. Again, by annealing, he found that hardness produced by plastic deformation can be removed on the strength of these experiments. Coulomb asserted that, in order to specify the mechanical characteristics of a material, we need two quantities, elastic modulus and the elastic limit, but we cannot change the elastic property of the material defined by the constant E (elastic modulus). The heat treatment changes the elastic limit but leaves the elastic property of the material unaltered. Coulomb frames the hypothesis that each elastic material has a certain characteristic arrangement of molecules which is not distributed by small elastic deformations. Beyond the elastic limit some permanent sliding of molecules takes place, resulting in an increase of cohesive forces, though the elastic property remains unaffected.

Coulomb made further progress in the theory of arches. In his time, it was known from experiments with models that typical failures of arches are such as those shown in Fig. 1.4. We can conclude from this figure that in investigating arch stability, it is not sufficient to consider only relative sliding of the wedges, but also the possibility of relative rotation must be checked.

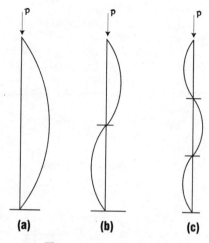

Fig. 1.3 Buckling curves.

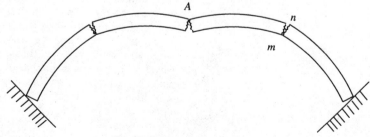

Fig. 1.4 Typical arch behavior.

History—Design Concepts and Material Behavior ■ 1-5

The memoirs of Coulomb presented in 1773 contained correct solutions for several important problems in strength of materials. However, it took engineers more than 40 years to understand them satisfactorily and use them in practical applications. Further progress was made in this science by *Navier* (1785–1836). He assumed that the neutral axis must divide the cross section such that the moment of tensile stresses about it must be equal to the moment of compressive stresses. He also proved that in materials following Hooke's law, the neutral axis must pass through the centroid of the cross section.

Navier was the first to evolve a general method of analyzing statically indeterminate problems in mechanics of materials. He states that such problems are indeterminate only so long as the bodies are considered absolutely rigid. By taking elasticity into consideration we can always add to the equations of statics a number of equations expressing the conditions of deformation, so that there will always be enough equations to allow evaluation of all unknown quantities. Considering, for example, a load P supported by several bars in one plane (Fig. 1.5), Navier states that if the bars are absolutely rigid the problem is indeterminate. He can take arbitrary values for the forces in all the bars except two and determine the forces in the last two by using equations of statics. But the problem becomes determinate if the elasticity of the bars is taken into consideration. If u and v are the horizontal and vertical components of the displacement of the point 0, he can express the elongations of the bars and forces in them as functions of u and v. Writing now two equations of statics, he finds u and v and then calculates the forces in all bars.

Navier made tests with thin spherical iron shells about 1 ft in diameter and 0.1 in thick. Subjecting them to internal pressures sufficient to produce rupture, he found that the ultimate strength of the material is approximately the same as it is when obtained from simple tensile tests. In his book on Strength of Materials, Navier presented satisfactory solutions to many structural problems.

Thomas Young (1773–1829) made his main contributions to strength of materials in his findings from experiments on tension and compression of bars. He introduced the modulus of elasticity for the first time. The definition of this quantity differs from that which we now use to specify Young's modulus. It states, "The modulus of elasticity of any substance is a column of the substance, capable of producing a pressure on its base which is to the weight causing a certain degree of compression as the length of the substance is to the diminution of its length."[1] Young had determined the weight of the modulus of steel from the frequency of vibration of a tuning fork and found it equal to 29×10^6 lb/in. Describing experiments on tension and compression of bars, Young draws the attention of his readers to the fact that longitudinal deformations

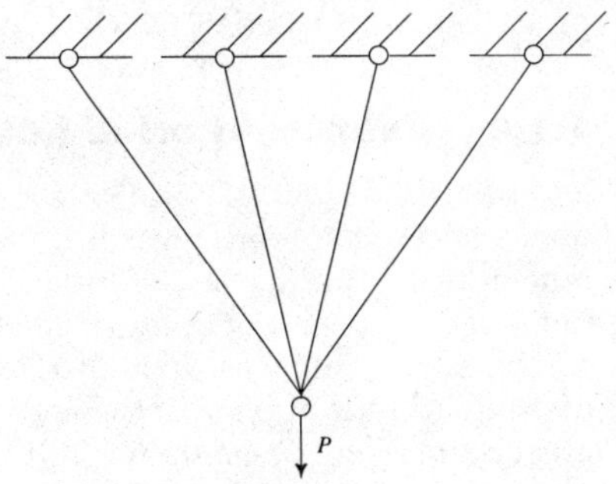

Fig. 1.5

are always accompanied by some change in the lateral dimensions. Introducing Hooke's law, he observes that it holds only up to a certain limit beyond which a part of the deformation is inelastic and constitutes permanent set.

In respect to shear forces, Young remarks that no direct test had been made to establish the relation between shearing forces and the deformations that they produce. He states, "It may be inferred, however, from the properties of twisted substances, that the force varies in the simple ratio of the distance of the particles from their natural position, and it must also be simply proportional to the magnitude of the surface to which it is applied." When circular shafts are twisted, Young points out that the applied torque is mainly balanced by shearing; stresses which act in the cross-sectional planes and are proportional to the distance from the axis of the shaft and to the angle of twist. He also notes that an additional resistance to torque, proportional to the cube of the angle of twist, will be furnished by the longitudinal stresses in the fibers, which will he bent to helices. Because of this, the outer fibers will be in tension and inner fibers in compression. Further, the shaft will be shortened during torsion "one-fourth as much as the external fibers could be extended if the length remained undiminished."

When considering inelastic deformations Young says, "A permanent alteration of form limits the strength of materials with regard to practical purposes, almost as much as fracture, since in general the force which is capable of producing this effect is sufficient, with a small addition, to increase it till fracture takes place." He concluded that where fracture failure is critical, working stresses should be kept much lower than the elastic limit of the material.

Young gave an interesting discussion of fracture of elastic bodies produced by impact. In this case, not the weight of the striking body but the amount of its kinetic energy must be considered. "Supposing the direction of the stroke to be horizontal, so that its effect may not be increased by the force of gravity," he concluded that "if the pressure of a weight of 100 lb (applied statically) broke a given substance after extending it through the space of an inch, the same weight would break it by striking it with the velocity that would be acquired by the fall of a heavy body from the height of half an inch, and a weight of one pound would break it by falling from a height of 50 inches." Young was the pioneer in analyzing stresses brought about by impact and gave a method of calculating them for perfectly elastic materials which follow Hooke's law up to fracture.

1.1.4 Historical Period between 1800 and 1899

Barré de Saint-Venant (1797–1886) was the first to examine the accuracy of the fundamental assumptions regarding bending: (1) The cross sections of a beam remain plane during the deformation. (2) The longitudinal fibers of a beam do not press up on each other during bending and are in a state of simple tension or compression. This was demonstrated by a beam subjected to two equal and opposite couples applied at the ends. Considering the pure bending of a rectangular beam (Fig. 1.6), he showed that the changes in length of fibers, and the corresponding lateral deformations, satisfy not only the above conditions but also the condition of continuity of deformation. He showed that the initially rectangular cross section changes its shape in Fig 1.6b owing to lateral contraction of the fibers on the convex side and expansion on the concave side.

The initially straight line ab becomes slightly bent and the corresponding radius of curvature is e/m, where m is Poisson's ratio and e is the radius of curvature of the axis of the bent bar. Because of this lateral deformation, the distances of neutral fibers a and b from upper and lower surfaces of

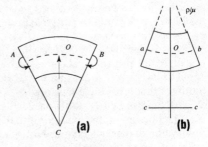

Fig. 1.6

the bar are also slightly altered. The upper and lower surfaces will be bent to antiplastic surfaces. This was the first time that the distortion of the shape of cross section of a bent bar was investigated.

Saint-Venant showed that the deflection of a cantilever can be calculated in an elementary way without integration of the appropriate differential equation, and introduced the method which now is called the area moment method. In the discussion of pure bending of beams (Fig. 1.6) Saint-Venant formulates the principle which now carries his name. He stated that the stress distribution for this case conforms to the rigorous solution only when the external forces applied at the ends are distributed over the end cross sections in the same manner as they are distributed over intermediate cross sections. He also states the solution obtained will be accurate enough for any other distribution of the forces at the ends, provided that the resultant force and the resultant couple of the applied forces always remain unchanged. He refers to some experiments which he carried out with rubber bars and says that those experiments showed that if a system of self-equilibrating forces is distributed on a small portion of the surface of a prism, a substantial deformation will be produced only in the vicinity of these forces. Figure 1.7 shows one of Saint-Venant's examples. The two equal and opposite forces which act on the rubber bar produce only a local deformation at the end, and the remainder is practically unaffected. Saint-Venant's principle is often used by engineers in analyzing stresses in structures.

Fig. 1.7

When allowable limits in structures have to be fixed, Saint-Venant always assumes that the elastic limit of a material is reached when the distances between the molecules are increased by the deformation beyond a certain value which is peculiar to any given material. Thus his formulas for calculating safe dimensions of structures are derived from maximum strain consid-

1-8 ■ Chapter One

erations. For instance, to cover the combined bending and torsion of shafts he establishes a formula for the maximum strain; this has been used by engineers ever since its appearance.

Saint-Venant was the first to show that pure shear is produced by tension in one direction and an equal compression perpendicular to it. Taking Poisson's ratio equal to 1/4, he concludes that the working stress in shear must be equal to eight-tenths of that for sample tension.

Karl Culmann (1821–1881) has as his principal accomplishment his systematic introduction of graphic methods into the analysis of all kinds of structures and the publication of his book on graphical statics. He incorporated many original graphical solutions in this book.[2]

Culmann considered protective geometry very important in the development of graphical statics and discussed the protective properties of systems of forces n in his introductory chapters. In dealing with applications, Culmann begins with parallel forces in one plane and shows how, by using the funicular polygon, the reactions at the supports of a beam can be determined. He demonstrated how to construct the bending-moment diagram and how to find the position of a moving load which produces the maximum value of the bending moment.

In his work on bending of beams, Culmann showed how the stresses at a point A in Fig. 1.8a can be analyzed graphically considering an infinitesimal element A_{mn} and denoting the stress components acting on the planes through A and perpen-

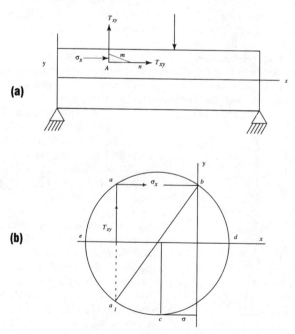

Fig. 1.8 Culmann's circle of stress.

dicular to the X and Y axes by 6x and T_{xy}. He demonstrated that the normal and tangential components of stress acting on any inclined plane mn are given by the coordinates of points on a circle of stress. To construct this circle, we have only to take the point a in Fig. 1.8b with coordinates T_{xy} and 6x and consider point a symmetrically placed with respect to a about the x axis. Then the line a, b represents the diameter of the circle of stress, and we can draw this circle.

From the equations of equilibrium of the element A_{mn}, Culmann showed that the stress components 6 and T acting on any plane mn. From this construction, he showed that the points d and e at the ends of the horizontal diameter define the magnitudes and the directions of the principal stresses at point A. He also proved that planes upon which the maximum shear acts bisect the angles between the principal planes and that the maximum shear is given by the magnitude of the radius of the circle of stress.

James Clerk Maxwell (1831–1879) showed how to analyze statically determinate and indeterminate trusses. He showed if we have a plane framework with n joints, we can write 2n equations of equilibrium. Three equations will usually be needed for calculating the reactions at the supports, and the remaining 2n-3 equations may be used to compute the forces in the members of the framework if the number of bars is equal to 2n-3. In the event the number of members is larger than 2n-3, the problem is statically indeterminate, so that the elastic properties of the structure must be taken into consideration and the deformation of the structure must be investigated for its solution. Regarding the method of solving this, Maxwell says, "I have therefore stated a general method of solving all such questions in the least complicated manner. The method is derived from the principle of conservation of energy and is referred to as Clapeyron's theorem." Maxwell's method of analysis can be explained by the example of this in Fig. 1.9a and b.

To calculate the deflection of the truss shown in Fig. 1.9: This truss is statically determinate, and we can readily find the forces in all members produced by given loads P1, P2, ... Let S_i be the force acting in any bar i; let L_i be the length of that bar and A_i be its cross sectional area. Then the elongation of the bar is $\Delta_i = S_i L_i / EA_i$. We now have the geometrical properties of finding the deflection of any joint, say A, knowing the elongations of all bars of the truss. Maxwell obtains the deflection at point A by summing up deflections of all members.

$$\Delta_a = \sum_{i=1}^{i=m} \frac{S_i s_i L_i}{EA_i}$$

In his work on deflections of trusses, Maxwell discovered the existence of a very important relation between the deflections of a truss produced by two different kinds of loading. Let us consider the two cases of loading shown in Fig. 1.10a and b. In the first case load P is applied at joint B, and it is required to find the deflection Δ_a at A. In the second, load P acts at A, and it is required to find the deflection, Δ_b.

Following the method explained above, we consider the two auxiliary cases shown in Fig.10c and d. Using the notation S_i and s_i for the forces in the ith bars of Fig. 1.10a and b, respectively, and S_i and s_i for the forces in Fig. 1.10b and c, equations for Δ_a and Δ_b become

1-10 ■ Chapter One

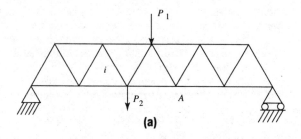

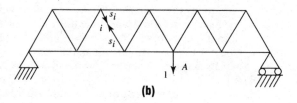

Fig. 1.9

$$\Delta_a = \Delta_b = \sum_{i=1}^{P_i=m} \frac{S_i s_i L_i}{EA_i}$$

Comparing Fig. 1.10c and b, it can be seen that $S_i = s_i P$ and, from Fig. 1.10d and a we find $S_i = s_i P$. Substituting this into the above equations,

$$\Delta_a = \Delta_b = \sum_{i=1}^{P_i=m} \frac{s_i P s_i L_i}{EA_i}$$

It can be seen from the two cases of loading shown in Fig. 1.10a and b that when the load is moved from joint B to joint A, the deflection of A moves to B. This is the reciprocity theorem as it was obtained by Maxwell in its simplest form.

Later, this theorem was generalized and became very important in the analysis of statically indeterminate structures.

Otto Mohr (1835–1918) designed some of the first steel trusses in Germany. Other important contributions were on the use of the funicular curve. In finding elastic deflections of beams, he derived the three moment equations for continuous beams, the supports of which are not on the same level. The first applications of influence lines were also published by him.

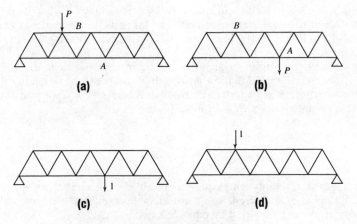

Fig. 1.10

Mohr used circles for representation of stresses and strength theory which can be adapted to various stress conditions. The results of this representation were in better agreement with experiments. He assumed that of all planes having the same magnitude of normal stress, the weakest one, on which failure is most likely to occur, is that with the maximum shearing stress. Under such circumstances, it is necessary to consider only the largest circle. Mohr calls it the *principal circle* and suggests that such circles should be constructed when experimenting for each stress condition in which failure occurs. Principal circles for cast iron tested to failure in tension, compression, and pure shear (torsion) are shown in Fig. 1.11.

If there is a sufficient number of such principal circles, an envelope of those circles can be drawn, and it can be assumed with sufficient accuracy that for any stress condition for which there are no experimental data, the corresponding lim-

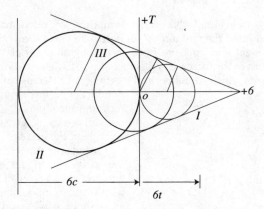

Fig. 1.11 Mohr's principal circles.

iting principal circle will also touch the envelope. For example, when considering cast iron, Mohr suggests that the envelope be taken as the two outer tangents to the circles I and II (Fig. 1.11) which correspond to experimental fracture in tension and compression.

The ultimate strength in shear is then found by drawing circle III, which has center at 0 and is tangential to the envelope. If $6t$ and $6c$ are the absolute values of the ultimate strength of material in tension and compression, we find from the figure that the ultimate strength in shear is

$$T_{ult} = \frac{6t6c}{6+6c}$$

Which agrees satisfactorily with experiments. Mohr's theory attracted great attention from engineers and physicists. A considerable amount of experimental work was done in connection with it by other researchers.

In Italy, *Alberto Castigliano* (1847–1884) gave his famous theorem in the following form: In case of trusses with ideal hinges, if external forces are applied at the hinges and the bars have prismatical form, the strain energy of such a system can be give in the form

$$V = \frac{½ S_i^2 L_i}{EA_i}$$

where V = strain energy
 L_i = length of each truss member
 S_i = force in each truss member
 A_i = area of each truss member
 E = modulus of elasticity

After having developed this general method of calculating deflections, Castigliano considered the two important particular cases shown in Fig. 1.12. He proved that if equal and opposite forces act along *ab* of an elastic system, the derivative dv/ds gives the increase of distance *ab* due to deformation of the system, for example, a truss which contains *a* and *b* as hinges. In the case of two forces perpendicular to line *ab* in the truss and forming a couple M (Fig. 1.12*b*), the derivative dv/dm gives the angle of rotation of the line *ab*.

Castigliano applied these results to the analysis of trusses with redundant bars and gave a proof of the principle of least work. In this analysis, he removed all redundant bars and replaced their action upon the rest of the system by

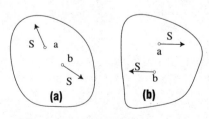

Fig. 1.12

forces as shown in Fig. 1.12b. The system is now statically determinate and strain energy V_1 is a function of the external forces P_i and of the unknown forces X_i acting in the redundant bars. On the strength of what was said in Fig. 1.12 he concluded that $-dV_1/dX_i$ represents the increase of the distance between the joints a and b. This increase is clearly equal to the elongation of bar ab in the redundant system in Fig. 1.13a; thus he obtained the relationship in Fig. 1.13, observing that the right-hand side of this equation represents the derivative with respect to X_i of the strain energy $X_i^2 L_i/2 E A_i$ of the redundant bar and denoting the total strain energy of the system with redundant bars included by V. He put the of strain energy equation into the form $dv/dx_i = 0$, which expresses the principle of least work. These results, which were obtained initially for hinged-truss analysis, were generalized by Castigliano to extend to an elastic solid of any shape. He developed expressions for the strain energy of bars under various kinds of deformation and used these expressions in numerous applications of his theory of the analysis of statically indeterminate problems of beams and arches.

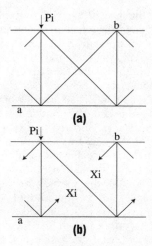

Fig. 1.13

Lord Rayleigh's principal contributions included his famous discussions of vibrations of strings, bars, membranes, plates, and shells in the first volume of his book *Theory of Sound*. He shows the advantages which an engineer can gain by adapting the notions of generalized forces and generalized coordinates. The introduction of these concepts and the use of the Bett-Rayleigh reciprocity theorem have brought about great simplification in handling redundant structures.

Rayleigh's work not only covers sound proper but also covers nonsonic vibration. Rayleigh demonstrated the benefits to be had by using normal coordinates and shows how by making velocities vanish, the solutions of static problems may be extracted from vibration analysis. Thus he obtains the deflections of bars, plates, and shells expressed in terms of normal functions; these ideas had been of great importance in engineering.

In finding frequencies of vibration of complicated systems, he obtained an approximate value by assuming a suitable form for the type of motion and so transforming the problem to that of the vibration of a simple oscillator. He described the steps which can be taken toward improving the approximation. This idea of calculating frequencies directly from an energy consideration, without solving differential equations, was later elaborated by Walter Ritz, and the Rayleigh Ritz method is now widely used in studying vibration.

Rayleigh used his approximate method in several complicated problems. In the vibration of strings, he showed how an approximate solution can be found where the mass of the string is not uniformly distributed along the length, and he investigated the effect on the frequency of a small concentrated mass attached to the string. In treating the longitudinal vibrations of bars, he estimated the error involved in neglecting the inertia of lateral motion of particles not situated on the axis. He showed that for a circular bar his result is in good agreement with results of a more elaborate theory (*The Theory of Sound*, vol. 1). Again, a correction is made to the simple theory of lateral vibration of rods to allow for the rotatory inertia.

An important contribution in the theory of vibration of thin shells was made by Rayleigh. He showed that two kinds of vibrations have to be considered: (1) extensional modes by which the middle surface of the shell undergoes stretching and (2) flexural or inextensional modes. In the first case, the strain energy of the shell is proportional to its thickness, while in the second it is proportional to the cube. Now all the strength of the principle that, for given displacements, the strain energy of the shell should be as small as possible, Rayleigh concluded that "when thickness is diminished without limit, the actual displacement will be one of pure bending, if such there be, consistent with the given conditions." Using this conclusion, he examined flexural vibrations of cylindrical, conical, and spherical shells and obtained the results which agree satisfactorily with experiments.

One of the most important contributions of Lord Rayleigh was the theory of elastic surface waves.[3] In the waves which bear his name, motion is propagated over the surface of an elastic medium in a manner similar to the rippling set up when the surface of still water is disturbed. As predicted by Rayleigh, these waves have been found to play an important part in earthquakes.

1.1.5 Progress Through 1900 to 1950

Development of Theory of Structures During 1900 to 1950 ■ Nineteenth century developments were principally for the analysis of trusses. However, to analyze fully rigid joint structures, the slope deflection method was developed by Axel Bendixen.[4] Considering the fixed-end element as shown in Fig. 1.14, the slope deflection equation can be used for bending moment M_{mn} acting at end m. $M_{mn} = 2K_{mn}(2o_{mn} + o_{nm}) + M_{mn0}$ of the bar mn. o_{mn} and o_{nm} are the angles of rotation of the ends, taken positive in the clockwise sense; $K_{mn} = EI_{mn}/L_{mn}$ is the so-called stiffness factor; and M_{mn0} is the fixed-end moment, that is, the moment which would act on the end m of the bar if its ends were fixed and only lateral loads P and Q were acting. These moments were taken positive if they act on the bar in the clockwise direction.

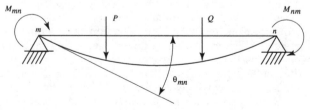

Fig. 1.14

History—Design Concepts and Material Behavior ■ 1-15

The method of slope deflection was further simplified by Hardy Cross in the United States.[5] He developed a method of directly distributing the unbalanced moments while unlocking one joint at a time and considering all other joints as fixed. This procedure of successive approximations was proposed in 1932 and found wide applications in the United States.

In the design of steel structures, it is sometimes necessary to consider not only loads at which yielding of the material sets in but also those loads under which complete collapse of the structure is brought about. Analysis shows that, if two structures are designed with the same factor of safety with respect to yielding, they may have very different factors of safety with respect to complete failure. For example, considering pure bending of beams and assuming that structural steel follows Hooke's law up to the yield point and that beyond that limit it stretches without strain hardening, we obtain the stress distributions shown in Fig. 1.15a and b for the two limiting conditions: (1) in the beginning of yielding and (2) complete collapse. The corresponding bending moments for a rectangular cross section (Fig. 1.15c) are

$$M_{yp} = 6yp\frac{bh^2}{6} \qquad M_{ult} = 6yp\frac{bh^2}{4}$$

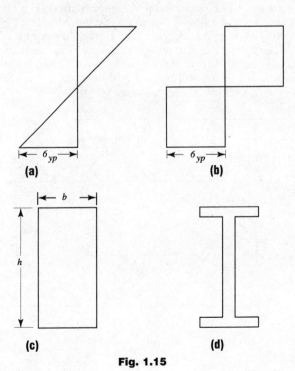

Fig. 1.15

Therefore, $M_{ult}/M_{yp} = 1.5$. For an I beam this ratio is much smaller than that of a rectangular beam. To allow for this fact, some engineers had recommended (Grüning[6]) cross-sectional dimensions of structural members should be chosen on the basis of ultimate strength. It has been shown that if we take the stress distribution shown in Fig 1.15b for the ultimate strength condition, the corresponding ultimate loads can be readily calculated. For instance, from a beam with built-in ends loaded at the middle (Fig. 1.15a) it can be concluded that complete collapse of the beam occurs when the bending moment reaches its ultimate value M_{ult} at the time cross sections a, b, and c. For any further loading, the condition will be the same as for two-hinged bars (Fig. 1.16b).

The magnitude of the ultimate load is then obtainable from the corresponding bending-moment diagram (Fig. 1.16a), from which we have

$$\tfrac{1}{4} P_{ult} = 2 M_{ult}$$

If we proceed in a similar manner for such highly statically indeterminate systems as the structures of office buildings and put hinges at all those cross sections at which bending moments reach their ultimate values, the problem of determining ultimate loads can be reduced to that of simple statics of rigid bodies, and the entire stress analysis of the system can be greatly simplified.

Further contribution to the analysis of the elastic stability of plates under various kinds of heading and edge conditions was made by S. Timoshenko. The books by S. Timoshenko on *Strength of Materials* and *Theory of Elasticity* published originally in Russian and later in English, were used throughout the world and are still being used as important reference.

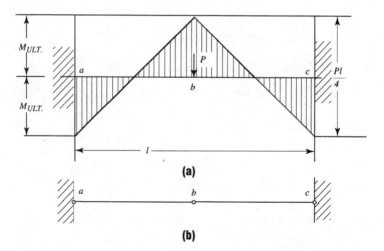

Fig. 1.16

In the early 1950s Lynn Beedle carried out research on the plastic behavior of steel beams at Lehigh University, and continued research including T.V. Galambos formed the basis of the current load factor design method adopted in American Institute of Steel Construction LRFD specification and *Manual of Steel Construction*.[7]

1.2 Design Concepts

1.2.1 Basic Concept of Plastic Analysis

The basis of analysis for indeterminate structures is that the structural material used should be ductile. Idealized characteristics of such a material are such that it initially behaves elastically, but after the yield point has been reached it continues to deform at a constant stress level known as yield stress f_y. This phenomenon of yielding at constant stress is known as plastic yielding. When the yield plateau is long compared with the elastic strain at yield, it becomes possible to use simple plastic theory to predict the ultimate load behavior of a frame.

Plastic analysis is a special case of limit states analysis wherein the limit state for strength must be achievement of plastic strength. This precludes having limit states based on instability, fatigue, or brittle fracture. In plastic analysis, the inherent ductility of steel is recognized and utilized in the analysis of statically indeterminate structures, such as continuous beams and rigid frames.

Achievement of plastic strength at one location in a statically indeterminate structure may not constitute achievement of maximum strength for the structure. After one location reaches its plastic strength, additional load is carried in different proportions throughout the structure until a second location of plastic strength is achieved. Once the structure has no further ability to carry an increased load, it is said to have reached a collapse mechanism. Even when plastic analysis is used for the strength requirement, serviceability requirements such as deflection are investigated at service load conditions.

1.2.2 Lower, Upper Bound, and Uniqueness Theorem

This theorem states that a load computed on the basis of a bending-moment distribution in which no moment exceeds the local value of M_p is equal to or less than the true ultimate limit load. Referring to the moment diagram in Fig. 1.17b, hinges have already formed at both ends so that fixing moments are equal to M_p. In the beginning the center moment is still elastic and has a value of $\frac{1}{2}M_p$. Then as the load intensity is increased, the center moment at point C approaches M_p. At this stage sufficient hinges have developed and the mechanism condition is satisfied. This approach is referred to as application of lower bound theorem.

Upper Bound Theorem ▪ In this theorem a load computed on the basis of an assumed mechanism is equal to or greater than the true plastic limit load. Consider the beam in Fig. 1.18a and let the simple assumption be made that the two hinges necessary for a hinged mechanism occur at the fixed end and near

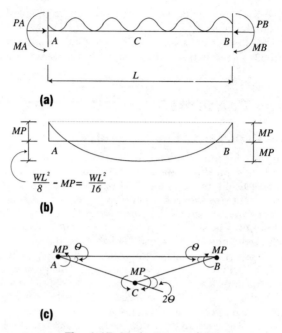

Fig. 1.17 Mechanism condition.

midspan at distance X. The resultant moment distribution is shown in Fig. 1.18b, and clearly both equilibrium and mechanism conditions are satisfied, though if the assumed location is changed to the center of span, the load-carrying capacity will be $12M_p/L^2$, which gives an upper limit to the plastic limit load which is actually $11.66M_p/L^2$. This approach is considered to form a basis for the upper bound theorem.

The following two conclusions can be drawn from the upper bound theorem. First the plastic limit load of a structure cannot be increased by removing material from any part. Second, after due consideration of all possible collapse mechanisms for a structure, the plastic limit load is equal to the lowest possible value.

Uniqueness Theorem ▪ This theorem states that a load computed on the basis of a bending-moment distribution which satisfies the conditions of equilibrium, plastic moment, and mechanism is the true plastic limit load. When a solution satisfies simultaneously the upper and lower bond conditions, the resulting estimate of collapse load must become the actual value.

The significance of this theorem is that if some mechanism is thought to be the actual one for collapse, then confirmation can be obtained by reference to the relevant bending-moment diagram. It must be emphasized that the ultimate check of any plastic analysis is to make sure that the plastic moment condition is not violated anywhere on the moment diagram. It follows from the uniqueness theorem

History—Design Concepts and Material Behavior ■ 1-19

that initial stresses, deformations, or settlement at supports have no effect on the plastic limit load.

1.2.3 LRFD Design Approach

Until recently engineers were basing the analysis and design of structures on a linear theory of elasticity. On the whole, the results have been satisfactory. The buildings and bridges have withstood the test of time. Why then should one be concerned with the LRFD method?

A part of the answer to this question is to point out some of the disadvantages of elastic design methods. On the one hand, ductile structural materials such as steel can withstand strains much larger than those encountered within the elastic limit. Design methods which are based on the elastic limit fail

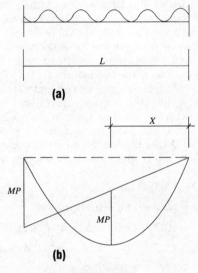

Fig. 1.18

to take advantage of the ability of such material to carry stresses above the yield stress (strain hardening). More important, this ductility in redundant structures permits a redistribution of stress beyond the elastic limit. These redistributed stresses can often carry very considerable additional loads. Therefore, from this viewpoint, elastic analysis is unduly conservative. On the other hand, elastic analysis of structures is generally concerned only with overall quantities such as moments and force resultants. Although these quantities are easily related to average stresses, actual stress distributions may exhibit considerable nonuniformity. It follows that, even though the average stress is below the elastic limit, highly stressed local regions may become plastic. Thus elastic analysis is frequently unrealistic.

Finally, elastic analysis of all but the simplest of structures is complicated. Simplifying assumptions must be made to the point where the relation between a practical solution and rigorous elasticity theory is often quite obscure. From a practical viewpoint elastically designed structures are safe because precautions are taken to err always on the conservative side. Obviously, the net result is a waste of material. Such waste should be generally deplored from an economic viewpoint. For structures such as aircraft, where weight is of prime importance, the results may be even more serious.

Trial and error, rule of thumb, and ad hoc regulations can of course reduce this waste, but resulting analysis can hardly be considered elastic. Further, since such an analysis would have little rational basis, a true estimation of the safety factor would become virtually impossible. The LRFD method offers some of the answers to the objections to elastic design. It takes full advantage of ductility, and the method is mathematically simple.

1-20 ■ Chapter One

Load Factors ■ Load factors increase the load imposed on structures to account for the uncertainties involved in estimating the magnitudes of dead or live loads. Generally the value for load factors used for dead loads is smaller than the one used for live loads because designers can estimate so much more accurately the magnitudes of dead loads. The LRFD method makes a designer more conscious of load variations.

ASCE Standard 7-93 recommends minimum loads to be used for design. The AISC LRFD specification gives load factors and load combinations that can be used with minimum loads in the ASCE Standard 7-93.[8]

The following six equations represent the load combinations given in the LRFD specification. Numbers of equations given on the right correspond to the AISC LRFD specification equation number.

$U = 1.4D$ (A4.1)

$U = 1.2D + 1.6L + 0.5 (L_r \text{ or } S \text{ or } R)$ (A4.2)

$U = 1.2D + 1.6 (L_r \text{ or } S \text{ or } R) + (0.5L \text{ or } 0.8W)$ (A4.3)

$U = 1.2D + 1.3W + 0.5L \text{ to } 0.5 (L_r \text{ or } S \text{ or } R)$ (A4.4)

$U = 1.2D \pm 1.0E + 0.5L + 0.25$ (A4.5)

$U = 0.9D \pm (1.3W \text{ or } 1.0E)$ (A4.6)

where U = ultimate load
D = dead load
L = live load
L_r = roof live load
R = initial rainwater or ice
S = snow load
W = wind loads
E = earthquake forces

Resistance Factors ■ To account for uncertainties in material strengths, dimensions, and workmanship, a resistance factor is introduced. The resistance factor is multiplied by nominal strength to arrive at the actual ultimate strength of the member. Typical resistance factors given in the AISC LRFD specification are shown in Table 1.1.

LRFD Specification and Reliability Index ■ The random nature of strength and load is recognized by load factors and resistance factors as discussed above. For safe structural strength of the structure, R is always greater than load effects

$$Q < R$$

Table 1.1 Resistance Factors Table

Resistance Factor	Condition where it is applicable
0.60	Bearing on concrete foundation
0.65	Bearing on bolts (other than A307)
0.75	Bolts in tension, plug, or slot welds, fracture in the net section of tension members
0.80	Shear on effective area of full-penetration groove welds, tension normal to effective area of partial-penetration groove welds
0.85	Columns, web crippling edge distance, and bearing capacity at holes
0.90	Beams in bending and shear, fillet welds with stress parallel to weld axis, groove welds base metal, yielding on gross section of tension members
1.00	Bearing on projected areas of pins, web yielding under concentrated loads. Slip resistance bolt shear values

Therefore, fundamental LRFD design criteria are stated as

$$\sum_{i=1}^{K} y_i Q_{ni} \leq QR_n$$

where $y_i Q_{ni}$ is a factored load effect and QR_n is factored resistance.

In the probabilistic concepts of LRFD it is recognized that load and resistance are both random variables. Since their magnitude at any given time cannot be predicted with certainty, certain data bases can be established through observations over long periods of time, which enables the designer to see how the values vary as well as whether there are any numbers that have a tendency to occur more frequently than others (see Fig. 1.19).

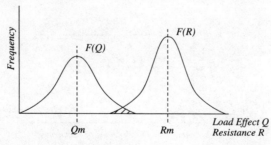

Fig. 1.19 Probability density function for load effect and strength.

The Reliability Index ■ The reliability index β (Fig. 1.20) has been used to measure reliability or safety of the structure. Whether it is related to the true probability of failure or is just a relative measure, its usefulness can be summarized as follows: When β is chosen as a constant value for the structure and all its elements, the probability of failure will be the same throughout regardless of the variability of the strengths and the loads.

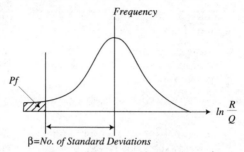

Fig. 1.20 The reliability index.

In the LRFD specification, safety is no longer empirical but rather is a rationally developed measure of structural performance. The LRFD specification is based on R/Q lognormal distribution. When complete distributions are known for R and Q, the probability values for failure take the following form for different β values (Fig. 1.21):

P.F. $= 2.3 \times 10^{-2}$ for $\beta = 2$

P.F. $= 1.4 \times 10^{-3}$ for $\beta = 3$

P.F. $= 3.2 \times 10^{-5}$ for $\beta = 4$

P.F. $= 2.9 \times 10^{-6}$ for $\beta = 5$

The magnitude of P.F. (probability of failure) is indicated by area under the "tail" distribution of $L_n(R/Q)$ for which $L_n(R/Q) \leq 0$.

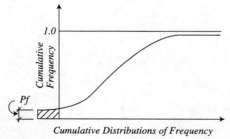

Fig. 1.21 Frequency vs. cumulative distribution.

1.3 Material Behavior

1.3.1 Material Properties

Carbon steel is an alloy of the metallic element iron and nonmetallic element carbon. Its composition is made of microscopically small crystals. Properties of steel can be varied by changing the ratio of carbon to iron and can be further modified by the addition of other alloying elements such as silicon, manganese, nickel, and copper. Steels containing any of the above elements in significant proportions are known as alloy steels. All steels also have a certain amount of small impurities such as sulfur or phosphorus which must be kept below specified limits to prevent adverse effects on weldability and embrittlements. American Society for Testing and Materials (ASTM) specifies chemical composition and other relevant properties for a number of steels for structural use.

The level of strength used in LRFD design is that of the yield plateau F_y. The length of the plastic plateau determines the ductility. The strain limit ε_{st} at the end of the plastic region for A36, A572, and A441 steels is approximately 12 times the strain at the initiation of yielding ε_y. In LRFD design actual stress-strain diagrams can be replaced by an idealized diagram representing steel as an elastic plastic material as shown in Fig. 1.22.

1.3.2 Idealized Stress-Strain Characteristics

The basis of LRFD design is that structural material will be ductile. Figure 1.22 shows the idealized stress-strain characteristics of steel. In the beginning the material behaves elastically, but after the yield point has been reached it continues to deform at a constant stress level known as yield stress (f_y).

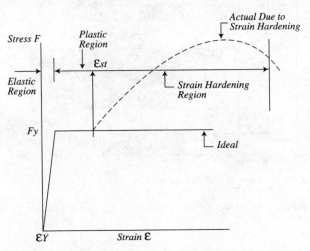

Fig. 1.22 Ideal stress-strain diagram.

1-24 ■ Chapter One

1.3.3 Plastic Modulus of Wide-Flange and Compound Sections

The bending moment which produces a plastic hinge in a member is known as full plastic moment. It is denoted by M_p. The plastic hinge is a condition of limiting moment resistance at the cross section of the beam. Increased load will produce greater strain, but moment remains constant at the plastic moment.

Table 1.2 Steel Properties

Availability of Shapes, Plates, and Bars According to ASTM Structural Steel Specifications

Steel Type	ASTM Designation	F_y Minimum Yield Stress (ksi)	F_u Tensile Stress[a] (ksi)	Shapes Group per ASTM A6					Plates and Bars									
				1[b]	2	3	4	5	To ½" incl.	Over ½" to ¾" incl.	Over ¾" to 1¼" incl.	Over 1¼" to 1½" incl.	Over 1½" to 2" incl.	Over 2" to 2½" incl.	Over 2½" to 4" incl.	Over 4" to 5" incl.	Over 5" to 6" incl.	Over 6" to 8" incl.	Over 8"
Carbon	A36	32	58–80																A
		36	58–80[c]	A	A	A	A	A	A	A	A	A	A	A	A	A	A	A	
	A529 Grade	42	60–85	A	A				A	A	A								
		50	70–100	A	A				A	A	A	A	A[d]						
High-Strength Low-alloy	A572 Grade	42	60	A	A	A	A	A	A	A	A	A	A	A					
		50	65	A	A	A	A	A	A	A	A	A	A	A					
		60	75	A	A	A			A	A	A								
		65	80	A	A				A	A									
Corrosion Resistant High-strength Low-alloy	A242	42	63											A	A	A	A		
		46	67									A	A						
		50	70	A	A	A			A	A	A								
	A588	42	63											A	A	A	A	A	
		46	67									A	A						
		50	70	A	A	A	A	A	A	A	A								
Quenched & Tempered Alloy	A852[e]	70	90–110						A	A	A	A	A	A	A				
Quenched & Tempered Low-Alloy	A514[e]	90	100–130											A	A	A	A		
	A514[e]	100	110–130						A	A	A	A	A	A					

[a] Minimum unless a range is shown.
[b] Includes bar-size shapes
[c] For shapes over 426 lb / ft minimum of 58 ksi only applies.
[d] Plates to 1 in. thick, 12 in. width; bars to 1½ in.
[e] Plates only.
[f] To improve the weldability of A529 steel, the specification of a maximum carbon equivalent (per ASTM Supplementary Requirement S78) is recommended.

■ Available
□ Not Available

History—Design Concepts and Material Behavior ■ 1-25

$$M_p = F_y Z \tag{1.1}$$

where Z = plastic section modulus, a geometric property of the cross section that can be found in the *AISC Manual of Steel Construction*.

When the full cross section of a wide flange becomes plastic, the resisting moment M_p is about 12 percent higher than the moment that causes the first yielding moment M_y:

Table 1.3 Steel Properties

	Availability of Steel Pipe and Structural Tubing According to ASTM Material Specifications							
			F_y Minimum Yield Stress (ksi)	F_u Minimum Tensile Stress (ksi)	Shape		
Steel	ASTM Specification	Grade			Round	Square & Rectangular	Availability
Electric-Resistance Welded	A53 Type E	B	35	60	■		Note 3
Seamless	Type S	B	35	60	■		Note 3
Cold Formed	A500	A	33	45	■		Note 1
		B	42	58	■		Note 1
		C	46	62	■		Note 1
		A	39	45		■	Note 1
		B	46	58		■	Note 2
		C	50	62		■	Note 1
Hot Formed	A501	—	36	58	■	■	Note 1
High-Strength Low-Alloy	A618	I	50	70	■	■	Note 1
		II	50	70	■	■	Note 1
		III	50	65	■		Note 1

Notes:
1. Available in mill quantities only; consult with producers.
2. Normally stocked in local steel service centers.
3. Normally stocked by local pipe distributors.

■ Available
□ Not Available

$$\frac{M_p}{M_y} = \frac{F_y Z}{F_y S} = \frac{Z}{S} = \text{shape factor} = 1.12 \tag{1.2}$$

For the rectangular section shown in Fig. 1.23 the stress distribution is shown by two rectangular blocks of magnitude $\pm f_y$ acting on two equal areas of $bd/2$ separated equally from the neutral axis.

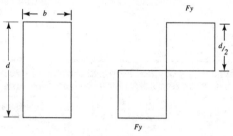

Fig. 1.23

Plastic moment = force × lever arm

$$M_p = \left(\frac{bd f_y}{2}\right)\frac{d}{2} = \frac{bd^2}{4} f_y$$

$$Z = \frac{bd^2}{4}$$

$$S = \frac{bd^2}{6}$$

Shape factor $\dfrac{Z}{S} = 1.5$

Shape factors for different types of members are given in Table 1.4.

1.3.4 Moment Curvature Graphs

For a plastic mechanism to develop fully, the first hinge formed must rotate at constant moment M_p until the last hinge develops. This means the first formed hinge needs to have rotation capacity. The moment curvature graphs in Fig. 1.24 show the ability of a beam to carry moment during rotation.

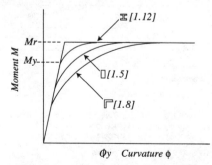

Fig. 1.24 Moment curvature graphs.

1.3.5 Effect of Cold Work on Tensile Properties

When the strain $\varepsilon = F_y/E$ at the first yield has been exceeded appreciably and the specimen is unloaded, if the load is applied again, stress-strain relationships are quite different from that observed during initial loading. Under normal elastic loading and unloading there is no residual strain.

When initial loading is beyond the yield point as in Fig. 1.25 to the point a, unloading will result to a strain point b and a permanent set ob will occur. The ductility capacity will be reduced from a strain of to the strain bf. If the specimen is reloaded, it will exhibit behavior as if the stress-strain origin were at point b. The plastic zone prior to strain hardening will also be reduced.

When the specimen is loaded until point c is reached, unloading follows the dashed line to point d. Now the origin of a new loading is a point d. The length of line cd will be greater, indicating that the yield point has increased. The increase in value of the yield point is known as the strain-hardening effect. When loading is resumed from point d, ductility is reduced from its original value prior to the initial loading. The process loading beyond the elastic range to cause a change in available ductility at atmospheric temperature is known as cold work. Inelastic deformation occurs when structural shapes are made by cold forming from plates at atmospheric temperature. Increase in yield strength occurs owing to cold working into the strain hardening range.

Table 1.4 Shape Factors for Different Types of Members

Member Profile	Shape Factor
Wide-flange sections	1.12
Solid rectangle	1.5
Solid round	1.7
Channels	1.2
Equal angles	1.82
Unequal angles	1.8
Circular hollow sections	1.35
Square hollow sections	1.2
Rectangular hollow sections	1.25

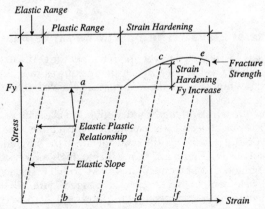

Fig. 1.25 Strain-hardening effects.

If steel is loaded in the plastic range and then unloaded after a period of time, the steel will have acquired different properties from those represented by points d, c, and e in Fig. 1.25 by strain-aging phenomena. Strain aging is shown in Fig. 1.26. It indicates an increase in yield point has occurred. The plastic zone has been restored at constant stress. It also gives the strain-hardening zone at elevated stress, even though the original shape of the stress-strain diagram is restored but the ductility is reduced. The new stress-strain diagram can be used for the analysis of cold-formed sections as long as the ductility remaining is adequate for the purpose for which it is being used.

The effect of cold work on ductility and strength can be eliminated by annealing or stress relieving. It involves elements to be heated above transformation range and allowing them to slowly cool. The recrystallization will occur and original properties are restored.

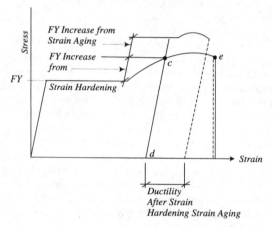

Fig. 1.26 Strain-hardening and strain-aging effects.

1.3.6 Effects of Elevated Temperature on Tensile Properties

Structural steel may be subjected to very high temperatures during fire and welding operations. When steel is subjected to temperatures over 200°F, the stress-strain curve becomes more rounded and nonlinear. No well-defined yield point is present at high temperatures.

The yield strength, tensile strength, and modulus of elasticity reduce as temperature increases. The rate of decrease is maximum in the temperature range from 800 to 1000°F. Each type of steel behaves differently owing to the different chemistry and microstructure of that steel. Figure 1.27 shows that for high-carbon steels such as A36 and A441 strain aging occurs at temperatures of 300 to 700°F. This increases the tensile strength and yield point in that temperature range. Figure 1.28 shows the variation of tensile strength against temperature rise, and Fig 1.29 shows elastic modulus variation against temperature rise. Tensile strength rises approximately 10 percent

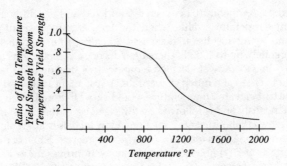

Fig. 1.27 Effect of temperature on yield strength.

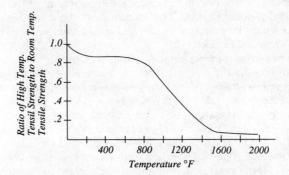

Fig. 1.28 Effect of temperature on tensile strength.

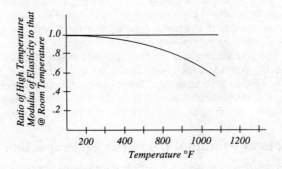

Fig. 1.29 Effect of temperature on modulus of elasticity.

above that at room temperature between 300 and 700°F. The yield point recovers to about its room temperature value when the temperature is about 500 to 600°F. However, strain aging results in decreased ductility. The modulus of elasticity decreases moderately up to 1000°F. Thereafter, decrease is rapid. See Fig 1.29.

Under long-time loading at elevated temperatures steel deforms rapidly at first, then continues to creep at a much slower rate. A systematic creep test for steel subjected to a constant elevated temperature would show that initial elongation occurs almost instantaneously and is followed by three stages. In the first stage elongation increases at a slower rate. In stage two elongation increases at an almost constant rate; in stage three elongation increases at an increasing rate. Some of the other high-temperature effects are: (1) notch impact resistance improves up to 150 to 200°F; (2) corrosion resistance of steels increases for temperatures up to 1000°F.

1.3.7 Residual Stresses

The rolled sections and fabricated member are known to have locked-in stresses which are called residual stresses. The magnitude of these stresses can be determined by removing a longitudinal section and measuring resulting strain. Usually longitudinal stress is measured.

Residual stress results from unequal cooling rates during cooling of hot-rolled structural members. In a wide-flange section, the center of the flange cools more slowly and develops tensile residual stress that is balanced by compressive stress elsewhere on the cross section (Fig. 1.30c). For welded sections residual tensile stresses develop near the weld, and compressive stresses elsewhere form the equilibrium (see Fig. 1.30d). For a welded I-shaped member the stress condition in the edge of the flanges before welding is shown in Fig. 1.30a.

In case of sheared edge plates, residual stresses at the edges vary through the plate thickness. Tensile stresses are present on one surface and compressive stresses on the opposite surface. Usually residual stress distribution is relatively constant along the length of the member. In some cases residual stress may occur at a particular location in a member, owing to localized plastic flow from fabrication operations like heat straightening or cold straightening. During application of loading to structural-steel members, residual stress causes some premature inelastic action. Yielding can occur in localized portions before the nominal stress reaches the yield point. The tension strength of the member is not usually affected by residual stress due to ductility of steel. However, excessive tensile residual stress in combination with other conditions can cause fracture. Buckling load will be decreased because of residual stresses from an ideal member. The influence of residual stress has been accounted for by AISC in general design criteria for compression members. In general for statically loaded members consideration of residual stresses is not necessary; the ultimate moment of compact members is not affected by residual stresses.

1.3.8 Effects of Principal Elements on Properties of Steel

Iron constitutes approximately 95 percent of the steel matrix. The other principal elements are listed below in alphabetical order.

History—Design Concepts and Material Behavior ■ 1-31

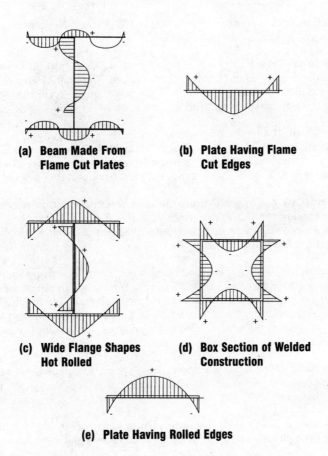

(a) Beam Made From Flame Cut Plates

(b) Plate Having Flame Cut Edges

(c) Wide Flange Shapes Hot Rolled

(d) Box Section of Welded Construction

(e) Plate Having Rolled Edges

Fig. 1.30 Residual stress distributions of typical conditions. *(Adapted from Brockenbrough and Merritt, Structural Steel Designer's Handbook, McGraw-Hill, 1994.)*

Aluminum (Al) ■ Aluminum is a good deoxidizer. It is used to obtain silicon-killed steel. Aluminum lowers the transition temperature and increases notch toughness; it also helps to form a fine-grained crystalline structure. If a large quantity of aluminum is added, finishing on plate becomes difficult.

Boron (B) ■ It is mainly used in quenched and tempered low-carbon constructional alloy steel. A trace of boron usually increases the strength of low-carbon steel.

Carbon (C) ■ Carbon is the one of the most important elements in steel. Increase of carbon content increases the yield strength and reduces the ductility and weldability. The increase of carbon by 0.01 percent increases yield strength by 0.5 ksi.

The maximum carbon content permitted is 0.30 percent. The general range of carbon content in steel is 0.15 to 0.3 percent.

Chromium (Cr) ■ It improves abrasion resistance, strength, and corrosion resistance. However, weldability of steel is reduced. Generally stainless steel has a large amount of chromium content, in the range of 8 percent. ASTM A588 steel usually has 0.10 to 0.90 percent chromium content.

Columbium (Cb) ■ Columbium enhances the yield strength, but the tensile strength increase is relatively small for carbon steel. The notch toughness in thick sections is substantially reduced. It improves corrosion resistance to some degree. Cb content in ASTM A572 steel varies between 0.005 and 0.15 percent.

Copper (Cu) ■ Copper is primarily used to improve atmospheric corrosion. Copper content typically varies from 0.20 to 0.35 percent. Copper is used as an anticorrosion component in steel grades A441 and A242. It is usually absorbed during the steelmaking process.

Hydrogen (H) ■ Hydrogen embrittles steel. Ductility can be improved with aging at room temperature. This will diffuse the hydrogen out of the steel. If the hydrogen content exceeds 0.0005 percent, internal cracks, bursts, and flaking will occur during cooling of the steel after the rolling. Normally in carbon steel flaking can be prevented by slow cooling after rolling. Slow cooling usually diffuses hydrogen out of steel.

Manganese (Mn) ■ It is used to increase strength, fatigue limit, notch toughness, and corrosion resistance. It usually lowers ductility. Manganese reduces the weldability of steel.

Molybdenum (Mo) ■ This has an effect similar to that of Mn. Molybdenum increases creep strength and can be useful for applications in high-temperature service.

Nickel (Ni) ■ It increases notch toughness, corrosion resistance, and strength and is used as a major element in stainless steel. However, ductility and weldability are reduced as its content increases.

Phosphorus (P) ■ It promotes internal segregation in the steel matrix. Phosphorus increases fatigue limit notch toughness and strength. However, it decreases weldability and ductility. It is usually limited to below 0.05 percent in steel.

Silicon (Si) ■ It increases notch toughness and strength and acts as a good deoxidizer. It reduces the weldability.

Sulfur (S) ■ Effects of sulfur are in many cases similar to those of phosphorus. Sulfur during steelmaking can form sulfide inclusions. This can lead to brittle failure by providing a stress riser from which fracture would initiate. Welding can cause hot cracking due to porosity of high-sulfur-content steel. Sulfur is also limited below 0.04 to 0.05 percent in steel.

Vanadium (V) ■ It increases creep strength and fracture toughness and helps to develop a crystalline structure. The amount of vanadium varies from 0.02 to 0.15 percent, depending upon the type of steel.

Weldability of Steel ■ For engineers it is necessary to determine whether the steel being used is weldable or not. The common method to determine this is called a carbon equivalent (CE) method.

$$CE = C + \frac{Mn + Si}{6} + \frac{Cr + Mo + V}{5} + \frac{Ni + Cu}{15} \leq 0.5 \qquad (1.3)$$

All elements are in percent of content. Carbon equivalent is related to the rate at which adjacent plate is cooled after welding without underbead cracks occurring. The higher the CE, the lower the allowable cooling rate. Preheating of steel and use of low-hydrogen welding electrodes can minimize the impact of the higher-carbon equivalent.

References

1. Young, Thomas, *A Course of Lectures of Natural Philosophy and Mechanical Arts*, vol. 2, London, 1807.

2. Culmann, K., *Die Graphische Statics*, Zurich, 1866.

3. Lord Rayleigh, "Theory of Elastic Surface Waves," *Proceedings London Mathematical Society*, vol. 17, 1887.

4. Bendixen, Axel, *Die Methode der Alpha-Glei Chungen zur Berechung von Rahmen Konstruktionen*, Berlin, 1914.

5. Cross, Hardy, "Moment Distribution, Procedure of Successive Approximations," *Transactions ASCE*, vol. 96, 1932.

6. Grüning, M., *Die Tragfähigkeit Statisch Unbestimmter Tragwerke*, Berlin, 1926.

7. American Institute of Steel Construction LFRD Specification and *Manual of Steel Construction*.

8. ASCE Standard 7-93, Recommended Minimum Loads to Be Used for Design.

Further Reading

Beedle, Lynn, *Plastic Design of Steel Frames*, Wiley, New York, 1958.

Brockenbrough, R., and F. Merritt, *Structural Steel Designers Handbook*, 2d. ed., McGraw-Hill, New York, 1994.

Ravindra, M. K., and T. V. Galambos, "Load and Resistance Factor Design for Steel," *ASCE Journal of the Structural Division*, vol. 104, no. ST109, pp. 1337–1353, Sept. 1978.

Geschwindener, Disque R., and R. Bjorhovde, *Load and Resistance Factor Design of Steel Structures*, Prentice-Hall, Englewood Cliffs, N.J., 1994.

Gaylord and Gaylord, *Structural Design Handbook*, McGraw-Hill, New York, 1996.

McCormac, Jack, *Structural Steel Design LRFD Method*, HarperCollins, New York, 1995.

Salmon, Charles, and John Johnson, *Steel Structures Design and Behavior*, 3d. ed., Harper & Row, New York, 1990.

2

Jack C. McCormac

DESIGN OF FLEXURAL MEMBERS*

2.1 Beam Behavior

2.1.1 Bending Moments

If gravity loads are applied to a fairly long, simply supported beam, the beam will bend downward, and its upper part will be placed in compression and will act as a compression member. The cross section of this "column" will consist of the portion of the beam cross section above the neutral axis. For the usual beam the "column" will have a much smaller moment of inertia about its y or vertical axis than about its x axis. If nothing is done to brace it perpendicular to the y axis, it will buckle laterally at a much smaller load than would otherwise have been required to produce a vertical failure. (You can verify these statements by trying to bend vertically a magazine held in a vertical position. The magazine will, just as a steel beam, always tend to buckle laterally unless it is braced in that direction.)

Lateral buckling will not occur if the compression flange of a member is braced laterally at frequent intervals. In this chapter the buckling moments of a series of compact ductile steel beams with different lateral bracing situations are considered. A *compact section* is one that has a sufficiently stocky profile so that it is capable of developing a fully plastic stress distribution before buckling.

---

*This chapter is an extension and adaptation of chapters in *Structural Steel Design-LRFD Method* by Jack McCormac, HarperCollins, New York, 1995.

In this chapter we will look at beams as follows:

1. First, the beams will be assumed to have continuous lateral bracing for their compression flanges.
2. Next, the beams will be assumed to be braced laterally at short intervals.
3. Finally, the beams will be assumed to be braced laterally at larger and larger intervals.

In Fig. 2.1 a typical curve showing the nominal resisting or buckling moments of one of these beams with varying unbraced lengths is presented.

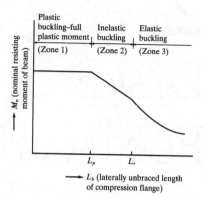

Fig. 2.1 Nominal moment as function of unbraced length of compression flange.

An examination of Fig. 2.1 will show that beams have three distinct ranges or zones of buckling depending on their lateral bracing situation. If we have continuous or closely spaced lateral bracing, the beams will buckle plastically and fall into what is classified as zone 1 buckling. As the distance between lateral bracing is increased further, the beams will begin to fail inelastically at smaller moments and fall into zone 2. Finally, with even larger unbraced lengths, the beams will fail elastically and fall into zone 3. A brief discussion of these three types of buckling is presented here, while the remainder of the chapter is devoted to a detailed discussion of each type, together with a series of numerical examples.

Plastic Buckling (Zone 1) ■ If we were to take a compact beam whose compression flange is continuously braced laterally, we would find that we could load it until its full plastic moment M_p is reached; further loading then produces a redistribution of moments. In other words, the moments in these beams can reach M_p and then develop a rotation capacity sufficient for moment redistribution.

If we now take one of these compact beams and provide closely spaced intermittent lateral bracing for its compression flanges, we will find that we can still load it until the plastic moment plus moment redistribution is achieved if the spacing between the bracing does not exceed a certain value called L_p here. (The value of L_p is dependent on the dimensions of the beam cross section and on its yield stress.) *Most beams fall in Zone 1.*

Inelastic Buckling (Zone 2) ■ If we now further increase the spacing between points of lateral bracing, the section may be loaded until some but not all of the compression fibers are stressed to F_y. The section will have insufficient rotation capacity to permit full moment redistribution and thus will not permit plastic analysis. In other words, in this zone we can bend the member until the yield strain is reached in some but not all of its compression elements before buckling occurs. This is referred to as *inelastic buckling*.

Design of Flexural Members ■ 2-3

As we increase the unbraced length we will find that the moment the section resists will decrease until finally it will buckle before the yield stress is anywhere reached. The maximum unbraced length at which we can still reach F_y at one point is the end of the inelastic range. It's shown as L_r in Fig. 2.1. Its value is dependent upon the properties of the beam cross section as well as on the yield and residual stresses of the beam. At this point, as soon as we have a moment which theoretically causes the yield stress to be reached anywhere (actually it's less than F_y because of residual stresses), the section will buckle.

Elastic Buckling (Zone 3) ■ If the unbraced length is greater than L_r, the section will buckle elastically before the yield stress is reached anywhere. As the unbraced length is further increased, the buckling moment becomes smaller and smaller. As the moment is increased in such a beam, the beam will deflect more and more transversely until a critical moment value M_{cr} is reached. At this time the beam cross section will twist and the compression flange will move laterally. The moment M_{cr} is provided by the torsional resistance and the warping resistance of the beam as is discussed in Section 2.1.4.

2.1.2 Plastic Buckling–Full Plastic Moment, Zone 1

Beam equations for plastic, inelastic, and elastic buckling are presented in the next few pages together with a discussion of each type and a series of numerical examples. After seeing some of the latter expressions the reader may become quite concerned that he or she is going to spend an enormous amount of time in formula substitution. This is not generally true, however, as the values needed are tabulated and graphed in simple form in the LRFD Manual.

When a steel section has a large shape factor, appreciable inelastic deformations may occur under service loads if the section is designed so that M_p is reached at the factored load condition. As a result LRFD Specification F1.1 limits the amount of such deformation for sections with shape factors larger than 1.5. This is done by limiting M_p to a maximum value of $1.5 M_y$.

If the unbraced length L_b of the compression flange of a compact I- or C-shaped section including hybrid members does not exceed L_p (if elastic analysis is being used) or L_{pd} (if plastic analysis is being used) then the member's bending strength about its major axis may be determined as follows:

$$M_n = M_p = F_y Z \leq 1.5 M_y$$
$$M_u = \phi_b M_n \quad \text{with} \quad \phi_b = 0.90$$
(LRFD Equation F1-1)

This part of the specification limiting M_n to $1.5 M_y$ for high shape factor sections such as WTs does not apply to hybrid sections with web yield stresses less than their flange yield stresses. Web yielding for such members does not result in significant inelastic deformations. For hybrid members the yield moment $M_y = F_{yf} S$.

For elastic analysis L_b may not exceed the value L_p to follow if M_n is to equal $F_y Z$.

$$L_p = \frac{300 r_y}{\sqrt{F_{yf}}}$$
(LRFD Equation F1-4)

For solid rectangular bars and box beams with A = cross-sectional area (in²) and J = torsional constant (in⁴)

$$L_p = \frac{3750 r_y}{M_p}\sqrt{JA}$$ (LRFD Equation F1-5)

For plastic analysis of doubly symmetric and singly symmetric I-shaped members with the compression flanges larger than their tension flanges (including hybrid members) loaded in the plane of the web L_b (which is defined as the laterally unbraced length of the compression flange at plastic hinge locations associated with failure mechanisms) may not exceed the value L_{pd} to follow if M_n is to equal $F_y Z$.

$$L_{pd} = \frac{3600 + 2200(M_1/M_2)}{F_y} r_y$$ (LRFD Equation F1-17)

In this expression M_1 is the smaller moment at the end of the unbraced length of the beam and M_2 is the larger moment at the end of the unbraced length and the ratio M_1/M_2 is positive when the moments cause the member to be bent in double curvature (∫) and negative if they bend it in single curvatures (∪). Only steels with F_y values (F_y is the specified minimum yield stress of the compression flange) of 65 ksi or less may be considered. Higher-strength steels may not be ductile.

There is no limit of the unbraced length for circular or square cross sections or for I-shaped beams bent about their minor axes. (If I-shaped sections are bent about their minor or y axes they will not buckle before the full plastic moment M_p about the y axis is developed.) Equation F1-18 of the LRFD specification also provides a value of L_{pd} for solid rectangular bars and symmetrical box beams.

For these sections to be compact the width-thickness ratios of the flanges and webs of I- and C-shaped sections are limited to the maximum values to follow, which are taken from Table B5.1 of the LRFD specification.

For flanges

$$\lambda_p = \frac{b}{t} \leq \frac{65}{\sqrt{F_y}}$$

For webs

$$\lambda_p = \frac{h}{t_w} \leq \frac{640}{\sqrt{F_y}}$$

In this last expression h is the distance from the web toe of the fillet in the top of the web to the web toe of the fillet in the bottom of the web (that is, twice the distance from the neutral axis to the inside face of the compression flange less the fillet or corner radius).

Design of Flexural Members ■ 2-5

2.1.3 Design of Beams, Zone 1

Included in the items that need to be considered in beam design are moments, shears, deflections, crippling, lateral bracing for the compression flanges, fatigue, and others. Beams will probably be selected that provide sufficient design moment capabilities ($\phi_b M_n$) and then checked to see if any of the other items are critical. The factored moment will be computed and a section having that much design moment capacity will be initially selected from the LRFD Manual.

A table is given in Section 4 of the LRFD Manual entitled "Load Factor Design Selection Table for Shapes Used as Beams." From this table steel shapes having sufficient plastic moduli to resist certain moments can quickly be selected. Two important items should be remembered in selecting shapes. These are:

1. These steel sections cost so many cents per pound and it is therefore desirable to select the lightest possible shape having the required plastic modulus (assuming that the resulting section is one that will reasonably fit into the structure). The table has the sections arranged in various groups having certain ranges of plastic moduli. The heavily typed section at the top of each group is the lightest section in that group and the others are arranged successively in the order of their plastic moduli. Normally the deeper sections will have the lightest weights giving the required plastic moduli, and they will be generally selected unless their depth causes a problem in obtaining the desired headroom, in which case a shallower but heavier section will be selected.

2. The plastic moduli values in the table are given about the horizontal axes for beams in their upright positions. If a beam is to be turned on its side the proper plastic modulus can be found in the tables giving dimensions and properties of shapes in Part 1 of the LRFD Manual. A W shape turned on its side may only be from 10 to 30 percent as strong as one in the upright position when subjected to gravity loads. In the same manner, the strength of a wood joist with the actual dimensions 2×10 in turned on its side would be only 20 percent as strong as in the upright position.

The examples to follow illustrate the design of compact steel beams whose compression flanges have full lateral support or bracing, thus permitting plastic analysis. For the selection of such sections the designer may enter the tables either with the required plastic modulus or with the factored design moment (if $F_y=36$ or 50 ksi).

Beam Weight Estimates ■ In each of the examples to follow, the weight of the beam is included in the calculation of the bending moment to be resisted, as the beam must support itself as well as the external loads. The estimates of beam weight are very close here because the author was able to perform a little preliminary paperwork before making his estimates. The reader is not expected to be able to glance at a problem and estimate exactly the weight of the beam required. A very simple method is available, however, with which beam weights can be quickly and accurately estimated. He or she can calculate the maximum factored bending moment, not counting the effect of the beam weight, and pick a section from the LRFD table. Then the weight of that section or a little bit more (since the beam's weight will increase the moment somewhat) can be used as the estimated beam weight. The results will almost always be very close to the weight of the member selected in the final design.

Example 2.1 ■ Select a beam section for the span and loading shown in Fig. 2.2 assuming full lateral support is provided for the compression flange by the floor slab above (that is, $L_b = 0$) and $F_y = 50$ ksi.

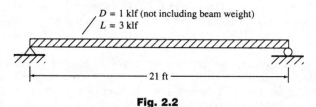

Fig. 2.2

Solution

Assume beam weight = 44 lb / ft

$$w_u = (1.2)(1.044) + (1.6)(3) = 6.05 \text{ klf}$$

$$M_u = \frac{(6.05)(21)^2}{8} = 333.5 \text{ ft-k}$$

$$Z_x \text{ required} = \frac{M_u}{\phi_b F_y} = \frac{(12)(333.5)}{(0.9)(50)} = 88.9 \text{ in}^3$$

Use W21×44

Example 2.2 ■ The 5-in reinforced-concrete slab shown in Fig. 2.3 is to be supported with steel W sections 8 ft 0 in on centers. The beams, which will span 20 ft, are assumed to be simply supported. If the concrete slab is designed to support a live load of 100 psf, determine the lightest steel section required to support the slab. It is assumed that the compression flange of the beam will be fully supported laterally by the concrete slab. The concrete weighs 150 lb/ft³. $F_y = 50$ ksi.

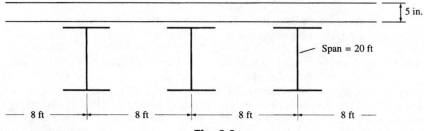

Fig. 2.3

Solution

$$\text{Dead loads: slab} = (\tfrac{5}{12})(150)(8) = 500 \text{ lb/ft}$$
$$\text{Estimated beam wt} = \underline{22}$$
$$\text{Total} = 522 \text{ lb/ft}$$
$$w_u = (1.2)(522) + (1.6)(8 \times 100) = 1906 \text{ lb/ft} = 1.906 \text{ klf}$$
$$M_u = \frac{(1.906)(20)^2}{8} = 95.3 \text{ ft-k}$$
$$Z_x \text{ required} = \frac{(12)(95.3)}{(0.9)(50)} = 25.4 \text{ in}^3$$
$$\underline{\text{Use W10} \times 22}$$

Holes in Beams ■ It is often necessary to have holes in steel beams. They are obviously required for the installation of bolts and sometimes for pipes, conduits, ducts, etc. If at all possible the latter type of hole should be completely avoided. When absolutely necessary they should be placed through the web if the shear is small and through the flange if the moment is small. Cutting a hole through the web of a beam does not reduce its section modulus greatly or its resisting moment; but a large hole in the web tremendously reduces the shearing strength of a steel section. When large holes are put in beam webs, extra plates are sometimes connected to the webs around the holes to serve as reinforcing against possible web buckling.

When large holes are placed in beam webs the strength limit states of the beams such as local buckling of the compression flange, the web, or the tee-shaped compression zone above or below the opening—or the moment-shear interaction or serviceability limit states may control the size of the member. A general procedure for estimating these effects and the design of any required reinforcement is available for both steel and composite beams.[1,2]

It is interesting to note that flexure tests of steel beams seem to show that their failure is based on the strength of the compression flange even though there may be bolt holes in the tension flange. The presence of these holes does not seem to be as serious as might be thought, particularly as compared with holes in a pure tension member. These tests show little difference in the strengths of beams with no holes and in beams with an appreciable number of bolt holes in either flange.

The LRFD specification does not require a reduction in flange area for bolts as long as the following expression is satisfied:

$$0.75 F_u A_{fn} \geq 0.9 F_y A_{fg} \qquad \text{(LRFD Equation B10-1)}$$

In this equation A_{fn} is the net flange area and A_{fg} is the gross flange area. Substituting into this expression we find that no deduction is necessary if the net flange area is equal to or greater than 74 percent of the gross flange area for A36 steel or 92 percent for A572 grade 50 steel. These values are shown in Table 2.1.

2-8 ■ Chapter Two

Table 2.1 When to Ignore Bolt Holes in Beam and Girder Flanges

Steel	No reduction if $A_{fn}/A_{fg} \geq$
A36 (F_u=58 ksi)	0.74
A572 grade 50 (F_u=65 ksi)	0.92
A588 (F_y=50 ksi, F_u=70 ksi)	0.86

Should $0.75F_u A_{fn}$ be less than $0.9F_y A_{fg}$, the LRFD specification requires that the flexural properties of the section be based on an effective tension flange area A_{fe}, determined as follows:

$$A_{fe} = \frac{5}{6}\frac{F_u}{F_y} A_{fn}$$ (LRFD Equation B10-3)

Bolt holes in the webs of beams are generally considered to be insignificant as they have almost no effect on Z calculations.

Some specifications, notably the bridge ones, and some designers have not adopted the idea of ignoring the presence of all or part of the holes in tension flanges. As a result they follow the more conservative practice of deducting 100 percent of all holes. For such a case the reduction in Z_x will equal the statical moment of the holes (in both flanges) taken about the neutral axis.

The usual practice is to assume that we have the same holes in both flanges whether or not they are actually present in the compression flange. (If we have bolt holes filled with bolts in the compression flange only, we forget the whole thing. This is because it is felt that the fasteners can adequately transmit compression through the holes by means of the bolts.) For a section with holes in the tension flange only, where a reduction in the tension flange area is required by the specification or is considered necessary, we make the same deduction from both flanges.

Should the application of LRFD Equation B10-1 show that $0.75F_u A_{fn} < 0.9F_y A_{fg}$ we will need to reduce Z_x. Its value will equal Z_x from the Manual tables minus the statical moment of $A_{fg} - A_{fe}$ for each flange taken about the neutral axis. The review of a beam with flange bolt holes is considered in Example 2.3.

Example 2.3 ■ Determine M_u for the fully braced A36 W24 × 176 shown in Fig. 2.4 for the following situations:

a. Using the LRFD specification and assuming two lines of 1-in bolts in each flange.

b. Using the LRFD specification and assuming four lines of 1-in bolts in each flange.

c. Same as (*b*) above, but conservatively assuming we must deduct all holes for computing beam properties.

Design of Flexural Members 2-9

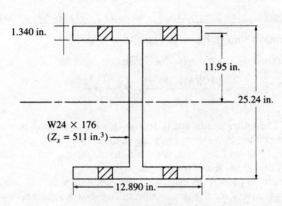

Fig. 2.4

Solution

a. With two holes in each flange (LRFD specification)

$$A_{fg} = (12.890)(1.340) = 17.27 \text{ in}^2$$
$$A_{fn} = 17.27 - (2)(1\tfrac{1}{8})(1.340) = 14.25 \text{ in}^2$$

For no reduction $0.75 F_u A_{fn} \geq 0.9 F_y A_{fg}$

$(0.75)(58)(14.25) = 619.9 \text{ k} > (0.9)(36)(17.27) = 559.5 \text{ k}$

∴ No reduction in flange area is required.

$$M_u = \phi M_m = \frac{(0.9)(36)(511)}{12} = \underline{\underline{1379.7 \text{ ft-k}}}$$

b. With four holes in each flange (LRFD specification)

$$A_{fg} = 17.27 \text{ in}^2$$
$$A_{fm} = 17.27 - (4)(1\tfrac{1}{8})(1.340) = 11.24 \text{ in}^2$$

For no reduction $0.75 F_u A_{fn} \geq 0.9 F_y A_{fg}$

$(0.75)(58)(11.24) = 488.9 \text{ k} < (0.9)(36)(17.27) = 559.5 \text{ k}$

∴ Flange area must be reduced

$$A_{fe} = \left(\frac{5}{6}\right)\left(\frac{58}{36}\right)(11.24) = 15.09 \text{ in}^2$$

Reduced $Z_x = 511 - (17.27 - 15.09)(11.95)(2) = 458.9 \text{ in}^3$

$$M_u = \phi M_n = \frac{(0.9)(36)(458.9)}{12} = \underline{\underline{1239.0 \text{ ft-k}}}$$

c. Reducing each flange area by four holes (not LRFD specification)

$$\text{Area of four holes} = (4)(1\tfrac{1}{8})(1.340) = 6.03 \text{ in}^2$$

$$\text{Reduced } Z_x = 511 - (6.03)(11.95)(2) = 366.9 \text{ in}^3$$

$$M_u = \phi M_n = \frac{(0.9)(36)(366.9)}{12} = \underline{\underline{990.6 \text{ ft-k}}}$$

These LRFD requirements apply to the design of hybrid beams and girders whose flanges consist of a stronger grade of steel than their webs. This is true as long as these members are not required to resist an axial force larger than $\phi_b(0.15 F_{yf})$ of the flanges times their gross areas.

Should a hole be present in only one side of a flange of a W section, there will be no axis of symmetry for the net section of the shape. A correct theoretical solution of the problem would be very complex. Rather than going through such a lengthy process over a fairly minor point, it seems logical to assume holes in both sides of the flange. The results obtained will probably be just as satisfactory as those obtained by a laborious theoretical method.

2.1.4 Introduction to Inelastic Buckling, Zone 2

If intermittent lateral bracing is supplied for the compression flange of a beam section such that the member can be bent until the yield strain is reached in some but not all of its compression elements before lateral buckling occurs, we have inelastic buckling. In other words the bracing is insufficient to permit the member to reach a full plastic strain distribution before buckling occurs.

Because of the presence of residual stresses, yielding will begin in a section at applied stresses equal to $F_y - F_r$ where F_y is the yield stress of the web and F_r equals the compressive residual stress assumed equal to 10 ksi for rolled shapes and 16.5 ksi for welded shapes. It should be noted that the definition of plastic moment $F_y Z$ in zone 1 is not affected by residual stresses because the sum of the compressive residual stresses equals the sum of the tensile residual stresses in the section and the net effect is theoretically zero.

If the unbraced length L_b of a compact I- or C-shaped section is larger than L_p, the beam will fail inelastically unless L_b is greater than a distance L_r (to be discussed) beyond which the beam will fail elastically before F_y is reached (thus falling into zone 3).

Bending Coefficients ■ In the formulas presented in these next few sections for inelastic and elastic buckling we will use a term C_b. This is a moment coefficient that is included in the formulas to account for the effect of different moment gradients on lateral-torsional buckling. In other words, lateral buckling may be appreciably affected by the end restraint and loading conditions of the member.

As an illustration the reader can see that the moment in the laterally unbraced beam of part *a* of Fig. 2.5 causes a worse compression flange situation than does the moment in the laterally unbraced beam of part *b*. For one reason the upper

Design of Flexural Members ■ 2-11

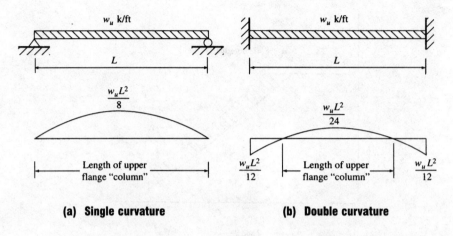

Fig. 2.5

flange of the beam in part *a* is in compression for its entire length, while in *b* the length of the "column," that is, the length of the upper flange that is in compression, is much less (thus a much shorter "column").

For the simply supported beam of part *a* of the figure, C_b is taken as 1.0, while for the beam of part *b* it is taken as larger than 1.0. The basic moment capacity equations for zones 2 and 3 were developed for laterally unbraced beams subject to single curvature with $C_b = 1.0$. Frequently, beams are not bent in single curvature with the result that they can resist more moment. We have seen this in Fig. 2.5. To handle this situation the LRFD specification provides moment or C_b coefficients larger than 1.0 which are to be multiplied by the computed M_n values. The results are higher moment capacities. The designer who conservatively says "I'll always use $C_b = 1.0$" is missing out on the possibility of significant savings in steel weight for some situations. *When using C_b values the designer should clearly understand that the moment capacity obtained by multiplying M_n by C_b may not be larger than the plastic M_n of zone I which is equal to F_yZ.* This situation is illustrated in Fig. 2.6.

The value of C_b is determined from the expression to follow in which M_{max} is the largest moment in an unbraced segment of a beam while M_A, M_B, and M_C are, respectively, the moments at the ¼ point, ½ point, and ¾ point in the segment.

$$C_b = \frac{12.5 M_{max}}{2.5 M_{max} + 3 M_A + 4 M_B + 3 M_C} \qquad \text{(LRFD Equation F1-3)}$$

C_b is equal to 1.0 for cantilevers or overhangs where the free end is unbraced. Some typical values of C_b calculated with the above equation are shown in Fig. 2.7 for various beam and moment situations.

2-12 ■ Chapter Two

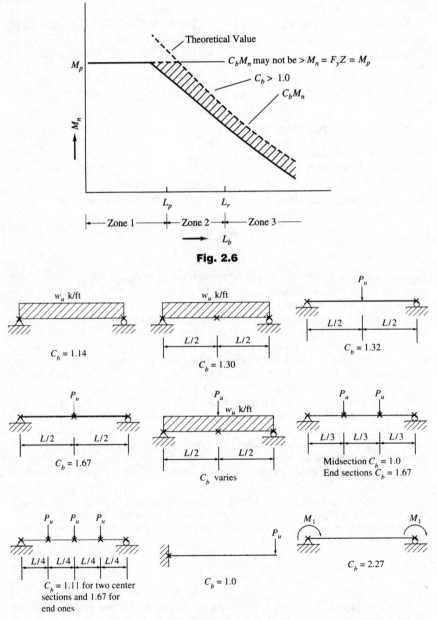

Fig. 2.6

Fig. 2.7 Sample C_b values.

2.1.5 Moment Capacities, Zone 2

As the unbraced length of the compression flange of a beam is increased beyond L_p the moment capacity of the section will become smaller and smaller. Finally, at an unbraced length L_r, the section will buckle elastically as soon as the yield stress is reached. Owing to the rolling operation, however, there is a residual stress in the section equal to F_r. Thus the elastically computed stress caused by bending can only reach $F_{yw} - F_r$. Assuming $C_b = 1.0$ the design moment capacity for a compact I- or C-shaped section bent about its x axis may be determined as follows if $L_b = L_r$:

$$M_u = \phi_b M_r = \phi_b S_x (F_{yw} - F_r)$$

L_r is a function of several of the section's properties such as its cross-sectional area, modulus of elasticity, yield stress, and warping and torsional properties. The very complex formulas needed for its computation are given in the LRFD specification (F_1) and space is not taken to show them here. Fortunately, numerical values have been determined for sections normally used as beams and are given in the Load Factor Design Selection Table.

Going backward from an unbraced length of L_r toward an unbraced length L_p we can see that buckling does not occur when the yield stress is first reached. We are in the inelastic range (zone 2) where there is some penetration of the yield stress into the section from the extreme fibers. For these cases when the unbraced length falls between L_p and L_r the moment capacity will fall approximately on a straight line between $M_u = \phi_b F_y Z$ at L_p and $\phi_b S_x (F_{yw} - F_r)$ at L_r. For intermediate values of the unbraced length the moment capacity may be determined by proportions or by substituting into the expression at the end of this paragraph. If C_b is larger than 1.0, the section will resist additional moment but not more than $\phi_b F_y Z = \phi_b M_p$.

$$\phi_b M_n = C_b \left[\phi_b M_p - BF(L_b - L_p) \right] \leq \phi_b M_p$$

in which BF is a factor given in the Load Factor Design Selection Table for each section, which enables us to do the proportioning with a simple formula.

Alternately, the value of M_n can be determined from the equation to follow and multiplied by ϕ_b to obtain M_u.

$$M_n = C_b \left[M_p - (M_p - M_r) \left(\frac{L_b - L_p}{L_r - L_p} \right) \right] \leq M_p \qquad \text{(LRFD Equation F1-2)}$$

Example 2.4 illustrates the determination of the moment capacities of a section with L_b between L_p and L_r, while Example 2.5 demonstrates the design of a beam in the same range.

Example 2.4 ■ Determine the moment capacity of a W24 × 62 with $F_y = 36$ ksi and again with $F_y = 50$ ksi if $L_b = 8.0$ ft and $C_b = 1.0$.

Solution if $F_y = 36$ ksi. From the Load Factor Design Selection Table for a W24 × 62

$$L_p = 5.8 \text{ ft}$$
$$L_r = 17.2 \text{ ft}$$
$$\phi_b M_r = 255 \text{ ft-k}$$
$$\phi_b M_p = 413 \text{ ft-k}$$
$$BF = 13.8 \text{ k}$$

Since $L_b > L_p < L_r$ the section is in zone 2 for inelastic buckling and $\phi_b M_n$ can be determined as follows:

$$\phi_b M_n = C_b [\phi_b M_p - BF(L_b - L_p)]$$
$$= 1.0[413 - (13.8)(8.0 - 5.8)] = \underline{382.6 \text{ ft-k}}$$

Or directly by proportions

$$\phi_b M_n = 255 + \left(\frac{17.2 - 8.0}{17.2 - 5.8}\right)(413 - 255) = 382.5 \text{ ft-k}$$

Solution if $F_y = 50$ ksi. From the Load Factor Design Selection Table

$$L_p = 4.9 \text{ ft}$$
$$L_r = 13.3 \text{ ft}$$
$$\phi_b M_r = 393 \text{ ft-k}$$
$$\phi_b M_p = 574 \text{ ft-k}$$
$$BF = 21.4 \text{ k}$$

Since $L_b > L_p < L_r$

$$\phi_b M_n = 1.0[574 - (21.4)(8.0 - 4.9)] = 507.7 \text{ ft-k}$$

Example 2.5 ■ Select the lightest available section for a factored moment of 290 ft-k if $L_b = 10.0$ ft. Use A36 steel and assume $C_b = 1.0$.

Solution. Enter the Load Factor Design Selection Table and notice that $\phi_b M_r$ for a W21 × 50 is 297 ft-k but L_p is 5.4 ft < L_b of 10.0 ft. Also $\phi_b M_r = 184$ ft-k and $BF = 10.5$ k.

$$\therefore \phi_b M_n = 1.0[297 - (10.5)(10.0 - 5.4)]$$
$$= 248.7 \text{ ft-k} < 290 \text{ ft-k} \quad \text{(NG)}$$

Moving up in the Load Factor Design Selection Table try a W24 × 55 (with L_p = 5.6 ft, L_r = 16.6 ft, $\phi_b M_r$ = 222 ft-k, $\phi_b M_p$ = 362 ft-k, and BF = 12.7 k).

$$\phi_b M_n = 1.0[362 - (12.7)(10.0 - 5.6)]$$
$$= 306.1 \text{ ft-k} > 290 \text{ ft-k}$$
$$\underline{\text{Use W24} \times 55} \quad \text{(OK)}$$

Note: A much easier solution is presented in Sec. 2.1.7.

2.1.6 Elastic Buckling, Zone 3

When a beam is not fully braced laterally, it may fail due to buckling laterally about the weaker axis between the points of lateral bracing. This will occur even though the beam is loaded so that it supposedly will bend about the stronger axis. The beam will bend initially about the stronger axis until a certain critical moment M_{cr} is reached. At that time it will buckle laterally about its weaker axis. As it bends laterally the tension in the other flange will try to keep the beam straight. As a result the buckling of the beam will be a combination of lateral bending and a twisting (or torsion) of the beam cross section. A sketch of this situation is shown in Fig. 2.8.

The critical moment or flexural-torsional moment M_{cr} in a beam will be made up of the torsional resistance (commonly called *St. Venant torsion*) and the warping resistance of the section. These are combined as follows:

$$M_{cr} = \sqrt{\left(\begin{array}{c}\text{torsional}\\\text{resistance}\end{array}\right)^2 + \left(\begin{array}{c}\text{warping}\\\text{resistance}\end{array}\right)^2}$$

Returning to the LRFD specification, if the unbraced length of the compression flange of a beam section is greater than L_r the section will buckle elastically

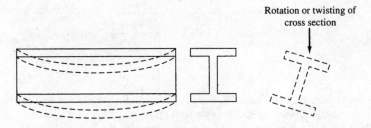

Fig. 2.8 Lateral torsional buckling of a simply supported beam.

2-16 ■ Chapter Two

before the yield stress is reached anywhere in the section. In Sec. F1.1.2b of the LRFD specification the classic equation for determining this flexural-torsional buckling moment called M_{cr} is presented. This expression follows:

$$M_{cr} = C_b \frac{\pi}{L_b} \sqrt{EI_y GJ + \left(\frac{\pi E}{L_b}\right)^2 I_y C_w} \qquad \text{(LRFD Equation F1-13)}$$

In this equation G is the shear modulus of elasticity of the steel = 11,200 ksi, J is a torsional constant (in^4), while C_w is the warping constant (in^6). The values of J and C_w are provided for rolled sections in the tables entitled "Torsion Properties" in Part 1 of the LRFD Manual.

This expression is applicable to compact doubly symmetric I-shaped members, channel sections loaded in the plane of their webs, and I-shaped singly symmetric sections with their compression flanges larger than their tension ones ($\top$). Expressions also are given in Sections F1.1.2b and F1.1.2c of the LRFD specification for M_{cr} in the elastic range for other sections such as solid rectangular bars, symmetric box sections, tees, and double angles.

It is not possible for lateral-torsional bucking to occur if the moment of inertia of the section about the bending axis is equal to or less than the moment of inertia out of plane. As a result the limit state of lateral-torsional buckling is not applicable for shapes bent about their minor axes, for shapes with $I_x \leq I_y$, or for circular or square shapes. Furthermore, yielding controls if the section is noncompact.

Example 2.6 illustrates the computation of $\phi_b M_{cr}$ for an elastic buckling situation.

Example 2.6 ■ Compute $M_u = \phi_b M_{cr}$ for a W18 × 97 consisting of A36 steel if the unbraced length L_b is 44 ft. Assume $C_b = 1.0$.

Solution. Noting $L_b = 44$ ft > $L_r = 38.1$ ft from Load Factor Design Selection Table (Part 4, LRFD Manual).

The following values for the W 18 × 97 are also obtained from the Manual: $I_y = 201$ in^4, $J = 5.86$ in^4, and $C_w = 15{,}800$ in^6.

$$\phi_b M_n = \phi M_{cr} = M_u = (0.9)(1.0)\left(\frac{\pi}{12 \times 44}\right)$$

$$\times \sqrt{(29 \times 10^3)(201)(11{,}200)(5.86) + \left(\frac{\pi 29 \times 10^3}{44 \times 12}\right)^2 (201)(15{,}800)}$$

$$= 3698.9 \text{ in-k} = 308.2 \text{ ft-k}$$

(This value may be checked with the LRFD Manual charts described in the next few paragraphs. There we will be able to read $\phi M_n = M_u = 308$ ft-k.)

The LRFD Specification (F1.2b) also presents the elastic buckling equation in an alternate form as follows:

Design of Flexural Members ■ 2-17

$$M_{cr} = \frac{C_b S_x X_1 \sqrt{2}}{L_b/r_y}\sqrt{1 + \frac{X_1^2 X_2}{2(L_b/r_y)^2}}$$ (LRFD Equation F1-13 Alternate Form)

In which

$$X_1 = \frac{\pi}{S_x}\sqrt{\frac{EGJA}{2}}$$ (LRFD Equation F1-8)

$$X_2 = 4\frac{C_w}{I_y}\left(\frac{S_x}{GJ}\right)^2$$ (LRFD Equation F1-9)

Values of X_1 and X_2 are shown for W shapes in the "Properties for W Shapes section of Part 1 of the Manual.

2.1.7 Design Charts

Fortunately the values of $\phi_b M_{cr} = \phi_b M_n$ for the sections normally used as beams are computed for different unbraced lengths and the results plotted as curves in Part 4 of the LRFD Manual. The values not only cover unbraced lengths in the elastic range but also the inelastic range, enabling us to very easily handle the problems falling in zones 1 and 2 as well as the ones in this section which fall in zone 3. The moments are plotted for F_y values of 36 ksi and 50 ksi and for $C_b = 1.0$.

The curve for a typical W section is shown in Fig. 2.9. For each of the shapes L_p is indicated with a solid circle (●) while L_r is shown with a hollow circle (○).

The charts were developed without regard to such things as shear, deflection, etc.—items that may occasionally control the design. The curves extend to unbraced lengths equal to 30 times the section depths. As such they cover almost

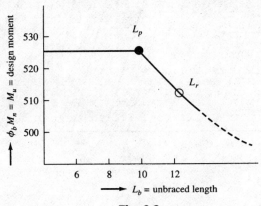

Fig. 2.9

2-18 ■ Chapter Two

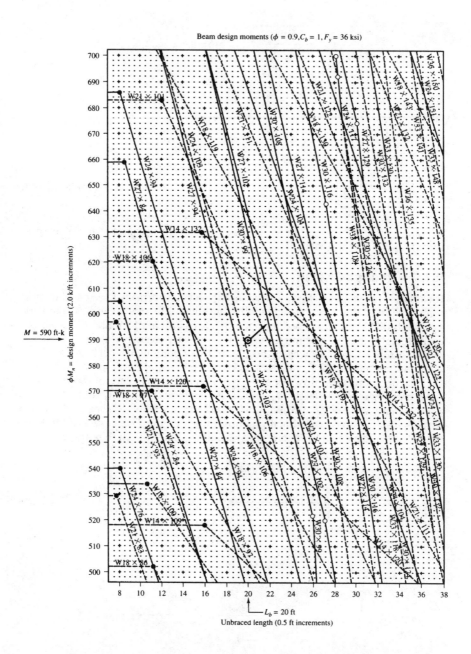

Fig. 2.10 Design moments for beams with different unbraced lengths.

Design of Flexural Members ■ 2-19

all of the unbraced lengths encountered in practice. If C_b is greater than 1.0 the values given will be magnified somewhat, as illustrated in Fig. 2.6.

To select a member it is only necessary to enter the chart with the unbraced length L_b and the factored design moment M_u. For an illustration let's assume F_y = 36 ksi and assume that we wish to select a beam with L_b = 20.0 ft for a moment M_u = 590 ft-k. We enter the charts in Part 4 entitled "Beam Design Moments" and go through the pages where F_y = 36 ksi until we find in the left-hand column $\phi_b M_n$ equal to 590. We proceed up from the bottom of the chart for an unbraced length = 20.0 ft until we intersect a horizontal line from the 590. Any section to the right and above this intersection point (✗) will have a greater unbraced length and a greater moment capacity. The appropriate page from the handbook is shown in Fig. 2.10 with the permission of the AISC.

Moving up and to the right we first encounter a W21 × 101. In this area of the charts this section is shown with a dashed line. This section will provide the necessary moment capacity but the dashed line indicates that it is in an uneconomical range. If we proceed farther upward and to the right, the first solid line that we encounter will represent the lightest available section. In this case it's a W30 × 99. Other illustrations of the use of these charts are presented in Examples 2.7 and 2.8.

Example 2.7 ■ Using A36 steel select the lightest available section for the beam of Fig. 2.11 which has lateral bracing provided for its compression flange only at its ends. Assume C_b = 1.0.

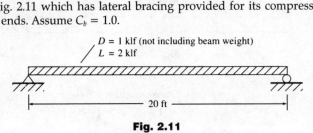

Fig. 2.11

Solution

$$\text{Assume beam weight} = 60 \text{ lb/ft}$$
$$w_u = (1.2)(1.060) + (1.6)(2) = 4.472 \text{ klf}$$
$$M_u = \frac{(4.472)(20)^2}{8} = 223.6 \text{ft-k}$$
$$\text{Noting } C_b = 1.0$$

Entering the beam design moments charts with L_b = 20 ft and M_u = 223.6 ft-k.

Use W18 × 60

For the example problem that follows C_b is greater than 1.0. For such a situation the reader should look back to Fig. 2.6. There he or she will see that the design moment strength of a section can go to $\phi_b C_b M_n$ when $C_b > 1.0$ but may under no circumstances exceed $\phi_b M_p = \phi_b F_y Z$.

2-20 ■ Chapter Two

To handle such a problem we calculate an effective moment as follows (the numbers being taken from Example 2.8):

$$M_{\text{effective}} = \frac{M_u}{C_b} = \frac{868.7}{1.67} = 520.2 \text{ ft-k}$$

Then we enter the charts with our unbraced length of 17 ft and with $M_{\text{effective}} = 520.2$ ft-k and there select a W27 × 84. We must, however, check to see that our M_u (868.7) does not exceed $\phi_b F_y Z$ for the section. In this case it does and we must keep going until we find the lightest section that has a $\phi_b M_n \geq 520.2$ ft-k at $L_b = 17$ ft and yet which has a $\phi_b F_y Z \geq 868.7$ ft-k.

Example 2.8 ■ Using A36 steel select the lightest available section for the situation shown in Fig. 2.12. Bracing is provided only at the ends and centerline of the members and thus $C_b = 1.67$.

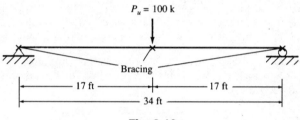

Fig. 2.12

Solution

Assume beam weight = 108 lb / ft

$$w_u = (1.2)(0.108) = 0.1296 \text{ klf}$$

$$M_u = \frac{(100)(34)}{4} + \frac{(0.1296)(34)^2}{8} = 868.7 \text{ ft-k}$$

Entering the Beam Design Moments charts with $M_{u \text{ effective}} = 868.7/1.67 = 520.2$ ft-k and $L_b = 17$ ft we find a W27 × 84 will do.

But $\phi_b M_n = 868.7$ ft-k may not exceed $\phi_b F_y Z$ of the section, and it does for a W27 × 84, as can be seen in the Load Factor Design Selection Table. Therefore, we continue looking in the table until we find the lightest section that has a $\phi_b F_y Z \geq 868.7$ ft-k.

Use W30 × 108

2.1.8 Design of Continuous Beams

Section A5 of the LRFD specification permits the design of beams analyzed either on the basis of elastic analysis with factored loads or by plastic analysis with the same factored loads. Design based on plastic analysis is permitted only for sections

Design of Flexural Members ■ 2-21

with yield stresses no greater than 65 ksi and is subject to some special requirements in Sections B5.2, C2, E1.2, F1.2d, H1, and I1 of the LRFD specification.

Both theory and tests show clearly that continuous ductile steel members meeting the requirements for compact sections with sufficient lateral bracing supplied for their compression flanges have the desirable ability of being able to redistribute moments caused by overloads. If plastic analysis is used, this advantage is automatically included in the analysis.

If elastic analysis is used, the LRFD handles the redistribution by a rule of thumb that approximates the real plastic behavior. The LRFD specification (A5) states that for continuous compact sections the design may be made on the basis of nine-tenths of the maximum negative moments caused by gravity loads which are maximum at points of support if the positive moments are increased by one-tenth of the average negative moments at the adjacent supports. *(The 0.9 factor is applicable only to gravity loads and not to lateral loads such as those caused by wind and earthquake. The factor can also be applied to columns that have axial stresses less than $0.15F_y$.)* This moment reduction does not apply to members consisting of A514 steel, hybrid girders, or to moments produced by loading on cantilevers.

Example 2.9 illustrates the design of a three-span continuous beam analyzed by (*a*) plastic analysis and (*b*) elastic analysis.

Example 2.9 ■ The beam shown in Fig. 2.13 is assumed to consist of A36 steel. (*a*) Select the lightest W section available using plastic analysis and assuming full lateral support is provided for its compression flanges. (*b*) Design the beam using elastic analysis with the factored loads and the 0.9 rule and assuming full lateral support is provided for flanges. (*c*) Design the beam using elastic analysis with the factored loads and the 0.9 rule assuming full lateral support is provided for the upper flange *but* only for the lower flange at support points.

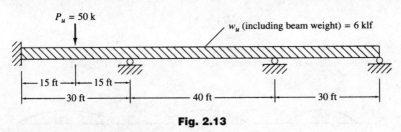

Fig. 2.13

Solution

a. Plastic analysis and design

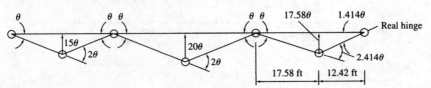

2-22 ■ Chapter Two

Span 1
$$\begin{cases} M_u 4\theta = (30w_u)\left(\frac{1}{2}\right)(15\theta) + (P_u)(15\theta) \\ M_u = 56.25w_u + 3.75P_u \\ M_u = (56.25)(6.0) + (3.75)(50) \\ M_u = 525 \text{ ft-k} \end{cases}$$

Span 2
$$\begin{cases} M_u 4\theta = (40w_u)\left(\frac{1}{2}\right)(20\theta) \\ M_u = 100w_u = (100)(6.0) \\ M_u = 600 \text{ ft-k} \leftarrow \end{cases}$$

Span 3
$$\begin{cases} M_u(3.414\theta) = (30w_u)\left(\frac{1}{2}\right)(17.58\theta) \\ M_u = 77.24w_u = (77.24)(6.0) \\ \qquad = 463.4 \text{ ft-k} \end{cases}$$

$$Z \text{ required} = \frac{M_u}{\phi_b F_y} = \frac{(12)(600)}{(0.9)(36)} = 222.2 \text{ in}^3$$

<u>Use W24 × 84</u>

<u>*b.* Elastic analysis and design—full lateral bracing both flanges</u>

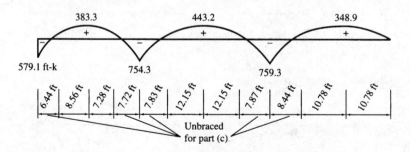

Fig. 2.14

Design of Flexural Members ■ 2-23

Maximum negative moment for design
$$= (0.9)(759.3) = -683.4 \text{ ft-k} \leftarrow$$
Maximum positive moment for design
$$= 443.2 + \frac{1}{10}\left(\frac{754.3 + 759.3}{2}\right) = +518.9 \text{ ft-k}$$
$$Z \text{ required} = \frac{(12)(683.4)}{(0.9)(36)} = 253.1 \text{ in}^3$$

Use W24 × 94

c. Elastic design—full lateral support supplied for top flange but only at support points for bottom flange

Referring to the moment diagram for the part *b* solution (Fig. 2.14) we see that the greatest unbraced length in the negative moment region is 8.44 ft and the reduced M_u is 683.4 ft-k.
From the Load Factor Design Selection Table with M_u = 683.4 ft-k we pick a W30 × 90. Its L_p = 8.7 ft > 8.44 ft.

Use W30 × 90

2.1.9 Shear

Generally, shear is not a problem in steel beams because the webs of rolled shapes are capable of resisting rather large shearing forces. Perhaps it is well, however, to list the most common situations where shear might he excessive. These are as follows.

1. Should large concentrated loads be placed near beam supports they will cause large external shears without corresponding increases in bending moments. A fairly common example of this type occurs in tall buildings where on a particular floor the upper columns are offset with respect to the columns below. The loads from the upper columns applied to the beams on the floor level in question will be quite large if there are many stories above.

2. Probably the most common shear problem occurs where two members (as a beam and a column) are rigidly connected together so their webs lie in a common plane. This situation frequently occurs at the junction of columns and beams (or rafters) in rigid frame structures.

3. Where beams are notched or coped as shown in Fig. 2.15, shear can be a problem. For this case shear forces must be calculated for the remain-

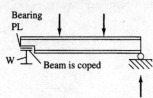

Fig. 2.15

ing beam depth. A similar discussion can be made where holes are cut in beam webs for ductwork or other items.
4. Theoretically, very heavily loaded short beams can have excessive shears, but practically this does not occur too often unless it is like case 1.
5. Shear may very well be a problem even for ordinary loadings when very thin webs are used as in plate girders or in light-gauge cold-formed steel members.

Shear strength expressions are given in LRFD Specification F2. In these expressions which follow, F_{yw} is the specified minimum yield stress of the web; h is the clear distance between the web toes of the fillets for rolled shapes, while for built-up welded sections it's the clear distance between flanges. For bolted built-up sections h is the distance between adjacent lines of bolts in the web. Different expressions are given for different h/t_w ratios depending on whether shear failures would be plastic, inelastic, or elastic.

1. Web yielding. Almost all rolled beam sections in the Manual fall into this classification. If $h/t_w \leq 418/\sqrt{F_{yw}} = 70$ for $F_y = 36$ ksi and 59 for $F_y = 50$ ksi

$$V_n = 0.6 F_{yw} A_w \qquad \text{(LRFD Equation F2-1)}$$

2. Inelastic buckling of web. If $418/\sqrt{F_{yw}} < h/t_w \leq 523/\sqrt{F_{yw}} = 87$ for $F_y = 36$ ksi and 74 for $F_y = 50$ ksi

$$V_n = 0.6 F_{yw} A_w (418/\sqrt{F_{yw}})/(h/t_w) \qquad \text{(LRFD Equation F2-2)}$$

3. Elastic buckling of web. If $523/\sqrt{F_{yw}} < h/t_w \leq 260$

$$V_n = (132{,}000 A_w)/(h/t_w)^2 \qquad \text{(LRFD Equation F2-3)}$$

For each of the situations given $V_u = \phi_c V_v$ with $\phi_v = 0.90$. The LRFD Appendix F2.2 gives expressions for the general design shear strength of webs with or without stiffeners. Appendix G3 also provides an alternative method for plate girders that utilizes tension field action.

In Example 2.10 the shear strength of a beam is computed.

Example 2.10 ■ A W24 × 55 ($d = 23.57$ in, $t_w = 0.395$ in, k = distance from extreme flange fiber to web toe of fillet = $1^5/_{16}$ in) consisting of 50 ksi steel is used for the beam and loading of Fig. 2.16. Check its adequacy in shear.

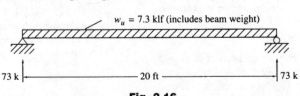

Fig. 2.16

Solution

$$h = 23.57 - (2)(1\tfrac{5}{16}) = 20.94 \text{ in}$$

$$\frac{h}{t_w} = \frac{20.94}{0.395} = 53.01 < \frac{418}{\sqrt{50}} = 59$$

$$\phi_v V_n = (0.90)(0.6)(F_{yw})(A_w)$$
$$= (0.90)(0.6)(50)(23.57 \times 0.395) = 251.4 \text{ k} > 73\text{k} \qquad \text{(OK)}$$

Note: In Part 4 of the LRFD Manual there is a large set of tables entitled "Beams W shapes maximum factored uniform load in kips for beams laterally supported." This title seems to the author to be quite inadequate because so much other information can be obtained from these tables, such as the design shear strengths or $\phi_v V_n$ values, and information concerning web yielding, web crippling, and other items to be discussed later.

Should V_u for a particular beam exceed the LRFD specified shear strength of the member, the usual procedure will be to select a little heavier section. If it is necessary, however, to use a much heavier section than required for moment, doubler plates (Fig. 2.17) may be welded to the beam web, or stiffeners may be connected to the webs in zones of high shear. Doubler plates must meet the width-thickness requirements for compact stiffened elements as prescribed in Section B.5 of the LRFD specification. In addition, they must be welded sufficiently to the member webs to develop their proportionate share of the load.

The LRFD specified shear strength of a beam or girder is based on the entire area of the web. Sometimes, however, a connection is made to only a small portion or depth of the web. For such a case the designer may decide to assume that the shear is spread over only part of the web depth for purposes of computing shear strength. Thus he or she may compute A_w as being equal to t_w times the smaller depth for use in the shear strength expression.

When beams that have their top flanges at the same elevations (the usual situation) are connected to each other, it is frequently necessary to cope one of them, as shown in Fig. 2.18. For such cases there is a distinct possibility of a block shear failure along the broken lines shown.

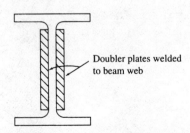

Fig. 2.17 Increasing shear strength of beam by using doubler plates.

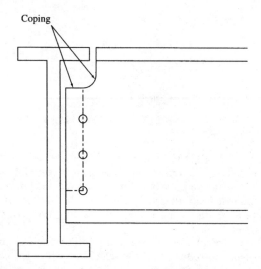

Fig. 2.18 Block shear failure possible along dashed line.

2.1.10 Unsymmetrical Bending

From mechanics of materials it is remembered that each beam cross section has a pair of mutually perpendicular axes known as the principal axes for which the product of inertia is zero. Bending that occurs about any axis other than one of the principal axes is said to be unsymmetrical bending. When the external loads are not in a plane with either of the principal axes or when loads are simultaneously applied to the beam from two or more directions, unsymmetrical bending is the result.

If a load is not perpendicular to one of the principal axes it may be broken into components which are perpendicular to those axes and the moments about each axis, M_{ux} and M_{uy}, determined as shown in Fig. 2.19.

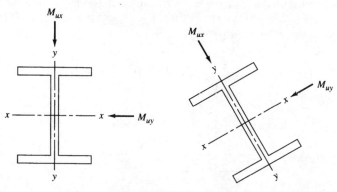

Fig. 2.19

When a section has one axis of symmetry, that axis is one of the principal axes and the calculations necessary for determining the moments are quite simple. For this reason unsymmetrical bending is not difficult to handle in the usual beam section, which is probably a W, S, M, or C. Each of these sections has at least one axis of symmetry and the calculations are appreciably reduced. A further simplifying factor is that the loads are usually gravity loads and probably perpendicular to the x axis.

Among the beams that must resist unsymmetrical bending are crane girders in industrial buildings and purlins, for ordinary roof trusses. The x axes of purlins are parallel to the sloping roof surface while the large percentage of their loads (roofing, snow, etc.) are gravity loads. These loads do not lie in a plane with either of the principal axes of the inclined purlins and the result is unsymmetrical bending. Wind loads are generally considered to act perpendicular to the roof surface and thus perpendicular to the x axes of the purlins with the result that they are not considered to cause unsymmetrical bending. The x axes of crane girders are usually horizontal, but the girders are subjected to lateral thrust loads from the moving cranes as well as to gravity loads.

To check the adequacy of members bent about both axes simultaneously the LRFD provides an equation in Section H1 of their specification. The equation to follow is for combined bending and axial tension or compression if $P_u/\phi P_n < 0.2$.

$$\frac{P_u}{2\phi P_n} + \left(\frac{M_{ux}}{\phi_b M_{nx}} + \frac{M_{uy}}{\phi_b M_{ny}}\right) \leq 1.0 \qquad \text{(LRFD Equation H1-1b)}$$

Since for the problem discussed here P_u is equal to zero, the formula reduces to

$$\frac{M_{ux}}{\phi_b M_{nx}} + \frac{M_{uy}}{\phi_b M_{ny}} \leq 1.0$$

This is an interaction or percentage equation. If M_{ux} is, say, 75 percent of the moment it could resist if bent about the x axis only ($\phi_b M_{nx}$), then M_{uy} can be no greater than 25 percent of what it could resist if bent about the y axis only ($\phi_b M_{ny}$).

Example 2.11 illustrates the design of a beam subjected to unsymmetrical bending. To illustrate the trial-and-error nature of the problem the author did not do quite as much scratchwork as in some of his earlier examples. The first design problems of this type which the reader attempts may quite well take several trials. Consideration needs to be given to the question of lateral support for the compression flange. Should the lateral support be of questionable nature the engineer should reduce the design moment resistance by means of one of the expressions previously given for that purpose.

Example 2.11 ■ A certain beam in its upright position is estimated to have a vertical bending moment M_{ux} = 160 ft-k and a lateral bending moment M_{uy} = 35 ft-k. These moments include the effect of the estimated beam weight. The loads are assumed to pass through the centroid of the section. Select a W24 shape of A36 steel that can resist these moments assuming full lateral support for the compression flange.

Solution

Try W24 × 62 ($\phi_b M_p = \phi_b M_{nx} = 413$ ft-k, $Z_y = 15.7$ in^3)

$$\phi_b M_{ny} = \frac{(0.9)(36)(15.7)}{12} = 42.39 \text{ ft-k}$$

$$\frac{M_{ux}}{\phi_b M_{nx}} + \frac{M_{uy}}{\phi_b M_{ny}} = \frac{160}{413} + \frac{35}{42.39} = 1.21 > 1.00 \quad \text{(NG)}$$

Try W24 × 68 ($\phi_b M_p = \phi_b M_{nx} = 478$ ft-k, $Z_y = 24.5$ in^3)

$$\phi_b M_{ny} = \frac{(0.9)(36)(24.5)}{12} = 66.15 \text{ ft-k}$$

$$\frac{160}{478} + \frac{35}{66.15} = 0.864 < 1.00 \quad \text{(OK)}$$

Use W24 × 68

It will be noted in the solution for Example 2.11 that though the procedure used will yield a section that will adequately support the moments given, the selection of the absolutely lightest section listed in the LRFD Manual could be quite lengthy because of the two variables Z_x and Z_y which affect the size. If we have a large M_{ux} and a small M_{uy} the most economical section will probably be quite deep and rather narrow, whereas if we have a large M_{uy} in proportion to M_{ux} the most economical section may be rather wide and shallow.

2.1.11 Ponding

It has been claimed that almost 50 percent of the lawsuits faced by building designers are concerned with roofing systems.[3] Ponding, a problem with many flat roofs, is one of the common subjects of such litigation. If water accumulates more rapidly on a roof than it runs off, ponding results because the increased load causes the roof to deflect into a dish shape that can hold more water, which causes greater deflections, and so on. This process continues until equilibrium is reached, or until collapse occurs. Ponding can be caused by increasing deflections, clogged roof drains, settlement of footings, warped roof slabs, and so on.

The best way to prevent ponding is to use appreciable roof slopes ($^1/_4$ in/ft or more) together with good drainage facilities. Supposedly more than two thirds of the flat roofs in the United States have slopes less than $^1/_4$ in/ft. The construction of roofs with slopes this large will increase building costs by only a few percent as compared to perfectly flat roofs. The supporting girders for flat roofs with long spans should definitely be cambered to reduce the possibility of ponding (as well as the sagging that is so disturbing to the people occupying a building).

Design of Flexural Members ■ 2-29

Section K2 of the LRFD specification states that, unless roof surfaces have sufficient slopes to areas of free drainage or sufficient individual drains to prevent water accumulation, the strength and stability of the roof systems during ponding conditions must be investigated. The very detailed work of Marino[4] forms the basis of the pending provisions of the LRFD specification. Many other useful references are available.[5-7]

The amount of water that can be retained on a roof depends on the flexibility of the framing. The specifications state that a roof system can be considered stable and not requiring further investigation if the following expressions are satisfied:

$C_p + 0.9C_s \leq 0.25$ (LRFD Equation K2-1)

$I_d \geq 25(S^4)10^{-6}$ (LRFD Equation K2-2)

where $C_p = 32 L_s L_p^4 / 10^7 I_p$

$C_s = 32 SL_s^4 / 10^7 I_s$

L_p = length of primary members, ft

L_s = length of secondary members, ft

S = spacing of secondary members, ft

I_p = moment of inertia of primary members, in^4

I_s = moment of inertia of secondary members, in^4

I_d = moment of inertia of the steel deck (if one is used) supported on the secondary members, in^4 / ft

Should steel roof decks be used, their I_d must at least equal the value given by Equation K2-2. If the roof decking is the secondary system (i.e., no secondary beams, joists, etc.), it should be handled with Equation K2-1.

Some other LRFD requirements in applying these expressions follow:

1. The moment of inertia I_s must be decreased by 15 percent for trusses and steel joists.
2. Steel decking is considered to be a secondary member supported directly by the primary members.
3. Stresses caused by wind or seismic forces do not have to be considered in the ponding calculations.

Should moments of inertia be needed for open-web joists. they can be computed from the member cross sections, or perhaps more easily backfigured from the resisting moments and allowable stresses given in the joist tables. (Since $M_R = FI/C$, we can compute $I = M_R c/F$.)

In effect these equations reflect *stress indexes* or percentage stress increases. For instance, here we are considering the percentage increase in stress in the steel members caused by ponding. If the stress in a member increases from $0.60F_y$ to $0.80F_y$ we say that the stress index U is

$$U = \frac{0.80F_y - 0.60F_y}{0.60F_y} = 0.33$$

The terms C_p and C_s are, respectively, the approximate stiffnesses of the primary and secondary support systems. LRFD Equation K2-1 ($C_p + 0.9C_s \leq 0.25$), which gives us an approximate stress index during ponding, is limited to a maximum value of 0.25. Should we substitute into this equation and obtain a value no greater than 0.25, ponding supposedly will not be a problem. Should the index be larger than 0.25, however, it will be necessary to conduct a further investigation. One method of doing this is presented in the LRFD Commentary and will be described later in this section.

Example 2.12 presents the application of LRFD Equation K2-1 to a roofing system.

Example 2.12 ■ Check the roof system shown in Fig. 2.20 for ponding, using the LRFD specification and A36 steel.

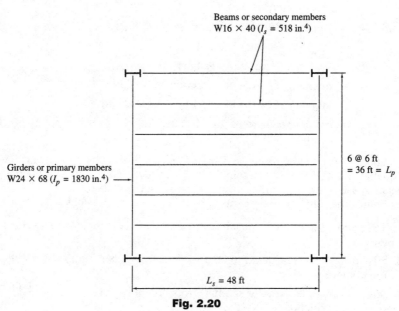

Fig. 2.20

Solution

$$C_p = \frac{32L_s L_p^4}{10^7 I_p} = \frac{(32)(48)(36)^4}{(10^7)(1830)} = 0.141$$

$$C_s = \frac{32SL_s^4}{10^7 I_s} = \frac{(32)(6)(48)^4}{(10^7)(518)} = 0.197$$

$$C_p + 0.9C_s = 0.141 + (0.9)(0.197) = 0.318 > 0.25$$

Design of Flexural Members ■ 2-31

Indicates insufficient strength and stability, and thus a more precise method of checking should be used.

The curves in the LRFD Commentary (K2) provide a design aid for use when we need to compute a more accurate flat-roof framing stiffness than is given by the specification provision that $C_p + 0.9C_s \leq 0.25$.

For this application the term f_o represents the computed bending stress due to the supported loads not including the ponding effect. The following stress indexes are computed for the primary and secondary members:

$$U_p = \left(\frac{F_y - f_o}{f_o}\right)_p$$

$$U_s = \left(\frac{F_y - f_o}{f_o}\right)_p$$

In these expressions f_o represents the stress due to $1.2D + 1.2R$ (where $D =$ nominal dead load and $R =$ nominal load due to rainwater or ice exclusive of ponding contribution). These loads should include any snow that is present, although most ponding failures have occurred during torrential summer rains.

We enter Fig. A-K2.1 in Section K of the LRFD Appendix with our computed U_p and move horizontally to the calculated C_s value of the secondary members. Then we go vertically to the abscissa scale and read the flexibility constant. If this value is more than our calculated C_p value computed for the primary members, the stiffness is sufficient. The same process is followed by Fig. A-K2.2, where we enter with our computed U_s and C_p values and pick from the abscissa the flexibility constant, which should be no less than our C_s value. This procedure is illustrated in Example 2.13.

Example 2.13 ■ Recheck the roof system considered in Example 2.12 using the LRFD curves. Assume that as ponding begins, f_o is 20 ksi in both girders and beams.

Solution. Checking girders:

$$U_p = \frac{F_y - f_o}{f_o} = \frac{36 - 20}{20} = 0.80$$

For $U_p = 0.80$ and $C_s = 0.197$, we read in Fig. A-K2.1 $C_p = 0.28 >$ our calculated C_p of 0.141.

∴ Girders OK

Checking beams:

$$U_s = \frac{F_y - f_o}{f_o} = \frac{36 - 20}{20} = 0.80$$

For $U_s = 0.80$ and $C_p = 0.141$, we read in Fig. A-K2.2 $C_s = 0.29 >$ our calculated C_s of 0.197.

∴ <u>Beams OK</u>

References

1. Darwin, D., *Steel and Composite Beams with Web Openings*, AISC Design Guide Series no. 2, American Institute of Steel Construction, Chicago, 1990.

2. ASCE Task Committee on Design Criteria for Composite Structures in Steel and Concrete, "Proposed Specification for Structural Steel Beams with Web Openings," D. Darwin, chairman, *Journal of Structural Engineering*, vol. 118, New York, December 1992.

3. Van Ryzin, Gary, "Roof Design: Avoid Ponding by Sloping to Drain," *Civil Engineering*, January 1980, pp. 77–81.

4. Marino, F. J., "Ponding of Two-way Roof System," *Engineering Journal*, 3d quarter, no. 3, pp. 93–100, 1966.

5. Burgett, L. B., "Fast Check for Ponding," *Engineering Journal*, 10, no. 1, 1st quarter, pp. 26–28, 1973.

6. Chinn, J., "Failure of Simply-Supported Flat Roofs by Ponding of Rain," *Engineering Journal*, no. 2, 2d quarter, pp. 38–41, 1965.

7. Ruddy, J. L., "Ponding of Concrete Deck Floors," *Engineering Journal*, 23, no. 2, 3d quarter, pp. 107–115, 1986.

3

Jack C. McCormac

DESIGN OF TENSION AND COMPRESSION MEMBERS*

3.1 Tension Members

3.1.1 Design Specifications

A ductile steel member without holes subject to a tensile load can resist without fracture a load larger than its gross cross-sectional area times its yield stress because of strain hardening. However, a tension member loaded until strain hardening is reached will lengthen a great deal before fracture, a fact that will in all probability take away the member's usefulness and may even cause failure of the structural system of which the member is a part.

If, on the other hand, we have a tension member with bolt holes, it can possibly fail by fracture at the net section through the holes. This failure load may very well be smaller than the load required to yield the gross section away from the holes. It is to be realized that the portion of the member length where we have a reduced cross-sectional area due to the presence of holes normally is very short compared with the total length of the member. Though the strain-hardening situation is quickly reached at the net section portion of the member, yielding there may not really be a limit state of significance because the overall change in length of the member due to yielding in this small part of the member length may be negligible.

---

*This chapter is an extension and adaptation of chapters in *Structural Steel Design-LRFD Method* by Jack McCormac, HarperCollins, New York, 1995.

As a result of the preceding information the LRFD specification (D1) states that the design strength of a tension member, $\phi_t P_n$, is to be the smaller of the values obtained by substituting into the following two expressions:

For the limit state of yielding in the gross section (which is intended to prevent excessive elongation of the member)

$$P_n = F_y A_s \qquad \text{(LRFD Equation D1-1)}$$
$$P_u = \phi_t F_y A_s \text{ with } \phi_t = 0.90$$

For fracture in the net section where bolt or rivet holes are present

$$P_n = F_u A_e \qquad \text{(LRFD Equation D1-2)}$$
$$P_u = \phi_t F_u A_e \text{ with } \phi_t = 0.75$$

In the preceding expression F_u is the specified minimum tensile stress and A_e is the effective net area that can be assumed to resist tension at the section through the holes. This area may be somewhat smaller than the actual net area A_n because of stress concentrations and other factors that are discussed in Sec. 3.1.4.

The design strengths presented here are not applicable to threaded steel rods or to members with pin holes (as in eyebars). These situations are discussed in Secs. 3.2.3 and 3.2.4.

It is not likely that stress fluctuations will be a problem in the average building frame because the changes in load in such structures usually occur only occasionally and produce relatively minor stress variations. Full design wind or earthquake loads occur so infrequently that they are not considered in fatigue design. Should there, however, be frequent variations or even reversals in stress, the matter of fatigue must be considered. This subject is presented in Chapter 8 of this handbook.

3.1.2 Net Areas

The presence of a hole obviously increases the unit stress in a tension member, even if the hole is occupied by a bolt. There is less area of steel to which the load can be distributed and there will be some concentration of stress along the edges of the hole.

Tension is assumed to be uniformly distributed over the net section of a tension member, although photoelastic studies show there is a decided increase in stress intensity around the edges of holes, sometimes equaling several times the stresses if the holes were not present. For ductile materials, however, a uniform stress distribution assumption is reasonable when the material is loaded beyond its yield stress. Should the fibers around the holes be stressed to their yield stress they will yield without further stress increase, with the result that there is a redistribution or balancing of stresses. At ultimate load it is reasonable to assume a uniform stress distribution. The importance of ductility on the strength of bolted tension members has been clearly demonstrated in tests. Tension members (with bolt holes) made from ductile steels have proved to be as much as one-fifth to one-sixth stronger than similar members made from brittle steels with the same strengths. We will learn in Chapter 8 of this handbook that it is possible for steel

Design of Tension and Compression Members ■ 3-3

to lose its ductility and become subject to brittle fracture. Such a condition can be created by fatigue type loads and by very low temperatures.

This initial discussion is applicable only for tension members subjected to relatively static loading. Should tension members be designed for structures subjected to fatigue type loadings, considerable effort should be made to minimize the items causing stress concentrations such as points of sudden change of cross section and sharp corners. In addition the members may have to be enlarged.

The term "net cross-sectional area" or simply "net area" refers to the gross cross-sectional area of a member minus any holes, notches, or other indentations. In considering the area of such items as these, it is important to realize that it is usually necessary to subtract an area a little larger than the actual hole. For instance, in fabricating structural steel which is to be connected with bolts, the holes are usually punched $1/16$ in larger than the diameter of the bolt. Furthermore, the punching of the hole is assumed to damage or even destroy $1/16$ in more of the surrounding metal; therefore, the diameter of the holes subtracted is $1/8$ in (3 mm) larger than the diameter of the bolt. The area of the holes subtracted is rectangular and equals the diameter of the hole times the thickness of the metal. (Should the holes be slotted, the usual practice is to add $1/16$ in to the actual width of the holes.)

It may be necessary to have an even greater latitude in meeting dimensional tolerances during erection and for high-strength bolts larger than $5/8$ in in diameter. For such a situation, holes larger than the standard ones may be used without reducing the performance of the connections. These oversized holes can be short-slotted or long-slotted.

Example 3.1 illustrates the calculations necessary for determining the net area of a plate type of tension member.

Example 3.1 ■ Determine the net area of the $3/8 \times 8$-in plate shown in Fig. 3.1. The plate is connected at its end with two lines of $3/4$-in bolts.

Solution.

$$\text{Net area} = A_n = \left(\frac{3}{8}\right)(8) - (2)\left(\frac{3}{4} + \frac{1}{8}\right)\left(\frac{3}{8}\right) = 2.34 \text{ in}^2 \quad (1510 \text{ mm}^2)$$

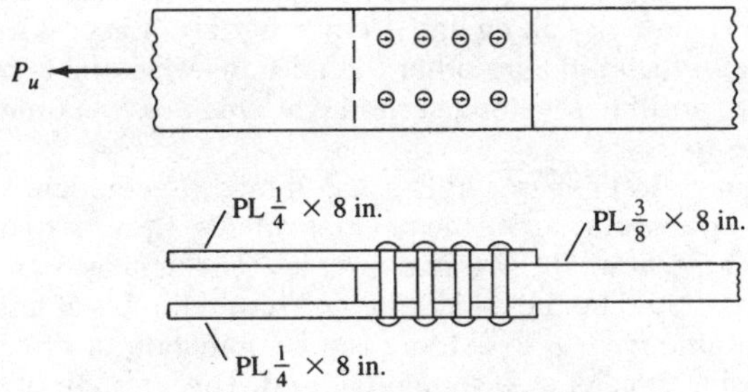

Fig. 3.1

The connections of tension members should be arranged so that no eccentricity is present. (An exception to this rule is permitted by the LRFD Specification J1.8 for certain bolted and welded connections). If this arrangement is possible the stress is assumed to be spread uniformly across the net section of a member. Should the connections have eccentricities, moments will be produced that will cause additional stresses in the vicinity of the connection. Unfortunately it is often quite difficult to arrange connections without eccentricity. Although specifications cover some situations, the designer may have to give consideration to eccentricities in some cases by making special estimates.

The lines of action of truss members meeting at a joint are assumed to coincide. Should they not coincide, eccentricity is present and secondary stresses are the result. The centers of gravity (cg's) of truss members are assumed to coincide with the lines of action of their respective forces. No problem is present in a symmetrical member as its center of gravity is at its centerline, but for unsymmetrical members the problem is a little more difficult. For these members the centerline is not the center of gravity, but the usual practice is to arrange the members at a joint so their gauge lines coincide. If a member has more than one gauge line, the one closest to the actual center of gravity of the member is used in detailing. Figure 3.2 shows a truss joint in which the cg's coincide.

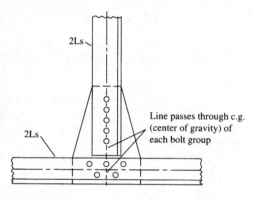

Fig. 3.2 Lining up cg's of members.

3.1.3 Effect of Staggered Holes

Should there be more than one row of bolt holes in a member it is often desirable to stagger them in order to provide as large a net area as possible at any one section to resist the load. In the preceding paragraphs tensile members have been assumed to fail transversely as along line AB in either Fig. 3.3a or 3.3b. Figure 3.3c shows a member in which a failure other than a transverse one is possible. The holes are staggered, and failure along section $ABCD$ is possible unless the holes are a large distance apart.

To determine the critical net area in Fig. 3.3c, it might seem logical to compute the area of a section transverse to the member (as AE) less the area of one hole and then the area along section $ABCD$ less two holes. The smallest value obtained along these sections would be the critical value. This method is at fault, however. Along the diagonal line from B to C there is a combination of direct stress and shear and a somewhat smaller area should be used. The strength of the member along section $ABCD$ is obviously somewhere between the strength obtained by

Design of Tension and Compression Members ■ 3-5

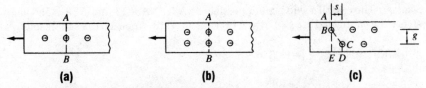

Fig. 3.3 Possible failure sections in plates.

using a net area computed by subtracting one hole from the transverse cross sectional area and the value obtained by subtracting two holes from section *ABCD*.

Tests on joints show that little is gained by using complicated theoretical formulas to consider the staggered hole situation, and the problem is usually handled with an empirical equation. The LRFD specification (B2) and other specifications use a very simple method for computing the net width of a tension member along a zigzag section[1]. The method is to take the gross width of the member regardless of the line along which failure might occur, subtract the diameter of the holes along the zigzag section being considered, and add for each inclined line the quantity given by the expression $s^2/4g$.

In this expression s is the longitudinal spacing (or pitch) of any two holes and g is the transverse spacing (or gauge) of the same holes. The values of s and g are shown in Fig. 3.3c. There may be several paths, any one of which may be critical at a particular joint. Each possibility should be considered and the one giving the least value should be used. The smallest net width obtained is multiplied by the plate thickness to give the net area A_n. Example 3.2 illustrates the method of computing the critical net area of a section which has three lines of bolts. (For angles, the gauge for holes in opposite legs is considered to be the sum of the gauges from the back of the angle minus the thickness of the angle.)

Example 3.2.■ Determine the critical net area of the ½-in-thick plate shown in Fig. 3.4 using the LRFD specification (Section B2). The holes are punched for ¾-in bolts.

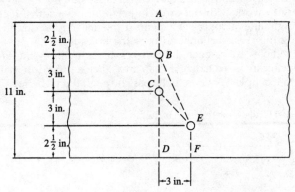

Fig. 3.4

3-6 ■ Chapter Three

Solution. The critical section could possibly be $ABCD$, $ABCEF$, or $ABEF$. Hole diameters to be subtracted are $3/4 + 1/8 = 7/8$ in. The net widths for each case are as follows:

$$ABCD = 11 - (2)\left(\frac{7}{8}\right) = 9.25 \text{ in}$$

$$ABCEF = 11 - (3)\left(\frac{7}{8}\right) + \frac{(3)^2}{(4)(3)} = 9.125 \text{ in (controls)}$$

$$ABEF = 11 - (2)\left(\frac{7}{8}\right) + \frac{(3)^2}{(4)(6)} = 9.625 \text{ in}$$

(The reader should note it is a waste of time to check route $ABEF$ for this plate. Two holes need to be subtracted for routes $ABCD$ and $ABEF$. As $ABCD$ is a shorter route it obviously controls over $ABEF$.)

$$A_n = (9.125)\left(\frac{1}{2}\right) = 4.56 \text{ in}^2 \qquad \text{Ans.}$$

Holes for bolts are normally punched in steel angles at certain standard locations. These locations or gauges are dependent on the angle-leg widths and on the number of lines of holes. Table 3.1, which is taken from Part 9 of the LRFD Manual, shows these gauges. It is unwise for the designer to require different gauges from those given in the table unless unusual situations are present, because of the appreciably higher fabrication costs that will result.

The $s^2/4g$ rule is merely an approximation or simplification of the complex stress variations which occur in members with staggered arrangements of bolts. Steel specifications can only provide minimum standards and designers will have to logically apply such information to complicated situations which the specifications could not cover in their attempts at brevity and simplicity.

The LRFD specification does not include a method to be used for determining the net widths of sections other than plates and angles. For channels, W sections, S sections, and others the web and flange thicknesses are not the same. As a result it is necessary to work with net areas rather than net widths. If the holes are placed in straight lines across such a member the net area can simply be obtained by subtracting the cross-sectional areas of the holes from the gross area of the member. If the holes are staggered it is necessary to multiply the $s^2/4g$ values by the applicable thickness to change it to an area. Such a procedure is illustrated for a W section in Example 3.3 where bolts pass through the web only.

Table 3.1 Usual Gauges for Angles in Inches

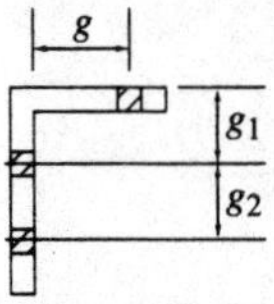

Leg	8	7	6	5	4	3½	3	2½	2	1¾	1½	1⅜	1¼	1
g	4½	4	3½	3	2½	2	1¾	1⅜	1⅛	1	⅞	⅞	¾	⅝
g_1	3	2½	2¼	2										
g_2	3	3	2½	1¾										

Design of Tension and Compression Members ■ 3-7

Example 3.3. ■ Determine the net area of the W12×16 (A_g = 4.71 in²) shown in Fig. 3.5 assuming the holes are for 1-in bolts.

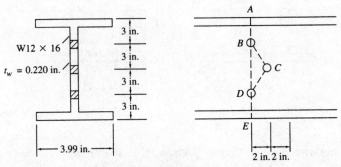

Fig. 3.5

Solution. Net areas:

$$ABDE = 4.71 - (2)\left(1\frac{1}{8}\right)(0.220) = 4.21 \text{ in}^2$$

$$ABCDE = 4.71 - (3)\left(1\frac{1}{8}\right)(0.220) + (2)\frac{(2)^2}{(4)(3)}(0.220) = 4.11 \text{ in}^2$$

Obviously there are situations not included in the LRFD specification or in the preceding discussion. For one example there is the case where bolt holes are used in the webs and flanges of W sections (with their different web and flange thicknesses) and staggered with respect to each other. For such cases the designer will have to use a reasonable procedure based on what he or she has learned in the last few pages.[2]

3.1.4 Effective Net Areas

When a member other than a flat plate or bar is loaded in axial tension until failure occurs across its net section, its actual tensile failure stress will probably be less than the coupon tensile strength of the steel *unless all of the various elements which make up the section are connected so stress is transferred uniformly across the section.*

If the forces are not transferred uniformly across a member cross section there will be a transition region of uneven stress running from the connection out into the member for some distance. This is the situation shown in Fig. 3.6a where a single angle tension member is connected by one leg only. At the connection more of the load is carried by the connected leg, and it takes the transition distance shown in part *b* of the figure for the stress to spread uniformly across the whole angle.

In the transition region the stress in the connected part of the member may very well exceed F_y and go into the strain-hardening range. Unless the load is

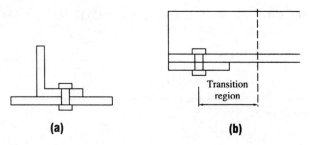

Fig. 3.6 Shear lag. (*a*) Angle connected by one leg only. (*b*) Stress in transition range > F_y.

reduced the member may fracture prematurely. The further we move out from the connection the more uniform the stress becomes. In the transition region the shear transfer has "lagged" and the phenomenon is referred to as *shear lag*.

In such a situation the flow of tensile stress between the full member cross section and the smaller connected cross section is not 100 percent effective. As a result the LRFD specification (B3) states that the effective net area A_e of such a member is to be determined by multiplying an area A (which is the net area or the gross area or the directly connected area as described in the next few pages) by a reduction factor U. The use of a factor such as U accounts for the nonuniform stress distribution in a simple manner. An explanation of the way in which U factors are determined follows.

$$A_e = AU \quad \text{(LRFD Equation B3-1)}$$

The angle shown in Fig. 3.7*a* is connected at its ends to only one leg. You can easily see that its area effective in resisting tension can be appreciably increased by shortening the width of the unconnected leg and lengthening the width of the connected leg as shown in Fig. 3.7*b*.

Investigators have found that a convenient measure of the effectiveness of a member such as an angle connected by one leg is the distance $\bar{x}$ measured from the plane of the connection to the centroid of the area of the whole section.[3,4] The smaller the value of $\bar{x}$ the larger the effective area of the member. The specification in effect reduces the length L of a connection with shear lag to a shorter effective

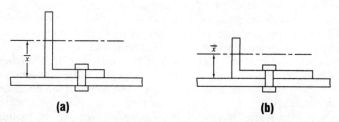

Fig. 3.7 Reducing shear lag by reducing length of unconnected leg and thus $\bar{x}$.

Design of Tension and Compression Members ■ 3-9

length L'. The value of U then equals L'/L or $1 - \bar{x}/L$. Several values of $\bar{x}$ are shown in Fig. 3.8. The remainder of this section is devoted to the determination of the effective areas of various bolted and welded tension members.

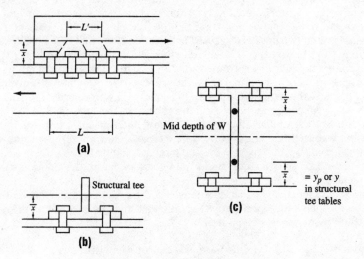

Fig. 3.8 Values of $\bar{x}$ for different shapes.

Bolted Members ■ Should a tension load be transmitted by bolts, A equals the net area A_n of the member and U is computed as follows:

$$U = 1 - \frac{\bar{x}}{L} \leq 0.9 \qquad \text{(LRFD Equation B3-2)}$$

The length L used in this expression is equal to the distance between the first and last bolts in the line. When there are two or more lines of bolts, it is the length of the line with the maximum number of bolts. Should the bolts be staggered, it is the out-to-out dimension between the extreme bolts. You will note that the longer the connection (L) becomes the larger U will become as will the effective area of the member. (On the other hand, the effectiveness of connectors is somewhat reduced if very long connections are used.) Insufficient data are available for the case in which only one bolt is used in each line. It is thought that a conservative approach for this case is to let $A_e = A_n$ of the connected element.

In order to calculate U for a W section connected by its flanges only, we will assume that the section is split into two structural tees. Then the value of $\bar{x}$ used will be the distance from the outside edge of the flange to the cg of the structural tee as shown in part c of Fig. 3.8. Parts b and c of Fig. C-B3 of the LRFD Commentary illustrate the recommended procedures for calculating $\bar{x}$ values for channel and I-shaped sections where the loads are transferred by means of bolts passing through the member webs only.

3-10 ■ Chapter Three

The LRFD specification permits the designer to use larger values of U than obtained from the equation if such values can be justified by tests or other rational criteria.

Section B3 of the LRFD Commentary provides suggested $\bar{x}$ values for use in the equation for U for several situations not addressed in the specification. Included are values for W and C sections bolted only through their webs. Also considered are single angles with two lines of staggered bolts in one of their legs. The basic idea for computing $\bar{x}$ for these cases is presented in the following paragraph.[5]

The channel of Fig. 3.9a is connected with two lines of bolts through its web. The "angle" part of this channel above the center of the top bolt is shown darkened in part b of the figure. This part of the channel is unconnected. For shear lag purposes we can compute the vertical distance from the center of the top bolt to the centroid of the "angle" above and the horizontal distance from the outside face of the web to the "angle" centroid. The larger value will represent the worst situation and will be the $\bar{x}$ used in the equation. It is felt that with this idea in mind the reader will be able to understand the values shown in the Commentary for other sections.

Example 3.4 illustrates the calculations necessary for determining the effective net area of a bolted W section connected only to its flanges. In addition the design strength of the member is computed.

Example 3.4 ■ Determine the tensile design strength of a W10×45 with two lines of ³⁄₄-in-diameter bolts in each flange using A572 grade 50 steel with $F_y = 50$ ksi and $F_u = 65$ ksi and the LRFD specification. There are assumed to be at least three bolts in each line 4 in on center, and the bolts are not staggered with respect to each other.

Solution. Using a W10×45 ($A_g = 13.3$ in², $d = 10.10$ in, $b_f = 8.020$ in, $t_f = 0.620$ in)

(a) $P_u = \phi_t F_y A_g = (0.90)(50)(13.3) = 598.5$ k

(b) $A_n = 13.3 - (4)\left(\dfrac{7}{8}\right)(0.620) = 11.13$ in² $= A$

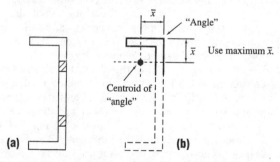

Fig. 3.9 Computing $\bar{x}$ for a channel bolted through its web.

Referring to the tables for half of a W10×45 (or to a WT5×22.5) we find $\bar{x}$ = 0.907 in. Then

$$U = 1 - \frac{\bar{x}}{L} = 1 - \frac{0.907}{8} = 0.89$$

$$A_e = UA = (0.89)(11.13) = 9.91 \text{ in}^2$$

$$P_u = \phi_t F_u A_e = (0.75)(65)(9.91) = 483.1 \text{ k}$$

Design strength P_u = 483.1 k

The 1986 LRFD specification presented a set of standard U values that could be used for bolted members instead of substituting into the $1-\bar{x}/L$ expression. These values, which are listed in Table 3.2, are still acceptable. (For Example 3.4 note that U from the table is 0.90 as b_f/d = 8.020/10.10 > 2/3.) The tabulated values are particularly useful for situations in which initial design size selections are being made and we have insufficient information to calculate U values. We will encounter this situation in Examples 3.8 and 3.9.

Welded Members ▪ When tension loads are transferred by welds the following rules from LRFD Specification B.3 are to be used to determine values for A and U. (A_e as for bolted connections = AU.)

1. Should the load be transmitted only by longitudinal welds to other than a plate member, or by longitudinal welds in combination with transverse welds, A is to equal the gross area of the member A_g.
2. Should a tension load be transmitted only by transverse welds, A is to equal the area of the directly connected elements and U is to equal 1.0.
3. Tests have shown that when flat plates or bars connected by longitudinal fillet welds are used as tension members, they may fail prematurely by shear lag at the corners if the welds are too far apart. Therefore, the LRFD specification states that when such situations are encountered the length of the welds may

Table 3.2 Permissible U Values for Bolted Connections

(a) W, M, or S shapes with flange widths not less than two-thirds the depth, and structural tees cut from these shapes, provided the connection is to the flanges and has no fewer than three fasteners per line in the direction of stress, U = 0.90

(b) W, M, or S shapes not meeting the conditions of subparagraph a, structural tees cut from these shapes, and all other shapes including built-up cross sections, provided the connection has no fewer than three fasteners per line in the direction of stress, U = 0.85

(c) All members having only two fasteners per line in the direction of stress, U = 0.75

not be less than the width of the plates or bars. The letter A represents the area of the plate and UA is the effective net area. For such situations the following values of U are to be used (LRFD Specification B3.2(d)):

When $l \geq 2w$ $\qquad$ $U=1.0$
When $2w > l \geq 1.5w$ $\qquad$ $U=0.87$
When $1.5w > l \geq w$ $\qquad$ $U=0.75$

where l = weld length, in
w = plate width (distance between welds), in

For combinations of longitudinal and transverse welds l is to be used equal to the length of the longitudinal weld because the transverse weld has little or no effect on the shear lag (that is, it does little to get the load into the unattached parts of the member).

In cases in which fillet welds are used to transmit tension loads to some but not all of the elements of a cross section the weld strength will control.

Examples 3.5 and 3.6 illustrate the calculation of the effective area and the design strengths of two welded members.

Example 3.5 ▪ The 1×6 in plate shown in Fig. 3.10 is connected to a 1×10 in plate with longitudinal fillet welds to support a tensile load. Determine the design strength P_u of the member if $F_y = 50$ ksi and $F_u = 65$ ksi.

Solution. Considering the smaller PL

(a) $P_u = \phi_t F_y A_g = (0.90)(50)(1 \times 6) = 270$ k

(b) $A = A_g = 1 \times 6 = 6$ in^2

$1.5w = 9$ in $> l = 8$ in $> w = 6$ in

$\therefore U = 0.75$

$A_e = AU = (6.0)(0.75) = 4.50$ in^2

$P_u = \phi_t F_u A_e = (0.75)(65)(4.50) = 219.4$ k ←

Design strength $P_u = 219.4$ k

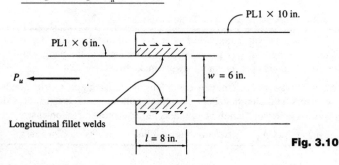

Fig. 3.10

Design of Tension and Compression Members ■ 3-13

Example 3.6 ■ Compute the design strength P_u for the angle shown in Fig. 3.11. It is welded on the end and sides of the 8-in leg only and $F_y = 50$ ksi and $F_u = 70$ ksi.

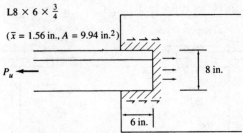

Fig. 3.11 Angle welded to 8-in leg only.

Solution. As only one leg of the angle is connected, a reduced effective area needs to be computed.

(a) $P_u = \phi_t F_y A_g = (0.9)(50)(9.94) = 447.3$ k

(b) $U = 1 - \dfrac{\bar{x}}{L} = 1 - \dfrac{1.56}{6.00} = 0.74$

$A_e = AU = (9.94)(0.74) = 7.36$ in^2

$P_u = \phi_t F_u A_e = (0.75)(70)(7.36) = 386.4$ k ←

Design strength $P_u = 386.4$ k

3.1.5 Connecting Elements for Tension Members

When splice or gusset plates are used as statically loaded tensile connecting elements their strength shall be determined as follows:
For yielding of welded or bolted connection elements

$\phi = 0.90$

$R_n = A_g F_y$ (LRFD Equation J5-1)

For fracture of bolted connection elements

$\phi = 0.75$

$R_n = A_n F_u$ with $A_n \leq 0.85 A_g$ (LRFD Equation J5-2)

The net area A_n to be used in the second of these expressions may not exceed 85 percent of A_g. Tests have shown for decades that bolted tension connection elements rarely have an efficiency greater than 85 percent, even if the holes represent a very small percentage of the gross area of the elements. In Example 3.7 the strength of a pair of tensile connecting plates is computed.

3-14 ■ Chapter Three

Example 3.7 ■ The tension member (F_y = 50 ksi and F_u = 65 ksi) of Example 3.4 is assumed to be connected at its ends with two $3/8 \times 12$-in plates as shown in Fig. 3.12. If two lines of $3/4$-in bolts are used in each plate, determine the design tensile force which the plates can transfer.

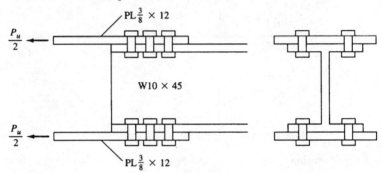

Fig. 3.12

Solution.

$$P_u = \phi_t F_y A_g = (0.9)(50)\left(2 \times \frac{3}{8} \times 12\right) = 405 \text{ k}$$

$$A_n \text{ of 2 plates} = \left(\frac{3}{8} \times 12 - \frac{7}{8} \times 2 \times \frac{3}{8}\right)2 = 7.69 \text{ in}^2$$

$$0.85 A_g = (0.85)\left(2 \times \frac{3}{8} \times 12\right) = 7.65 \text{ in}^2 = A_n$$

$$P_u = \phi_t F_u A_n = (0.75)(65)(7.65) = 372.9 \text{ k} \leftarrow$$

$$\underline{\underline{P_u = 372.9 \text{ k}}}$$

3.2 Design Topics

3.2.1 Selection of Sections

In this section the selection of members to support given tension loads is described. Although the designer has considerable freedom in the selection, the resulting members should have the following properties: (1) compactness, (2) dimensions that fit into the structure with reasonable relation to the dimensions of the other members of the structure, and (3) connections to as many parts of the sections as possible to minimize shear lag.

The choice of member type is often affected by the type of connections used for the structure. Some steel sections are not very convenient to bolt together with the required gusset or connection plates, while the same sections may be welded together with little difficulty. Tension members consisting of angles, channels, and W or S sections will probably be used when the connections are made with bolts, while plates, channels, and structural tees might be used for welded structures.

Design of Tension and Compression Members ■ 3-15

Various types of sections are selected for tension members in the examples to follow, and in each case where bolts are used some allowance is made for holes. Should the connections be made entirely by welding, no holes have to be added to the net areas to give the required gross area. *The reader should realize, however that very often welded members may have holes punched in them for temporary bolting during field erection before the permanent field welds are made. These holes need to be considered in design.* It is also to be remembered that in LRFD Equation D1-2 ($P_n = F_u A_e$) the value of A_e may be less than A_g even though there are no holes, depending on the arrangement of welds and on whether all of the parts of the members are connected.

The slenderness ratio of a member is the ratio of its unsupported length to its least radius of gyration. Steel specifications give preferable maximum values of slenderness ratios for both tension and compression members. The purpose of such limitations for tension members is to ensure the use of sections with sufficient stiffness to prevent undesirable lateral deflections or vibrations. Although tension members are not subject to buckling under normal loads, stress reversal may occur during shipping and erection and perhaps due to wind or earthquake. The specifications recommend that slenderness ratios be kept below certain maximum values in order that some minimum compressive strengths be provided in the members. For tension members other than rods LRFD Specification B7 recommends maximum slenderness ratios of 300. Members whose design is controlled by tension loads, but which may be subjected to some compression due to other loading conditions, are not required to meet the preferable maximum slenderness ratio requirement for compression members, which is 200. (For slenderness ratios greater than 200, design compressive stresses will be very small—in fact, less than 5.33 ksi for all grades of steel.)

It should be noted that out-of-straightness does not affect the strength of tension members very much because the tension loads tend to straighten the members. (The same statement cannot be made for compression members.) For this reason the LRFD specification is a little more liberal in its consideration of tension members, including those subject to some compressive forces due to transient loads such as wind or earthquake.

The *recommended* maximum slenderness ratio of 300 is not applicable to tension rods. Maximum L/r values for rods are left to the designer's judgment. If a maximum value of 300 was specified for them, they would seldom be used because of their extremely small radii of gyration.

Example 3.8 illustrates the design of a bolted tension member with a W section. The design strength P_u is the lesser of (a) $\phi_t F_y A_g$ or (b) $\phi_t F_u A_e$.

1. To satisfy the first of these expressions the minimum gross area must be at least equal to the following:

$$\min A_g = \frac{P_u}{\phi_t F_y} \tag{1}$$

2. To satisfy the second expression the minimum value of A_e must be at least

$$\min A_e = \frac{P_u}{\phi_t F_u}$$

3-16 ■ Chapter Three

And since $A_e = UA_n$ for a bolted member the minimum value of A_n is

$$\min A_n = \frac{\min A_e}{U} = \frac{P_u}{\phi_t F_u U}$$

Then the minimum A_g for the second expression must at least equal the minimum value of A_n plus the estimated hole areas.

$$\min A_g = \frac{P_u}{\phi_t F_u U} + \text{estimated hole areas} \tag{2}$$

The designer can substitute into Eqs. 1 and 2, taking the larger value of A_g so obtained for an initial size estimate. It is, however, well to notice that the maximum preferable slenderness ratio L/r is 300. From this value it is easy to compute the least preferable value of r for a particular design, that is, the value of r for which the slenderness ratio will be exactly 300. It is undesirable to consider a section whose least r is less than this value because its slenderness ratio would exceed the preferable maximum value of 300.

$$\min r = \frac{L}{300} \tag{3}$$

In Example 3.8 a W section is selected for a given set of tensile loads. For this first application of the tension design formulas the author has narrowed the problem down to one series of W shapes so the reader can concentrate on the application of the formulas and not become lost in considering W8s, W10s, W 14s, and so on. Exactly the same procedure can be used for trying these other series as is used here for the W12.

Example 3.8 ■ Select a 30-ft-long W12 section of A572 grade 50 steel to support a tensile service dead load $P_D = 130$ k and a tensile service live load $P_L = 110$ k. As shown in Fig.3.13, the member is to have two lines of bolts in each flange for $7/8$-in bolts (at least three in a line 4 in on center).

Solution. Considering the two ordinary load conditions

$$P_u = 1.4 P_D = (1.4)(130) = 182 \text{ k}$$
$$P_u = 1.2 P_D + 1.6 P_L = (1.2)(130) + (1.6)(110) = 332 \text{ k} \leftarrow$$

Computing the minimum A_g required

1. $\min A_g = \dfrac{P_u}{\phi_t F_y} = \dfrac{332}{(0.90)(50)} = 7.38 \text{ in}^2$

2. $\min A_g = \dfrac{P_u}{\phi_t F_u} + \text{estimated hole areas}$

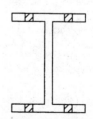

Fig. 3.13 Cross section of member for Example 3.8.

Design of Tension and Compression Members ■ 3-17

Assume $U = 0.90$ from Table 3.2 and assume flange thickness is about 0.380 in after looking at W12 sections in the LRFD Manual which have areas of 7.38 in² or more.

$$\min A_g = \frac{332}{(0.75)(65)(0.90)} + (4)(1.00)(0.380) = 9.0 \text{ in}^2 \leftarrow$$

3. Preferable min $r = \dfrac{L}{300} = \dfrac{(12)(30)}{300} = 1.2$ in

Try W12×35 ($A_g = 10.3$ in², $d = 12.50$ in, $b_f = 6.56$ in, $t_f = 0.520$ in, $r_y = 1.54$ in)

Checking

1. $P_u = \phi_t F_y A_g = (0.90)(50)(10.3) = 463.5 \text{ k} > 332 \text{ k}$ \hfill (OK)
2. $\bar{x}$ for half of W12×35 or that is a WT6 ×17.5 = 1.30 in
   $L = (2)(4) = 8$ in

$$U = 1 - \frac{\overline{X}}{L} = 1 - \frac{1.30}{8} = 0.84 \text{ (from Table 3.2 } U = 0.85)$$

$A_n = 10.3 - (4)(1.00)(0.520) = 8.22 \text{ in}^2$

$P_u = \phi_t F_u A_e = (0.75)(6.5)(0.84 \times 8.22) = 336.6 \text{ k} > 332 \text{ k}$ \hfill (OK)

3. $\dfrac{L}{r} = \dfrac{(12)(30)}{1.54} = 234 < 300$ \hfill (OK)

Use W12×35

3.2.2 Built-up Tension Members

Sections D2 and J3.5 of the LRFD specification provide a definite set of rules describing how the different parts of built-up tension members are to be connected together.

1. When a tension member is built up from elements in continuous contact with each other, such as a plate and a shape, or two plates, the longitudinal spacing of connectors between those elements must not exceed 24 times the thickness of the thinner plate, or 12 in if the member is to be painted or if it is not to be painted and not to be subjected to corrosive conditions.
2. Should the member consist of unpainted weathering steel elements in continuous contact and be subject to atmospheric corrosion, the maximum permissible connector spacings are 14 times the thickness of the thinner plate, or 7 in.
3. Should a tension member be built up from two or more shapes separated by intermittent fillers, the shapes must be connected to each other at intervals such that the slenderness ratio of the individual shapes between the fasteners does not exceed 300.

3-18 ■ Chapter Three

4. The distance from the center of any bolts to the nearest edge of the connected part under consideration may not be larger than 12 times the thickness of the connected part, or 6 in.

Example 3.9 illustrates the review of a tension member that is built up from two channels that are separated from each other. This example does not include the design of the tie plates or tie bars needed to hold the channels together as shown in Fig. 3.14b. These plates which are used to connect the parts of built-up members on their open sides result in more uniform stress distribution among the various parts. Section D-2 of the LRFD specification provides empirical rules for their design. (Perforated cover plates may also be used.) The rules are based on many decades of experience with built-up tension members.

In Fig. 3.14 the location of the bolts, that is, the usual gauge for these channels, is shown as being $1^3/_4$ in from the back of the channels. The LRFD Manual does not provide the usual gauges except for angles, and those are given in Part 9. For other rolled shapes such as Cs, Ws, and Ss, it is necessary to refer to a manufacturer's catalog or to old AISC Steel Manuals. Though rather standard values of these gauges are used by the steel industry for the various rolled shapes, they are not specified in the LRFD Manual in order to give steel fabricators more freedom in placing the holes.

In Fig. 3.14 the distance between the lines of bolts connecting the tie plates to the channels can be seen to equal 8.50 in. The LRFD specification (D2) states that the length of tie plates may not be less than two thirds the distance between the lines of connectors. Furthermore, their thickness may not be less than one-fiftieth of this distance.

The minimum permissible width of tie plates (not mentioned in the specification) is the width between the lines of connectors plus the necessary edge distance on each side to keep the bolts from splitting the plate. For this example this minimum edge distance is taken as $1^1/_2$ in from Table J3.4 of the LRFD specification. The plate dimensions are rounded off to agree with the plate sizes available from the steel mills as given in the Bars and Plates section of Part I of the LRFD Manual. It is much cheaper to select standard thicknesses and widths rather than picking odd ones that will require cutting or other operations.

The LRFD Specification (D2) provides a maximum spacing between tie plates by stating that the L/r of each individual component of a built-up member running along by itself between tie plates should preferably not exceed 300. If the designer substitutes into this expression ($L/r = 300$), the least r of an individual component of the built up member the value of L may be computed. This will be the maximum spacing of the tie plates preferred by the LRFD specification for this member.

Example 3.9 ■ Two C12×30s (see Fig. 3.14) have been selected to support a dead tensile working load of 170 k and a 275-k live tensile working load. The member is 30 ft long, consists of a steel with $F_y = 50$ ksi and $F_u = 65$ ksi, and has one line of at least three $7/_8$-in bolts in each channel flange 3 in on center. Using the LRFD specification, determine whether the member is satisfactory and design the necessary tie plates. Assume centers of bolt holes are $1^3/_4$ in from the backs of the channels.

Design of Tension and Compression Members ■ 3-19

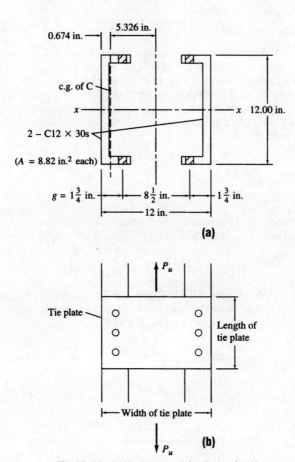

Fig. 3.14 Built-up section for Example 3.9.

Solution. Using C12×30s (A_g=8.82 in² each, t_f=0.501 in, I_x=162 in⁴ each, I_y=5.14 in⁴ each, y axis 0.674 in from back of C, r_y=0.763 in).

Load to be resisted

$P_u = (1.2)(170) + (1.6)(275) = 644$ k

Design strengths

$P_u = \phi_t F_y A_g = (0.90)(50)(2 \times 8.82) = 793.8$ k > 644 k (OK)

$A_n = [8.82 - (2.0)(0.501)]2 = 15.64$ in²

$U = 0.85$ from Table 3.2

$P_u = \phi_t F_u A_n U = (0.75)(65)(15.64)(0.85) = 648.1$ k > 644 k (OK)

3.2.3 Rods and Bars

When rods and bars are used as tension members they may be simply welded at their ends, or they may be threaded and held in place with nuts. The LRFD nominal tensile design stress for threaded rods is given in their Table J3.2 and equals $\phi 0.75 F_u$ and is to be applied to the gross area of the rod A_D computed with the major thread diameter, that is, the diameter to the outer extremity of the thread. The area required for a particular tensile load can then be calculated from the following expression:

$$A_D = \frac{P_u}{\phi 0.75 F_u} \text{ with } \phi = 0.75$$

In Table 8-7 of the LRFD Manual entitled "Threading Dimensions for High-Strength and Non-High-Strength Bolts" properties of standard threaded rods are presented. Example 3.10 which follows illustrates the selection of a rod using this table. It will be noted that the LRFD specification (Part 6, Section J1.7) states that the factored load P_u used for connection design may not be less than 10 k except for lacing, sag rods, or girts.

Example 3.10 ■ Using a steel with $F_u = 65$ ksi and the LRFD specification, select a standard threaded rod of A36 steel to support a tensile working dead load of 10 k and a tensile working live load of 20 k.

Solution.

$P_u = (1.2)(10) + (1.6)(20) = 44 \text{ k} \leftarrow$

$$A_D = \frac{P_u}{\phi 0.75 F_u} = \frac{44}{(0.75)(0.75)(65)} = 1.20 \text{ in}^2$$

Use $1\frac{1}{4}$ - in - diameter rod with 7 threads per in ($A_D = 1.23 \text{ in}^2$)

As shown in Fig. 3.15, upset rods sometimes are used, where the rod ends are made larger than the regular rod and the threads are placed in the upset ends. Threads obviously reduce the cross-sectional area of a rod. If a rod is upset and the threads are placed in that part of the rod, the result will be a larger cross-sectional area at the root of the thread than we would have if the threads were placed in the regular part of the rod.

Table J3.2 in the LRFD specification states that the nominal tensile strength of the threaded portion of the upset ends is equal to $0.75 F_u A_D$, where A_D is the cross-sectional area of the rod at its major thread diameter. This value must be larger than the nominal body area of the rod (before upsetting) times F_y.

Upsetting permits the designer to use the entire area of the regular part of the bar for strength calculations.

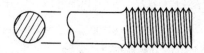

Fig. 3.15 A round upset rod.

Design of Tension and Compression Members ■ 3-21

Nevertheless, the use of upset rods probably is not economical and should be avoided unless a large order is being made.

One situation in which tension rods are sometimes used is in steel frame industrial buildings with purlins running between their roof trusses to support the roof surface. These types of buildings will also frequently have girts running between the columns along the vertical walls. (Girts are horizontal beams used on the sides of buildings, usually industrial, to resist lateral bending due to wind. They also are often used to support corrugated or other types of siding.) Sag rods may be required to provide support for the purlins parallel to the roof surface and vertical support for the girts along the walls. For roofs with steeper slopes than 1 vertically to 4 horizontally, sag rods are often considered necessary to provide lateral support for the purlins particularly where the purlins consist of steel channels. Steel channels are commonly used as purlins, but they have very little resistance to lateral bending. Although the resisting moment needed parallel to the roof surface is small, an extremely large channel is required to provide such a moment. The use of sag rods for providing lateral support to purlins made from channels usually is economical because of the bending weakness of channels about their y axes. For light roofs (as where trusses support corrugated steel roofs) sag rods will probably be needed at the one-third points if the trusses are more than 20 ft on centers. Sag rods at the midpoints are usually sufficient if the trusses are less than 20 ft on centers. For heavier roofs such as those made of slate, cement tile, or clay tile, sag rods will probably be needed at closer intervals. The one-third points will probably be necessary if the trusses are spaced at greater intervals than 14 ft and the midpoints will be satisfactory if truss spacings are less than 14 ft. Some designers assume that the load components parallel to the roof surface can be taken by the roof, particularly if it consists of corrugated steel sheets, and that tie rods are unnecessary. This assumption, however, is open to some doubt and definitely should not be followed if the roof is very steep.

Designers have to use their own judgment in limiting the slenderness values for rods, as they will usually be several times the limiting values mentioned for other types of tension members. A common practice of many designers is to use rod diameters no less than 1/500th of their lengths to obtain some rigidity, even though design calculations may permit smaller sizes.

It is usually desirable to limit the minimum size of sag rods to ⅝ in because smaller rods than these are often injured during construction. The threads on smaller rods are quite easily injured by overtightening, which seems to be a frequent habit of construction workers.

3.2.4 Pin-Connected Members

Until the early years of the twentieth century nearly all truss bridges in the United States were pin-connected, but today pin-connected bridges are used infrequently because of the advantages of bolted and welded connections. One trouble with the old pin-connected trusses was the wearing of the pins in the holes, which caused looseness of the joints.

An eyebar is a special type of pin-connected member whose ends where the pin holes are located are enlarged as shown in Fig. 3.16. Though just about obsolete today, eyebars at one time were very commonly used for the tension members of bridge trusses.

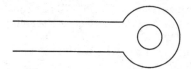

Fig. 3.16 An eyebar.

Pin-connected eyebars are used occasionally today as tension members for long-span bridges and as hangars for some types of bridges and other structures where they are normally subjected to very large dead loads. As a result the eyebars are usually prevented from rattling and wearing under live loads.

Eyebars are generally made not by forging them, but by thermally cutting them from plates. As stated in the LRFD Commentary (D3), extensive testing has shown that thermally cut members result in more balanced designs. The heads of eyebars are specially shaped so as to provide optimum stress flow around the holes. These proportions are based on long experience and testing with forged eyebars, and the results are rather conservative for today's thermally cut members.

The LRFD specification (D3) provides detailed requirements for pin-connected members as to strength and proportions of the pins and plates. The design strength of such a member is the lowest value obtained from the following equations. Reference is made to Fig. 3.17.

1. *Tensile strength on the effective net area. See Fig. 3.17a.*

$\phi = \phi_t = 0.75$

$P_n = 2tb_{eff}F_u$ (LRFD Equation D3-1)

In which t = plate thickness and b_{eff} = $2t + 0.63$, but may not exceed the distance from the hole edge to the edge of the part measured perpendicular to the line of force.

2. *Shear design strength on the effective area. See Fig. 3.17b.*

$\phi = \phi_{sf} = 0.75$

$P_n = 0.6A_{sf}F_u$ (LRFD Equation D3-2)

In which $A_{sf} = 2t(a + d/2)$, where a is the shortest distance from the edge of the pin hole to the member edge measured parallel to the force.

3. *Strength of surfaces in bearing. See Fig. 3.17c.*

$\phi = 0.75$

$R_n = 1.8F_y A_{pb}$ (LRFD Equation J8-1)

In which A_{pb} = projected bearing area = dt. Notice that LRFD Equation J8-1 applies to milled surfaces, pins in reamed, drilled, or bored holes, and ends of fitted bearing stiffeners. (LRFD specification J8 also provides other equations for determining the bearing strength for expansion rollers and rockers.)

4. *Tensile strength of the gross section. See Fig. 3.17d.*

$\phi_b = 0.90$

$P_n = F_y A_g$ (LRFD Equation D1-1)

Design of Tension and Compression Members ■ 3-23

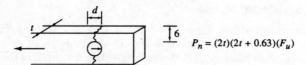

(a) Tensile strength on net section

(b) Shear design strength on effective area

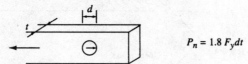

(c) Bearing strength of surface (This is bearing on projected rectangular area behind bolt)

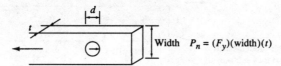

(d) Tensile strength of gross section

Fig. 3.17 Strength of pin-connected tension members.

The LRFD specification (D3) says that thicknesses $<\frac{1}{2}$ in for both eyebars and pin-connected plates are only permissible when external nuts are provided to tighten the pin plates and filler plates into snug contact. The bearing design strength of such plates is provided in LRFD specification J8.

In addition to the other requirements mentioned, LRFD Specification D3 specifies certain proportions between the pins and the eyebars. These values are based on long experience in the steel industry, and experimental work by B. G. Johnston.[6] It has been found that when eyebars and pin-connected members are made from steels with yield stresses greater than 70 kips per square inch (ksi) there may be a possibility of "dishing" (a complicated inelastic stability failure where the head of the eyebar tends to curl laterally into a dishlike shape). For this reason the LRFD specification requires stockier member proportions for such situations (hole diameter not to exceed five times the plate thickness and the width of the eyebar reduced accordingly).

3.2.5 Block Shear

The design strength of a tension member is not always controlled by $\phi_t F_y A_g$ or $\phi_t F_u A_e$ or by the strength of the bolts or welds with which the member is connected. It may instead be controlled by its *block shear* strength as described in this section.

The failure of a member may occur along a path involving tension on one plane and shear on a perpendicular plane as shown in Fig. 3.18, where several possible block shear failures are shown. For these situations it is possible for a "block" of steel to tear out.

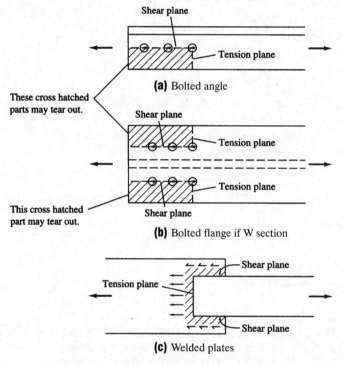

Fig. 3.18 Block shear.

When a tensile load applied to a particular connection is increased the fracture strength of the weaker plane will be approached. That plane will not fail then because it is restrained by the stronger plane. The load can be increased until the fracture strength of the stronger plane is reached. During this time the weaker plane is yielding. The total strength of the connection equals the fracture strength of the stronger plane plus the yield strength of the weaker plane.[7] Thus it is not realistic to add the fracture strength of one plane to the fracture strength of the other plane to determine the block shear resistance of a particular member. *You can see that block shear is a tearing or rupture situation and not a yielding situation.*

Design of Tension and Compression Members ■ 3-25

The member shown in Fig. 3.19a has a large shear area and a small tensile area; thus the primary resistance to a block shear failure is shearing and not tensile. The LRFD specification states that it is logical to assume that when a shear fracture occurs on this large shear-resisting area the small tensile area has yielded.

Part b of Fig. 3.19 shows a free body of the block that tends to tear out of the angle of part a. You can see in this sketch that the block shear is caused by the bolts bearing on the back of the bolt holes.

In part c of Fig. 3.19 a member is shown which, so far as block shear goes, has a large tensile area and a small shear area. The LRFD feels that for this case the primary resisting force against a block shear failure will be tensile and not shearing. Thus a block shear failure cannot occur until the tensile area fractures. At that time it seems logical to assume that the shear area has yielded.

Based on the preceding discussion the LRFD specification (J4.3) states that the block shear design strength of a particular member is to be determined by (1) computing the tensile fracture strength on the net section in one direction and adding to that value the shear yield strength on the gross area on the perpendicular segment and (2) computing the shear fracture strength on the gross area subject to tension and adding it to the tensile yield strength on the net area subject to shear on the perpendicular segment. The expression to use is the larger rupture term.

Test results show that this procedure gives good results. Furthermore, it is consistent with the calculations previously used for tension members where gross areas are used for one limit state of yielding ($\phi_t F_y A_g$) and net area for the fracture limit state ($\phi_t F_y A_g$). The LRFD specification (J4.3) states that the block shear rupture design strength is to be determined as follows:

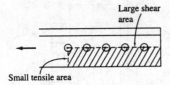

(a) Shear fracture and tension yielding

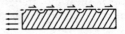

(b) Free body of "block" that tends to shear out in angle of part (a)

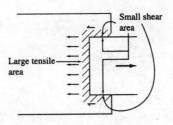

(c) Tensile fracture and shear yielding

Fig. 3.19 Block shear.

3-26 ■ Chapter Three

1. If $F_u A_{nt} \geq 0.6 F_u A_{nv}$ we will have shear yielding and tension fracture and thus should use the equation to follow:

$$\phi R_n = \phi\left[0.6 F_y A_{gv} + F_u A_{nt}\right]$$ (LRFD Equation J4-3a)

2. If $0.6 F_u A_{nc} > F_u A_{nt}$ we will have tension yielding and shear fracture and thus should use the following equation:

$$\phi R_n = \phi\left[0.6 F_u A_{nv} + F_y A_{gt}\right]$$ (LRFD Equation J4-3b)

in which $\phi = 0.75$

A_{gv} = gross area subjected to shear
A_{gt} = gross area subjected to tension
A_{nv} = net area subjected to shear
A_{nt} = net area subjected to tension

Examples 3.11 and 3.12 illustrate the determination of the block shear strengths for two members. In the study of connections we will find that it is absolutely necessary to check the block shear of beam ends where they are coped or cut back.

Example 3.11 ■ The A572 grade 50 tension member shown in Fig. 3.20 is connected with three $^3/_4$-in bolts. Determine the block shearing strength of the member and its tensile strength.

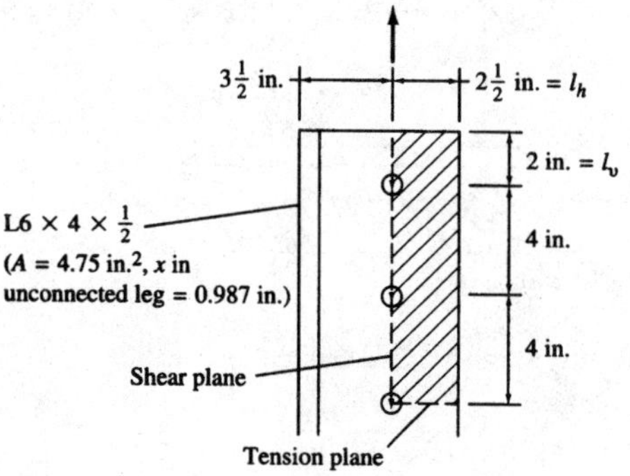

L6 × 4 × $\frac{1}{2}$
($A = 4.75$ in.^{2}, x in unconnected leg = 0.987 in.)

Fig. 3.20

Solution.

$$A_{gv} = (10)\left(\frac{1}{2}\right) = 5.0 \text{ in}^2$$

$$A_{gt} = (2.5)\left(\frac{1}{2}\right) = 1.25 \text{ in}^2$$

$$A_{nv} = \left(10 - 2.5 \times \frac{7}{8}\right)\left(\frac{1}{2}\right) = 3.91 \text{ in}^2$$

$$A_{nt} = \left(2.50 - \frac{1}{2} \times \frac{7}{8}\right)\left(\frac{1}{2}\right) = 1.03 \text{ in}^2$$

$$F_u A_{nt} = (65)(1.03) = 66.9 \text{ k} < 0.6 F_u A_{nv} = (0.6)(65)(3.91) = 152.5 \text{ k}$$

∴ Use LRFD Equation J4-3b

$$\phi R_n = 0.75[(0.6)(65)(3.91) + (50)(1.25)] = 161.2 \text{ k}$$

Tensile strength of angle

(a) $P_u = \phi_t F_y A_g = (0.9)(50)(4.75) = 213.7$ k ←

(b) $A_n = 4.75 - (1)(\frac{7}{8})(\frac{1}{2}) = 4.31$ in^2 = A

$U = 1 - \dfrac{0.987}{8} = 0.88$

$A_e = UA = (0.88)(4.31) = 3.79$ in^2

$P_u = \phi_t F_u A_e = (0.75)(65)(3.79) = 184.8$ k

P_u for member = smaller of $\phi R_n = 161.2$ k or of $P_u = 184.8$ k

$\underline{\underline{P_u = 161.2 \text{ k}}}$

Tables are available in Part 8 of the LRFD Manual with which the block shear strengths of coped W beams can be determined. In Table 8-47(a) values of $\phi F_u A_{nt}$ are tabulated per inch of material thickness, and then in Table 8-47(b) values of $\phi(0.6 F_y A_{gv})$ per inch of material thickness are given.

Example 3.12 ■ Determine the block shear design strength of the A36 welded member shown in Fig. 3.21.

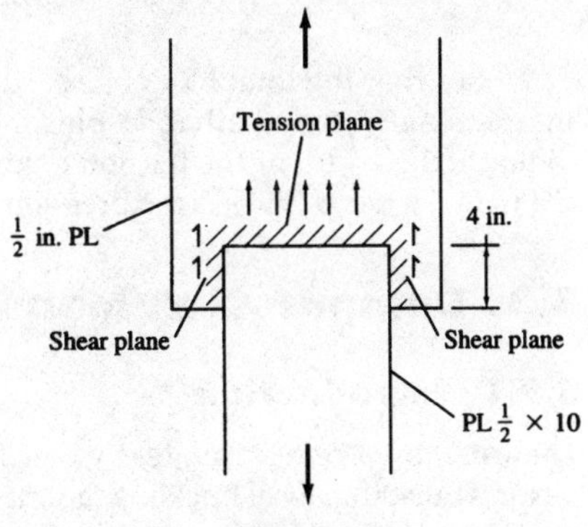

Solution.

$A_{gv} = \left(\dfrac{1}{2}\right)(8) = 4.0$ in^2

$A_{gt} = \left(\dfrac{1}{2}\right)(10) = 5.0$ in^2

$A_{nv} = \left(\dfrac{1}{2}\right)(8) = 4.0$ in^2

$A_{nt} = \left(\dfrac{1}{2}\right)(10) = 5.0$ in^2

$F_u A_{nt} = (58)(5.0) = 290$ k $> 0.6 F_u A_{nv} = (0.6)(58)(4.0) = 139.2$ k

∴ Must use LRFD Equation J4-3a

$\phi R_n = 0.75[(0.6)(36)(4.0) + (58)(5.0)] = 282.3$ k

Fig. 3.21

Tensile strength of plate

$P_u = \phi F_y A_g = (0.9)(36)\left(\dfrac{1}{2} \times 10\right) = 162.0$ k ←

$\underline{\text{Design strength of plate} = 162.0 \text{ k}}$

3-28 ■ Chapter Three

Sometimes cases are encountered where it is not altogether clear what sections should be considered for block shear calculations. For such situations designers will have to use their own judgment. One such case is shown in Fig. 3.22. In part *a* of the figure it is first assumed that web tear-out will occur along the lines *abcdef*. An alternate tear-out possibility for the same member along lines *abdef* is shown in part *b* of the figure. For this connection it is assumed that the load to be resisted is distributed equally among the five bolts. Thus, when web tear-out is considered for case *b*, we will assume only $4/5 P_u$ is carried by the section in question, because one of the bolts is outside of the tear-out area.

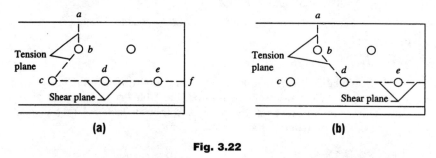

Fig. 3.22

Notice that the total block shear strength of the member will equal the block shear strength along path *abdef* plus the strength of bolt C, as it also must fail. To compute the width of the tension planes *abc* and *abd* for these cases, it seems logical to make use of the $s^2/4g$ expression presented in Sec. 3.1.3.

3.3 Compression Members

3.3.1 Introduction

There are three general modes by which axially loaded columns can fail. These are flexural buckling, local buckling, and torsional buckling. These modes of buckling are briefly defined below.

1. *Flexural buckling* (also called Euler buckling) is the primary type of buckling discussed in this chapter. Members are subject to flexure or bending when they become unstable.

2. *Local buckling* occurs when some part or parts of the cross section of a column are so thin that they buckle locally in compression before the other modes of buckling can occur. The susceptibility of a column to local buckling is measured by the width thickness ratios of the parts of its cross section. This topic is addressed in Sec. 3.3.3.

3. *Torsional buckling* may occur in columns that have certain cross-sectional configurations. These columns fail by twisting (torsion) or by a combination of torsional and flexural buckling. This topic is addressed in Sec. 3.3.15.

The longer a column becomes for the same cross section the greater becomes its tendency to buckle and the smaller becomes the load it will support. The tendency of a member to buckle is usually measured by its *slenderness ratio*, which has previously been defined as the ratio of the length of the member to its least radius of gyration. The tendency to buckle is also affected by such factors as the types of end connections, eccentricity of load application, imperfection of column material, initial crookedness of columns, residual stresses from manufacture, etc.

Slight imperfections in tension members and beams can be safely disregarded as they are of little consequence. On the other hand, slight defects in columns may be of major significance. A column that is slightly bent at the time it is put in place may have significant bending moments equal to the column load times the initial lateral deflection. (Figure 3 of Section 7-11 of the LRFD Commentary on the Code of Standard Practice for Steel Buildings and Bridges, located in Part 6 of the LRFD Manual, shows that the maximum out-of-straightness permitted in columns is $L/1000$ where L is the distance between braced points. Section E2 of the Commentary on the LRFD specification states that average out-of-straightness values of $L/1500$ were used in developing the LRFD column formulas.)

Obviously, a column is a more critical member in a structure than is a beam or tension member because minor imperfections in materials and dimensions mean a great deal. This fact can be illustrated by a bridge truss that has some of its members damaged by a truck. The bending of tension members probably will not be serious as the tensile loads will tend to straighten those members; but the bending of any compression members is a serious matter, as compressive loads will tend to magnify the bending in those members.

The spacing of columns in plan establishes what is called a *bay*. For instance, if the columns are 20 ft on center in one direction and 25 ft in the other direction the bay size is 20 ft × 25 ft. Larger bay sizes increase the user's flexibility in space planning. As to economy, a detailed study by John Ruddy[8] indicates that when shallow spread footings are used, bays with length-to-width ratios of about 1.25 to 1.75 and areas of about 1000 sq ft are the most cost-efficient. When deep foundations are used, his study shows that larger bay areas are more economical.

3.3.2 End Restraint and Effective Lengths of Columns

End restraint and its effect on the load-carrying capacity of columns is a very important subject indeed. Columns with appreciable end restraint can support considerably more load than those with little end restraint as at hinged ends.

To successfully use the LRFD equations for practical columns the value of L should be the distance between points of inflection in the buckled shape. This distance is referred to as the *effective length* of the column. For a pinned-end column (whose ends can rotate but cannot translate) the points of inflection or zero moment are located at the ends a distance L apart. For columns with different end conditions the effective lengths may be entirely different. In steel specifications the effective length of a column is referred to as KL where K is the *effective length factor*. K is the number that must be multiplied by the length of the column to obtain its effective length. Its size depends on the rotational restraint supplied at the ends of the column and upon the resistance to lateral movement provided.

The concept of effective lengths is simply a mathematical method of taking a column, whatever its end and bracing conditions, and replacing it with an equivalent pinned-end braced column. A complex buckling analysis could be made for a frame to determine the critical stress in a particular column. The K factor is determined by finding the pinned-end column with an equivalent length that provides the same critical stress. The K factor procedure is a method of making simple solutions for complicated frame buckling problems.

Columns with different end conditions have entirely different effective lengths. For this initial discussion it is assumed that no sidesway or joint translation is possible. Sidesway or joint translation means that one or both ends of a column can move laterally with respect to each other. Should a column be connected with frictionless hinges as shown in part a of Fig. 3.23, its effective length would be equal to the actual length of the column and K would equal 1.0. If there were such a thing as a perfectly fixed-ended column, its points of inflection (or points of zero moment) would occur at its one-fourth points and its effective length would equal $L/2$ as shown in part b of Fig. 3.23. As a result its K value would equal 0.50.

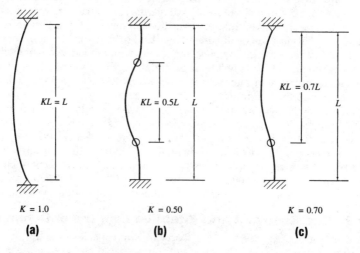

Fig. 3.23 Effective lengths for columns in braced frames (sidesway prevented).

Obviously, the smaller the effective length of a particular column, the smaller its danger of lateral buckling and the greater its load-carrying capacity. In part c of Fig. 3.23 a column is shown with one end fixed and one end pinned. The K for this column is theoretically 0.70.

Actually there are neither perfect pin connections nor any perfect fixed ends, and the usual column falls in between the two extremes. This discussion would seem to indicate that column effective lengths always vary from an absolute minimum of $L/2$ to an absolute maximum of L, but there are exceptions to this rule. An example is given in Fig. 3.24a where a simple bent is shown. The base of each

Design of Tension and Compression Members ■ 3-31

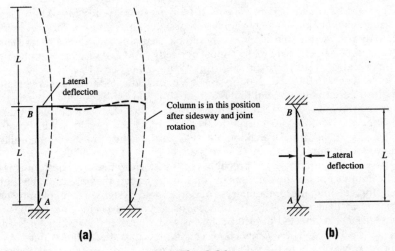

Fig. 3.24

of the columns is pinned and the other end is free to rotate and move laterally (called sidesway). Examination of this figure will show that the effective length will exceed the actual length of the column as the elastic curve will theoretically take the shape of the curve of a pinned-end column of twice its length and K will theoretically equal 2.0. Notice in part b of the figure how much smaller the lateral deflection of column AB would be if it were pinned both top and bottom so as to prevent sidesway.

Structural steel columns serve as parts of frames, and these frames are sometimes *braced* and sometimes *unbraced*. A braced frame is one for which sidesway or joint translation is prevented by means of bracing, shear walls, or lateral support from adjoining structures. An unbraced frame does not have any of these types of bracing supplied and must depend on the stiffness of its own members to prevent lateral buckling. For braced frames K values can never be greater than 1.0, but for unbraced frames the K values will always be greater than 1.0 because of sidesway.

Table C-C2.1 of the Commentary on the LRFD specification gives recommended effective length factors when ideal conditions are approximated. This table is reproduced here as Table 3.3 with the permission of the AISC. Two sets of K values are provided in the table, one being the theoretical values and the other being the recommended design values based on the fact that perfectly pinned and fixed conditions are not possible. If the ends of the column of Fig. 3.23b were not quite fixed, the column would be a little freer to bend laterally and its points of inflection would be farther apart. The recommended design K given is 0.65, while the theoretical value is 0.5. As no column ends are perfectly fixed or perfectly hinged, the designer may wish to interpolate between the values given in the table, the interpolation to be based on his or her judgment of the actual restraint conditions.

3-32 ■ Chapter Three

The values in Table 3.3 are very useful for preliminary designs. When using this table we will almost always apply the design values and not the theoretical values. In fact, the theoretical values should be used only for those very rare situations where fixed ends are really almost perfectly fixed and/or when simple supports are almost perfectly frictionless (this means almost never).

You will note in the table that for cases $a, b, c,$ and e the design values are greater than the theoretical values, but that is not the situation for cases d and f, where the values are the same. The reason for this in each of these two latter cases is that if the pinned conditions are not perfectly frictionless the K values will become smaller, not larger. Thus, by making the design values the same as the theoretical ones we are staying on the safe side.

The K values in Table 3.3 are probably very satisfactory to use for designing isolated columns, but for the columns in continuous frames they are probably only satisfactory for making preliminary or approximate designs. Such columns are restrained at their ends by their connections to various beams and the beams themselves are connected to other columns and beams at their other ends and thus also restrained. These connections can appreciably affect the K values. As a result, for most situations the values in Table 3.3 are not sufficient for final designs. They are usually satisfactory for preliminary designs and for situations in which sidesway is prevented by bracing. Should the columns be part of a continuous

Table 3.3 Column Effective Lengths

	(a)	(b)	(c)	(d)	(e)	(f)
Buckled shape of column is shown by dashed line						
Theoretical K value	0.5	0.7	1.0	1.0	2.0	2.0
Recommended design value when ideal conditions are approximated	0.65	0.80	1.2	1.0	2.10	2.0
End condition code		Rotation fixed and translation fixed Rotation free and translation fixed Rotation fixed and translation free Rotation free and translation free				

Source: Load and Resistance Factor Design Specification for Structural Steel Buildings, December 1, 1993 (Chicago: AISC, 1994), p. 6-184 in the LRFD Manual.

frame subject to sidesway, however, it is often advantageous to make a more detailed analysis as described in this section. To a lesser extent this is also desirable for columns in frames braced against sidesway.

Perhaps a few explanatory remarks should be made at this point, defining sidesway as it pertains to effective lengths. For this discussion sidesway refers to a type of buckling. In statically indeterminate structures sidesway occurs where the frames deflect laterally due to the presence of lateral loads or unsymmetrical vertical loads or where the frames themselves are unsymmetrical. Sidesway also occurs in columns whose ends can move transversely when they are loaded until buckling occurs.

Should frames with diagonal bracing or rigid shear walls be used, the columns will be prevented from sidesway and provided with some rotational restraint at their ends. For these situations, pictured in Fig. 3.25, the K factors will fall somewhere between cases a and d of Table 3.3.

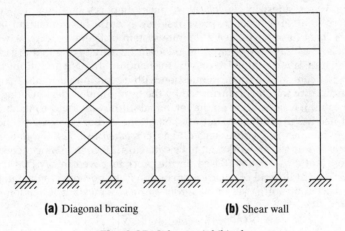

(a) Diagonal bracing **(b)** Shear wall

Fig. 3.25 Sidesway inhibited.

The LRFD specification (C2) states that $K = 1.0$ should be used for columns in frames with sidesway inhibited unless an analysis shows that a smaller value can be used. $K = 1.0$ is often quite conservative, and an analysis made as described here may result in some appreciable savings.

The effective length of a column is a property of the whole structure of which the column is a part. In many existing buildings it is probable that the masonry walls provide sufficient lateral support to prevent sidesway. When light curtain walls are used, however, as they often are in modern buildings, there is probably little resistance to sidesway. Sidesway is also present in tall buildings in appreciable amounts unless a definite diagonal bracing system or shear walls are used. For these cases it seems logical to assume that resistance to sidesway is primarily provided by the lateral stiffness of the frame alone.

Theoretical mathematical analyses may be used to determine effective lengths, but such procedures are usually too lengthy and perhaps too difficult for the average designer. The usual procedure is to use either Table 3.3, interpolating between the idealized values as the designer feels appropriate, or the alignment charts that are described in this section.

The charts shown in Fig. 3.26 present a practical method for estimating K values.[2,3] They were developed from a slope-deflection analysis of the frames that includes the effect of column loads. One chart was developed for columns braced against sidesway and one for columns subject to sidesway. Their use enables the designer to obtain good K values without struggling through lengthy trial-and-error procedures with the buckling equations.

To use the alignment charts it is necessary to have preliminary sizes for the girders and columns framing into the column in question before the K factor can be determined for that column. In other words, before the chart can be used we have to either assume some member sizes or carry out a preliminary design.

When we say sidesway is *inhibited* we mean there is something present other than just columns and girders to prevent sidesway or the horizontal translation of the joints. That means we have a definite system of lateral bracing or we have shear walls. If we say that sidesway is *uninhibited* we are saying that resistance to horizontal translation is supplied only by the bending strength and stiffness of the girders and beams of the frame in question with its continuous joints.

The resistance to rotation furnished by the beams and girders meeting at one end of a column is dependent on the rotational stiffnesses of those members. The moment needed to produce a unit rotation at one end of a member if the other end of the member is fixed is referred to as its *rotational stiffness*. From our structural analysis studies this works out to be equal to $4EI/L$ for a homogeneous member of constant cross section. Based on the preceding we can say that the rotational restraint at the end of a particular column is proportional to the ratio of the sum of the column stiffnesses to the girder stiffnesses meeting at that joint.[11]

$$G = \frac{\sum(4EI/L) \text{ for columns}}{\sum(4EI/L) \text{ for girders}} = \frac{\sum(I_c/L_c)}{\sum(I_g/L_g)}$$

To determine a K value for a particular column, the following steps are taken:

1. Select the appropriate chart (sidesway prevented or sidesway uninhibited).
2. Compute G at each end of the column and label the values G_A and G_B as desired.
3. Draw a straight line on the chart between the G_A and G_B values and read K where the line hits the center K scale.

When G factors are being computed for a rigid frame structure (rigid in both directions) the torsional resistance of the girders is generally neglected in the calculations. With reference to Fig. 3.27 it is assumed that we are calculating G for the

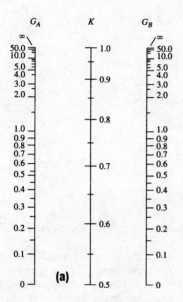

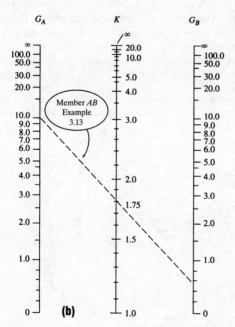

The subscripts A and B refer to the joints at the two ends of the column section being considered. G is defined as

$$G = \frac{\Sigma \dfrac{I_c}{L_c}}{\Sigma \dfrac{I_g}{L_g}}$$

in which Σ indicates a summation of all members rigidly connected to that joint and lying in the plane in which buckling of the column is being considered, I_c is the moment of inertia and L_c the unsupported length of a column section, and I_g is the moment of inertia and L_g the unsupported length of a girder or other restraining member. I_c and I_g are taken about axes perpendicular to the plane of buckling being considered.

For column ends supported by but not rigidly connected to a footing or foundation, G is theoretically infinity, but, unless actually designed as a true friction free pin, may be taken as '10' for practical designs. If the column end is rigidly attached to a properly designed footing, G may be taken as 1.0. Smaller values may be used if justified by analysis.

From American Institute of Steel Construction, *Manual of Steel Construction Load & Resistance Factor Design*, vol. I, 2d ed. (Chicago: AISC, 1994), p. 6-186.

Fig. 3.26 Alignment charts for effective lengths of columns in continuous frames. (*a*) Sidesway prevented. (*b*) Sidesway uninhibited.

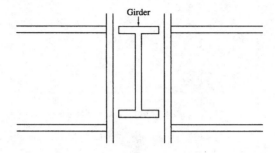

Fig. 3.27

joint shown for buckling in the plane of the paper. For such a case the torsional resistance of the girder shown, which is perpendicular to the plane being considered, is probably neglected.

The effective lengths of each of the columns of a frame are estimated with the alignment charts in Example 3.13. (When sidesway is possible, it will be found that the effective lengths are always greater than the actual lengths as is illustrated in this example. When frames are braced in such a manner that sidesway is not possible, K will be less than 1.0.) An initial design has provided preliminary sizes for each of the members in the frame. After the effective lengths are determined, each column can be redesigned. Should the sizes change appreciably, new effective lengths can be determined, the column designs repeated, and so on. Several tables are used in the solution of this example. These should be self-explanatory after the clear directions given on the alignment chart are examined.

Example 3.13 ■ Determine the effective lengths of each of the columns of the frame shown in Fig. 3.28 if the frame is not braced against sidesway. Use the alignment charts of Fig. 3.26b.

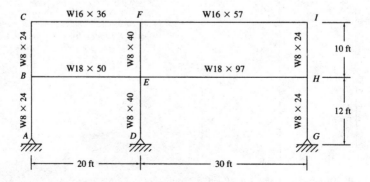

Fig. 3.28

Design of Tension and Compression Members ■ 3-37

Solution. Stiffness Factors:

Member	Shape	I	L	I/L
AB	W8×24	82.8	144	0.575
BC	W8×24	82.8	120	0.690
DE	W8×40	146	144	1.014
EF	W8×40	146	120	1.217
GH	W8×24	82.8	144	0.575
HI	W8×24	82.8	120	0.690
BE	W18×50	800	240	3.333
CF	W16×36	448	240	1.867
EH	W18×97	1750	360	4.861
FI	W16×57	758	360	2.106

G factors for each joint:

Joint	$\Sigma(I_c/L_c)/\Sigma(I_g/L_g)$	G
A	See Fig. 3-26(b)	10.0
B	$\dfrac{0.575 + 0.690}{3.333}$	0.380
C	$\dfrac{0.690}{1.867}$	0.370
D	See Fig. 7-2(b)	10.0
E	$\dfrac{1.014 + 1.217}{3.333 + 4.861}$	0.272
F	$\dfrac{1.217}{1.867 + 2.106}$	0.306
G	See Fig. 3-26(b)	10.0
H	$\dfrac{0.575 + 0.690}{4.861}$	0.260
I	$\dfrac{0.690}{2.106}$	0.328

Column K factors from chart (Fig. 3.26b):

Column	G_A	G_B	K*
AB	10.0	0.380	1.75
BC	0.380	0.370	1.13
DE	10.0	0.272	1.73
EF	0.272	0.306	1.11
GH	10.0	0.260	1.73
HI	0.260	0.328	1.11

*It is a little difficult to read the charts to the 3 places shown by the author. He has used a larger copy of Fig. 3.26 for his work. For all practical design purposes, however, the K values can be read to two places which can easily be accomplished with this figure.

For most buildings the values of K_x and K_y should be examined separately. The reason for such individual study lies in the different possible framing conditions in the two directions. Many multistory frames consist of rigid frames in one direction and conventionally connected frames with sway bracing in the other. In addition the points of lateral support may often be entirely different in the two planes.

There is available a set of rather simple equations for computing effective length factors. On some occasions the designer may find these expressions very convenient to use as compared to the alignment charts just described. Perhaps the most useful situation is for computer programs. You can see that it would be rather inconvenient to stop occasionally in the middle of a computer design to read K factors from the charts and input them to the computer. The equations, however, can easily be included in the programs, eliminating the necessity of using alignment charts.[12]

The alignment chart of Fig. 3.26b for frames with sidesway uninhibited always indicates that $K \geq 1.0$. In fact calculated K factors of 2.0 to 3.0 are common, and even larger values are occasionally obtained. To many designers such large factors seem completely unreasonable. If the designer obtains seemingly high K factors, he or she should carefully review the numbers used to enter the chart (that is, the G values) as well as the basic assumptions used in preparing the charts. These assumptions are discussed in detail in Sec. 3.3.9.

3.3.3 Stiffened and Unstiffened Elements

Up to this point the author has only considered the overall stability of members and yet it is entirely possible for the thin flanges or webs of a column or beam to buckle locally in compression well before the calculated buckling strength of the whole member is reached. When thin plates are used to carry compressive stresses they are particularly susceptible to buckling about their weak axes due to the small moments of inertia in those directions.

The LRFD specification (Section B5) provides limiting values for the width-thickness ratios of the individual parts of compression members and for the parts of beams in their compression regions. The reader is quite well aware of the lack of stiffness of thin pieces of cardboard or metal or plastic with free edges. If, however, one of these elements is folded or restrained, its stiffness is appreciably increased. For this reason two categories are listed in the LRFD Manual: *stiffened elements* and *unstiffened elements*.

An unstiffened element is a projecting piece with one free edge parallel to the direction of the compression force, while a stiffened element is supported along the two edges in that direction. These two types of elements are illustrated in Fig. 3.29. In each case the width b and the thickness t of the elements in question are shown.

Depending on the ranges of different width thickness ratios for compression elements and depending on whether the elements are stiffened or unstiffened, the elements will buckle at different stress situations.

For establishing width thickness ratio limits for the elements of compression members, the LRFD specification divides members into three classifications as follows: compact sections, noncompact sections, and slender compression elements. These classifications, which decidedly affect the design compression stresses to be used for columns, are discussed in the paragraphs to follow.

Compact Sections ■ A compact section is one that has a sufficiently stocky profile so that it is capable of developing a fully plastic stress distribution before buckling. The term *plastic* means stressed throughout to the yield stress. For a

Design of Tension and Compression Members ■ 3-39

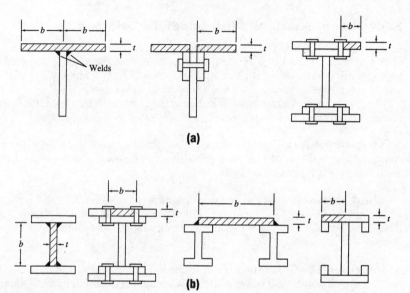

Fig. 3.29 Stiffened and unstiffened elements. (*a*) Unstiffened elements. (*b*) Stiffened elements.

compression member to be classified as compact its flanges must be continuously connected to its web or webs and the width-thickness ratios of its compression elements may not be greater than the limiting ratios λ_p given in Table B5.1 of Part 6 of the LRFD Manual.

Noncompact Sections ■ A noncompact section is one for which the yield stress can be reached in some but not all of its compression elements before buckling occurs. It is not capable of reaching a fully plastic stress distribution. In LRFD Table B5.1 the noncompact sections are those which have width-thickness ratios greater than λ_p but not greater than λ_r.

Slender Compression Elements ■ A slender element with a cross section that does not satisfy the width thickness requirements of Table B5.1 can still be used as a column, but the procedure for doing so is very complex. Furthermore the reduction in design stress is severe. As a result it is usually more economical to thicken members to take them out of the slender range.

Almost all of the W, M, and S shapes listed in the LRFD Manual are compact for 36 and 50 ksi yield stress steels. A few of them are noncompact (and are so indicated in the column and beam tables of the Manual). None of them are classified as being slender for these two yield stresses.

Should the width thickness limits for noncompact sections be exceeded, Appendix B5.3 of the LRFD specification must be consulted. The formulas presented there are so complex and tedious to apply that we would be wise to almost never use members falling in this classification.

3.3.4 Long, Short, and Intermediate Columns

Long steel columns will fail at loads that are proportional to the bending rigidity of the column (EI) and independent of the strength of the steel. For instance, a long column constructed with a 36 ksi yield stress steel will fail at just about the same load as one constructed with a 100 ksi yield stress steel.

Columns are sometimes classed as being long, short, or intermediate. A brief discussion of each of these classifications is presented in the paragraphs to follow.

Long Columns ■ The Euler formula predicts very well the strength of long columns where the axial buckling stress remains below the proportional limit. Such columns will buckle *elastically*.

Short Columns ■ For very short columns the failure stress will equal the yield stress and no buckling will occur. (For a column to fall into this class it would have to be so short as to have no practical application. Thus no further reference is made to them here.)

Intermediate Columns ■ For intermediate columns some of the fibers will reach the yield stress and some will not. The members will fail by both yielding and buckling and their behavior is said to be *inelastic*. Most columns fall into this range. (For the Euler formula to be applicable for such columns it would have to be modified according to the reduced modulus concept or the tangent modulus concept to account for the presence of residual stresses.)

In Sec. 3.3.5 formulas are presented with which the LRFD estimates the strength of columns in these different ranges.

3.3.5 Column Formulas

The LRFD specification provides one equation (it's the Euler equation) for long columns with elastic buckling and an empirical parabolic equation for short and intermediate columns. With these equations a critical or buckling stress F_{cr} is determined for a compression member. Once this stress is computed for a particular compression member it is multiplied by the cross-sectional area of the member to obtain the member's nominal strength. The design strength of the member can then be determined as follows:

$$P_n = A_g F_{cr} \quad \text{(LRFD Equation E2-1)}$$

$$P_u = \phi_c A_g F_{cr} \text{ with } \phi = 0.85$$

One LRFD equation for F_{cr} is for inelastic buckling and the other is for elastic buckling. In both equations λ_c in easily remembered form is $\sqrt{F_y/F_e}$ where F_e is the Euler stress or $\pi^2 E/(KL/r)^2$. Substituting this value for F_e we get the form of λ_c given in the LRFD specification.

$$\lambda_c = \frac{KL}{r\pi}\sqrt{\frac{F_y}{E}} \quad \text{(LRFD Equation 2-4)}$$

Design of Tension and Compression Members ■ 3-41

Both equations for F_{cr} include the estimated effects of residual stresses and initial out-of-straightness of the members. The inelastic formula that follows is an empirical or test result formula.

$$F_{cr} = (0.658^{\lambda_c^2})F_y \text{ for } \lambda_c \leq 1.5 \qquad \text{(LRFD Equation E2-2)}$$

The other equation is for elastic or Euler buckling and is the familiar Euler equation multiplied by 0.877 to estimate the effect of out-of-straightness.

$$F_{cr} = \left(\frac{0.877}{\lambda_c^2}\right)F_y \text{ for } \lambda_c > 1.5 \qquad \text{(LRFD Equation E2-3)}$$

These equations are represented graphically in Fig. 3.30.

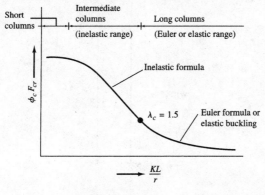

Fig. 3.30

3.3.6 Maximum Slenderness Ratios

In Part 6, Section B7, the LRFD states that compression members *preferably* should be designed with KL/r ratios not exceeding 200. The reader might note from LRFD Tables 3-36 and 3-50 that design stresses $\phi_c F_{cr}$ for KL/r values of 200 are both 5.33 ksi. Should slenderness ratios larger than these be used the $\phi_c F_{cr}$ values will be very small and will require the user to substitute into the column formulas provided in Sec. 3.3.5.

3.3.7 Example Problems

In this section three simple numerical column problems are presented. In each case the design strength of a column is calculated. In Example 3.14a the author determines the strength of a W section. The value of K is determined as described in Sec. 3.3.2, the effective slenderness ratio is computed, and the design stress of the member $\phi_c F_{cr}$ is selected from the appropriate LRFD table and multiplied by the cross-sectional area of the column.

3-42 ■ Chapter Three

It will be noted that the LRFD Manual in its Part 3 has further simplified the calculations required by computing the column design strength $\phi_c F_{cr} A_g$ for each of the steel shapes normally used as columns for F_y values of 36 and 50 ksi for most of the commonly used effective lengths or KL values given in feet. The use of these tables is illustrated in Example 3.14b.

Example 3.14 ■

a. Using the column design stress values shown in Table 3-50, Part 6 of the LRFD Manual, determine the design strength $(P_u = \phi_c P_n)$ of the $F_y = 50$ ksi axially loaded column shown in Fig. 3.31.

b. Repeat the problem using the column tables of Part 2 of the Manual.

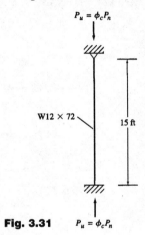

Fig. 3.31 $P_u = \phi_c P_n$

Solution.

a. Using a W12×72 (A=21.1 in², r_x=5.31 in, r_y=3.04 in)

$K = 0.80$ from Table 3.3

Obviously $(KL/r)_y > (KL/r)_x$ and thus controls

$$\left(\frac{KL}{r}\right)_y = \frac{(0.80)(12 \times 15)}{3.04} = 47.37$$

$\phi_c F_{cr} = 36.07$ ksi from Table 3-50, Part 6, LRFD Manual

$P_u = \phi_c P_n = \phi_c F_{cr} A_g = (36.07)(21.1) = 761.1$ k

b. Entering column tables Part 2 of Manual with $K_y L_y$ in feet.

$K_y L_y = (0.80)(15) = 12$ ft

$P_u = \phi_c P_n = 761$ k

Design of Tension and Compression Members ■ 3-43

In Example 3.15 the author illustrates the computations necessary to determine the design strength of a built-up column section. Several special requirements for built-up column sections are described in Secs. 3.3.11 to 3.3.14.

Example 3.15 ■ Determine the design strength $\phi_c P_n$ of the axially loaded column shown in Fig. 3.32 if $KL = 19$ ft and 50 ksi steel is used.

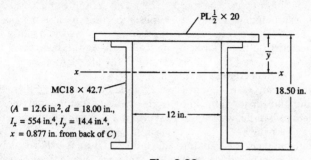

MC18 × 42.7

($A = 12.6$ in.^{2}, $d = 18.00$ in.,
$I_x = 554$ in.^{4}, $I_y = 14.4$ in.^{4},
$x = 0.877$ in. from back of C)

Fig. 3.32

Solution.

$$A = (20)(\tfrac{1}{2}) + (2)(12.6) = 35.2 \text{ in}^2$$

$$\bar{y} \text{ from top} = \frac{(10)(0.25) + (2)(12.6)(9.50)}{35.2} = 6.87 \text{ in}$$

$$I_x = (2)(554) + (25.2)(2.63)^2 + (\tfrac{1}{12})(20)(\tfrac{1}{2})^3 + (10)(6.62)^2 = 1721 \text{ in}^4$$

$$I_y = (2)(14.4) + (12.6)(6.877)^2(2) + (\tfrac{1}{12})(\tfrac{1}{2})(20)^3 = 1554 \text{ in}^4$$

$$\text{Least } r = r_y = \sqrt{\frac{1554}{35.2}} = 6.64 \text{ in}$$

$$\left(\frac{KL}{r}\right)_y = \frac{(12)(19)}{6.64} = 34.34$$

$$\phi_c F_{cr} = 38.99 \text{ ksi}$$

$$\phi_c P_n = P_u = (38.99)(35.2) = \underline{1372.4 \text{ k}}$$

To determine the design compression stress to be used for a particular column it is theoretically necessary to compute both $(KL/r)_x$ and $(KL/r)_y$. The reader will notice, however, that for most of the steel sections used for columns r_y will be much less than r_x. As a result only $(KL/r)_y$ is calculated for most columns and used in the applicable column formulas.

For some columns, particularly the long ones, bracing is supplied perpendicular to the weak axis, thus reducing the slenderness or the length free to buckle in that direction. This may be accomplished by framing braces or beams into the sides of a column. For instance, horizontal members called *girts* running parallel to the exterior walls of a building frame may be framed into the sides of columns. The result is stronger columns and ones for which the designer needs to calculate both $(KL/r)_x$ and $(KL/r)_y$. The larger ratio obtained for a particular column indicates the weaker direction and will be used for calculating the design stress $\phi_c F_{cr}$ for that member.

Bracing members must be capable of providing the necessary lateral forces without buckling themselves. The forces to be taken are quite small and are often conservatively estimated to equal 0.02 times the column design loads. These members can be selected as are other compression members. A bracing member must be connected to other members that can transfer the horizontal force by shear to the next restrained level. If this is not done, little lateral support will be provided for the original column in question.

If the lateral bracing were to consist of a single bar or rod ($\pm$) it would not prevent twisting and torsional buckling of the column. As torsional buckling is a difficult problem to handle, we should provide lateral bracing that prevents lateral movement and twist.[13]

Steel columns may also be built into substantial masonry walls in such a manner that they are substantially supported in the weaker direction. The designer, however, should be quite careful in assuming complete lateral support parallel to the wall, because a poorly built wall will not provide 100 percent lateral support.

Example 3.16 illustrates the calculations necessary to determine the design strength of a column with two unbraced lengths.

Example 3.16

a. Using Table 3-50 of Part 6 of the LRFD Manual, determine the design strength $\phi_c P_n$ of the 50 ksi axially loaded W14×90 shown in Fig. 3.33. Because of its considerable length this column is braced perpendicular to its weak or y axis at the points shown in the figure. These connections are assumed to permit rotation of the member but to prevent translation or sidesway.

b. Repeat part *a* using the column tables of Part 2 of the Manual.

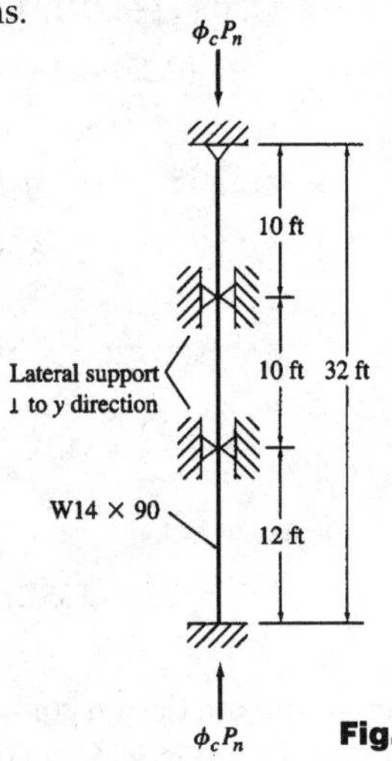

Fig. 3.33

Solution.

a. Using W14×90 (A = 26.5 in², r_x = 6.14 in, r_y = 3.70 in)
Determining effective lengths

$$K_x L_x = (0.80)(32) = 25.6 \text{ ft}$$
$$K_y L_y = (1.0)(10) = 10 \text{ ft} \leftarrow$$
$$K_y L_y = (0.80)(12) = 9.6 \text{ ft}$$

Computing slenderness ratios

$$\left(\frac{KL}{r}\right)_x = \frac{(12)(25.6)}{6.14} = 50.03 \leftarrow$$

$$\left(\frac{KL}{r}\right)_y = \frac{(12)(10)}{3.70} = 32.43$$

$\phi_c F_{cr} = 35.39$ ksi

$\phi_c P_n = (35.39)(26.5) = \underline{937.8 \text{ k}}$

b. Noting from part *a* solution that there are two different KL values

$K_x L_x = 25.6$ ft

$K_y L_y = 10$ ft

Controlling $K_y L_y$ for use in tables is either 10 ft or $K_x L_x / r_x/r_y$

$\dfrac{r_x}{r_y}$ for W14×90 from column tables = 1.66

$\dfrac{K_x L_x}{r_x / r_y} = \dfrac{25.6}{1.66} = 15.42$ ft $\leftarrow$

From column tables with $K_y L_y = 15.42$ ft we find by interpolation

$\phi_c P_n = \underline{937.8 \text{ k}}$

3.3.8 Design of Axially Loaded Columns

The design of columns with formulas involves a trial-and-error process. The design stress $\phi_c F_{cr}$ is not known until a column size is selected and vice versa. Once a trial section is assumed, the r values for that section can be obtained and substituted into the appropriate column equation to determine its design stress. Examples 3.17 and 3.18 illustrate this procedure.

The designer may assume a design stress, divide that stress into the factored column load to give an estimated column area, select a column section which has approximately that area, determine its design stress, multiply that stress by the cross-sectional area of the section to obtain the member's design strength to see if the section selected is over- or underdesigned, and if appreciably so, try another size. The reader may feel that he or she does not have sufficient background or knowledge to make reasonable initial design stress assumptions. If this person will read the information contained in the next few paragraphs, however, he or she will immediately be able to make excellent estimates.

The effective slenderness ratio (KL/r) for the average column of 10- to 15-ft length will generally fall between about 40 and 60. If for a particular column a KL/r

3-46 ■ Chapter Three

somewhere in this approximate range is assumed and substituted into the appropriate column equation (in the LRFD this usually means looking into the tables where the design stresses have already been calculated for KL/r values from 0 to 200), the result will usually be a very satisfactory design stress estimate.

In Example 3.17 a column with $KL = 10$ ft is selected using the LRFD formulas. An effective slenderness ratio of 50 is assumed, the design stress for that value is determined from Table 3-50 of Part 6 of the Manual, and the resulting stress is divided into the factored column load to obtain an estimated column area. After a trial section is selected with approximately that area, its actual slenderness ratio and design strength are determined. The first estimated size in Example 3.17, though quite close, is a little too small, but the next larger section in that series of shapes is tried and found to be satisfactory.

To estimate the effective slenderness ratio for a particular column the designer may estimate a value a little higher than 40 to 60 if the column is appreciably longer than the 10- to 15-ft range and vice versa. A very heavy factored column load, say in the 750- or 1000-k range or higher, will require a rather large column for which the radii of gyration will be larger and the designer may estimate a little smaller value of KL/r. For lightly loaded bracing members he or she may estimate high slenderness ratios, perhaps over 100.

Example 3.17 ■ Using $F_y = 36$ ksi select the lightest W14 available for the service column loads $P_D = 100$ k and $P_L = 160$ k. $KL = 10$ ft.

Solution.

$$P_u = (1.2)(100) + (1.6)(160) = 376 \text{ k}$$

Assume $KL/r = 50$

$\phi_c F_{cr}$ from Table 3-36 (Part 6 of Manual) = 26.83 ksi

$$A \text{ required} = \frac{376}{26.83} = 14.01 \text{ in}^2$$

Try W14×48 ($A = 14.1 \text{ in}^2$, $r_x = 5.85$ in, $r_y = 1.91$ in)

Obviously $(KL/r)_y > (KL/r)_x$ and $\therefore$ controls

$$(KL/r)_y = \frac{(12)(10)}{1.91} = 62.83$$

$\phi_c F_{cr}$ from Table 3-36 = 24.86 ksi

$$\psi_c \iota_n = (24.86)(14.1) = 350 \text{ k} < 376 \text{ k} \quad \text{(NG)}$$

$\therefore$ try next larger W14.

Try W14×53 ($A = 15.6 \text{ in}^2$, $r_x = 5.89$ in, $r_y = 1.92$ in)

$$\left(\frac{KL}{r}\right)_y = \frac{(12)(10)}{1.92} = 62.5$$

$\phi_c F_{cr} = 24.91$ ksi

$\phi_c P_n = (24.91)(15.6) = 388.6$ k > 376 k (OK)

Use W14×53

For Example 3.18, Part 3 of the LRFD Manual is used to select various column sections from tables without the necessity of using a trial-and-error process. These tables provide axial design strengths ($\phi_c P_n$) for various practical effective lengths of the steel sections commonly used as columns (W's, pipes, tubes, pairs of angles, and structural tees). The values are given with respect to the least radius of gyration for steels with $F_y = 36$ ksi and 50 ksi (except that the pipe strengths are given for 36 ksi steel only, while the strengths for the square and rectangular tubes are for 46 ksi steel).

For most columns consisting of single steel shapes the effective slenderness ratio with respect to the y axis $(KL/r)_y$ is larger than the effective slenderness ratio with respect to the x axis $(KL/r)_x$. As a result the controlling or smaller design stress is for the y axis. Because of this, the LRFD tables provide design strengths of columns with respect to their y axes. We will learn in the pages to follow how to use the AISC tables and handle situations in which $(KL/r)_x$ is larger than $(KL/r)_y$.

The resulting tables are very simple to use. The designer takes $K_y L_y$ in feet, enters the table in question from the left-hand side, and moves horizontally across the table. Under each section is listed the design strength $\phi_c P_n$ for that KL and the steel yield stress. (The 50 ksi column values are shaded in the tables.) As an illustration it is assumed that we have a factored design load $P_u = \phi_c P_n = 1200$ k, $K_y L_y = 12$ ft, and we want to select the lightest available W14 section using 50 ksi steel. We enter the tables with $KL = 12$ ft in the left column and read from left to right for 50 ksi steel the numbers 7240, 8530, 7760, 7030 k, and so on until several pages later where the consecutive values 1220 and 1110 k are found. The 1110 k value is not sufficient, and we go back to the 1220 k which falls under the W14×109. A similar procedure can be followed for the other available shapes.

Example 3.18 which follows, illustrates the selection of W sections as well as pipes and tubes. It is possible to support a given column load with a standard pipe column; or with an extra strong pipe column (× strong) which has a smaller diameter but thicker walls, thus is heavier and more expensive; or with a double extra strong pipe column (×× strong) which has an even smaller diameter and even thicker walls and heavier weight.

Example 3.18 ■ Using the LRFD column tables of Part 3 of the Manual.

a. Select the lightest W section available for the loads, steel, and KL of Example 3.17.

b. Select the lightest standard, extra strong, and double extra strong pipe columns for the situation of part *a* of this problem.

c. Select the lightest square and rectangular tubes satisfactory for the situation of part *a*, except use an F_y of 46 ksi.

Solution.

    a. Enter tables with $K_y L_y = 10$ ft and $P_u = \phi_c P_n = 376$ k
        Lightest suitable section in each W series

$$W14 \times 53 (\phi_c P_n = 389 \text{ k})$$
$$W12 \times 53 (\phi_c P_n = 422 \text{ k})$$
$$W10 \times 49 (\phi_c P_n = 392 \text{ k}) \leftarrow$$
$$W8 \times 58 (\phi_c P_n = 441 \text{ k})$$
$$\underline{\text{Use } W10 \times 49}$$

    b. Pipe columns

$$\text{Pipe 12 std } (\phi_c P_n = 429 \text{ k}) \text{ wt} = 49.56 \text{ lb/ft}$$
$$\text{Pipe 10} \times \text{strong } (\phi_c P_n = 465 \text{ k}) \text{ wt} = 54.74 \text{ lb/ft}$$
$$\text{Pipe 6} \times \times \text{ strong } (\phi_c P_n = 399 \text{ k}) \text{ wt} = 53.16 \text{ lb/ft}$$

    c. Square and rectangular tubing ($F_y = 46$ ksi)

$$8 \times 8 \times \tfrac{3}{8} (\phi_c P_n = 392 \text{ k}) \text{ wt} = 37.60 \text{ lb/ft}$$
$$12 \times 10 \times \tfrac{1}{4} (\phi_c P_n = 390 \text{ k}) \text{ wt} = 36.03 \text{ lb/ft}$$

An axially loaded column is laterally restrained in its weak direction in Fig. 3.34. Example 3.19 illustrates the design of such a column with its different unsupported lengths in the x and y directions. The reader can easily solve this problem by trial and error. A trial section can be selected as described earlier in this section, the slenderness values $(KL/r)_x$ and $(KL/r)_y$ computed, $\phi_c F_{cr}$ determined for the larger value and multiplied by A_g to obtain $\phi_c P_n$. Then, if necessary, another size can be tried, and so on.

For this discussion it is assumed that K is the same in both directions. Then if we are to have equal strengths about the x and y axes the following relation must hold:

$$\frac{L_x}{r_x} = \frac{L_y}{r_y}$$

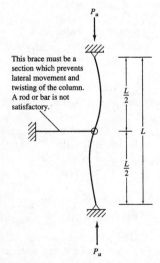

Fig. 3.34 A column laterally restrained at middepth in its weaker direction.

For L_y to be equivalent to L_x we would have

$$L_x = L_y \frac{r_x}{r_y}$$

If $L_y(r_x/r_y)$ is less than L_x then L_x controls; if greater than L_x then L_y controls.

Based on the preceding information the LRFD Manual provides a method with which a section can be selected from its tables with little trial and error work when the unbraced lengths are different. The designer enters the appropriate table with K_yL_y, selects a shape, takes the r_x/r_y value given in the table for that shape, and multiplies it by L_y. If the result is larger than K_xL_x, then K_yL_y controls and the shape initially selected is the correct one. If the result of the multiplication is less than K_xL_x then K_xL_x controls and the designer will reenter the tables with a larger K_yL_y equal to $K_xL_x/(r_x/r_y)$ and select the final section.

Example 3.19 illustrates the two procedures described here for selecting a W section which has different effective lengths in the x and y directions.

Example 3.19 ▪ Select the lightest satisfactory W12 for the following conditions: $F_y = 50$ ksi, $P_u = 900$ k, $K_xL_x = 26$ ft, and $K_yL_y = 13$ ft.

a. By trial and error
b. Using LRFD tables

Solution.

a. Using trial and error

Assume $\dfrac{KL}{r} = 50$

$\phi_c F_{cr} = 35.40$ ksi

A required $= \dfrac{900}{35.40} = 25.42$ in^2

Try W12×87 ($A = 25.6$ in^2, $r_x = 5.38$ in, $r_y = 3.07$ in)

$\left(\dfrac{KL}{r}\right)_x = \dfrac{(12)(26)}{5.38} = 57.99 \leftarrow$

$\left(\dfrac{KL}{r}\right)_y = \dfrac{(12)(13)}{3.07} = 50.81$

$\phi_c F_{cr} = 33.23$ ksi

$\phi_c P_n = (33.23)(25.6) = 850.6$ k < 900 k (NG)

A subsequent check of the next larger W section (W12×96) shows it will work.

Use W12×96

b. Using LRFD tables

Enter tables with $K_y L_y = 13$ ft

Try W12×87 $\left(\dfrac{r_x}{r_y} = 1.75\right)$ with $\phi_c P_n$ based on $K_y L_y$

$$(K_y L_y)\left(\dfrac{r_x}{r_y}\right) = (13)(1.75) = 22.75 < K_x L_x$$

Therefore, $K_x L_x$ controls.

Reenter tables with new $K_y L_y = \dfrac{K_x L_x}{r_x / r_y} = \dfrac{26}{1.75} = 14.86$

Use W12×96

3.3.9 Stiffness Reduction Factors

The alignment charts were developed on the basis of a set of idealized conditions which are seldom if ever completely met in a real structure. A complete list of these assumptions is shown in Section C2 of the Commentary on the LRFD specification. Among these are: column behavior is purely elastic, all columns buckle simultaneously, all members have constant cross sections, all joints are rigid, and so on.

If the actual conditions are different from these assumptions, unrealistically high K factors may be obtained from the charts, and overconservative designs may result. A large percentage of columns will fall in the inelastic range, but the alignment charts were prepared assuming elastic failure. This situation, previously discussed in Sec. 3.3.5, is illustrated in Fig. 3.35. For such cases the alignment chart K values are too conservative and should be corrected as described in this section.

In the elastic range the stiffness of a column is proportional to EI where $E = 29,000$ ksi, while in the inelastic range its stiffness is more accurately proportional to $E_T I$ where E_T is a reduced or tangent modulus.

The buckling strength of columns in framed structures was shown in the alignment charts to be related to

$$G = \dfrac{\text{column stiffness}}{\text{girder stiffness}} = \dfrac{\Sigma(EI/L) \text{ columns}}{\Sigma(EI/L) \text{ girders}}$$

If the columns behave elastically, the modulus of elasticity will be canceled from the preceding expression for G. If the column behavior is inelastic, however, the column stiffness factor will be smaller and will equal $E_T I/L$. As a result

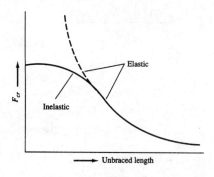

Fig. 3.35

Design of Tension and Compression Members ■ 3-51

the G factor used to enter the alignment chart will be smaller, and the K factor selected from the chart will be smaller.

Though the alignment charts were developed for elastic column action, they may be used for an inelastic column situation if the G value is multiplied by a correction factor called the *stiffness reduction factor* (SRF). This reduction factor equals the tangent modulus over the elastic modulus (E_T/E) and is approximately equal to $F_{cr\,inelastic}/F_{cr\,elastic} \approx (P_u/A)/F_{cr\,elastic}$. Values of this correction factor are shown for various P_u/A values in Table 3.4 which is Table 3.1 in the LRFD Manual. A direct design method for considering inelastic buckling is presented in the Manual. It involves the following steps:

1. Calculate P_u and select a trial column size.
2. Calculate P_u/A and pick the SRF from Table 3.4. (If P_u/A is less than the values given in the table the column is in the elastic range and no reduction needs to be made.)
3. The value of $G_{elastic}$ is computed and multiplied by the SRF and K is picked from the chart.
4. The effective slenderness ratio KL/r is computed, and $\phi_c F_{cr}$ obtained from the Manual is multiplied by the column area to obtain P_u. If this value is appreciably different from the value computed in step 1, another trial column size is attempted and the four steps are repeated.

Table 3.4 Stiffness Reduction Factors (SRF) for Columns

P_u/A ksi	F_y 36 ksi	F_y 50 ksi	P_u/A ksi	F_y 36 ksi	F_y 50 ksi
42	—	0.03	26	0.38	0.82
41	—	0.09	25	0.45	0.85
40	—	0.16	24	0.52	0.88
39	—	0.21	23	0.58	0.90
38	—	0.27	22	0.65	0.93
37	—	0.33	21	0.70	0.95
36	—	0.38	20	0.76	0.97
35	—	0.44	19	0.81	0.98
34	—	0.49	18	0.85	0.99
33	—	0.53	17	0.89	1.00
32	—	0.58	16	0.92	↓
31	—	0.63	15	0.95	
30	0.05	0.67	14	0.97	
29	0.14	0.71	13	0.99	
28	0.22	0.75	12	1.00	
27	0.30	0.79	11	↓	

— Indicates not applicable

Source: American Institute of Steel Construction, *Manual of Steel Construction Load & Resistance Factor Design*, 2d ed., vol. I, AISC, Chicago, 1994, p. 3–7. Reprinted with the permission of AISC.

3-52 ■ Chapter Three

Example 3.20 which follows illustrates these steps for the design of a column in a frame subject to sidesway. *In this example, note that the author has only considered in-plane behavior and only bending about the x axis. As a result of inelastic behavior, the effective length factor is appreciably reduced.*

Example 3.20 ■ Select a W12 section for column AB of the frame shown in Fig. 3.36 assuming (*a*) elastic column behavior and (*b*) inelastic column behavior. $P_u = 1290$ k and A36 steel is used. The columns above and below AB are assumed to be approximately the same size as AB. Consider only in-plane behavior.

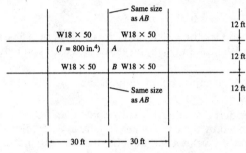

Fig. 3.36

Solution.
   a. Assuming column in elastic range and selecting a trial section based on an estimated $K_y L_y = 12$ ft
     Try W12 × 170 ($A = 50.0$ in^2, $I_x = 1650$ in^4, $r_x = 5.74$ in)

$$G_A = G_B = \frac{\Sigma(I_c/L_c)}{\Sigma(I_g/L_g)} = \frac{(2)(1650/12)}{(2)(800/30)} = 5.16$$

     $K = 2.26$ from Fig. 7-1 alignment chart

$$\left(\frac{KL}{r}\right)_x = \frac{(2.26)(12 \times 12)}{5.74} = 56.7$$

     $\phi_c F_{cr} = 25.83$ ksi
     $P_u = (25.83)(50.0) = 1291.5$ k > 1290 k   (OK)
              Use W12×170

   b. Inelastic solution
     Try the next lightest section W12×152 ($A = 44.7$ in^2, $I_x = 1430$ in^4, $r_x = 5.66$ in)

$$\frac{P_u}{A} = \frac{1290}{44.7} = 28.86 \text{ ksi}$$

     SRF = 0.151 from Table 3-1 in Manual
        ∴ Column is in inelastic range

Design of Tension and Compression Members ■ 3-53

$$G_A = G_B = \frac{\Sigma(I_c/L_c)}{\Sigma(I_g/L_g)}(SRF)$$

$$= \frac{(2)(1430/12)}{(2)(800/30)}(0.151) = 0.675$$

$K = 1.24$ from Fig 3.21*b* alignment chart

$$\frac{KL}{r} = \frac{(1.24)(12 \times 12)}{5.66} = 31.55$$

$\phi_c F_{cr} = 29.06$ ksi

$P_u = (29.06)(44.7) = 1299$ k > 1290 k

Use W12×152

3.3.10 Single-Angle Compression Members

You will note that the author has not discussed the design of single-angle compression members up to this point. The AISC has long been concerned about the problems involved in loading such members fairly concentrically. It can be done rather well if the ends of the angles are milled and if the loads are applied through bearing plates. In practice, however, single-angle columns are often used in such a manner that rather large eccentricities of load applications are present. The sad result is that it is somewhat easy to greatly underdesign such members.

In Part 6 of the LRFD Manual a special specification is provided for the design of single-angle compression members. Though this specification includes information for tensile, shear, compressive, flexural, and combined loadings, the present discussion is concerned only with the compression case. Some rather complicated formulas are presented in the specification for computing the axial design strengths of single angles.

These formulas were developed to account for three different limit states that might occur in single-angle columns. These are flexural buckling, local buckling of thin angle legs, and flexural torsional buckling. Using these expressions the design strengths of concentrically loaded single angles are computed for 36 ksi and 50 ksi steels for a range of *KL* values. These strengths are shown at the end of Part 3 of the LRFD Manual.

Just before the single-angle column tables are presented in Part 3 of the Manual, a very practical example problem is presented in which a single-angle strut loaded with an estimated eccentricity is analyzed.

3.3.11 Built-up Columns

Compression members may be constructed with two or more shapes built up into a single member. They may consist of parts in contact with each other, such as cover-plated sections I or they may consist of parts in near contact with each other, such as pairs of angles ⊤⊤ that may he separated by a small distance from each other equal to the thickness of the end connections or gusset plates between them. They may consist of parts that are spread well apart, such as pairs of channels][or four angles]⋅[, and so on.

Two-angle sections probably are the most common type of built-up member. (For example, they frequently are used as the members of light trusses.) When a pair of angles are used as a compression member they need to be fastened together so they will act as a unit. Welds may be used at intervals (with a spacer bar between the parts if the angles are separated) or they may be connected with "stitch bolts." When the connections are bolted, washers or "ring fills" are placed between the parts to keep them at the proper spacing if the angles are to be separated.

For long columns it may be convenient to use built-up sections where the parts of the columns are spread out or widely separated from each other. Before heavy W sections were made available, such sections were very commonly used in both buildings and bridges. Today these types of built-up columns are commonly used for crane booms and for the compression members of various kinds of towers. The widely spaced parts of these types of built-up members must be carefully laced or tied together.

3.3.12 Built-up Columns with Components in Contact with Each Other

Should a column consist of two equal size plates, as shown in Fig. 3.37, and should those plates not be connected together, each plate will act as a separate column and each will resist approximately half of the total column load. In other words, the total moment of inertia of the column will equal two times the moment of inertia of one plate. The two "columns" will act the same and have equal deformations, as shown in part b of the figure.

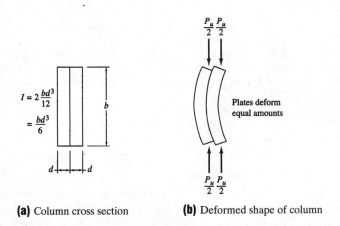

(a) Column cross section (b) Deformed shape of column

Fig. 3.37 Column consisting of two plates not connected to each other.

Design of Tension and Compression Members ■ 3-55

Should the two plates be connected together sufficiently to prevent slippage on each other, as shown in Fig. 3.38, they will act as a unit. Their moment of inertia may be computed for the whole built-up section as shown in the figure and will be four times as large as it was for the column of Fig. 3.37, where slipping between the plates was possible. The reader should also notice that the plates of the column of Fig. 3.38 will deform different amounts as the column bends laterally.

Should the plates be connected in a few places, it would appear that the strength of the resulting column would be somewhere in between the two cases just described.

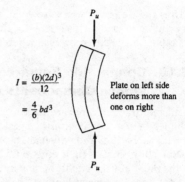

Fig. 3.38 Column consisting of two plates fully connected to each other.

Reference to Fig. 3.37b shows that the greatest displacement between the two plates tends to occur at the ends and the least displacement tends to occur at middepth. As a result connections placed at column ends which will prevent slipping between the parts have the greatest strengthening effect, while those unlaced at middepth have the least effect.

Should the plates be fastened together at their ends with slip-resistant connectors, those ends will deform together and the column will take the shape shown in Fig. 3.39. As the plates are held together at the ends the column will bend in an S shape as shown in the figure.

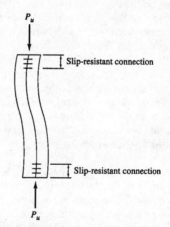

Fig. 3.39 Column consisting of two plates connected at its ends only.

If the column were to bend in the S shape shown, its K factor would theoretically equal 0.5 and its KL/r value would be the same as the one for the continuously connected column of Fig. 3.38.[14]

$$\frac{KL}{r} \text{ for the column of Fig. 3.38} = \frac{(1)(L)}{\sqrt{\frac{4}{6}bd^3 / 2bd}} = 1.732L$$

$$\frac{KL}{r} \text{ for the end-fastened column of Fig. 3.39} = \frac{(0.5)(L)}{\sqrt{\frac{1}{6}bd^3 / 2bd}} = 1.732L$$

Thus the design stresses are equal for the two cases and the columns would carry the same loads. This is true for the particular case described here, but is not applicable for the common case where the parts of Fig. 3.39 begin to separate.

3.3.13 Connection Requirements for Built-up Columns Whose Components Are in Contact with Each Other

Several requirements concerning built-up columns are presented in LRFD specification E4. When such columns consist of different components which are in contact with each other and which are bearing on base plates or milled surfaces, they must be connected at their ends with bolts or welds. If welds are used the weld lengths must at least equal the maximum width of the member. If bolts are used they may not be spaced longitudinally more than four diameters on center and the connection must extend for a distance at least equal to $1\frac{1}{2}$ times the maximum width of the member.

The LRFD specification also requires the use of welded or bolted connections between the end ones described in the last paragraph. These must be sufficient to provide for the transfer of calculated stresses. If it is desired to have a close fit over the entire faying surfaces between the components, it may be necessary to place the connectors even closer than is required for shear transfer.

When the component of a built-up column consists of an outside plate, the LRFD specification provides specific maximum spacings for fastenings If intermittent welds are used along the edges of the components or if bolts are provided along all gauge lines at each section, their maximum spacing may not be greater than $127/\sqrt{F_y}$ times the thickness of the thinner outside plate or 12 in. Should these fasteners be staggered on each gauge line, however, they may not be spaced farther apart on each gauge line than $190/\sqrt{F_y}$ times the thickness of the thinner part or 18 in.

In the following discussion the letter a represents the distance between connectors and r_i is the least radius of gyration of an individual component of the column. If compression members consisting of two or more shapes are used they must be connected together at intervals such that the effective slenderness ratio Ka/r_i of each of the component shapes between the connectors is not larger than $\frac{3}{4}$ times the governing or controlling slenderness ratio of the whole built-up member. The end connections must be made with welds or slip-critical bolts with clean mill scale or blasted cleaned faying surfaces with class A coatings. (A class A coating is one that has a mean slip coefficient not less than 0.33. See Section 5(b) in Specification for Structural Joints Using A325 or A490 Bolts in Part 6 of the LRFD Manual.) Class A surfaces are typical in structural steel work. Snug-tight bolts may be used for interior bolts.

The design strength of compression members built up from two or more shapes in contact with each other will be determined with the usual applicable LRFD equations (E2-1, E2-2, and E2-3) with one exception. Should the column tend to buckle in such a manner that relative deformations in the different parts cause shear forces in the connectors between the parts, it will be necessary to modify the KL/r value for that axis of buckling. This modification is required by Section E4 of the LRFD specification.

Reference is made here to the cover-plated column of Fig. 3.40. If this section tends to buckle about its y axis the connectors between the W shape and the plates are not subjected to any calculated load. If, on the other hand, it tends to buckle about its x axis the connectors are subjected to shearing forces. The flanges of the W section and the cover plates will have different stresses and thus different deformations. The result will be shear in the connection between these parts and $(KL/r)_x$ will have to be modified by LRFD Equation E4-1 or E4-2 as described below. Equation E4-1 is based upon test results that supposedly account for shear deformations in the connectors. Equation 4-2 is based upon theory and was checked by means of tests.

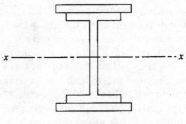

Fig. 3.40

a. For intermediate connectors that are snug-tight bolted:

$$\left(\frac{KL}{r}\right)_m = \sqrt{\left(\frac{KL}{r}\right)_0^2 + \left(\frac{a}{r_i}\right)^2} \qquad \text{(LRFD Equation E4-1)}$$

Note that it is important to remember that the design strength of a built-up column will be reduced if the spacing of connectors is such that one of the components of the column can buckle before the whole column buckles.

b. For intermediate connectors that are welded or have fully tensioned bolts as required for slip-critical joints:

$$\left(\frac{KL}{r}\right)_m = \sqrt{\left(\frac{KL}{r}\right)_0^2 + 0.82\frac{\alpha^2}{1+\alpha^2}\left(\frac{a}{r_{ib}}\right)^2} \qquad \text{(LRFD Equation E4-2)}$$

In these two equations

$(KL/r)_o$ = unmodified slenderness of the whole built-up member acting as a unit
$(KL/r)_m$ = modified slenderness of built-up member
a = distance between connectors, in
r_i = minimum radius of gyration of individual component, in
r_{ib} = radius of gyration of individual component relative to its centroidal axis parallel to the member axis of buckling, in
h = distance between centroids of individual components perpendicular to the member axis of buckling, in
α = separation ratio = $h/2r_{ib}$

For the case in which the column tends to buckle about an axis such as to cause shear in the connection between the column parts, it will be necessary to compute a modified slenderness ratio $(KL/r)_m$ for that axis and to check to see whether that value will cause a change in the design strength of the member. If it does, it may be necessary to revise sizes and repeat the steps just described.

Example 3.21 illustrates the design of a column consisting of a W section with cover plates bolted to its flanges as shown in Fig. 3.41. Even though snug-tight bolts are used for this column, you should realize that LRFD specification E4 states that the end bolts must be slip-critical or the ends must be welded. This is required so the parts of the built-up section will not slip with respect to each other and thus will act as a unit in resisting loads. (As a practical note, the usual steel company required to tighten the end bolts to a slip-critical condition will probably just go ahead and tighten them all to that condition.)

As this type of built-up section is not shown in the column tables of the LRFD Manual, it is necessary to use a trial-and-error design procedure. An effective slenderness ratio is assumed. Then, $\phi_c F_{cr}$ for that slenderness ratio is determined and divided into the column design load to estimate the column area required. The area of the W section is subtracted from the estimated total area to obtain the estimated cover plate area. Cover plate sizes are then selected to provide the required estimated area.

Next, the properties of the trial section are calculated. You will note that it is necessary to compute a modified value of $(KL/r)_x$. In this case the section tends to buckle about its y axis—thus longitudinal shear is not developed between the W section and the cover plates.

Example 3.21 ■ It is desired to design a column for $P_u = 2375$ k using $F_y = 50$ ksi and $KL = 14$ ft. A W12×120 (for which $\phi_c P_n = 1220$ k from the Part 3 tables of the Manual) is on hand. Design cover plates to be snug-tight bolted at 6 in spacings to the W section shown in Fig. 3.41 to enable the column to support the required load.

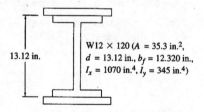

Fig. 3.41 W section used as column with cover plates.

Solution.

Assume $KL/r = 50$

$$\phi_c F_{cr} = 35.40 \text{ ksi}$$

$$A \text{ required} = \frac{2375}{35.40} = 67.09 \text{ in}^2$$

$$- A \text{ of } W12 \times 120 = \underline{-35.30}$$

Estimated A of 2 plates = 31.79 in^2 or 15.90 in^2 each

Try 1 PL1×16 each flange

$$A = 35.30 + (2)(1)(16) = 67.30 \text{ in}^2$$

$$I_x = 1070 + (2)(16)(7.06)^2 = 2665 \text{ in}^4$$

$$r_x = \sqrt{\frac{2665}{67.30}} = 6.29 \text{ in}$$

$$\left(\frac{KL}{r}\right)_x = \frac{(12)(14)}{6.29} = 26.71$$

$$I_y = 345 + (2)\left(\frac{1}{12}\right)(1)(16)^3 = 1027.7 \text{ in}^4$$

$$r_y = \sqrt{\frac{1027.7}{67.30}} = 3.91 \text{ in}$$

$$\left(\frac{KL}{r}\right)_y = \frac{(12)(14)}{3.91} = 42.97$$

$$P_u = (37.14)(67.30) = 2500 \text{ k} > 2375 \text{ k} \tag{OK}$$

Use W12×120 with 1 cover plate 1×16 each flange. (Note: Many other plate sizes could have been selected.)

3.3.14 Built-up Columns with Components Not in Contact with Each Other

The open sides of compression members that are built up from plates or shapes may be connected together with continuous cover plates with perforated holes for access purposes, or they may be connected together with lacing and tie plates.

The purpose of the perforated cover plates and the lacing or lattice work is to hold the various parts parallel and the correct distance apart and to equalize the stress distribution between the various parts. The reader will understand the necessity for lacing if he or she considers a built-up member consisting of several sections which supports a heavy compressive load. Each of the parts will tend to individually buckle laterally unless they are tied together to act as a unit in supporting the load. In addition to lacing, it is necessary to have tie plates (also called stay plates or batten plates) as near the ends of the member as possible and at intermediate points if the lacing is interrupted. Parts *a* and *b* of Fig. 3.42 show arrangements of tie plates and lacing. Other possibilities are shown in parts *c* and *d* of the same figure.

If continuous cover plates perforated with access holes are used to tie the members together, the LRFD specification E4 states that (*a*) they must comply with the limiting width thickness ratios specified for compression elements in Section B5.1 of the LRFD specification; (*b*) the ratio of the access hole length (in the direction of stress) to the hole width may not exceed 2; and (*c*) the clear distance between the holes in the direction of stress may not be less than the transverse distance between the nearest lines of connecting fasteners or welds. Stress concentrations and secondary bending stresses are usually neglected, but lateral shearing forces must be checked as they are for other types of lattice work. (The unsupported width of such plates at access holes is assumed to contribute to the design strength $\phi_c P_n$ of the member if the conditions as to sizes, width-thickness ratios, etc., described in the LRFD specification are met.)

Dimensions of tie plates and lacing are usually controlled by specifications. Section E4 of the LRFD specification states that tie plates shall have a thickness at least equal to one fiftieth of the distance between the connection lines of welds or other fasteners.

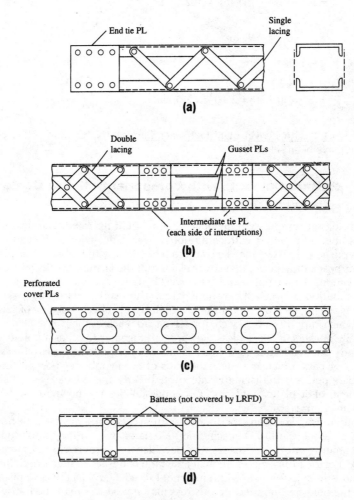

Fig. 3.42 Lacing and perforated cover plates.

Design of Tension and Compression Members ■ 3-61

Lacing may consist of flat bars, angles, channels, or other rolled sections. These pieces must be so spaced that the individual parts being connected will not have L/r values between connections which exceed the governing value for the entire built-up member. (The governing value is KL/r for the whole built-up section.) Lacing is assumed to be subjected to a shearing force normal to the member equal to not less than 2 percent of the compression design strength $\phi_c P_n$ of the member. The LRFD column formulas are used to design the lacing in the usual manner. Slenderness ratios are limited to 140 for single lacing and 200 for double lacing. Double lacing or single lacing made with angles should preferably be used if the distance between connection lines is greater than 15 in.

3.3.15 Flexural-Torsional Buckling of Compression Members

Axially loaded compression members can theoretically fail in three different fashions: by flexural buckling, by torsional buckling, or by flexural-torsional buckling.

Flexural buckling (also called Euler buckling) is the situation considered up to this point in our column discussions where we have computed slenderness ratios for the principal column axes and determined $\phi_c F_{cr}$ for the highest ratios so obtained. Doubly symmetrical column members (such as W sections) are subject only to flexural buckling and torsional buckling.

As torsional buckling can be very complex, it is very desirable to prevent its occurrence. This may be done by careful arrangements of the members and by providing bracing to prevent lateral movement and twisting. If sufficient end supports and intermediate lateral bracing are provided, flexural buckling will always control. The column design strengths given in the LRFD column tables for W, M, S, tube, and pipe sections are based on flexural buckling.

Open sections such as Ws, Ms, and channels have little torsional strength, but box beams have a great deal. Thus if a torsional situation is encountered, it may be well to use box sections or to make box sections out of W sections by adding welded side plates ☐. Another way in which torsional problems can be reduced is to shorten the lengths of members that are subject to torsion.

For a singly symmetrical section such as a tee or double angle, Euler buckling may occur about the x or y axis. For equal-leg single angles Euler buckling may occur about the z axis. For all these sections flexural-torsional buckling is definitely a possibility and may control. (It will always control for unequal-leg single-angle columns.) The values given in the LRFD column load tables for double-angle and structural tee sections were computed for buckling about the weaker of the x or y axis and for flexural-torsional buckling.

Usually symmetrical members such as W sections are used as columns. Torsion will not occur in such sections if the lines of action of the lateral loads pass through their shear centers. The shear centers of the commonly used doubly symmetrical sections occur at their centroids. This is not necessarily the case for other sections such as channels and angles. Shear center locations for several types of sections are shown in Fig 3.43. Also shown in the figure are the coordinates x_0 and y_0 for the shear center of each section with respect to its centroid. These values are needed to solve the flexural-torsional formulas, presented later in this section.

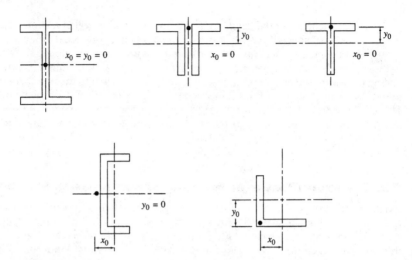

Fig. 3.43 Shear center locations for some common column sections.

Even though loads pass through shear centers, torsional buckling still may occur. If you load any section through its shear center no torsion will occur—but one still computes torsional buckling strength for these members, i.e., buckling load does not depend on the nature of the axial or transverse loading but, rather, on the cross-section properties, column length, and support conditions.

The average designer does not consider the torsional buckling of symmetrical shapes or the flexural-torsional buckling of unsymmetrical shapes. Their usual feeling is that these conditions don't control the critical column loads, or at least don't affect them very much. Should we, however, have unsymmetrical columns or even symmetrical columns made up of thin plates, we will find that torsional buckling or flexural torsional buckling may significantly reduce column capacities.

In Appendix E of the LRFD specification a long list of formulas is presented for computing the flexural-torsional strength of column sections. The values given for column design strengths ($\phi_c P_n$ values) for double angles, single angles, and tees in Part 3 of the LRFD Manual make use of these formulas.

For flexural-torsion $P_u = \phi_c P_n = \phi_c A_g F_{cr}$ with $\phi_c = 0.85$ and F_{cr} to be determined from the formulas to follow from the specification. A list of definitions that are needed for using these formulas also is provided.

If $\lambda_e \sqrt{Q} \leq 1.5$

$$F_{cr} = Q(0.658^{Q\lambda_e^2})F_y \qquad \text{(LRFD Equation A-E3-2)}$$

If $\lambda_e \sqrt{Q} > 1.5$

$$F_{cr} = \left(\frac{0.877}{\lambda_e^2}\right)F_y \qquad \text{(LRFD Equation A-E3-3)}$$

Design of Tension and Compression Members ■ 3-63

in which $Q = 1.0$ for elements meeting the width thickness ratios λ_r of LRFD section B5.1 and if not calculated as described in LRFD Appendixes E3 and B5.3.

$$\lambda_e = \sqrt{\frac{F_y}{F_e}} \qquad \text{(LRFD Equation A-E3-4)}$$

F_e = critical flexural - torsional elastic buckling stress

For doubly symmetric shapes

$$F_e = \left[\frac{\pi^2 E C_w}{(K_z L)^2} + GJ\right] \frac{1}{I_x + I_y} \qquad \text{(LRFD Equation A-E3-5)}$$

For singly symmetric shapes where y is the axis of symmetry

$$F_e = \frac{F_{ey} + F_{ez}}{2H}\left[1 - \sqrt{1 - \frac{4 F_{ey} F_{ez} H}{(F_{ey} + F_{ez})^2}}\right] \qquad \text{(LRFD Equation A-E3-6)}$$

For unsymmetrical section F_e is the lowest root of the following cubic equation

$$(F_e - F_{ex})(F_e - F_{ey})(F_e - F_{ez}) - F_e^2(F_e - F_{ey})\left(\frac{x_0}{\bar{r}_0}\right)^2$$
$$- F_e^2(F_e - F_{ex})\left(\frac{y_0}{\bar{r}_0}\right)^2 = 0 \qquad \text{(LRFD Equation A-E3-7)}$$

A perhaps more convenient form of LRFD Equation A-E3-7 follows:

$$HF_e^3 + \left[\frac{1}{\bar{r}_0^2}(y_0 F_{ex} + x_0^2 F_{ey}) - (F_{ex} + F_{ey} + F_{ez})\right]F_e^2$$
$$+ (F_{ex}F_{ey} + F_{ex}F_{ez} + F_{ey}F_{ez})F_e - F_{ex}F_{ey}F_{ez} = 0$$

K_z = effective length factor for torsional buckling
G = shear modulus (ksi)
C_w = warping constant (in^6)
J = torsional constant (in^4)

$$\bar{r}_0^2 = x_0^2 + y_0^2 + \frac{I_x + I_y}{A} \qquad \text{(LRFD Equation A - E3 - 8)}$$

$$H = 1 - \left(\frac{x_0^2 + y_0^2}{\bar{r}_0^2}\right) \qquad \text{(LRFD Equation A - E3 - 9)}$$

$$F_{ex} = \frac{\pi^2 E}{(KL/r)_x^2} \qquad \text{(LRFD Equation A - E3 - 10)}$$

$$F_{ey} = \frac{\pi^2 E}{(KL/r)_y^2} \qquad \text{(LRFD Equation A-E3-11)}$$

$$F_{ez} = \left[\frac{\pi^2 E C_w}{(K_z L)^2} + GJ\right]\frac{1}{A\bar{r}_0^2} \qquad \text{(LRFD Equation A-E3-12)}$$

The values of C_w, J, $\bar{r}_0$, and H are provided for many sections in the "Flexural-Torsional Properties" tables of Part 1 of the Manual.

3.4 Tension and Compression Members with Flexure

3.4.1 Introduction

Structural members that are subjected to a combination of bending and axial force are far more common than the reader may realize. Columns that are part of a steel building frame must nearly always resist sizable bending moments in addition to the usual compressive loads. It is almost impossible to erect and center loads exactly on columns, even in a testing lab, and in an actual building the reader can see that it is even more impossible. Even if building loads could be perfectly centered at one time they would not stay in one place. Furthermore, columns may be initially crooked or have other flaws with the result that lateral bending is produced. The beams framing into columns are quite commonly supported with framing angles or brackets on the sides of the columns. These eccentrically applied loads produce moments. Wind and other lateral loads cause columns to bend laterally, and the columns in rigid frame buildings are subjected to moments even when the frame is supporting gravity loads alone. The members of bridge portals must resist combined forces as do building columns. Among the causes of the combined forces are heavy lateral wind loads, vertical traffic loads whether symmetrical or not, and the centrifugal effect of traffic on curved bridges.

The previous experience of the reader has probably been to assume truss members axially loaded only. Purlins for roof trusses, however, are frequently placed in between truss joints, causing the top chords to bend. Similarly, the bottom chords may be bent by the hanging of light fixtures, ductwork, and other items between the truss joints. All horizontal and inclined truss members have moments caused by their own weights, while all truss members—whether vertical or not—are subjected to secondary bending forces. Secondary forces are developed because the members are not connected with frictionless pins, as assumed in the usual analysis, the member centers of gravities or those of their connectors do not exactly coincide at the joints, etc.

Moments in tension members are not as serious as those in compression members because tension tends to reduce lateral deflections while compression increases them. Increased lateral deflection in turn results in larger moments, which result in larger lateral deflections, etc. It is hoped that members in such situations are quite stiff so as to keep the additional lateral deflections from becoming excessive.

Design of Tension and Compression Members ■ 3-65

3.4.2 Members Subject to Axial Tension and Flexure

A few types of members subject to both bending and axial tension are shown in Fig 3.44.

In Section H1 of the LRFD specification the following interaction equations are given for symmetric shapes subjected simultaneously to bending and axial tensile forces. These equations are also applicable to members subjected to bending and axial compression forces as will be described in the next few sections.

If $\dfrac{P_u}{\phi_t P_n} \geq 0.2$

$$\frac{P_u}{\phi_t P_n} + \frac{8}{9}\left(\frac{M_{ux}}{\phi_b M_{nx}} + \frac{M_{uy}}{\phi_b M_{ny}}\right) \leq 1.0 \qquad \text{(LRFD Equation H1-1a)}$$

If $\dfrac{P_u}{\phi_t P_n} < 0.2$

$$\frac{P_u}{2\phi_t P_n} + \left(\frac{M_{ux}}{\phi_b M_{nx}} + \frac{M_{uy}}{\phi_b M_{ny}}\right) \leq 1.0 \qquad \text{(LRFD Equation H1-1b)}$$

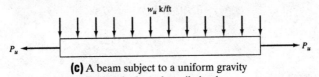

(a) A hanger subject to an off-center tensile load

(b) A hanger subject to an axial tensile load and a lateral load (as wind) or other lateral moment

(c) A beam subject to a uniform gravity load and a lateral tensile load

Fig. 3.44 Some members subject to axial tension and flexure.

The terms in these equations have previously been defined where P_u and M_u are the required tensile and flexural strengths, P_n and M_n are the nominal tensile and flexural strengths, and the resistance factors ϕ_t and ϕ_b are determined as before for axially loaded members. Usually only a first-order analysis (that is, not including any secondary forces as described in the next section) is made for members subject to bending and axial tension. However, the analyst may—in fact, is encouraged to—make second-order analyses for these members and use the results in his or her designs. Examples 3.22 and 3.24 illustrate the use of the interaction equations to review members subjected simultaneously to bending and axial tension, while Example 3.23 illustrates the design of such a member.

Example 3.22 ■ A W12×35 tension member with no holes consisting of 50 ksi steel is subjected to a factored tensile force P_u of 80 k and a factored bending moment M_{uy} of 35 ft-k. Is the member satisfactory if $L_b < L_p$?

Solution. Using a W12×35 ($A = 10.3$ in^3, $Z_y = 11.5$ in^3)

$$\phi_t P_n = \phi_t F_y A_g = (0.9)(50)(10.3) = 463.5 \text{ k}$$

$$\frac{P_u}{\phi_t P_n} = \frac{80}{463.5} = 0.173 < 0.2$$

∴ Use LRFD Equation H1-1b

$$\phi_b M_{ny} = \phi_b F_y Z_y = \frac{(0.9)(50)(11.5)}{12} = 43.12 \text{ ft-k}$$

$$\frac{P_u}{2\phi_t P_n} + \left(\frac{M_{ux}}{\phi_b M_{nx}} + \frac{M_{uy}}{\phi_b M_{ny}}\right) = \frac{80}{(2)(463.5)} + \left(0 + \frac{35}{43.12}\right)$$

$$= 0.898 < 1.0 \quad \text{(OK)}$$

Example 3.23 ■ A W10×30 tensile member with no holes consisting of 50 ksi steel with $L_b = 12.0$ ft is subjected to a factored tensile force P_u of 120 k and to the factored moments $M_{ux} = 90$ ft-k and $M_{uy} = 0$. If $C_b = 1.0$, is the member satisfactory?

Solution. Using a W10×30 ($A = 8.84$ in^2, $L_p = 4.8$ ft, $L_r = 14.5$ ft)

$$\phi_t P_n = (0.9)(50)(8.84) = 397.8 \text{ k}$$

$$\frac{P_u}{\phi_t P_n} = \frac{120}{397.8} = 0.302 > 0.2$$

∴ Use LRFD Equation H1-1a

Noting $L_b > L_p$, from the Load Factor Design Selection Table in the LRFD Manual

Design of Tension and Compression Members ■ 3-67

$$\phi_b M_p = 137 \text{ ft-k}$$
$$\phi_b M_r = 97.2 \text{ ft-k}$$
$$BF = 4.13$$
$$\phi_b M_n = C_b[\phi_b M_p - BF(L_b - L_p)]$$
$$= 1.0[137 - 4.13(12.0 - 4.8)] = 107.26 \text{ ft-k}$$

$$\frac{P_u}{\phi_t P_n} + \frac{8}{9}\left(\frac{M_{ux}}{\phi_b M_{nx}} + \frac{M_{uy}}{\phi_b M_{ny}}\right) = \frac{120}{397.8} + \frac{8}{9}\left(\frac{90}{107.26} + 0\right)$$
$$= 1.048 > 1.0 \quad \text{(NG)}$$

Example 3.24 ■ Select an 8-ft-long W10 to support a factored tensile load of 110 k applied with an eccentricity of 5 in with respect to the x axis. The A36 member is to be welded and braced laterally only at its supports. Assume C_b is equal to 1.0.

Solution. Try a W10×26 ($A = 7.61$ in^2, L_p=5.7 ft, L_r=18.5 ft, M_{ux}=(5)(110)=550 in-k = 45.8 ft-k, $\phi_t P_n = (0.9)(36)(7.61) = 246.6$ k.

$$\frac{P_u}{\phi_t P_n} = \frac{110}{246.6} = 0.446 > 0.2$$

∴ Use LRFD Equation H1-1a

Noting that $L_b > L_p < L_r$, we determine $\phi_b M_n$ as follows, with reference to the Load Factor Design Selection Table in the Manual:

$$\phi_b M_p = 84.5 \text{ ft-k}$$
$$\phi_b M_r = 54.4 \text{ ft-k}$$
$$BF = 2.34$$
$$\phi_b M_{nx} = C_b[\phi_b M_p - BF(L_b - L_p)]$$
$$= 1.0[84.5 - (2.34)(8.0 - 5.7)] = 79.1 \text{ ft-k}$$

$$\frac{P_u}{\phi_t P_n} + \frac{8}{9}\left(\frac{M_{ux}}{\phi_b M_{nx}} + \frac{M_{uy}}{\phi_b M_{ny}}\right) = \frac{110}{246.6} + \frac{8}{9}\left(\frac{45.8}{79.1} + 0\right)$$
$$= 0.961 < 1.0 \quad \text{(OK)}$$

Subsequent check of W10×22 shows it will not do.

Use W10×26

3-68 ■ Chapter Three

3.4.3 First-Order and Second-Order Moments for Members Subject to Axial Compression And Bending

When a beam-column is subjected to moment along its unbraced length it will be displaced laterally in the plane of bending. The result will be an increased or secondary moment equal to the axial compression load times the lateral displacement or eccentricity. In Fig. 3.45 we can see that the member moment is increased by an amount $P_u\delta$. This moment will cause additional lateral deflection, which will in turn cause a larger column moment, which will cause a larger lateral deflection, and so on until equilibrium is reached.

If a frame is subject to sidesway where the ends of the columns can move laterally with respect to each other, additional secondary moments will result. In Fig. 3.46 the secondary moment produced due to sidesway is equal to $P_u\Delta$.

The moment M_1 is assumed by the LRFD specification to equal M_{nt} (which is the required flexural strength of a member assuming ther is no lateral translation plus the moment due to $P_u\delta$. The moment M_2 is assumed to equal M_{lt} (the moment due to the lateral translation of the frame) plus the moment due to $P_u\Delta$.

The required total flexural strength of a member must equal at least the sum of the first-order and second-order moments. Several methods are available for determining this required strength, ranging from very simple approximations to very rigorous procedures.

The LRFD specification C.1 states that we can either (1) make a second-order analysis to determine the maximum factored load strength required or (2) use a first-order elastic analysis and amplify the moments obtained with some amplification factors called B_1 and B_2 and described in the paragraphs to follow.

Should the designer make a second-order analysis, he or she should realize that it must account for the interaction of the factored load effects. That is, we must consider combinations of factored loads acting at the same time. We cannot correctly make separate analyses and superimpose the results.

In this section the author presents the approximate method of analysis given in the LRFD Manual. We will make two first-order analyses—one an analysis

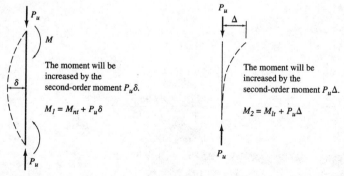

Fig. 3.45 Moment amplification of a column that is braced against sidesway.

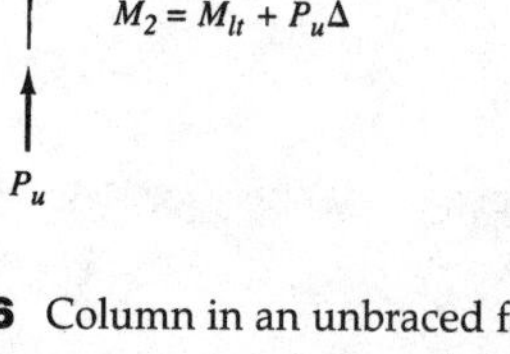

Fig. 3.46 Column in an unbraced frame.

where the frame is assumed to be braced so that it cannot sway. We will call these moments M_{nt} and will multiply them by a magnification factor called B_1 to account for the P-δ effect (see Fig. 3.45). Then we will analyze the frame again, allowing it to sway. We will call these moments M_{lt} and will multiply them by a magnification factor called B_2 to account for the P-Δ effect (see Fig. 3.46). The final moment in a particular member will equal

$$M_u = B_1 M_{nt} + B_2 M_{lt}$$ (LRFD Equation C1-1)

Instead of using the LRFD empirical procedure described here, the designer may and is encouraged to use a theoretical second-order analysis, provided he or she meets the requirements of Sections C1 and C2 of the specification. These requirements pertain to axial deformations, maximum axial forces permitted in members, bracing, K factors, and so on.

3.4.4 Magnification Factors

The magnification factors are B_1 and B_2. With B_1 the analyst attempts to estimate the $P_u\delta$ effect for a column whether the frame is braced or unbraced against sidesway. With B_2 he or she attempts to estimate the $P_u\Delta$ effect in unbraced frames.

These factors are theoretically applicable when the connections are fully restrained or when they are completely unrestrained. The LRFD Manual indicates that the determination of secondary moments in between these two classifications for partially restrained moment connections is beyond the scope of their specification.

In the expression for B_1 that follows, C_m is a term that will be defined in the next section, P_u is the required axial strength of the member, and P_{e1} is the member's Euler buckling strength equal to $A_g F_y/\lambda_c^2$ for a braced frame. In this expression $\lambda_c = (KL/r\pi)\sqrt{F_y/E}$ as given by LRFD Equation E2-4. I and KL are both taken in the plane of bending determined in accordance with LRFD Specification C2.1 for a *braced frame*.

Substituting this value of λ_c into the expression for P_{e1} and replacing A_g with I/r^2, it becomes

$$P_{e1} = \frac{A_g F_y}{\lambda_c^2} = \frac{\frac{I}{r^2} F_y}{\left(\frac{kL}{2\pi}\sqrt{\frac{F_y}{E}}\right)^2}$$

$$P_{e1} = \frac{\pi^2 EI}{(KL)^2}$$

In a similar fashion P_{e2} is the Euler buckling strength $\pi^2 EI/(KL)^2$ with K determined in the plane of bending for the *unbraced frame*.

In Table 8 of the LRFD Specification Appendix the value of P_{e2}/A_g can be obtained directly. Multiplying it by A_g will provide the value of P_{e2}. Should we compute a B_1 or B_2 value that is less than 1.0 we must use 1.0. (If we are supposedly magnifying moments, we don't want to multiply them by a number less than 1.0.)

The expression to follow for B_1 was derived for a member braced against sidesway. It will be used only to magnify the M_{nt} moments (those moments computed assuming there is no lateral translation of the frame).

$$B_1 = \frac{C_m}{1 - P_u/P_{e1}} \geq 1.0 \qquad \text{(LRFD Equation C1-5)}$$

The horizontal deflection of a multistory building due to wind or seismic load is called *drift*. It is represented by Δ in Figs. 3.46 and 3.47. In the formula to follow

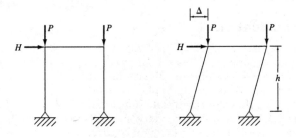

Fig. 3.47 Drift in building frame.

the term Δ_{oh} used. It represents the lateral deflection of the story in question calculated with service loads with respect to the story below. Drift is measured with the so-called *drift index* Δ_{oh}/h where Δ_{oh} is the lateral deflection and h is the height or distance to the lower level. For the comfort of occupants of a building the index usually is limited at working or service loads to a value between 0.0015 and 0.0030, and at factored loads to about 0.0040.

The designer may use either of the following expressions for B_2. In the first of these expressions ΣP_u represents the total required axial strength of all the columns on the floor in question, Δ_{oh}/L represents the story drift index, and ΣH is the sum of all the story horizontal forces producing Δ_{oh}. The value of P_{e2} is defined as before for P_{e1}, except that the effective length factor K is determined in the plane of bending for *an unbraced frame*.

The values shown for ΣP_u and ΣP_{e2} are for all of the columns on the floor in question. This is considered to be necessary because the B_2 term is used to magnify column moments for sidesway. For sidesway to occur in a particular column, it is necessary for all of the columns on the floor to sway simultaneously. The ΣH value used in the first of the B_2 expressions represents the sum of the lateral loads acting above the floor being considered. To compute the ratio $\Delta_{oh}/\Sigma H$ we may use either factored or unfactored loads.

$$B_2 = \frac{1}{1 - \Sigma P_u (\Delta_{oh} / \Sigma HL)} \qquad \text{(LRFD Equation C1-4)}$$

or

$$B_2 = \frac{1}{1 - \Sigma P_u / \Sigma P_{e2}} \qquad \text{(LRFD Equation C1-5)}$$

We must remember that the amplification factor B_2 is only applicable to moments caused by forces that cause sidesway and is to be computed for an entire story. To use the B_2 value given by LRFD Equation C1-5 we must select initial member sizes (that is so we can compute a value for P_{e2}). Should you be designing a building frame where you are limiting the drift index Δ_{oh}/L to a certain maximum value, you can compute B_2 with LRFD Equation C1-4 before you design the member. In this way you can set a drift limit in advance so that secondary bending is insignificant.

To calculate the values of ΣP_u and ΣP_e some designers will calculate the values for the columns in the one frame under consideration (or for that single line of columns perpendicular to the wind). This, however, is a rather bad practice unless all the other frames are exactly the same as the one under study.

Of the two expressions given for B_2, the first one (C1-4), which involves a drift index, is better suited for design office practice. If we assume drifts at factored loads as large as about 0.0040, we are probably being quite conservative. That is, the real structures may not drift this much.

3.4.5 Moment Modification or C_m Factors

In the last two sections the subject of moment magnification due to lateral deflections was introduced, and the factors B_1 and B_2 were presented with which the moment increases could be estimated. In the expression for B_1 a term C_m, called the *modification factor*, was included. The magnification factor B_1 was developed for the largest possible lateral displacement. On many occasions the displacement is not that large and B_1 overmagnifies the column moment. As a result, the moment needs to be reduced or modified with the C_m factor. You can see that this is not the case in Fig. 3.48, where we have a column bent in single curvature with equal end moments such that the column bends laterally by an amount δ at middepth. The maximum total moment occurring in the column will clearly equal M plus the increased moment $P_u \delta$. As a result no modification is required and $C_m = 1.0$.

An entirely different situation is considered in Fig. 3.49 where the end moments tend to bend the member in reverse curvature. The initial maximum moment occurs at one of the ends, and we shouldn't increase it by a value $P_u \delta$ that occurs some distance out in the column because we will be overdoing the moment magnification. The purpose of the modification factor is to modify or reduce the magnified moment when the variation of the moments in the column is such that B_1 is made too large. If we didn't use a modification factor we would end up with the same total moments in the columns of both Figs. 3.48 and 3.49, assuming the same dimensions and initial moments and load.

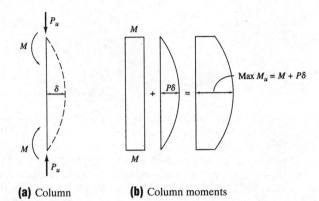

Fig. 3.48 Moment magnification for column bent in single curvature.

Modification factors are based on the rotational restraint at the member ends and on the moment gradients in the members. The LRFD specification (C1) includes two categories of C_m as described in the next few paragraphs.

In category 1 the members are prevented from joint translation or sidesway and they are not subject to transverse loading between their ends. For such members the modification factor is based on an elastic first-order analysis.

$$C_m = 0.6 - 0.4\frac{M_1}{M_2} \qquad \text{(LRFD Equation C1-3)}$$

In this expression M_1/M_2 is the ratio of the smaller moment to the larger moment at the ends of the unbraced length in the plane of bending under consideration. The ratio is negative if the moments cause the member to bend in single curvature and positive if they bend the members in reverse or double curvature. As previously described, a member in single curvature has larger lateral

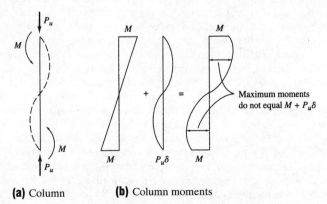

Fig. 3.49 Moment magnification for column bent in double curvature.

Design of Tension and Compression Members ■ 3-73

deflections than a member bent in reverse curvature. With larger lateral deflections the moments due to the axial loads will be larger.

Category 2 applies to members that are subjected to transverse loading between their joints in the plane of loading. The compression chord of a truss with a purlin load between its joints is a typical example of this category. The LRFD specification states that the value of C_m is to be determined either by rational analysis or by using one of the values to follow.

a. For members with restrained ends $C_m = 0.85$.

b. For members with unrestrained ends $C_m = 1.0$.

If either one of these two values is used, or if we interpolate between them where we feel our conditions warrant, the results will probably be quite reasonable.

Instead of using these values for transversely loaded members, the values of C_m for category 2 may be determined for various end conditions and loads from Table 3.5, which is a reproduction of Table C-C1.1 of the Commentary on the LRFD Specification. In the expressions given in the table P_u is the factored column axial load and P_{e1} is the elastic buckling load for a braced column for the axis about which bending is being considered.

Table 3.5 Modification Factors for Beam Columns Subject to Transverse Loads Between Joints

Case	ψ	C_m
(simply supported with axial load P_u)	0	1.0
(fixed one end, simply supported other)	−0.4	$1 - 0.4 \dfrac{P_u}{P_{e1}}$
(fixed both ends with distributed load)	−0.4	$1 - 0.4 \dfrac{P_u}{P_{e1}}$
(simply supported with point load at midspan)	−0.2	$1 - 0.2 \dfrac{P_u}{P_{e1}}$
(fixed one end, point load at $L/2$)	−0.3	$1 - 0.3 \dfrac{P_u}{P_{e1}}$
(fixed both ends with point load)	−0.2	$1 - 0.2 \dfrac{P_u}{P_{e1}}$

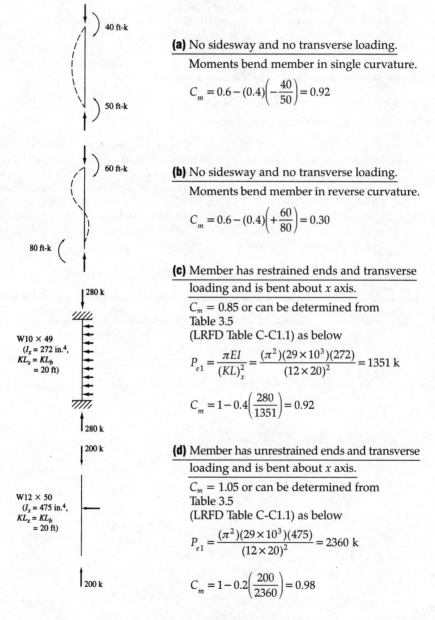

Fig. 3.50 Example modification or C_m factors.

Design of Tension and Compression Members ■ 3-75

$$P_{e1} = \frac{\pi^2 EI}{(KL)^2}$$

In Table 3.5 note that some members have restrained ends and some do not.
Sample values of C_m are calculated for four beam columns and shown in Fig. 3.50.

3.4.6 Review of Beam Columns in Braced Frames

The same interaction equations are used for members subject to axial compression and bending as were used for members subject to axial tension and bending. However, some of the terms involved in the equations are defined somewhat differently. For instance, P_u and P_n refer to compressive forces rather than tensile forces, ϕ_c is 0.85 for axial compression, and ϕ_b is 0.9 for bending.

To analyze a particular beam column or a member subject to both bending and axial compression we need to make both a first-order and a second-order analysis to obtain the bending moments. The first-order moment is usually obtained by making an elastic analysis and consists of the moments M_{nt} (these are the moments in beam columns caused by gravity loads) and the moments M_{lt} (these are the moments in beam columns due to the lateral loads, that is, due to lateral translation).

Examples 3.25 to 3.27 illustrate the application of the interaction equations to beam-columns that are members of braced frames. Thus, only B_1 will be computed, as B_2 is not applicable. *It is to be remembered that C_m was developed for braced frames and thus must be used in these three examples for calculating B_1.*

Example 3.25 ■ A 12-ft W12×96 (A36 steel) is used as a beam-column in a braced frame. It is bent in single curvature with equal and opposite end moments and is not subjected to intermediate transverse loads. Is the section satisfactory if $P_u = 500$ k and the first-order moment $M_{ntx} = 125$ ft-k?

Solution. Using a W12×96 ($A = 28.2$ in^2, $I_x = 833$ in^4, $L_p = 12.9$ ft, $\phi_b M_b = 397$ ft-k)

For a braced frame $K = 1.0$
∴ $K_x L_x = K_y L_y = (1.0)(12) = 12$ ft
$\phi_c P_n = 770$ k from LRFD column tables

$$\frac{P_u}{\phi_c P_n} = \frac{500}{770} = 0.649 > 0.2$$

∴ Must use LRFD Equation H1-1a

As the only moment is M_{ntx} there is no lateral translation of the frame, that is $M_{lt} = 0$.
∴ $M_{ux} = B_1 M_{ntx}$

$$C_m = 0.6 - 0.4\left(\frac{M_1}{M_2}\right) = 0.6 - 0.4\left(-\frac{125}{125}\right) = 1.0$$

$$P_{e1} = \frac{\pi^2 EI_x}{\left(K_x L_x\right)^2} = \frac{(\pi^2)(29 \times 10^3)(833)}{(12 \times 12)^2} = 11{,}498 \text{ k}$$

3-76 ■ Chapter Three

Notice this value can be determined from the bottom of the column tables where $P_e(KL^2)/10^4$ is given for both the x and y axes. (*Note*: the value of P_e can also be determined from Table 8 in Part 6 of the Manual. First K_xL_x/r_x is computed and the P_e/A_g is picked from the table. Finally, $P_e = A_g$ times the table value.)

$$B_1 = \frac{C_m}{1-\frac{P_u}{P_{e1}}} = \frac{1.0}{1-\frac{500}{11,498}} = 1.045$$

$$M_{ux} = (1.045)(125) = 130.6 \text{ ft-k}$$

Applying LRFD Equation H1-1a

Since $L_b = 12$ ft $< L_p = 12.9$ ft, $\phi_b M_{nx} = \phi_b M_p = 397$ ft-k

$$\frac{P_u}{\phi_c P_n} + \frac{8}{9}\left(\frac{M_{ux}}{\phi_b M_{nx}} + \frac{M_{uy}}{\phi_b M_{ny}}\right)$$

$$= \frac{500}{770} + \frac{8}{9}\left(\frac{130.6}{397} + 0\right) = 0.942 < 1.0$$

∴ Section is satisfactory.

Example 3.26 ■ A 14-ft W14×120 (A36 steel) is used as a beam-column in a braced frame. It is bent in single curvature with equal and opposite moments. Its ends are restrained and it is not subjected to intermediate transverse loads. Is the section satisfactory if $P_u = 180$ k and if it has the first-order moments $M_{ntx} = 150$ ft-k and $M_{nty} = 100$ ft-k?

Solution. Using a W14×120 ($A = 35.3$ in², $L_p = 15.6$ ft, $I_x = 1380$ in⁴, $I_y = 495$ in⁴, $Z_x = 212$ in³, $Z_y = 102$ in³.)

For a braced frame $K_x L_x = K_y L_y = (1.0)(14) = 14$ ft

$\phi_c P_n = 971$ k from LRFD column tables

$$\frac{P_u}{\phi_c P_n} = \frac{180}{971} = 0.185 < 0.2$$

∴ Must use LRFD Equation H1-1b

As we have only the moments M_{ntx} and M_{nty} there is no lateral translation of the frame and thus $M_{ltx} = M_{lty} = 0$.

$M_{ux} = B_1 M_{ntx}$ and $M_{uy} = B_1 M_{nty}$

$C_m = 0.6 - 0.4\left(-\dfrac{150}{150}\right) = 1.0$

P_{e1x} from column table $= \dfrac{(10)^4(39{,}300)}{(12 \times 14)^2} = 13{,}924$ k

$B_{1x} = \dfrac{C_m}{1 - \dfrac{P_u}{P_{e1x}}} = \dfrac{1.0}{1 - \dfrac{180}{13{,}924}} = 1.013$

$M_{ux} = (1.013)(150) = 152$ ft-k

P_{e1y} from column table $= \dfrac{(10)^4(14{,}100)}{(12 \times 14)^2} = 4996$ k

$B_{1y} = \dfrac{C_m}{1 - \dfrac{P_u}{P_{e1y}}} = \dfrac{1.0}{1 - \dfrac{180}{4996}} = 1.037$

$M_{uy} = (1.037)(100) = 103.7$ ft-k

Since $L_b = 14$ ft $< L_p = 15.6$ ft

$\phi_b M_{nx} = \phi_b F_y Z = \dfrac{(0.9)(36)(212)}{12} = 572.4$ ft-k

$\phi_b M_{ny} = \phi_b F_y Z = \dfrac{(0.9)(36)(102)}{12} = 275.4$ ft-k

Applying LRFD Equation H1-1b

$\dfrac{P_u}{2\phi P_n} + \left(\dfrac{M_{ux}}{\phi_b M_{nx}} + \dfrac{M_{uy}}{\phi_b M_{ny}}\right)$

$= \dfrac{180}{(2)(971)} + \left(\dfrac{152}{572.4} + \dfrac{103.7}{275.4}\right)$

$= 0.735 < 1.0$ (OK but overdesigned)

Example 3.27 ■ For the truss shown in Fig. 3.51a a W8×31 is used as a continuous top chord member from joint L_0 to joint U_3. If the member consists of A36 steel, does it have sufficient strength to resist the factored loads shown in part b of the figure. Part b shows the portion of the chord from L_0 to U_1 and the 16-k load represents the effect of a purlin. It is assumed that lateral support is provided for this member at its ends and centerline.

3-78 ■ Chapter Three

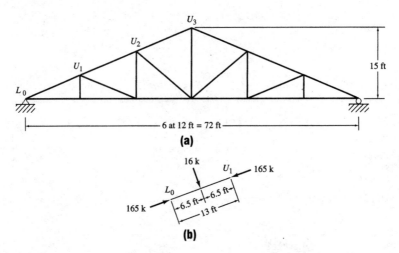

Fig. 3.51 A truss whose top chord is subject to intermediate loads.

Solution. Using a W8×31 ($A = 9.13$ in², $I_x = 110$ in⁴, $r_x = 3.47$ in, $r_y = 2.02$ in, $L_p = 8.4$ ft, $Z_x = 30.4$ in³)

For a frame braced against sidesway $K = 1.0$

$$\left(\frac{KL}{r}\right)_x = \frac{(1.0)(12 \times 13)}{3.47} = 44.96$$

$$\left(\frac{KL}{r}\right)_y = \frac{(1.0)(12 \times 6.5)}{2.02} = 38.61$$

$\phi_c F_{cr} = 27.50$ ksi (from LRFD Table 3-36)

$\phi_c P_n = (27.50)(9.13) = 251$ k

$$\frac{P_u}{\phi_c P_n} = \frac{165}{251} => 0.2$$

∴ Must use LRFD Equation H1-1a

For a braced frame M_{lt} (the moment due to lateral loads) = 0

$$P_{e1x} = \frac{(10)^4(3150)}{(12 \times 13)^2} \text{ from column table} = 1294 \text{ k}$$

From Table 3.5

For determining the values of C_m and M_{nt} the author assumes that the member end support conditions are an average of the two situations pictured below

For $C_m = 1 - 0.2\dfrac{P_u}{P_{e1}} = 1 - 0.2\left(\dfrac{165}{1294}\right) = 0.974$

For $C_m = 1 - 0.3\dfrac{P_u}{P_{e1}} = 1 - 0.3\left(\dfrac{165}{1294}\right) = 0.962$

Average $C_m = 0.97$

Computing moment

For $M_{nt} = \dfrac{(16)(13)}{4} = 52$ ft-k

For $M_{nt} = \dfrac{(3)(16)(13)}{16} = 39$ ft-k

Avg $M_{nt} = \dfrac{52 + 39}{2} = 45.5$ ft-k

Computing P_{e1x} and B_{1x}

$$B_{1x} = \dfrac{C_{mx}}{1 - \dfrac{P_u}{P_{e1x}}} = \dfrac{0.97}{1 - \dfrac{165}{1294}} = 1.112$$

$M_{ux} = (1.112)(45.5) = 50.6$ ft-k

Since $L_b = 6.5$ ft $< L_p = 8.4$ ft

$$\phi_b M_{nx} = \phi_b F_y Z_x = \dfrac{(0.9)(36)(30.4)}{12} = 82.1 \text{ ft-k}$$

Applying LRFD Equation H1-1a

$$\dfrac{P_u}{\phi P_n} + \dfrac{8}{9}\left(\dfrac{M_{ux}}{\phi_b M_{nx}} + \dfrac{M_{uy}}{\phi_b M_{ny}}\right)$$

$$= \dfrac{165}{251} + \dfrac{8}{9}\left(\dfrac{50.6}{82.1} + 0\right) = 1.21 > 1.0 \quad \text{(NG)}$$

3.4.7 Review of Beam-Columns in Unbraced Frames

The maximum primary moments in unbraced frames almost always occur at the column ends. As you can see in Fig. 3.46 the maximum sidesway moments always occur at the member ends and the total moment for a particular column is determined by adding its primary end moment to its sidesway moment. Therefore, it is not necessary to use a modification factor and C_m is not used in the B_2 expressions.

Example 3.28 ■ A 10-ft W12×106 (F_y = 50 ksi) is used as a beam-column in an unbraced frame. It is bent in reverse curvature with equal and opposite moments and is not subject to intermediate transverse loads. Is the section satisfactory if P_u = 250 k, M_{ntx} = 60 ft-k, M_{nty} = 40 ft-k, M_{ltx} = 100 ft-k, and M_{lty} = 80 ft-k? The total factored gravity load ΣP_u above this level has been calculated to equal 5000 k. Assume that ΣP_{ex} = 40,000 k and ΣP_{ey} = 20,000 k. $K_x = K_y$ = 1.2.

Solution. Using a W12×106 (A=31.2 in², I_x=933 in⁴, I_y=301 in⁴, L_p=11.0 ft, Z_x=164 in³, Z_y=75.1 in³.)

$$K_x L_x = K_y L_y = (1.2)(10) = 12 \text{ ft}$$

$$\phi_c P_n = 1130 \text{ k from LRFD column tables}$$

$$\frac{P_u}{\phi_c P_n} = \frac{250}{1130} = 0.221 > 0.2$$

∴Must use LRFD Equation H1-1a

As we have the moments M_{ntx} and M_{nty} as well as M_{ltx} and M_{lty}

$$M_{ux} = B_1 M_{ntx} + B_2 M_{ltx}$$
$$M_{uy} = B_1 M_{nty} + B_2 M_{lty}$$

For reverse curvature and equal end moments

$$C_{mx} = C_{my} = 0.6 - 0.4(+1.0) = 0.2$$

$$P_{e1x} = \frac{\pi^2 E I_x}{(K_x L_x)^2} = \frac{(\pi^2)(29 \times 10^3)(933)}{(12 \times 12)^2} = 12{,}877 \text{ k}$$

$$B_{1x} = \frac{C_m}{1 - \frac{P_u}{P_{e1x}}} = \frac{0.2}{1 - \frac{250}{12{,}877}} = 0.204 < 1.0 \quad \text{Use 1.0}$$

$$P_{e1y} = \frac{\pi^2 E I_y}{(K_y L_y)^2} = \frac{(\pi^2)(29 \times 10^3)(301)}{(12 \times 12)^2} = 4154 \text{ k}$$

$$B_{1y} = \frac{C_m}{1-\frac{P_u}{P_{e1y}}} = \frac{0.2}{1-\frac{250}{4154}} = 0.213 < 1.0 \quad \underline{\text{Use } 1.0}$$

$$B_{2x} = \frac{1}{1-\frac{\Sigma P_u}{\Sigma P_{e2x}}} = \frac{1}{1-\frac{5000}{40{,}000}} = 1.143$$

$$B_{2y} = \frac{1}{1-\frac{\Sigma P_u}{\Sigma P_{e2y}}} = \frac{1}{1-\frac{5000}{20{,}000}} = 1.333$$

$$M_{ux} = (1.0)(60) + (1.143)(100) = 174.3 \text{ ft-k}$$
$$M_{uy} = (1.0)(40) + (1.333)(80) = 146.6 \text{ ft-k}$$

Since $L_b = 10$ ft $< L_p = 11.0$ ft

$$\phi_b M_{nx} = \phi_b F_y Z_x = \frac{(0.9)(50)(164)}{12} = 615 \text{ ft-k}$$

$$\phi_b M_{ny} = \phi_b F_y Z_y = \frac{(0.9)(50)(75.1)}{12} = 281.6 \text{ ft-k}$$

Applying LRFD Equation H1-1a

$$\frac{P_u}{\phi P_n} + \frac{8}{9}\left(\frac{M_{ux}}{\phi_b M_{nx}} + \frac{M_{uy}}{\phi_b M_{ny}}\right) = \frac{250}{1130} + \frac{8}{9}\left(\frac{174.3}{615} + \frac{146.6}{281.6}\right) = 0.936 < 1.0 \quad \text{(OK)}$$

3.4.8 Design of Beam-Columns—Braced or Unbraced

The design of beam-columns involves a trial-and-error procedure. A trial section is selected by some process and is then checked with the appropriate interaction equation. If the section does not satisfy the equation or if it's too much on the safe side (that is, if it's overdesigned) another section is selected and the interaction equation is applied again.

A common method used for selecting sections to resist both moments and axial loads is the *equivalent axial load or effective axial load method*. In this method the axial load (P_u) and the bending moment (M_{ux} and/or M_{uy}) are replaced with a fictitious concentric load P_{ueq} equivalent to the actual design axial load plus the design moment.

It is assumed for this discussion that it is desired to select the most economical section to resist both a moment and an axial load. By a trial-and-error procedure it is eventually possible to select the lightest section. Somewhere, however, there is a fictitious axial load that will require the same section as the one required for

3-82 ■ Chapter Three

the actual moment and the actual axial load. This fictitious load is called the equivalent axial load or the effective axial load P_{ueq}.

Equations are used to convert the bending moment into an estimated equivalent axial load P'_u which is added to the actual design axial load P_u. The total of $P_u + P'_u$ is the equivalent or effective axial load P_{ueq}, and it is used to enter the concentric column tables of the LRFD Manual for choice of a trial section. In the approximate formula for P_{ueq} which follows, m is a factor given in Table 3.6 (Table 3-2 in Part 3 of the LRFD Manual).

$$P_{ueq} = P_u + M_{ux}m + M_{uy}mu$$

To apply this expression a value of m is taken from the first approximation section of Table 3.6 and u is assumed equal to 2. *In applying the equation the moments* M_{ux} *and* M_{uy} *must be used in ft-k.* The equation is solved for P_{ueq} and a column is selected from the concentrically loaded column tables for that load. Then the equation for P_{ueq} is solved again using a revised value of m from the subsequent approximations part of the table and the value of u is taken from the column tables for the column initially selected. Another shape is selected and the process is continued until m and u stabilize (that is, until the column size selected does not change).

Finally, it is necessary to check the trial column size with the appropriate LRFD interaction equation (H1-1a) or (H1-1b). The equivalent axial load equation yields sections that are generally on the conservative side. For this reason the designer may select a section by the equivalent axial load method and then try the interaction equation on a section one or two sizes smaller. This procedure may very well provide significant savings in steel weight.

Limitations of P_{ueq} Formula. ■ The application of the equivalent axial load formula and Table 3.6 results in economical beam-column designs unless the moment becomes quite large in comparison with the axial load. For such cases the members selected will be capable of supporting the loads and moments but may very well be rather uneconomical. The tables for concentrically loaded columns of Part 3 of the Manual are limited to the W14s, W12s, and shallower sections, but when the moment is large in proportion to the axial load there will often be a much deeper and appreciably lighter section such as a W27 or W30 which may be a good bit lighter and which will satisfy the appropriate interaction equations.

As there are so many different loading situations that can occur, the author presents quite a few example problems (3.29 to 3.33) in the pages to follow. After a section is selected by the approximate P_{ueq} formula, it is necessary to check it with the appropriate interaction equation. As we have been doing just this in the last several examples (3.25 to 3.28), the checking part of the examples to follow is somewhat abbreviated in some cases.

Example 3.29 ■ Select a 14-ft 50 ksi beam column to support the following: $P_u = 600$ k, first-order moment $M_{ux} = 200$ ft-k, and $M_{uy} = 0$. The member is to be bent in single curvature with no transverse loading and the frame is to be braced. Use $K = 1.0$ and $C_m = 0.85$.

Design of Tension and Compression Members ■ 3-83

Table 3.6 Preliminary Beam-Column Design
$F_y = 36$ ksi, $F_y = 50$ ksi

	Values of m											
F_y	36 ksi						50 ksi					
KL, ft	10	12	14	16	18	20	22 and over	10	12	14	16	18	20	22 and over
	1st approximations											
All shapes	2.0	1.9	1.8	1.7	1.6	1.5	1.3	1.9	1.8	1.7	1.6	1.4	1.3	1.2
	Subsequent approximations											
W4	3.1	2.3	1.7	1.4	1.1	1.0	0.8	2.4	1.8	1.4	1.1	1.0	0.9	0.8
W5	3.2	2.7	2.1	1.7	1.4	1.2	1.0	2.8	2.2	1.7	1.4	1.1	1.0	0.9
W6	2.8	2.5	2.1	1.8	1.5	1.3	1.1	2.5	2.2	1.8	1.5	1.3	1.2	1.1
W8	2.5	2.3	2.2	2.0	1.8	1.6	1.4	2.4	2.2	2.0	1.7	1.5	1.3	1.2
W10	2.1	2.0	1.9	1.8	1.7	1.6	1.4	2.0	1.9	1.8	1.7	1.5	1.4	1.3
W12	1.7	1.7	1.6	1.5	1.5	1.4	1.3	1.7	1.6	1.5	1.5	1.4	1.3	1.2
W14	1.5	1.5	1.4	1.4	1.3	1.3	1.2	1.5	1.4	1.4	1.3	1.3	1.2	1.2

Source: This table is from a paper in AISC *Engineering Journal* by Uang, Wattar, and Leet, 1990.

Solution. For first approximation $m=1.7$ (from Table 3.6) and $u=2.0$.

$$P_{ueq} = P_u + M_{ux}m + M_{uy}mu = 600 + (200)(1.7) + 0 = 940 \text{ k}$$

Entering column tables with $P_u = 940$ k and $K_xL_x = K_yL_y = (1.0)(14) = 14$ ft
Try W14×90 ($m = 1.4$ from Table 3.6 and $u = 1.94$ from column tables)

$$P_{ueq} = 600 + (200)(1.4) + 0 = 880 \text{ k}$$

Returning to the column tables we find that the W14×90 ($\phi_c P_n = 969$ k) still is required. Checking it with the appropriate interaction equation:

$$\left(\frac{KL}{r}\right)_x = \frac{12 \times 14}{6.14} = 27.36$$

$$P_{e1} \text{ from column table} = \frac{(10)^4(28,600)}{(12 \times 14)^2} = 10,133 \text{ k}$$

$$B_1 = \frac{C_m}{1 - \dfrac{P_u}{P_{e1}}} = \frac{0.85}{1 - \dfrac{600}{10,133}} < 1.0 \quad \therefore \underline{\text{Use 1.0}}$$

$$\therefore M_{ux} = (1.0)(200) = 200 \text{ ft-k}$$

$\phi_b M_{nx} = 587$ ft-k from Load Factor Design Selection Table as $L_b < L_p$

3-84 ■ Chapter Three

$$\frac{P_u}{\phi_c P_n} = \frac{600}{969} > 0.2$$

∴ Must use LRFD Equation H1-1a

$$\frac{P_u}{\phi_c P_n} + \frac{8}{9}\left(\frac{M_{ux}}{\phi_b M_{nx}} + \frac{M_{uy}}{\phi_b M_{ny}}\right) = \frac{600}{969} + \frac{8}{9}\left(\frac{200}{587}\right) = 0.922 < 1.0$$

Use W14×90 OK

Example 3.30 ■ A 12-ft beam-column of A36 steel is to be used in a braced frame for the following: $P_u = 200$ k; first-order moment $M_{ux} = 150$ ft-k and $M_{uy} = 100$ ft-k. Assuming it is to have restrained ends and to be subject to intermediate transverse loads, select a W12 section. Assume $K = 1.0$.

Solution. 1st approximation making use of Table 3.6 ($m = 1.9$, $u = 2.0$)

$$P_{ueq} = P_u + M_{ux}m + M_{uy}m = 200 + (150)(1.9) + (100)(1.9)(2.0) = 865 \text{ k}$$

Entering column tables with $P_u = 865$ k and $K_x L_x = K_y L_y = (1)(12) = 12$ ft.

Try W12×120 ($m = 1.7$ from Table 3.6 and $u = 2.12$ from column tables)

$$P_{ueff} = 200 + (150)(1.7) + (100)(1.7)(2.12) = 815 \text{ k}$$

Try W12×106 ($m = 1.7$, $u = 2.12$)

$$P_{ueq} = 200 + (150)(1.7) + (100)(1.7)(2.12) = 815 \text{ k}$$

The column load tables still call for a W12×106.

Checking a W12×106 ($A = 31.2$ in^2, $L_p = 13.0 > L_b$ $\phi_b M_{nx} = 443$ ft-k, $Z_y = 75.1$ in^3, $\phi_b M_{ny} = (0.9)(36)(75.1)/12 = 202.8$ ft-k, $r_x = 5.47$ in, $r_y = 3.11$ in, $\phi_c P_n = 853$ k)

$C_{mx} = C_{my}$ for members with restrained ends subject to transverse loads between ends = 0.85.

$$P_{ex1} = \frac{(10)^4(26,700)}{(12 \times 12)^2} = 12,876 \text{ k}$$

$$B_{1x} = \frac{C_{mx}}{1 - \dfrac{P_u}{P_{e1x}}} = \frac{0.85}{1 - \dfrac{200}{12,876}} = 0.86 \quad \text{Use 1.0}$$

$$M_{ux} = (1.0)(150) = 150 \text{ ft-k}$$

$$P_{ey1} = \frac{(10)^4(8640)}{(12 \times 12)^2} = 4167 \text{ k}$$

$$B_{1y} = \frac{C_{my}}{1-\dfrac{P_u}{P_{e1y}}} = \frac{0.85}{1-\dfrac{200}{4167}} = 0.89 \quad \underline{\text{Use 1.0}}$$

$$M_{uy} = (1.0)(100) = 100 \text{ ft-k}$$

$$\frac{P_u}{\phi_c P_n} = \frac{200}{853} = 0.234 > 0.2$$

∴ Must use LRFD Equation H1-1a

$$\frac{P_u}{2\phi P_n} + \frac{8}{9}\left(\frac{M_{ux}}{\phi_b M_{nx}} + \frac{M_{uy}}{\phi_b M_{ny}}\right) = \frac{200}{853} + \frac{8}{9}\left(\frac{150}{443} + \frac{100}{202.8}\right) = 0.974 < 1.00$$

<u>Use W12×106</u>

Example 3.31 ■ Select a 10-ft W12 beam-column (A36 steel) for a braced symmetrical frame with P_u = 200 k; first-order moments M_{ltx} = 120 ft-k, M_{lty} = 80 ft-k, $M_{ntx} = M_{nty}$ = 0. The computed ΣP_u for all of the columns on this level is 3600 k while ΣP_{ex2} has been calculated to equal 50,000 k and ΣP_{ey2} to equal 25,000 k. Assume $K_x = K_y$ = 1.2. The column is to be bent in single curvature with no intermediate loads.

Solution. Determining M_{ux} and M_{uy} from LRFD Equation C1-1

$$M_u = B_1 M_{nt} + B_2 M_{lt}$$

$$B_{2x} = \frac{1.0}{1-\dfrac{\Sigma P_u}{\Sigma P_{ex2}}} = \frac{1.0}{1-\dfrac{3600}{50,000}} = 1.078$$

$$B_{2y} = \frac{1.0}{1-\dfrac{\Sigma P_u}{\Sigma P_{ey2}}} = \frac{1.0}{1-\dfrac{3600}{25,000}} = 1.168$$

The total moments are

$$M_{ux} = (1.078)(120) = 129.4 \text{ ft-k}$$
$$M_{uy} = (1.168)(80) = 93.4 \text{ ft-k}$$
$$KL = (1.2)(10) = 12 \text{ ft}$$

For first approximation m = 1.9 and u = 2.0

$$P_{ueq} = 200 + (129.4)(1.9) + (93.4)(1.9)(2.0) = 800.8 \text{ k}$$

<u>Try W12×106</u> (m = 1.7, u = 2.12)

3-86 ■ Chapter Three

$P_{ueq} = 200 + (129.4)(1.7) + (93.4)(1.7)(2.12) = 756.6$ k

Try W12×96 ($m = 1.7, u = 2.10$)

$P_{ueq} = 200 + (129.4)(1.7) + (93.4)(1.7)(2.10) = 753.4$ k

The column tables still call for a W12×96. Checking a W12×96 ($\phi_c P_n = 770$ k, $\phi_b M_{nx} = 397$ ft-k since $L_b < L_p$, $Z_y = 67.5$ in³, $\phi_b M_{ny} = (0.9)(36)(67.5)/12 = 182.2$ ft-k)

$$\frac{P_u}{\phi_c P_n} = \frac{200}{770} = 0.260 > 0.2$$

∴ Must use LRFD Equation H1-1a

$$\frac{P_u}{\phi P_n} + \frac{8}{9}\left(\frac{M_{ux}}{\phi_b M_{nx}} + \frac{M_{uy}}{\phi_b M_{ny}}\right) = \frac{200}{770} + \frac{8}{9}\left(\frac{129.4}{397} + \frac{93.4}{182.8}\right) = 1.006 \approx 1.00$$

Use W12×96 (OK)

Example 3.32 ■ Select a W14 beam-column of A36 steel with $K_x = 1.2 = K_y$ for the following: $P_u = 350$ k, first-order moment $M_{lty} = 100$ ft-k due to wind, with all other moments equal to 0. The 14-ft member is to be used in a symmetrical unbraced frame with an allowable story drift index $\Delta_{oh}/L = 0.0020$ as a result of a total service or unfactored load of 100 k. The total factored gravity load above has been calculated to equal 5000 k.

Solution.

$K_x L_x = K_y L_y = (1.2)(14) = 16.8$ ft

Having set a drift index=0.002 we can backfigure B_2 with LRFD Formula C1-4.

$$B_2 = \frac{1}{1 - \Sigma P_u\left(\dfrac{\Delta_{oh}}{\Sigma HL}\right)} = \frac{1}{1 - 5000\left(\dfrac{0.0020}{100}\right)} = 1.111$$

∴ $B_2 M_{lty} = (1.111)(100) = 111.1$ ft-k

Trying a W14 with $m = 1.36$ and $u = 2.0$

$P_{ueq} = P_u + M_{ux} m + M_{uy} mu = 350 + 0 + (111.1)(1.36)(2.0) = 652.2$ k

Try W14×90 ($m = 1.36$ and $u = 2.02$)

$P_{ueq} = 350 + 0 + (111.1)(1.36)(2.02) = 655.2$ k

Column tables still call for a W14×90. Checking a W14×90 ($Z_y = 75.6$ in³)

$\phi_c P_n = 693.6$ k from column tables by interpolation

$$\frac{P_u}{\phi_c P_n} = \frac{350}{693.6} > 0.2$$

∴ Must use LRFD Equation H1-1a

$$\phi_b M_{ny} = \frac{(0.9)(36)(75.6)}{12} = 204.1 \text{ ft-k}$$

$$\frac{P_u}{\phi P_n} + \frac{8}{9}\left(\frac{M_{ux}}{\phi_b M_{nx}} + \frac{M_{uy}}{\phi_b M_{ny}}\right) = \frac{350}{693.6} + \frac{8}{9}\left(0 + \frac{111.1}{204.1}\right) = 0.989 < 1.0 \quad \text{(OK)}$$

Use W14×90

Example 3.33 ■ Design a 12-ft W12 (A36) for a symmetrical unbraced frame to support the following: $P_u = 280$ k, $M_{ntx} = 60$ ft-k, $M_{nty} = 40$ ft-k, $M_{ltx} = 100$ ft-k, and $M_{lty} = 70$ ft-k. The member is to be subjected to intermediate transverse loads and its ends are restrained. It is to have an allowable drift index = 0.0020 due to the total horizontal service or unfactored loads of 120 k perpendicular to the x axis and 80 k perpendicular to the y axis. Assume $K_x = K_y = 1.2$ and $\Sigma P_u = 5000$ k.

Solution. We do not know the value of P_{e2x}, P_{e2y}, B_{1x}, or B_{1y}. Assume $B_{1x} = B_{1y} = 1.0$ and compute the values of B_{2x} and B_{2y} as follows using LRFD Equation C1-4.

$$B_{2x} = \frac{1}{1 - \Sigma P_u\left(\frac{\Delta_{oh}}{\Sigma HL}\right)} = \frac{1}{1 - 5000\left(\frac{0.002}{120}\right)} = 1.09$$

$$B_{2y} = \frac{1}{1 - 5000\left(\frac{0.002}{80}\right)} = 1.14$$

$$M_{ux} = (1.00)(60) + (1.09)(100) = 169 \text{ ft-k}$$
$$M_{uy} = (1.00)(40) + (1.14)(70) = 119.8 \text{ ft-k}$$

Selecting a trial section

$KL = (1.2)(12) = 14.4$ ft

$m = 1.58$ and $u = 2.0$

$P_{ueq} = 280 + (169)(1.58) + (119.8)(1.58)(2.0) = 925.6$ k

Try W12×120 ($m = 1.58$, $u = 2.12$)

$P_{ueq} = 280 + (169)(1.58) + (119.8)(1.58)(2.12) = 948.3$ k

Tables call for W12×136, but try a W12×120 ($\phi_c P_n = 920$ k, $L_p = 13.0$ ft, $\phi_b M_{nx} = 502$ ft-k, $Z_y = 85.4$ in^3.)

$$\frac{P_u}{\phi_c P_n} = \frac{280}{920} = 0.304 > 0.2$$

∴ Must use LRFD Equation H1-1a

$$\phi_b M_{ny} = \frac{(0.9)(36)(85.4)}{12} = 230.6 \text{ ft-k}$$

$$\frac{P_u}{\phi_c P_n} = \frac{8}{9}\left(\frac{M_{ux}}{\phi_b M_{nx}} + \frac{M_{uy}}{\phi_b M_{ny}}\right) = \frac{280}{920} + \frac{8}{9}\left(\frac{169}{502} + \frac{119.8}{230.6}\right) = 1.07 > 1.00 \quad \text{(NG)}$$

Use W12×136

References

1. Cochrane, V.H. "Rules for Riveted Hole Deductions in Tension Members," *Engineering News Record*, Nov. 16, 1922, pp. 847–848.

2. McCormac, J.C., *Structural Steel Design LRFD Method*, 2d ed., Harper Collins, New York, 1995, pp. 75–76.

3. Gaylord, E.H., Jr., and Gaylord, C.N. and J.E. Stallmeyer, *Design of Steel Structures*, 3d ed., McGraw-Hill, New York, 1992, pp. 143–145.

4. Munse, W.H., E. Chesson, Jr., "Riveted and Bolted Joints: Net Section Design," *Journal of the Structural Division ASCE*, 89, ST1, February 1963.

5. Easterling, W.S., and L.G. Giroux, "Shear Lag Effects in Steel Tension Members," *Engineering Journal*, AISC, no. 3, 3d quarter, 1993, pp. 77–89.

6. Johnston, B.G., "Pin Connected Plate Links," *Transactions ASCE*, 104, 1939.

7. Burgett, L.B., "Fast Check for Block Shear," *Engineering Journal*, AISC, 29, no. 4, 4th quarter, pp. 125–127, 1992.

8. Ruddy, J.L., "Economics of Low-Rise Steel Framed Structures," *Engineering Journal*, AISC, 20, no. 3, 3d quarter, pp. 107–118, 1983.

9. Julian, O.G., and L.S. Lawrence, *Notes on J and L Monograms for Determination of Effective Lengths*, 1959.

10. Structural Stability Research Council, *Guide to Stability Design Criteria for Metal Structures*, 4th ed., T.V. Galambos, ed. Wiley, New York, 1988.

11. Segui, W.T., *LRFD Steel Design*, PWS Publishing Company, Boston, 1994, pp. 96–97.

12. Dumonteil, P., "Simple Equations for Effective Length Factors," *Engineering Journal*, AISC, 29, no. 3, 3d quarter, 1992, pp. 111–115.

13. Yura, J.A., "Elements for Teaching Load and Resistance Factor Design," AISC, New York, August 1987, p. 20.

14. Yura, J.A., "Elements for Teaching Load Resistance Factor Design," AISC, Chicago, July 1987, pp. 17–19.

4

Louis F. Geschwindner

TORSION

4.1 Introduction

Although torsional moment is one of the four components of structural member response, with bending moment, shear force, and axial force, it remains the least understood of these actions. This may be due to the fact that, because typical structural members with open cross section are weak in torsion, torsion is usually avoided, or to the fact that member response to torsion appears to be significantly more complex than response to the other actions. This chapter addresses the designer's needs in order to approach torsion with the same confidence as is normally associated with the other structural responses.

In order to understand the design aspects of torsion, it will be helpful to identify those physical conditions which bring torsion into play for a structure. The easiest case of torsion to understand is one that does not occur in the normal structural design realm but rather in the field of machine design; that of the crank. As shown in Fig. 4.1, a force applied to the crank handle at some distance from the longitudinal axis of the crankshaft applies a moment about that longitudinal axis. It is this moment which is referred to as a torque. One of the first questions to be answered is why a torque must be treated differently than a bending moment. The answer is simple: Structural members respond differently to moments applied about

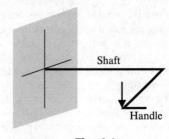

Fig. 4.1

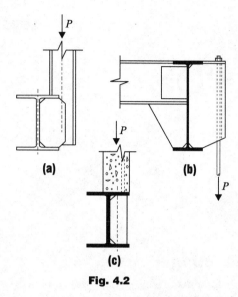

Fig. 4.2

their longitudinal axis than they do to moments applied about their transverse axes. Thus torsion becomes a topic of its own.

There are numerous ways that a torsional moment may be applied to a structural member. Figure 4.2 shows three fairly obvious examples where a load is applied at some distance to the side of the structural member, causing it to twist. The wide-flange member in Fig. 4.2a is loaded through a compression member that has a centerline to the right of the centerline of the supporting member. The actual method of attachment will have a significant impact on the overall response, but for now it will simply be assumed to apply an eccentric load. The similar example in Fig. 4.2b shows an eccentric load induced through an off-center hanger. Again, the actual connection and the connection to other members will influence the total response, but for now the significance is the eccentricity of the load. Figure 4.2c illustrates a similar case where a wall applies a load at an eccentricity from the supporting member. In each of these cases, a change in the detailing might have been able to avoid the application of torsion to the supporting member and thus reduce the complexity of design as well as any negative impact of the resulting torsional stresses.

The framing plan shown in Fig. 4.3 illustrates a significantly more common situation where torsion may play a role. The attachment of floor beams to spandrel beams, as shown along member AB in the figure, has the potential to induce a torsion into the member AB at points 1 and 2 where the floor beams are attached to the spandrel. Again, the actual details of the connection will greatly influence the amount of torsion to be carried by the spandrel. Two problems could result from the different interpretations in this case. The induced torsion might be ignored, resulting in a spandrel member that is significantly overstressed for the actual case

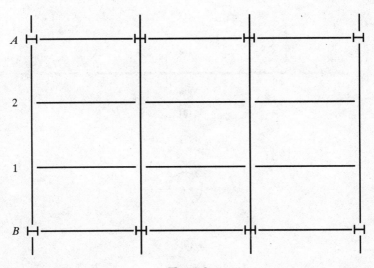

Fig. 4.3

or the full torsion applied by the beam reaction at some eccentricity could be applied to the spandrel, resulting in a beam significantly larger than actually necessary. This example points out one of the most significant reasons to have an understanding of torsion, so that it may be ignored or included in a design as appropriate, not arbitrarily.

Response to Torsion ■ A member AB subjected to an applied torsion at each end is shown in Fig. 4.4a. Note that in this representation, the double-headed vector represents a torsional moment applied about the x axis according to the right-hand rule. If the member is a circular shaft as shown in Fig. 4.4b, the shaft will rotate counterclockwise (relative to point A) through the angle θ as shown. This condition, the circular shaft and the applied torsion, represents what is generally known as "pure torsion." It is the simplest of the cases where torsion is a factor and will be discussed in more detail later.

If the member of interest is a wide flange, it will warp as it twists. The member shown in section in Fig. 4.5a is subjected to the same torsion as was applied to the member in Fig. 4.4. Again looking at section 1-1 of the member AB from Fig. 4.4a, the counterclockwise rotation θ takes place as shown in Fig. 4.5b. In addition to this rotation about the member longitudinal axis, a deformation takes place as shown in Fig. 4.5c. This deformation is called warping and, for the open shapes normally used in structures, may play a significant role in structural response.

Another wide-flange member is shown in plan in Fig. 4.6. This member is loaded through an applied torsional moment at the midlength of the member. Note that the middle section of the member has overturned, or twisted, while the ends of the member have maintained their original vertical webs. In addition, the

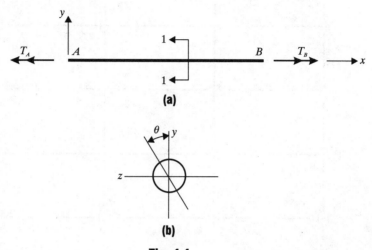

Fig. 4.4

warping of the member is seen at the ends where the tips of the flanges have displaced along the longitudinal axis of the member. It should be obvious from this example that there must be some specific end conditions which permit this type of member response. It is also clear that, as with bending moment, shear force, and axial force, these end conditions will influence the response of the member at each section along its length.

Support Definitions ■ The terminology used to define support conditions with respect to torsion is similar to that used for the other member actions: free, pinned, and fixed. Unfortunately, these terms may seem to have a somewhat different meaning when applied to torsion. A torsionally free end is permitted to rotate about the member longitudinal axis and is also permitted to warp as shown

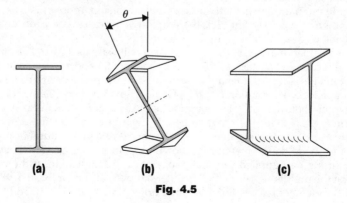

Fig. 4.5

Torsion ■ 4-5

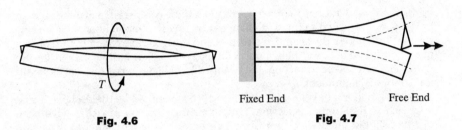

Fixed End — Fig. 4.7 — Free End

Fig. 4.6

in Fig. 4.7. A pinned end is not permitted to rotate; however, it is permitted to warp. The ends of the member shown in Fig. 4.6 are an example of a pinned end. This is quite close to the end condition that would exist for a beam with a double-angle web connection. As was the case for the fixed end in a bending situation, the fixed end in torsion does not permit any deformation. Thus the fixed end will neither rotate nor will it warp. The torsionally fixed end is more difficult to produce in an actual application than is the flexurally fixed end. It is important, as will be shown later, that the proper end conditions be used when analyzing member torsion. As with support conditions for bending, attaining a true pinned or fixed support for torsion is difficult and some judgment may need to be applied.

Avoiding or Limiting Torsion ■ Cases in actual practice where torsion comes into play are relatively limited and, when they do occur, may actually result in only a secondary effect. It is most desirable to design the structural loading and support conditions in such a way as to avoid torsion completely. This may be accomplished most easily by bringing load to a member through the point known as the shear center. The determination of the shear center will be addressed later. For now it is sufficient to recognize that every member, regardless of what shape the cross section may take, has a reference point through which a transverse load may be applied without inducing torsion into the member. When this is not possible, additional bracing may be required to take the torsion out of a member through other direct action as shown in Fig. 4.8, where the kicker is used for that purpose.

Perhaps the most significant factor in limiting the torsion induced into a member comes about as a result of the stiffness of the members actually

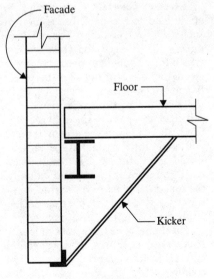

Fig. 4.8

4-6 ■ Chapter Four

bringing the load into the member in question. If the compression member shown in Fig. 4.2a is attached to the wide flange shown with a moment connection, the torsional rotation of the beam cross section and the flexural rotation of the column must be the same. Since the column will likely be significantly stiffer in flexure than the beam will be in torsion, the beam will not be required to resist as significant a torsional moment.

A combination of avoiding and limiting torsion should be the designer's first approach to designing for torsion. Only after these two approaches have been pursued should the design approaches presented in the remainder of this chapter become necessary

Shape Implications ■ As is the case for the other force actions, certain shapes are most appropriate to resist torsion. Shapes required to resist torsion are categorized as either closed or open sections. Generally, closed sections such as tubes, boxes, and solid shafts are the most efficient while open sections such as wide flanges, channels, and angles are the least efficient. The hollow circular tube is the most efficient shape for torsion resistance and will be used to develop the basic equations of pure torsion. These equations will then be modified to include the response of open shapes.

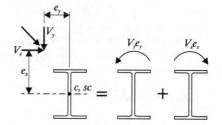

Fig. 4.9

Shear Center ■ In order to apply load to a beam cross section in such a way as to avoid torsion, the load must pass through the shear center, which for doubly symmetric shapes is located at the centroid. For the wide-flange shape shown in Fig. 4.9, with the resultant of the loads in the two principal axes passing through the centroid, and thus the shear center, there will be no torsion. Looking at the torsional moments which result from each component taken individually, the torsion applied by V_x at the eccentricity e_x is balanced by the torsion applied by V_y at the eccentricity e_y.

For shapes which are not symmetric about both major axes, the shear center and centroid will not coincide. The channel in Fig. 4.10 is one example. If a vertical load V is applied to a beam

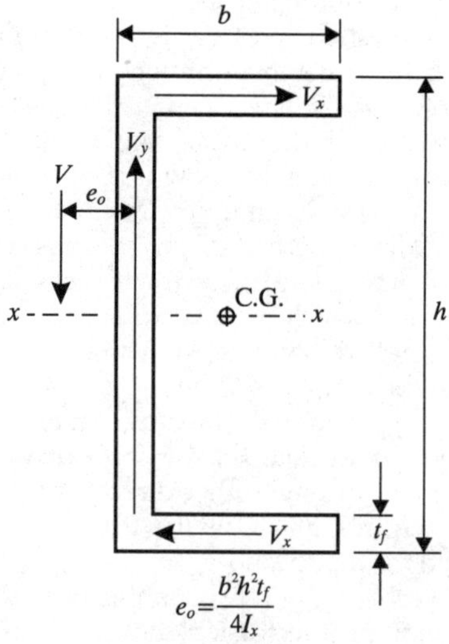

$$e_o = \frac{b^2 h^2 t_f}{4 I_x}$$

Fig. 4.10

with this channel section, it will result in a vertical shear V_y in the web and equal and opposite horizontal shear forces V_x in the flanges as shown. In order for there to be no resulting torsion, the load V must be applied at the eccentricity e_0, thus defining the location of the shear center. A rigorous derivation for the location of the shear center for various shapes may be found in numerous references such as Muvdi and McNabb. The location of the shear center for selected shapes and the necessary equations to determine the numerical values are given in Fig. 4.11.

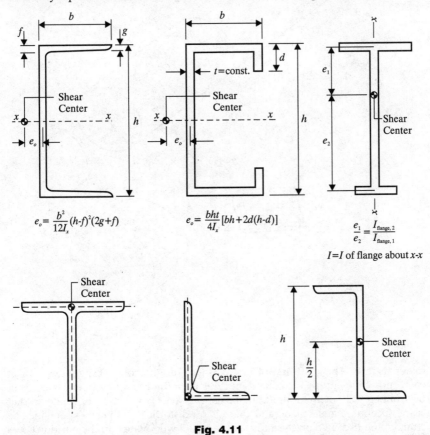

Fig. 4.11

4.2 Pure Torsion

In order to develop an understanding of torsional stress and deformation, the case of pure torsion in a circular shaft will first be addressed. Pure torsion is that condition where the stress at any point may be represented as pure shear. The shaft of Fig. 4.12 will be used for the development of the basic equations of stress and deformation. The complete derivation is found in most basic strength of

4-8 ■ Chapter Four

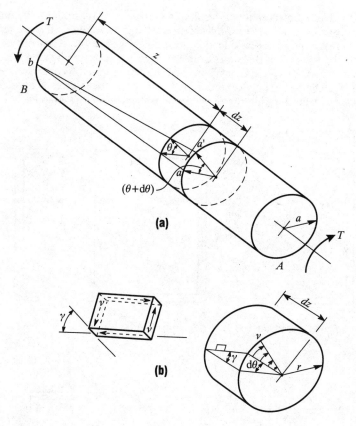

Fig. 4.12

materials texts. The shaft AB of Fig. 4.12 is loaded with concentrated torques T at each end while end B is fixed and end A is permitted to rotate freely. As a result of symmetry about the longitudinal axis, the shaft is assumed to rotate so that plane sections remain plane and there is no distortion of the cross section. The straight line ab drawn on the surface of the shaft will rotate into the position $a'b$ as shown in Fig. 4.12. Taking the differential segment of length dz from the shaft, the rate of twist may be expressed

$$\text{Rate of twist} = \frac{d\theta}{dz} \tag{4.1}$$

For the differential slice shown in Fig. 4.12b, the radial line on the section rotates through the angle $d\theta$ while the line on the surface rotates through the angle γ, thus $rd\theta = \gamma dz$. The rectangular element on the surface is distorted through the angle γ and Hooke's law yields the shear stress on each face as

$$v = G\gamma \quad (4.2)$$

where G is the shear modulus or the modulus of rigidity. Based on the assumptions made above and the geometry of the figure, it is seen that the angle γ and thus the shear stress v vary with the radial distance r. Since the shear stress around a ring of thickness dr is uniform, the total force exerted on this ring may be written as $v(2\pi r)dr$. The torsional resistance provided by this ring force is then the force times the arm r, which yields the incremental torsion

$$dT = v(2\pi r)rdr \quad (4.3)$$

Integrating over the entire cross section, with a radius a, yields the relation between the applied torque and the shear stress as

$$T = \int_0^a 2v\pi r^2 dr \quad (4.4)$$

Since the shear stress varies linearly with the radial distance, $v = v_{max}(r/a)$ and the torque becomes

$$T = \frac{v_{max}}{a} \int_0^a 2\pi r^3 dr \quad (4.5)$$

where the integral is the torsional cross section property, which for a circular section is the polar moment of inertia J. From the geometry of Fig. 4.12b, the distortion is given as $rd\theta = \gamma dz$. When this is combined with the above equations, the ratio of twist becomes

$$\text{Rate of twist} = \frac{d\theta}{dz} = \frac{T}{GJ} \quad (4.6)$$

The term GJ is the torsional rigidity and the equation is the basic differential equation of pure torsion, similar to the well-known flexure equation.

When the shaft of interest is rectangular in shape, the analysis becomes more complex. If the ends of the shaft are unrestrained, the warping that will take place is unrestrained so the rate of twist of the shaft remains constant. Shear stresses will develop parallel to the faces of the rectangle as shown in Fig. 4.13a and b. These components of shear stress combine to provide a shear flow as shown in Fig. 4.13c. An approximate analysis presented by Salmon and Johnson[2] shows that the torsional constant may be taken as $bt^3/3$ in the limit for thin rectangles, where b is the width (largest dimension) and t is the thickness (smallest dimension). When rectangles are combined, the shear flow becomes even more complex as shown in Fig. 4.14. For shapes composed of a series of rectangles, as for the wide flange shown in Fig. 4.15, the total torsional resistance may be taken as the sum of the individual contributions. Thus $T = T_{web} + 2T_{flange}$, which yields

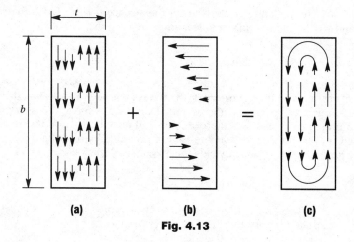

Fig. 4.13

$$T = \frac{b_w t_w^3}{3}G\frac{d\theta}{dz} + 2\frac{b_f t_f^3}{3}G\frac{d\theta}{dz} = \Sigma\frac{bt^3}{3}G\frac{d\theta}{dz} \qquad (4.7)$$

Therefore, a reasonable approximation of the torsional section property for shapes composed of a series of rectangles is given by $\Sigma bt^3/3$.

The maximum shear stress due to pure torsion, in each component rectangle, occurs at the midpoint of the long side of the rectangle. It may be determined as

$$v_{f\max} = \frac{t_f T}{\Sigma bt^3/3} \quad \text{and} \quad v_{w\max} = \frac{t_w T}{\Sigma bt^3/3} \qquad (4.8)$$

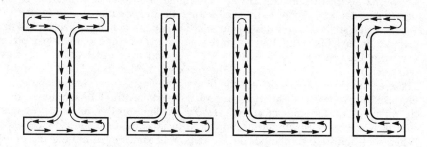

Fig. 4.14

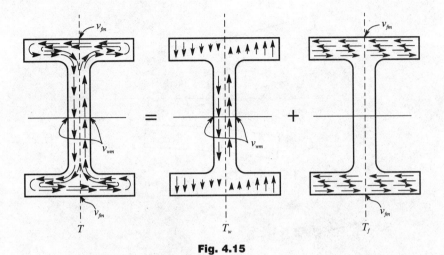

Fig. 4.15

4.3 Warping Torsion

Open sections, such as the wide flange and channel, rotate in pure torsion when the ends of the member are completely free to distort through warping, as discussed previously and shown in Figs. 4.5 and 4.6. However, if there is restraint of warping as shown at the support of the cantilever in Fig. 4.7, additional forces will develop as a result of that restraint and the member undergoes nonuniform torsion. A closer look at the restraint forces is given in Fig. 4.16b, where the forces F must be developed to restrain the flanges from the distortion that would have occurred had the support permitted warping as in Fig. 4.16a. The actual distribution of these tension and compression forces is not important at this time, but the resulting shearing forces that develop in the flanges are of particular interest. The differential elements of the two flanges of the member shown in Fig. 4.16b show these shear forces. They result in the transverse shear forces H shown in the flanges of the rotated section from Fig. 4.17b. These shear forces are very much like the forces that would develop if the flanges were to bend about their strong axis as a rectangular beam. Note also that the two flanges are bending in opposite directions. Although there is the possibility of bending in the web as well, this component is normally so small that it is reasonable to neglect its contribution as being only slightly conservative. The flange forces times the distance between them represent a force couple which is in the same direction as the resisting torsion. Thus the restraint of warping reduces the stress that must be resisted by pure torsion and uses the bending of the warping elements to contribute to the total torsion resistance. These stresses are normally referred to as warping stresses, although a better term might be resistance to warping stresses.

4-12 ■ Chapter Four

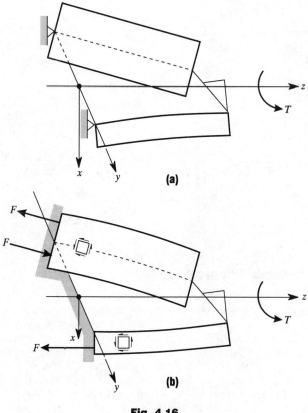

Fig. 4.16

Considering again the cantilever of Fig. 4.16b, it is clear from equilibrium that the torsion is constant along the member. It is also clear that, due to the warping constraint, this total torsion is composed of two parts, pure torsion and warping torsion. This may be written as

$$T = T_{st.v} + T_w \qquad (4.9)$$

where the pure torsion, also known as St. Venant torsion, is $T_{st.v}$, and warping torsion is T_w. An additional complexity enters the problem since the rate of twist varies along the member even though the torque does not. It is seen in Fig 4.16b that the rate of twist at the support is zero while it has a finite value at the free end. Although this will make the determination of rotation and stress more difficult, the solutions are not outside the capabilities of the typical engineer.

The pure torsional stress may be obtained by rewriting Eq. 4.6 with this new notation such that

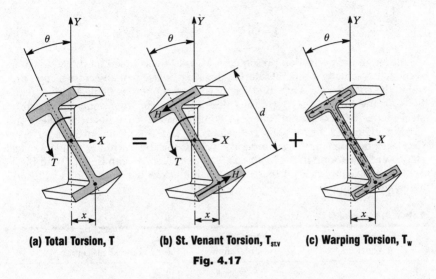

(a) Total Torsion, T   **(b) St. Venant Torsion, $T_{ST.V}$**   **(c) Warping Torsion, T_w**

Fig. 4.17

$$T_{st.v} = GJ \frac{d\theta}{dz} \qquad (4.10)$$

The expression for warping resistance will be obtained by considering the flanges as rectangular beams using the normal flexure relations so that

$$H = \frac{dM_f}{dz} = -EI_f \frac{d^3 x}{dz^3} \qquad (4.11)$$

where M_f is the bending moment in the flange that yields the shear H and I_f is the moment of inertia of the flange about its major axis. If the influence of the web is ignored, I_f may be taken as $I_y/2$ for the full cross section. From the geometry shown in Fig. 4.17b, the lateral displacement of the flange x may be related to the torsional rotation through $x = \theta d/2$. Making these substitutions into Eq. 4.11 yields

$$\frac{d^3\theta}{dz^3} = -\frac{T_w}{EI_y d^2/4} \qquad (4.12)$$

The section-related terms in the denominator of the right-hand side of Eq. 4.12 may readily be combined into a term referred to as the warping constant C_w. Substituting for the warping constant and rearranging Eq. 4.12 yields

$$T_w = -EC_w \frac{d^3\theta}{dz^3} \qquad (4.13)$$

Substitution of Eqs. 4.10 and 4.13 into Eq. 4.9 yields the complete differential equation for nonuniform torsion.

$$T = GJ\frac{d\theta}{dz} - EC_w\frac{d^3\theta}{dz^3} \qquad (4.14)$$

Although the derivation presented for Eq. 4.14 was based on the wide-flange section, it is actually a general solution that may be used for any cross section provided the correct values are used for the section properties. Table 4.1 gives equations for the torsional section properties of interest for four common structural steel shapes. More accurate equations, which include the influence of fillets and sloping elements, have been published by the American Institute of Steel Construction.[3] In addition, numerical values for the currently available steel shapes are also found in the *Manual of Steel Construction*, both the ASD and LRFD versions.[4,5] To simplify Eq. 4.14 for solution, both sides will be divided by EC_w and a new term defined.

Table 4.1 Torsional Properties

O = shear center	J = torsional constant $\qquad C_w$ = warping constant
G = centroid	I_p = polar moment of inertia about shear center

Shape	Properties
I-section: width b, height h, flange thickness t_f, web thickness t_w; O,G at center, $h/2$	$J = \tfrac{1}{3}(2bt_f^3 + ht_w^3)$ $C_w = \dfrac{I_f h^2}{2} = \dfrac{t_f b^3 h^2}{24} = \dfrac{h^2 I_y}{4}$ $I_p = I_x + I_y$
Channel: height h, width b, flange thickness t_f, web thickness t_w, shear center O at distance q, centroid G at $\bar{x}$	$J = \tfrac{1}{3}(2bt_f^3 + ht_w^3)$ $C_w = \dfrac{t_f b^3 h^2}{12}\left(\dfrac{3bt_f + 2ht_w}{6bt_f + ht_w}\right) = \dfrac{h^2}{4}(I_y + A\bar{x}^2 - q\bar{x}A)$ $q = \dfrac{th^2 b^2}{4I_x}$
Angle: legs b and h, thicknesses t_1, t_2; O at corner, G inside	$J = \tfrac{1}{3}(bt_1^3 + ht_2^3)$ $C_w = \dfrac{1}{36}(b^3 t_1^3 + h^3 t_2^3)$ $\approx$ zero for small t
Tee: flange width b, thickness t_f, stem height h, thickness t_w; O at flange, G below	$J = \tfrac{1}{3}(bt_f^3 + ht_w^3)$ $C_w = \dfrac{1}{36}\left(\dfrac{b^3 t_f^3}{4} + h^3 t_w^3\right)$ $\approx$ zero for small t

$$a^2 = \frac{EC_w}{GJ} \tag{4.15}$$

Eq. 4.14 is then rewritten as

$$\frac{T}{EC_w} = \frac{1}{a^2}\frac{d\theta}{dz} - \frac{d^3\theta}{dz^3} \tag{4.16}$$

As would be the case for solution of the differential equation of bending, both the boundary conditions and the loading pattern will influence the distribution of stress and the rotation. In order to look at an example, the simple case of the cantilever with torsion applied at the free end as previously discussed and illustrated in Fig. 4.16b will be addressed. For this case, where T is constant, the solution to the differential equation (Eq. 4.16) yields

$$\theta = \frac{Tz}{JG} + A\sinh\frac{z}{a} + B\cosh\frac{z}{a} + C \tag{4.17}$$

where A, B, and C are constants that will be determined by applying the appropriate boundary conditions and sinh and cosh are the hyperbolic functions. The boundary conditions at the support, $z = 0$ will be $\theta = 0$ and $d\theta/dz = 0$ and at the free end, $z = l$, $d^2\theta/dz^2 = 0$. Through application of these boundary conditions, the constants of integration may be obtained and the twist angle and its derivatives will follow. Carrying out those manipulations yields

$$\theta = \frac{T}{JG}\left\{z + \frac{\sinh[(l-z)/a] - \sinh(l/a)}{(1/a)\cosh(l/a)}\right\} \tag{4.18}$$

$$\frac{d\theta}{dz} = \frac{T}{JG}\left\{1 - \frac{\cosh[(l-z)/a]}{a\cosh(l/a)}\right\} \tag{4.19}$$

$$\frac{d^2\theta}{dz^2} = \frac{T}{JG}\left\{\frac{\sinh[(l-z)/a]}{\cosh(l/a)}\right\} \tag{4.20}$$

$$\frac{d^3\theta}{dz^3} = -\frac{T}{JG}\left\{\frac{\cosh[(l-z)/a]}{a^2\cosh(l/a)}\right\} \tag{4.21}$$

Combining Eqs. 4.10 and 4.19 yields

$$T_{st.v} = T\left\{1 - \frac{\cosh[(l-z)/a]}{\cosh(l/a)}\right\} \tag{4.22}$$

and Eqs. 4.13 and 4.21 yield

$$T_w = T\left\{\frac{\cosh[(l-z)/a]}{\cosh(l/a)}\right\} \tag{4.23}$$

4-16 ■ Chapter Four

It is important to recognize that Eq. 4.17 is applicable only if the torsion is constant and Eqs. 4.18 through 4.23 are applicable only for the combination of constant torsion and the boundary conditions given in Fig. 4.16.

Example 4.1 ■ Using the equations developed for the cantilever of Fig. 4.16b, determine the proportion of pure torsion and warping torsion for a W14×61 cantilever of length 12 ft. The constants for this shape are taken from AISC[5] as $J = 2.20$ in^4, $C_w = 4710$ in^6, $a = 74.5$ in. Using Eq. 4.22, determine the St. Venant torsion as

$$T_{st.v} = T\left\{1 - \frac{\cosh[(144-z)/74.5]}{\cosh(144/74.5)}\right\}$$

Equation 4.23 is used to determine the portion of T carried as warping torsion which yields

$$T_w = T\left\{\frac{\cosh[(144-z)/74.5]}{\cosh(144/74.5)}\right\}$$

The rotation of the member may be determined from Eq. 4.18 as

$$\theta = \frac{T}{2.2(11,200)}\left\{z + \frac{\sinh[(144-z)/74.5] - \sinh(144/74.5)}{(1/74.5)\cosh(144/74.5)}\right\}$$

A plot of twist angle and the distribution of the total torsion is given in Fig. 4.18.

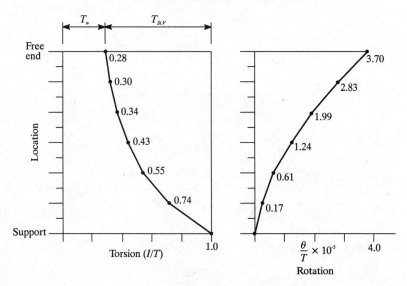

Fig. 4.18

The solutions for Eq. 4.16 have been presented for 12 combinations of boundary conditions and loading cases in AISC.[3] These cases are defined in Table 4.2, which is taken from the reference. Plots of the rotation and its three derivatives obtained as a solution to Eq. 4.16 for cases defined in Table 4.1 as cases 1, 2, and 3 are presented in Fig. 4.19. It is suggested that these curves be used only for an understanding of the analysis and that numerical data be obtained from the more accurately presented curves of AISC.[3]

4.4 Combined Stresses

Once the proportion of torsion resisted as pure torsion and warping torsion has been determined, the resulting stresses may be evaluated. In addition, the torsion stresses may be combined with the stresses that result from the other loading components as required.

Pure Torsion Shear Stress ■
The shear stress resulting from pure torsion given by Eqs. 4.8 may be restated, with the use of the torsional section property as

$$v_{st.v} = \frac{t T_{st.v}}{J} \quad (4.24)$$

The distribution of these stresses is shown in Fig. 4.20a for a wide-flange member.

Warping Torsion Shear Stress ■
The shear component of the warping torsion resistance is the result of the force in the flange that was used in the development of Eq. 4.13. This force is assumed to act in the same way that shear from bending on a rectangular section acts. Thus a parabolic distribution as shown in Fig. 4.20b represents the shear stress distribution from warping torsion. The maximum stress may be calculated through the normal shear stress equation as

$$v_w = \frac{3H}{2 b_f t_f} \quad (4.25)$$

Since $H = T_w/(d - t_f)$, Eq. 4.25 may be rewritten as

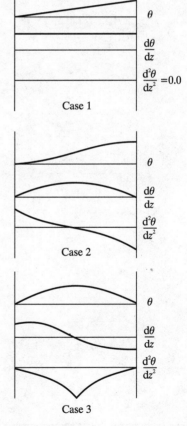

Fig. 4.19

4-18 ■ Chapter Four

Table 4.2 Case Chart Listing—Prismatic Beams (AISC 1996)

Various Torsional Loadings and End Conditions

Case	Torsional Loading	Torsional End Restraints Left End	Right End	Remarks
1	M ... M (free ends)	Free $\phi=\phi''=0$	Free $\phi''=0$	Concentrated torques at ends of member with free ends
2	M ... M (fixed ends)	Fixed $\phi=\phi'=0$	Fixed $\phi'=0$	Concentrated torques at ends of member with fixed ends
3	M at aL, $L(1-a)$ (pinned)	Pinned $\phi=\phi''=0$	Pinned $\phi=\phi''=0$	Concentrated torque at $\alpha=0.1, 0.3,$ or 0.5 on member with pinned ends
4	m distributed (pinned)	Pinned $\phi=\phi''=0$	Pinned $\phi=\phi''=0$	Uniformly distributed torque on member with pinned ends
5	$M=m(z/L)$, m (pinned)	Pinned $\phi=\phi''=0$	Pinned $\phi=\phi''=0$	Linearly varying torque on member with pinned ends
6	M at aL, $L(1-a)$ (fixed)	Fixed $\phi=\phi'=0$	Fixed $\phi=\phi'=0$	Concentrated torque at $\alpha=0.1, 0.3,$ or 0.5 on member with fixed ends
7	m distributed (fixed)	Fixed $\phi=\phi'=0$	Fixed $\phi=\phi'=0$	Uniformly distributed torque on member with fixed ends
8	$M=m(z/L)$, m (fixed)	Fixed $\phi=\phi'=0$	Fixed $\phi=\phi'=0$	Linearly varying torque on member with fixed ends
9	M at aL (fixed-free)	Fixed $\phi=\phi'=0$	Free $\phi''=0$	Concentrated torque at $\alpha=0.1, 0.3, 0.5, 0.7, 0.9,$ or 1.0 on member with fixed and free ends
10	m partial along aL	Fixed $\phi=\phi'=0$	Free $\phi''=0$	Partial uniformly distributed torque along $\alpha=0.1, 0.3, 0.5, 0.7, 0.9,$ or 1.0 with fixed end and free end
11	$M=m(z/L)$, m (free-fixed)	Free $\phi''=0$	Fixed $\phi=\phi'=0$	Linearly varying torque on member with free and fixed ends
12	m distributed (fixed-pinned)	Fixed $\phi=\phi'=0$	Pinned $\phi=\phi''=0$	Uniformly distributed torque on member with fixed and pinned ends

Note: The end restraints apply to torsional conditions only, $\phi=\theta$

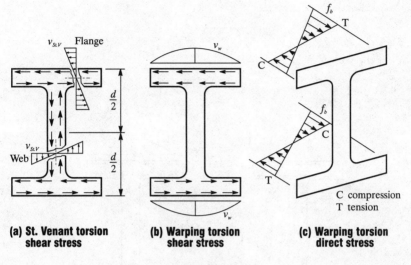

(a) St. Venant torsion shear stress   **(b) Warping torsion shear stress**   **(c) Warping torsion direct stress**

Fig. 4.20

$$v_w = \frac{3T_w}{2b_f t_f (d - t_f)} \qquad (4.26)$$

Warping Torsion Normal Stress ■ Since warping of the section causes the flanges to bend about their strong axis, additional stresses develop as shown in Fig. 4.20c. These are tension and compression stresses which will combine both positively and negatively with any normal bending stresses that would occur as a result of other loadings. The flange moment may be determined by integrating Eq. 4.11 and substituting for x as was done for the derivation of Eq. 4.12, to yield

$$M_f = -\frac{EI_f d}{2} \frac{d^2\theta}{dz^2} = -\frac{EC_w}{d} \frac{d^2\theta}{dz^2} \qquad (4.27)$$

Using the flexure formula to determine bending stress yields

$$f_b = \frac{6M_f}{t_f b_f^2} = -\frac{6EC_w}{dt_f b_f^2} \frac{d^2\theta}{dz^2} \qquad (4.28)$$

These flange bending stresses are distributed as shown in Fig. 4.20c.

Example 4.2 ■ Using the results from Example 4.1, determine the stress distribution at the free end and at the support. For the free end, the resulting stresses are shown in Fig. 4.21a and b. Shear stress from pure torsion may be determined from Eq. 4.24 for the flange as

$$v_{\text{st.v}} = \frac{0.645(0.72T)}{2.2} = 0.211T$$

and for the web as

$$v_{\text{st.v}} = \frac{0.375(0.72T)}{2.2} = 0.123T$$

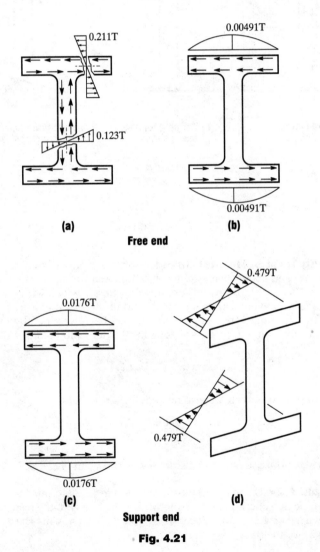

(a)   (b)

Free end

(c)   (d)

Support end

Fig. 4.21

Shear from warping torsion is determined through Eq. 4.25 where

$$H = \frac{0.28T}{13.89 - 0.645} = 0.0211T$$

which yields

$$v_w = \frac{3(0.0211T)}{2(9.995)(0.645)} = 0.00491T$$

At the fixed end the results are shown in Fig. 4.21c and d. Since $T_{st.v} = 0$, the only shear stress is developed due to warping and may be found through the application of Eq. 4.25 again.

$$H = \frac{T}{13.89 - 0.645} = 0.0755T$$

and

$$v_w = \frac{3(0.0755T)}{2(9.995)(0.645)} = 0.0176T$$

The direct stress will be greatest at the support since that is the location of maximum warping torsion. Combining Eq. 4.28 with Eq. 4.20 yields

$$f_b = \frac{6(74.5)T}{13.89(0.645)(9.995)^2}(0.959) = 0.479T$$

Stress Combinations ■ When torsion occurs in combination with other actions, the resulting stresses must be combined to determine the maximum normal and shear stress conditions. For wide flange shapes, the normal stress due to bending is maximum on the outer face of the flanges while the maximum for torsion is maximum on the tips of the flanges as was shown in Fig. 4.20c. Thus the maximum normal bending stress and maximum normal warping torsion stress will always combine positively at one point on each flange, regardless of the sign of the bending moment and torsion. Shear stresses in the flanges of the wide flange shape due to warping, St. Venant torsion, and bending will also all combine with the same sign at some point on the flange, independent of the sign of the moment and torsion. For the web of the wide flange, the St. Venant torsion shear stress and the bending shear stress will also combine with the same sign at some point, regardless of the direction of the torsion and bending moment. This makes the determination of the maximum combined stresses a straightforward calculation for wide flange shapes under the action of torsion and bending. If axial forces are present, their direction must be accounted for in the calculation, but they too will combine to maximize one of the resulting stress conditions.

For shapes other than the wide flange, the direction of the resulting stresses must be determined and properly combined. Reference to AISC[3] will provide further guidance in the determination of these stresses.

4-22 ■ Chapter Four

Example 4.3 ■ For the beam shown in Fig. 4.22a, determine the resulting stresses. The support conditions are pinned for bending and torsion; thus, as far as torsion is concerned, this may be seen as having the same boundary conditions as the beam considered in Examples 4.1 and 4.2 when one half of the beam is considered. For bending at midspan

$$f_b = \frac{1080}{92.2} = 11.71 \text{ ksi}$$

Direct shear is constant over the half span; thus shear stress is given as

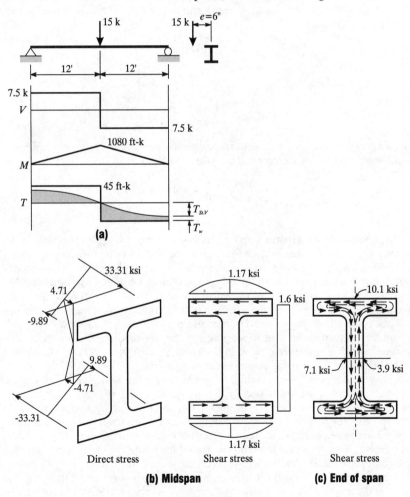

Fig. 4.22

$$v_{web} = \frac{7.5(51.1)}{640(0.375)} = 1.60 \text{ ksi} \quad \text{and} \quad v_{flange} = \frac{7.5(21)}{640(0.645)} = 0.38 \text{ ksi}$$

For torsion at the support (equivalent to the free end of Example 4.1 and 4.2)

$T_w = 0.28(45) = 12.6$ ksi

$T_{st.v} = 0.72(45) = 32.4$ ksi

$v_w = 0.00491(45) = 0.221$ ksi

$v_{st.v(flange)} = 0.211(45) = 9.495$ ksi

$v_{st.v(web)} = 0.123(45) = 5.535$ ksi

At midspan (equivalent to the support for the member of Examples 4.1 and 4.2) $T_w = 45$ in-kips and $T_{st.v} = 0$. Thus in the flange

$f_{bw} = 0.479(45) = 21.6$ ksi and $v_w = 0.0176(45) = 0.792$ ksi

Combined stresses at midspan yield

Direct stress $= 11.71 \pm 21.6 = -9.89, +33.31$ ksi

Flange shear stress $= 0.0792 + 0.38 = 1.17$ ksi

Web shear stress $= 1.60$ ksi

Combined stresses at the support yield

Flange shear $= 9.495 + 0.221 + 0.38 = 10.1$ ksi

Web shear $= 1.6 \pm 5.535 = +7.1, -3.9$ ksi

4.5 Bending Analogy

The determination of stresses due to combined bending and torsion for a given loading and cross section has been shown to be fairly straightforward, although solution of a differential equation is required. In the design stage the section has not yet been determined and it would be helpful to have a somewhat more direct method to obtain an estimate of the required section properties. An easily applied approach has been presented by Salmon and Johnson[2] that will be outlined here.

In most cases where torsion is to be included with bending in the design, it is the direct stress in the flange which has the predominant impact in selecting a section. In order to estimate this stress it will be useful to convert the torsional moment to a force couple as shown in Fig. 4.23. Thus the stress in the flange is determined as a combination of the stress due to bending about the strong axis and the stress due to an equivalent bending about the weak axis which represents the torsion. It can be shown that this approach is conservative and in most practical instances too conservative to be a useful approach to design. It is best used if

4-24 ■ Chapter Four

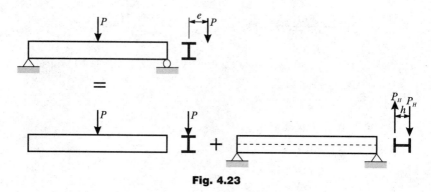

Fig. 4.23

the distance between torsional supports is small. If this approach was used to determine the stresses for the loading and support conditions from the member in Example 4.3, an indication of the conservativeness will be seen.

Example 4.4 ■ Determine the maximum direct and shear stresses for the conditions of Example 4.3 using the bending analogy. Compare these results to those obtained through the solution of the differential equations as found in Example 4.3. Using the beam analogy, the bending in one flange due to a 90 in-kip torsion is

$$M_f = \frac{[90/(13.89-0.645)](24)}{4} = 40.8 \text{ kip-ft}$$

the normal stress is

$$f_b = \frac{40.8(12)}{21.5/2} = 45.5 \text{ ksi}$$

Torsional shear stresses in the flange and web are given through

$$v_{flange} = \frac{(90/2)(0.645)}{2.2} = 13.19 \text{ ksi}$$

$$v_{web} = \frac{(90/2)(0.375)}{2.2} = 7.67 \text{ ksi}$$

The shear stress due to warping is treated as shear from lateral bending so that

$$v_w = \frac{1.5(3.4)}{(9.995)(0.645)} = 0.79 \text{ ksi}$$

The maximum combinations of stress result in

	Bending analogy	Differential equations
Nominal stress =	11.71 + 45.5 = 57.2 ksi	33.3 ksi at midspan
Web shear =	1.6 + 7.67 = 9.27 ksi	7.1 ksi at support
Flange shear =	0.38 + 13.19 + 0.79 = 14.36 ksi	10.1 ksi at support

As can be seen from these results, the approach is overly conservative, particularly for the normal stress calculations. Since the bending analogy assumes that the shear force in the flange is distributed as for a simple beam, the moment is significantly exaggerated when compared to that which would come from the actual distribution of the shear force, accounting for the distribution of warping and St. Venant torsion.

In order to correct for the conservatism of the bending analogy, Salmon and Johnson[2] have presented a method to modify the direct stress resulting from torsion. They provide tables for selected loading and support conditions which provide values for the modifier β. For the case in question here, $\beta = 0.496$ from their tables. Thus the resulting stress, given as:

$$f_b = \frac{\beta TL}{2(d-t_f)S_y}$$

may be determined by multiplying the direct stress determined previously by 0.496 which yields a direct stress of $(0.496)(45.5) = 22.57$ ksi which is quite close to the value of 21.6 ksi previously determined in Example 4.3.

Although it is shown that the calculation of stress may be approached in this fashion, the real usefulness of this approach is in the selection of a trial section. In those cases, tables of β values will be helpful for the variety of boundary conditions to be considered. This approach is discussed later when design problems are addressed.

4.6 Serviceability

The rotation of a member due to torsion is of concern as it impacts on the structural and nonstructural elements that are attached to the member undergoing torsional rotation. Since there are no uniform published criteria for limitation of the angle of rotation of a member, the designer is cautioned to evaluate the particular situation under consideration and to establish a limitation on the rotation that will be compatible with the response limitations of the attached elements. The rotation may be determined through the solution of Eq. 4.16 and the displacements of the attached elements subsequently determined. When serviceability is a critical factor, however, it may be most appropriate to redesign the details of the structure to eliminate the torsional loading completely.

4.7 LRFD Design

The AISC LRFD specification[5] considers torsion and torsion combined with flexure, shear, and axial force in Chapter H. Although Chapter H provides interaction equations for flexure and axial forces, it does not provide interaction equations for use when torsion is involved. Instead it takes the approach that has been used historically, that of combining elastic stresses that result from a factored load analysis and limiting the maximum value to a specified level. That limitation is given as

a. For the limit state of yielding under normal stress:

$$f_{un} \leq \phi F_y$$
$$\phi = 0.9$$
(4.29)

b. For the limit state of yielding under shear stress:

$$f_{uv} \leq 0.6\phi F_y$$
$$\phi = 0.9$$
(4.30)

c. For the limit state of buckling:

$$f_{un} \text{ or } f_{uv} \leq \phi_c F_{cr}$$
$$\phi_c = 0.85$$
(4.31)

Example 4.5 ■ Using the stress results from Example 4.3, check for compliance with the AISC LRFD criteria of Eqs. 4.29 through 4.31. The steel used has $F_y = 50$ ksi.

a. For normal stress, $f_{un} = 0.9(50) = 45.0$ ksi > 33.31 ksi; thus it is adequate
b. For shear stress, $f_{uv} = 0.9(50) = 45.0$ ksi > 7.1 ksi; thus it is also adequate

Thus the W14×61 $F_y = 50$ ksi member is acceptable for the combined stress conditions.

4.8 Simplified Equations

The previously developed equations provide an approach to analysis for torsion that is built upon the solution to the differential equation given as Eq. 4.16 for the particular loading and boundary conditions specified. In order to make the transition to other loads and boundary conditions, it is helpful to reform the stress equations in terms of member rotation θ and its derivatives. Then general solutions such as those found in AISC[3] may be used in an analysis.

Pure Torsion Shear Stress ■ Combining Eq. 4.10 and Eq. 4.24 yields

$$v_{st.v} = Gt\frac{d\theta}{dz}$$
(4.32)

Warping Torsion Shear Stress ■ Combining Eq. 4.13 and Eq. 4.26 yields

$$v_w = \frac{-3C_w E}{2b_f(d-t_f)t_f}\frac{d^3\theta}{dz^3} = \frac{-ES_{ws}}{t_f}\frac{d^3\theta}{dz^3}$$
(4.33)

where the newly introduced term S_{ws} is provided in AISC.[5]

Warping Torsion Direct Stress ■ Rewriting Eq. 4.28 yields

$$f_b = \frac{6C_w E}{dt_f b_f^2}\frac{d^2\theta}{dz^2} = EW_{ns}\frac{d^2\theta}{dz^2}$$
(4.34)

where W_{ns} is also provided in AISC.[5]

Torsion ■ 4-27

Example 4.6 ■ For the beam and loading shown in Fig. 4.24, determine the combined shear and direct stresses and compare them to the LRFD specified limits. The beam is identified as case 4 in Table 4.1. The factored loads are

$$w_u = 1.2(0.36 + 0.14) + 1.6(1.5) = 3.0 \text{ k/ft}$$

and the ultimate bending moment is $M_u = 3(24)^2/8 = 216$ ft-kips. For torsion, since the self-weight of the member does not contribute to torsion, $w_u = 1.2(0.36) + 1.6(1.5) = 2.83$ k/ft. This yields an applied ultimate torsion $m = 18.4$ in-kips/ft.

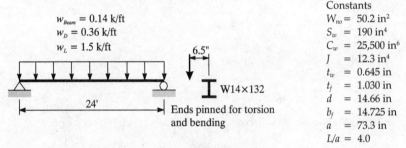

$w_{Beam} = 0.14$ k/ft
$w_D = 0.36$ k/ft
$w_L = 1.5$ k/ft

Constants
$W_{no} = 50.2$ in^2
$S_w = 190$ in^4
$C_w = 25,500$ in^6
$J = 12.3$ in^4
$t_w = 0.645$ in
$t_f = 1.030$ in
$d = 14.66$ in
$b_f = 14.725$ in
$a = 73.3$ in
$L/a = 4.0$

Fig. 4.24

For the W14×132 member using A36 steel, the following values shown in Fig. 4.24 are obtained from AISC.[5] For loading case 4, AISC[3] yields the following maximum values for rotation and its derivatives:

At midspan (L/2)

$$\theta \frac{GJ}{m} \frac{1}{2a^2} = 0.63$$

$$\frac{d^2\theta}{dz^2} \frac{GJ}{m} = -0.74$$

At span ends (0 and L)

$$\frac{d\theta}{dz} \frac{GJ}{m} \frac{L}{10a^2} = 0.42$$

$$\frac{d^3\theta}{dz^3} \frac{GJ}{m} a = -0.96$$

At the ends of the member, warping torsion shear yields

$$v_w = \frac{-ES_{ws}}{t_f} \frac{d^3\theta}{dz^3} = \frac{-ES_{ws}}{t_f}(-0.96)\left(\frac{m}{GJa}\right) = 0.81 \text{ ksi}$$

and pure torsion shear in the web is

$$v_{st.v} = Gt\frac{d\theta}{dz} = G(0.645)(0.42)\left(\frac{10a^2m}{GJL}\right) = 6.28 \text{ ksi}$$

4-28 ■ Chapter Four

and in the flange

$$v_{\text{st.v}} = Gt\frac{d\theta}{dz} = 6.28\left(\frac{1.03}{0.645}\right) = 10.03 \text{ ksi}$$

At midspan, shear stresses are zero while direct stress due to warping is given as

$$f_b = EW_{ns}\frac{d^2\theta}{dz^2} = EW_{ns}(-0.74)\left(\frac{m}{GJ}\right) = -11.13 \text{ ksi}$$

Stresses due to bending and shear due to bending are

$$f_b = \frac{216(12)}{209} = 12.4 \text{ ksi}$$

$$v_{\text{flange}} = \frac{36.0(49.9)}{1530(14.725)} = 0.08 \text{ ksi}$$

$$v_{\text{web}} = \frac{36.0(117.0)}{1530(0.625)} = 4.27 \text{ ksi}$$

These results are combined in Table 4.3. A comparison with the code-specified values results in

$$f_{un} = 0.9(36.0) = 32.4 \text{ ksi} > 23.5 \text{ ksi}$$

$$f_{uv} = 0.6(0.9)(36.0) = 19.44 \text{ ksi} > 14.3 \text{ ksi}$$

which indicates that this beam is adequate to carry the imposed load.

Table 4.3 Results for Example 4.6

		Direct/Normal						
		f_b	f_{bw}	Σ	v_t	v_w	v_b	Σ
Midspan	Flange	12.4	11.13	23.5	0	0	0	0
	Web				0	0	0	0
End	Flange	0	0		6.28	0.81	0.08	7.17
	Web				10.03	—	4.27	14.3

Example 4.7 ■ For the beam and loading shown in Fig. 4.25, determine the combined shear and direct stresses under the left load and compare them to the LRFD specified limits. This beam is the same section as was used in Example 4.6 and is loaded so that the bending moment and maximum torsion are the same as in that example. The beam is identified as case 3 in Table 4.1. The factored loads are given in the figure. The ultimate bending moment is $M_u = 30(7.2) = 216$ ft-kips and the torsion is $T_u = 220.8$ in-kips.

Torsion ■ 4-29

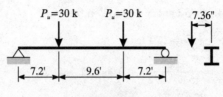

Fig. 4.25

For the W14×132 member used, the values shown in Fig. 4.24 are used. For loading case 3, AISC[3] yields the following maximum values for rotation and its derivatives:

At the left load due to the left load	At the left load due to the right load	Combined
$\theta \dfrac{GJ}{T}\dfrac{1}{L} = 0.095$	$\theta \dfrac{GJ}{T}\dfrac{1}{L} = 0.068$	0.163
$\dfrac{d^2\theta}{dz^2}\dfrac{GJa}{T} = -0.455$	$\dfrac{d^2\theta}{dz^2}\dfrac{GJa}{T} = -0.085$	−0.54
$\dfrac{d\theta}{dz}\dfrac{GJ}{T} = 0.155$	$\dfrac{d\theta}{dz}\dfrac{GJ}{T} = -0.20$	−0.045
$\dfrac{d^3\theta}{dz^3}\dfrac{GJ}{T}a^2 = -0.54$	$\dfrac{d^3\theta}{dz^3}\dfrac{GJ}{T}a^2 = 0.10$	−0.44

At the load points, warping torsion shear stress is

$$v_w = \frac{-ES_{ws}}{t_f}\frac{d^3\theta}{dz^3} = \frac{-ES_{ws}}{t_f}(-0.44)\left(\frac{T}{GJa^2}\right) = 0.69 \text{ ksi}$$

Pure torsion shear in the web is

$$v_{st.v} = Gt\frac{d\theta}{dz} = G(0.645)(-0.045)\left(\frac{T}{GJ}\right) = 0.51 \text{ ksi}$$

and in the flange is

$$v_{st.v} = Gt\frac{d\theta}{dz} = 0.51\left(\frac{1.03}{0.645}\right) = 0.81 \text{ ksi}$$

Direct stress due to warping is given as

$$f_b = EW_{ns}\frac{d^2\theta}{dz^2} = EW_{ns}(-0.54)\left(\frac{T}{GJa}\right) = -15.65 \text{ ksi}$$

Chapter Four

Stresses due to bending and shear due to bending are

$$f_b = \frac{216(12)}{209} = 12.4 \text{ ksi}$$

$$v_{flange} = \frac{30.0(49.9)}{1530(14.725)} = 0.066 \text{ ksi}$$

$$v_{web} = \frac{30.0(117.0)}{1530(0.625)} = 3.67 \text{ ksi}$$

At the left load due to the left load	At the left load due to the right load	Combined
$\theta \dfrac{GJ}{T}\dfrac{1}{L} = 0$	0.0	0.0
$\dfrac{d^2\theta}{dz^2} \dfrac{GJa}{T} = 0$	0.0	0.0
$\dfrac{d\theta}{dz} \dfrac{GJ}{T} = 0.4$	−0.246	0.154
$\dfrac{d^3\theta}{dz^3} \dfrac{GJ}{T} a^2 = -0.3$	0.055	−0.245

At the ends, warping torsion shear stress is

$$v_w = \frac{-ES_{ws}}{t_f} \frac{d^3\theta}{dz^3} = \frac{-ES_{ws}}{t_f}(-0.245)\left(\frac{T}{GJa^2}\right) = 0.384 \text{ ksi}$$

Pure torsion shear in the web is

$$v_{st.v} = Gt\frac{d\theta}{dz} = G(0.645)(0.154)\left(\frac{T}{GJ}\right) = 1.74 \text{ ksi}$$

and in the flange is

$$v_{st.v} = Gt\frac{d\theta}{dz} = 1.74\left(\frac{1.03}{0.645}\right) = 2.77 \text{ ksi}$$

Direct stress due to warping at the end is zero, as is stress due to bending. Stresses due to shear from bending are

$$v_{flange} = \frac{30.0(49.9)}{1530(14.725)} = 0.066 \text{ ksi}$$

$$v_{web} = \frac{30.0(117.0)}{1530(0.625)} = 3.67 \text{ ksi}$$

These results are combined in Table 4.4. A comparison with the code-specified values results in

$$f_{un} = 0.9(36.0) = 32.4 \text{ ksi} > 28.05 \text{ ksi}$$
$$f_{uv} = 0.6(0.9)(36.0) = 19.44 \text{ ksi} > 5.41 \text{ ksi}$$

As can be seen from these two examples, the use of the solutions found in AISC[3] makes the determination of stresses due to torsion a straightforward exercise.

Table 4.4 Results for Example 4.7

		Direct/Normal						
		f_b	f_{bw}	Σ	v_t	v_w	v_b	Σ
Midspan	Flange	12.4	15.65	28.05	0.81	0.69	0	1.5
	Web				0.51	0	0	0.51
End	Flange	0	0	0	2.77	0.384	0.066	3.22
	Web				1.74	—	3.67	5.41

References

1. Muvdi, B. B., and J.W. McNabb, *Engineering Mechanics of Materials*, Macmillan, New York, 1980.

2. Salmon, C. G., and J. E. Johnson, *Steel Structures–Design and Behavior*, 3d ed., HarperCollins, New York, 1990.

3. American Institute of Steel Construction, *Torsional Analysis of Structural Steel Members*, P. A. Seaburg and C. J. Carter, AISC, Chicago, Ill., 1996.

4. American Institute of Steel Construction, *Manual of Steel Construction—ASD*, 9th edition, AISC, Chicago, Ill., 1989.

5. American Institute of Steel Construction, *Manual of Steel Construction—LRFD*, 2d edition, AISC, Chicago, Ill., 1994.

5

Stephen J. Y. Tang (1919–1993)
Ian R. Chin
Jerome W. Rasgus
Richard F. Rowe

BUILDING DESIGN LOADING CRITERIA AND GENERAL CONSIDERATIONS*

5.1 Planning Building Structures

The design of the structural system for a building may be visualized as consisting of the following steps:

1. Programming, or definition of the problem. Analysis of the functional requirements dictated by the building use, location, character, etc.
2. Evaluation of the loads. Determination of the type, magnitude, and possible combinations of the loads which act on the building
3. Planning, or preliminary design. Determination of the type, material, layout, and scale of the system
4. Investigation of the structural performance or behavior under load. Both qualitative and quantitative determination of the external and internal forces generated by the loads
5. Definitive design of the system. Selection of the size, shape, quality, etc., of all the parts of the structure
6. Integration of the structure in the building. Relation of the structure to architectural details, mechanical equipment, building occupancy requirements, etc.
7. Construction detailing. Study of the methods, details, and sequence of fabrication and assembly of the structure

*This section is an adaptation of Section 21 in Gaylord and Gaylord *Structural Engineering Handbook*, McGraw-Hill, New York, 1996.

The planning, or preliminary design, must be done with an intelligent anticipation of the problems which will arise in the later stages of design and construction. In many respects, this makes it the most difficult as well as the most important step. A poor job at this stage may void all the following work and make a completely new start necessary. Since the preliminary design must be based largely on experience and judgment, some amount of reworking must be anticipated where the designer lacks experience or the job is of a complex or unique nature. In fact, except for the very simple and ordinary structures, even the experienced designer will usually study several alternative systems, materials, and layouts before the final scheme is set.

5.1.1 Selection of Structural Scheme

The selection of the structural system for a particular building involves the comparison of the unique requirements of the building with the general characteristics of the various systems available. In most designs, the relative merit of several systems will need to be carefully weighed in order to make the best possible selection. The feasibility of various systems must first be established with respect to fire resistance, exposure, appearance, height restrictions, construction time constraints, size of elements, spans, loads, adaptability to openings and special loads, attachment of nonstructural elements, etc. They must be compared with respect to the total cost of the finished structure, including the cost of materials, labor for fabrication and erection, transportation to the site, temporary forms or bracing required, and any maintenance required during the life of the building. The extent to which a particular system reduces or increases the cost of other elements of the construction must be considered. Thus, the ability of the structure to develop the required fire resistance with or without applied protection, to facilitate the inclusion or attachment of nonstructural elements, etc., must be considered in any comparative study of systems. The relative depth of a floor system may be critical as it affects the required height of the building in multistory construction.

Complete and fair economic comparisons of structurally equivalent systems are very difficult to make. In addition to considerations of total cost of the structure in place and cost of the complete building, other constantly changing relationships influence the comparative cost. Some of these are: local availability of materials; competition of suppliers; development of new materials which make established ones less competitive; changing building code criteria which make better—and usually more economical—designs possible in certain materials; familiarity of local builders with new systems, which reflects itself in fewer contingencies in bids; etc.

Intelligent selection of the structural scheme can only be made after the criteria for the particular building have been thoroughly itemized. This entails considerable judgment and the cooperation of all those involved in the planning. Because of the unique or complex nature of a problem, or the lack of experience of any or all of the members of the design team, it may be highly unlikely that the best solution can be arrived at in a single attempt. In fact, except for repeated similar projects, the typical design effort usually involves partial development of alternative schemes, in order to establish the optimal choice.

5.1.2 Spatial Requirements

The architectural design will also determine the clear spans and clear heights required. Structural considerations may influence these decisions but usually only by way of establishing practical limitations. The determination of the modular system of the building, if any, may require some coordination of architectural and structural requirements.

Limitations on the size of the structural members may be imposed by the architectural design. Reduction of column sizes may be critical in conserving space, especially in the lower floors of multistory buildings. Limitations on the depth of horizontal elements of the structural frame may be important in reducing the building height and volume, resulting in savings on exterior walls, stairs, piping, ducts, wiring, etc. In some cases structural efficiency may be sacrificed if it results in savings in other aspects of the building construction.

Special building uses may dictate the shape and layout of the structure and in turn influence the choice of the type of structural system. Thus, in the design of an auditorium the requirements for an efficient acoustical form and good sight lines may set the shape of the building and indicate the use of a particular system. For a gymnasium or large assembly arena, requiring a long clear span and high open space, an efficient long-span system, such as a truss, arch, dome, or suspension system, will be most appropriate.

5.1.3 Lateral Load Systems

Lateral load pressure on exterior vertical surfaces of the building is transmitted to the roof and floors and through them to the shear walls, rigid frames, vertical bracing, wind towers, super frames, or whatever combination of these elements is provided. Transfer of the lateral force through unbraced roofs and floors may involve considerable judgment with regard to the diaphragm capabilities of the deck and the adequacy of the assembly details.

Floors and roofs may be stiffened to increase their resistance to in-plane distortion. Intersecting beams may be designed for Vierendeel truss action, horizontal bracing may be provided, and single panel strips of the floor system may be stiffened so that loads are transferred to them by other beams in column action. Walls can act as shear panels if they are designed and constructed to act as part of the lateral system.

Provision must be made for both overturning and sliding. Codes specify that the resisting dead-load moment must be 1.5 to 2 times the lateral load overturning moment, or else anchorage must be provided. Horizontal resistance is usually provided by frictional resistance of the footings and slabs on ground, and by passive earth pressure on foundation walls. The lateral resistance may be increased by specifying a highly compacted, granular backfill.

5.1.4 Deflection

Deformation is an important issue in design. Major considerations have to do with deflection, vibration, and with the somewhat nebulous but very real quality of solidity. Deflections may be objectionable for various reasons. Excessive deflections may result in curvatures or misalignments perceptible to the eye. Lesser deflections may result in fracture of architectural elements, such as plaster or masonry. Where neither of these is critical, deflection may still be objectionable if it results in the transfer of load to nonstructural elements, such as window frames or interior partitions. The latter situation is often handled by detailing connections which allow for movements. This requires close coordination of the architectural and structural detailing.

Table 5.1 Comfort Criteria for Building Occupants

Acceleration, %g	Motion
Under 0.5	Not perceptible
0.5–1.5	Threshold of perceptibility
1.5–5	Annoying
5–15	Very annoying
Over 15	Unbearable

Vibration of building structures may result from various sources and may be objectionable because it is perceptible to the occupants, because the use of the building involves delicate work or sensitive machinery, or because it causes actual physical damage, such as fracture of window glass.

Limits and thresholds of tolerable physiological and psychological response to movements are difficult to determine. It is generally agreed that, under normal conditions vertical deflection to $1/600$ of the span and lateral deflection of $1/360$ to $1/1000$ of the building height are reasonable and acceptable. However, rate of movement and frequency of vibration play an important part. Table 5.1 can be used as a guide.

5.1.5 Structural Materials

Considerations to be made in the selection and use of structural materials are:

Structural properties—strength, stress-strain characteristics, time-dependent behavior, temperature effects on strength and stiffness, etc.

Physical properties—form, weight, ductility, hardness, durability, fire resistance, coefficient of expansion, resistance to corrosion, rot, insect attack, etc.

General properties—cost, availability, reliability of quality control, forming limits, general workability, etc.

5.1.6 Fire Resistance

Fire-resistant ratings required for the various parts of the building, including the members of the structural system, will be established by the building location and use and will usually be specified in detail by the local building code.

Fire resistive ratings are expressed in hours and refer to the length of time the member can be subjected to a standard fire test without failing. Failure may consist of any of the following:

1. Transmission failure. Rise of temperature of, or passage of hot gases to, the side opposite the fire. This type of failure may be due to conduction or to formation of holes or cracks in the members.

2. Structural failure at high temperature. The element must retain sufficient strength to support the building and its contents during the fire. Failure may be

Building Design Loading Criteria and General Considerations ■ 5-5

one of actual collapse of the element or of excessive deformation resulting in failure of other parts of the structure.

3. Failure when subjected to a stream of water. The structure should retain its strength when subjected to a hose stream during or after the fire.

It should be noted that standard fire ratings refer only to the action of a member during a fire and in no way certify the usefulness of the structure after the fire. Fire ratings are established, in most cases, by fire tests on full-scale elements. Fire-resistive ratings of some typical construction systems are shown in Table 5.2.

Table 5.2 Hourly Fire-Resistive Ratings of Typical Construction Elements

	Elements	Ratings and criteria
	Beams	
	Steel section, poured concrete cover, stone concrete	1 to 4 h, depending on thickness of cover, type of tie or wire wrap around beam, specific type of aggregate
	Steel section, lath and plaster cover	1 to 4 h, depending on type of plaster, thickness of plaster, type of lath, space between lath and beam
	Steel section, sprayed fiber protection	3 to 4 h, depending on type of fiber, thickness of covering
	Steel section, enclosed between floor and ceiling	0 to 4 h, depending on type of ceiling, space between ceiling and beam
	Steel section, fireproof paint cover	1 to 2 h, depending on thickness and fire test
	Steel section, water and antifreeze filled, auxiliary mechanical pump system	4 h with constant circulating water
	Steel section with deflector and interior fireproofing	Rating to be determined by fire test, 1 to 4 h
	Steel section, masonry cover	0 to 4 h, depending on thickness and material

Table 5.2 Hourly Fire-Resistive Ratings of Typical Construction Elements (*Continued*)

Elements	Ratings and criteria
Columns	
Steel section, poured concrete cover	1 to 4 h, depending on thickness of cover, type of aggregate, and size of column
Steel section, lath and plaster cover	1 to 4 h, depending on thickness of plaster, type of plaster, type of lath, space between lath and column, size of steel section
Steel section, sprayed fiber protection	1 to 4 h, depending on type of fiber, thickness of cover, details of the application process
Steel section masonry cover	1 to 4 h, depending on thickness of cover
Steel section, fireproof paint cover (paint expands 20 or 30 times original thickness)	1 to 2 h, depending on thickness and fire test
Steel section with oversized flange and web so that actual required section temp. < 1000°F (537.7°C)	1 to 4 h, depending on fire test
Steel section, water and antifreeze filled, auxiliary mechanical pump system	4 h, with constant circulating water
Floor and roof decks	
Steel deck, concrete fill, hung ceiling of lath and plaster	3 to 4 h, depending on thickness and type of concrete fill, space between deck and ceiling, type and thickness of plaster
Steel deck, concrete fill, sprayed fiber on underside of deck	2 to 4 h, depending on thickness and type of concrete, thickness and type of fiber cover
Concrete slab on steel deck, composite action, negative reinforcing in concrete, no protection on underside	Up to 4 h, depending on overall thickness of slab, type of aggregate, cover on negative reinforcing
Reinforced-concrete slab, no protection	1 to 4 h, depending on overall thickness of slab, cover on reinforcing, type of aggregate
Steel open-web joists, enclosed between deck and ceiling	0 to 4 h, depending on type of deck, type of ceiling, space (if any) between ceilings and joists

Building Design Loading Criteria and General Considerations ■ 5-7

Since there are many materials and building systems, coupled with numerous design approaches such as liquid-filled steel columns and beams, flame-shielded exterior surfaces of exposed spandrel girders, and fire-protection paints and coatings, etc., many building codes permit architects and engineers to certify the fire-resistance quality and rating of the submitted material and system by procuring a Fire Underwriters label.

Fire separation and compartmentation to minimize the spread of fire may require special types of walls and certain structural materials and systems. Another concept of fire containment is to introduce positive pressure mechanically in the corridors, stairways, vertical mechanical and elevator shafts, and certain designated areas.

A variety of automatic fire-prevention and fire-detection devices and systems, such as smoke and fire sensor systems, alarm and communication systems, and wet and dry sprinkler systems, are used in conjunction with fire-resistant structural systems to provide the required rating for the building.

Design for specific fire ratings will usually require careful coordination of the architectural and structural detailing. Where plastering or other covering of the structure is provided, it may, with proper detailing, be utilized as fire insulation for the bare structure. This is especially important with framed structures of metal.

5.1.7 Deterioration

Building structures must resist deterioration due to exposure to various corrosive or wearing actions (Table 5.3). Metal structures may be painted, galvanized, or otherwise protected. However, certain metals, such as aluminum alloys and some highly corrosion-resistant steels, may be left unfinished in many exposure conditions, as they form their own protective finish because of initial oxidation which inhibits further deterioration. Nonaccessible thin metal elements should be protected from electrolysis, rusting, and fumes.

Concrete structures may be treated with sealers or have their surfaces artificially hardened with compatible chemicals. Exposed concrete structures should be air-entrained to increase their freeze/thaw resistance and should be designed and constructed with good-quality concrete; epoxy-coated, galvanized steel, or stainless-steel reinforcing bars with a minimum concrete cover of $1\frac{1}{2}$ in to increase their long-term durability. If the exposed surface has a sandblast or bushhammer finish, the concrete should be stronger and the concrete cover of reinforcing steel should be increased.

The necessity for any of these treatments may be reduced or eliminated where architectural details provide protection for the structure. Coordination of the architectural and structural detailing is required in order to determine the extent of protection required.

5.1.8 Provision for Environmental-Control Systems

Integrated planning of the environmental-control system and the structural and architectural systems helps to minimize cost. One may elect to use larger openings in structural elements, and deeper beams or joists, at additional structural cost, to accommodate a mechanical circulation pattern. On the other hand, one may use

Table 5.3 Various Types of Deterioration of Some Structural Materials and Their Prevention

Material	Type of deterioration	Prevention or remedy
Concrete	Freezing, thawing cycles	Avoid crack development as much as possible Use air-entraining additives Apply and maintain membrane system to minimize moisture penetration
	Corrosive gases, liquids	Use good-quality concrete and reinforcement, and proper concrete cover Coat or artificially harden surface
	Wear	Coat, cover, or artificially harden surface Overdesign for some loss Use metal nosings and edgings Use high-quality concrete
	Cracking, shrinking	Use mesh, reinforcing steel to minimize cracking Use expandable-type concrete to minimize shrinking Use densifier and other admixtures to minimize cracking in poured-in-place concrete Use denser and low-slump concrete for precast-concrete members For prestressed, precast members use initial prestressing force to minimize cracking Use control joints to minimize cracking Use epoxy-coated, galvanized or stainless steel reinforcement where chlorides are present and to maximize durability. Use proper concrete cover
Metals	Oxidation	Paint, galvanize, etc., to protect surface Use nonprogressive oxidizing alloys Use specified minimum thicknesses for safety against dimensional loss Certain steels possess self-protective rust coating that prevents further rusting (used as architectural exposed finish) Aluminum has self-protecting film that prevents further oxidation Avoid chloride environment
	Corrosive gases, liquids Electrolysis	As for oxidation. Plastic coatings effectively used Avoid contact of dissimilar metals, especially where water and chlorides are present
	Wear	Coat, cover, or harden surface

a more expensive environmental control system, such as a high-velocity or high-pressure system with smaller ducts, in order to maintain the floor-construction depth. In high-rise buildings a mechanical floor is sometimes located at midheight to reduce the required floor area for vertical circulation. This may save enough floor area, depending on the number of floors, to compensate for the extra floor. This floor also can be used as a fire-resistance compartment in helping to solve fire-resistance problems.

5.1.9 Limitations of Various Systems

For each particular type of structural system, executed in various materials, practical ranges may be established for the size of elements, spans, load capacities, modular units, etc. In addition, certain functional limitations may be inherent in the system, such as practicable rise-span ratios in arch construction. Each system lends itself to certain geometric configurations which limit its applicability to particular architectural solutions. Systems vary in the degree to which they adapt to variations of shape or dissymmetry and to the accommodation of openings, special loads, etc. Some systems may be varied or blended with other systems to facilitate special conditions. Thus an arch system may use suspended tendons or struts to support a horizontal surface; a linear, rectangular frame may utilize trussing or local rigid frame action to achieve spatial stability or increased stiffness; a suspension membrane may be stiffened for reverse action as a vault or dome to reduce flutter; etc.

Each structural system has unique fabrication and erection problems. Some of these are

1. Availability of materials, skilled labor, special equipment, etc. This is especially critical with respect to new systems, special materials, and highly complex or unique systems.

2. Speed of erection. This factor may influence the rejection of certain systems. The time rate of progress of other aspects of the construction relative to that of the structure should be considered. Prefabrication may speed field erection.

3. Maximum size of elements for handling, transporting, and erecting. Several factors must be considered, such as the degree to which the structure is prefabricated or shop-assembled, the distance from the shop to the site, the equipment available for handling, transporting, and erection, and the accessibility of the site.

4. Location of erection joints and sequence of erection. Construction joints should be located where they interfere least with the structural action, or where they can be designed for the least or simplest stresses.

5. Erection loads and stresses. Special conditions usually exist during the construction due to construction sequence, the location of construction joints, handling of elements, loads caused by the weight of construction equipment or materials, etc. In some cases the structure itself may be designed for these loads, usually at an increase in allowable stresses due to the temporary nature of the condition. In other situations, temporary supports, bracing, or shoring may be required. For some systems these considerations may be an important aspect of the structural design.

5.2 Loads

5.2.1 Dead Load

Dead load includes the weight of the structure plus any permanently attached elements of the construction. Table 5.4 gives approximate densities of various building materials. Table 5.5 gives the average weights of various typical elements.

Evaluation of the dead load on any individual structural element is sometimes difficult to make. Allowances are often made for openings in walls, and the weight of items such as doors, windows, hardware, trim, electrical fixtures and wiring, ductwork, plumbing fixtures and piping, and railings.

Although the structure is usually designed for the dead load of the completed building, for certain types of structures it may be necessary to investigate several stages of loading during the construction with respect to the changing effect on the stress distribution and stability of the structure (especially composite and pre-

Table 5.4 Densities of Building Materials

Material	Weight, pcf	Material	Weight, pcf
Steel, rolled	490	Glass, common	156
Aluminum alloy, 6061-T6, structural wrought	170	Glass, plate	161
		Plaster, cement and sand	100
Iron, cast	450	Plaster, gypsum and fiber	30
Bronze	550	Earth, clay, dry	65
Copper, cast or rolled	556	Earth, clay, wet	110
Lead	710	Earth, sand and gravel, dry, loose	100
Wood, eastern spruce	28	Earth, sand and gravel, wet, loose	125
Wood, redwood	28	Earth, sand and gravel, dry, packed	115
Wood, Douglas fir, coastal	34		
Wood, southern longleaf pine	41	Earth, sand and gravel, wet, packed	135
Wood, white oak	47		
Wood, red cedar	23	Earth, glacial till, mixed grain size, dry	130
Wood, hard maple	42		
Concrete, sand and gravel aggregate, structural grade	145	Earth, glacial till, mixed grain size, wet	145
Concrete, expanded shale aggregate, structural grade	90–115	Earth, loam, loose, dry	75
		Earth, loam, moist, packed	95
Concrete, perlite aggregate, insulating fill	31	Plastic, soft (vinyl)	85
		Plastic, hard, acrylate (Lucite)	90
Brick, soft, common	100	Fiberboard, hard pressed (Masonite)	45
Brick, hard-burned, common	120		
Brick, pressed, finish	140	Fiberboard, soft, structural, deck (Tectum)	25
Limestone	165		
Marble	170	Fiberboard, soft, structural, wall (Celotex)	20
Granite	165		

Table 5.5 Weights of Elements of Building Construction
All weights given are in psf and are average

Walls:		Floor and roof construction:	
4-in brick, soft common	35	Wood joists, 2×8, 16 in o.c.,	
4-in brick, hard common	40	wood subfloor	6
4-in brick, pressed finish	45	Wood joists, 2×12, 16 in o.c.,	
4-in hollow concrete block,		wood subfloor	8
stone aggregate (heavy)	30	Metal deck, 22 gauge, $1\frac{1}{2}$ in	
6-in hollow concrete block,		deep, 1 in insulation board	3
stone aggregate (heavy)	40	Metal deck, 18 gauge, $1\frac{1}{2}$ in	
8-in hollow concrete block,		deep, 1 in insulation board	4
stone aggregate (heavy)	55	Metal deck, 20 gauge, $1\frac{1}{2}$ in	
10-in hollow concrete block,		deep, stone concrete fill, 3 in	
stone aggregate (heavy)	68	thick overall	38
12-in hollow concrete block,		Metal deck, 20 gauge, $1\frac{1}{2}$ in	
stone aggregate (heavy)	80	deep perlite concrete fill,	
4-in hollow concrete block, light-		$2\frac{1}{2}$ in thick overall	10
weight aggregate	20	Wood deck, solid, per inch	$2\frac{1}{2}$
6-in hollow concrete block,		Reinforced-concrete slab, stone	
lightweight aggregate	30	aggregate, per inch	$12\frac{1}{2}$
8-in hollow concrete block,		Reinforced-concrete slab, shale	
lightweight aggregate	38	or slag aggregate, per inch	$9\frac{1}{2}$
10-in hollow concrete block,		Gypsum Pyrofill concrete on	
lightweight aggregate	45	formboard, bulb tee frame,	
12-in hollow concrete block,		3-in thick	10–12
lightweight aggregate	55	Wood W-trusses, 20–30 ft span,	
3-in gypsum block	10	24 ft o.c., $\frac{1}{2}$-in plywood	
4-in gypsum block	14	sheathing	8–10
2-in solid plaster on metal lath		Finish materials:	
and studs	20	Hung ceiling, metal lath, sand	
4-in glass block	20	plaster (portland cement)	10–25
2×4 studs, plaster board and		Hung ceiling, metal lath, gyp-	
plaster, 2 sides	15	sum and fiber plaster	8
2×4 studs, $\frac{1}{2}$-in gypsum dry		Hung ceiling, metal runners	
wall, 2 sides	8	with fiber or metal tile	5–8
Glass wall, small single glazed		Siding, wood	1–2
panes, light mullions,	5–10	Siding, asbestos shingles	$1\frac{1}{2}$–2
Glass wall, large plate, heavy		Siding, 20-gauge corrugated	
mullions	10–15	metal	2
Movable metal partitions	5–10	Roofing, built-up, tar and felt,	
Plaster 1 side of wall	4–5	gravel topping	5–8
Miscellaneous elements:		Roofing, shingles, asphalt or	
Skylight, $\frac{3}{8}$-in wire glass, light		wood	2–3
frame	8	Roofing, slate shingles	10–18
Insulation, loose or rigid, per		Roofing, tile, clay	10–20
inch	$\frac{1}{2}$–1	Roofing, copper or monel metal	$1\frac{1}{2}$–$2\frac{1}{2}$
Stairs, steel risers, treads, string-			
ergs terrazzo fill	30–35		
Stairs, steel treads and stringers,			
open risers, no fill	15		
Stairs, wood	8–10		

stressed systems). Temporary supports, bracing, reinforcing, or even modifications in the design may be necessary to provide for the sequence of loadings which occur during construction.

5.2.2 Live Load

Live loads are usually specified by local or model building codes. Table 5.6 lists minimum floor live loads for building occupancies. Live loads for some buildings, particularly custom-use and special buildings, that are established by the owner are often greater than code-required live load and should be reduced only with the knowledge and approval of the owner. Conformance of the minimum loads shown in this table with the governing building code and requirements of the owner for the project should be checked.

Some types of live loads may be practically permanent in nature, although subject to removal or relocation. Movable partitions, hung ceilings, and building equipment are in this category. They should be considered as dead load when evaluating settlement of clay foundations, creep of concrete, sag of timber, etc.

For structural elements that support large areas of floor or multiple floors, live loads are often reduced based on the probability of full loading on the area or floors (Table 5.6). Caution and judgment should be exercised in reducing specified live loads for high live-load to dead-load ratios and when deflections are critical.

Table 5.7 lists minimum roof live loads specified by the Uniform Building Code (UBC).[1] The same minimums are also specified in ASCE 7-93[2] but with formulas to compute loads for tributary areas between 200 and 600 ft² and between slopes of 4:12 and 12:12. The formulas are linear, so the same values can be obtained by linear interpolation in Table 5.7. These minimum roof loads do not include snow loads.

5.2.3 Snow Loads

The following snow loads for roofs on structures were obtained from ASCE 7-93.[2] Figure 5.1 gives ground snow loads p_g for the United States. Large-scale maps of this figure are given in Ref. 2. These maps were prepared by placing the 2 percent annual-probability-of-being-exceeded values (50-year mean recurrence interval) and the maximum observed values during the period of record (through the 1979–1980 winter) for each of more than 9000 recording stations. Snow load on a roof is usually less than the ground snow load.

Flat roof (less than 1 in/ft slope) ■ The snow load p_f on an unobstructed flat roof with a slope less than 1 in/ft (5 degrees) is calculated in pounds per square foot as

$$p_f = 0.7 C_e C_t I p_g \text{ in the contiguous United States}$$

$$p_f = 0.6 C_e C_t I p_g \text{ in Alaska}$$

where C_e = exposure factor
C_t = thermal factor
I = importance factor

Table 5.6 Minimum Live Loads for Floors[a,b]

Use or occupancy	Uniform load[c]	Concentrated load
Access floor systems		
Office	50	2000[d]
Computer use	100	2000[d]
Armories	150	0
Assembly areas[e] auditoriums, and balconies therewith		
Fixed seating areas	50	0
Movable seating and other areas	100	0
Stage areas and enclosed platforms	125	0
Cornices, marquees, and residential balconies	60[i]	0
Exit facilities: public corridors serving an occupancy load of 10 persons or more, exterior exit balconies, stairways, fire escapes, and similar uses	100	0[f]
Garages		
General storage and/or repair	100	[g]
Private or pleasure-type motor vehicle storage	50	[g]
Hospitals: wards and rooms	40	1000[d]
Libraries		
Reading rooms	60	1000[d]
Stack rooms	125	1500[d]
Manufacturing		
Light	75	2000[d]
Heavy	125	3000[d]
Offices	50	2000[d]
Printing plants		
Press rooms	150	2500[d]
Composing and linotype rooms	100	2000[d]
Residential: private dwellings, apartments, and hotel guest rooms	40	0[f]
Exterior balconies	60[i]	0
Decks	40[i]	0
Rest rooms[h]		
Reviewing stands, grand stands, and bleachers	100	0
School classrooms	40	1000[d]
Sidewalks and driveways for public access	250	[g]
Storage		
Light	125	
Heavy	250	
Stores		
Retail	100	3000[d]
Pedestrian bridges and walkways	100	

SOURCE: Reproduced from the 1994 edition of the Uniform Building Code, copyright 1994, with permission of the publishers, the International Conference of Building Officials.

[a]When it can be determined that the actual live load will be greater than the value shown, the actual live loads shall be used in the design of such buildings or parts thereof, and special provisions shall be made for machine or apparatus loads.

[b]Floors in office buildings and in other buildings where partition loads are subject to change shall be designed to support, in addition to all other loads, a uniformly distributed dead load of 20 psf. Access floor systems may be designed to support, in addition to all other loads, a uniformly distributed load of 10 psf.

[c]Except for places of public assembly and except for live loads greater than 100 psf, the design live load may be reduced on any member, including flat slabs, supporting more than 150 ft^2 by $0.08(A-150)$ percent, where A = area supported, but not to exceed 40 percent for members receiving load from only one level, 60 percent for other members, or $23.1\,(1+D/L)$ percent, where D = dead load, psf, and L = unit live load, psf. The reduction shall not exceed 40 percent in garages for storage of private pleasure cars having a capacity of not more than nine passengers per vehicle. No reduction shall be made for storage live loads exceeding 100 psf except that design live loads on columns may be reduced 20 percent.

[d]Provision shall be made for this load acting on any space $2\frac{1}{2}$-ft square wherever it produces stresses greater than those caused by the specified uniform load on an otherwise unloaded area.

[e]Includes dance halls, drill rooms, gymnasiums, playgrounds, plazas, terraces, and similar occupancies generally accessible to the public

[f]Individual stair treads shall be designed to support a 300-lb concentrated load placed to produce maximum stress. Stair stringers may be designed for the uniform load.

[g]In areas where vehicles are used or stored, provision shall be made for concentrated loads consisting of two or more loads spaced 5 ft on center, without uniform load, each load to be 40 percent of the gross weight of the maximum-size vehicle to be accommodated. Parking garages for storage of private or pleasure-type vehicles shall have a floor system designed for a concentrated wheel load of not less than 2000 lb without uniform load. In both cases the condition of concentrated or uniform load producing the greater stresses shall govern.

[h]Not less than the load for the occupancy with which they are associated, but need not exceed 50 psf.

[i]Check snow loads due to drift and use if greater than design live loads.

Table 5.7 Minimum Live Loads* for Roofs

Roof slope		Tributary area for member, ft²		
Shed or gable	Arch or dome	0–200	201–600	Over 600
$0 \leq$ slope $< 4/12$	$h/l < 1/8$	20	16	12
$4/12 \leq$ slope $< 12/12$	$1/8 \leq h/l < 3/8$	16	14	12
$12/12 \leq$ slope	$3/8 \leq h/l$	12	12	12

SOURCE: Reproduced, in part, from the 1994 edition of the Uniform Building Code, copyright 1994 with permission of the publishers, the International Conference of Building Officials.[1]
*On horizontal projection, psf.

The following exposure factors C_e are prescribed

0.8 for windy area with roof exposed on all sides with no shelter afforded by terrain, higher structures, or trees

0.9 for windy areas with little shelter available

1.0 for locations in which snow removal by wind cannot be relied on to reduce roof loads because of terrain, higher structures, or several trees nearby

1.1 for areas that do not experience much wind and where terrain, higher structures, or several trees shelter the roofs

1.2 for densely forested areas that experience little wind, with roof located tight in among conifers

The following thermal factors C_t are prescribed:

1.0 for heated structures

1.1 for structures kept just above freezing

1.2 for unheated structures

The following importance factors I are prescribed:

1.0 for all buildings and structures except those listed below.

1.1 for buildings and structures where the primary occupancy is one in which more than 300 people congregate in one area

1.2 for buildings and structures designated as essential facilities including, but not limited to hospitals and other medical facilities having surgery or other emergency treatment areas; fire or rescue and police stations; structures and equipment in communication centers and other facilities required for emergency response; power stations and other utilities required in an emergency; structures having critical national defense capabilities; and designated shelters for hurricanes

0.8 for buildings and structures that represent a low hazard to human life in the event of failure, such as agricultural buildings, certain temporary facilities, and minor storage facilities

Low Sloped Roof ■ According to ASCE, the low-slope roof snow load p_f should apply to shed, hip, and gable roofs with slopes from 5 degrees to less than

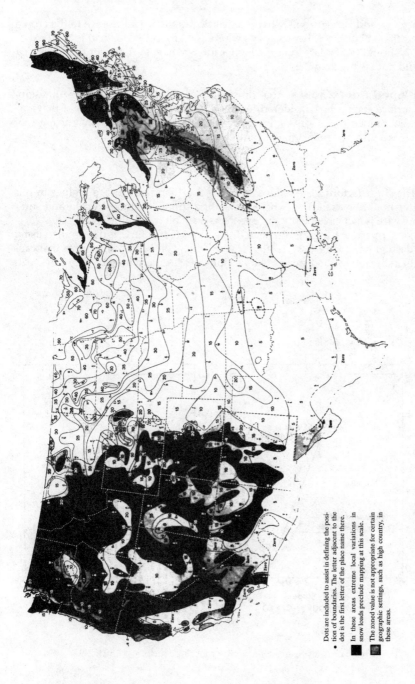

Fig. 5.2 Ground snow load, psf, 50-year reoccurrence interval (ASCE 7-93).

5-16 ■ Chapter Five

15 degrees and curved roofs where the vertical angle from the eave to the crown is less than 10 degrees. For locations where the ground snow load p_g is 20 psf or less, p_f should not be less than Ip_g. For locations where p_g is greater than 20 psf, p_f should not be less than $20I$.

Sloped Roof Snow ■ According to ASCE, all snow loads acting on a sloping surface should be considered to act on the horizontal projection of that surface. The sloped-roof snow load p_s should be obtained by multiplying the flat-roof snow load p_f by the roof slope factor C_s:

$$p_s = p_f C_s$$

The slope factor C_s is obtained from Fig 5.2. "Slippery surface" values in this figure should be used only where the sliding surface is unobstructed and sufficient space is available below the eaves to accept all the sliding snow.

For slippery, warm roofs on heated structures ($C_t = 1.0$), C_s is determined using the dashed line in Fig. 5.2a. For other warm roofs on heated structures that cannot be relied on to shed snow load by sliding, C_s is determined by using the solid line in Fig. 5.2a.

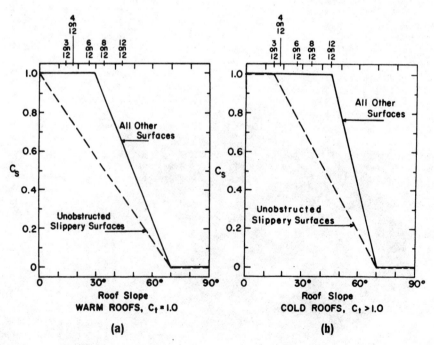

Fig. 5.2 Graphs for determining roof slope factor C_s (ASCE 7-93).

Building Design Loading Criteria and General Considerations 5-17

For slippery cold roofs on unheated structures ($C_t = 1.2$) or on structures just kept above freezing ($C_t = 1.1$), C_s determined using the dashed line in Fig. 5.2b. For other cold roofs on unheated structures or on structures just kept above freezing that cannot be relied on to shed snow load by sliding, C_s is determined by using the solid line in Fig. 5.2b.

For curved roofs, C_s is determined from the appropriate curve in Fig. 5.2 by basing the slope on the vertical angle from the eave to the crown. The eave of a curved roof should be considered as the point at which the slope of roof exceeds 70 degrees. Portions of curved roofs having a slope exceeding 70 degrees should be considered free from snow.

For multiple folded plate, sawtooth, and barrel vault roofs, no reduction in snow load should be applied because of slope (that is, $C_s = 1.0$ regardless of slope, and therefore $p_s = p_f$).

Unbalanced Snow Loads ■ For hip and gable roofs with a slope less than 15 degrees or exceeding 70 degrees, unbalanced snow loads need not be considered. For slopes between 15 and 70 degrees, the roof structure should be designed with an unbalanced uniform snow load on the lee side of $1.5 \, p_s/C_e$ and the windward side free of snow. In addition, the roof structure of all hip and gable roofs should be designed with the uniform balanced snow load p_s.

ASCE 7-93 gives unbalanced snow load requirements for curved, multiple folded plate, sawtooth, and barrel vault roofs.

Drifts on Lower Roofs of a Structure ■ Drifting of snow can occur in the wind shadow of higher portions of a structure or of an adjacent structure. According to ASCE, the surcharge load due to drifting is a triangular cross section that is superimposed on the balanced snow load, as shown in Fig 5.3. When considering drifting, it is assumed that all snow has blown off the upper roof near its eave. Notations in Fig. 5.3 and in applicable equations are as follows:

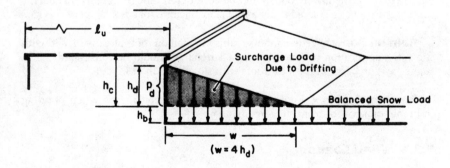

Fig. 5.3 Configuration of drift on lower roofs (ASCE 7-93).

h_b = height of balanced snow load on lower roof, in feet, that is, p_f or p_s divided by density of snow, γ

h_c = clear height from top of balanced snow load to closest point on adjacent upper roof, to top of parapet, or to top of a projection on the roof, in feet

h_d = height of drift, in feet as determined from Fig. 5.4, but not greater than h_c

l_u = length of the roof upwind of the drift, in feet

p_d = maximum intensity of drift surcharge load = $h_d \gamma$ in pounds per square foot

w = width of snow drift = $4h_d$, in feet. If ω is greater than the width of the lower roof, truncate the drift at the far edge of the roof

γ = snow density in pounds per cubic foot, that is, $0.13p_g + 14$ but not more than 35 psf

Drift loads need not be considered when h_c/h_d is less than 0.2 or in areas where the ground snow load p_g is less than 10 pounds per square foot.

The above snow drifting methodology can also be used to calculate surcharge snow loads caused by drifting on a roof within 20 ft of a higher structure or terrain feature that could cause snow to accumulate on it. However, the maximum intensity of the drift surcharge load p_d should be reduced by the factor $1-S/20$ to account for separation s between the structures. When S is greater than 20 ft, drift loads from an adjacent structure or terrain feature need not be considered.

Roof Projections ■ The above snow drifting methodology should be used to calculate drift loads on all sides of roof obstructions that are longer than 15 ft. Drifts created by parapet walls should be computed using half the drift height from Fig. 5.4 (that is, $0.5\ h_d$) with l_u equal to the length of the roof upwind of the parapet.

Sliding Snow ■ Sliding of snow from a sloped roof onto a lower roof increasing the load on the lower roof should be considered. ASCE suggests that such a surcharge load be calculated using a roof slope factor C_s of a nonslippery surface.

Rain on Snow ■ Where intense rains may fall on roofs already loaded with snow, the application of a rain-on-snow surcharge should be considered. ASCE suggests a surcharge of 5 psf for roofs with slopes less than 1/2 in/ft.

Ponding Loads ■ Roof deflections should be considered when ponding from rain on snow or from snow meltwater.

5.2.4 Wind Loads

Wind pressures on a structure depend on the wind velocity and the shape, size, dynamic response, and exposure of the structure.

Building Design Loading Criteria and General Considerations ■ 5-19

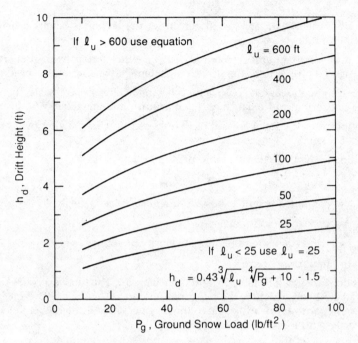

Fig. 5.4 Graphs and equations for determining drift weight, h_d.

According to ASCE 7-93[2], the velocity pressure, psf, is given by

$$q = 0.00256V^2$$

where V = fastest mile wind speed, mph

Wind speed varies with height and with the terrain, and the velocity pressure q_z at a height z above ground is given by

$$q_{xz} = 0.00256 K_z V^2$$

where K_z = velocity pressure exposure coefficient
V = basic wind speed, mph

Wind pressure on a structure also depends on the dynamic response of the structure to wind and can be determined by

$$q_{eff} = G_z q = 0.00256 G_z K_z V^2$$

where q_{eff} = effective pressure
G_z = gust response factor

Values of K_z and G_z are given in ASCE 7-93[2] for the following four exposure categories:

A. Large city centers with at least 50 percent of the buildings having a height exceeding 70 ft.
B. Urban and suburban areas, wooded areas, or other terrain with numerous closely spaced obstructions the size of single-family dwellings or larger.
C. Open terrain with scattered obstructions having heights generally less than 30 ft. This category includes flat lands, open country, and grasslands.
D. Flat, unobstructed coastal areas directly exposed to wind flowing over large bodies of water.

According to the UBC,[1] design wind pressure is given by

$$P = C_e C_q q_s I_w$$

where C_e = combined height, exposure, and gust factor coefficient
C_q = pressure coefficient
q_s = wind stagnation pressure = velocity pressure
I_w = importance factor

Values of C_e are given in Table 5.8 and C_q in Table 5.9. The three exposure categories in Table 5.8 correspond to the categories B, C, and D of ASCE 7-93. Note that the values of C_e are very nearly equal to the product $K_z G_z$ for the higher velocity of the velocity ranges in the table for those categories.

Table 5.8 Combined Height, Exposure, and Gust Factor Coefficient C_e

Height above average level of adjoining ground, ft*	Exposure D	Exposure C	Exposure B
0–15	1.39	1.06	0.62
20	1.45	1.13	0.67
25	1.50	1.19	0.72
30	1.54	1.23	0.76
40	1.62	1.31	0.84
60	1.73	1.43	0.95
80	1.81	1.53	1.04
100	1.88	1.61	1.13
120	1.93	1.67	1.20
160	2.02	1.79	1.31
200	2.10	1.87	1.42
300	2.23	2.05	1.63
400	2.34	2.19	1.80

*Values for intermediate heights above 15 ft may be interpolated.

SOURCE: Adapted from the 1994 edition of the Uniform Building Code, copyright 1994, with permission of the publishers, the International Conference of Building Officials.[1]

Table 5.9 Pressure Coefficients C_q

Structure or part thereof	Description	C_q factor

1. Primary frames and systems	Method 1 (normal force method)
	Walls:
	  Windward wall
	  Leeward wall
	Roofs:[a]
	  Wind perpendicular to ridge
	    Leeward roof or flat roof
	    Windward roof
	      Less than 2:12 (16.7%)
	      Slope 2:12 (16.7%) to less than 9:12 (75%)
	
	      Slope 9:12 (75%) to 12:12 (100%)
	      Slope > 12:12 (100%)
	  Wind parallel to ridge and flat roofs
	
	Method 2 (projected area method)
	  On vertical projected area
	    Structures 40 ft (12,192 mm) or less in height
	    Structures over 40 ft (12,192 mm) in height
	  On horizontal projected area[d]
		
		0.8 inward
		0.5 outward
		
		0.7 outward
		
		0.7 outward
		0.9 outward or
		  0.3 inward
		0.4 inward
		0.7 inward
		0.7 outward
		
		1.3 horizontal any direction
		1.4 horizontal any direction
		
		0.7 upward

2. Elements and components not in areas of discontinuity[b]	Wall elements
	  All structures
	  Enclosed and unenclosed structures
	  Partially enclosed structures
	  Parapet walls
	
	Roof elements[c]
	  Enclosed and unenclosed structures
	    Slope < 7:12 (58.3%)
	    Slope 7:12 (58.3%) to 12:12 (100%)
	
	  Partially enclosed structures
	    Slope < 2:12 (16.7%)
	    Slope 2:12 (16.7%) to 7:12 (58.3%)
	
	    Slope > 7:12 (58.3%) to 12:12 (100%)
		
		1.2 inward
		1.2 outward
		1.6 outward
		1.3 inward or outward
		
		1.3 outward
		1.3 outward or inward
		
		1.7 outward
		1.6 outward or
		  0.8 inward
		1.7 outward or inward

3. Elements and components in areas of discontinuities[b,d,e]	Wall corners[f]
	
	Roof eaves, rakes, or ridges without overhangs[f]
	  Slope < 2:12 (16.7%)
	  Slope 2:12 (16.7%) to 7:12 (58.3%)
	  Slope > 7:12 (58.3%) to 12:12 (100%)
	For slopes less than 2:12 (16.7%)
	  Overhangs at roof eaves, rakes, or ridges, and canopies
		1.5 outward or 1.2 inward
		
		2.3 upward
		2.6 outward
		1.6 outward
		
		0.5 added to values above

4. Chimneys, tanks, and solid towers	Square or rectangular Hexagonal or octagonal Round or elliptical	1.4 any direction 1.1 any direction 0.8 any direction

5. Open-frame towers[g,h]	Square and rectangular
	  Diagonal
	  Normal
	Triangular
		
		4.0
		3.6
		3.2

6. Tower accessories (such as ladders, conduit, lights, and elevators)	Cylindrical members
	  2 in (51 mm) or less in diameter
	  Over 2 in (51 mm) in diameter
	Flat or angular members
		
		1.0
		0.8
		1.3

7. Signs, flagpoles, lightpoles, minor structures[h]		1.4 any direction

SOURCE: Adapted from the 1994 edition of the Uniform Building Code, copyright 1994, with permission of the publishers, the International Conference of Building Officials.[1]

Table 5.9 Pressure Coefficients C_q (Continued)

[a] For one story or the top story of multistory partially enclosed structures, an additional value of 0.5 shall be added to the outward C_q. The most critical combination shall be used for design. For definition of open structures, see Section 1613.

[b] C_q values listed are for 10-square-foot (0.93 m²) tributary areas of 100 square feet (9.29 m²), the value of 0.3 may be subtracted from C_q, except for areas at discontinuities with slopes less than 7 units vertical in 12 units horizontal (58.3% slope) where the value of 0.8 may be subtracted from C_q. Interpolation may be used for tributary areas between 10 and 100 square feet (0.93 m² and 9.29 m²). For tributary areas greater than 1000 square feet (92.9 m²), use primary frame values.

[c] For slopes greater than 12 units vertical in 12 units horizontal (100% slope), use wall element values.

[d] Local pressures shall apply over a distance from the discontinuity of 10 feet (3048 mm) or 0.1 times the least width of the structure, whichever is smaller.

[e] Discontinuities at wall corners or roof ridges are defined as discontinuous breaks in the surface where the included interior angle measures 170 degrees or less.

[f] Load is to be applied on either side of discontinuity but not simultaneously on both sides.

[g] Wind pressures shall be applied to the total normal projected area of all elements on one face. The forces shall be assumed to act parallel to the wind direction.

The wind stagnation pressure normally used in design is computed for the wind speed with a mean recurrence period of 50 years (Fig. 5.5) unless terrain features and local records indicate higher values. Recurrence periods of 100 and 25 years can be obtained by multiplying the 50-year velocity by 1.07 and 0.95, respectively.[2]

The UBC requires the minimum importance factor $I=1$ for all buildings except for essential facilities which must be safe and usable after a windstorm and for hazardous facilities containing toxic or exposure chemicals or substances. For

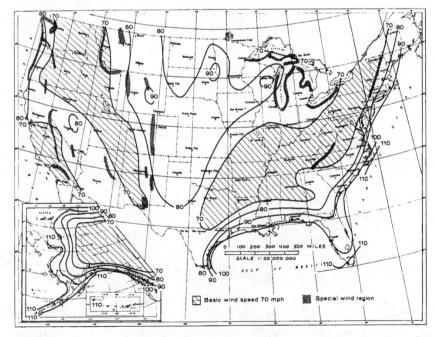

Fig. 5.5 Fastest-mile wind speed 33 ft above ground, 50-year mean recurrence interval.

these facilities, the minimum $I = 1.15$. Since velocity pressure is proportional to velocity squared, this has the effect of designing for a 100-year return velocity ($1.07^2 \approx 1.15$). Essential facilities include hospitals and other medical facilities having surgery or emergency treatment areas, fire and police stations, and municipal government disaster operation.

Main Resisting Systems ■ The UBC allows the design wind pressure for the design of main windforce resisting systems to be computed by either of two procedures. In what is called the normal force method, wind pressures are assumed to act simultaneously normal to all exterior surfaces. This procedure is required for gabled rigid frames and may be used for any structure. In what is called the projected area method, which may be used for any structure less than 200 ft high (except for gabled frames), horizontal pressure is assumed to act on the full vertical projection of the structure and the vertical pressure is assumed to act simultaneously on the full horizontal projected area. The applicable pressure coefficients are given in Table 5.9, where it is noted that the numerical sum of the pressure coefficients on the windward and leeward walls equals the pressure coefficient 1.3 to be used in calculating the pressure on the vertical projected area of buildings 40 ft or less in height. This means, of course, that the pressures for the design of the main frame by the two methods are equal for this case.

Flexible Structures ■ ASCE 7-93 requires that gust factors for flexible structures defined as those with a ratio of height to least horizontal dimension greater than 5 and a fundamental structural frequency of vibration less than 1 Hz, be obtained by "rational analysis" and gives a procedure for calculating them. UBC-1994 has a similar provision in that it requires buildings sensitive to dynamic effects such as those with a ratio of height to least horizontal dimension greater than 5, structures sensitive to wind-excited oscillation, and buildings over 400 ft tall to be designed in accordance with "approved national standards."

Local Pressures ■ Wind pressures at building corners and at ridges, eaves, cornices, and 90° corners of roofs may be considerably higher than average pressures. Suction at these points is important in respect to fastening of elements such as cladding. Local-pressure coefficients are given in Table 5.9 and in ASCE 7-93.

Tanks, Chimneys, Towers, Signs, and Miscellaneous Structures ■ Local-pressure coefficients are given in Table 5.9 and in ASCE 7-93.

Overturning Moment ■ According to UBC,[1] the base overturning moment for the entire structure or for any of its primary lateral resisting elements must not exceed two-thirds of the dead-load resisting moment. The combined effects of uplift and overturning for an entire structure may be reduced by one-third if the height to width ratio is 0.5 or less in the wind direction and the height is 60 ft or less.

5.2.5 Seismic Loads

Figure 5.6 shows seismic-risk zones of the United States as defined by the UBC.[1] This code specifies two procedures for calculating seismic lateral forces: (*a*) a static lateral-force procedure and (*b*) a dynamic lateral-force procedure.

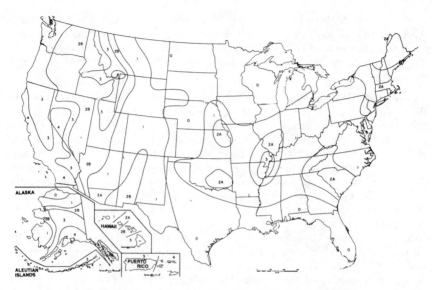

Fig. 5.6 Seismic risk zone map of the United States (Ref. 1).

Static Lateral-Force Procedure ▪ This procedure may be used for all structures in zone 1 and for those of standard occupancy (Table 5.10) in zone 2, for regular structures under 240 ft in height with lateral-load systems listed in Table 5.11, and for irregular structures not more than 5 stories or 65 ft high. (Various irregularities which disqualify a structure from being classed as regular are described in Ref. 1.) The total base shear V in a given direction is

$$V = \frac{ZICW}{R_w}$$

where Z = seismic-zone factor = 0.075, 0.15, 0.20, 0.30, and 0.40 for zones 1, 2A, 2B, 3, and 4, respectively (Fig. 5.6)

I = occupancy importance factor (Table 5.10)

C = coefficient from equation below

W = total dead load plus (a) weight of permanent equipment, (b) at least 25 percent of floor live load in storage and warehouse occupancies, (c) at least 10 psf if partition loads are used in floor design, (d) snow load if it exceeds 30 psf*

R_w = coefficient from Tables 5.11 and 5.12

The coefficient C is given by

$$C = \frac{1.25S}{T^{2/3}}$$

---

*If siting, configuration, and duration of load warrant, snow load may be reduced up to 75 percent if approved by the building official.

where T = fundamental period of vibration, s, of structure in direction considered
 S = site coefficient defined below

However, C need not exceed 2.75. Furthermore, the minimum value of C/R_w is 0.075 except for those provisions where code-prescribed forces are scaled up by $3R_w/8$.

Site coefficients $S = S_1, S_2, S_3,$ and S_4 in the equation above are the following:

S_1 = 1 for soil profiles with rocklike materials, characterized by shear-wave velocities greater than 2500 ft/s or other suitable means of classification, or for stiff or dense soil conditions where the soil depth is less than 200 ft.

S_2 = 1.2 for soil profiles with dense or stiff soil conditions where the depth exceeds 200 ft.

S_3 = 1.5 for soil profiles 40 ft or more in depth and containing more than 20 ft of soft to medium-stiff clay, but not more than 40 ft of soft clay.

S_4 = 2 for soil profiles containing more than 40 ft of soft clay.

If soil properties are not well enough known to determine the profile, S_2 or S_3, whichever gives the higher base shear, is to be used.

The value of T may be approximated by

$$T = C_1 h_n^{3/4}$$

where C_1 = 0.035 for steel moment-resisting frames, 0.030 for reinforced-concrete moment-resisting frames, 0.020 for all other buildings
 h_n = height to uppermost level in main portion of structure, ft

Alternatively, $C_1 = 0.1/\sqrt{A_c}$ may be used for structures with concrete or masonry shear walls, in which A_c is given by

$$A_c = \Sigma A_e \left[0.2 + \left(\frac{D_e}{h_n} \right)^2 \right]$$

Table 5.10 UBC Occupancy Importance Factors for Design for Seismic Loads

Occupancy	I	I_p
Essential facilities*	1.25	1.50
Hazardous facilities†	1.25	1.50
Special occupancy‡	1.0	1.0
Standard occupancy§	1.0	1.0

SOURCE: Adapted from the 1994 edition of the Uniform Building Code, copyright 1994, with permission of the publishers, the International Conference of Building Officials.[1]

*Same as snow-load importance factor.

†Structures housing, supporting, or containing sufficient quantities of toxic or explosive substances which would be dangerous to the safety of the general public if released.

‡Various types of structures with capacities of 50 up to 5000 persons. Also jails and detention facilities. See Ref. 1.

§All structures having occupancies or functions not listed above.

Table 5.11 Values of R_w

Basic structural system[a]	Lateral-load resisting system	R_w	H, ft[f]
Bearing wall systems[b]	Light-framed walls with shear panels		
	Plywood walls for structures three stories or less	8	65
	All other light framed walls	6	65
	Shear walls, concrete or masonry	6	160
	Light steel-framed bearing walls with tension-only bracing	4	65
	Braced frames where bracing carries gravity loads		
	Steel	6	160
	Concrete[g]	4	—
	Heavy timber	4	65
Building-frame system[c]	Steel EBF	10	240
	Light-framed walls with shear panels		
	Plywood walls for structures three stories or less	9	65
	All other light-framed walls	7	65
	Shear walls		
	Concrete	8	240
	Masonry	8	160
	Concentric-braced frames		
	Steel	8	160
	Concrete[g]	8	—
	Heavy timber	8	65
	Special concentrically braced steel frames	9	240
Moment resisting frame systems	SMRFs		
	Steel	12	No limit
	Concrete	12	No limit
	Masonry moment-resistant frames	9	160
	Concrete IMRFs[g]	8	—
	OMRFs		
	Steel	6	160
	Concrete[g]	5	—
Dual systems	Shear walls		
	Concrete with SMRF	12	No limit
	Concrete with steel OMRF	6	160
	Concrete with concrete IMRF[g]	9	160
	Masonry with SMRF	8	160
	Masonry with steel OMRF	6	160
	Masonry with concrete IMRF[g]	7	—
	Steel EBF with steel SMRF	12	No limit
	Steel EBF with steel OMRF	6	160
	Concentric braced frames		
	Steel with steel SMRF	10	No limit
	Steel with steel OMRF	6	160
	Concrete with concrete SMRF[g]	9	—
	Concrete with concrete IMRF[g]	6	—
	Special concentrically braced frames		
	Steel with steel SMRF	11	No limit
	Steel with steel OMRF	6	160

SOURCE: Adapted from the 1994 section of the Uniform Building Code, copyright 1994, with permission of the publishers, the International Conference of Building Officials

KEY: EBF = eccentric braced frame.
   SMRF = special moment resisting frame.
   IMRF = intermediate moment resisting frame.
   OMRF = ordinary moment resisting frame.

[a]See Ref. 2 for combinations of structural systems.

[b]A bearing wall system is a system without a complete vertical load-carrying space frame, in which bearing walls or bracing systems provide support for all or most of the gravity loads while resistance to lateral load is provided by shear walls or braced frames.

[c]A building frame system is one with an essentially complete space frame to support gravity loads while resistance to lateral load is provided by shear walls or braced frames.

[d]A moment-resisting frame system is one with an essentially complete space frame to support gravity loads. It resists lateral load primarily by flexure of its members.

[e]A dual system is one with an essentially complete space frame to support the gravity loads with resistance to lateral load provided by a specially detailed moment-resisting space frame, concrete or steel, capable of resisting at least 25 percent of the base shear and shear walls or braced frames. The two systems are designed to resist the lateral load in proportion to their rigidities.

[f]Height limit H applies for seismic zones 3 and 4, except that for regular, unoccupied structures not accessible to the general public H may exceed the stated limit by up to 50 percent.

[g]Prohibited in seismic zones 3 and 4, but see *Nonbuilding Structures* for IMRF exception.

Building Design Loading Criteria and General Considerations ■ 5-27

where A_e = minimum cross-sectional shear area, ft^2, in any horizontal plane in the first story of a shear wall

D_e = length, ft, of a first-story shear wall in the direction of the applied forces.

In lieu of the equation for T above, the UBC also allows T to be calculated, based on the structural properties of the resisting elements of the structure.

Vertical distribution of lateral forces is discussed in following paragraphs. The UBC-prescribed distribution is essentially the same.

Dynamic Lateral-Force Procedure ■ This requires an analysis based on representations of the group motion. In lieu of response spectra developed for the site the designer may use response spectra given in the code. If the base shear in a given direction determined by the dynamic procedure is less than V, it must be increased to 90 percent of that value for regular buildings and 100 percent for irregular buildings and must not be less than 80 percent of V using the period T. Should the base shear by the dynamic procedure be more than V, it may be reduced to the applicable value, 90 or 100 percent noted above. Response parameters, such as deflections and moments, are adjusted proportionately.

Nonbuilding Structures ■ These are defined in the UBC as all self-supporting structures, other than buildings, which carry gravity load and must resist earthquakes. If the structure has a structural system similar to those described in Table 5.11 for buildings, lateral force design procedures are the same as for buildings. However, the intermediate moment-resisting space frame may be used in zones 3 and 4 for the standard and special occupancies of Table 5.10 if the structure is less than 50 ft tall and $R_w = 4$ is used in the analysis.

If the structure is rigid (defined as one with a period T less than 0.06 s), the lateral force, which is to be distributed according to the distribution of mass, is given by

$$V = 0.5ZIW$$

Table 5.12 Values of R_r for Nonbuilding Structures

Type of structure	R_w
Tanks, vessels, or pressurized spheres on braced or unbraced legs	3
Cast in place concrete silos and chimneys having walls continuous to the foundation	5
Distributed mass cantilever structures such as stacks, chimneys, silos, and skirt-supported vertical vessels	4
Trussed towers (freestanding or guyed), guyed stacks and chimneys	4
Inverted pendulum-type structures	3
Cooling towers	5
Bins and hoppers on braced or unbraced legs	4
Storage racks	5
Signs and billboards	5
Amusement structures and monuments	3
Self-supporting structures not otherwise covered	4

SOURCE: Adapted from the 1994 edition of the Uniform Building Code, copyright 1994, with permission of the publishers, the International Conference of Building Officials.[1]

Nonbuilding structures with structural systems not covered in Table 5.11 must be designed to resist seismic lateral forces not less than V using R_w from Table 5.12. However, the ratio C/R_w used for design must not be less than 0.5.

Parts of Structure ■ According to the UBC,[1] parts and portions of structures, and permanent nonstructural components and their attachments, must be designed to resist a total lateral force F_p given by

$$F_p = ZI_p C_p W_p$$

where W_p = weight of part

Z = same as for the building

I_p = same as for the building, except use 1.5 for anchorage of machinery and equipment for life safety systems and for tanks and vessels containing quantities of highly toxic or explosive substances sufficient to be hazardous to safety of general public

For rigid items (defined as those having a fixed-base period equal to or less than 0.06 s) use $C_b = 2$ for unbraced (cantilevered) parapets, exterior and interior ornamentations and appendages, signs, and billboards. Also use $C_b = 2$ for a chimney, stack, trussed tower, or tank on legs if it is supported on or projects as an unbraced cantilever above the roof more than half its height; otherwise use $C_p = 0.75$.

Use $C_p = 0.75$ for exterior walls (except cantilevered parapets) above the ground floor; interior bearing and nonbearing walls and partitions; masonry or concrete fences over 6 ft high; penthouses not framed by an extension of the building frame; connections for prefabricated structural elements other than walls (zones 2, 3, and 4 only); mechanical, plumbing, and electrical equipment and machinery and associated piping; tanks and vessels plus contents; storage racks including contents; anchorage for permanent floor-supported cabinets and bookstacks (including contents) more than 5 ft high; anchorage for suspended ceilings and light fixtures; and access floor systems.

For nonrigid or flexibly supported items on a structure above grade, in lieu of a detailed analysis the value of C_p is taken at twice the value for the corresponding rigid item but need not exceed 2. For elements or components supported at or below ground level, C_p may be taken as two-thirds the value for the corresponding nonrigid item.

5.2.6 Load Factors and Load Combinations*

The required strength of the structure and its elements must be determined from the appropriate critical combination of factored loads. The most critical effect may occur when one or more loads are not acting. The following load combinations and the corresponding load factors should be investigated:

1.4D

1.2D + 1.6L + 0.5 (L_r or S or R)

1.2D + 1.6 (L_r or S or R) + (0.5L[†] or 0.8W)

*Adapted from AISC LRFD Specification for structural steel and buildings, Dec. 1, 1993.
†The load factor L should equal 1.0 for garages, areas occupied as places of public assembly, and all areas where the live load is greater than 100 psf.

Building Design Loading Criteria and General Considerations ■ 5-29

$1.2D + 1.3W + 0.5L^† + 0.5 (L_r \text{ or } S \text{ or } R)$

$1.2D \pm 1.0E + 0.5L^† + 0.2S$

$0.9D \pm (1.3W \text{ or } 1.0E)$

D = dead load due to the weight of the structural elements and the permanent features on the structure

L = live load due to occupancy and movable equipment

L_r = roof live load

W = wind load

S = snow load

E = earthquake load determined in accordance with Part I of the AISC *Seismic Provisions for Structural Steel Buildings*

R = load due to initial rainwater or ice exclusive of the ponding contribution

5.3 Miscellaneous Considerations

5.3.1 Openings and Voids

Some structural elements contain natural voids. Trusses facilitate passage of elements of a size limited only by the scale of the truss. The Vierendeel truss is most accommodating in this respect. Hollow units such as metal deck or metal sandwich panels may receive wiring, piping, or even circulating air in their voids.

Beams and girders accept small openings in their webs without seriously impairing their strength. Location, as well as size, is important. Shear considerations usually dictate the location of web openings near midspan, while for bending the optimal location is near the neutral axis of the cross section. Flanges of steel beams may be punctured or notched in zones of low moment. Round openings, or square openings with rounded corners, minimize stress concentrations. Reinforcing should be provided around large web openings.

It is always best to anticipate the need for openings in steel members and detail them for shop fabrication. Indiscriminate cutting of structural members in the field is a practice which should not be allowed. Where errors or late changes make cutting unavoidable, specifications should be given as to where and how cutting may be done, and careful field supervision should be provided.

5.3.2 Thermal and Seismic Movement

Exterior elements and surfaces of buildings are subjected to a range of temperature which can be as much as 170°F. The resulting expansion and contraction presents a number of problems.

1. Movement of the entire building structure. Since the substructure is kept at a relatively constant temperature by the ground, there is a buildup of differential length between the lower and upper portions of the building. This can cause crit-

ical conditions of shear in end walls or sidesway deflection of end columns. Where separate wings of the building are joined, as in L- or H- shaped plans, one wing may pull away from or push against the other. The solution to this problem is usually to provide expansion joints (Fig. 5.7).

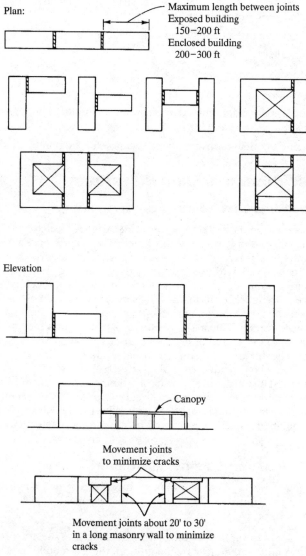

Fig. 5.7 Location of expansion joints in building.

Building Design Loading Criteria and General Considerations ■ 5-31

An expansion joint usually requires careful design and construction attention. The expansion continuity must be broken without losing the weather seal and structural integrity. This often calls for sliding or sealing elements and duplication of structural elements. Details of some typical expansion joints are shown in Fig 5.8.

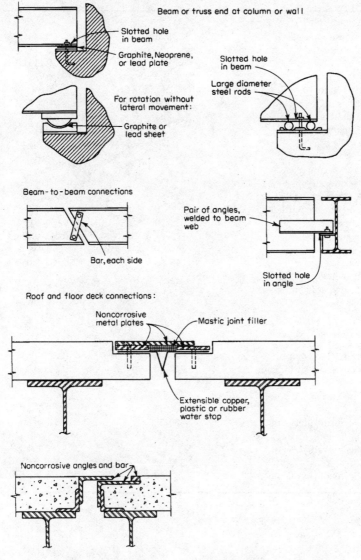

Fig. 5.8 Schematic expansion joint details.

Table 5.13 Coefficients of Linear Expansion—Typical Building Materials

Material	Coefficient of linear expansion per unit of length, per °F, value given × 10^{-7}	Approximate total length change in 10 ft for thermal change of 100°F, in
Metals:		
Aluminum	128	0.154
Bronze	101	0.121
Copper	98	0.118
Iron, cast, gray	60	0.072
Iron, wrought	67	0.080
Lead	159	0.191
Steel, mild (structural)	65	0.078
Steel, stainless, 18-8	96	0.115
Zinc, rolled	173	0.208
Stone and masonry:		
Ashlar masonry	35	0.042
Brick masonry	35–50	0.042–0.060
Cement mortar, portland	70	0.084
Concrete, stone	55–70	0.066–0.084
Concrete, perlite	43–61	0.052–0.073
Limestone	40	0.048
Marble	45–55	0.054–0.066
Plaster cement and sand	90	0.108
Plaster, gypsum and fiber	85	0.102
Sandstone	55	0.066
Slate	45	0.054
Timber:		
Fir, parallel to grain	20–30	0.024–0.036
Fir, perpendicular to grain	200–300	0.240–0.360
Miscellaneous:		
Glass	45	0.054
Plastic, lucite, plexiglass	450–500	0.540–0.600
Plastic, nylon	2000	2.40
Plastic, polyethylene	1000	1.200
Plastic, styrene	330–450	0.396–0.540
Plastic, foam (styrofoam)	400	0.480

Building Design Loading Criteria and General Considerations ■ 5-33

Seismic shocks may produce large-scale sawing motions at the joints and result in the working out or rupture of the joint filler or sealing material. If possible, such joints should be carefully detailed so that these materials can be easily replaced.

2. Differential movement between internal and external structure. Where any portion of the structure is exposed, there is the possibility of critical differential movement due to the expansion and contraction of the exposed portion while the interior is maintained at a relatively constant temperature. These movements may be accommodated in expansion joints in the structure and walls.

3. Differential movement between structure and skin. Where the structure is protected by considerable thickness of surface materials, a critical temperature differential may develop between the surface materials and the structure to which they are attached. In this case, expansion jointing should be located only in the surface elements. This may be easily done if the elements are in separate units and can be joined to allow the accumulation of movement in one unit. Where the surfacing elements are continuous—as with masonry or stucco—expansion joints must be provided at relatively short intervals.

4. Differential movement between dissimilar materials. Where materials with considerably different expansion rates are attached, critical distortions may develop over relatively short distances. Copper, bronze, aluminum, and stainless steel all have expansion rates higher than structural steel, and facings of these materials should be allowed to expand free of any supporting structure of steel. Similarly, reinforcing elements of steel encased by light mullions of aluminum or bronze should be allowed to slip longitudinally. Long elements of steel supporting masonry or plaster may accumulate critical differential expansion, unless adequate expansion jointing is provided in either, or both, elements.

Table 5.13 gives the coefficient of linear expansion for most common building-construction materials. Some materials, such as wood and paper, have different rates of expansion in different directions. Some expand differently in different directions because of their configuration, e.g., corrugated metal deck, which tends to accumulate movement in a direction parallel to the corrugations but absorbs motion in the other direction by slight flexing of the corrugations.

References

1. Uniform Building Code, International Conference of Building Officials, Whittier, Calif., 1994.

2. American Society of Civil Engineers (ASCE), *Minimum Design Loads for Buildings and Other Structures*, ASCE 7-93, New York, N.Y., 1994.

6
J. Y. Richard Liew
W. F. Chen

LRFD–Limit Design of Frames

6.1 General Principles

This chapter outlines the general principles that apply to the design of single-story and multistory steel building frames.

6.1.1 Introduction

Structural engineers are responsible for the design of the overall framework to ensure stability for safety. The design of all structural parts and components should be compatible even where some of the details of parts and components are not designed by the same structural engineer.

A structure should be designed to transmit dead, live, and wind loads in a direct way to the foundations. The structural arrangement should be stable so that the structure will not overturn or collapse under the actions of external loading and due to misuse or accidental damage to any one of the structural elements. Consideration should also be given to the erection procedure and stability during construction.

6.1.2 Stability

Single-Story Frames. ▪ Lateral stability to single-story frames is generally provided in two directions approximately at right angles to each other. This may be achieved by a rigid-frame system or vertical braced bays in conjunction with plan bracing or a diaphragm-braced system. For a diaphragm-braced system, either plywood or light gauge steel deck can be used to transmit the lateral forces.

However, when the structure length-to-width ratio exceeds 3, the diaphragm system requires a large amount of fasteners and thus becomes uneconomical. In such cases, plan bracing in the form of lattice trusses proves to be economical. The lateral force can be transmitted from the side cladding through the plan bracing and then to the vertical braces. However, when the buildings are long and narrow and when bracing is not allowed in the vertical plane, the best system to resist lateral load is the rigid frame system. This system can be placed in both directions of the structures to provide overall stability.

Multistory Frames ■ Unbraced Rigid Frames. A rigid unbraced frame is characterized by rigid joints between frame members. The frame does not require an additional bracing system, and it can resist all the design forces, including gravity as well as lateral forces. Columns in an unbraced frame are subjected to both axial load and moments and will experience lateral translation. The columns can provide stability for gravity columns whose simple connections to the frame do not provide resistance to lateral loads. However, the leaning effects from the gravity columns must be included in the design of the moment-frame columns. For tall building frames, unbraced rigid frames can be combined with vertical braced truss to produce a better system to reduce frame drift. Although rigid connections are more expensive to fabricate and take longer time to erect, rigidly framed systems have some benefits:

1. The connections are more ductile and they tend to perform better in earthquakes.
2. From the architectural and functional points of view, it can be advantageous not to have any bracing in the structure.

Braced Frames. For frames in which the members are pin-connected, lateral stability is provided by placing vertical and horizontal bracing within the structure in two directions approximately at right angles to each other. Bracing can generally be provided by means of structural walls enclosing the stairs, lifts, service ducts, etc. Additional lateral stiffness can also be provided by bracing within other external or internal walls. The bracing should preferably be distributed throughout the structure so that the combined shear center for lateral resistance is located approximately with the center of shear resultant on plan. Where this is not possible, torsional forces may be induced, and they must be considered when calculating the load carried by each braced system. Braced bays should be effective throughout the full height of the building. If it is essential for bracing to be discontinuous at one level, provision must be made to transfer the forces to other braced bays.

Forms of Bracing ■ Bracing mainly consists of (1) vertical bracing and (2) horizontal bracing. Vertical bracing may consist of triangulated steel trusses, moment resisting frame, reinforced-concrete shear walls preferably not less than 120 mm in thickness, masonry walls preferably not less than 150 mm in thickness adequately tied to the steel frames, or any combination of lateral force resisting elements. Temporary bracing should be provided during erection before the permanent bracing systems are installed.

Horizontal bracing may consist of triangulated steel trusses, concrete floors or roofs, and profiled metal decking either in the form of roof sheeting or as formwork to a composite floor. Since the floor may not be fully loaded during construction, the profiled decking, if adequately connected, provides the diaphragm action and lateral bracing to floor beams during construction.

6.1.3 Other Design Considerations

All members of a structure should be effectively tied together in the longitudinal, transverse, and vertical directions. Members whose failure would cause collapse of a part of or the entire structure should be avoided. Where this is not possible, alternative load paths should be identified or the member in question strengthened.

Movement joints should be provided to minimize the effects of movements arising from temperature variations and relative displacement of adjacent structures. The movement of joints should be effective enough to divide the structure into a number of independent sections. The joints should pass through the whole structure above ground level in one plane. The structure should be framed on both sides of the joint, and each structure should be independent and designed to be stable without relying on the stability of the adjacent structure.

Joints may also be required where there is a significant change in the type of foundation, soil condition, plan configuration, or the height of the structure to accommodate differential settlement. Where detailed calculations are not made, joints to permit movement up to 1 in (25 mm) should normally be provided at 50-m centers along the length of the frame. In the design of multistory building frames, a gap should generally be provided between steelworks and masonry cladding to allow for shortening of columns under high gravity loads.

6.1.4 Design Limit States

Under the load and resistance factor design (LRFD) format, a design is said to be satisfactory if it can be demonstrated that the structure satisfies both the ultimate strength and serviceability limit states. The strength limit states under the format of LRFD may be represented by the following equation:

$$\phi R_n \geq \Sigma \gamma_i Q_{ni} \qquad (6.1)$$

where Q_{ni} and γ_i = nominal load effects and their corresponding load factors

R_n = nominal resistance

ϕ = resistance factor corresponding to R_n.

The required strength on the right side of the inequality is the summation of the various load effects Q_{ni} multiplied by their respective load factor γ_i. The design strength, on the left side, is the nominal strength or resistance R_n multiplied by the resistance factor ϕ. The load combinations and load factors to be used in design for the limit states of strength and stability are shown in Table 6.1. The resistance factors, which are dependent on the types of members, are given in Table 6.2.

6-4 ■ Chapter Six

Table 6.1 Load Factors and Load Combinations

1	$1.4D$
2	$1.2D + 16L + 0.5(L_r \text{ or } S \text{ or } R)$
3	$1.2D + 1.6(L_r \text{ or } S \text{ or } R) + (0.5L \text{ or } 0.8W)$
4	$1.2D + 1.3W + 0.5L + 0.5(L_r \text{ or } S \text{ or } R)$
5	$1.2D + 1.5E + 0.5L + 0.2S$
6	$0.9D - (1.3W \text{ or } 1.5E)$

where: D = dead load due to the weight of the structural elements and the permanent features on the structure including finishes, fixtures, and fixed partitions
L = live load due to occupancy and movable equipment
L_r = roof live load
W = wind load
S = snow load
E = earthquake load (determined in accordance with Part 1 of the AISC seismic provisions for structural-steel buildings)
R = load due to initial rainwater on ice exclusive of the ponding contribution

The load factors and their combinations should be used to produce the most onerous condition. When appropriate, temperature effects should be considered with load combinations 1, 2, 3, and 4 in Table 6.1.

For serviceability limit states design, the LRFD specification lists five topics: (1) deflection, vibration, and drift, (2) thermal expansion and contraction, (3) connection slip, (4) camber, and (5) corrosion. The two most common parameters that affect the design serviceability of a steel building frame are deflections and accelerations.

Building deflections are usually measured in terms of a nondimensional parameter call drift. Interstory drift is the ratio of interstory deflection to story height.

Table 6.2 Resistance Factors for Different Member Type and Limit States

Member type and limit states	Resistance factors, ϕ
Tension member, limit state: yielding	0.90
Tension member, limit state: fracture	0.75
Pin-connected member, limit state: tension	0.75
Pin-connected member, limit state: shear	0.75
Pin-connected member, limit state: bearing	1.00
Column, all limit states	0.85
Beams, all limit states	0.90
High-strength bolts, limit state: tension	0.75

High-strength bolts, limit state: shear
  A307 bolts
  Others
	
	0.60
	0.65

Frame drift is the ratio of the lateral deflection at the top of the frame to the frame height. For steel building frames, commonly used values for first-order frame drift range from $H/400$ to $H/600$ under a 50-year wind, where H is the frame height. If the drift is calculated for a different return period, the drift limit should be adjusted appropriately (Iyengar et al.[9]). Maximum interstory drift from first-order analysis is normally limited to the range of $h_s/300$ to $h_s/400$, where h_s is the story height.

Service beam live-load deflections may be limited to $L/360$ when the beam carries brittle finishers such as plaster ceilings. Otherwise the beam deflection limit may be increased to $L/240$, where L is the span length of the beam. A maximum absolute deflection limit must be prescribed depending on the constructional materials and the manufacturers which supply those materials. This absolute limit is in the range of ¾ to 1 in (Fisher and West[7]).

Table 6.3 summarizes the deflection limits, which are intended to be the reasonable values for general applications. For some multistory frames, other values than those shown may be more appropriate. In particular for multistory buildings a ratio of height of story/500 may be more suitable where the cladding cannot accommodate larger movements. Coordination is required between the deflected structure and cladding components to ensure that the limits are appropriate in the particular circumstances.

Another parameter that is related to the calculation of building drift is the second-order $P\text{-}\Delta$ effect. $P\text{-}\Delta$ effect occurs because of the presence of gravity loads acting on the displaced structural configuration. A steel building of typical dimensions and designed to a first-order drift of 1/500 may experience second-order deflections in the range of 10 to 30 percent of the first-order deflection. Therefore, if second-order analysis is used directly to assess the serviceability performance of a building frame, the limit for lateral drift and interstory deflection may be reduced (Chen et al.[5]).

Table 6.3 Recommended Deflection Limits for Building Frames

Beam deflections from unfactored live loads	
Beams carrying plaster or brittle finish	$L_b/360$ (with maximum of ¼ to 1 in)
Other beams	$L_b/240$
Column deflections from unfactored live and wind loads	
Column in single story frames	$h_c/300$
Column in multistory frames	$h_s/300 \sim h_s/400$
For frame supporting cladding which is sensitive to large movement	$h_s/500$
Frame drift under 50 years wind load	
Frame drift	$H/400 \sim H/600$

L_b = span length of beam; h_c = column height; h_s = height of story; H = frame height.

6.1.5 Live-Load Reduction

Subject to certain limitations, ASCE 7-88[3] suggests that members having an influence area of 400 ft² or more may be designed for a reduced live load determined by applying the following equation:

$$L = L_0 \left(0.25 + \frac{15}{\sqrt{A_I}} \right) \quad (6.2)$$

where L = reduced design live load per square foot of area supported by the member

L_0 = unreduced design live load per square foot of area supported by the member

A_I = influence area, sq ft

The influence area is four times the tributary area for a column, two times the tributary area for a beam, and equal to the panel area for a two-way slab. The reduced design live load should not be less than 50 percent of the unit live load L_0 for members supporting one floor nor less than 40 percent of the unit live load L_0 otherwise.

6.1.6 Beam-Column Interaction Equations

Member forces obtained from global analysis must be checked against the interaction equations to ensure that they satisfy member strength and stability. The AISC-LRFD specification[2] provides the following equations for checking steel beam-columns subject to combined axial force and bending about the major and minor axes:

$$\frac{P_u}{\phi_c P_n} + \frac{8}{9} \left(\frac{M_{ux}}{\phi_b M_{nx}} + \frac{M_{uy}}{\phi_b M_{ny}} \right) \leq 1.0 \quad \text{for} \quad \frac{P_u}{\phi_c P_n} \geq 0.2 \quad (6.3)$$

$$\frac{P_u}{2\phi_c P_n} + \left(\frac{M_{ux}}{\phi_b M_{nx}} + \frac{M_{uy}}{\phi_b M_{ny}} \right) \leq 1.0 \quad \text{for} \quad \frac{P_u}{\phi_c P_n} < 0.2 \quad (6.4)$$

where P_u = required axial strength

P_n = nominal axial compressive strength evaluated using Eq. 6.5

M_{ux}, M_{uy} = required flexural strength about the strong and weak axes, respectively, within the unbraced length of the member

M_{nx}, M_{ny} = nominal flexural strengths about the strong and weak axes, respectively

ϕ_c = column resistance factor (= 0.85)

ϕ_b = beam resistance factor (= 0.90)

For the evaluation of P_n, the LRFD specification has adopted the following column strength formulas:

$$P_n = \begin{cases} P_y(0.685^{\lambda_c^2}) & \text{for } \lambda_c \leq 1.5 \\ P_y\left(\dfrac{0.877}{\lambda_c^2}\right) & \text{for } \lambda_c > 1.5 \end{cases} \quad (6.5)$$

where P_y = yield strength

λ = column slenderness ratio = $\dfrac{KL}{\pi r}\sqrt{\dfrac{F_y}{E}}$

6.1.7 Second-Order Effects

A second-order elastic analysis may be used to compute the required axial and flexural strengths in a member for direct substitution into Eq. 6.3 or 6.4. Alternatively, a first-order elastic analysis may be used, but the first-order forces must be amplified using B_1 and B_2 moment-amplification factors. The LRFD specification suggests the following procedure to estimate the second-order elastic moment M_u in lieu of a direct second-order elastic analysis:

$$M_u = B_1 M_{nt} + B_1 M_{1t} \quad (6.6)$$

where M_{nt} = first-order elastic moments in the member based on "no lateral translation" (NT) analysis (see Fig. 6.1a)

M_{1t} = maximum first-order elastic moments in the member based on "lateral translation" (LT) analysis (see Fig. 6.1b)

B_1 = P-δ moment-amplification factor which accounts for the amplification of the first-order NT moments associated with member curvature effects. B_1 is expressed as

$$B_1 = \dfrac{C_m}{1 - \left(P_u / P_{e1}\right)} \geq 1 \quad (6.7)$$

where P_{e1} = member buckling load defined as $\pi^2 EI/(KL)^2$, where K is an effective length factor in the plane of bending based on the assumption that sidesway is prevented

P_u = required axial compressive strength for member under consideration

C_m = a coefficient based on elastic first-order analysis assuming no lateral translation of the frame whose value is taken as

$$C_m = 0.6 - \dfrac{0.4 M_1}{M_2} \quad (6.8)$$

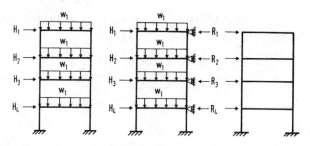

Original Frame = (a) Frame for M_{nt} + (b) Frame for M_{lt}

Fig. 6.1 AISC LRFD approach for calculating M_{nt} and M_{lt}.

for compression members subjected to end moments only; where M_1 and M_2 are the smaller and larger end moments obtained from the first-order NT analysis, respectively. M_1/M_2 is positive when the member is bent in reverse curvature and negative when it is bent in single curvature.

For compression members having transverse loading between their supports, the C_m value may be determined by rational analysis (Chen and Lui[4]) or by the following values:

$$C_m = 0.85$$

for members whose ends are restrained, and

$$C_m = 1.00$$

for members whose ends are unrestrained.

The B_2 factor in Eq. 6.6 is a $P\text{-}\Delta$ moment-amplification factor which amplifies the first-order end moments associated with lateral translation of the column. This term is expressed by the LRFD specification as

$$B_2 = \frac{1}{1 - \Sigma P_u \left(\Delta_{oh} / \Sigma HL \right)} \quad (6.9)$$

or

$$B_2 = \frac{1}{1 - \left(\Sigma P_u / \Sigma P_{e2} \right)} \quad (6.10)$$

where ΣP_u = summation of the axial load of all columns in a story

Δ_{oh} = lateral interstory deflection obtained from first-order elastic analysis

ΣH = sum of the story horizontal forces producing Δ_{oh}

L = story height

P_{e2} = member buckling load defined as $\pi^2 EI/(KL)^2$ in which K is an effective length factor in the plane of bending for the unbraced frame.

The factor B_2 applies to moments due to forces producing sidesway and is calculated for the entire story. When a frame is designed to limit Δ_{oh}/L to a predetermined value, the B_2 factor from Eq. 6.9 may be found in advance without knowing the member sizes. Thus Eq. 6.9 allows an estimation of second-order effect without actually knowing the member sizes.

6.1.8 Column Effective Length

Several rational methods have been proposed to estimate the effective length of the columns in an unbraced frame with sufficient accuracy (Liew et al.[12,13]). These methods may be classified broadly as (see Fig. 6.2)

1. Alignment chart method
2. Story buckling analysis
3. System buckling analysis

Although attempts were made to formulate general interaction equations without using the K factor, it was found that this was almost impossible if the interaction equations were to be versatile enough for a wide range of slenderness ratios and load combinations (Liew et al.[12,13]). However, the authors have devised a way of eliminating the effective length factor for beam-column capacity checks using a

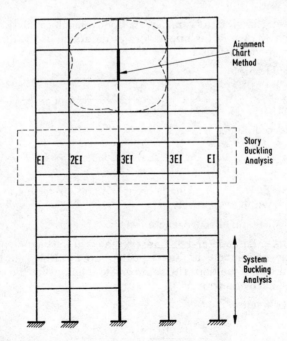

Fig. 6.2 Different methods of calculating column effective length factor.

notional load procedure (Liew et al.[16] Chen et al.[5]). The basic concept of the notional load method is discussed in Sec. 6.1.10.

Alignment Chart Method ▪ The specification describes a procedure by which the effective length factor is determined from the alignment charts that correspond to the sidesway-inhibited and sidesway-permitted cases. However, one must realize the underlying assumptions of which the alignment charts given in AISC LRFD[2] are derived:

1. All members are elastic and have constant cross sections. These members are connected by rigid joints.
2. For braced frames, rotations at opposite ends of beams are equal in magnitude, producing single-curvature bending. For unbraced frames, rotations at opposite ends of beams are equal in magnitude, producing double-curvature bending.
3. The stiffness parameters $L\sqrt{P/EI}$ for all columns are equal, where P is the axial force, L is the column length, and EI is the bending stiffness.
4. Joint restraint is distributed to the column above and below the joint in proportion to I/L of the columns.
5. All columns buckle simultaneously.
6. Axial forces in the connecting beams are small and have no significant effect on their stiffness.

The alignment chart K factors are derived from a stability analysis of a frame subassemblage as shown in the upper story of the frame in Fig. 6.2. The resulting equations from the analysis are:

For braced frames:

$$\frac{G_A G_B}{4}\frac{\pi^2}{K^2} + \frac{G_A + G_B}{2}\left[1 - \frac{\pi/K}{\tan(\pi/K)}\right] + \frac{2}{\pi/K}\tan\left(\frac{\pi}{2K}\right) - 1 = 0 \qquad (6.11)$$

and for unbraced frames:

$$\frac{G_A G_B (\pi/K)^2 - 36}{6(G_A + G_B)} - \frac{\pi/K}{\tan(\pi/K)} = 0 \qquad (6.12)$$

where $G = \Sigma(I_c/L_c)/\Sigma(I_b/L_b)$ with subscripts A and B referring to column end A and B, respectively

K = column effective length factor

The solutions to Eqs. 6.11 and 6.12 are presented in the alignment charts in Fig. 6.3. Approximate solutions to these equations are available, and they provide a reasonably accurate solution. The approximate solutions for the effective length factors are (Geschwindner et at.[8]):

For braced frames:

$$K = \frac{3G_A G_B + 1.4(G_A + G_B) + 0.64}{3G_A G_B + 2.0(G_A + G_B) + 1.28} \qquad (6.13)$$

For unbraced frames:

$$K = \sqrt{\frac{1.6 G_A G_B + 4.0(G_A + G_B) + 7.5}{G_A + G_B + 7.5}} \qquad (6.14)$$

The approximate solutions are accurate within 2 percent of the exact solutions from Eqs. 6.11 and 6.12.

Story Buckling Analysis ▪ One significant drawback of the alignment chart method is that it does not account for the fact that the stronger columns in a story can provide lateral resistance to weaker columns in a story mode as shown in Fig. 6.2. A more accurate way to obtain column effective length factor is to perform a story buckling analysis which assumes that (Liew et al.[12,13]):

1. The sum of the gravity load that causes sway buckling of a story is equal to the sum of the individual buckling loads of columns that provide story sidesway resistance.
2. The individual column buckling loads are equal to the buckling load determined using the alignment chart K factor.

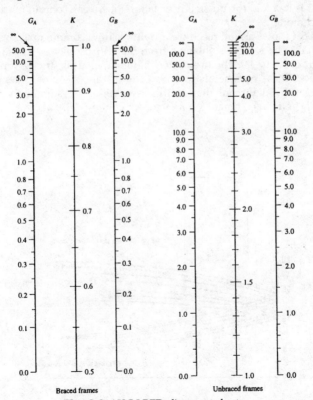

Fig. 6.3 AISC LRFD alignment charts.

6-12 ■ Chapter Six

This results in a modified effective length factor K_i' equation which has the form (AISC LRFD[2])

$$K_i' = \sqrt{\frac{I_i}{P_{ui}} \frac{\Sigma P_u}{\Sigma \left(I_i/K_i^2\right)}} \geq \sqrt{\frac{5}{8}} K_i \qquad (6.15)$$

where K_i' = effective length factor with story stability effect for ith rigid column
 I_i = moment of inertia in plane of bending of ith rigid column
 K_i = effective length factor for ith rigid column based on alignment chart for unbraced frame
 P_{ui} = required axial compressive strength for ith rigid column
 ΣP_u = required axial compressive strength of all columns in a story

The equation includes the effect of leaning columns and can be applied for columns with different stiffness parameter $L\sqrt{P/EI}$ in a story. The limiting K value in Eq. 6.15 is to avoid the possibility of failure of a weak column in the sidesway-prevented mode.

Figure 6.4 shows an unbraced frame with columns of unequal lengths defined as $\gamma = L_{AB}/L_{CD}$. The K factors obtained from Eq. 6.15 are plotted in the figure for γ ranging from 1.0 to 3.0. The use of the alignment chart is strictly applicable only when the stiffness parameters of both columns are the same (i.e., $\gamma = 1.0$). The direct use of the alignment chart in this case will underestimate the K factor for column CD and overestimate the K factor for column AB. Therefore, the ultimate

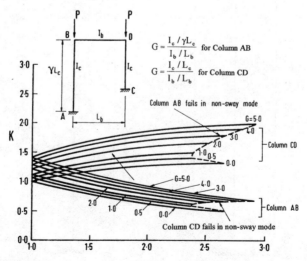

Fig. 6.4 K factors for columns of unequal length in a portal frame.

strength of column CD will be overpredicted whereas the strength for column AB will be underpredicted. The dashed curve in Fig. 6.4 is due to the failure of the weaker column in the sidesway-prevented mode.

System Buckling Analysis ▪ The system buckling load is a common basis for calculation of column effective length and estimating second-order amplification of forces and displacements. For nonrectangular and irregular framework involving a mixture of elements providing restraint against sidesway (see the lower stories of Fig. 6.2), a system buckling analysis is necessary for accurate determination of the effective length factors to be used in the P_n term of the beam column-interaction formulas. The effective length factor can be calculated using the equation

$$K_i = \sqrt{\frac{P_e}{P_{cr}}} = \sqrt{\frac{\pi^2 EI/L^2}{P_{cr}}} \qquad (6.16)$$

where P_e = Euler buckling load of column

P_{cr} = axial force in column at incipient buckling of the frame

An example for application of system buckling analysis is the lower portion of the framework shown in Fig. 6.2, in which the concept of "story buckling" is absent.

6.1.9 Preliminary Selection of Member Sizes

The design of beam-columns is a trial-and-error process. The selection of trial sections for use as beam-columns is facilitated by rewriting the beam-column interaction equations, Eqs. 6.3 and 6.4, into the so-called *equivalent axial load* form as

$$P_u + m_x M_{ux} + m_y U M_{uy} \le \phi_c P_n \qquad \text{for } \frac{P_u}{\phi_c P_n} \ge 0.2 \qquad (6.17)$$

$$\frac{P_u}{2} + \frac{9}{8}\left(m_x M_{ux} + m_y U M_{uy}\right) \le \phi_c P_n \qquad \text{for } \frac{P_u}{\phi_c P_n} < 0.2 \qquad (6.18)$$

where $m_x = \dfrac{8}{9}\left(\dfrac{\phi_c P_n}{\phi_b M_{nx}}\right)$

$m_y U = \dfrac{8}{9}\left(\dfrac{\phi_c P_n}{\phi_b M_{ny}}\right)$

Numerical values for m and U are provided in AISC *Manual of Steel Construction* (AISC LRFD[2]). The advantage of using Eqs. 6.17 and 6.18 for initial selection of member sizes is that the terms on the left-hand side of the inequality equations can be treated as an equivalent axial load term. This allows the designer to take advantage of the column tables provided in the manual for selecting trial sections.

6.1.10 Notional Load Approach for Assessment of Column Stability

The LRFD approach for beam-column design requires the use of effective length factor for the determination of the strength of members in frames. An alternate way to beam-column design is to consider the effects of end restraints in sway columns using the actual column length with an applied notional horizontal load. For sway columns, the notional load accounts for various imperfection effects that influence beam-column strength (Liew et al.[10]).

The notional load analysis involves the application of a set of notional loads at each and every story of a frame. The magnitude of the notional load H acting at every story may be expressed as

$$H = \alpha \Sigma P_u \qquad (6.19)$$

where ΣP_u = sum of all gravity loads acting on the story
α = 0.005, notional load parameter

The basic concept of the method and the detailed calibration of the notional load parameter have been reported in Liew et al.[10] The objective of the calibration is to determine a unique value for which can be universally adopted with the assurance that the deduced story or member strength and stability are accurate and conservative when compared with the "exact" solutions for a wide range of sway frames.

For multistory frames, the notional lateral load may be reduced because not all the columns in a building frame are arranged with out-of-plumbness leaning in the same direction. The expression for story notional lateral load H may be expressed as

$$H = \alpha K_c K_s \Sigma P_u \qquad (6.20)$$

in which

$$K_c = \sqrt{\frac{1}{2} + \frac{1}{n_c}} \leq 1.0 \qquad (6.21)$$

$$K_s = \sqrt{\frac{1}{5} + \frac{1}{n_s}} \leq 1.0 \qquad (6.22)$$

where n_c = number of columns in the plane of the story
n_s = number of stories in the plane of the frame
ΣP_u = sum of gravity loads acting on the story

The coefficients K_c and K_s in Eqs. 6.21 and 6.22 account for the fact that a large number of columns in a story and a large number of stories in a frame would reduce the influence of initial imperfection effects on frame or story stability;

therefore, the notional lateral load may be reduced accordingly. It should be noted that the story notional loads should be applied in conjunction with the design lateral load to cause additional frame deformation.

For braced frames where lateral stability is provided by adequate attachment to shear walls, or other systems that provide lateral force resistance, the member initial imperfection effects often have a greater influence on the stability of individual members and the overall frame than the sway imperfection effects. The member imperfection effects are particularly important for beam-column members subjected to high axial forces, e.g., $P > 0.25\pi^2 EI/L^2$. In this case, the effects of member imperfections and the associated P-δ effects must be considered in the analysis of the entire system. One way to account for these local P-δ effects is to apply a notional force at the maximum moment point of the member. This notional force should be counterbalanced by the two end reaction forces to maintain equilibrium of forces at the individual member level.

Schematic diagrams shown in Fig. 6.5 demonstrate how the notional load approach may be applied to irregular and nonrectangular frameworks. The story

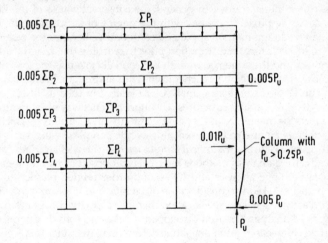

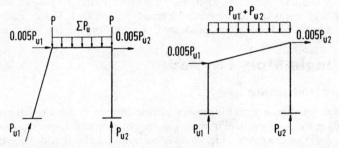

Fig. 6.5 Notional loads for irregular and nonrectangular frameworks.

notional load parameter α is taken as 0.5 percent, whereas the member notional load parameter is 1 percent as given in Fig. 6.5. The principal advantage of the notional load approach is its simplicity. As the actual length ($K = 1$) is used in the notional load procedure to determine the nominal strength P_n for use in the beam-column interaction equations, there is no need to compute the values of G factors and hence K factors. In addition, the notional approach no longer requires the distinction between sway and nonsway frames as in the case of the effective length approach.

6.1.11 Advanced Inelastic Analysis

Recent research by Liew et al.[14,15] has shown that elastic analysis procedures based on specification member capacity checks are limited in their ability to provide true assessment of the maximum strength behavior of highly redundant frameworks. The design procedure employed by the member interaction equations does not represent the true behavior of a frame after inelastic redistribution of forces occurs. Therefore, its application is truly valid for frames that are essentially elastic. The research also found that the LRFD elastic analysis/design approach is conservative for most conventional frames. In general, the conservativeness of the LRFD elastic analysis/design approach increases with the degree of structural redundancy. The elastic analysis/design method should be conservative because the reserved capacities due to inelastic force redistribution prior to reaching the design load level have been ignored, and the system's strength is essentially taken as the load at which the most critically loaded member reaches its individual strength. In summary, any specification checks based on elastic analysis are limited in their ability to estimate the true strength of redundant structures. The more rational way of estimating the interactive effects of member and system strength and stability—which implicitly includes member imperfections and also avoids complex issues such as effective length concept—is through *advanced inelastic analysis* (Liew et al.[16], Liew and Chen[15]).

Two advanced analysis methods have been proposed by the authors: (1) notional load plastic hinge analysis and (2) second-order refined plastic hinge analysis. Both approaches take into consideration all significant nonlinear effects in the global analysis so that separate specification member capacity checks are not required. The advanced analysis has been developed to take into consideration the flexibility of semirigid connections and panel-zone deformation in frames (Liew et al.,[14,15] Liew and Chen[10,11]). Practical procedure for the design of semirigid frames ranging from member-by-member approach to more sophisticated computer-based advanced analysis/design approach have been developed and verified. The details of these studies can be found in the recent books by Chen and Toma[6] and Chen et al.[5]

6.2 Single-Story Frames

6.2.1 Introduction

This section offers advice on the general principles to be applied when preparing a scheme for single-story steel frames. The emphasis is to examine various structural schemes that are practicable, sensibly economic, and not unduly sensitive to changes that are likely to be imposed as the design develops into the final stage.

LRFD—Limit Design of Frames ■ 6-17

Residential frames are simpler to construct, as the span lengths are usually small in comparison with industrial buildings which can range from the simplest warehouse-type structure to a sophisticated framing system that requires long-span construction. The principal requirement for the majority of single-story frames is to construct four walls and a roof so as to provide a structural envelope to the structure. The structural envelope is mostly made of profile steel sheets, reinforced concrete, precast concrete, or masonry.

In the design of the structural frameworks it is essential to ascertain correctly the loads applied to the structure and to predict the load paths from the wind to the cladding, to the purlins and side rails, to the girders and columns, and finally to the foundations. Loads should be carried to the foundation by the shortest and most direct routes.

For ease of construction, simplicity implies (among other matters) repetition, avoidance of complex joint details, with straightforward temporary works and minimal requirements for unorthodox sequencing to achieve the intended behavior of the completed structure. Sizing of structural members should be based on the longest and largest areas of roof or floor. The same sections should be assumed for similar but less onerous cases. This saves design and construction time and is of advantage to produce a cost-effective structure. Simple structural schemes are quick to design and easy to build.

General structural arrangements in a building plan should include bracing, type of roof and wall cladding, beam and column sizes, and typical joint details. Other nonstructural issues such as fire and corrosion protection must also be addressed so as to provide a serviceable design.

6.2.2 Loading

The loads are transferred from the sheeting onto the purlins and rails, which in turn are supported by a primary structure. These loads can be obtained from the loading codes such as ASCE 7-88.[3] They will include:

1. Dead loads, such as self-weight of cladding or slab, secondary beams, and primary structural members
2. Service loads, etc.
3. Vertical and horizontal live loads
4. Snow loads
5. Wind loads
6. Earthquake loads, in some areas

When the length of the structure is not greater than, say, 50 to 100 m, it is normal practice not to consider the thermal effects produced by the change of temperature. For industrial buildings, wind can cause pressure, suction, and drag loads on the cladding. Figure 6.6 illustrates the various components of load to be considered in the design of single-story industrial portal frames.

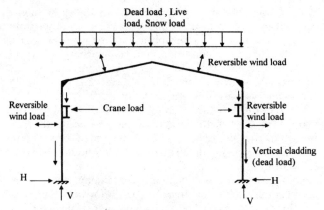

Fig. 6.6 Frame loads and reactions.

6.2.3 Material Selection

Three groups of structural steel are available in the United States for steel frame construction: (1) carbon steels: ASTM A36 and A529, (2) high-strength low-alloy steel: ASTM A440, A441, A572, A242, and A588, and (3) heat-treated low-alloy steel: ASTM A514. The most commonly used type and grade of structural steel for frame construction is A36 carbon steel. The most common choice where high-strength steel is desired is ASTM A572 grade 50. A36 steel is used generally for hot-rolled beams and columns. High strength steel is invariably economical because of lighter members unless local buckling, instability, deflection, and vibration are likely to be critical.

6.2.4 Structural Forms and Framing

The most common forms of single-story frames are:

1. Flat-roof portal frames used mainly for residential and small industrial buildings
2. Pitched-roof portal frames using hot-rolled beams or castellated beams for medium to large industrial buildings
3. Lattice frames using latticed girders or space frame roofs which provide long span capability

These frames may be provided with fixed-base columns, but the foundations need to be designed for moment in addition to axial and shear forces. The following frame spacing and spans are likely to be economic:

Framing system	Spacing, m	Span, m
Flat-roof portals	4.5–6.0	Up to 12
Pitched-roof portals	4.5–9.0	Up to 40
Lattice frames	6.0–9.0	>20

General design guidelines of the framing are given in the following:

1. For portal frames, provide longitudinal stability against horizontal forces by placing vertical bracing in the sidewalls deployed symmetrically wherever possible.
2. For lattice frames, provide stability against lateral forces in two directions approximately at right angles to each other by arranging suitably braced bays deployed symmetrically wherever possible.
3. Provide bracing in the roof plane of all single-story construction to transfer horizontal loads to the vertical bracing.
4. Provide bracing to the bottom of members of trusses or lattice girders if needed to cater for reversal of forces in these members because of wind uplift.
5. Consider the provision of movement joints for buildings whose plan dimensions exceed 50 m, approximately, depending on the temperature range.
6. Purlins should be supported at node points for lattice girders and trusses. If this cannot be achieved in practice, the effects of local bending at the chord members need to be taken into consideration.

Finally, a choice has to be made between the various framing schemes and their bracing arrangements. This should be based on the required appearance of the structure, span length, imposed load, the extent of support required in the roof for ceilings and services, local fabrication practice, cranage, length and height restriction for transportation, and overall economy.

6.2.5 Portal Frames

Portal frames are single-story, single- or multibay frames with flat or pitched roofs as shown in Fig. 6.7. The two common forms of portal construction are pinned-base frames and fixed-base frames. Frames that are designed based on pinned bases are heavier than those designed for fixed bases. However, the frame cost is offset by the provision of simple foundations which are designed only for axial force and shear force.

Structural Forms ■ Figure 6.7a shows the simplest framing solution which can be used to provide structural support to single-story buildings. Where flat roof construction is acceptable, this form of framing is used predominantly in spans of up to 12 m with frame spacing between 5 and 7 m. The frame comprises standard hot-rolled sections having simple or moment-resisting joints.

For long-span construction, a heavier section is required to control deflection, particularly if the beam is designed as simply supported. For gravity loading the column is in compression with a small bending moment at the top of the column due to the eccentricity of the beam connection. The beam acts in bending owing to the applied gravity loads, the compression flange being restrained either by purlins, which support the roof sheet, or by slab that spans between the main frames. Flat roofs are more difficult to weatherproof, since deflections of the beams induce ponding of rainwater on the roof which tends to penetrate the laps of roof cladding and the seams of roofing fabric. To counteract this problem, the

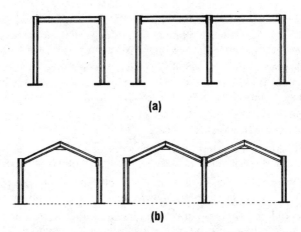

Fig. 6.7 Various types of portal frames: (*a*) flat-roof portals, (*b*) pitched portal.

beam may be cambered to provide the required fall across the roof, or the roof cladding may be laid to a predetermined fall so as to facilitate drainage of water off the roof.

Pitched-roof portal frames as shown in Fig. 6.7*b* may be used in spans of up to 40 m with frame spacing on the order of 6 to 9 m. The falls required to the roof are provided by the cladding carried on purlins which, in turn, are supported by the main frame members. The minimum slope for drainage is on the order of $1/8$ to $1/4$ in/ft, although a 1/10 roof slope is more commonly used for architectural requirement. The beam deflections become critical at low slopes, and design consideration must be given to the large resultant horizontal forces at the column bases.

Bracing ▪ For frames in which the structural members are pin-connected, resistance to lateral loads is achieved by the use of plan bracing and vertical bracing. Plan bracing is usually arranged by having diagonal members to form a truss in the roof plane to transfer horizontal loads to the vertical bracing. Lateral load is transmitted from the side cladding to the bracing in the roof plane, to the vertical braced plane, and then to the foundation. Typical arrangements of plan and vertical bracings are shown in Figs. 6.8 and 6.9 for flat-roof and pitched-roof frames, respectively. The bracing is generally designed as a pinpointed frame, in keeping with the simple joints used in the main frame. Resistance to lateral loading in the plane of the main frame can also be achieved either by the use of rigid connections at the column-to-beam joint (Fig. 6.10*a*) or by designing the columns as fixed-base cantilevers (Fig. 6.10*b*). Other combinations of joint details for beam-to-beam, column-to-beam, and column-to-foundation connections are shown in Fig. 6.10*c* and *d*. Since single-story buildings are normally light and low in profile, wind and seismic forces are relatively low. Rigid frame action can be easily achieved by the rigid frame action in the plane of the frame. Rigid beam-to-column connections reduce

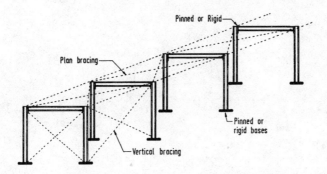

Fig. 6.8 Bracing system for flat-roof portal frame.

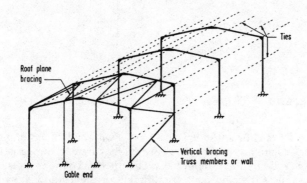

Fig. 6.9 Bracing arrangement of pitched-roof portal frame.

the beam deflection and rigid column base joints reduce frame drift, and thus the final design requires less weight of structural steel for the frame. However, resistance to lateral load in the direction perpendicular to the main frames is not inherent and hence must be provided. Buildings that employ portal frame construction often have brickwork cladding in the vertical plane. With proper detailing, the brickwork can be designed to provide the sway bracing.

Portal Frames with Fabricated Sections ▪ In many circumstances, pitched-roof portal frames are constructed with haunches at the eaves and apex. The haunches improve, rather significantly, the stiffness of frame because of its large section properties in the regions of high bending moment. The haunch length, which is approximately 10 percent of the span length, can be formed by cutting from the same rolled section as the rafter. The depth of the haunch at the column face is about the depth of the rafter used. Portal frames can be also fabricated with tapered members. Frames of this type are common in the United States and are being used more frequently in Europe.

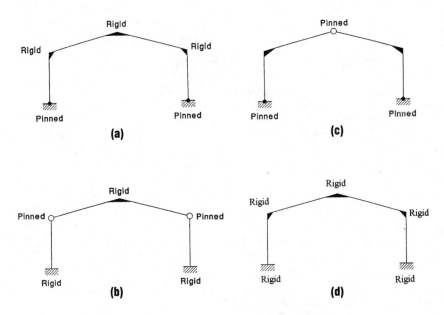

Fig. 6.10 Portal frames with various kinds of joint combinations.

Multibay Portal Frames ■ Portal frames with multibay configurations of 20- to 30-m spans are not uncommon. For multibay frames, the design is almost invariably governed by the exterior bays. The internal columns are subjected to little bending unless the bay spans are different or the loading is asymmetrical. Therefore, these columns can be sized the same as the columns for the single-span case. The rafter size multibay be kept constant throughout unless different span lengths are used in the may frame. The eave deflections of a pitched-roof multibay frame should be checked carefully, as the horizontal deflections will be cumulative from span to span.

Pin-ended props may be used as the internal columns for multibay frame construction. However, these columns do not participate in lateral resistance of the structure; instead they are leaning on the other columns that contribute to the lateral stiffness of the structure (Fig. 6.11a). A leaning column tends to destabilize the whole structure because any lateral displacement causes the prop to induce an additional lateral load to adjacent columns that participate in lateral load resistance. As a result of this leaning effect, columns that participate in lateral load resistance must be designed using larger effective length factor (see Sec. 6.1.8). Valley beams may be used, instead, to replace internal columns to gain extra free space. It is common to use this construction in Portal frames with valley columns omitted at alternate bays as shown in Fig. 6.11b. The valley beams provide no stabilizing effect to the structure. They may be modeled as sliding supports, or otherwise in accordance with the detailing requirement.

LRFD—Limit Design of Frames ■ 6-23

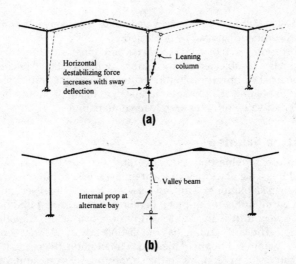

Fig. 6.11 Multibay portal frames: (*a*) frame with internal prop, (*b*) frame with valley beam.

Analysis and Design ■ Manual methods and charts are available for the analysis and design of simple portal frames with prismatic members (Owen and Knowles[17]). The analysis of portal frames with tapered members may be carried out using a computer. The tapered members can be discretized into several beam-column elements; each element is considered to be a prismatic member and is assigned with the average section properties over its length. The procedures for analysis and design of portal frames are described as follows:

1. Perform elastic buckling analysis and obtain the buckling load factor λ_{cr}
2. Calculate column effective length factor from

$$K_i = \sqrt{\frac{P_e}{P_{cr}}} = \sqrt{\frac{\pi^2 EI_c / L_c^2}{P_{cr}}} \qquad (6.23)$$

3. Calculate moment amplification factor A_F associated with lateral translation from

$$A_F = \frac{1}{1 - \left(1/\lambda_{cr}\right)} \qquad (6.24)$$

where $\lambda_{cr} = P_{cr}/P_u$ is the buckling load factor

P_u = design factored load

P_{cr} = elastic buckling load of the structure for the same load combination.

The moment amplification is recommended only for frames with $5 < \lambda_{cr} < 10$. If $\lambda_{cr} < 5$, a direct second-order analysis is more appropriate to obtain the required moment for design. For $\lambda_{cr} > 10$ sidesway instability is unlikely to be critical (SCI[18]), and the second-order effects can be neglected.

The B_2 expression given by AISC LRFD is valid only for rectangular framing in which the beam is not subjected to any compression force. Since pitched-roof portals have large axial compression in the inclined beams, separate buckling analysis is required to obtain the moment-amplification factor. It should be noted that there is a different value of λ_{cr} for each load combination.

6.2.6 Lattice Frames

Lattice frames can provide both attractive and structural efficient solutions for industrial buildings, e.g., factories and warehouses, leisure buildings, sport buildings, and commercial buildings. Latticed truss will have a larger second moment of area and section modulus for in-plane bending than a corresponding hot-rolled joist of similar weight. It is the lightest form of construction, though it requires more fabrication. The lattice frame has the distinct advantage that services can be accommodated within the depth of the girder. In addition, the reduction in weight of the girder can result in economy in the supporting structure and foundations.

Structural Forms ■ Lattice roof structures are able to meet a medium- to long-span requirement. They are particularly economic for span length greater than 20 m. Main frames are normally spaced at 6- to 9-m centers. These spacings generally provide economic solutions for the purlin and side-rail arrangements. The members of the roof truss can be arranged to provide the required cladding slope and to support internal services within the roof zone. The top chord of the lattice truss follows the required slope of the roof, but the bottom chord can be either parallel to the roof slope or horizontal as shown in Fig. 6.12. A horizontal bottom chord provides a larger service zone but results in bracing members of varying lengths. A bottom chord parallel to the roof slope results in bracing of the same length, ease of fabrication, smaller dimension for transport, and increase in clear internal height at midspan. A multispan roof can be constructed from a series of simple truss spans. The roof may be supported at valley and/or at the apex by girders spanning two or more bays in the longitudinal direction as shown in Fig. 6.13. To maximize the floor areas to be covered with minimum internal columns, the alternate columns in the valley can be omitted.

Analysis of the lattice frames can be carried out by hand calculation or computer. In the hand calculation method, it is essential to assume all joints are pin-connected and the girder end support conditions are such that the truss is statically determinate. In computer calculation, it is often assumed that connections between web and chord members are pinned for calculation of axial forces in the members.

Bracing ■ The top (compressive) chord of the lattice structure needs to be restrained laterally. This restraint is provided by the purlins, which provide lateral stability to the compressive chord. The forces in the purlins are transmitted to the foundation via a diagonal bracing system. The bracing system is usually provided in each end bay of the building. In addition to providing lateral stability, the

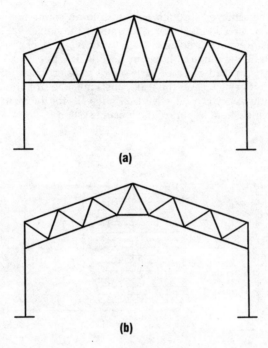

Fig. 6.12 Lattice frames with (a) horizontal bottom chord, (b) parallel chords.

bracing system also resists wind load applied on the gable end of the building, and it provides stability during erection of the main frames.

Under the action of load reversal due to wind, bottom chord tension members have to resist compressive forces. Similar lateral members, linked into the bracing system, must be provided. Many single-story lattice frame arrangements have inherent in-plane resistance to sidesway by providing moment-resisting joints at the column-to-truss member connections. Moment connections help to reduce deflection and member size. In addition, they provide resistance to lateral in-plane load such as those from side winds and crane movements. If in-plane sta-

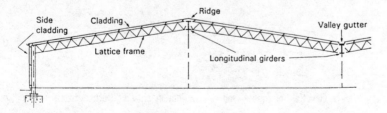

Fig. 6.13 Multibay lattice frames with longitudinal girders and internal props.

bility cannot be achieved through truss-to-column connections, roof bracing in the longitudinal direction is required to transfer the horizontal load to the gable frames which are again braced against sidesway. A typical bracing arrangement of an industrial building frame is shown in Fig. 6.14. It is possible to use the cladding and wall panels instead of braced trusses to transfer longitudinal loads to the foundations, but in such cases the wall panels must be sized based on stress skin design. When masonry is used as vertical sidewalls for the main frames, it is feasible to use that element as part of the bracing system.

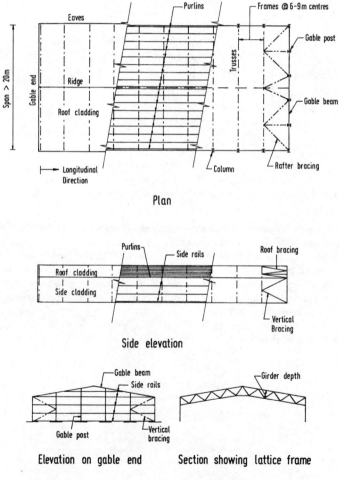

Fig. 6.14 Structural layout of a typical lattice frame building.

The bracing can be single diagonals or cross members. If the former system is adopted, the members are designed for compressive and tensile loads. When cross members are used, the members in tension may be assumed to be effective, and those in compression are designed to satisfy the slenderness criteria of $L/r = 200$.

Bracing may be located either at midlength of the building (Fig. 6.15a, 6.15b) or at its ends (Fig. 6.16a, b). Bracing at the ends helps to provide stability during erection, since it provides a stable structure at one end of the building when the erection begins. Its disadvantage is that it prohibits the free movement of the frame in which thermal stresses may be incurred as the frame is constrained at both ends. If the braces are put at the midlength, this disadvantage disappears, since the building is able to expand or contract freely. However, the horizontal forces have to "travel" from their point of application in the gable end to the brace bay, thus causing compression in the purlins. This "unfavorable" situation can be avoided by providing plan bracing at both ends, as the applied lateral load is taken completely by the braced frame at the end. The lateral load travels to the bracing members, then to the columns of the braced frames, and finally to the foundations. A shorter load path normally implies a more efficient load transfer system. A typical example of such a bracing arrangement is shown in Fig. 6.16.

Design of Lattice Trusses ■ The members in the lattice girders or trusses should be designed using the following criteria:

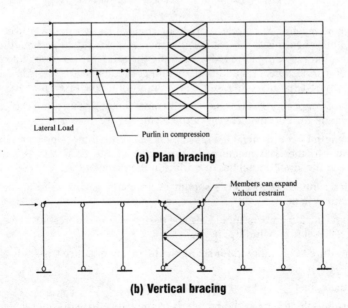

Fig. 6.15 Vertical bracing at the midlength of the building.

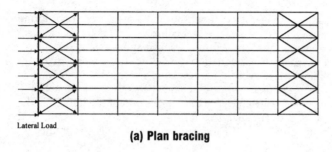

(a) Plan bracing

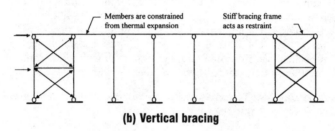

(b) Vertical bracing

Fig. 6.16 Vertical bracing at the gable ends.

1. Members meeting at a node should be arranged so that their centroidal axes coincide. When this is not possible the members should be designed to resist the resulting bending moments caused by the axial forces acting at eccentricities of connections. In addition, bending moments arising from loading between node points, other than self-weight, should be taken into account.
2. Fixity of connections and rigidity of members may be taken into account for calculating the effective lengths of the members.
3. The length of chord members may be taken as the distance between the connections to the web members in the plane of the truss and the distance between the longitudinal ties or purlins in the plane of the roof cladding.
4. All ties connecting to chords should be properly anchored to an adequate restraint system.
5. Bottom chord members should also be designed for compression force due to load reversal from wind uplift.

The procedure to be adopted for sizing the members of lattice trusses is given as follows:

1. Calculate the total factored load on the roof.
2. Determine the forces in the members of the lattice truss for all relevant load combinations by analysis.

LRFD—Limit Design of Frames ■ 6-29

3. For compression members calculate the member effective lengths L_e and design the section using the section table in Part 2 of the AISC LRFD Manual.
4. For tension members, design the section by reference to AISC LRFD, Sec. 2.
5. The deflection of girder and trusses should be checked to ensure that serviceability with particular reference to roof drainage is not impaired.

Column Design

Columns Braced in Both Directions

1. Calculate the factored axial load P_u on the column from the roof and side cladding.
2. Calculate the factored wind loading on the sidewalls and on the roof.
3. Calculate the factored horizontal component W of the wind force on the roof for use in the design of bracing members.
4. Calculate the total factored sidewall wind loads on the columns and calculate the corresponding maximum factored moments arising from wind.
5. Calculate the factored moments on the columns arising from the live load and dead load by elastic analysis and add these to the factored wind moments.
6. Compute the required moment by amplifying the column end moment using the B_1 factor.
7. Select a section and assume column effective length factor $K=1.0$ or less (depends on the column end restraint); check the column capacity using Eq. 6.5.

Columns Braced in One Direction Only

1. Calculate the factored axial load P_u on the column from the roof, and from the side cladding.
2. Calculate the factored wind loading on the sidewalls and on the roof.
3. Calculate the factored horizontal component of the wind force on the roof.
4. Calculate the total factored sidewall wind loads on the column.
5. Calculate the maximum factored moments on the columns arising from wind, dead, and live loads by elastic analysis assuming that the columns and lattice girders or trusses act as frames in the unbraced direction.
6. Amplify the column end moments using the B_1 and B_2 factors as in Sec. 6.1.7.
7. Select a section and check the design of the column using the interaction equations given by Eq. 6.3 or 6.4.

6.2.7 Example: Single-Story Multibay Braced Frame

Figure 6.17a shows a single-story structure in which moment connections are provided at the roof level for columns A1, B1, A4, and B4. All other column connections are pin-jointed. The in-plane stability and lateral load resistance are provided by the X bracing as shown. The out-of-plane stability and lateral load resistance

6-30 ■ Chapter Six

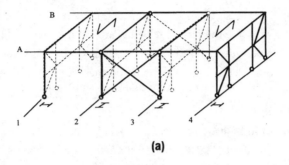

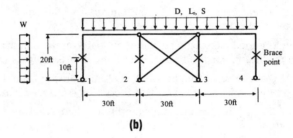

Fig. 6.17 (a) Three-dimensional braced frame, (b) two-dimensional braced frame.

are provided by the braced frames at grid lines 1 and 4 and that the columns are braced to prevent buckling at the midheight in the out-of-plane direction. Design the columns and the roof beam for the given load shown in Fig. 6.17b. Check that the frame drift under unfactored wind loads is limited to height/400. The beams are assumed to be fully restrained in the out-of-plane direction.

Uniformly Distributed Load

Dead load $D = 1.8$ kips/ft   Roof live load $L_r = 0.3$ kip/ft
Snow load $S = 0.6$ kip/ft   Wind load $W = 0.6$ kip/ft

Load Combinations 1

$1.2D + 1.6(L_r \text{ or } S) + 0.8W = 2.16$ kips/ft $+ 0.96$ kip/ft $+ 0.48$ kip/ft
$\qquad = 3.12$ kips/ft $+ 0.48$ kip/ft

Load Combinations 2

$1.2D + 0.5(L_r \text{ or } S) + 1.3W = 2.16$ kips/ft $+ 0.3$ kip/ft $+ 0.78$ kip/ft
$\qquad = 2.46$ kips/ft $+ 0.78$ kip/ft

LRFD—Limit Design of Frames ■ 6-31

Design Assumptions ■ The following assumptions are made for the frame shown in Fig. 6.17b:

1. The frame is fully braced in both planes, i.e., sidesway is prevented.
2. The beams are continuously restrained in the lateral direction.
3. Column is braced at the midheight in the weak-axis directions. The braces act to resist buckling only and do not participate in resisting bending and/or compression.

Preliminary Sizing of Beams ■ The beam at the center bay is considered to be most critical, because it is subjected to axial compression and lateral load under the wind load combination. The beam is simply supported at both ends and is continuously restrained against lateral-torsional buckling. Serviceability deflection may control the design of this beam, which has a span length of 30 ft. Serviceability limit under unfactored live load

$$\Delta_{max} = \frac{span}{360} = \frac{30 \times 12}{360} = 1.0 \text{ in}$$

For uniformly distributed load

$$\Delta_{max} = \frac{5wL^4}{384EI} = 1 \text{ in}$$

Service load $w = 0.6$ kip/ft

$$\text{Required stiffness } I_x = \frac{5wL^4}{384E} = \frac{5 \times (0.6/12) \times (30 \times 12)^4}{384 \times 29,000} = 377 \text{ in}^4$$

Select W24×62 for the roof beam. The section properties are

$$I_x = 1550 \text{ in}^4 > 377 \text{ in}^4$$

$$Z_x = 153 \text{ in}^3, \ d = 23.74 \text{ in}, \ t_w = 0.43 \text{ in}$$

Preliminary Sizing of Columns ■ Assuming that the internal and external columns are the same size, and by inspection, the internal column is likely to be more critical as it supports twice the gravity load in comparison with the exterior column. Hence in preliminary design, we consider only the factored gravity load (1.2D+1.6S) for sizing the internal columns which are pin-supported at both ends.

$$P_u = (2.16 + 0.96) \times 30 = 93.6 \text{ kips}$$

Since the frame is braced at both directions, the effective length factors $K_x = K_y = 1.0$.

With bracing at midheight, the effective length in the y-y axis is $K_y L_y = 10$ ft.

Effective length in the x-x axis is $K_x L_x = 20$ ft.

From the connection point of view a W12 column section is required in order to connect the column to the beam section of W24×62.

6-32 ■ Chapter Six

Select W 12×65 section for the column.
From p. 3-24 of the AISC LRFD[2] manual, we have

$$K_y L_y = 10 \text{ ft}, \quad \phi_c P_n = 538 \text{ kips} > 93.6 \text{ kips}$$

Equivalent effective length for x-x axis is

$$(K_x L_x)_{eqv} = \frac{20}{1.75} = 11.4$$

$$\phi_c P_n = 525 \text{ kips} > 93.6 \text{ kips}$$

Hence buckling about the x-x axis is more critical.

Preliminary Sizing of Bracing Members ■ Factored wind load at the beam level is

$$W = \frac{0.78 \times 20}{2} = 7.8 \text{ kips}$$

Assuming the compression bracing is not effective in resisting the wind load, the wind load is resisted by the tension brace only. The tension force in the bracing is

$$T = \frac{7.8}{\cos(33.7°)} = 9.4 \text{ kips}$$

$$\text{Net area required in tension} = A_{req} = \frac{T}{\phi \rho_y} = \frac{9.4}{0.9 \times 36} = 0.29 \text{ in}^2$$

For a tension member, maximum member slenderness ratio is $(L/r)_{max} = 300$. Since L = length of bracing = 36.1 ft, the minimum radius of gyration to satisfy the slenderness ratio limit is

$$r_{min} = 36.1 \times \frac{12}{300} = 1.444 \text{ in}$$

Select 2L5×5×5/16

Area $A = 6.05 \text{ in}^2 > 0.29 \text{ in}^2$ OK

Minimum radius of gyration $r = 1.57 \text{ in} > 1.444 \text{ in}$ OK

Analysis ■ The bending moments and forces obtained from the first-order elastic analysis under the load combination $1.2D+1.6L$ are shown in the following figures.

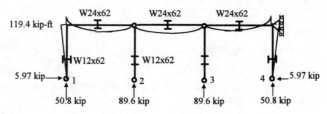

From first order elastic analysis under factored wind load 1.3W, the forces are shown below.

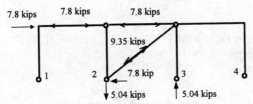

Detailed Design Checks

1. Roof Beams. Check the beam between columns 2 and 3 for load combination $1.2D+1.6L_r+0.8W$. The load acting on the beam is shown in the following figure:

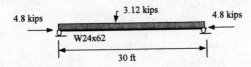

$P_u = 4.8$ kips, $M_{nt} = 3.12 \times 30^2 / 8 = 351$ kip-ft

For W24×62 section, $L=30$ ft/s, $r_x=9.23$ in, $F_y=36$ ksi, $E=29,000$ ksi

$$\lambda_c = \frac{30 \times 12}{9.23\pi} \sqrt{\frac{36}{29,000}} = 0.4374 < 1.5$$

$$P_{e1} = A_g F_y / \lambda_c^2 = \frac{18.2 \times 36}{0.4374^2} = 3425 \text{ kips}$$

$C_m = 1$

$$B_1 = \frac{C_m}{1-\left(P_u/P_{e1}\right)} = \frac{1.0}{1-\left(4.8/3425\right)} = 1.0014 > 1.0$$

$M_u = B_1 M_{nt} = 1.0014 \times 351 = 351.5$ kips

Since $\lambda_c \leq 1.5$

$$P_n = P_y(0.658^{\lambda_c^2}) = (36 \times 18.2) \times (0.658^{0.4374^2}) = 604 \text{ kips}$$

Using the design table in p. 4-18 in the AISC LRFD Manual[2]

$\phi_b M_{nx} = \phi_b M_{px} = 413$ kip-ft

since $P_u/\phi_c P_n < 0.2$, check the following interaction equation in accordance with Eq. 6.4 as follows:

$$\frac{P_u}{2\phi_c P_n} + \left(\frac{M_{ux}}{\phi_b M_{nx}} + \frac{M_{uy}}{\phi_b M_{ny}}\right) = \frac{4.8}{2(0.85 \times 604)} + \left(\frac{351.5}{413} + 0\right) = 0.856 < 1.0 \quad \text{OK}$$

Check the beam for the load combination $1.2D+0.5L_r+1.3W$. The load acting on the beam is shown in the following figure:

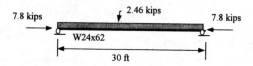

$$P_u = 7.8 \text{ kips}, \quad M_{nt} = \frac{2.46 \times 30^2}{8} = 276.8 \text{ kip-ft}$$

$$B_1 = \frac{C_m}{1-\left(P_u/P_{e1}\right)} = \frac{1.0}{1-\left(7.8/3425\right)} = 1.0023 > 1.0$$

$$M_u = B_1 M_{nt} = 1.0023 \times 276.8 = 277.4 \text{ kips}$$

since $P_u/\phi_c P_n < 0.2$, check the following interaction equation in accordance with Eq. 6.4:

$$\frac{P_u}{2\phi_c P_n} + \left(\frac{M_{ux}}{\phi_b M_{nx}} + \frac{M_{uy}}{\phi_b M_{ny}}\right) = \frac{7.8}{2(0.85 \times 604)} + \left(\frac{277.8}{413} + 0\right) = 0.68 < 1.0 \quad \text{OK}$$

W24×62 is adequate.

2. Leaning Columns at Grid Lines 2 and 3

For load combination $1.2D+1.6L_r+0.8W$

Axial load $P_u = 89.6 + \left(\frac{4.8}{7.8}\right)(5.04) = 92.7$ kips

For load combination $1.2D+0.5L_r+1.3W$

Axial load $P_u = \left(\frac{2.46}{3.12}\right)(89.6) + 5.04 = 75.7$ kips

Maximum required axial compressive strength $P_u = 92.7$ kips

Effective length for $y-y$ axis is $K_y L_y = 10$ ft

Equivalent effective length for the $x-x$ axis is $(K_x L_x)_{eqv} = \frac{20}{1.75} = 11.4$ ft

(Control!)

For W12×65:

$\phi_c P_n = 525$ kips

$\dfrac{P_u}{\phi_c P_n} = \dfrac{92.7}{525} = 0.177 < 1.0$ hence section W12×65 is adequate.

3. Exterior Columns

Governing load combination is $1.2D + 1.6L_r + 0.8W$

$P_u = 50.8$ kips, $M_{nt} = 119.4$ kip-ft

For W12×65, $L_x = 20$ ft, $L_y = 10$ ft

For buckling about y-y axis, $K_y L_y = 10$ ft

For buckling about x-x axis,

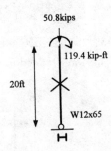

$G_{top} = \dfrac{1170/(20 \times 12)}{1.5(1550/(30 \times 12))} = 0.75$

(Note that the beam stiffness in unbraced frames is modified for the far end pin condition by a factor of 1.5)

$G_{bottom} = 10$ (pinned end)

From alignment chart for braced frame, $K_x = 0.84$.

Equivalent effective length in the x-x axis is $(K_x L_x)_{eqv} = 0.84 \times 20/1.75 = 9.6$ ft $< K_y L_y$; therefore y-y axis controls!

$\phi_c P_n = 538$ kips (p. 3-24, AISC LRFD Manual[2])

$\phi_b M_n = 261$ kip-ft (page 4-19, AISC LRFD Manual[2])

$P_{e1} = \dfrac{\pi^2 E I_x}{(KL_x)^2} = \dfrac{\pi^2 (29{,}000)(533)}{(0.84 \times 20 \times 12)^2} = 3754$ kips

since $M_1 = 0$,

$C_m = \dfrac{0.6 - 0.4 M_1}{M_2} = 0.6$

$B_1 = \dfrac{C_m}{1 - \left(P_u / P_{e1}\right)} = \dfrac{0.6}{1 - (50.8/3754)} = 0.608$

Since B_1 must be greater than 1.0, therefore, it is taken as 1.0

$$M_u = B_1 M_{nt} = 119.4 \text{ kips}$$

Check for interaction formula

$$\frac{P_u}{\phi_c P_n} = \frac{50.8}{538} = 0.094$$

which is less than 0.2; therefore the following interaction equation may be applied:

$$\frac{P_u}{2\phi_c P_n} + \left(\frac{M_{ux}}{\phi_b M_{nx}} + \frac{M_{uy}}{\phi_b M_{ny}}\right) = \frac{50.8}{2(538)} + \left(\frac{119.4}{261} + 0\right) = 0.50 < 1.0 \quad \text{OK}$$

W12×65 is adequate for the column.

4. Sway Deflection Due to Service Wind Load, 1.0W

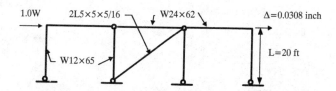

Allowable sway limit $= \dfrac{L}{400} = \dfrac{20 \times 12}{400} = 0.6$ in

Deflection under service wind load from first-order elastic analysis is 0.0308 in, which is less than the allowable limit of 0.6 in.

6.2.8 Example: Single-Story Multibay Unbraced Frame with Leaning Columns

The frame shown in Fig. 6.18 is similar to that in Sec. 6.2.7 except that the in-plane stability and lateral load resistance are provided by the rigid frame system at the exterior bays. The intermediate bracing at the midheight of the column is removed. Design the frame for in-plane stability and strength and check that the frame drift under unfactored wind load is limited to height/300. The loading is the same as that in the example given in Sec. 6.2.7.

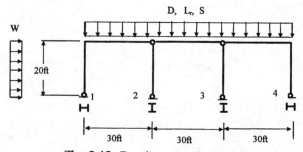

Fig. 6.18 Two-dimensional unbraced frame.

Uniformly Distributed Load

Dead load $D=1.8$ kips/ft   Roof live load $L_r=0.3$ kips/ft
Snow load $S=0.6$ kip/ft   Wind load $W=0.6$ kip/ft

Preliminary Sizing of Members ▪ Section sizes similar to the example frame in Sec. 6.2.7 may be used as initial sizing for this frame. The beam is found to be satisfactory but the columns are undersized. After one more trial and error, the final member sizes are selected as shown in the following table.

Section		Area	Second moment of area		Radius of gyration	
		A	I_x	I_y	r_x	r_y
Beams	W24×62	18.2	1550	34.5	9.23	1.38
Columns	W18×71	20.8	1170	60.3	7.5	1.7

Analysis ▪ Elastic first-order analysis of the frame for different load combinations is carried out. The bending moments and forces are shown in Fig. 6.19a to d for NT and LT analyses.

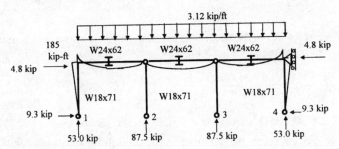

Fig. 6.19a Nontranslation (NT) analysis for load case 1: $1.2D + 1.6S + 0.8W$.

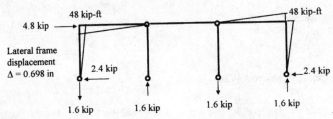

Fig. 6.19b Lateral-translation (LT) analysis for wind load of $0.8W$ under load case 1.

6-38 ■ Chapter Six

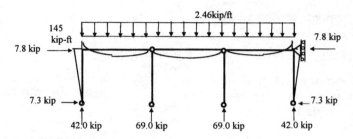

Fig. 6.19c Nontranslation (NT) analysis for load case 2: 1.2D + 0.5S + 1.3W.

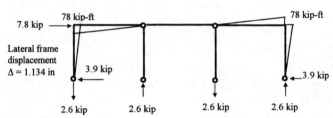

Fig. 6.19d Lateral-translation (LT) analysis for wind load of 1.3W under load case 2.

Check Interior Column W18×71. The interior column is a leaning column; therefore, weak-axis buckling controls. Compute the nondimensional slenderness ratio

$$\lambda_{cy} = \frac{K_y L_y}{\pi r_y}\sqrt{\frac{F_y}{E}} = \frac{1.0 \times 20 \times 12}{\pi \times 1.70}\sqrt{\frac{36}{29,000}} = 1.58 > 1.5$$

From the AISC LRFD Manual, the column axial resistance is $\phi P_n = 222$ kips.

The required axial strength can be obtained from the gravity load combination in Fig. 6.19a as

$$P_u = 87.5 \text{ kips}$$

Check the column strength

$$\frac{P_u}{\phi_c P_n} = \frac{87.5}{222} = 0.394 < 1.0 \quad \text{OK}$$

Check exterior column W18×71

For load case 1: 1.2D+1.6S+0.8W

The required force for load case 1: 1.2D+1.6S+0.8W is shown in Fig. 6.20. Calculate the effective length factor based on the story buckling concept from Eq. 6.15:

$$K'_i = \sqrt{\frac{I_i}{P_{ui}} \frac{\Sigma P_u}{\Sigma \left(I_i/K_i^2\right)}} \geq \sqrt{\frac{5}{8}K_i}$$

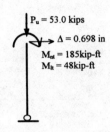

Fig. 6.20 Required forces and displacement for load case 1.

I_i=1170 in^4, P_{ui}=53 kips, ΣP_u = 3.12×90=280.8 kips. K_i is the effective length factor from the alignment chart.
For exterior column,

$$G_{top} = \frac{1170/(20 \times 12)}{0.5(1550/(30 \times 12))} = 2.3$$

(Note that the beam stiffness in the unbraced frame is modified for far end pin condition by a factor of 0.5.)

$G_{bottom} = 10$ (pinned end)

From the alignment chart for the unbraced frame, K_x=2.2.

For the leaning column $K_i = \infty$ should be taken to compute the summation term in Eq. 6.15.

$$K'_i = \sqrt{\frac{1170}{53} \frac{280.8}{2 \times (1170/2.2^2)}} = 3.58 \geq 1.17$$

Nondimensionalized column slenderness ratios

$$\lambda_{cx} = \frac{K_x L_x}{\pi r_x}\sqrt{\frac{F_y}{E}} = \frac{3.58 \times 20 \times 12}{\pi \times 7.5}\sqrt{\frac{36}{29,000}} = 1.29 < 1.5$$

$$\lambda_{cy} = \frac{K_y L_y}{\pi r_y}\sqrt{\frac{F_y}{E}} = \frac{1.0 \times 20 \times 12}{\pi \times 1.70}\sqrt{\frac{36}{29,000}} = 1.58 > 1.5$$

therefore, weak axis buckling controls, and ϕP_n=222 kips.

$$\frac{P_u}{\phi_c P_n} = \frac{53}{222} = 0.24 > 0.2$$

Equation 6.3 should be used for the member capacity check.

P_u=53 kips

M_{nt}=185 kip-ft

M_{1t}=48 kip-ft

Δ_{oh}=0.698 in

For L_b=20 ft, $\phi_b M_n$=285 kip-ft (page 4-130, AISC LRFD Manual[2])

(Note that C_b is assumed to be 1.0 as a conservative estimation.)

$$M_{ux} = B_1 M_{nt} + B_2 M_{lt}$$

$$P_{e1} = \frac{\pi^2 \times 29{,}000 \times 1170}{(3.58 \times 20 \times 12)^2} = 454 \text{ kips}$$

$$B_1 = \frac{C_m}{1-\left(P_u/P_{e1}\right)} = \frac{0.6}{1-\left(53/454\right)} = 0.68 < 1.0$$

Hence $B_1 = 1.0$.

From Eq. 6.9:

$$B_2 = \frac{1}{1 - \sum P_u\left(\Delta_{oh}/\Sigma HL\right)} = \frac{1}{1 - 280.8\left(0.698/4.8 \times 20 \times 12\right)} = 1.205$$

$$M_{ux} = 1.0 \times 185 + 1.205 \times 48 = 243 \text{ kip-ft}$$

Check the interaction equation from Eq. 6.3:

$$\frac{P_u}{\phi_c P_n} + \frac{8}{9}\frac{M_{ux}}{\phi_b M_{nx}} = \frac{53}{222} + \frac{8}{9}\frac{243}{285} = 0.996 < 1.0 \quad \text{OK}$$

For load case 2: $1.2D + 0.5S + 1.3W$
The required force in the column is shown in Fig. 6.21.

$P_u = 42.0$ kips

$M_{nt} = 145.0$ kip-ft

$M_{lt} = 78.0$ kip-ft

$\Delta = 1.134$ in

$\Sigma P_u = 2.46 \times 90 = 221.4$ kips

$\Sigma H = 7.8$ kips

$B_1 = 1.0$

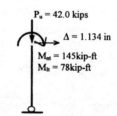

Fig. 6.21 Required forces and displacement for load case 2.

$$B_2 = \frac{1}{1 - \sum P_u\left(\Delta_{oh}/\Sigma HL\right)} = \frac{1}{1 - 221.4\left(1.134/7.8 \times 20 \times 12\right)} = 1.155$$

$$M_{ux} = 1.0 \times 145 + 1.155 \times 78 = 235 \text{ kip-ft}$$

$$\frac{P_u}{\phi_c P_n} = \frac{42}{222} = 0.188 < 0.2$$

Check the interaction equation from Eq. 6.3:

$$\frac{P_u}{2\phi_c P_n} + \frac{M_{ux}}{\phi_b M_{nx}} = \frac{42}{2 \times 222} + \frac{235}{285} = 0.92 < 1.0 \quad \text{OK}$$

W18×71 column section is adequate.

Check Sway Deflection Due to Service Wind Load, 1.0W

Allowable sway limit = $L/300$ = 20×12/300 = 0.8 in.

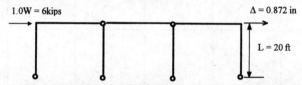

Deflection under service wind load from first-order elastic analysis is $\Delta = 0.872$ in, which is slightly greater than the allowable limit of 0.8 in. Therefore, serviceability check is not satisfactory. Larger column or beam size should be used.

6.3 Multistory Frames

6.3.1 Introduction

Multistory steel frames consist of columns and beams which are connected to form a three-dimensional structure. The frame can be stabilized by a bracing system (braced frames) or can be stabilized by itself (unbraced frames). Most building frames require bracing systems to ensure stability against lateral load. A common approach is to provide a gravity framing system with one or more lateral bracing systems attached to it. This type of framing system is called simple braced frames, which are found to be cost-effective for multistory buildings of moderate height.

For gravity frames, the beams and columns are pin-connected and the frames are not capable of resisting any lateral loads. The stability of the entire structure must be provided by attaching the gravity frames to some forms of bracing systems. The lateral loads are resisted by the bracing systems while the gravity loads are resisted by both the gravity frame and the bracing system. In most cases, the bracing system's response to lateral forces is sufficiently stiff such that second-order effects may be neglected for the design of such frames. This kind of structure may be classified as nonsway frames. Figure 6.22 shows the principal components—gravity frame and bracing system—of such a structure.

For moment-resisting frames, the beams and columns are rigidly connected to provide moment resistance at joints, and they may be used to resist lateral forces in the absence of any bracing systems. However, rigid joints are rather expensive to fabricate. In addition, it takes longer to erect a rigid frame than a simple frame.

A cost-effective framing system can be achieved by minimizing the number of moment joints, replacing field welding by field bolting, and combining the moment resisting frames with appropriate bracing systems to increase frame stiffness against sidesway.

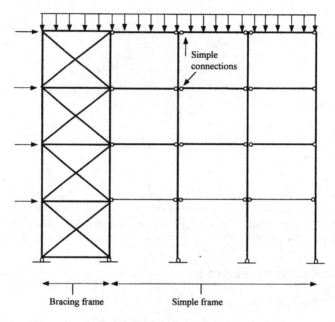

Fig. 6.22 Simple braced frames.

In the design of multistory frames, the applied loads should be transmitted to the foundation by the shortest and most direct routes. For ease of construction, the structural schemes should be simple enough, which implies repetition of member and joints, adoption of standard structural details, straightforward temporary works, and minimal requirements for interrelated erection procedure to achieve the intended behavior of the completed structure.

Sizing of structural members may be based on the longest spans and largest attributed roof and/or floor areas. The same sections should be used for similar but less onerous cases. Simple structural schemes are quick to design and easy to erect. They also provide a good "benchmark" for further refinement. Scheme drawings for multistory frame construction should include (1) general arrangement of the structure including column and beam layout, bracing frames, and floor systems, (2) critical and typical member sizes, (3) typical cladding and bracing details, (4) typical and unusual connection details, and (5) proposals for fire and corrosion protection.

This section offers advice on the general principles to be applied when preparing a structural scheme for multistory frames. The aim is to establish a structural scheme that is practicable, sensibly economic, and not unduly sensitive to the various changes that are likely to be imposed as the overall design develops. The following sections examine the design procedure and construction considerations that are specific to simple frames, braced frames, and moment-resisting frames, and demonstrate the design approaches which should be adopted in accordance with AISC LRFD.[2]

6.3.2 Gravity Frames

Gravity frames, which are sometimes referred to as *simple frames*, assume that the beam and girder connections transfer only vertical shear reactions without bending moment. Gravity frames need to be attached to a bracing system which is designed to resist lateral loads and to provide lateral stability to the part of the structure resisting gravity load (see Fig. 6.22). This section focuses on the design of gravity framing while the design of bracing system is given in Sec. 6.3.3.

General Guides. ■ The following points should be observed in the design of gravity frames:

1. Provide lateral stability to simple framing by arranging suitable braced bays or core walls deployed symmetrically in orthogonal directions, or wherever possible, to resist lateral forces.
2. Adopt a simple arrangement of slabs, beams, and columns so that loads can be transmitted to the foundations by the shortest and most direct load paths.
3. Tie all the columns effectively in orthogonal directions at every story. This may be achieved by the provision of beams or ties that are placed as close as practicable to the columns and to the edges of the slabs.
4. Select a floor construction that provides adequate lateral restraint to the beams and adequate diaphragm action to transfer the lateral load to the bracing system.

Structural Layout. ■ In simple framing, greater economy can be achieved through a repetition of similarly fabricated components. For example, a regular column grid is significantly less expensive than a nonregular grid for a given floor area; orthogonal arrangements of beams and columns, as opposed to skewed arrangements, provide the most cost-effective layout. In addition, greater economies can be achieved when the column grids in plan are rectangular, in which the secondary beams should span in the longer direction and the primary beams in the shorter, as shown in Fig. 6.23. This arrangement reduces the number

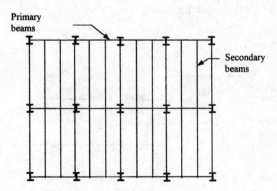

Fig. 6.23 Rectangular grid layout.

of beam-to-beam connections and the number of individual members per unit area of supported floor.

In simple frames, the beams are assumed to be simply supported between columns. The cost-effective span length will be dependent on the applied load, the type of beam system employed, and the restrictions on structural floor depth. The floor-to-floor height in a multistory building is influenced by the restrictions on overall building height and the requirements for services above and/or below the floor slab. Figure 6.24 shows the practical span ranges for these different systems when used in office building construction. Naturally, flooring systems involving the use of structural steel members that act compositely with the concrete slab achieve the longest spans.

Analysis. ■ The analysis and design of a simple braced frame must recognize the following points:

1. The members intersecting at a joint are pin-connected.
2. The columns are not subjected to any direct moment transferred through the connection. The design axial force in the column is proportional to the floor loading and its tributary areas. Live-load reduction should be applied, if possible, to calculate the required axis strength in the columns.
3. The structure is statically determinate. The internal forces and moments are therefore determined from a consideration of statics.
4. Gravity frames must be attached to a bracing system so as to provide lateral stability to the part of the structure resisting gravity load. Gravity frames may be designed as a nonsway frame, and the second-order moments associated with frame drift can be ignored.
5. The leaning effects of gravity columns must be considered in the design of the frames that participate in sidesway resistance.

Design of Beams. ■ Based on the above assumptions, beams can be designed as simply supported between the supports. The beam moments and shears are independent of beam size, and therefore the initial sizing may be

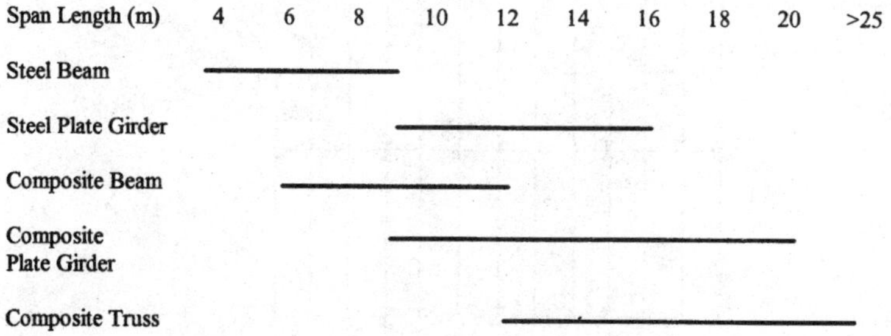

Span Length (m)	4	6	8	10	12	14	16	18	20	>25
Steel Beam	—	—	—							
Steel Plate Girder				—	—	—				
Composite Beam		—	—	—	—					
Composite Plate Girder					—	—	—	—	—	
Composite Truss							—	—	—	—

Fig. 6.24 Effective span lengths for different beam options.

obtained from the load tables in Part 4 of the AISC LRFD Manual for steel beams or Part 5 of the AISC LRFD Manual for composite beams.

Most conventional types of floor slab construction will provide adequate lateral restraint to the compression flange of the beam. Consequently, the beams may be designed as laterally restrained beams without the moment resistance being reduced by lateral-torsional buckling effects. Under the service loading (unfactored loads), the total central deflection of the beam or the deflection of the beam due to unfactored live load (with proper cambering for dead load) should satisfy the deflection limits (see Table 6.3). On some occasions, it may be necessary to check the dynamic sensitivity of the beams. When assessing the deflection and dynamic sensitivity of secondary beams, the deflection of the supporting beams must also be included. Whether it is the strength deflection or dynamic sensitivity which controls the design will depend on the span-to-depth ratio of the beam. Figure 6.24 gives typical span ranges for beams in office buildings for which these design criteria (strength, deflection, vibration) may be dominant.

Design of Gravity Columns. ■ The required axial forces in the columns can be derived from the cumulative reaction forces from those beams that frame into the columns. Column members having an influence area of 400 ft^2 or more may be designed for a reduced live load as discussed in Sec. 6.1.5. For a braced frame, the column node points are prevented from lateral translation. A conservative estimate of column effective length KL for buckling considerations is $1.0L$, where L is the story height. However, in cases where the columns above and below the story under consideration are underutilized in terms of load resistance, the restraining effects offered by these members may result in an effective length of less than $1.0L$ for the column under consideration.

Such a situation arises where the column is continuous through the restraint points and the columns above and/or below the restraint points are of different length. An example of such cases is the continuous column shown in Fig. 6.25 in which column AB is longer than column BC and hence column AB is restrained by column BC at the restraint point at B. A buckling analysis shows that the critical buckling load for the continuous column is $P_{cr} = 5.89EI/L^2$, which gives rise to an effective length factor of $K = 0.862$ for column AB and $K = 1.294$ for column BC. Column BC has a larger effective length factor because it provides restraint to column AB, whereas column AB has a smaller effective length factor because it is restrained by column BC during buckling. Figure 6.26 summarizes the reductions in effective length which may be considered for columns in a frame with different story heights having various values of a/L ratios.

Simple Connections. ■ Simple connections should be designed and detailed to prevent excessive transfer of moment between the beams and columns. Such connections should comply with the classification requirement for a "nominally pinned connection" in terms of both strength and stiffness. A computer program for connection classification has been made available in a book by Chen et al.[5], and their design applications for semirigid frames are discussed in Liew et al.[14], and Liew and Chen.[10]

Effective Length in Braced Column

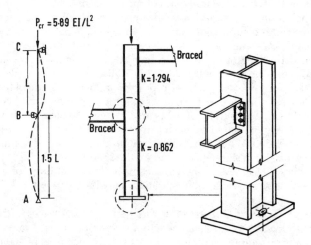

Fig. 6.25 Effective length factor of column with an intermediate restraint.

Column	Frame	a/L				
		0.2	0.4	0.6	0.8	1.0
EI		0.76	0.82	0.88	0.94	1.0
EI		0.57	0.65	0.75	0.87	1.0
EI		0.74	0.79	0.84	0.91	1.0

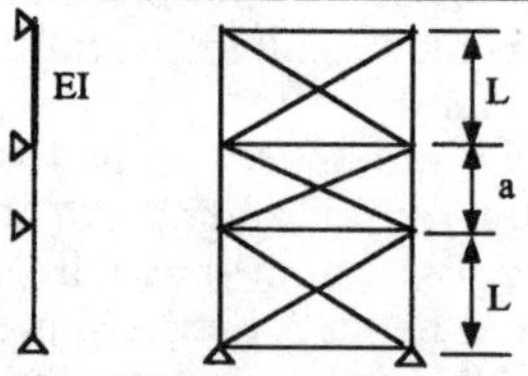

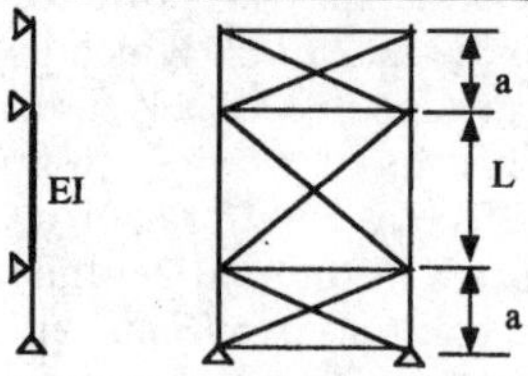

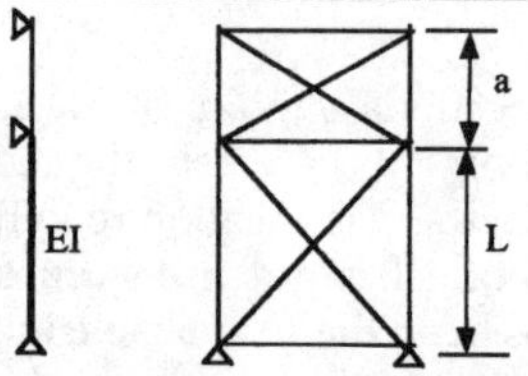

Fig. 6.26 Effective length factor of continuous braced columns.

Simple beam-to-column connections are designed to resist the vertical shear from the beam reaction. Depending on the connection details adopted, it may also be necessary to consider an additional bending moment resulting from the eccentricity of the bolt line from the supporting face. Figure 6.27 shows the typical beam-to-beam and beam-to-column connections that satisfy the assumption for pinned connections. They are (a) single or double web-angle connections, (b) header-plate connections, and (c) single-plate connections, among others. A comprehensive collection of semirigid connections including pinned connections in the form of databank and computer diskette containing the information required for connection classification has been made available in a book by Chen and Toma.[6]

Web-angle connections consist of single or double angle, either bolted or welded to both the column and the beam web. A single-plate connection uses the plate instead of angle, and the plate is welded to the column flange. It requires less

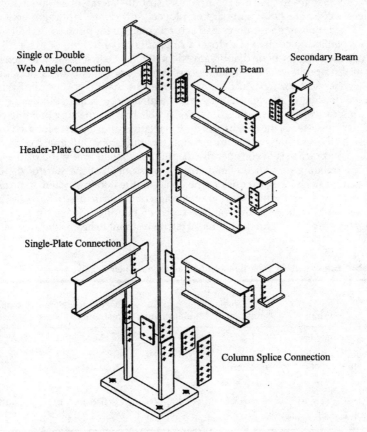

Fig. 6.27 Simple beam-to-beam and beam-to-column connections.

material than angle connection. A header-plate connection consists of an end plate, whose length is less than the beam depth, welded to the beam web and bolted to the column. A header-plate connection is used mainly to transfer the beam reaction to the column and is classified as a connection in type 3 framing in accordance with the AISC ASD Specification (AISC[1]).

Flooring System ■ All stability systems use the floor diaphragm action to transfer lateral loads from their point of application to the vertical bracing elements. The lateral shear at any story is distributed in proportion to the stiffness of the bracing systems. However, the designer should ensure that the flooring system is capable and rigid enough to transmit the lateral forces.

The adequacy of the floor to act as a diaphragm depends very much on its type. Precast concrete floor planks offer limited resistance to the racking effects of diaphragm action if they are acting independently. In such cases, supplementary bracing systems in plan are required to distribute lateral forces adequately. Where precast concrete floor units are employed, sufficient diaphragm action can be achieved by using a reinforced structural concrete topping, so that all individual floor planks are combined to form a single floor diaphragm.

The rigidity of floor diaphragms depends on their tendency to deflect under load. Composite concrete floors incorporating permanent metal decking behave like concrete flooring which provides excellent diaphragm action. By fixing the metal decking to the floor beams, an adequate floor diaphragm can be achieved during the construction stage. It is conceivable that in a narrow building consisting of bracing systems, the diaphragm floor is relatively flexible as compared to the stiffness of the bracing system. In general, for multistory building construction, the common practice is to consider the diaphragm as rigid unless the plan aspect ratio of the building exceeds 3 (Taranath[19]). It is essential, at the start of the design of structural steelworks, to consider the details of the flooring system to be used since these have a significant effect on the design of the structure. Table 6.4 summarizes the salient features of the various types of flooring systems in terms of their diaphragm actions, and the degree of lateral restraint to beams during construction.

Table 6.4 Details of Typical Flooring Systems and Their Relative Merits

Floor system	Typical span length, m	Typical depth, mm	Degree of lateral restraint to beams	Degree of diaphragm action*
In situ concrete	3–6	150–250	Very good	Very good
Steel deck with in situ concrete	2.5–3.6 unshored > 3.6	110–180	Very good	Very good
Precast concrete	3–6	100–200	Satisfactory	Satisfactory
Prestressed concrete	6–9	100–200	Good	Good

*Note: Special attention has to be given for plan aspect ratio greater than 3.

6.3.3 Bracing Systems

The main purpose of a bracing system is to provide lateral stability to the entire structure. It has to be designed to resist all possible kinds of lateral loading due to external forces, e.g., wind forces, earthquake forces, and "leaning forces" from the gravity frames. The wind or the equivalent earthquake forces on the structure, whichever is greater, should be assessed and divided into the number of bracing bays resisting the lateral forces in each direction.

Structural Forms ▪ Steel braced systems are open in a form of vertical truss which behaves like cantilever elements under lateral loads, developing tension and compression in the column chords. Shear forces are resisted by the bracing members. The truss diagonalization may take various forms, as shown in Fig. 6.28. The design of such structures must take into account the manner in which the frames are erected, the distribution of lateral forces, and their sideway resistance.

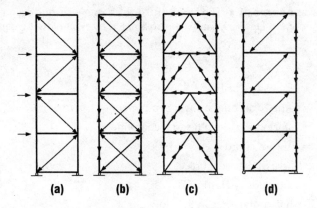

Fig. 6.28 (a) Diagonal bracing, (b) cross bracing, (c) K bracing, (d) eccentric bracing.

In a cross-braced system (Fig. 6.28b), the brace members are usually designed to resist tension only (see the example in Sec. 6.2.7). Consequently, light sections such as structural angles and channels, or tie-rods can be used to provide a very stiff overall structural response. The advantage of the cross-braced system is that the beams are not subjected to significant axial force, as the lateral forces are mostly taken up by the bracing members.

In the single-braced forms, where a single diagonal brace is used (Fig. 6.28a), it must be capable of resisting both tensile and compressive axial forces caused by the alternate wind load. In addition, gravity forces may tend to dominate the axial forces in the diagonal braces and due consideration must be given in the design of such members. AISC LRFD recommends that the slenderness ratio of the tension bracing member (L/r) must not be greater than 300 to prevent the self-weight deflection of the brace limiting its compressive resistance.

The K trusses are common since the diagonals do not participate extensively in carrying gravity load and can thus be designed for wind axial forces without gravity axial force being considered as a major contribution. K-braced frame is more efficient in preventing sidesway than cross-braced frame for equal steel areas of braced members used. This type of system is preferred for longer bay width because of the shorter length of the braces. K-braced frame is found to be more efficient if the apexes of all the braces are pointing in the upward direction (Fig. 6.28c).

For eccentrically braced frame, the centerline of the brace is positioned eccentrically to the beam-column joint, as shown in Fig. 6.28d. The system relies, in part, on flexure of the short segment of the beam between the brace-beam joist and the beam-column joint. The forces in the braces are transmitted to the column through shear and bending of the short beam segment. This particular arrangement provides a more flexible overall response. Nevertheless, it is more effective against seismic loading because it allows for energy dissipation due to flexural and shear yielding of the short beam segment.

Drift Assessment ■ The story drift Δ of a single-story diagonally braced frame, as shown in Fig. 6.29, can be approximated by the following equation:

$$\Delta = \Delta_s + \Delta_f = \frac{HL_d^3}{A_d EL^2} + \frac{Hh^3}{A_c EL^2} \qquad (6.25)$$

where Δ = interstory drift
Δ_s = story drift due to shear component
Δ_f = story drift due to flexural component
A_c = area of chord
A_d = area of diagonal brace
E = modulus of elasticity
H = horizontal force in the story
h = story height
L = length of braced bay
L_d = length of diagonal brace

The shear component of the frame drift Δ_s in Eq. 6.25 is caused mainly by the straining of the diagonal brace. The deformation associated with girder compression has been neglected in the calculation of Δ_s because the axial stiffness of the girder is very much larger than the stiffness of the brace. The elongation of the diagonal braces gives rise to shear deformation of the frame. The shear deformation is a function of the brace length L_d and the length ratio of the diagonal brace to bay width L_d/L. A shorter brace length with a smaller brace-to-bay length ratio will produce a smaller lateral drift.

The flexural component of the frame drift Δ_f is due to tension and compression of the windward and leeward columns. The extension of the windward column and shortening of the leeward column cause flexural deformation of the frame,

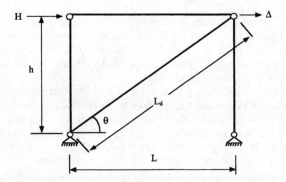

Fig. 6.29 Lateral displacement of a diagonally braced frame.

which is a function of the area of the column and the ratio of the height to bay length h/L. For a slender bracing frame with large h/L ratio, the flexural component can contribute significantly to the overall story drift. Low-rise braced frames deflect predominantly in shear mode while high-rise braced frames tend to deflect more in flexural mode. This is because the story shear H, which is cumulative from story to story becomes larger at the lower story of a high-rise building frame.

Design Considerations ▪ Frames with braces connecting columns obstruct locations of access openings such as windows and doors; thus they should be placed where such access is not required, for example, around elevators and service and stair wells. The location of the bracing systems within the structure will influence the efficiency with which the lateral forces can be resisted. The most appropriate position for the bracing systems is in the periphery of the building (Fig. 6.30a) since this arrangement provides greater torsional resistance. Bracing frames should be situated where the shear center for lateral resistance is approximately equal to the center of shear resultant on plan. Where this is not possible, torsional forces will be induced, and they must be considered when calculating the load carried by each braced system.

When core braced systems are used, they are normally located in the center of the building (Fig. 6.30b). The torsional stability is then provided by the torsional rigidity of the core brace. For tall building frames, a minimum of three braced bents are required to provide transitional and torsional stability. These bents should be carefully arranged so that their planes of action do not meet at one point. Figure 6.30c shows two examples of such practice which should be avoided. The flexibilities of different bracing systems must be taken into account in the analysis, since the stiffer braces will attract a larger share of the applied lateral load. For tall and slender frames, the bracing system itself can be a sway frame, and a second-order analysis is generally required to evaluate the required forces for ultimate strength and serviceability checks.

Lateral loads produce transverse shears, overturning moments, and sidesway. The stiffness and strength demands on the lateral system increase dramatically with height. The shear increases linearly, the overturning moment as a second power and sway as a fourth power of the height of the building. Therefore, apart from providing the strength to resist lateral shear and overturning moments, the dominant design consideration (especially for tall buildings) is to develop adequate lateral stiffness to control frame drift.

For serviceability verification, it requires that both the interstory drifts and the lateral deflections of the structure as a whole must be limited. The limits depend on the sensitivity of the structural elements to shear deformations. Recommended limits for typical multistory frames are given in Table 6.3. When considering the ultimate limit state, the bracing system must be capable of transmitting the factored lateral loads safely down to the foundations. Braced bays should be effective throughout the full height of the building. If it is essential for bracing to be

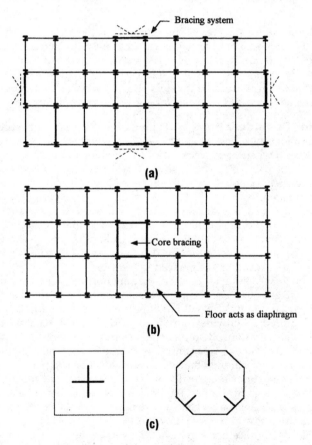

Fig. 6.30 Locations of bracing systems.

discontinuous at one level, provision must be made to transfer the forces to other braced bays. Where this is not possible, torsional forces may be induced.

The design of the internal members in a steel bracing system is similar to the design of lattice girders as discussed in Sec. 6.2.6. The horizontal member in a latticed bracing system serves also as a floor beam. This member will be subjected to bending due to gravity loads and axial compression due to wind and leaning force from adjacent gravity columns. The resistance of the element should therefore be checked as a beam-column using Eqs. 6.3 and 6.4 based on the appropriate load combinations.

6.3.4 Unbraced Rigid Frames

Structural Concept. ■ In cases where bracing systems would disturb the functioning of the building, rigidly jointed moment-resisting frames may be adopted to provide lateral stability to the framework, as illustrated in Fig. 6.31. The efficiency of development of lateral stiffness is dependent on bay span, number of bays in the frame, number of frames, and the available depth in the floors for the frame girders. Bay dimensions in the range of 6 to 9 m and structural height up to 20 to 30 stories are commonly used. For tall building frames, deeper girders are required to control drift, thus the design becomes uneconomical.

When a rigid unbraced frame is subjected to lateral load, the horizontal shear in a story is resisted predominantly by the bending of columns and beams. These deformations cause the frame to deform in a shear mode. The design of these frames is controlled therefore by the bending stiffnesses of individual members. The deeper the member, the more efficiently the bending stiffness can be developed. A small part of the frame sidesway is caused by the overturning of the entire frame, resulting in shortening and elongation of the columns at opposite sides of the frame. For unbraced rigid frames up to 20 to 30 stories, the overturning moment contributes to about 10 to 20 percent of the total sway, whereas shear racking accounts for the remaining 80 to 90 percent (Fig. 6.31*b*). However, the

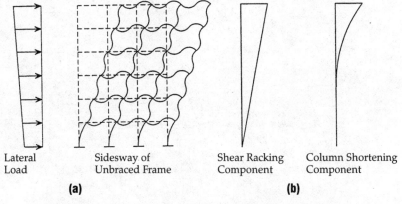

Fig. 6.31 Sidesway characteristics of rigid unbraced frame.

story drift due to overall bending tends to increase with height, while that due to shear racking tends to decrease.

Drift Assessment. ▪ Since shear racking accounts for most of the lateral sway, the design of such frames should direct toward minimizing the sidesway due to shear. Figure 6.32 shows a free-body diagram of a typical story bounded between the point of contraflexure in the column above and below the story. It is assumed that the horizontal shear forces above and below the story are the same. The shear displacement Δ_i of the ith story can be approximately computed from the following expression:

$$\Delta_i = \frac{V_i h_i^2}{12E}\left(\frac{1}{\Sigma(I_{ci}/h_i)} + \frac{1}{\Sigma(I_{bi}/L_i)}\right) \quad (6.26)$$

where Δ_i = shear deflection of ith story
E = modulus of elasticity
I_c, I_b = second moment of area for columns and beams, respectively
h_i = height of ith story
L_i = length of girder in ith story
V_i = total horizontal shear force in ith story
$\Sigma(I_{ci}/h_i)$ = sum of column stiffness in ith story
$\Sigma(I_{bi}/L_i)$ = sum of girder stiffness in ith story

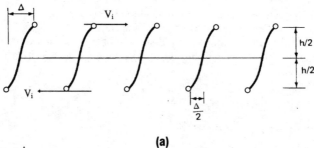

(a)

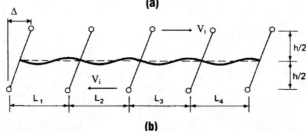

(b)

Fig. 6.32 Story drift due to (a) bending of columns, (b) bending of girders.

LRFD—Limit Design of Frames ■ 6-55

Examination of Eq. 6.26 shows that sidesway deflection caused by story shear is influenced by the sum of column and beam stiffnesses in a story, as shown in Fig. 6.32a and b, respectively. The total deflection of the frame can be obtained by the summation of the deflections for each story. For multistory frames, girder spans are generally larger than the story height; hence the moment of inertia of the girders needs to be larger to match the column stiffness, as both of these members contribute equally to the story drift. As the beam span increases, considerably deeper beam sections will be required to control frame drift.

Since the gravity forces in columns are cumulative, larger column sizes are needed on lower stories as the frame height increases. Similarly, story shear forces are cumulative, and therefore, larger beam properties in lower stories are required to control lateral drift. Because of limitations in available depth, heavier beam members will need to be provided at lower floors. This is the basic source of inefficiency in unbraced rigid frames which involves considerable premium for taller buildings.

Apart from the beam span, height-to-width ratio of the frame plays an important role in the design of such structures. Wider building frames allow a larger number of bays, i.e., larger values for story summation terms for $\Sigma(I_{ci}/h_i)$ and $\Sigma(I_{bi}/L_i)$ in Eq. 6.26 with consequent reduction in frame drift. Rigid frames with closed spaced columns which are connected by deep beams are very effective in resisting sidesway. This kind of framing system is suitable for use in the exterior planes of the building.

Rigid Joints. ■ Fully welded rigid joints are rather costly to fabricate. To minimize labor cost and to speed up site erection, field bolting instead of field welding should be used. Figure 6.33a to d shows a few types of bolted or welded moment connections currently used in practice. Beam-to-column flange connections can be shop-fabricated by welding of a beam stub to an end plate or directly to a column. The beam can then be erected by field bolting the end plate to the column flanges or splicing beams (Fig. 6.33c and d).

An additional parameter to be considered for the design of unbraced rigid frames is the modeling of connections. When an unbraced frame is subjected to lateral load, additional shear forces are induced in the column web panel as illustrated in Fig. 6.34b. The shear force is induced by the unbalanced moments from the adjoining beams, causing the joint panel to deform in shear. The deformation is attributed to the large flexibility of the unstiffened column web.

The factored shear force in the column web F along plane A-B in Fig. 6.35 may be computed as

$$F = \frac{M_{u1}}{0.95 d_{b1}} + \frac{M_{u2}}{0.95 d_{b2}} - V_u \qquad (6.27)$$

where M_{u1}, M_{u2} = factored beam moments
d_{b1}, d_{b2} = beam depths
V_u = factored column shear force

In LRFD, the design shear resistance of the panel zone ϕR_v must be greater than the factored shear force F. In such cases, no reinforcement in the web panel

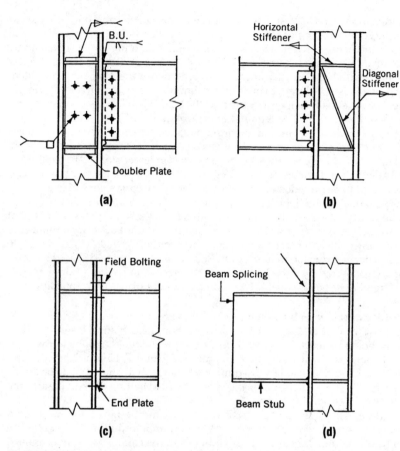

Fig. 6.33 Rigid connections: (*a*) bolted and welded connection with doubler plate, (*b*) bolted and welded connection with diagonal stiffener, (*c*) bolted end-plate connection, (*d*) beam stub welded to column.

is required. If ϕR_v is smaller than ΣF, the web needs to be reinforced by either doubler plate or diagonal stiffeners, as shown in Fig. 6.33*a* and *b*.

In evaluating the shear resistance of a panel zone, two approaches are recommended in AISC LRFD.[2] The first approach is based on the first-yield criterion for the panel web considering the reduction of shear resistance due to the presence of axial force. The second approach is based on the assumption that the panel joint can develop significant postyield shear strength beyond initial yielding. The second approach recommends that the panel-zone shear strength be increased by a factor that is related to the stiffness of the flange elements. These two approaches are discussed in the following section.

LRFD—Limit Design of Frames ■ 6-57

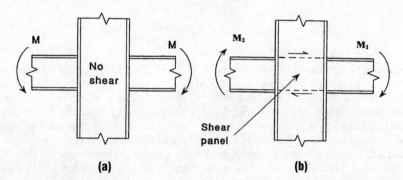

Fig. 6.34 Forces acting on a panel joint: (*a*) frame subject to gravity load, (*b*) frame subject to lateral load.

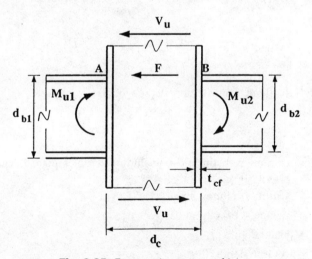

Fig. 6.35 Forces acting on a panel joint.

Design of Panel Joints ■ Using the first-yield criterion for panel web considering the reduction of shear resistance due to the presence of axial force, the shear resistance R_v of a panel zone is determined as

$$R_v = 0.6 F_y d_c t_w \qquad \text{for } P_u \leq 0.4 P_y \qquad (6.28a)$$

$$R_v = 0.6F_y d_c t_w \left(1.4 - \frac{P_u}{P_y}\right) \quad \text{for } P_u > 0.4P_y \qquad (6.28b)$$

where P_u = design axial force
P_y = squash load
d_c = column depth
t_w = total thickness of panel zone

Equations 6.28a and 6.28b should be used when the effect of panel zone deformation on frame stability is not considered in the analysis. Since most of the general structural analysis software cannot account for panel joint flexibility, Eqs. 6.28a and b thus limit the panel-zone behavior to the elastic range. Therefore, comparable results can be obtained by using a conventional method of analysis that does not account for panel-zone deformations. If adequate joint ductility is provided and the frame analysis considers inelastic panel-zone deformations, then an additional inelastic shear resistance of the panel zone may be recognized by considering the postyield shear strength of the panel zone. The shear resistance of the panel zone may be written as:

$$R_v = 0.6F_y d_c t_w (FCF) \quad \text{for } P_u \leq 0.75P_y \qquad (6.29a)$$

$$R_v = 0.6F_y d_c t_w \left(1.9 - \frac{1.2P_u}{P_y}\right)(FCF) \quad \text{for } P_u > 0.75P_y \qquad (6.29b)$$

where FCF is the flange contribution factor, which is defined as

$$FCF = \frac{1 + 3b_{cf} t_{cf}^2}{d_b d_c t_w} \qquad (6.30)$$

where t_w = column web thickness (including doubler plate thickness when required)
b_{cf} = width of column flange
t_{cf} = thickness of column flange
d_b = beam depth
d_c = column depth
F_y = yield strength of column web
P_y = $F_y A$, axial yield strength of the column
A = column cross-sectional area

If the beam-to-column panel zones are ductile and the global frame analysis considers the inelastic panel-zone deformations, then Eqs. 6.29a and b may be used to account for the additional inelastic shear resistance given by the flange contribution factor defined in Eq. 6.30. This new provision enables the design of panel-zone joints with substantially reduced doubler plate requirements. Consequently, the frame is more flexible and the inelastic panel-zone deformations must be considered in the drift calculations for rigid frames.

The nonlinear behavioral responses of a beam-column model with joint elements included can be quite different from those of a beam-column model without them. It is essential that structural analysis explicitly models the behavior of panel-zone joints for a better prediction of the postyield behavior of moment-resisting sway frames. However, it has been shown that panel-zone deformations need not be accounted for in the analysis when the ratios of girder depth to story height d_b/h and column depth to girder span d_c/L_b are less than or equal to the following limits (Liew and Chen[11]):

$$\frac{d_b}{h} \leq \frac{1}{8} \quad \frac{d_c}{L_b} \leq \frac{1}{20} \tag{6.31}$$

If the above requirements are satisfied, the conventional structural analysis based on the member centerline to centerline distance is accurate enough to compensate for the effects of elastic panel-zone deformations. Otherwise a more rigorous second-order analysis, which considers panel-zone deformations, may be required for an accurate assessment of the frame response (Liew and Chen[11]).

Analysis and Design of Unbraced Frames ▪ Rigid multistory frames are statically indeterminate; the required design forces can be determined using either:

1. Elastic analysis
2. Second-order plastic hinge analysis

While elastic methods of analysis can be used for all kind of sections, plastic analysis is applicable only where the frame members are of compact sections so as to enable the development of plastic hinges.

In a first-order elastic analysis, the additional secondary actions due to the deformation of the structure are ignored. This assumption is valid only in the following cases:

1. Where the frame is braced and not subjected to sidesway
2. Where an indirect allowance for second-order effects is made through the use of B_1 and B_2 factors

The first-order elastic analysis is a convenient approach. Most design offices possess computer software capable of performing this method of analysis on large and highly indeterminate structures. As an alternative, hand calculations can be performed on appropriate subframes within the structure (see Fig. 6.36) comprising a significantly reduced number of members. However, when conducting the analysis of an isolated subframe it is important that:

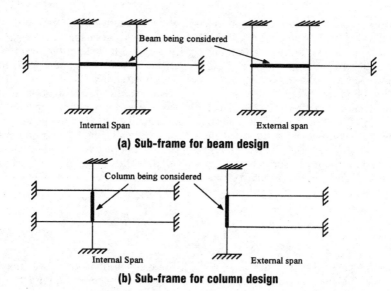

Fig. 6.36 Sub-frames for gravity load analysis.

1. The subframe is indeed representative of the structure as a whole.
2. The selected boundary conditions are appropriate.
3. Account is taken of the possible interaction effects between adjacent subframes.
4. Second-order effects are allowed for through the use of column effective length and moment amplification factors.

Plastic hinge analysis generally requires more sophisticated computer programs, which enable second-order effects to be taken into account. Computer software is now available through recent publications made available by Chen et al.[4,5] For building structures in which the required rotations are not calculated, all members containing plastic hinges should have plastic cross sections.

A basic procedure for the design of an unbraced frame is as follows:

1. Obtain approximate member size based on gravity load analysis of subframes as shown in Fig. 6.36 If sidesway deflection is likely to control (e.g., slender frames), use Eq. 6.26 to estimate the member sizes.
2. Determine wind moments from the analysis of the entire frame subjected to lateral load. A simple portal wind analysis may be used in lieu of the computer analysis.
3. Check member capacity for the combined effects of factored lateral load plus gravity loads.
4. Check beam deflection and frame drift.

5. Redesign the members and perform final analysis and design check (a second-order elastic analysis is preferable in the final stage).

Repeating the analysis to correspond to changed section sizes is unavoidable for highly redundant frames. Iteration of steps 1 to 5 gives results that will converge to the economical design satisfying the various design constraints imposed on the analysis.

6.3.5 Combined Systems

Figure 6.37 compares the sway characteristic of a 20-story frame subjected to the same lateral forces, but with different structural schemes, namely, (1) unbraced rigid

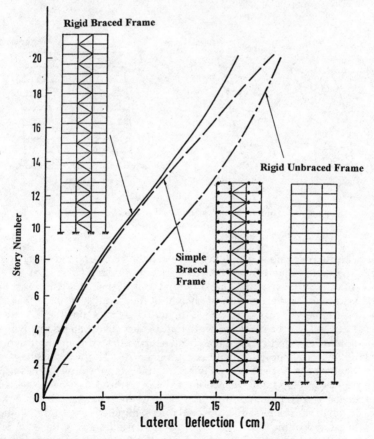

Fig. 6.37 Sway characteristics of rigid braced frame, simple braced frame, and rigid unbraced frame.

frame, (2) simple braced frame, and (3) rigid-braced frame. The simple braced frame helps to control lateral drift at the lower stories, but the overall frame drift increases toward the top of the frame. The unbraced rigid frame, on the other hand, shows an opposite characteristic for sidesway in comparison with the simple braced frame. The combination of rigid frame and bracing frame (called rigid-braced frame) provides overall improvement in reducing frame drift; the benefit becomes more pronounced toward the top of the frame. The trussed frame is restrained by the rigid frame at the upper part of the building while at the lower part, the rigid frame is restrained by the truss frame. This is because the slope of frame sway displacement is relatively smaller than that of the truss at the top while the proportion is reversed at the bottom. The interacting forces between the truss frame and rigid frame, as shown in Fig. 6.38, enhance the combined truss-frame stiffness to a level larger than the summation of individual frame and truss stiffnesses.

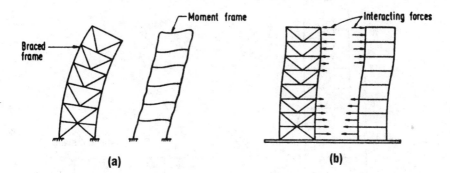

Fig. 6.38 Interaction between braced frame and moment frame.

Other forms of combined systems for tall building construction are categorized in Fig. 6.39 in accordance with their efficiency in resisting lateral load. The efficiency of a structural system is measured in terms of their ability to resist higher lateral load which increases with the height of the frame. It is noted that frames involving cantilever truss action have higher efficiencies, but the overall effectiveness depends on the height-to-width ratio. Interactive systems involving both moment frame and vertical truss are effective up to 40 stories and represent most steel building forms for tall structures. Outrigger truss and belt truss help to further enhance the lateral stiffness by engaging the exterior simple frames with the core braces to develop cantilever actions. Exterior framed tube systems with closely spaced exterior columns connected by deep girders mobilize the three-dimensional action to resist lateral and torsional forces. Bundled tubes improve the efficiency of exterior frame tubes by providing internal stiffening to the exterior tube concept. Finally, by providing diagonal braces to the exterior framework, a superframe is formed and can be used for ultra-tall megastructures.

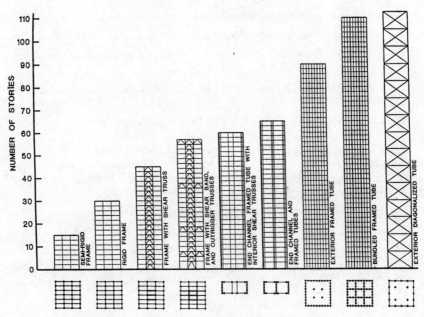

Fig. 6.39 Categorization of tall building frame systems.

6.3.6 Examples on Unbraced Frames

Two-Story Unbraced Frame. ■ A two-story frame shown in Fig. 6.40 has a story height of 3.7 m (12 ft) and bay width of 7.4 m (24 ft). The frame is load with vertical distributed load and lateral wind load applied at the story level. The most critical load combination has been found to be $1.2D + 0.5L + 1.3W$, and the factored loads are shown in Fig. 6.40. Design the frame for the wind load combination using the following assumptions:

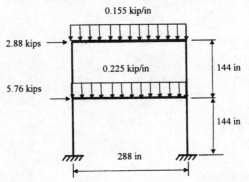

Fig. 6.40 A story subassemblage under load combination: $1.2D + 0.5L + 1.3W$.

1. The beam-to-column connections are assumed to be rigid.
2. The column size should be kept constant, and A36 steel is used for all sections.
3. The beams are continuously restrained from lateral-torsional buckling by the floor slab.
4. The columns are braced at midheight and at the story level in the out-of-plane direction.
5. The column bases are rigidly attached to the foundations.

1. Preliminary Sizing of Members. The sizes of member sections are chosen in such a manner that the width of the beam flange is less than or equal to that of the column flange. This is due to the need of detailing the beam-to-column connections.

(a) For the roof beam, assuming simple support under gravity load:

$$M_{max} = \frac{wL^2}{8} = \frac{0.155 \times 288^2}{8} = 1607 \text{ kip-in}$$

Required section modulus $Z_x = \frac{M}{F_y} = \frac{1607}{36} = 44.6 \text{ in}^3$

Try W16× 31 (Z_x=54 in³) for the roof beam.

(b) For the floor beam, assuming simple support under gravity load:

$$M_{max} = \frac{wL^2}{8} = \frac{0.225 \times 288^2}{8} = 2333 \text{ kips-in}$$

Required section modulus $Z_x = \frac{M}{F_y} = \frac{2333}{36} = 64.8 \text{ in}^3$

Try W21×44 (Z_x=95.4 in³) for the floor beam.

(c) For columns, calculate the gravity load acting on the lower column:

$P_u = (0.155 + 0.225) \times 144 = 54.7$ kips

Assume effective length factor K_x=2.0 for sway column:

$$K_x L_x = 2.0 \times \frac{144}{12} = 24 \text{ ft}$$

For weak-axis buckling, use K_y=1.0, and since the column is braced at midheight, L_y=72 in.

$$K_y L_y = 1.0 \times \frac{72}{12} = 6 \text{ ft}$$

Assume r_x/r_y=1.7.

Equivalent $\frac{K_x L_x}{r_x/r_x} = \frac{24}{1.7} = 14.1 \text{ ft}$ (Control!)

Approximated moment in the lower column is

$$M_u = \frac{2.88}{2} \times 288 + \frac{2.88}{2} \times 144 = 829 \text{ kip-in} = 69.1 \text{ kip-ft}$$

Equivalent gravity load on the column from p. 3-12 of AISC LRFD Manual[2] (see Sec. 6.1.9):

$P_{ueq} = 54.7 + 1.8(69.1) = 179$ kips

From p. 3-27 of AISC LRFD Manual:[2]

Try W10×33 with $\phi P_n = 200$ kips.

2. Analysis. From first-order elastic analysis under the wind load combination, the member forces from nontranslation and translation analysis are shown in Fig. 6.41a and b, respectively.

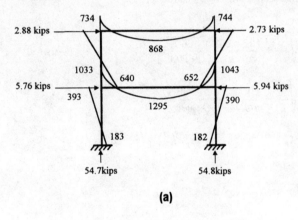

(a)

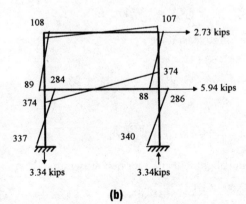

(b)

Fig. 6.41 (a) Nontranslation analysis for load combination 1.2D + 0.5L + 1.3W. (b) Translation analysis for load combination 1.2D + 0.5L + 1.3W.

3. Beam-Column Capacity Checks

For columns W10×33:

A=9.71 in^2, I_x=170 in^4, I_y=36.6 in^4, Z_x=38.8 in^3, Z_y=14.0 in^3, r_x=4.19 in, r_y= 1.94 in

$$G_T = \frac{\sum(EI/L)_c}{\sum(EI/L)_b} = \frac{170/144 + 170/144}{843/288} = 0.807$$

$$G_B = \frac{\sum(EI/L)_c}{\sum(EI/L)_b} = \infty \text{ for rigid base}$$

From alignment chart for sway frame
$K_x = 1.13$

$$\lambda_{cx} = \frac{K_x L_x}{\pi r_x}\sqrt{\frac{F_y}{E}} = \frac{1.13 \times 144}{\pi(4.19)}\sqrt{\frac{36}{29,000}} = 0.436 < 1.5 \qquad \text{(Control!)}$$

$$\lambda_{cy} = \frac{K_y L_y}{\pi r_y}\sqrt{\frac{F_y}{E}} = \frac{1.0 \times 72}{\pi(1.94)}\sqrt{\frac{36}{29,000}} = 0.416 < 1.5$$

$$P_n = 0.658^{\lambda_c^2} F_y A = 0.658^{(0.436)^2}(36)(9.71) = 323 \text{ kips}$$

$$L_b = \frac{144/2}{12} = 6 \text{ ft}$$

$L_p = 8$ ft, from p. 4-134, AISC LRFD Manual[2]

$L_b < L_p$

$M_n = M_p = Z_x F_y = 38.8 \times 36 = 1397$ kip-in

Calculate the required axial force P_u:

$P_u = 54.8 + 3.34 = 58.1$ kips

Compute the required second-order moment from

$$M_u = B_1 M_{nt} + B_2 M_{lt}$$

$$B_1 = 0.6 - 0.4\left(\frac{M_1}{M_2}\right) = 0.6 - 0.4\left(\frac{182}{390}\right) = 0.41$$

$K_x = 0.62$ from alignment chart for nonsway frame

$$P_{e1} = \frac{\pi^2 \times 29,000 \times 170}{(0.62 \times 144)^2} = 6104 \text{ kips}$$

and the axial force from nontranslation analysis is (Fig. 6.41a)

$P_u = 54.8$ kips

$B_1 = \dfrac{0.41}{1 - 54.8/6104} = 0.41 < 1.0$ use 1.0

$B_2 = \dfrac{1}{1 - \left(\Sigma P_u / \Sigma P_{e2}\right)}$

where $\Sigma P_u = 54.7 + 54.8 = 109$ kips

$$\Sigma P_{e2} = \dfrac{\pi^2 \times 29{,}000 \times 170}{(1.13 \times 144)^2} \times 2 = 3675 \text{ kips}$$

$$B_2 = \dfrac{1}{1 - (109/3675)} = 1.03$$

The amplified moment is given as

$M_u = 10 \times 390 + 1.03 \times 286.4 = 685$ kip-ft at the top of the column   (Control)

$M_u = 10 \times 182 + 1.03 \times 340 = 532$ kip-ft at the bottom of the column

Determine the appropriate interaction equation to be used:

$$\dfrac{P_u}{\phi_c P_n} = \dfrac{58.1}{0.85 \times 323} = 0.21 > 0.2$$

Use Eq. 6.3, which yields

$$\dfrac{P_u}{\phi_c P_n} + \dfrac{8}{9}\left(\dfrac{M_{ux}}{\phi_b M_{nx}} + \dfrac{M_{uy}}{\phi_b M_{ny}}\right) = 0.21 + \dfrac{8}{9}\left(\dfrac{685}{0.9 \times 397}\right) = 0.69 < 1.0 \quad \text{OK}$$

4. Check the Roof Beam

For W16×31, $I_x = 375$ in^4, $Z_x = 54$ in^3

Neglect the axial-force effect since it is small, check the beam for

$$\dfrac{M_{ux}}{\phi_b M_{nx}} \le 1.0$$

$M_{ux} = 1.0(744) + (1.03)(107) = 855$ kip-in

$M_{nx} = Z_x F_y = 54 \times 36 = 1944$ kip-in

$$\dfrac{M_{ux}}{\phi_b M_{nx}} = \dfrac{855}{0.90 \times 1944} = 0.49 \le 1.0 \quad \text{OK!}$$

5. Check the Floor Beam

For W21×44, $I_x = 843$ in^4, $Z_x = 95.4$ in^3

Neglect the axial-force effect

6-68 ■ Chapter Six

$$M_{ux} = 1.0(1043) + (1.03)(374) = 1428 \text{ kip-in}$$

$$M_{nx} = Z_x F_y = 95.4 \times 36 = 3434 \text{ kip-in}$$

$$\frac{M_{ux}}{\phi_b M_{nx}} = \frac{1428}{0.90 \times 3434} = 0.46 \leq 1.0 \quad \text{OK!}$$

6. Serviceability Checks. The overall frame drift and interstory drift under unfactored wind load should be limited to $H/300$. The beam deflection under unfactored live load should be limited to span/360.

Multistory Unbraced Frame ■ An exterior column (column AB) from an intermediate level of a three-bay multistory rigid frame is shown in Fig. 6.42. The columns are braced at the story level in the out-of-plane direction. Assuming that the same section will be used for the level above and below column AB, determine whether a W14×109 A36 column is adequate to resist the gravity load and wind load combinations. A first-order elastic analysis of the frame for gravity plus wind loads results in the forces shown in Fig. 6.43a, while the results for gravity load combination are shown in Fig. 6.43b.

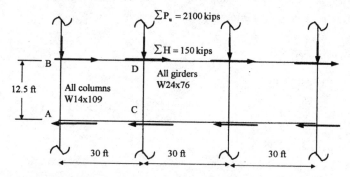

Fig. 6.42 A story subassemblage under load combination: $1.2D + 0.5L + 1.3W$.

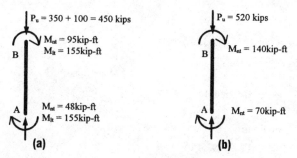

Fig. 6.43 Element and forces for column AB. (a) $1.2D + 0.5L + 1.3W$. (b) $1.2D + 1.6L$.

For beam section W24×76, I_x=2100 in⁴, and for column section W14×109, I_x=1240 in⁴.

$$G_{top} = G_{bot} = \frac{2 \times 1240/12.5}{2100/30} = 2.83$$

From the alignment chart for the unbraced frame in Fig. 6.3, the effective length factor for column AB is

K=1.80

With r_x/r_y ratio as 1.67, an equivalent effective length for the x-x axis is determined as

$$(K_x L_x)_{eq} = \frac{1.8 \times 12.5}{1.67} = 13.4 \text{ ft}$$

and for the y-y axis

$K_y L_y$=12.5 ft

Therefore, the x-x axis controls, and from the column table in the AISC Manual, for KL=13.4 ft,

$\phi_c P_n$=888 kips

Check column AB for load combination 1: $1.2D+0.5L+1.3W$.

Figure 6.43a shows the column end moments obtained from a combination of nontranslation and lateral translation analysis. These moments are

M_{nt}=95 kip-ft, M_{lt}=155.0 kip-ft

The nontranslation moments must be amplified by the B_1 factor as follows:

$$C_m = 0.6 - 0.4 \frac{48.0}{95.0} = 0.4$$

From the alignment chart in Fig. 6.3

$K_x = 0.88$ for $G_{top} = G_{bot} = 2.83$

$K_x L = 0.88 \times 12.5 = 11.0$ ft

$$P_{e1} = \frac{\pi^2 \times 29{,}000 \times 1240}{(11.0 \times 12)^2} = 20{,}369 \text{ kips}$$

and the axial force from nontranslation analysis is (Fig. 6.43a)

P_u=350 kips

$$B_1 = \frac{0.4}{1 - 350/20{,}639} = 0.41 < 1.0$$

Hence

B_1=1.0

The lateral translation moment must be amplified by the B_2 factor. Using the following expression for B_2,

$$B_2 = \frac{1}{1-\left(\Sigma P_u / \Sigma P_{e2}\right)}$$

where $\Sigma P_u = 2100$ kips

For column AB

$$P_{e2} = \frac{\pi^2 \times 29{,}000 \times 1240}{(1.8 \times 12.5 \times 12)^2} = 4868 \text{ kips}$$

For column CD, the G factors at the top and bottom of the column are

$$G_{top} = G_{bot} = \frac{2 \times 1240/12.5}{2 \times 2100/30} = 1.42$$

The corresponding effective length factor for column CD is $K=1.46$. P_{e2} for column CD is

$$P_{e2} = \frac{\pi^2 \times 29{,}000 \times 1240}{(1.42 \times 12.5 \times 12)^2} = 7823 \text{ kips}$$

The summation of story buckling loads is

$\Sigma P_{e2} = 2(4868 + 7823) = 25{,}382$ kips

Thus

$$B_2 = \frac{1}{1-\left(2100/25{,}382\right)} = 1.09$$

The amplified moment is given as

$M_u = 1.0 \times 95 + 1.09 \times 155.0 = 264$ kip-ft

The unbraced length of the compression flange for pure bending is 12.5 ft, which is less than $L_p = 15.5$ ft for the column section. Therefore, the moment capacity of the section is $\phi_b M_p = 518$ kips. Determine the appropriate interaction equation to be used:

$$\frac{P_u}{\phi_c P_n} = \frac{450}{888} = 0.507 > 0.2$$

Use Eq. 6.3, which yields

$$\frac{P_u}{\phi_c P_n} + \frac{8}{9}\left(\frac{M_{ux}}{\phi_b M_{nx}} + \frac{M_{uy}}{\phi_b M_{ny}}\right) = 0.507 + \frac{8}{9}\left(\frac{264}{518}\right) = 0.96 < 1.0 \qquad \text{OK}$$

Check the column for load combination 2: $1.2D + 1.6L$.

For $C_m = 0.4$, $P_{e1} = 15{,}774$ kips, and $P_u = 520$ kips

$B_1 = 1.0$

Since there is no lateral translation under the gravity load combination, $M_{1t}=0$, and the required moment for design is

$M_u = 1.0 \times 140 = 140$ kip-ft

Check the interaction equation:

$$\frac{P_u}{\phi_c P_n} + \frac{8}{9}\left(\frac{M_{ux}}{\phi_b M_{nx}} + \frac{M_{uy}}{\phi_b M_{ny}}\right) = \frac{520}{888} + \frac{8}{9}\left(\frac{140}{518}\right) = 0.83 < 1.0 \qquad \text{OK}$$

The W14×109 section for column AB is adequate.

6.3.7 Example: Multistory Braced Frame

The building frame shown in Fig. 6.44 is eight stories high. The frames are braced against lateral translation by two bracing frames as shown Fig. 6.44b. The interior frames are designed to resist gravity load only, whereas the bracing frames are designed to resist both gravity and lateral loads. Design the gravity frames and bracing frames using the following assumptions:

1. All beams and braces are pin-connected to the columns.
2. All gravity loads are resisted by the beams and columns.
3. Alternate wind loads are resisted by the bracing frames. Only the braces in tension are effective in resisting the wind forces.
4. Same beam, column, and brace sizes are to be used for the entire frame.
5. The partition load is considered to be part of the live load.
6. All steel sections are A36 steel.
7. All floor beams are fully braced against out-of-plane displacement.

Loading

Roof loads: Dead load=45 psf; snow load=45 psf

Floor loads: Dead load=55 psf; live load=45 psf; partition=20 psf

Wind load: Pressure=30 psf

1. Design of Columns in the Gravity Frame

Considering the load distribution in a typical floor, the tributary area for the interior column per floor is

$A_T = 35 \times 25 = 875 \text{ft}^2/\text{floor}$

The dead load acting on column $C,4$ is

$DL = 0.045 \times 875 + 0.055 \times 875 \times 7 = 376.25$ kips

The unreduced live load is

$LL = (0.05 + 0.02)(875)(7) = 428.75$ kips

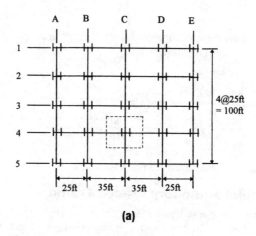

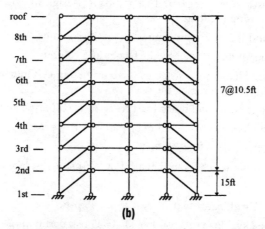

Fig. 6.44 Frame configurations for Example 6.3.7: (*a*) Floor plan, (*b*) front elevation.

The influence area for column C_14 at the lowest story is

$$A_T = 875 \times 4 \times 7 = 24{,}500 \text{ ft}^2$$

The reduced live load can be computed from Eq. 6.2 as follows:

$$L_r = L_0\left(0.25 + \frac{15}{\sqrt{A_I}}\right) = 428.75\left(0.25 + \frac{15}{\sqrt{24{,}500}}\right) = 148.28 \text{ kips}$$

The reduced live load should not be less than 40 percent of the unit live load L_0:

LRFD—Limit Design of Frames ■ 6-73

$0.4L_0 = 0.4 \times 428.75 = 171.5$ kips

Therefore, the minimum reduced live load is $L_r = 171.5$ kips.

Snow load on the roof top is

$SL = 0.03 \times 875 = 26.25$ kips

Calculate the axial force on column C,4 based on the gravity load combinations given in Table 6.1 as follows:

$1.4DL = 1.4 \times 376.25 = 527$ kips

$1.2DL + 1.6LL + 0.5SL = 1.2(376.25) + 1.6(171.5) + 0.5(26.25) = 739$ kips

$1.2DL + 1.6SL + 0.5LL = 1.2(376.25) + 1.6(26.25) + 0.5(171.5) = 579$ kips

Therefore, the required axial strength for design is $P_u = 739$ kips. The column is not subjected to any moment, because it is a leaning column. The column effective length factors about the strong and weak axis are $K_x = K_y = 1.0$. The column strength is controlled by weak-axis buckling, and the corresponding effective length is

$K_y L_y = 1.0(15) = 15$ ft

From page 3-20 of the AISC LRFD Manual[2] select W14×99 column which provides axial resistance of

$\phi P_u = 787$ kips > 739 kips

2. Design of Beams in the Gravity Frame

Influence area of the beam is

$A_T = 35 \times (25 \times 2) = 1750$ ft^2/floor

The uniformly distributed dead load on the beam is

$DL = 0.055 \times 25 = 1.375$ kips/ft

The corresponding live load can be computed as follows:

$LL = (0.05 + 0.02)(25) = 1.75$ kips/ft

The reduced live load on the beam is

$$L_r = L_0 \left(0.25 + \frac{15}{\sqrt{A_I}} \right) = 1.75 \left(0.25 + \frac{15}{\sqrt{1750}} \right) = 1.06 \text{ kips/ft}$$

The reduced live load is subjected to a minimum of

$L_r = 0.4L_0 = 0.4 \times 1.75 = 0.7$ kip/ft

Therefore, the reduced live load to be used for design is $L_r = 1.06$ kips/ft. The governing load combination is:

$1.2DL + 1.6LL = 1.2(1.375) + 1.6(1.06) = 3.346$ kips/ft

6-74 ■ Chapter Six

The maximum moment of a simply supported beam occurs at the midspan:

$$M_u = \frac{1}{8}wL^2 = \frac{1}{8}(3.346)(35)^2 = 512.4 \text{ kip-ft}$$

From page 4-18 of the AISC LRFD Manual[2] select the lightest section W24×76 which provides a flexural resistance of

$\phi M_p = 540$ kip-ft > 512.4 kip-ft

Compute the serviceability deflection based on unfactored live load (without reduction for live load) of

$w = (0.05 + 0.02) \times 25 = 1.75$ kip/ft

The midspan beam deflection is

$$\Delta_{max} = \frac{5}{384}\frac{wL^4}{EI} = \frac{5}{384}\frac{(1.75/12) \times (35 \times 12)^4}{29,000(2100)} = 0.97 \text{ in}$$

Assuming the beams carry plaster or brittle finishes, the allowable deflection is (Table 6.3)

$$\Delta_{allow} = \frac{L}{360} = \frac{35 \times 12}{360} = 1.17 \text{ in} > 0.97 \text{ in} \qquad \text{OK}$$

Hence W24×76 is satisfactory. However, the beam may need to be precambered to counter the deflection due to dead load.

3. Design of Bracing Frame (Frame 4, A-B)

Dead load at column A,4:

Roof level: $DL = 0.045(25)(25/2) = 14.1$ kips

Floor level: $DL = 0.055(25)(25/2) = 17.2$ kips/floor

Dead load at column B,4:

Roof level: $DL = 0.045(25)(25/2 + 35/2) = 33.8$ kips

Floor level: $DL = 0.055(25)(25/2 + 35/2) = 41.3$ kips/floor

Live load at column A,4:

$LL = (0.055 + 0.02)(25)(25/2) = 21.9$ kips

Influence area $A_1 = 25(25 \times 2)(7) = 8750$ ft²

Reduced live load is

$$L_r = 21.9\left(0.25 + \frac{15}{\sqrt{8750}}\right) = 8.99 \text{ kips}$$

with a minimum value of

$0.4L_0 = 0.4 \times 21.9 = 8.75$ kips

Therefore, the reduced live load is $L_r = 8.99$ kips.

Live load at column B,4:

$L = (0.05 + 0.02)(25)(25/2 + 25/2) = 52.5$ kips

Influence area $A_I = (25 \times 2)(25 + 35)(7) = 21,000$ ft^2

Reduced live load is

$$L_r = 52.5\left(0.25 + \frac{15}{\sqrt{21,000}}\right) = 18.6 \text{ kips}$$

Minimum live load is

$0.4L_0 = 0.4 \times 52.5 = 21.0$ kips

Therefore, the reduced live load is $L_r = 21.0$ kips.

Note that the live load reduction scheme is for the design of the column at the first floor. For columns above the first floor, the reduced live load may be calculated based on the respective influence area, which is different from floor to floor.

Snow load at column A,4:

$SL = 0.03(25)(25/2) = 9.375$ kips

Snow load at column B,4:

$SL = 0.03(25)(25/2 + 35/2) = 22.5$ kips

Compute the wind forces acting at the story level based on wind pressure of 30 psf.

Wind force at the second floor level is

$WL = 0.03(15/2 + 10.5/2)(25) = 9.56$ kips

Wind force at the third- to seventh-floor level is

$W = 0.03(10.5/2)(25) = 7.88$ kips

Wind force at the roof level is

$W = 0.03(10.5/2)(25) = 3.94$ kips

For load combination $1.2DL + 1.6LL + 0.5SL$, compute the gravity forces on the columns as follows:

Column A,4 at the roof level:

$1.2(14.1) + 1.6(0) + 0.5(9.375) = 21.6$ kips

Column A,4 at the typical floor level:

$1.2(17.2) + 1.6(8.99) + 0.5(0) = 35.0$ kips

Column B,4 at the roof level:

$1.2(33.8) + 1.6(0) + 0.5(22.5) = 51.8$ kips

Column $B,4$ at the typical floor level:

$1.2(41.3) + 1.6(21.0) + 0.5(0) = 83.2$ kips

For load combination $1.2DL + 1.3WL + 0.5LL$ compute the vertical and horizontal forces on the frame:

Vertical load on column $A,4$ ($1.2D + 0.5L$):

Roof level: $1.2(14.1) + 0.5(0) = 16.9$ kips

Floor level: $1.2(17.2) + 0.5(8.99) = 25.1$ kips

Vertical load at column $B,4$ ($1.2D + 0.5L$):

Second-floor level: $1.3(9.36) = 12.4$ kips

Third to eighth floor: $1.2(17.2) + 0.5(8.99) = 25.1$ kips

Factored wind force ($1.3W$) acting on the story levels:

Second-floor level: $1.3(9.56) = 12.4$ kips

Third to eighth floor: $1.3(7.88) = 10.2$ kips

Roof level: $1.3(3.94) = 5.12$ kips

Vertical load on column $B,4$ ($1.2D + 0.5L$):

Roof level: $1.2(33.8) + 0.5(0) = 40.6$ kips

Floor level: $1.2(41.3) + 0.5(21.0) = 60.1$ kips

Figure 6.45 shows the design forces acting on the bracing frame. The forces given in brackets represent the values obtained from the wind load combination.

Preliminary Sizing of Columns ■ The overturning moment at the second-floor level caused by the wind loads can be estimated as follows:

$M_T = (5.12)(88.5) + 10.5(78 + 67.5 + 57 + 46.5 + 36 + 25.5) + (12.4)(15) = 3899$ kip-ft

The overturning force acting on the opposite sides of the columns is

$$F = \frac{3899}{25} = 156 \text{ kips}$$

The critical forces acting on the lower portion of the bracing frame can be approximated by the system of forces as shown in Fig. 6.46. The forces shown within brackets represent the values obtained from the wind load combination.

The axial force acting on the column at line B under load case 1: $1.2DL + 1.6LL$ is

$P_u = 634$ kips

The axial force under load case 2: $1.2DL + 1.3WL + 0.5LL$ is

$P_u = 156 + 461 = 617$ kips

Therefore, load case 1 controls, and the required axial strength is $P_u = 634$ kips.

The column effective length is $(KL)_y = 15$ ft. From page 3-20 of the AISC LRFD Manual, select W14×90, which gives an axial resistance of

LRFD—Limit Design of Frames 6-77

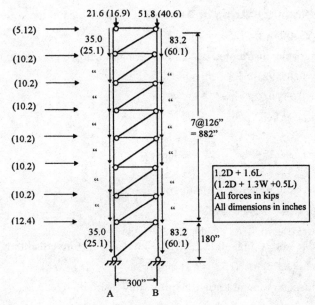

Fig. 6.45 Bracing frame subject to (1) $1.2D + 1.6L +$ and (2) $1.2D + 1.3W + 0.5L$.

$\phi P_n = 716$ kips > 634 kips   OK

Preliminary Sizing of Braces ▪ Assuming the diagonal brace resists the lateral wind force, the tensile force T, in the diagonal brace can be computed by equating its horizontal resultant force with the lateral force at the first story:

$$T\left(\frac{25}{29}\right) = 78.7$$
$$\Rightarrow T = 91 \text{ kips}$$

Compute the required area based on the above tensile force as follows:

$$A_{req} = \frac{T}{\phi F_y} = \frac{91}{0.9 \times 36} = 2.8 \text{ in}^2$$

From page 1-160 of the AISC LRFD Manual, select L4×4×7/16, which gives a gross area of

$$A = 3.31 \text{ in}^2$$

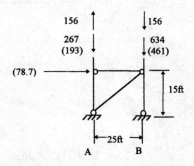

Fig. 6.46 Factored forces acting on lower portion of the bracing frame.

Preliminary Sizing of Beams ■ Compute the influence area on the beam

$A_I = 25 \times (25 \times 2) = 1250$ ft^2

The uniformly distributed dead and live loads acting on the beam are

$DL = (0.055)(25) = 1.375$ kips/ft

$LL = (0.05 + 0.02)(25) = 1.75$ kips/ft

The reduced live load is

$$L_r = 1.75\left(0.25 + \frac{15}{\sqrt{1250}}\right) = 1.18 \text{ kips/ft}$$

which is subjected to a minimum of

$0.4L_0 = 0.4 \times 1.75 = 0.7$ kip

Therefore, the reduced live load is $L_r = 1.18$ kips/ft.

Under load combination 1: $1.2DL + 1.6LL$, the factored distributed load is

$W = 1.2(1.375) + 1.6(1.18) = 3.54$ kips/ft

The midspan moment under the uniformly distributed load is

$$M_u = \frac{1}{8}wL^2 = \frac{1}{8}(3.54)(25)^2 = 276 \text{ kip-ft}$$

From page 4-19 of the AISC LRFD Manual, the lightest section is W21×50, which gives

$\phi M_p = 297$ kip-ft

Under load combination 2: $1.2DL + 1.3WL + 0.5LL$, the forces acting on the beam are shown below:

The midspan moment under the uniformly distributed load is

$$M_u = \frac{1}{8}wL^2 = \frac{1}{8}(2.24)(25)^2 = 175 \text{ kip-ft}$$

Since the beam is fully braced against out-of-plane buckling, compute the effective length factor for in-plane bending:

$(KL)_x = 1.0 \times 25 = 25$ ft

From page 3-12 of AISC LRFD Manual, $m = 1.3$, and the equivalent axial strength is

$P_{u,\,eq} = P_u + mM_u = 78.7 + 1.3(175) = 306$ kips

Select W21×50, which satisfies this load combination.

LRFD—Limit Design of Frames ■ 6-79

Summary of Trial Sections for Bracing Frame:

Columns: W14×90
Beams: W21×50
Braces: L4×4×7/16

Structural Analysis ■ An elastic first-order analysis is carried out for the two load combinations, and the internal forces at the lower story of the bracing frame are shown in Fig. 6.47a and b.

Checking of Columns: W14×90
From page 3-20 of the AISC LRFD Manual, $\phi P_n = 716$ kips

From p. 4-126 of the AISC LRFD Manual, $L_p = 15.4$ ft $> L_b = 15$ ft; hence

$\phi_b M_n = \phi_b M_p = 5087$ kip-in

For load combination: $1.2DL + 1.6LL + 0.5SL$
The required axial strength is $P_u = 634.2$ kips

$$C_m = 0.6 - 0.4 \frac{M_1}{M_2} = 0.6 - 0.4(0) = 0.6$$

$$P_{e1} = \frac{\pi^2 EI}{(KL)^2} = \frac{\pi^2 (29,000)(999)}{(1.0 \times 15 \times 12)^2} = 8825 \text{ kips}$$

$$B_1 = \frac{C_m}{1 - P_n / P_{e1}} = \frac{0.6}{1 - 634.2 / 8825} = 0.65 < 1.0$$

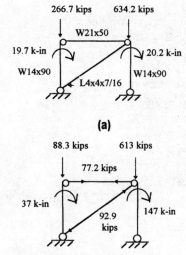

Fig. 6.47 (a) First-order elastic analysis under $1.2D + 1.6L + 0.5W$. (b) First-order elastic analysis under $1.2D + 1.3L + 0.5W$.

Use $B_1=1.0$.
   The required flexural strength is

   $M_u=1.0\times 20.2=20.2$ kip-in

Checking the beam-column interaction equation:

$$\frac{P_u}{\phi P_n}=\frac{634.2}{716}=0.89>0.2$$

$$\frac{P_u}{\phi_c P_n}+\frac{8}{9}\frac{M_u}{\phi_b M_{nx}}=0.89+\frac{8}{9}\frac{20.2}{5087}=0.890<1.0 \quad OK$$

For load combination: $1.2D+1.3W+0.5L$.
   The required axial load strength is $P_u=613$ kips. Compute the various terms for use in the beam-column capacity checks:

$C_m=0.6$

$P_{e1}=8825$ kips

$$B_1=\frac{C_m}{1-P_n/P_{e1}}=\frac{0.6}{1-613/8825}=0.64<1.0$$

Use $B_1=1.0$.
   The required flexural strength is

   $M_u=1.0\times 147=147$ kip-in

Checking the beam-column interaction equation

$$\frac{P_u}{\phi_c P_n}+\frac{8}{9}\frac{M_u}{\phi_b M_{nx}}=\frac{613}{716}+\frac{8}{9}\frac{147}{5087}=0.88<1.0 \quad OK$$

Column size W14×90 is satisfactory.

Checking of Beams: W21×50
For load combination: $1.2D+1.6L$

$$\frac{M_u}{\phi_b M_p}=\frac{276}{297}=0.93<1.0 \quad OK!$$

For load combination: $1.2D+1.3W+0.5L$
   The required axial load strength and flexural strength are

   $P_u=77.2$ kips

   $M_u=175$ kip-ft

For strong-axis bending, the axial and flexural resistances are

   $\phi P_n=419.1$ kips

   $\phi_b M_n=\phi_b M_p=3564$ kip-in

Checking the beam-column interaction equation

$$\frac{P_u}{\phi P_n} = \frac{77.2}{419.1} = 0.184 < 0.2$$

$$\frac{P_u}{2\phi_c P_n} + \frac{M_u}{\phi_b M_{nx}} = \frac{0.184}{2} + \frac{175 \times 12}{3564} = 0.681 < 1.0 \qquad \text{OK}$$

Hence W21×50 is satisfactory.

Checking of Braces: L4×4×7/16. The controlling load combination is $1.2D+1.3W+0.5L$, and the required tensile force in the brace is $T_u=92.9$ kips. The tensile force resistance of the brace is computed as

$\phi T_n = 0.9(3.31)(36) = 107$ kips

Checking the axial strength of the brace

$$\frac{T_u}{\phi T_n} = \frac{92.9}{107} = 0.87 < 1.0 \qquad \text{OK}$$

Hence L4×4×7/16 is satisfactory.

Serviceability Deflections ■ The midspan deflection of the beam is computed based on the unfactored live load:

$w = (0.05 + 0.02)(25) = 1.75$ kips/ft

The midspan deflection is

$$\Delta_{max} = \frac{5}{384} \frac{wL^4}{EI} = \frac{5}{384} \frac{(1.75/12) \times (25 \times 12)^4}{29,000(984)} = 0.54 \text{ in}$$

The deflection should not exceed the following limit:

$$\Delta_{allow} = \frac{L}{360} = \frac{25 \times 12}{360} = 0.833 \text{ in} > 0.54 \text{ in} \qquad \text{OK}$$

Lateral frame drift due to unfactored wind load obtained from computer analysis is

$\Delta_{max} = 2.0$ in

The allowable frame drift is

$$\Delta_{allow} = \frac{H}{400} = \frac{88.5 \times 12}{400} = 2.66 \text{ in} > 2.0 \text{ in} \qquad \text{OK}$$

4. Summary of Design. Figure 6.48 shows the member sizes in the braced frame. The same floor beam size can be repeated for all stories except the roof level, in which a smaller beam size can be selected. Smaller column and brace sizes may be selected for the upper stories. Members at the upper stories can be determined by the same design procedure as described in this example.

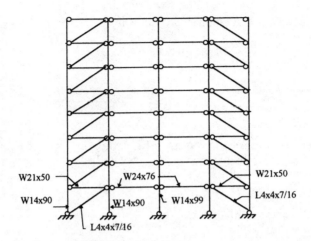

Fig. 6.48 Final member sizes of braced frame.

References

1. AISC, Allowable Stress Design and Plastic Design Specifications for Structural Steel Buildings, 9th ed., American Institute of Steel Construction, Chicago, Ill., 1989.

2. AISC, Load and Resistance Factor Design Specification for Structural Steel Buildings, American Institute of Steel Construction, 2d ed., Chicago, Ill., 1993.

3. ASCE 7-88, *Minimum Design Loads for Buildings and Other Structures*, ASCE Standard, American Society of Civil Engineers, 1990.

4. Chen, W. F. and E. M. Lui, *Stability Design of Steel Frames*, CRC Press, Boca Raton, Fl., 1991.

5. Chen, W. F., Y. Goto, and J.Y.R. Liew, *Stability Design of Semi-Rigid Frames*, Wiley-Interscience, New York, 1996.

6. Chen, W. F. and S. Toma, *Advanced Analysis of Steel Frames*, CRC Press, Boca Raton, Fl., 1994.

7. Fisher, J. M., and M. A. West, *Serviceability Design Considerations for Low-Rise Buildings*, American Institute of Steel Construction, Chicago, Ill., 1990.

8. Geschwindner, L. F., R. O. Disque, and R. Bjorhovde, *Load and Resistance Factored Design of Steel Structures*, Prentice-Hall, Englewood Cliffs, N.J., 1994.

9. Iyengar, S. H., W. F. Baker, and R. Sinn, "Multi-Story Buildings," in *Constructional Steel Design, An International Guide*, chap. 6.2, P.J. Dowling et al.(eds.), Elsevier, England, 1992, pp. 645–670.

10. Liew, J. Y. R., and W. F. Chen, "Implications of Using Refined Plastic Hinge Analysis for Load and Resistance Factor Design," *Journal of Thin-Walled Structures*, Elsevier, London, UK, vol. 20, nos. 1–4, pp. 17–47, 1994.

11. Liew, J. Y. R., and W. F. Chen, "Analysis and Design of Steel Frames Considering Panel Joint Deformations," *Journal of Structural Engineering, ASCE*, vol. 121, no. 10, pp. 1531–1540, October 1995.

12. Liew, J. Y. R., D. W. White, and W. F. Chen, "Beam-Column Design in Steel Frameworks—Insight on Current Methods and Trends," *Journal of Constructional Steel Research*, vol. 18, pp. 259–308, 1991.

13. Liew, J. Y. R, D. W. White, and W. F. Chen, "Beam-Columns," in *Constructional Steel Design, An International Guide*, chap. 5.1, P. J. Dowling et al. (eds.), Elsevier, England, 1992, pp. 105–132.

14. Liew J. Y. R, D. W. White, and W. F. Chen, "Limit-States Design of Semi-Rigid Frames Using Advanced Analysis: Part 1: Connection Modeling and Classification, Part II: Analysis and Design," *Journal of Constructional Steel Research*, Elsevier, London, vol. 26, no. 1, 1993, pp. 1–57.

15. Liew, J. Y. R., D. W. White, and W. F. Chen, "Second-Order Refined Plastic Hinge Analysis for Frame Design: Parts 1 and 2," *Journal of Structural Engineering, ASCE*, vol. 119, no. 11, 3196–3237, November 1993.

16. Liew, J. Y. R., D. W. White, and W. F. Chen, "Notional Load Plastic Hinge Method for Frame Design," *Journal of Structural Engineering, ASCE*, vol. 120, no. 5, pp. 1434–1454, May 1994.

17. Owens, G. W. and P. R. Knowles, *Steel Designers' Manual*, 5th ed., Black Scientific Publications, London, 1992.

18. SCI, "Plastic Design of Single-Story Pitched-Roof Portal Frames to Eurocode 3," *Technical Report*, SCI Publication 147, The Steel Construction Institute, London, 1995.

19. Taranath, B. S., *Structural Analysis and Design of Tall Buildings*, McGraw-Hill, N.Y., London, 1988.

7

William A. Thornton
Thomas Kane

CONNECTIONS

7.1 Introduction

In this chapter, the term connection is used in a general sense to include all types of joints in structural steel made with fasteners or welds. Emphasis, however, is placed on connections, such as beam column connections, main-member splices, vertical bracing connections, hanger-type connections, moment connections, and truss connections. Connections are an intimate part of a steel structure, and their proper treatment is essential for a safe and economic structure. An intuitive knowledge of how a system will transmit loads (the art of load paths) and an understanding of structural mechanics (the science of equilibrium and limit states) are necessary to achieve connections which are both safe and economic.

This material is based on the specifications of the American Institute of Steel Construction (AISC), Load and Resistance Factor Design Specification for Structural Steel Buildings, 1993. Standard specifications for structural steel for buildings contain general criteria governing the design of bolted connections. They cover such essentials as permissible fastener size, sizes of holes, arrangements of fasteners, size and attachment of fillers, and installation methods. Structural steel fabricators prefer that job specifications state that "shop connections shall be made with bolts or welds" rather than restricting the type of connection that can be used. This allows the fabricator to make the best use of available equipment and to offer a more competitive price.

7.2 Bolted Connections

7.2.1 Types of Bolts

For general purposes, A325 and A490 high-strength bolts may be specified. Each type of bolt can be identified by the ASTM designation and the manufacturer's mark on the bolt head and nut (Fig. 7.1). The cost of A490 bolts is 15 to 20 percent greater than that of A325 bolts.

Carbon-steel bolts should not be used in connections subject to fatigue. In building construction, snug-tight bearing-type connections can be used for most cases, including connections subject to stress reversal due to wind or seismic loading. The American Institute of Steel Construction (AISC) requires that fully tensioned high-strength bolts or welds be used for the following joints.

Column splices in multistory framing, if it is more than 200 ft high, or when it is between 100 and 200 ft high and the smaller horizontal dimension of the framing is less than 40 percent of the height, or when it is less than 100 ft high and the smaller horizontal dimension is less than 25 percent of the height.

Connections, in framing more than 125 ft high, on which column bracing is dependent and connections of all beams or girders to columns.

Crane supports, as well as roof-truss splices, truss-to-column joints, column splices and bracing, and crane supports, in framing supporting cranes with capacity exceeding 5 tons.

Connections for supports for impact loads, loads causing stress reversal, or running machinery.

The height of framing should be measured from curb level (or mean level of adjoining land when there is no curb) to the highest point of roof beams for flat roofs or to mean height of gable for roofs with a rise of more than $2\frac{2}{3}$ in 12. Penthouses may be excluded.

The AISC imposes special requirements on use of welded splices and similar connections in heavy sections. This includes ASTM A6 group 4 and 5 shapes and splices in built-up members with plates over 2 in thick subject to tensile stresses due to tension or flexure. Charpy V-notch tests are required, as well as special fabrication and inspection procedures. Where feasible, bolted connections are preferred to welded connections.

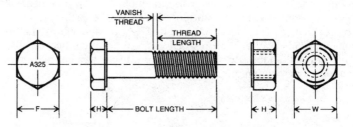

Fig. 7.1 High-strength structural-steel bolt and nut.

Job specifications often require that "main connections shall be made with bolts conforming to the Specification for Structural Joints Using ASTM A325 and A490 Bolts." This specification, approved by the Research Council on Structural Connections (RCSC) of the Engineering Foundation, establishes bolt, nut, and washer dimensions, minimum fastener tension, and requirements for design and installation.

As indicated in Table 7.1, many sizes of high-strength bolts are available. Most standard connection tables, however, apply primarily to $3/4$- and $7/8$-in-diameter bolts. Shop and erection equipment is generally set up for these sizes, and workers are familiar with them.

"Secondary connections may be made with unfinished bolts conforming to the Specifications for Low Carbon Steel ASTM A307" is an often-used specification. (Unfinished bolts also may be referred to as *machine*, *common*, or *ordinary bolts*.) When this specification is used, secondary connections should be carefully defined to preclude selection by ironworkers of the wrong type of bolt for a connection. A307 bolts generally have no identification marks on their square, hexagonal, or countersunk heads (Fig. 7.2), as do high-strength bolts. Use of high-strength bolts where A307 bolts provide the required strength merely adds to the cost of a structure. High-strength bolts cost at least 10 percent more than machine bolts.

A disadvantage of A307 bolts is the possibility that the nuts may loosen. This may be eliminated by use of lock washers. Alternatively, locknuts can be used or threads can be jammed, but either is more expensive than lock washers.

7.2.2 Fully Tensioned and Snug-Tight Bolt

Calibrated-Wrench Method ■ When a calibrated wrench is used, it must be set to cut off tightening when the required tension has been exceeded by 5 percent. The wrench should be tested periodically (at least daily on a minimum of three bolts of each diameter being used). For the purpose, a calibrating device that gives the bolt tension directly should be used. In particular, the wrench should be calibrated when bolt size or length of air hose is changed. When bolts are tightened, bolts previously tensioned may become loose because of compression of the connected parts. The calibrated wrench should be reapplied to bolts previously tightened to ensure that all bolts are tensioned to the prescribed values.

Table 7.1 Thread Lengths for High-Strength Bolts

Bolt diameter, in	Nominal thread, in	Vanish thread, in	Total thread, in
$1/2$	1.00	0.19	1.19
$5/8$	1.25	0.22	1.47
$3/4$	1.38	0.25	1.63
$7/8$	1.50	0.28	1.78
1	1.75	0.31	2.06
$1 1/8$	2.00	0.34	2.34
$1 1/4$	2.00	0.38	2.38
$1 3/8$	2.25	0.44	2.69
$1 1/2$	2.25	0.44	2.69

7-4 ■ Chapter Seven

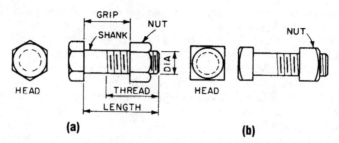

Fig. 7.2 Unfinished (machine) or common bolts.

Turn-of-the-Nut Method ■ When the turn-of-the-nut method is used, tightening may be done by impact or hand wrench. This method involves three steps:

1. *Fit-up of connection.* Enough bolts are tightened a sufficient amount to bring contact surfaces together. This can be done with fit-up bolts, but it is more economical to use some of the final high-strength bolts.

2. *Snug tightening of bolts.* All high-strength bolts are inserted and made snug-tight (tightness obtained with a few impacts of an impact wrench or the full effort of a person using an ordinary spud wrench). While the definition of snug-tight is rather indefinite, the condition can be observed or learned with a tension-testing device.

3. *Nut rotation from snug-tight position.* All bolts are tightened by the amount of nut rotation specified in Table 7.2. If required by bolt-entering and wrench-operation clearances, tightening, including by the calibrated-wrench method, may be done by turning the bolt while the nut is prevented from rotating.

Table 7.2 Number of Nut or Bolt Turns from Snug-Tight Condition for High-Strength Bolts*

	Slope of outer faces to bolted parts		
Bolt length (Fig. 7.1)	Both faces normal to bolt axis	One face normal to bolt axis and the other sloped[†]	Bolt faces sloped[†]
Up to 4 diameters	1/3	1/2	2/3
Over 4 diameters but not more than 8 diameters	1/2	2/3	5/6
Over 5 diameters but not more than 12 diameters[‡]	2/3	5/6	1

*Nut rotation is relative to the bolt regardless of whether the nut or bolt is turned. For bolts installed by 1/2 turn and less, the tolerance should be ± 30°. For bolts installed by 2/3 turn and more, the tolerance should be ± 45°. This table is applicable only to connections in which all material within the grip of the bolt is steel.
[†]Slope is not more than 1:20 from the normal to the bolt axis, and a beveled washer is not used.
[‡]No research has been performed by RCSC to establish the turn-of-the-nut procedure for bolt lengths exceeding 12 diameters. Therefore, the required rotation should be determined by actual test in a suitable tension-measuring device that simulates conditions of solidly fitted steel.

Direct-Tension-Indicator Tightening ■ Two types of direct-tension-indicator devices are available: washers and twist-off bolts. The hardened-steel load-indicator washer has dimples on the surface of one face of the washer. When the bolt is torqued, the dimples depress to the manufacturer's specification requirements, and proper torque can be measured by the use of a feeler gage. Special attention should be given to proper installation of flat hardened washers when load-indicating washers are used with bolts installed in oversize or slotted holes and when the load-indicating washers are used under the turned element.

The twist-off bolt is a bolt with an extension to the actual length of the bolt. This extension will twist off when torqued to the required tension by a special torque gun. A representative sample of at least three bolts and nuts for each diameter and grade of fastener should be tested in a calibration device to demonstrate that the device can be torqued to 5 percent greater tension than that required.

When the direct-tension indicator involves an irreversible mechanism such as yielding or fracture of an element, bolts should be installed in all holes and brought to the snug-tight condition. All fasteners should then be tightened, progressing systematically from the most rigid part of the connection to the free edges in a manner that will minimize relaxation of previously tightened fasteners prior to final twist-off or yielding of the control or indicator element of the individual devices. In some cases, proper tensioning of the bolts may require more than a single cycle of systematic tightening.

Washer Requirements ■ The RCSC specification requires that design details provide for washers in connections with high-strength bolts as follows:

1. A hardened beveled washer should be used to compensate for the lack of parallelism where the outer face of the bolted parts has a greater slope than 1:20 with respect to a plane normal to the bolt axis.

2. For A325 and A490 bolts for slip-critical connections and connections subject to direct tension, hardened washers are required as specified in items 3 through 7 below. For bolts permitted to be tightened only snug-tight, if a slotted hole occurs in an outer ply, a flat hardened washer or common plate washer should be installed over the slot. For other connections with A325 and A490 bolts, hardened washers are not generally required.

3. When the calibrated wrench method is used for tightening the bolts, hardened washers should be used under the element turned by the wrench.

4. For A490 bolts tensioned to the specified tension, hardened washers are used under the head and nut in steel with a specified yield point less than 40 ksi.

5. A hardened washer conforming to ASTM F436 is used for A325 or A490 bolts 1 in or less in diameter tightened in an oversized or short slotted hole in an outer ply.

6. Hardened washers conforming to F436 but at least $5/16$ in thick are used, instead of washers of standard thickness, under both the head and nut of A490 bolts more than 1 in in diameter tightened in oversized or short slotted holes in an outer ply. This requirement is not met by multiple washers even though the combined thickness equals or exceeds $5/16$ in.

7. A plate washer or continuous bar of structural-grade steel, but not necessarily hardened, at least $5/16$ in thick and with standard holes, is used for an A325 or A490 bolt 1 in or less in diameter when it is tightened in a long slotted hole in an outer ply. The washer or bar should be large enough to cover the slot completely after installation of the tightened bolt. For an A490 bolt more than 1 in in diameter in a long slotted hole in an outer ply, a single hardened washer (not multiple washers) conforming to F436 but at least $5/16$ in thick, is used instead of a washer or bar of structural-grade steel.

The requirements for washers specified in items 4 and 5 above are satisfied by other types of fasteners meeting the requirements of A325 or A490 and with a geometry that provides a bearing circle on the head or nut with a diameter at least equal to that of hardened F436 washers. Such fasteners include "twist-off" bolts with a splined end that extends beyond the threaded portion of the bolt. During installation, this end is gripped by a special wrench chuck and is sheared off when the specified bolt tension is achieved.

The RCSC specification permits direct tension-indicating devices, such as washers incorporating small, formed arches designed to deform in a controlled manner when subjected to the tightening force. The specification also provides guidance on use of such devices to assure proper installation.

7.2.3 Bearing Type and Slip-Resistant Connections

High-strength bolts may be used in either slip-critical or bearing-type connections subject to various limitations. Bearing-type connections have higher allowable loads and should be used where permitted. Also, bearing-type connections may be either fully tensioned or snug-tight, subject to various limitations. Snug-tight bolts are much more economical to install and should be used where permitted.

Bearing versus Slip-Critical Joints ■ Connections made with high-strength bolts may be slip-critical (material joined being clamped together by the tension induced in the bolts by tightening them) or bearing-type (material joined being restricted from moving primarily by the bolt shank). In bearing-type connections, bolt threads may be included in or excluded from the shear plane. Different stresses are allowed for each condition. The slip-critical connection is the most expensive, because it requires that the faying surfaces be free of paint, grease, and oil. Hence this type of connection should be used only where required by the governing design specification, e.g., where it is undesirable to have the bolts slip into bearing or where stress reversal could cause slippage (Sec. 7.2.1). Slip-critical connections, however, have the advantage in building construction that when used in combination with welds, the fasteners and welds may be considered to share the stress (Sec. 7.2.4).

Threads in Shear Planes ■ The bearing-type connection with threads in shear planes is frequently used. Since location of threads is not restricted, bolts can be inserted from either side of a connection. Either the head or the nut can be the element turned. Paint is permitted on the faying surfaces.

The bearing-type connection with threads excluded from shear planes is the most economical high-strength bolted connection, because fewer bolts generally are needed for a given capacity. But this type should be used only after careful consideration of the difficulties involved in excluding the threads from the shear planes. The location of the thread runout depends on which side of the connection the bolt is entered and whether a washer is placed under the head or the nut. This location is difficult to control in the shop but even more so in the field. The difficulty is increased by the fact that much of the published information on bolt characteristics does not agree with the basic specification used by bolt manufacturers (American National Standards Institute B 18.2.1).

Total nominal thread lengths and vanish thread lengths for high-strength bolts are given in Table 7.1. It is common practice to allow the last $1/8$ in of vanish thread to extend across a single shear plane.

In order to determine the required bolt length, the value shown in Table 7.3 should be added to the grip (i.e., the total thickness of all connected material, exclusive of washers). For each hardened flat washer that is used, add $5/32$ in and for each beveled washer, add $5/16$ in. The tabulated values provide appropriate allowances for manufacturing tolerances and also provide for full thread engagement with an installed heavy hex nut. The length determined by the use of Table 7.3 should be adjusted to the next longer $1/4$-in length.

7.2.4 Bolts in Combination with Welds

In new work, ASTM A307 bolts or high-strength bolts used in bearing-type connections should not be considered as sharing the stress in combination with welds. Welds, if used, should be provided to carry the entire stress in the connection. High-strength bolts proportioned for slip-critical connections may be considered as sharing the stress with welds.

In welded alterations to structures, existing rivets and high-strength bolts tightened to the requirements for slip-critical connections are permitted for carrying stresses resulting from loads present at the time of alteration. The welding needs to be adequate to carry only the additional stress.

If two or more of the general types of welds (groove, fillet, plug, slot) are combined in a single joint, the effective capacity of each should be separately computed with reference to the axis of the group in order to determine the allowable capacity of the combination.

Table 7.3 Lengths to Be Added to Grip

Nominal bolt size, in	Addition to grip for determination of bolt length, in
$1/2$	$11/16$
$5/8$	$7/8$
$3/4$	1
$7/8$	$1 1/8$
1	$1 1/4$
$1 1/8$	$1 1/2$
$1 1/4$	$1 5/8$
$1 3/8$	$1 3/4$
$1 1/2$	$1 7/8$

7.2.5 Standard, Short Slotted, and Long Slotted Holes for Bolts

In general, a connection with a few large-diameter fasteners costs less than one of the same capacity with many

7-8 ■ Chapter Seven

small-diameter fasteners. The fewer the fasteners the fewer the number of holes to be formed and the less installation work. Larger-diameter fasteners are particularly favorable in connections where shear governs, because the load capacity of a fastener in shear varies with the square of the fastener diameter. For practical reasons, however, $3/4$- and $7/8$-in-diameter fasteners are preferred.

Standard specifications require that holes for bolts be $1/16$ in larger than the nominal fastener diameter. In computing the net area of a tension member, the diameter of the hole should be taken $1/16$ in larger than the hole diameter.

Standard specifications also require that the holes be punched or drilled. Punching usually is the most economical method. To prevent excessive damage to material around the hole, however, the specifications limit the maximum thickness of material in which holes may be punched full size. These limits are summarized in Table 7.4.

In *buildings*, holes for thicker material may be either drilled from the solid or subpunched and reamed. The die for all subpunched holes and the drill for all subdrilled holes should be at least $1/16$ in smaller than the nominal fastener diameter.

Oversize holes can be used in slip-critical connections, and the oversize hole can be in all the plies connected. The oversize holes are $3/16$ in larger than the bolt diameter for bolts $5/8$- to $7/8$-in-diameter. For bolts 1 in in diameter, the oversize hole is $1/4$ in larger and for bolts $1 1/8$ in in diameter and greater, the oversize hole will be $5/16$ in larger.

Short slotted holes can be used in any or all of the connected plies. The load has to be applied 80 to 100° normal to the axis of the slot, and the design strength is greater than the factored nominal load. Short slots can be used without regard to the direction of the applied load when the design slip resistance is greater than the factored nominal load. The short slots for $5/8$ to $7/8$ in in diameter bolts are $1/16$ in larger in width and $1/4$ in larger in length than the bolt diameter. For bolts 1 in in diameter, the width is $1/16$ in larger and the length $5/16$ in larger and for bolts $1 1/8$ in in diameter and greater, the slot will be $1/16$ in larger in width and $3/8$ in larger in length.

Long slots have the same requirement as the short slotted holes, except that the long slot has to be in one of the connected parts at the faying surface of the connection. The width of all long slots for bolts is $1/16$ in greater than the bolt diameter, and the length of the long slots is $5/8$-in-diameter bolt $15/16$ in greater, $3/4$-in-diameter bolts $1 1/8$ greater, $7/8$-in-diameter bolt $1 5/16$ greater, 1-in-diameter bolt $1 1/2$ greater, and $1 1/8$ in diameter and larger $2 1/2$ × diameter of bolt. When finger shims are fully inserted between the faying surfaces of load-transmitting parts of the connections, this will not be considered as a long slot

7.2.6 Edge Distances and Spacing of Bolts

Minimum distances from centers of fasteners to any edges are given in Table 7.5. The AISC specifications for

Table 7.4 Maximum Material Thickness (in) for Punching Fastener Holes*

Type of steel	AISC
A36 steel	$d+1/8$[†]
High-strength steels	$d+1/8$[†]
Quenched and tempered steels	$1/2$[‡]

* Unless subpunching or subdrilling and reaming are used.
[†] d = fastener diameter, in.
[‡] A514 steel.

Table 7.5 Minimum Edge Distances (in) for Fastener Holes in Steel Buildings

Fastener diameter, in	At sheared edges	At rolled edges of plates, shapes, or bars or gas-cut edges*
$1/2$	$7/8$	$3/4$
$5/8$	$1 1/8$	$7/8$
$3/4$	$1 1/4$	1
$7/8$	$1 1/2$†	$1 1/8$
1	$1 3/4$†	$1 1/4$
$1 1/8$	2	$1 1/2$
$1 1/4$	$2 1/4$	$1 5/8$
Over $1 1/4$	$1 3/4 d$‡	$1 1/4 d$‡

* All edge distances in this column may be reduced $1/8$ in when the hole is at a point where stress does not exceed 25 percent of the maximum allowed stress in the element.
† These may be $1 1/4$ in at the ends of beam connection angles.
‡ d = fastener diameter, in.

Source: From AISC Specification for Structural Steel Buildings.

structural steel for buildings have provisions for minimum edge distance: The distance from the center of a standard hole to an edge of a connected part should not be less than the applicable value from Table 7.5.

Maximum edge distances are set for sealing and stitch purposes. AISC specifications limit the distance from center of fastener to nearest edge of parts in contact to 12 times the thickness of the connected part, with a maximum of 6 in. For unpainted weather steel, the maximum is 7 in or 14 times the thickness of the thinner plate. For painted or unpainted members not subject to corrosion, the maximum spacing is 12 in or 24 times the thickness of the thinner plate. Pitch is the distance (inches) along the line of principal stress between centers of adjacent fasteners. It may be measured along one or more lines of fasteners. For example, suppose bolts are staggered along two parallel lines. The pitch may be given as the distance between successive bolts in each line separately. Or it may be given as the distance, measured parallel to the fastener lines, between a bolt in one line and the nearest bolt in the other line.

Gage is the distance (inches) between adjacent lines of fasteners along which pitch is measured or the distance (inches) from the back of an angle or other shape to the first line of fasteners.

The minimum distance between centers of fasteners should be at least three times the fastener diameter. (The AISC specification, however, permits $2 \, 2/3$ times the fastener diameter.)

Limitations also are set on maximum spacing of fasteners, for several reasons. In built-up members, *stitch fasteners*, with restricted spacings, are used between components to ensure uniform action. Also, in compression members such fasteners are required to prevent local buckling.

Designs should provide ample clearance for tightening high strength bolts. Detailers who prepare shop drawings for fabricators generally are aware of the necessity for this and can, with careful detailing, secure the necessary space. In tight situations, the solution may be staggering of holes (Fig. 7.3), variations from standard gages (Fig. 7.4), use of knife-type connections, or use of a combination of shop welds and field bolts. Minimum clearances for tightening high-strength bolts are indicated in Fig. 7.5 and Table 7.6.

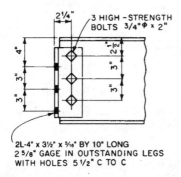

Fig. 7.3 Staggered holes provide clearance for high-strength bolts.

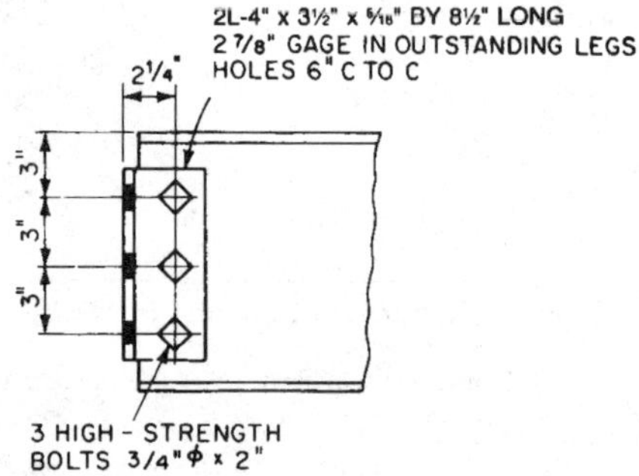

Fig. 7.4 Increasing the gage in framing angles.

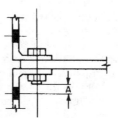

Fig. 7.5 The usual minimum clearances.

Table 7.6 Clearances for High-Strength Bolts

Bolt diameter, in	Nut height, in	Usual min clearance, in A	Min clearance for twist-off bolts, in A	
			Small tool	Large tool
$5/8$	$5/8$	1	$1 5/8$	
$3/4$	$3/4$	$1 1/4$	$1 5/8$	$1 7/8$
$7/8$	$7/8$	$1 3/8$	$1 5/8$	$1 7/8$
1	1	$1 7/16$		$1 7/8$
$1 1/8$	$1 1/8$	$1 9/16$		
$1 1/4$	$1 1/4$	$1 11/16$		

7.2.7 Installation

All parts of a connection should be held tightly together during installation of fasteners. Drifting done during assembling to align holes should not distort the metal or enlarge the holes. Holes that must be enlarged to admit fasteners should be reamed. Poor matching of holes is cause for rejection.

For connections with high-strength bolts, surfaces, when assembled, including those adjacent to bolt heads, nuts, and washers, should be free of scale, except tight mill scale. The surfaces also should be free of defects that would prevent solid seating of the parts, especially dirt, burrs, and other foreign material. Contact surfaces within slip-critical joints should be free of oil, paint, lacquer, and rust inhibitor.

Each high-strength bolt should be tightened so that when all fasteners in the connection are tight it will have the total tension (kips) for its diameter. Tightening should be done by the turn-of-the-nut method or with properly calibrated wrenches. High-strength bolts usually are tightened with an impact wrench. Only where clearance does not permit its use will bolts be hand-tightened.

Requirements for joint assembly and tightening of connections are given in the Specification for Structural Joints Using ASTM A325 or A490 Bolts, Research Council on Structural Connections of the Engineering Foundation.

In buildings, connections carrying calculated stresses, except lacing, sag bars, and girts, should be designed to support at least 6 kips of service load.

Long Grips ■ In buildings, if A307 bolts in a connection carry calculated stress and have grips exceeding five diameters, the number of these fasteners used in the connection should be increased 1 percent for each additional $1/16$ in in the grip. A *filler* is a plate inserted in a splice between a gusset or splice plate and stress-carrying members to fill a gap between them. Requirements for fillers included in the AISC specifications for structural steel for buildings are as follows.

In welded construction, a filler $1/4$ in or more thick should extend beyond the edge of the splice plate and be welded to the part on which it is fitted (Fig. 7.7). The welds should be able to transmit the splice-plate stress, applied at the surface of the filler, as an eccentric load. The welds that join the splice plate to the filler

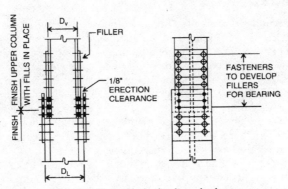

Fig. 7.6 Typical bolted splice of columns.

7-12 ■ Chapter Seven

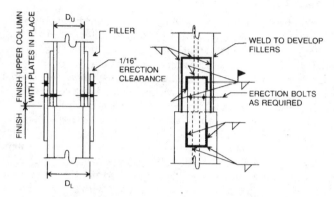

Fig. 7.7 Typical welded splice of columns.

should be able to transmit the splice-plate stress and should have sufficient length to prevent overstress of the filler along the toe of the welds. A filler less than $1/4$ in thick should have edges flush with the splice-plate edges. The size of the welds should equal the sum of the filler thickness and the weld size necessary to resist the splice-plate stress.

In bearing connections with bolts carrying computed stress passing through fillers thicker than $1/4$ in, the fillers should extend beyond the splice plate (Fig. 7.6). The filler extension should be secured by sufficient bolts to distribute the load on the member uniformly over the combined cross section of member and filler. Alternatively, an equivalent number of bolts should be included in the connection. Fillers $1/4$ to $3/4$ in thick need not be extended if the allowable shear stress in the bolts is reduced by the factor $0.4(t - 0.25)$, where t is the total thickness of the fillers but not more than $3/4$ in thick.

7.3 Welds

Welded connections often are used because of simplicity of design, fewer parts, less material, and decrease in shop handling and fabrication operations. Frequently, a combination of shop welding and field bolting is advantageous. With connection angles shop welded to a beam, field connections can be made with high-strength bolts without the clearance problems that may arise in an all-bolted connection.

Welded connections have a rigidity that can be advantageous if properly accounted for in design. Welded trusses, for example, deflect less than bolted trusses, because the end of a welded member at a joint cannot rotate relative to the other members there. If the end of a beam is welded to a column, the rotation there is practically the same for column and beam.

A disadvantage of welding, however, is that shrinkage of large welds must be considered. It is particularly important in large structures where there will be an accumulative effect.

Connections ■ 7-13

Properly made, a properly designed weld is stronger than the base metal. Improperly made, even a good-looking weld may be worthless. Properly made, a weld has the required penetration and is not brittle.

Prequalified joints, welding procedures, and procedures for qualifying welders are covered by AWS D1.1, Structural Welding Code—Steel, American Welding Society (1996). Common types of welds with structural steels intended for welding when made in accordance with AWS specifications can be specified by note or by symbol with assurance that a good connection will be obtained.

In making a welded design, designers should specify only the amount and size of weld actually required. Generally, a $5/16$-in weld is considered the maximum size for a single pass.

The cost of fit-up for welding can range from about one-third to several times the cost of welding. In designing welded connections, therefore, designers should consider the work necessary for the fabricator and the erector in fitting members together so they can be welded.

7.3.1 Types of Welds

The main types of welds used for structural steel are fillet, groove, plug, and slot. The most commonly used weld is the fillet. For light loads, it is the most economical, because little preparation of material is required. For heavy loads, groove welds are the most efficient, because the full strength of the base metal can be obtained easily. Use of plug and slot welds generally is limited to special conditions where fillet or groove welds are not practical.

More than one type of weld may be used in a connection. If so, the allowable capacity of the connection is the sum of the effective capacities of each type of weld used, separately computed with respect to the axis of the group.

Tack welds may be used for assembly or shipping. They are not assigned any stress-carrying capacity in the final structure. In some cases, these welds must be removed after final assembly or erection.

Fillet welds have the general shape of an isosceles right triangle (Fig. 7.8). The size of the weld is given by the length of leg. The strength is determined by the

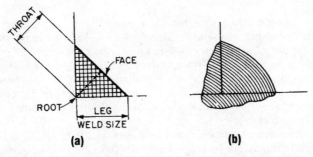

Fig. 7.8 Fillet weld. (*a*) Theoretical cross section. (*b*) Actual cross section.

throat thickness, the shortest distance from the root (intersection of legs) to the face of the weld. If the two legs are unequal, the nominal size of the weld is given by the shorter of the legs. If welds are concave, the throat is diminished accordingly, and so is the strength.

Fillet welds are used to join two surfaces approximately at right angles to each other. The joints may be lap (Fig. 7.9) or tee or corner (Fig. 7.10). Fillet welds also may be used with groove welds to reinforce corner joints. In a skewed tee joint, the included angle of weld deposit may vary up to 30° from the perpendicular, and one corner of the edge to be connected may be raised, up to $3/16$ in. If the separation is greater than $1/16$ in, the weld leg should be increased by the amount of the root opening.

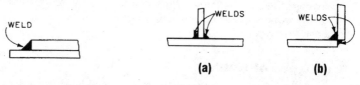

Fig. 7.9 Welded lap joint.   **Fig. 7.10** (*a*) Tee joint. (*b*) Corner joint.

Groove welds are made in a groove between the edges of two parts to be joined. These welds generally are used to connect two plates lying in the same plane (butt joint), but they also may be used for tee and corner joints.

Standard types of groove welds are named in accordance with the shape given the edges to be welded: square, single vee, double vee, single bevel, double bevel, single U, double U, single J, and double J (Fig. 7.11). Edges may be shaped by flame cutting, arc-air gouging, or edge planing. Material up to $5/8$ in thick, however, may be groove-welded with square-cut edges, depending on the welding process used.

Groove welds should extend the full width of parts joined. Intermittent groove welds and butt joints not fully welded throughout the cross section are prohibited.

Groove welds also are classified as complete-penetration and partial-penetration welds.

In a *complete-penetration weld*, the weld material and the base metal are fused throughout the depth of the joint. This type of weld is made by welding from both sides of the joint or from one side to a backing bar or backing weld. When the joint is made by welding from both sides, the root of the first-pass weld is chipped or gouged to sound metal before the weld on the opposite side, or back pass, is made. The throat dimension of a complete-penetration groove weld, for stress computations, is the full thickness of the thinner part joined, exclusive of weld reinforcement.

Partial-penetration welds generally are used when forces to be transferred are small. The edges may not be shaped over the full joint thickness, and the depth of weld may be less than the joint thickness (Fig. 7.11). But even if edges are fully

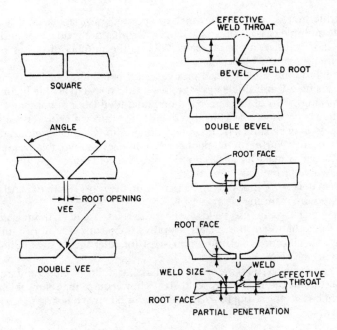

Fig. 7.11 Groove welds.

shaped, groove welds made from one side without a backing strip or made from both sides without back gouging are considered partial-penetration welds. They often are used for splices in building columns carrying axial loads only. In bridges, such welds should not be used where tension may be applied normal to the axis of the welds.

Plug and slot welds are used to transmit shear in lap joints and to prevent buckling of lapped parts. In buildings, they also may be used to join components of built-up members. (Plug or slot welds, however, are not permitted on A514 steel.) The welds are made, with lapped parts in contact, by depositing weld metal in circular or slotted holes in one part. The openings may be partly or completely filled, depending on their depth. Load capacity of a plug or slot completely welded equals the product of hole area and allowable stress. Unless appearance is a main consideration, a fillet weld in holes or slots is preferable.

Economy in Selection ■ In selecting a weld, designers should consider not only the type of joint but also the type of weld that would require a minimum amount of metal. This would yield a saving in both material and time.

While strength of a fillet weld varies with size, volume of metal varies with the square of the size. For example, a $1/2$-in fillet weld contains four times as much metal per inch of length as a $1/4$-in weld but is only twice as strong. In general, a smaller but longer fillet weld costs less than a larger but shorter weld of the same capacity.

Furthermore, small welds can be deposited in a single pass. Large welds require multiple passes. They take longer, absorb more weld metal, and cost more. As a guide in selecting welds, Table 7.7 lists the number of passes required for some frequently used types of welds.

Double-V and double-bevel groove welds contain about half as much weld metal as single-V and single-bevel groove welds, respectively (deducting effects of root spacing). Cost of edge preparation and added labor of gouging for the back pass, however, should be considered. Also, for thin material, for which a single weld pass may be sufficient, it is uneconomical to use smaller electrodes to weld from two sides. Furthermore, poor accessibility or less favorable welding position (Sec. 7.3.4) may make an unsymmetrical groove weld more economical, because it can be welded from only one side.

When bevel or V grooves can be flame-cut, they cost less than J and U grooves, which require planning or arc-air gouging.

For a given size of fillet weld, cooling rate is faster and restraint is greater with thick plates than with thin plates. To prevent cracking due to resulting internal stresses, specifications set minimum sizes for fillet welds depending on plate thickness (Table 7.8).

To prevent overstressing of base material at a fillet weld, standard specifications also limit the maximum weld size. They require that allowable stresses in adjacent base material not be exceeded when a fillet weld is stressed to its allowed capacity.

Table 7.7 Number of Passes for Welds

Weld size,* in	Fillet welds	Single-bevel groove welds (backup weld not included)		Single-V groove welds (backup weld not included)		
		30° bevel	45° bevel	30° open	60° open	90° open
$3/16$	1					
$1/4$	1	1	1	2	3	3
$5/16$	1					
$3/8$	3	2	2	3	4	6
$7/16$	4					
$1/2$	4	2	2	4	5	7
$5/8$	6	3	3	4	6	8
$3/4$	8	4	5	4	7	9
$7/8$		5	8	5	10	10
1		5	11	5	13	22
$1 1/8$		7	11	9	15	27
$1 1/4$		8	11	12	16	32
$1 3/8$		9	15	13	21	36
$1 1/2$		9	18	13	25	40
$1 3/4$		11	21			

* Plate thickness for groove welds.

Table 7.8 Minimum Fillet-Weld Sizes and Plate-Thickness Limits

Size of fillet welds,* in		Minimum plate thickness for fillet welds on each side of the plate, in	
Buildings† AWS D1.1	Maximum plate thickness, in§	36-ksi steel	50-ksi steel
1/8¶	1/4		
3/16	1/2	0.43	0.31
1/4	3/4	0.57	0.41
5/16	Over 3/4	0.75	0.52

*Weld size need not exceed the thickness of the thinner part joined, but AISC requires that care be taken to provide sufficient preheat to ensure weld soundness.
†When low-hydrogen welding is employed. AWS D1.1 permits the thinner part joined to be used to determine the minimum size of fillet weld.
§Plate thickness is the thickness of the thicker part joined.
¶Minimum weld size for structures subjected to dynamic loads is 3/16 in.

A limitation also is placed on the maximum size of fillet welds along edges. One reason is that edges of rolled shapes are rounded, and weld thickness consequently is less than the nominal thickness of the part. Another reason is that if weld size and plate thickness are nearly equal, the plate corner may melt into the weld, reducing the length of weld leg and the throat. Hence standard specifications require the following: *Along edges of material less than 1/4 in thick, maximum size of fillet weld may equal material thickness. But along edges of material 1/4 in or more thick, the maximum size should be 1/16 in less than the material thickness.*

Weld size may exceed this, however, if drawings definitely show that the weld is to be built out to obtain full throat thickness. AWS D1.1 requires that the minimum effective length of a fillet weld be at least four times the nominal size, or else the weld must be considered not to exceed 25 percent of the effective length.

Subject to the preceding requirements, intermittent fillet welds may be used in buildings to transfer calculated stress across a joint or faying surfaces when the strength required is less than that developed by a continuous fillet weld of the smallest permitted size. Intermittent fillet welds also may be used to join components of built-up members in buildings.

Intermittent welds are advantageous with light members where excessive welding can result in straightening costs greater than the cost of welding. Intermittent welds often are sufficient and less costly than continuous welds (except girder fillet welds made with automatic welding equipment).

Weld lengths specified on drawings are effective weld lengths. They do not include distances needed for start and stop of welding.

To avoid the adverse effects of starting or stopping a fillet weld at a corner, welds extending to corners should be returned continuously around the corners in the same plane for a distance of at least twice the weld size. This applies to side and top fillet welds connecting brackets, beam seats, and similar connections, on the plane about which bending moments are computed. End returns should be indicated on design and detail drawings.

Fillet welds deposited on opposite sides of a common plane of contact between two parts must be interrupted at a corner common to both welds.

If longitudinal fillet welds are used alone in end connections of flat-bar tension members, the length of each fillet weld should at least equal the perpendicular distance between the welds. The transverse spacing of longitudinal fillet welds in end connections should not exceed 8 in unless the design otherwise prevents excessive transverse bending in the connections.

In material $5/8$ in or less thick, the thickness of plug or slot welds should be the same as the material thickness. In material more than $5/8$ in thick, the weld thickness should be at least half the material thickness but not less than $5/8$ in.

Diameter of hole for a plug weld should be at least the depth of the hole plus $5/16$ in, but the diameter should not exceed $2\,1/4$ times the thickness of the member.

Thus the hole diameter in $3/4$-in plate could be a minimum of $3/4 + 5/16 = 1\,1/16$ in. Depth of metal would be at least $5/8$ in > ($1/2 \times 3/4 = 3/8$ in).

Plug welds may not be spaced closer center to center than four times the hole diameter.

Length of slot for a slot weld should not exceed 10 times the part thickness. Width of slot should be at least depth of hole plus $5/16$ in, but the width should not exceed $2\,1/4$ times the part thickness.

Thus width of slot in $3/4$-in plate could be a minimum of $3/4 + 5/16 = 1\,1/16$ in. Weld metal depth would be at least $5/8$ in > ($1/2 \times 3/4 = 3/8$ in). The slot could be up to $10 \times 3/4 = 7\,1/2$ in long.

Slot welds may be spaced no closer than four times their width in a direction transverse to the slot length. In the longitudinal direction, center-to-center spacing should be at least twice the slot length.

7.3.2 Welding Symbols

These should be used on drawings to designate welds and provide pertinent information concerning them. The basic parts of a weld symbol are a horizontal line and an arrow:

Extending from either end of the line, the arrow should point to the joint in the same manner as the electrode would be held to do the welding.

Welding symbols should clearly convey the intent of the designer. For the purpose, sections or enlarged details may have to be drawn to show the symbols, or notes may be added. Notes may be given as part of welding symbols or separately. When part of a symbol, the note should be placed inside a tail at the opposite end of the line from the arrow:

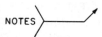

NOTES

Type and length of weld are indicated above or below the line. If noted below the line, the symbol applies to a weld on the arrow side of the joint, the side to which the arrow points. If noted above the line, the symbol indicates that the

Connections ■ 7-19

other side, the side opposite the one to which the arrow points (not the far side of the assembly), is to be welded.

A fillet weld is represented by a right triangle extending above or below the line to indicate the side on which the weld is to be made. The vertical leg of the triangle is always on the left.

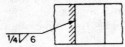

The preceding symbol indicates that a $\frac{1}{4}$-in fillet weld 6 in long is to be made on the arrow side of the assembly. The following symbol requires a $\frac{1}{4}$-in fillet weld 6 in long on both sides.

If a weld is required on the far side of an assembly, it may be assumed necessary from symmetry, shown in sections or details, or explained by a note in the tail of the welding symbol. For connection angles at the end of a beam, far-side welds generally are assumed:

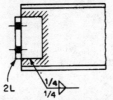

Length of weld is not shown on the symbol in this case because the connection requires a continuous weld the full length of each angle on both sides of the angle. Care must be taken not to omit length unless a continuous full-length weld is wanted. "Continuous" should be written on the weld symbol to indicate length when such a weld is required. In general, a tail note is advisable to specify welds on the far side, even when the welds are the same size.

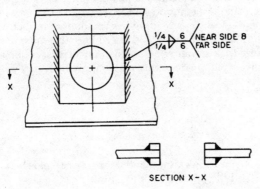

SECTION X-X

For many members, a stitch or intermittent weld is sufficient. It may be shown as

$$\swarrow \overline{1/4 \triangleright 2\text{-}10}$$

This symbol calls for $1/4$-in fillet welds on the arrow side. Each weld is to be 2 in long. Spacing of welds is to be 10 in center to center. If the welds are to be staggered on the arrow and other sides, they can be shown as

$$\swarrow \overline{\underset{1/4 \triangleright 2\text{-}10}{1/4 \triangleright 2\text{-}10}}$$

Usually, intermittent welds are started and finished with a weld at least twice as long as the length of the stitch welds. This information is given in a tail note:

$$\swarrow \overline{\underset{1/4 \triangleright 2\text{-}10}{1/4 \triangleright 2\text{-}10}} \langle 4\text{" at ends}$$

When the welding is to be done in the field rather than in the shop, a triangular flag should be placed at the intersection of arrow and line:

This is important in ensuring that the weld will be made as required. Often, a tail note is advisable for specifying field welds.

A continuous weld all around a joint is indicated by a small circle around the intersection of line and arrow:

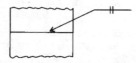

Such a symbol would be used, for example, to specify a weld joining a pipe column to a base plate. The all-around symbol, however, should not be used as a substitute for computation of actual weld length required. Note that the type of weld is indicated below the line in the all-around symbol, regardless of shape or extent of joint.

The preceding devices for providing information with fillet welds also apply to groove welds. In addition, groove-weld symbols also must designate material preparation required. This often is best shown on a cross section of the joint.

A square-groove weld (made in thin material) without root opening is indicated by

Length is not shown on the welding symbol for groove welds because these welds almost always extend the full length of the joint.

A short curved line below a square-groove symbol indicates weld contour. A short straight line in that position represents a flush weld surface. If the weld is not to be ground, however, that part of the symbol is usually omitted. When grinding is required, it must be indicated in the symbol.

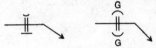

The root-opening size for a groove weld is written in within the symbol indicating the type of weld. For example, a ⅛-in root opening for a square-groove weld is specified by

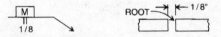

And a ⅛-in root opening for a bevel weld, not to be ground, is indicated by

In this and other types of unsymmetrical welds, the arrow not only designates the arrow side of the joint but also points to the side to be shaped for the groove weld. When the arrow has this significance, the intention often is emphasized by an extra break in the arrow.

The angle at which the material is to be beveled should be indicated with the root opening:

A double-bevel weld is specified by

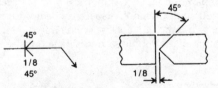

A single-V weld is represented by

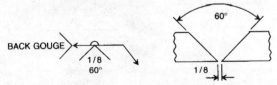

A double-V weld is indicated by

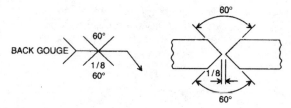

Summary ▪ Standard symbols for various types of welds are summarized in Fig. 7.12. The symbols do not indicate whether backing, spacer, or extension bars are required. These should be specified in general notes or shown in detail drawings. Preparation for J and U welds is best shown by an enlarged cross section. Radius and depth of preparation must be given.

Fig. 7.12 Summary of welding symbols.

In preparing a weld symbol, insert size, weld-type symbol, length of weld, and spacing, in that order from left to right. The perpendicular leg of the symbol for fillet, bevel, J, and flare-bevel welds should be on the left of the symbol. Bear in mind also that arrow-side and other-side welds are the same size unless otherwise noted. When billing of detail material discloses the identity of the far side with the near side, the welding shown for the near side also will be duplicated on the far side. Symbols apply between abrupt changes in direction of welding unless governed by the all-around symbol or dimensioning shown.

Where groove preparation is not symmetrical and complete, additional information should be given on the symbol. Also it may be necessary to give weld-penetration information, as in Fig. 7.13. For the weld shown, penetration from either side must be a minimum of $3/16$ in. The second side should be back-gouged before the weld there is made.

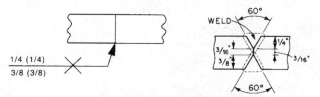

Fig. 7.13 Penetration information is given on the welding symbol in (a) for the weld shown in (b). Penetration must be at least $3/16$ in. Second side must be back-gouged before the weld on that side is made.

Welds also may be a combination of different groove and fillet welds. While symbols can be developed for these, designers will save time by supplying a sketch or enlarged cross section. It is important to convey the required information accurately and completely to the workers who will do the job. Actually, it is common practice for designers to indicate what is required of the weld and for fabricators and erectors to submit proposed procedures.

7.3.3 Welding Material

Weldable structural steels permissible in buildings are listed with required electrodes in Table 7.9. Welding electrodes and fluxes should conform to AWS 5.1, 5.5,

Table 7.9 Matching Filler-Metal Requirements for Complete-Penetration Groove Welds in Building Construction

	Welding process			
Base metal*	Shielded metal-arc	Submerged-arc	Gas metal-arc	Flux-cored arc
A36[†], A53 grade B A500 grades A and B A501, A529, and A570 grades 30 through 50	AWS A5.1 or A5.5[§] E60XX E70XX E70XX-X	AWS A5.17 or A5.23[§] F6XX-EXXX F7XX-EXXX or F7XX-EXX-XX	AWS A5.18 ER70S-X	AWS A5.20 E6XT-X E7XT-X (Except −2, −3, −10, −GS)
A242[‡], A441, A572 grade 42 and 50, and A588[‡] (4 in and under)	AWS A5.1 or A5.5[§] E7015, E7016, E7018, E7028 E7015-X, E7016-X, E7018-X	AWS A5.17 or A5.23[§] F7XX-EXXX F7XX-EXX-XX	AWS A5.18 ER70S-X	AWS A5.20 E7XT-X (Except −2, −3, −10, −GS)
A572 grades 60 and 65	AWS A5.5[§] E8015-X, E8016-X E8018-X	AWS A5.23[§] F8XX-EXX-XX	AWS A5.28[§] ER 80S-X	AWS A5.29[§] E8XTX-X
A514 over 2½ in thick	AWS A5.5[§] E10015-X, E10016-X, E10018-X	AWS A5.23[§] F10XX-EXX-XX	AWS A5.28[§] ER 100S-X	AWS A5.29[§] E10XTX-X
A514 2½ in thick and under	AWS A5.5[§] E11015-X, E11016-X, E11018-X	AWS A5.23[§] F11XX-EXX-XX	AWS A5.28[§] ER 110S-X	AWS A5.29[§] E11XTX-X

* In joints involving base metals of different groups, low-hydrogen filler-metal requirements to the lower-strength group may be used. The low-hydrogen processes are subject to the technique requirements applicable to the higher-strength group.

[†] Only low-hydrogen electrodes may be used for welding A36 steel more than 1 in thick for dynamically loaded structures.

[‡] Special welding materials and procedures (e.g., E80XX-X low-alloy electrodes) may be required to match the notch toughness of base metal (for applications involving impact loading or low temperature) or for atmospheric corrosion and weathering characteristics.

[§] Deposited weld metal should have a minimum impact strength of 20 ft-lb at 0°F when Charpy V-notch specimens are required.

5.17, 5.18, 5.20, 5.23, 5.25, 5.26, 5.28, or 5.29 or applicable provisions of AWS D1.1. Weld metal deposited by electroslag or electrogas welding processes should conform to the requirements of AWS D1.1 for these processes. For welded connections in buildings, the electrodes or fluxes given in Table 7.9 should be used in making complete-penetration groove welds.

7.3.4 Welding Positions

The position of the stick electrode relative to the joint when a weld is being made affects welding economy and quality. In addition, AWS specification D1.0 and D1.5 prohibit use of some welding positions for some types of welds. Careful designing should eliminate the need for welds requiring prohibited welding positions and employ welds that can be efficiently made.

The basic welding positions are as follows:

Flat, with face of weld nearly horizontal. Electrode is nearly vertical, and welding is performed from above the joint.

Horizontal, with axis of weld horizontal. For groove welds, the face of weld is nearly vertical. For fillet welds, the face of weld usually is about 45° relative to horizontal and vertical surfaces.

Vertical, with axis of weld nearly vertical. (Welds are made upward.)

Overhead with face of weld nearly horizontal. Electrode is nearly vertical, and welding is performed from below the joint.

Where possible, welds should be made in the flat position. Weld metal can be deposited faster and more easily. Generally the best and most economical welds are obtained. In a shop, the work usually is positioned to allow flat or horizontal welding. With care in design, the expense of this positioning can be kept to a minimum. In the field, vertical and overhead welding sometimes may be necessary. The best assurance of good welds in these positions is use of proper electrodes by experienced welders.

The AWS specifications require that only the flat position be used for submerged-arc welding, except for certain sizes of fillet welds. Single-pass fillet welds may be made in the flat or the horizontal position in sizes up to $5/16$ in with a single electrode and up to $1/2$ in with multiple electrodes. Other positions are prohibited.

When groove-welded joints can be welded in the flat position, submerged-arc and gas metal-arc processes usually are more economical than the manual shielded metal-arc process.

Designers and detailers should detail connections to ensure that welders have ample space for positioning and manipulating electrodes and for observing the operation with a protective hood in place. Electrodes may be up to 18 in long and $3/8$ in in diameter.

In addition, adequate space must be provided for deposition of the required size of fillet weld. For example, to provide an adequate landing c (in) for the fillet weld of size D (in) in Fig. 7.14, c should be at least $D + 5/16$. In building column splices, however, $c = D + 3/16$ often is used for welding splice plates to fillers.

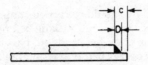

Fig. 7.14 Minimum landing for a fillet weld.

7.3.5 Weld Procedures

Welds should be qualified and should be made only by welders, welding operators, and tackers qualified as required in AWS D1.1 for buildings. Welding should not be permitted under any of the following conditions:

When the ambient temperature is below 0°F.

When surfaces are wet or exposed to rain, snow, or high wind.

When welders are exposed to inclement conditions.

Surfaces and edges to be welded should be free from fins, tears, cracks, and other defects. Also, surfaces at and near welds should be free from loose scale, slag, rust, grease, moisture, and other material that may prevent proper welding. AWS specifications, however, permit mill scale that withstands vigorous wire brushing, a light film of drying oil, or antispatter compound to remain. But the specifications require all mill scale to be removed from surfaces on which flange-to-web welds are to be made by submerged-arc welding or shielded metal-arc welding with low-hydrogen electrodes.

Parts to be fillet-welded should be in close contact. The gap between parts should not exceed $^3/_{16}$ in. If it is $^1/_{16}$ in or more, fillet-weld size should be increased by the amount of separation. The separation between faying surfaces for plug and slot welds and for butt joints landing on a backing should not exceed $^1/_{16}$ in. Parts to be joined at butt joints should be carefully aligned. Where the parts are effectively restrained against bending due to eccentricity in alignment, an offset not exceeding 10 percent of the thickness of the thinner part joined, but in no case more than $^1/_8$ in, is permitted as a departure from theoretical alignment. When correcting misalignment in such cases, the parts should not be drawn in to a greater slope than $^1/_2$ in in 12 in.

For permissible welding positions, see Sec. 7.3.4. Work should be positioned for flat welding whenever practicable.

In general, welding procedures and sequences should avoid needless distortion and should minimize shrinkage stresses. As welding progresses, welds should be deposited so as to balance the applied heat. Welding of a member should progress from points where parts are relatively fixed in position toward points where parts have greater relative freedom of movement. Where it is impossible to avoid high residual stresses in the closing welds of a rigid assembly, these welds should be made in compression elements. Joints expected to have significant shrinkage should be welded before joints expected to have lesser shrinkage, and restraint should be kept to a minimum. If severe external restraint against shrinkage is present, welding should be carried continuously to completion or to a point that will ensure freedom from cracking before the joint is allowed to cool below the minimum specified preheat and interpass temperature.

In shop fabrication of cover-plated beams and built-up members, each component requiring splices should be spliced before it is welded to other parts of the member. Up to three subsections may be spliced to form a long girder or girder section.

With too rapid cooling, cracks might form in a weld. Possible causes are shrinkage of weld and heat-affected zone, austenite-martensite transformation, and entrapped hydrogen. Preheating the base metal can eliminate the first two causes. Preheating reduces the temperature gradient between weld and adjacent base metal, thus decreasing the cooling rate and resulting stresses. Also, if hydrogen is present, preheating allows more time for this gas to escape. Use of low-hydrogen electrodes, with suitable moisture control, also is advantageous in controlling hydrogen content.

High cooling rates occur at arc strikes that do not deposit weld metal. Hence arc strikes outside the area of permanent welds should be avoided. Cracks or blemishes resulting from arc strikes should be ground to a smooth contour and checked for soundness.

Table 7.10 Requirements of AWS D1.1 for Minimum Preheat and Interpass Temperature (°F) for Welds in Buildings*

Thickness of thickest part at point of welding, in	Shielded metal-arc with other than low-hydrogen electrodes ASTM A36†, A53 grade B, A501, A529 A570 all grades	Shielded metal-arc with low-hydrogen electrodes; submerged-arc, gas metal-arc, or flux-cored arc ASTM A36, A53 grade B, A242 A441, A501, A529 A570 all grades A572 grades 42 50, A588	Shielded metal-arc with low-hydrogen electrodes; submerged-arc, gas metal-arc, or flux-cored arc ASTM A572 grades 60 and 65	Shielded metal-arc with low-hydrogen electrodes; submerged-arc, with carbon or alloy steel wire neutral flux, gas metal-arc, or flux-cored arc ASTM A514
To ¾	0‡	0‡	50	50
Over ¾ to 1½	150	50	150	125
Over 1½ to 2½	225	150	225	175
Over 2½	300	225	300	225

* In joints involving combinations of base metals, preheat as specified for the higher-strength steel being welded.

† Use only low-hydrogen electrodes when welding A36 steel more than 1 in thick for dynamically loaded structures.

‡ When the base-metal temperature is below 32°F, the base metal should be preheated to at least 70°F and the minimum temperature maintained during welding.

To avoid cracks and for other reasons, standard specifications require that under certain conditions, before a weld is made the base metal must be preheated. Table 7.10 lists typical preheat and interpass temperatures. The tables recognize that as plate thickness, carbon content, or alloy content increases, higher preheats are necessary to lower cooling rates and to avoid microcracks or brittle heat-affected zones.

Preheating should bring to the specified preheat temperature the surface of the base metal within a distance equal to the thickness of the part being welded, but not less than 3 in of the point of welding. This temperature should be maintained as a minimum interpass temperature while welding progresses.

Preheat and interpass temperatures should be sufficient to prevent crack formation. Temperatures above the minimums in Table 7.10 may be required for highly restrained welds.

For A514, A517, and A852 steels, the maximum preheat and interpass temperature should not exceed 400°F for thicknesses up to $1\frac{1}{2}$ in, inclusive, and 450°F for greater thicknesses. Heat input during the welding of these quenched and tempered steels should not exceed the steel producer's recommendation. Use of stringer beads to avoid overheating is advisable.

Peening sometimes is used on intermediate weld layers for control of shrinkage stresses in thick welds to prevent cracking. It should be done with a round-nose tool and light blows from a power hammer after the weld has cooled to a temperature warm to the hand. The root or surface layer of the weld or the base metal at the edges of the weld should not be peened. Care should be taken to prevent scaling or flaking of weld and base metal from overpeening.

When required by plans and specifications, welded assemblies should be stress-relieved by heat treating. (See AWS D1.1 for temperatures and holding times required.) Finish machining should be done after stress relieving.

Tack and other temporary welds are subject to the same quality requirements as final welds. For tack welds, however, preheat is not mandatory for single-pass welds that are remelted and incorporated into continuous submerged-arc welds. Also, defects such as undercut, unfilled craters, and porosity need not be removed before final submerged-arc welding. Welds not incorporated into final welds should be removed after they have served their purpose, and the surface should be made flush with the original surface.

Before a weld is made over previously deposited weld metal, all slag should be removed, and the weld and adjacent material should be brushed clean.

Groove welds should be terminated at the ends of a joint in a manner that will ensure sound welds. Where possible, this should be done with the aid of weld tabs or runoff plates. AWS D1.1 does not require removal of weld tabs for statically loaded structures but does require it for dynamically loaded structures. The ends of the welds then should be made smooth and flush with the edges of the abutting parts.

After welds have been completed, slag should be removed from them. The metal should not be painted until all welded joints have been completed, inspected, and accepted. Before paint is applied, spatter, rust, loose scale, oil, and dirt should be removed.

7.3.6 Weld Quality

A basic requirement of all welds is thorough fusion of weld and base metal and of successive layers of weld metal. In addition, welds should not be handicapped by craters, undercutting, overlap, porosity, or cracks. (AWS D1.1 gives acceptable tolerances for these defects.) If craters, excessive concavity, or undersized welds occur in the effective length of a weld, they should be cleaned and filled to the full cross section of the weld. Generally, all undercutting (removal of base metal at the toe of a weld) should be repaired by depositing weld metal to restore the original surface. Overlap (a rolling over of the weld surface with lack of fusion at an edge), which may cause stress concentrations, and excessive convexity should be reduced by grinding away of excess material (see Figs. 7.15 and 7.16). If excessive porosity, excessive slag inclusions, or incomplete fusion occur, the defective portions should be removed and rewelded. If cracks are present, their extent should be determined by acid etching, magnetic-particle inspection, or other equally positive means. Not only the cracks but also sound metal 2 in beyond their ends should be removed and replaced with the weld metal. Use of a small electrode for this purpose reduces the chances of further defects due to shrinkage. An electrode not more than $5/32$ in in diameter is desirable for depositing weld metal to compensate for size deficiencies.

AWS D1.1 limits convexity C to the values in Table 7.11.

Weld-quality requirements should depend on the job the welds are to do. Excessive requirements are uneconomical. Size, length, and penetration are always

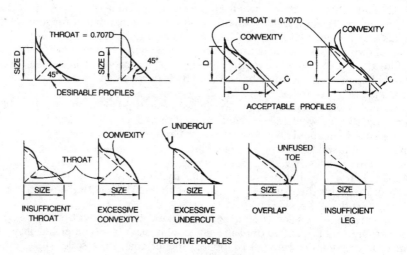

Fig. 7.15 Profiles of fillet welds.

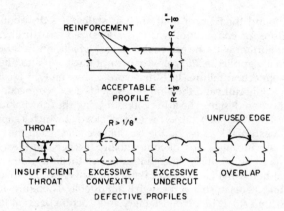

Fig. 7.16 Profiles as groove welds.

important for a stress-carrying weld and should completely meet design requirements. Undercutting, on the other hand, should not be permitted in main connections, such as those in trusses and bracing, but small amounts might be permitted in less important connections, such as those in platform framing for an industrial building. Type of electrode, similarly, is important for stress-carrying welds but not so critical for many miscellaneous welds. Again, poor appearance of a weld is objectionable if it indicates a bad weld or if the weld will be exposed where aesthetics is a design consideration, but for many types of structures, such as factories, warehouses, and incinerators, the appearance of a good weld is not critical. A sound weld is important. But a weld entirely free of porosity or small slag inclusions should be required only when the type of loading actually requires this perfection.

Table 7.11 AWS D1.1 Limits on Convexity of Fillet Welds

Measured leg size or width of surface bead, in	Maximum convexity, in
$5/16$ or less	$1/16$
Over $5/16$ but less than 1	$1/8$
1 or more	$3/16$

Welds may be inspected by one or more methods: visual inspection; nondestructive tests, such as ultrasonic, x-ray, dye penetration, and magnetic particle; and cutting of samples from finished welds. Designers should specify which welds are to be examined, extent of the examination, and methods to be used.

7.4 Design of Connections

7.4.1 Philosophy

Connection design is both an art and a science. The science involves equilibrium, limit states, and the lower-bound theorem of limit analysis. The art involves the determination of the most efficient load paths for the connection and is necessary because most connections are statically indeterminate.

The lower-bound theorem of limit analysis states: If a distribution of forces within a structure (or connection, which is a localized structure) can be found which is in equilibrium with the external load and which satisfies the limit states, then the externally applied load is less than, or at most equal to, the load which would cause connection failure. In other words, any solution for a connection which satisfies equilibrium and the limit states yields a safe connection. This is the science of connection design. The art involves finding the internal force distribution (or load paths) which maximizes the external load at which a connection fails. This maximized external load is also the true failure load when the internal force distribution results in satisfaction of compatibility (no gaps and tears) within the connection in addition to satisfying equilibrium and the limit states.

It should be noted that, strictly speaking, the lower-bound theorem applies only to yield limit states in structures which are ductile. Therefore, in applying it to connections, limit states involving stability and fracture (lack of ductility) must be considered to preclude these modes of failure.

7.4.2 General Procedure

Determine the external (applied) loads and their lines of action. Make a preliminary layout, preferably to scale. The connection should be as compact as possible to conserve material and to minimize interferences with utilities, equipment, and access. Decide on where bolts and welds will be used and select bolt type and size. Decide on a load path through the connection. For a statically determinate connection, there is only one, but for indeterminate connections there are many possibilities. Use judgment, experience, and published information to arrive at the best load path. Now provide sufficient strength, stiffness, and ductility, using the limit states identified for each part of the load path, to make the connection adequate to carry the given loads. Complete the preliminary layout, check specification required spacings, and finally check to ensure that the connection can be fabricated and erected.

7.4.3 Types of Connections

There are three basic forces to which connections are subjected. These are axial force, shear force, and moment. Many connections are subject to two or more of these simultaneously. Connections are usually classified according to the major load type to be carried, such as shear connections, which carry primarily shear; moment connections, which carry primarily moment; and axial force connections, such as splices, bracing and truss connections, and hangers, which carry primarily axial force. Subsequent sections of this chapter deal with these three basic types of connections.

7.4.4 Axial Force Connections

Bracing Connections ▪ The lateral force resisting system in buildings may consist of a vertical truss. This is referred to as a braced frame, and the connections of the diagonal braces to the beams and columns are the bracing connections. For the bracing system to be a true truss, the bracing connections should be concentric; that is, the gravity axes of all members at any joint should intersect at a single point.

Connections   7-31

If the gravity axes are not concentric, the resulting couples must be considered in the design of the members. The examples of this section are of concentric type.

Example 1 ■ Consider the bracing connection of Fig. 7.17. The brace load is 855 kips, the beam shear is 10 kips, and the beam axial force is 411 kips. The horizontal component of the brace force is 627 kips, which means that 627 – 411 = 216 kips is transferred to the opposite side of the column from the brace side. There must be a connection on this side to "pick up" this load, i.e., provide a load path.

The design of this connection involves the design of four separate connections. These are (1) the brace to gusset connection, (2) the gusset to column connection, (3) the gusset to beam connection, and (4) the beam to column connection. A fifth connection is the connection on the other side of the column, which will not be considered here.

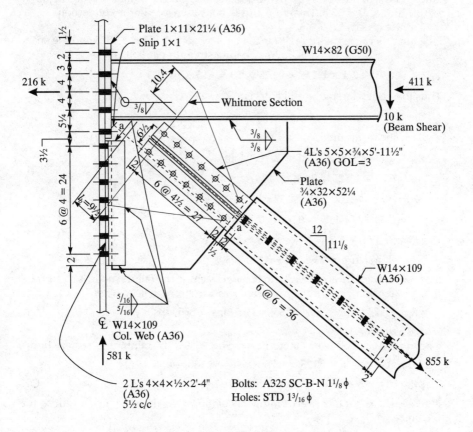

Fig. 7.17 Example 1, bracing connection design.

1. Brace to Gusset. This part of the connection is designed first because it provides a minimum size for the gusset plate which is then used to design the gusset to column and gusset to beam connections. Providing an adequate load path involves the following limit states:

a. Bolts (A325SC-B-N-1⅛-in-diameter STD holes)

The above notation indicates that the bolts are slip-critical, the surface class is B, and threads are not excluded from the shear planes. The slip-critical design strength per bolt is

$$\phi r_{str} = 1 \times 1.13 \times 0.5 \times 56 = 31.6 \text{ kips}$$

The bearing design strength is

$$\phi r_v = 0.75 \times \frac{\pi}{4} \times 1.125^2 \times 48 = 35.8 \text{ kips}$$

Since 31.6 < 35.8, use 31.6 kips as the design strength. The number of bolts required is 855/(31.6 × 2) = 13.5. Therefore, use 14 bolts each side of the connection.

b. Brace checks

i. Bearing: Because the edge distance $L_e = 2 > 1.5 \times 1.125 = 1.6875$ and the spacing $s = 4.5 > 3 \times 1.125 = 3.375$, the bearing strength per bolt is

$$\phi r_p = 0.75 \times 2.4 \times 1.125 \times 0.525 \times 58 = 61.7 \text{ kips}$$

Thus the design strength of 14 bolts is

$$\phi R_p = 14 \times 61.2 = 864 \text{ kips} > 855 \text{ kips} \qquad \text{OK}$$

ii. Block shear rupture:

$$A_{gv} = (2 + 6 \times 6) \times 0.525 \times 2 = 39.9 \text{ in}^2$$

$$A_{gt} = 6.75 \times 0.525 = 3.54 \text{ in}^2$$

$$A_{nv} = 39.9 - 6.5 \times 1.25 \times 0.525 \times 2 = 31.4 \text{ in}^2$$

$$A_{nt} = 3.54 - 1 \times 1.25 \times 0.525 = 2.88 \text{ in}^2$$

Shear fracture = 31.4 × 0.6 × 58 = 1093 kips

Tension fracture = 2.88 × 58 = 167 kips

Since shear fracture is greater than tension fracture, the failure mode is shear fracture and tension yield; thus the design fracture strength is

$$\phi R_{bs} = 0.75 (1093 + 3.54 \times 36) = 915 \text{ kips} > 855 \text{ kips} \qquad \text{OK}$$

c. Gusset checks

i. Bearing: Again, edge distance and spacing exceed the minimums required, so

$$\phi r_p = 0.75 \times 2.4 \times 1.125 \times 0.75 \times 58 = 88.1 \text{ kips/bolt}$$

$$\phi R_p = 14 \times 88.1 = 1233 \text{ kips} > 855 \text{ kips} \qquad \text{OK}$$

ii. Block shear rupture: Performing calculations which are similar to those for the brace,

$$\phi R_{bs} = 949 > 855 \text{ kips} \qquad \text{OK}$$

iii. Whitmore section: Since the brace load can be compression, this check is used to check for gusset buckling. Figure 7.17 shows the "Whitmore section" length, which is normally $l_w = (27 \tan 30) \times 2 + 6.5 = 37.7$ in, but the section passes out of the gusset and into the beam web at its upper side. Because of the fillet weld of the gusset to the beam flange, this part of the Whitmore section is not ineffective; i.e., load can be passed through the weld to be carried on this part of the Whitmore section. The effective length of the Whitmore section is thus

$$l_{we} = (37.7 - 10.4) + 10.4 \times \frac{0.510}{0.75} \times \frac{50}{36} = 27.3 + 10.1 = 37.1 \text{ in}$$

The gusset buckling length is, from Fig. 7.17, $l_b = 9.5$ in., and the slenderness ratio is

$$\frac{kl_b}{r} = \frac{0.5 \times 9.5 \times \sqrt{12}}{0.75} = 21.9$$

In the above formula the theoretical factor of 0.5 for fixed base is used rather than the usually recommended value of 0.65 for columns, because of the conservatism of this buckling check as determined by Gross (1990) from full scale tests. From the AISC LRFD Specification Table 3-36, $Kl_b/r = 21.9$, the design buckling strength is

$$\phi F_{cr} = 29.85 \text{ ksi}$$

and the Whitmore section buckling strength is thus

$$\phi R_{wb} = 29.85 \times 37.1 \times 0.75 = 831 \text{ kips} < 855 \text{ kips}$$

The design buckling load of 831 kips is slightly less (2.9 percent) than the required strength, but this is deemed acceptable because of the very conservative nature of the buckling check.

  d. Brace to gusset connection angles
  i. Gross and net area: The gross area required is $855 / (0.9 \times 36) = 26.4$ in^2
  Try 4L's $5 \times 5 \times ¾$, $A_{gt} = 6.94 \times 4 = 27.8$ in^2 OK
  The net area is $A_{nt} = 27.8 - 4 \times 0.75 \times 1.25 = 24.1$ in^2
  The effective net area is the lesser of $0.85 A_{gt}$ or UA_{nt}, where $U = 1 - 1.51/27 = 0.94$ use $U = 0.9$. Thus $0.85 A_{gt} = 0.85 \times 27.8 = 23.6$ and $UA_{nt} = 0.9 \times 24.1 = 21.7$ and then $A_e = 21.7$. Therefore, the net tensile design strength is $\phi R_t = 0.75 \times 58 \times 21.7 = 944$ kips > 855 kips OK
  ii. Bearing:

$$\phi R_p = 0.75 \times 2.4 \times 58 \times 0.75 \times 1.125 \times 14 = 1233 \text{ kips} > 855 \text{ kips} \quad \text{OK}$$

  iii. Block shear rupture (tearout): The length of the connection on the gusset side is the shorter of the two and is therefore the more critical. Per angle,

$$A_{nv} = (29 - 6.5 \times 1.25) \times 0.75 = 15.66 \text{ in}^2$$

$$A_{nt} = (2 - 0.5 \times 1.25) \times 0.75 = 1.03 \text{ in}^2$$

Since $0.6 F_u A_{nv} > F_u A_{nt}$,

$$\phi R_{bs} = 0.75 \times (0.6 \times 58 \times 15.66 + 36 \times 2 \times 0.75) \times 4 = 1800 \text{ kips} > 855 \text{ kips} \quad \text{OK}$$

7-34 ■ Chapter Seven

This completes the design checks for the brace to gusset connection. All elements of the load path, which consists of the bolts, the brace web, the gusset, and the connection angles, have been checked. The remaining connection interfaces require a method to determine the forces on them. Research has shown that the best method for doing this is the uniform force method (UFM). The force distributions for this method are shown in Fig. 7.18.

From the design of the brace to gusset connection, a certain minimum size of gusset is required. This is the gusset shown in Fig. 7.17. Usually this gusset size, which is a preliminary size, is sufficient for the final design. From Figs. 7.17 and 7.18, the basic data are

$$\tan\theta = \frac{12}{11.125} = 1.079$$

$$e_B = \frac{14.31}{2} = 7.155$$

$$e_C = 0$$

The quantities α and β locate the centroids of the gusset edge connections, and in order for no couples to exist on these connections, α and β must satisfy the following relationship given in Fig. 7.18b:

$$\alpha - \beta \tan\theta = e_C \tan\theta - e_C$$

Thus $\alpha - 1.079\beta = 7.155 \times 1.079 - 0 = 7.720$.

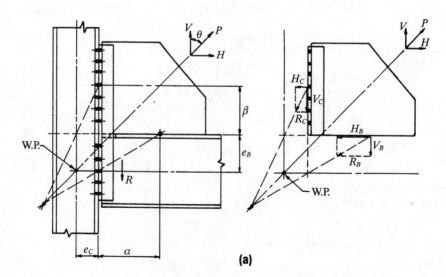

Fig. 7.18a The uniform force method.

From the geometry given in Fig. 7.17, a 7-row connection at 4-in pitch will give $\beta = 17.5$ in. Then $\alpha = 7.720 + 1.079 \times 17.5 = 26.6$ in and the horizontal length of the gusset is $(26.6 - 1) \times 2 + 1 = 52.2$ in. Choose a gusset length of $52\frac{1}{4}$ in. With $\alpha = 26.6$ and $\beta = 17.5$,

$$r = \sqrt{(\alpha + e_C)^2 + (\beta + e_B)^2} = \sqrt{(26.6 + 0)^2 + (17.5 + 7.155)^2} = 36.27$$

and

$$V_C = \frac{\beta}{r}P = \frac{17.5}{36.27} \times 855 = 413 \text{ kips}$$

$$H_C = \frac{e_C}{r}P = 0 \text{ kips}$$

$$H_B = \frac{\alpha}{r}P = \frac{26.6}{36.27} \times 855 = 627 \text{ kips}$$

$$V_B = \frac{e_b}{r}P = \frac{7.155}{36.27} \times 855 = 169 \text{ kips}$$

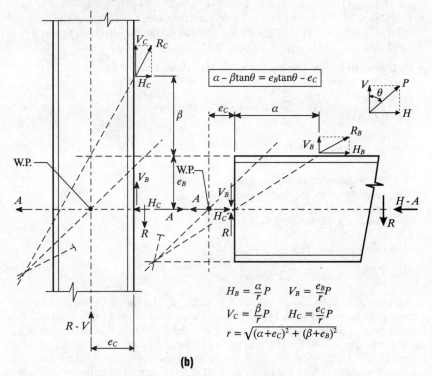

(b)

Fig. 7.18b Force distributions for the uniform force method.

7-36 ■ Chapter Seven

2. Gusset to Column. The loads are 413 kips shear and 0 kips axial.
 a. Bolts and clip angles
Bolts: A325SC-B-N $1^1/_8 \phi$; STD holes
Clip angles: Try L's $4 \times 4 \times \frac{1}{2}$
The shear per bolt is

$$V = 413/14 = 29.5 \text{ kips} < 31.6 \text{ kips} \qquad \text{OK}$$

The bearing strength of the clip angles is

$$\phi R_p = 0.75 \times 2.4 \times 58 \times 0.5 \times 1.125 \times 14 = 822 \text{ kips} > 413 \text{ kips} \qquad \text{OK}$$

The bearing strength of the W14 × 109 column web is

$$\phi R_p = \frac{822 \times 0.525}{0.5} = 863 \text{ kips} > 413 \text{ kips} \qquad \text{OK}$$

The net shear (or block shear rupture) strength of the clips is

$$\phi R_{bs} = 0.75 \times 0.6 \times 58 \,(28 - 7 \times 1.25) \times 0.5 \times 2 = 502 \text{ kips} > 413 \text{ kips} \qquad \text{OK}$$

 b. Fillet weld of clip angles to gusset. The length of this clip angle weld is 28 in. From AISC LRFD Manual, vol. II, Table 8-42, $l = 28$, $kl = 3.0$, $k = 0.107$, $al = 4 - xl = 4 - 0.009 \times 28 = 3.75$ and $a = 0.134$. By interpolation, $c = 1.71$, and the required fillet weld size is $D = 413/(1.73 \times 28 \times 2) = 4.26$, so the required fillet weld size is $^5/_{16}$, and no proration is required because of the $^3/_4$-in-thick gusset.

3. Gusset to Beam. The loads are 627 kips shear and 169 kips axial.
 a. Gusset stresses

$$f_v = \frac{627}{0.75 \times 51.25} = 16.3 \text{ ksi} < 0.9 \times 0.6 \times 36 = 19.4 \text{ ksi} \qquad \text{OK}$$

$$f_a = \frac{169}{0.75 \times 51.25} = 4.40 \text{ ksi} < 0.9 \times 36 = 32.4 \text{ ksi} \qquad \text{OK}$$

 b. Weld of gusset to beam bottom flange
The resultant force per inch of weld is

$$f_r = \sqrt{16.3^2 + 4.40^2} \times \frac{0.75}{2} = 6.33 \text{ kips/in}$$

The required weld size is

$$D = \frac{6.33}{1.392} \times 1.4 = 6.37$$

which indicates that a $^7/_{16}$-in fillet weld is required. The factor 1.4 is a ductility factor from the work of Richard (1986). Even though the stress in this weld is calculated as being uniform, it is well known that there will be local peak stresses, especially in the area where the brace to gusset connection comes close to the gusset to beam weld. An indication of high stress in this area is also indicated by the Whitmore section cutting into the beam web. Also, as discussed later, frame action will give rise to distortion forces which modify the force distribution given by the UFM.

Connections ■ 7-37

As an alternate method to determine the size of this weld, LRFD Table 8-38 can be used. The resultant force $P_u = \sqrt{627^2 + 169^2} = 649$ kips and the angle between the resultant and the weld axis is $\tan^{-1}(169/627) = 15.1°$. From Table 8-38 for 15° inclination of load, and $k = a = 0.0$, $c = 2.97$. Thus $D = (649/2.97 \times 51.25) \times 1.4 = 5.97$, or a ³⁄₈-in fillet weld is required. Figure 7.17 shows the ³⁄₈-in fillet weld.

c. Checks on the beam web. The 627-kips shear is passed into the beam through the gusset to beam weld. All of this load is ultimately distributed over the full cross section of the W14×82, 411 kips passes to the right, and 216 kips is transferred across the column. The length of web required to transmit 627 kips of shear is l_{web}, where $627 = 0.9 \times 0.6 \times 50 \times 0.510 \times l_{web}$. Thus

$$l_{web} = \frac{627}{0.9 \times 0.6 \times 50 \times 0.510} = 45.5 \text{ in}$$

which is reasonable. Note that this length can be longer than the gusset to beam weld but probably should not exceed about half the beam span.

The vertical component can cause beam web yielding and crippling.

i. Web yielding: The web yield design strength is

$\phi R_{wy} = 1 \times 0.66 \times 50(51.25 + 2.5 \times 1.625) = 1825$ kips > 169 kips OK

ii. Web crippling: The web crippling design strength is

$$\phi R_{wcp} = 0.75 \times 135 \times t_w^2 \left[1 + 3\left(\frac{N}{d}\right)\left(\frac{t_w}{t_f}\right)^{1.5}\right] \sqrt{\frac{F_y t_f}{t_w}}$$

$$= 0.75 \times 135 \times 0.510^2 \left[1 + 3\left(\frac{51.25}{14.31}\right)\left(\frac{0.510}{0.855}\right)^{1.5}\right] \sqrt{\frac{50 \times 0.855}{0.510}}$$

$$= 1435 \text{ kips} > 167 \text{ kips} \qquad\qquad\qquad\qquad\qquad\qquad\qquad\qquad \text{OK}$$

The above two checks on the beam web seldom control but should be checked "just in case." The web crippling formula used is that for locations not near the beam end because the beam to column connection will effectively prevent crippling near the beam end. The physical situation is closer to that at some distance from the beam end rather than that at the beam end.

4. Beam to Column. The loads are 216 kips axial, the specified transfer force, and a shear which is the sum of the nominal minimum beam shear of 10 kips and the vertical force from the gusset to beam connection of 169 kips. Thus, the total shear is $10 + 169 = 179$ kips.

a. Bolts and end plate. As established earlier in this example, the bolt design strength in shear is $\phi r_{str} = 31.6$ kips. In this connection, since the bolts also see a tensile load, there is an interaction between tension and shear which must be satisfied. If V is the factored shear per bolt, the design tensile strength is

$$\phi r'_t = 1.13T_b\left(1 - \frac{V}{\phi r_{str}}\right) \leq 0.75 \times 90 A_b$$

where T_b is the bolt pretension of 56 kips for A325 $1^1/_8$-in-diameter bolts and A_b is the bolt nominal area $= \pi/4 \times 1.125^2 = 0.994$ in^2. For $V = 179/10 = 17.9$ kips < 31.6 kips (OK), $\phi r'_t = 1.13 \times 56 (1 - 17.9/31.6) = 27.4$ kips and $0.75 \times 90 \times 0.994 = 67.1$ kips. Thus $\phi r'_t = 27.4$ kips and $\phi R_t = 10 \times 27.4 = 274$ kips > 216 kips OK. Checking the interaction for an N-type bearing connection,

$$\phi r_t = \phi(117 A_b - 1.9V) \leq \phi(90) A_b$$

Thus $\phi r'_t = 0.75(117 \times 0.994 - 1.9 \times 17.9) = 61.7$ kips and $0.75 \times 90 \times 0.994 = 67.1$ kips, so $\phi r'_t = 61.7$ kips > 27.4 kips means that bearing does not control.

To determine the end plate thickness required, the critical dimension is the distance b from the face of the beam web to the center of the bolts. For $5^1/_2$ cross-centers $b = (5.5 - 0.5)/2 = 2.5$ in. To make the bolts above and below the flanges approximately equally critical, they should be placed no more than $2^1/_2$ in above and below the flanges. Figure 7.17 shows them placed at 2 in. Let the end plate be 11 in wide. Then $a = (11 - 5.5)/2 = 2.75 < 1.25 \times 2.5 = 3.125$ OK. The edge distance at the top and bottom of the end plate is 1.5 in, which is more critical than 2.75 in, and will be used in the following calculations. The notation for a and b follows that of the AISC LRFD Manual (1994), as does the remainder of this procedure.

$$b' = b - \frac{d}{2} = 2.5 - \frac{1.125}{2} = 1.9375$$

$$a' = a + \frac{d}{2} = 1.5 + \frac{1.125}{2} = 2.0625$$

$$\rho = \frac{b'}{a'} = 0.94$$

$$\beta = \frac{1}{\rho}\left(\frac{\phi r'_t}{T} - 1\right)$$

where T = factored tension per bolt = 21.6 kips

$$\beta = \frac{1}{0.94}\left(\frac{27.4}{21.6} - 1\right) = 0.286$$

$$\alpha' = \min\left[\frac{1}{\delta}\left(\frac{\beta}{1-\beta}\right), 1\right]$$

where $\delta = 1 - d'/p = 1 - 1.1875/4 = 0.70$

In the above expression p is the tributary length of end plate per bolt. For the bolts adjacent to the beam web, this is obviously 4 in. For the bolts adjacent to the flanges, it is also approximately 4 in for p since at $b = 2.0$ in, a 45° spread from the center of the bolt gives $p = 4$ in. Note also that p cannot exceed one-half of the width of the end plate.

$$\alpha' = \min\left[\frac{1}{0.70}\left(\frac{0.286}{1-0.286}\right),1\right] = 0.572$$

The required end plate thickness is

$$t_{\text{req'd}} = \sqrt{\frac{4.44Tb'}{pF_y(1+\delta\alpha')}} = \sqrt{\frac{4.44\times 21.6 \times 1.9375}{4.0\times 36\times(1+0.70\times 0.572)}} = 0.960 \text{ in}$$

Use a 1-in end plate, 11 in wide and $14\frac{1}{4} + 2 + 2 + 1\frac{1}{2} + 1\frac{1}{2} = 21.25$ in long.

b. Weld of beam to end plate. All of the shear of 179 kips exists in the beam web before it is transferred to the end plate by the weld of the beam to the end plate. The shear capacity of the beam web is

$$\phi R_v = 0.9 \times 0.6 \times 50 \times 0.510 \times 14.31 = 197 \text{ kips} > 179 \text{ kips} \qquad \text{OK}$$

The weld to the end plate that carries this shear is the weld to the beam web plus the weld around to about the k_1 distance inside the beam profile and $2k_1$ on the outside of the flanges. This length is thus $2(d - 2t_f) + 4[k_1 - (t_w/2)] + 4k_1 = 2 \times (14.31 - 2 \times 0.855) + 4 \times (1 - 0.510/2) + 4 \times 1 = 32.2$ in. The force in this weld per inch due to shear is

$$f_v = \frac{179}{32.2} = 5.56 \text{ kips/in}$$

The length of weld that carries the axial force of 216 kips is the entire profile weld whose length is $4 \times 10.13 - 2 \times 0.510 + 2 \times 14.31 = 68.1$ in. The force in this weld per inch due to axial force is

$$f_a = \frac{216}{68.1} = 3.17 \text{ kips/in}$$

Also, where the bolts are close together, a "hot spot" stress should be checked. The most critical bolt in this regard is the one at the center of the W14×82. The axial force in the weld local to these bolts is

$$f'_a = \frac{2\times 21.6}{8} = 5.4 \text{ kips/in}$$

The controlling resultant force in the weld is thus

$$f_R = \sqrt{5.56^2 + 5.4^2} = 7.75 \text{ kips/in}$$

and the required weld size is

$$D = \frac{7.75}{1.392} = 5.57 \quad \text{or} \quad \frac{3}{8} \text{ fillet weld}$$

As a final check, make sure that the beam web can deliver the axial force to the bolts. The tensile load for 2 bolts is $2 \times 21.6 = 43.2$ kips, and 4 in of the beam web must be capable of delivering this load, i.e., providing a load path. The tensile capacity of 4 in of beam web is $4 \times 0.510 \times 0.9 \times 50 = 91.8$ kips > 43.2 kips   OK

Some Observations on the Design of Gusset Plates ■ It is a tenet of all gusset plate design that it must be able to be shown that the stresses on any cut section of the gusset do not exceed the yield stresses on this section. Now, once the resultant forces on the gusset horizontal and vertical sections are calculated by the UFM, the resultant forces on any other cut section, such as section *a-a* of Fig. 7.17, are easy to calculate (see the appropriate free-body diagram incorporating this section, as shown in Fig. 7.19, where the resultant forces on section *a-a* are shown), but the determination of the stresses is not. The traditional approach to the determination of stresses, as mentioned in many books [(Blodgett (1966), Gaylord and Gaylord (1972), Kulak, Fisher, and Struik (1987)], and papers [(Whitmore (1952), Vasarhelyi (1971)], is to use the formulas intended for long, slender members, i.e., $f_a = P/A$ for axial stress, $f_b = Mc/I$ for bending stress, and $f_v = V/A$ for shear stress. It is well known that these are not correct for gusset plates (Timoshenko, 1970). They are recommended only because there is seemingly no alternative. Actually, the uniform force method, coupled with the Whitmore section and the block shear fracture (tearout) limit state, is an alternative, as will be shown subsequently.

Applying the slender member formulas to the section and forces of Fig. 7.19, the stresses and stress distribution of Fig. 7.20 result. The stresses are calculated as

Shear: $$f_v = \frac{291}{0.75 \times 42} = 9.24 \text{ ksi}$$

Axial: $$f_a = \frac{314}{0.75 \times 42} = 9.97 \text{ ksi}$$

Bending: $$f_b = \frac{7276 \times 6}{0.75 \times 42^2} = 33.0 \text{ ksi}$$

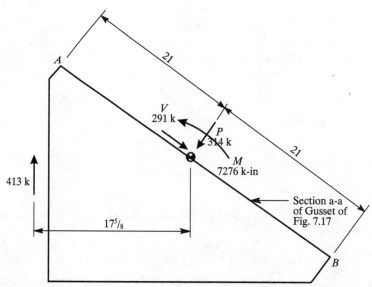

Fig. 7.19 Free-body diagram of portion of gusset cut at section *a-a* of Fig. 7.17.

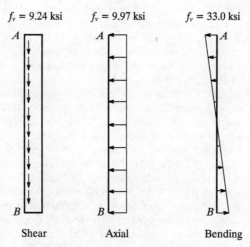

Fig. 7.20 Traditional cut section stresses.

These are the basic "elastic"* stress distributions. The peak stress occurs at point A and is

Shear: $f_V = 9.24$ ksi

Normal: $f_a + f_b = 9.97 + 33.0 = 43.0$ ksi

The shear yield stress (design strength) is $\phi F_V = \phi(0.6 F_y) = 0.9(0.6 \times 36) = 19.4$ ksi. Since 9.24 < 19.4, the section has not yielded in shear. The normal yield stress (design strength) is $\phi F_n = \phi F_y = 0.9(36) = 32.4$ ksi. Since 43.0 > 32.4, the yield strength has been exceeded at point A. At this point, it appears that the design is unsatisfactory (i.e., not meeting AISC requirements). But consider that the normal stress exceeds yield over only about 11 in of the 42-in-long section starting from point A. The remaining 42 − 11 = 31 in have not yet yielded. This means that failure has not occurred because the elastic portion of the section will constrain unbounded yield deformations; i.e., the deformation is "self-limited." Also, the stress of 43.0 ksi is totally artificial! It cannot be achieved in an elastic perfectly plastic material with a design yield point of 32.4 ksi. What *will* happen is that when the design yield point of 32.4 ksi is reached, the stresses on the section will redistribute until the design yield point is reached at *every* point of the cross section. At this time, the plate will fail by unrestrained yielding if the applied loads are such that higher stresses are required for equilibrium.

---

* Actually the shear stress is not elastic because it is assumed uniform. The slender beam theory elastic shear stress would have a parabolic distribution with a peak stress of $9.24 \times 1.5 = 13.9$ ksi at the center of the section.

To conclude on the basis of 43.0 ksi at point A that the plate has failed is thus false. What must be done is to see if a redistributed stress state on the section can be achieved which nowhere exceeds the design yield stress. Note that if this can be achieved, all AISC requirements will have been satisfied. AISC specifies that the design yield stress should not be exceeded, but it does *not* specify the formulas used to determine this.

The shear stress f_v and the axial stress f_a are already assumed uniform. Only the bending stress f_b is nonuniform. To achieve simultaneous yield over the entire section, the bending stress must be adjusted so that when combined with the axial stress, a uniform normal stress is achieved. To this end, consider Fig. 7.21. Here the bending stress is assumed uniform but of different magnitudes over the upper and lower parts of the section. Note that this can be done because M of Fig. 7.19, although shown at the centroid of the section, is actually a free vector which can be applied anywhere on the section or indeed anywhere on the free-body diagram. This being the case, there is no reason to assume that the bending stress distribution is symmetrical about the center of the section. Considering the distribution shown in Fig. 7.21, because the stress from A to the center is too high, the zero point of the distribution can be allowed to move down the amount e toward B. Equating the couple M of Fig. 7.19 to the statically equivalent stress distribution of Fig. 7.21 and taking moments about point D,

$$M = \frac{t}{2}\left[f_1(a+e)^2 + f_2(a-e)^2\right]$$

where t is the gusset thickness. Also, from equilibrium

$$f_1(a+e)t = f_2(a-e)t$$

The above two equations permit a solution for f_1 and f_2 as

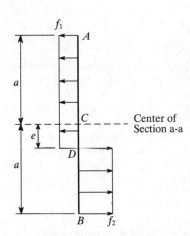

Fig. 7.21 Admissable bending stress distribution of section a-a.

$$f_1 = \frac{M}{at(a+e)}$$

$$f_2 = \frac{M}{at(a-e)}$$

For a uniform distribution of normal stress,

$$f_1 + f_a = f_2 - f_a$$

from which e can be obtained as

$$e = \frac{1}{2}\left[\sqrt{\left(\frac{M}{atf_a}\right)^2 + 4a^2} - \frac{M}{atf_a}\right]$$

Substituting numerical values,

$$e = \frac{1}{2}\left[\sqrt{\left(\frac{7276}{(21)(0.75)(9.97)}\right)^2 + 4(21)^2} - \frac{7276}{(21)(0.75)(9.97)}\right] = 8.10 \text{ in}$$

Thus,

$$f_1 = \frac{7276}{(21)(0.75)(21+8.10)} = 15.9 \text{ ksi}$$

$$f_2 = \frac{7276}{(21)(0.75)(21-8.10)} = 35.8 \text{ ksi}$$

and the normal stress at point A is

$$f_{nA} = f_1 + f_a = 15.9 + 9.97 = 25.9 \text{ ksi}$$

and at point B

$$f_{nB} = f_2 - f_a = 35.8 - 9.97 = 25.9 \text{ ksi}$$

Now the entire section is uniformly stressed. Since

$$f_v = 9.24 \text{ ksi} < 19.4 \text{ ksi}$$

$$f_n = 25.9 \text{ ksi} < 32.4 \text{ ksi}$$

at all points of the section, the design yield stress is nowhere exceeded and the connection is satisfactory.

It was stated above that there is an alternative to the use of the inappropriate slender beam formulas for the analysis and design of gusset plates. The above analysis of the special section a-a demonstrates the alternative which results in a

7-44 ■ Chapter Seven

true limit state (failure mode or mechanism) rather than the fictitious calculation of "hot spot" point stresses which, since their associated deformation is totally limited by the remaining elastic portions of the section, cannot correspond to a true failure mode or limit state. The uniform force method performs exactly the same analysis on the gusset horizontal and vertical edges, and on the associated beam to column connection. It is capable of producing forces on all interfaces which give rise to uniform stresses. Each interface is designed to just fail under these uniform stresses. Therefore, true limit states are achieved at every interface. For this reason, the uniform force method achieves a good approximation to the greatest lower-bound solution (closest to the true collapse solution) in accordance with the lower-bound theorem of limit analysis.

The uniform force method is a complete departure from the so-called traditional approach to gusset analysis using slender beam theory formulas. It has been validated against all known full-scale gusseted bracing connection tests (Thornton, 1991, 1995B). It does not require the checking of gusset sections such as that studied in this section (section a-a of Fig. 7.19). The analysis at this section was done to prove a point. But the uniform force method does include a check in the brace to gusset part of the calculation which is closely related to the special section a-a of Fig. 7.19. This is the shear rupture or "tearout" section of Fig. 7.22 [Hardash and Bjorhovde, 1985), (Richard, 1983)] which is included in section J4 of the AISC LRFD specification (AISC, 1994). Following the notation of the AISC LRFD commentary, the tearout capacity is calculated as follows:

The gross shear area is

$$A_{vg} = (27 + 2.0) \times 0.75 \times 2 = 43.5 \text{ in}^2$$

The net shear area is

$$A_{vn} = 43.5 - 6.5 \times 1.25 \times 0.75 \times 2 = 31.3 \text{ in}^2$$

The gross tension area is

$$A_{tg} = 6.5 \times 0.75 = 4.875 \text{ in}^2$$

The net tension area is

$$A_{tn} = 4.875 - 1 \times 1.25 \times 0.75 = 3.9375 \text{ in}^2$$

The shear fracture design strength is

$$\phi F_{vf} = \phi(0.6 F_u \times A_{vn}) = 0.75 \times 0.6 \times 58 \times 31.3 = 817 \text{ kips}$$

The tension fracture design strength is

$$\phi F_{tf} = \phi(F_u \times A_{tn}) = 0.75 \times 58 \times 3.9375 = 171 \text{ kips}$$

The controlling limit state is the one with the larger fracture component. Therefore, associated with the shear fracture, calculate tension yield as

$$\phi F_y = \phi(F_y \times A_g) = 0.75 \times 36 \times 4.875 = 132 \text{ kips}$$

The shear fracture (tearout) design strength is thus

$$\phi F_{to} = 817 + 132 = 949 \text{ kips}$$

Since 949 kips > 855 kips, the gusset plate is satisfactory for tearout.

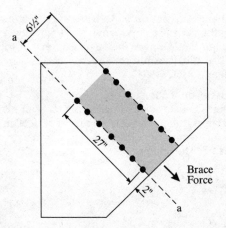

Fig. 7.22 Block shear rupture and its relation to gusset section a-a.

Comparing the tearout limit state to the special section a-a limit state, a reserve capacity in tearout = [(949 − 855)/949]100 = 9.9 percent is found, and the reserve capacity of the special section [(32.4 − 25.9)/32.4]100 = 20 percent which shows that tearout gives a conservative prediction of the capacity of the closely related special section.

A second check on the gusset performed as part of the UFM is the Whitmore section check. From Fig. 7.17, the Whitmore section area is

$$A_w = (37.7 - 10.4) \times 0.75 + 10.4 \times 0.510 \times \frac{50}{36} = 27.8 \text{ in}^2$$

The Whitmore section design strength in tension is

$$\phi F_w = \phi(F_y \times A_w) = 0.9(36 \times 27.8) = 900 \text{ kips}$$

The reserve capacity of the Whitmore section in tension is [(900 − 855)/900] × 100 = 5 percent, which again gives a conservative prediction of capacity when compared to the special section a-a.

With these two limit states, block shear rupture (tearout) and Whitmore, the special section limit state is closely bounded and rendered unnecessary. The routine calculations associated with tearout and Whitmore are sufficient in practice to eliminate the consideration of any sections other than the gusset to column and gusset to beam sections.

Example 2 ■ This connection is shown in Fig. 7.23. The member on the right of the joint is a "collector" that adds load to the bracing truss. The design in this example is for seismic loads. The brace consists of 2 MC12 × 45's with toes 1½ in apart. The gusset thickness is thus chosen to be 1½ in and is then checked. The completed design is shown in Fig. 7.23. In this case, because of the high specified

7-46 ■ Chapter Seven

beam shear of 170 kips, it is proposed to use a special case of the uniform force method which sets the vertical component of the load between the gusset and the beam, V_B to zero. Figure 7.24 shows the resultant force distribution. This method is called "Special Case 2" of the uniform force method and is discussed in the AISC books, AISC (1992) and AISC (1994).

1. Brace to Gusset Connection

a. **Weld.** The brace is field welded to the gusset with fillet welds. Because of architectural constraints, the gusset size is to be kept to 30 in horizontally to $24\frac{1}{2}$ in vertically. From the geometry of the gusset and brace, about 17 in of fillet weld can be accommodated. The weld size is

$$D = \frac{855}{4 \times 17 \times 1.392} = 9.03$$

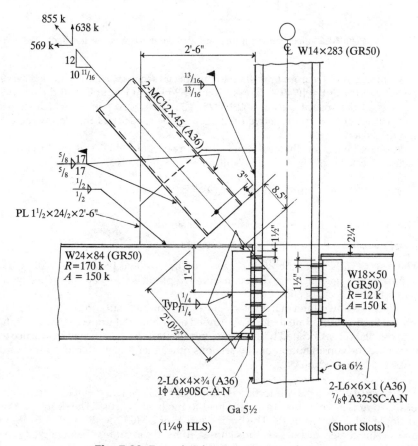

Fig. 7.23 Example 2, bracing connection design.

A ⅝-in fillet weld is indicated, but the flange of the MC 12×45 must be checked to see if an adequate load path exists. The average thickness of 0.700 in occurs at the center of the flange, which is 4.012 in wide. The thickness at the toe of the flange, because of the usual inside flange slope of $2/12$ or $16\tfrac{2}{3}$ percent, is $0.700 - 2/12 \times 2.006 = 0.366$ in (see Fig. 7.25). The thickness at the toe of the fillet is $0.366 + 2/12 \times 0.625 = 0.470$ in. The design shear rupture strength of the MC 12 flange at the toe of the fillet is

$$\phi R_v = 0.75 \times 0.6 \times 58 \times 0.470 \times 17 \times 4 = 834 \text{ kips}$$

The design tensile rupture strength of the toe of the MC flange under the fillet is

$$\phi R_t = 0.75 \times 36 \times \left(\frac{0.366 + 0.470}{2}\right) \times 0.625 \times 4 = 28 \text{ kips}$$

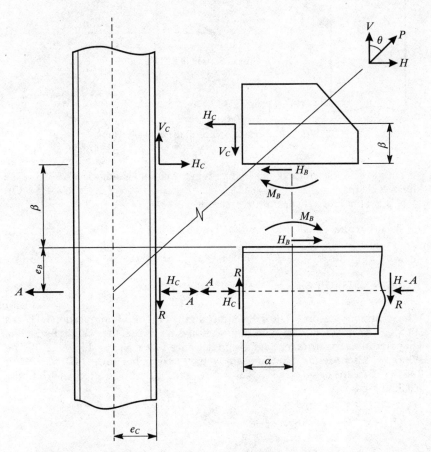

Fig. 7.24 Force distribution for special case 2 of the uniform force method.

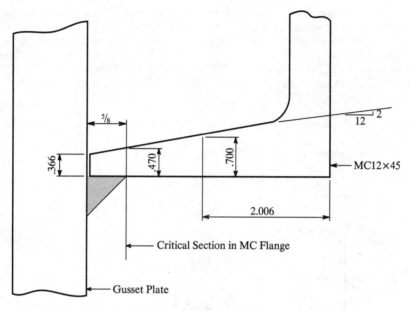

Fig. 7.25 Critical section at toe of fillet weld.

Thus the total strength of the load path in the channel flange is 834 + 28 = 862 kips > 855 kips   OK

   b. Gusset to brace shear rupture (tearout)

Shear fracture: $\phi R_v = 0.75 \times 0.6 \times 58 \times 1.5 \times 17 \times 2 = 1331$ kips

Tension fracture: $\phi R_t = 0.75 \times 58 \times 1.5 \times 12 = 783$ kips

$\phi R_{bs} = 1331 + 0.75 \times 36 \times 1.5 \times 12 = 1817$ kips > 855 kips   OK

   c. Whitmore section.   The theoretical length of the Whitmore section is (17 tan 30) × 2 + 12 = 31.6 in. The Whitmore section extends into the column by 5.40 in. The column web is stronger than the gusset since 1.29 × 50/36 = 1.79 > 1.5 in. The Whitmore also extends into the beam web by 6.80 in, but since 0.470 × 50/36 = 0.653 < 1.5 in, the beam web is not as strong as the gusset. The effective Whitmore section length is

$$l_{w\,\text{eff36}} (31.6 - 6.80) + 6.80 \times \frac{0.470}{1.5} \times \frac{50}{36} = 27.8 \text{ in}$$

The effective length is based on $F_y = 36$ and the gusset thickness of 1.5 in.

Since the brace force can be tension or compression, compression will control. The slenderness ratio of the unsupported length of gusset is

$$\frac{Kl}{r} = \frac{0.5 \times 8.5 \times \sqrt{12}}{1.5} = 9.8$$

From LRFD table 3-36, the buckling strength is

$\phi F_a = 30.4$ ksi

and the buckling strength of the gusset is

$\phi R_{wb} = 27.8 \times 1.5 \times 30.4 = 1268 > 855$ kips   OK

This completes the brace to gusset part of the design. Before proceeding, the distribution of forces to the gusset edges must be determined. From Fig. 7.23, $e_B = 24.10/2 = 12.05$, $e_c = 8.37$, $\bar{\beta} = 12.25$, $\bar{\alpha} = 15.0$, $\theta = \tan^{-1}(10.6875/12) = 41.6°$.

$V_C = P\cos\theta = 855 \times 0.747 = 638$ kips

$$H_C = \frac{V_C e_C}{e_B + \bar{\beta}} = \frac{638 \times 8.37}{12.05 + 12.25} = 220 \text{ kips}$$

$H_B = P\sin\theta - H_C = 855 \times 0.665 - 220 = 349$ kips

$M_B = H_B e_B = 349 \times 12.05 = 4205$ kip-in

Note that in this Special Case 2, the calculations can be simplified as shown above. The same results can be obtained formally with the UFM by setting $\beta = \bar{\beta} = 12.25$ and proceeding as follows: With $\tan\theta = 0.8906$, $\alpha - 0.8906$, $\beta = 12.05 \times 0.8906 - 8.37 = 2.362$. Setting $\beta = \bar{\beta} = 12.25$, $\alpha = 13.27$. Since $\bar{\alpha} = 15.0$, there will be a couple M_B on the gusset to beam edge. Continuing

$$r = [(13.27 + 8.37)^2 + (12.25 + 12.05)^2]^{\frac{1}{2}} = 32.54$$

$$\frac{P}{r} = 26.27$$

$$H_B = \frac{\alpha}{r}P = 349 \text{ kips}$$

$$H_C = \frac{e_C}{r}P = 220 \text{ kips}$$

$$V_B = \frac{e_B}{r}P = 317 \text{ kips}$$

$$V_C = \frac{\beta}{r}P = 322 \text{ kips}$$

$$M_B = |V_B(\alpha - \bar{\alpha})| = 548 \text{ kip-in}$$

This couple is clockwise on the gusset edge. Now, introducing Special Case 2, in the notation of the AISC LRFD Manual (1994), vol. II, p. 11-22, set $\Delta V_B = V_B = 317$ kips. This reduces the vertical force between the gusset and beam to zero, increases the gusset to column shear V_C to $317 + 322 = 639$ kips, and creates a counterclockwise couple on the gusset to beam edge of $\Delta V_B \bar{a} = 317 \times 15.0 = 4755$ kip-in. The total couple on the gusset to beam edge is thus $M_B = 4755 - 548 = 4207$ kip-in. It can be seen that these gusset interface forces are the same as those obtained from the simpler method.

2. Gusset to Column. The loads are 638 kips shear and 220 kips axial.
a. Gusset stresses:

$$f_v = \frac{638}{1.5 \times 24.5} = 17.4 \text{ ksi} < 0.9 \times 0.6 \times 36 = 19.4 \text{ ksi} \qquad \text{OK}$$

$$f_a = \frac{220}{1.5 \times 24.5} = 5.98 \text{ ksi} < 0.9 \times 36 = 32.4 \text{ ksi} \qquad \text{OK}$$

b. Weld of gusset to column flange. Using AISC LRFD Table 8-38, $P_u = \sqrt{638^2 + 220^2} = 675$ kips and the angle from the longitudinal weld axis is $\tan^{-1}(220/638) = 19°$, so using the table for 15° with $k = a = 0.0, c = 2.97$. Thus

$$D = \frac{675}{2.97 \times 24.5} \times 1.4 = 12.98$$

which indicates that a $^{13}/_{16}$ fillet is required. The ductility factor 1.4 is used because the weld is assumed to be uniformly loaded, but research shows that the ratio of peak to average stresses is about 1.4 (Richard, 1986).

c. Checks on column web.
i. Web yielding (under normal load H_C):

$$\phi R_{wy} = 1.0 \times 0.66 \times 50 \times 1.290 \,(24.5 + 5 \times 2^{3}/_{4}) = 1628 \text{ kips} > 220 \text{ kips} \qquad \text{OK}$$

ii. Web crippling (under normal load H_C):

$$\phi R_{wcp} = 0.75 \times 135 \times 1.290^2 \left[1 + 3 \left(\frac{24.5}{16.74} \right) \left(\frac{1.290}{2.070} \right)^{1.5} \right] \sqrt{\frac{50 \times 2.070}{1.290}}$$

$$= 4769 \text{ kips} > 220 \text{ kips} \qquad \text{OK}$$

iii. Web shear: The horizontal force H_C is transferred to the column by the gusset to column connection, and back into the beam by the beam to column connection. The situation is similar to that shown in Fig. 7.21, with $f_1(a + e)t = f_2(a - e)t = H_C$. Thus the column web sees $H_C = 220$ kips as a shear. The column shear capacity is

$$\phi R_v = 0.9 \times 0.6 \times 50 \times 1.290 \times 16.74 = 583 \text{ kips} > 220 \text{ kips} \qquad \text{OK}$$

3. Gusset to Beam. The loads are 349 kips shear and a 4205 kip-in couple.
a. Gusset Stresses

$$f_v = \frac{349}{1.5 \times 30} = 7.76 \text{ ksi} < 19.4 \text{ ksi} \qquad \text{OK}$$

$$f_a = 0$$

$$f_b = \frac{4205 \times 4}{1.5 \times 30^2} = 12.5 \text{ ksi} < 32.4 \text{ ksi} \qquad \text{OK}$$

b. Weld of gusset to beam flange

$$f_R = \sqrt{7.76^2 + 12.5^2} \times \frac{1.5}{2} = 11.0 \text{ kips/in}$$

$$f_{ave} = \sqrt{7.76^2 + 6.25^2} \times \frac{1.5}{2} = 7.47 \text{ kips/in}$$

Since $11.0/7.47 = 1.47 > 1.4$, the weld size based on the peak force in the weld f_R effectively includes a ductility factor; therefore

$$D = \frac{11.0}{1.392} = 7.9$$

A $\frac{1}{2}$ fillet weld is indicated.

An alternate method for calculating the weld size required is to use Table 8-38 of the AISC LRFD, vol. II, p. 8-163, special case $k = 0$, $P_u = 349$, and $al = 4205/349 = 12.05$ in; thus $a = 12.05/30 = 0.40$ and $c = 2.00$ and the required weld size is

$$D = \frac{349}{2.0 \times 30} = 5.8$$

A $\frac{3}{8}$ fillet is indicated. This method does not give an indication of peak and average stresses, but it will be safe to use the ductility factor. Thus the required weld size would be

$$D = 5.8 \times 1.4 = 8.12$$

Thus a $\frac{9}{16}$ (8.12) fillet is indicated, which is about the same size (within 3 percent) as the $\frac{1}{2}$ (7.9) required by the classical method. The $\frac{1}{2}$ fillet weld will be OK in this case.

c. Checks on beam web.
i. Web yield: Although there is no axial component, the couple $M_B = 4205$ kip-in is statically equivalent to equal and opposite vertical shears at a lever arm of one-half the gusset length or 15 in. The shear is thus

$$V_s = \frac{4205}{15} = 280 \text{ kips}$$

This shear is applied to the flange as a transverse load over 15 in of flange. It is convenient for analysis purposes to imagine this load doubled and applied over the contact length $N = 30$ in. The design web yielding strength is

$$\phi R_{wy} = 1.0 \times 0.66 \times 50 \times 0.470(30 + 5 \times 1\tfrac{9}{16}) = 586 \text{ kips} > 280 \times 2 = 560 \text{ kips} \qquad \text{OK}$$

ii. Web crippling:

$$\phi R_{wcp} = 0.75 \times 135 \times 0.470^2 \left[1 + 3\left(\frac{30}{24.10}\right)\left(\frac{0.470}{0.770}\right)^{1.5}\right]\sqrt{\frac{50 \times 0.770}{0.470}}$$

$$= 563 \text{ kips} > 560 \text{ kips}$$

OK

iii. Web Shear

$\phi P_v = 0.9 \times 0.6 \times 50 \times 0.470 \times 24.10 = 306$ kips > 280 kips   OK

The maximum shear due to the couple is centered on the gusset 15 in from the beam end. It does not reach the beam to column connection where the beam shear is 170 kips. Because of the total vertical shear capacity of the beam and the gusset acting together, there is no need to check the beam web for a combined shear of V_s and R of $280 + 170 = 450$ kips.

4. Beam to Column. The shear load is 170 kips and the axial force is $H_C \pm A = 220 + 150$ kips. Since the W18×50 is a collector, it adds load to the bracing system. Thus the axial load is $220 + 150 = 370$ kips. However, the AISC book on connections, AISC (1992), addresses this situation and states that because of frame action (distortion), which will always tend to reduce H_C, it is reasonable to use the larger of H_C and A as the axial force. Thus the axial load would be 220 kips in this case. The connection will be designed in this way and then the method will be justified.

a. Bolts and clips. The bolts are A490 SC-A-N 1 in in diameter in oversize $1\frac{1}{4}$-in-diameter holes. Thus, for shear

$\phi r_{str} = 0.85 \times 1.13 \times 0.33 \times 64 = 20.3$ kips / bolt

and for tension

$\phi r_t = 0.75 \times 113 \times 0.7854 = 66.6$ kips / bolt

The clips are L's $4 \times 4 \times \frac{3}{4}$ with seven rows of bolts. For shear

$\phi R_v = 20.3 \times 14 = 284$ kips > 170 kips   OK

For tension, the bolts and clips are checked together for prying action. Since all of the bolts are subjected to tension simultaneously, there is interaction between tension and shear. The reduced tensile capacity is

$$\phi r'_t = 1.13 \times 64 \left(1 - \frac{170/14}{20.3}\right) = 29.1 \text{ kips / bolt}$$

Since 29.1 kips $> 220 / 14 = 15.7$ kips, the bolts are OK for tension. The bearing-type interaction expression should also be checked but it will not control. Prying action is now checked using the method and notation of the AISC LRFD Manual.

$$b = \frac{5.5 - 0.470}{2} - 0.75 = 1.765$$

$$a = \frac{8 + 0.470 - 5.5}{2} = 1.485$$

Check $1.25b = 1.25 \times 1.765 = 2.206$. Since $2.206 > 1.485$, use $a = 1.485$.

$$b' = 1.765 - \frac{1.0}{2} = 1.265$$

$$a' = 1.485 + \frac{1.0}{2} = 1.985$$

$$\rho = 0.637$$

$$\delta = 1 - \frac{1.25}{3} = 0.583$$

$$t_c = \sqrt{\frac{4.44 \times 29.1 \times 1.265}{3 \times 36}} = 1.23$$

$$\alpha' = \frac{1}{0.583 \times 1.637}\left[\left(\frac{1.23}{0.75}\right)^2 - 1\right] = 1.77$$

Since $\alpha' > 1$, use $\alpha' = 1.0$. The design strength per bolt including prying is

$$T_d = 29.1\left(\frac{0.75}{1.23}\right)^2 (1 + 0.583 \times 1.0) = 17.1 \text{ kips} > 15.70 \text{ kips} \qquad \text{OK}$$

In addition to the prying check, the clips should also be checked for gross and net shear. These will not control in this case.

b. *Weld of clips to beam web.* The weld is a C-shaped weld with length $l = 21$ in, $kl = 3.5$ in, $k = 3.5/21 = 0.167$. From AISC LRFD Manual, Table 8-42, $xl = 0.0220 \times 21 = 0.462$, so $al = 6 - 0.462 = 5.538$, and $a = 5.538/21 = 0.264$. Since $\tan^{-1} 220/170 = 52.3°$, use the chart on p. 8-190 for 45°. By interpolation, $C = 1.92$. A 1/4 fillet weld has a capacity of $\phi R_w = 1.92 \times 4 \times 2 \times 21 = 323$ kips. In order to support this weld, the web thickness required is $0.9 \times 0.6 \times 50 \times t_w \geq 1.392 \times 4 \times 2$. Thus $t_{w\,\text{req'd}} \geq 0.41$ in. Since the actual web thickness is 0.470 in, the weld is fully effective and has the calculated capacity. Thus, since 323 kips $> \sqrt{220^2 + 170^2} = 278$ kips, the 1/4 fillet weld is OK.

c. *Bending of the column flange.* Because of the axial force, the column flange can bend just as the clip angles. A yield-line analysis derived from Mann and Morris (1979) can be used to determine an effective tributary length of column flange per bolt. The yield lines are shown in Fig. 7.26. From Fig. 7.26,

$$p_{\text{eff}} = \frac{(n-1)p + \pi\bar{b} + 2\bar{a}}{n}$$

where

$$\bar{b} = \frac{5.5 - 1.290}{2} = 2.105$$

7-54 ■ Chapter Seven

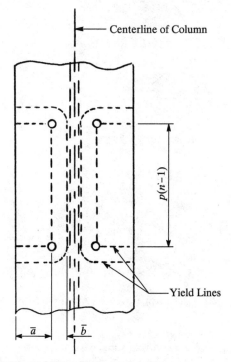

Fig. 7.26 Yield lines for flange bending.

$$\bar{a} = \frac{16.110 - 5.5}{2} = 5.305$$

$p = 3$

$n = 7$

Thus $p_{\text{eff}} = \frac{6 \times 3 + \pi \times 2.105 + 2 \times 5.305}{7} = 5.032$

Using p_{eff} in place of p, and following the AISC procedure,

$b = \bar{b} = 2.105$

$b' = 2.105 - \frac{1.0}{2} = 1.605$

$a = \min\left\{\frac{4 + 4 + 0.470 - 5.5}{2}, 5.305, 1.25 \times 2.105\right\} = \min.\{1.485, 5.305, 2.63\} = 1.485$

$$a' = 1.485 + 0.5 = 1.985$$

$$\rho = \frac{b'}{a'} = 0.81$$

$$\delta = 1 - \frac{1.0625}{5.032} = 0.79$$

Note that standard holes are used in the column flange.

$$t_c = \sqrt{\frac{4.44 \times 29.1 \times 1.605}{5.032 \times 50}} = 0.91$$

$$\alpha' = \frac{1}{0.79 \times 1.81}\left[\left(\frac{0.91}{2.070}\right)^2 - 1\right] = -0.564$$

Since $\alpha' < 0$, use $\alpha' = 0$.

$T_d = 29.1$ kips / bolt > 15.7 kips / bolt   OK

When $\alpha' < 1$, the bolts, and not the flange, control the strength of the connection.

Frame Action ■ The method of bracing connection design presented here, the uniform force method, is an equilibrium-based method. Every proper method of design for bracing connections, and in fact for every type of connection, must satisfy equilibrium. The set of forces derived from the uniform force method, as shown in Fig. 7.18, satisfy equilibrium of the gusset, the column, and the beam with axial forces only. Such a set of forces is said to be "admissible." But equilibrium is not the only requirement that must be satisfied to establish the true distribution of forces in a structure or connection. Two additional requirements are the constitutive equations which relate forces to deformations, and the compatibility equations which relate deformations to displacements.

If it is assumed that the structure and connection behave elastically (an assumption as to constitutive equations) and that the beam and column remain perpendicular to each other (an assumption as to deformation-displacement equations), then an estimate of the moment in the beam due to distortion of the frame (frame action) (Thornton, 1991) is given by

$$M_D = 6 \frac{P}{Abc} \frac{I_b I_c}{\frac{I_b}{b} + \frac{2I_c}{c}} \frac{b^2 + c^2}{bc}$$

where D subscript D denotes distortion, and

I_b = moment of inertia of beam = 2370 in^4
I_c = moment of inertia of column = 3840 in^4
P = brace force = 855 kips
A = brace area = 26.4 in^2

7-56 ■ Chapter Seven

b = length of beam to inflection point (assumed at beam midpoint) = 175 in

c = length of columns to inflection points (assumed at column midlengths) = 96 in

With $2I_c/c = 80$ and $I_b/b = 13.5$,

$$M_D \frac{6 \times 855 \times 2370 \times 3840}{26.4 \times 175 \times 96 \times (13.5 + 80)} \frac{175^2 + 96^2}{175 \times 96} = 2670 \text{ kip-in}$$

This moment M_D is only an estimate of the actual moment that will exist between the beam and column. The actual moment will depend on the strength of the beam to column connection. The strength of the beam to column connection can be assessed by considering the forces induced in the connection by the moment M_D as shown in Fig. 7.27. The distortion force F_D is assumed to act as

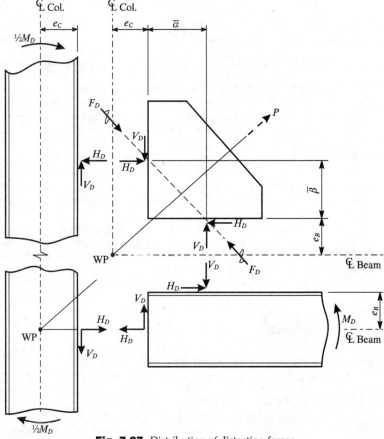

Fig. 7.27 Distribution of distortion forces.

shown through the gusset edge connection centroids. If the brace force P is a tension, the angle between the beam and column tends to decrease, compressing the gusset between them, so F_D is a compression. If the brace force P is a compression, the angle between the beam and column tends to increase and F_D is a tension. Figure 7.27 shows how the distortion force F_D is distributed throughout the connection. From Fig. 7.27, the following relationships exist between F_D, its components H_D and V_D), and M_D:

$$F_D = \sqrt{H_D^2 + V_D^2}$$

$$\overline{\beta} H_D = \overline{\alpha} V_D$$

$$H_D = \frac{M_D}{\overline{\beta} + e_B}$$

For the elastic case with no angular distortion

$$H_D = \frac{2670}{12.25 + 12.05} = 110 \text{ kips}$$

$$V_D = \frac{\overline{\beta}}{\overline{\alpha}} H_D = \frac{12.25}{15} \times 110 = 89.8 \text{ kips}$$

It should be remembered that these are just estimates of the distortion forces. The actual distortion forces will be dependent also upon the strength of the connection. But it can be seen that these estimated distortion forces are not insignificant. Compare, for instance, H_D to H_C. H_C is 220 kips tension when H_D is 110 kips compression. The net axial design force would then be $220 - 110 = 110$ kips rather than 220 kips.

The strength of the connection can be determined by considering the strength of each interface, including the effects of the distortion forces. The following interface forces can be determined from Figs. 7.18 and 7.27.

For the gusset to beam interface:

T_B (tangential force) $= H_B + H_D$

N_B (normal force) $= V_B - V_D$

For the gusset to column interface:

$T_C = V_C + V_D$

$N_C = H_C - H_D$

For the beam to column interface:

$T_{BC} = |V_B - V_D| + R$

$N_{BC} = |H_C - H_D| \pm A$

The only departure from a simple equilibrium solution to the bracing connection design problem was in the assumption that frame action would allow the beam to column connection to be designed for an axial force equal to the maximum of H_c and A, or max (220, 150) = 220 kips. Thus, the design shown in Fig. 7.23 has its beam to column connection designed for N_{BC} = 220 kips and T_{BC} = 170 kips. Hence

$$N_{BC} = |220 - H_D| + 150 = 220$$

means that $H_D = 150$ kips and

$$V_D = \frac{12.25}{15} \times 150 = 122.5 \text{ kips}$$

From

$$T_{BC} = |V_B - 122.5| + 170 = 170$$

$$V_B = 122.5 \text{ kips}$$

Note that in order to maintain the beam to column loads of 170 kips shear and 220 kips tension, the gusset to beam shear V_B must increase from 0 to 122.5 kips. Figure 7.28 shows the transition from the original load distribution to the final distribution

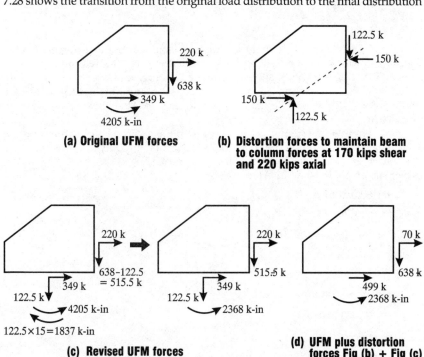

Fig. 7.28 Admissable combining of UFM and distortion forces.

as given in Fig. 7.28d. Note also that N_{BC} could have been set as $17.1 \times 14 = 239$ kips, rather than 220 kips, because this is the axial capacity of the connection at 170 kips shear. The N_{BC} value of 220 kips is used to cover the case when there is no excess capacity in the beam to column connection. Now the gusset to beam and gusset to column interfaces will be checked for the redistributed loads of Fig. 7.28d.

Gusset to beam
a. Gusset stresses

$$f_v = \frac{499}{1.5 \times 30} = 11.1 \text{ ksi} < 19.4 \text{ ksi} \qquad \text{OK}$$

$$f_b = \frac{2368 \times 4}{1.5 \times 30^2} = 7.02 \text{ ksi} < 32.4 \text{ ksi} \qquad \text{OK}$$

b. Weld of gusset to beam flange

$$f_R = \sqrt{11.1^2 + 7.02^2} \times \frac{1.5}{2} = 9.85$$

$$D = \frac{9.85}{1.392} = 7.08$$

A $\frac{1}{2}$-in fillet weld is indicated, which is what was provided. No ductility factor is used here because the loads include a redistribution.

Gusset to Column. This connection is OK without calculations because the loads of Fig. 7.28d are no greater than the original loads of Fig. 7.28a.

Discussion. From the foregoing analysis, it can be seen that the AISC suggested procedure for the beam to column connection, where the actual normal force

$$N_{BC} = |H_C - H_D| \pm A$$

is replaced by

$$N_{BC} = \max (H_C, A)$$

is justified.

It has been shown that the connection is strong enough to carry the distortion forces of Fig. 7.28b, which are larger than the elastic distortion forces.

In general, the entire connection could be designed for the combined uniform force method forces and distortion forces, as shown in Fig. 7.28d for this example. This set of forces is also admissible. The UFM forces are admissible because they are in equilibrium with the applied forces. The distortion forces are in equilibrium with zero external forces. Under each set of forces, the parts of the connection are also in equilibrium. Therefore, the sum of the two loadings is admissible because each individual loading is admissible. A safe design is thus guaranteed by the lower-bound theorem of limit analysis. The difficulty is in determining the distortion forces. The elastic distortion forces could be used, but they are only an estimate of the true distortion forces. The distortion forces depend as much on the properties of the connection, which are inherently inelastic and affect the maintenance of the angle between the members, as on the properties and lengths of the members of the frame. For this example, the distortion forces are $[(150 - 110)/110] \times 100 = 36$ percent greater than the elastic distortion forces. In full-scale tests by Gross (1990) as reported by Thornton (1991), the distortion forces were about $2\frac{1}{2}$ times the elastic distortion forces while the overall frame remained elastic. Because of the difficulty in establishing values for the distortion forces, and because the uniform force method has been shown to be conservative when they are ignored (Thornton, 1991, 1995B), they are not included in bracing connection design, except implicitly as noted here to justify replacing $|H_C - H_D| \pm A$ with max (H_C, A).

Truss Connections ■ The uniform force method (UFM) as originally formulated can be applied to trusses as well as to bracing connections. After all, a vertical bracing system is just a truss. But bracing systems generally involve orthogonal members whereas trusses, especially roof trusses, often have a sloping top chord. In order to handle this situation, the UFM has been generalized as shown in Fig. 7.29 to include nonorthogonal members. As before, α and β locate the centroids of the gusset edge connections and must satisfy the constraint shown in the box on Fig. 7.29. This can always be arranged when designing a connection, but in checking a given connection designed by some other method, the constraint may not be satisfied. The result is gusset edge couples which must be considered in the design.

Example ■ As an application of the UFM to a truss, consider the situation of Fig. 7.30. This is a top chord connection in a large aircraft hangar structure. The truss is cantilevered from a core support area. Thus the top chord is in tension. The design shown in Fig. 7.30 was obtained by generalizing the KISS method (Thornton, 1995B) shown in Fig. 7.31 for orthogonal members to the nonorthogonal case. The KISS method is the simplest admissible design method for truss and bracing connections. On the negative side, however, it generates large, expensive, and unsightly connections. The problem with the KISS method is the couples required on the gusset edges to satisfy equilibrium of all parts. In the Fig. 7.30 version of the KISS method, the truss diagonal horizontal and vertical components are placed at the gusset edge centroids as shown. The couples 15,860 kip-in on the top edge and 3825 kip-in on the vertical edge are necessary for equilibrium of the gusset, top chord, and truss vertical, with the latter two experiencing only axial forces away from the connection. It is these couples which require the $3/4$-in chord doubler plate, the $7/16$-in fillets between the gusset and chord, and the 38-bolt $7/8$-in end plate on the vertical edge.

The design shown in Fig. 7.32 is also obtained by the KISS method with the brace force resolved into tangential components on the gusset edges. Couples still result but are much smaller than in Fig. 7.30. The resulting connection requires no chord doubler plate, $5/16$-in fillets of the gusset to the chord, and a 32-bolt $3/4$-in end plate on the vertical edge. This design is much improved over that of Fig. 7.30.

When the uniform force method of Fig. 7.29 is applied to this problem, the resulting design is as shown in Fig. 7.33. The vertical connection has been reduced to only 14 bolts and a $5/8$-in end plate.

The designs of Figs. 7.30, 7.32, and 7.33 are all satisfactory designs for some admissible force system. For instance, the design of Fig. 7.30 will be satisfactory for the force systems of Figs. 7.32 and 7.33, and the design of Fig. 7.32 will be satisfactory for the force system of Fig. 7.33. How can it be determined which is the "right" or "best" admissible force system to use? The lower-bound theorem of limit analysis provides an answer. This theorem basically says that for a given connection configuration, i.e., Fig. 7.30 or Fig. 7.32 or Fig. 7.33, the statically admissible force distribution which maximizes the capacity of the connection is closest to the true force distribution. As a converse to this, for a given load, the smallest connection satisfying the limit states is closest to the true required connection. Of the three admissible force distributions given in Figs. 7.30, 7.32, and 7.33, the distribution of Fig. 7.33, based on the UFM, is the "best" or "right" distribution.

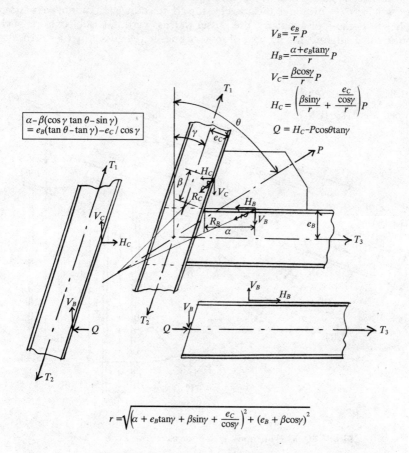

$$V_B = \frac{e_B}{r} P$$
$$H_B = \frac{\alpha + e_B \tan\gamma}{r} P$$
$$V_C = \frac{\beta\cos\gamma}{r} P$$
$$H_C = \left(\frac{\beta\sin\gamma}{r} + \frac{\frac{e_C}{\cos\gamma}}{r}\right) P$$
$$Q = H_C - P\cos\theta\tan\gamma$$

$$\alpha - \beta(\cos\gamma \tan\theta - \sin\gamma) = e_B(\tan\theta - \tan\gamma) - e_C/\cos\gamma$$

$$r = \sqrt{\left(\alpha + e_B\tan\gamma + \beta\sin\gamma + \frac{e_C}{\cos\gamma}\right)^2 + (e_B + \beta\cos\gamma)^2}$$

Fig. 7.29 Generalized uniform force method.

7-62 ■ Chapter Seven

A Numerical Example ■ To demonstrate the calculations required to design the connections of Figs. 7.30, 7.32, and 7.33, for the statically admissible forces of these figures, consider, for instance, the UFM forces and the resulting connection of Fig. 7.33.

The geometry of Fig. 7.33 is arrived at by trial and error. First, the brace to gusset connection is designed, and this establishes the minimum size of gusset. For calculations for this part of the connection, see Sec. 7.4.4. Normally, the gusset is squared off as shown in Fig. 7.32, which gives 16 rows of bolts in the gusset to truss vertical connection. The gusset to top chord connection is pretty well constrained by geometry to be about 70 in long plus about $13\frac{1}{2}$ in for the cutout. Starting from the configuration of Fig. 7.32, the UFM forces are calculated from the formulas of Fig. 7.29, and the design is checked. It will be found that Fig. 7.32 is a satisfactory design via the UFM, even though it fails via the KISS method forces of Fig. 7.30. Although the gusset to top chord connection cannot be reduced in length because of geometry, the gusset to truss vertical is subject to no such constraint. Therefore, the number of rows of bolts in the gusset to truss vertical is

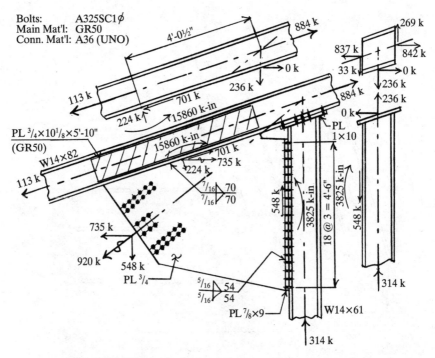

Fig. 7.30 KISS method—gusset forces are brace components.

sequentially reduced until failure occurs. The last achieved successful design is the final design as shown in Fig. 7.33.

The calculations for Fig. 7.33 and the intermediate designs and initial design of Fig. 7.32 are performed in the following manner. The given data for all cases are

$P = 920$ kips

$e_B = 7$ in

$e_c = 7$ in

$\gamma = 17.7°$

$\theta = 36.7°$

The relationship between α and β is

$\alpha - \beta(0.9527 \times 0.7454 - 0.3040) = 7(0.7454 - 0.3191) - 7/0.9527$

$\alpha - 0.4061\beta = -4.363$

This relationship must be satisfied for there to be no couples on the gusset edges. For the configuration of Fig. 7.33 with 7 rows of bolts in the gusset to truss

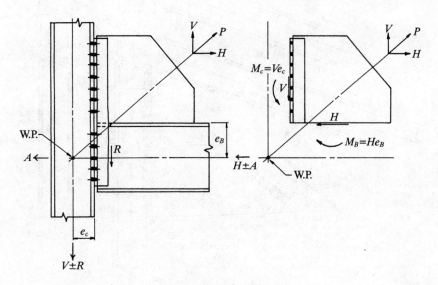

Fig. 7.31 The KISS method.

vertical connection (which is considered the gusset to beam connection of Fig. 7.29), $\bar{a} = 18.0$ in. Then,

$$\beta = \frac{18 + 4.363}{0.4061} = 55.07 \text{ in}$$

From Fig. 7.33, the centroid of the gusset to top chord (which is the gusset to column connection of Fig. 7.29) is $\bar{\beta} = 13.5 + 70/2 = 48.5$ in. Since $\bar{\beta} \neq \beta$, there will be a couple on this edge unless the gusset geometry is adjusted to make $\bar{\beta} = \beta = 55.07$. In this case, we will leave the gusset geometry unchanged and work with the couple on gusset to top chord interface.

Rather than choosing $\bar{a} = 18.0$ in, we could have chosen $\bar{\beta} = 48.5$ and solved for $\bar{a} \neq a$. In this case, a couple will be required on the gusset to truss vertical interface unless gusset geometry is changed to make $\bar{a} = a$.

Of the two possible choices, the first is the better one because the rigidity of the gusset to top chord interface is much greater than that of the gusset to truss vertical interface. This is so because the gusset is direct welded to the center of the top chord flange and is backed up by the chord web, whereas the gusset to truss vertical involves a flexible end plate and the bending flexibility of the flange of the

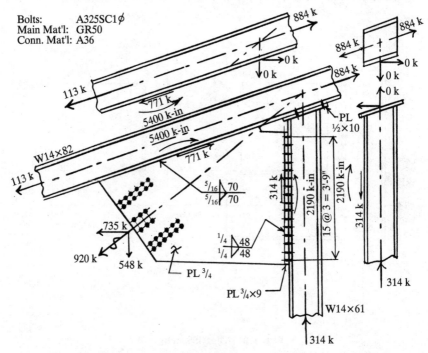

Fig. 7.32 KISS method—brace components are tangent to gusset edges.

truss vertical. Thus any couple required to put the gusset in equilibrium will tend to migrate to the stiffer gusset to top chord interface. With $\alpha = 18.0, \beta = 55.07$

$$r = \left[\left(18.0 + 7 \times 0.3191 + 55.07 \times 0.3040 + \frac{7}{0.9527}\right)^2 + (7 + 55.07 \times 0.9527)^2\right]^{1/2}$$

$= 74.16$ in

and from the equations of Fig. 7.29

$V_C = 648$ kips

$H_C = 298$ kips

$V_B = 87$ kips

$H_B = 250$ kips

For subsequent calculations, it is necessary to convert the gusset to top chord forces to normal and tangential forces as follows: The tangential or shearing component is

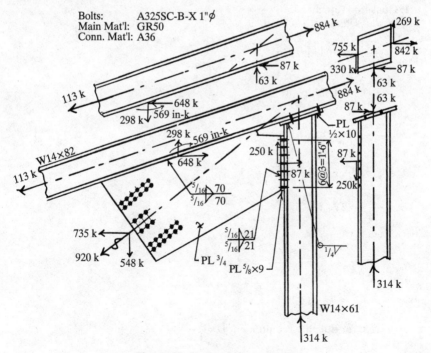

Fig. 7.33 Uniform force method.

$$T_C = V_C \cos\gamma + H_C \sin\gamma = (\beta + e_C \tan\gamma)\frac{P}{r}$$

The normal or axial component is

$$N_C = H_C \cos\gamma - V_C \sin\gamma = \frac{e_C P}{r}$$

The couple on the gusset to top chord interface is then

$$M_C = |N_C(\beta - \bar{\beta})|$$

Thus

$$T_C = (55.07 + 7 \times 0.3191)\frac{920}{74.16} = 711 \text{ kips}$$

$$N_C = 7 \times \frac{920}{74.16} = 86.6 \text{ kips}$$

$$M_C = 86.6 \times (55.07 - 48.5) = 569 \text{ kip-in}$$

Each of the connection interfaces will now be designed.

1. Gusset to top chord
a. Weld. Weld length is 70 in

$$f_v = \frac{711}{2 \times 70} = 5.08 \text{ kips/in}$$

$$f_a = \frac{86.6}{2 \times 70} = 0.61 \text{ kip/in}$$

$$f_b = \frac{569 \times 2}{70^2} = 0.23 \text{ kip/in}$$

$$f_R = \sqrt{(5.08)^2 + (0.61 + 0.23)^2} = 5.1 \text{ kips/in}$$

$$D = \frac{5.1}{1.392} = 3.7/16$$

Check ductility:

$$f_{ave} = \sqrt{5.08^2 + \left(0.61 + \frac{0.23}{2}\right)^2} = 5.1 \text{ kips/in}$$

Since $1.4 \times f_{ave} = 7.1 > 5.1$,

Size weld for ductility requirement $D = \dfrac{7.1}{1.392} = 5.1$

Use $5/16$ fillet weld (2 percent violation of ductility requirement OK)

b. Gusset stress

$$f_v = \frac{711}{0.75 \times 70} = 13.5 \text{ ksi} < 19.4 \text{ ksi} \qquad \text{OK}$$

$$f_a + f_b = \frac{86.6}{0.75 \times 70} + \frac{569 \times 4}{0.75 \times 70^2} = 2.27 \text{ ksi} < 32.4 \text{ ksi} \qquad \text{OK}$$

c. Top chord web yield. The normal force between the gusset and the top chord is $T_c = 86.6$ kips, and the couple is $M_c = 569$ kip-in. The contact length N is 70 in. The couple M_c is statically equivalent to equal and opposite normal forces $V_s = M_c/(N/2) = 569/35 = 16.2$ kips. The normal force V_s acts over a contact length of $N/2 = 35$ in. For convenience, an equivalent normal force acting over the contact length N can be defined as

$$N_{C\text{ equiv.}} = N_C + 2 \times V_s = 86.6 + 2 \times 16.2 = 119 \text{ kips}$$

Now, for web yielding

$$\phi R_{wy} = 1.0 \times (5k + N) F_{yw} t_w = 1.0 \times (5 \times 1.625 + 70) \times 50 \times 0.510$$
$$= 1992 \text{ kips} > 119 \text{ kips} \qquad \text{OK}$$

d. Top chord web crippling

$$\phi R_{wcp} = 0.75 \times 135 \times t_w^2 \left[1 + 3\left(\frac{N}{d}\right)\left(\frac{t_w}{t_f}\right)^{1.5}\right] \sqrt{\frac{F_{yw} t_f}{t_w}}$$

$$= 0.75 \times 135 \times 0.510^2 \left\{1 + 3\left[\frac{70}{14.31}\left(\frac{0.510}{0.855}\right)^{1.5}\right]\right\} \sqrt{\frac{50 \times 0.855}{0.510}}$$

$$= 1871 \text{ kips} > 119 \text{ kips} \qquad \text{OK}$$

In the web crippling check, the formula used is that for a location greater than $d/2$ from the chord end because $\bar{\beta} = 13.5 + 70/2 = 48.5$ in $> {}^{14.31}/_2 = 7.2$ in. $\bar{\beta}$ is the position of the equivalent normal force.

The checks for web yield and crippling could have been dismissed by inspection in this case but were completed to illustrate the method. Another check that should be made when there is a couple acting on a gusset edge is to ensure that the transverse shear induced on the supporting member, in this case the top chord W14×82, can be sustained. In this case, the induced transverse shear is $V_s = 16.2$ kips. The shear capacity of the W 14×82 is $0.510 \times 14.31 \times 0.9 \times 0.6 \times 50 = 197$ kips > 16.2 kips, OK. Now consider for contrast, the couple of 15,860 kip-in shown in Fig. 7.30. For this couple, $V_s = 15,860/35 = 453$ kips > 197 kips, so a ³⁄₄-in doubler plate of GR50 steel is required as shown in Fig. 7.30.

2. Gusset to Truss Vertical
a. Weld

$$f_v = \frac{250}{2 \times 21} = 5.95 \text{ kips/in}$$

$$f_a = \frac{87}{2 \times 21} = 2.07 \text{ kips/in}$$

$$f_R = \sqrt{5.95^2 + 2.07^2} = 6.30 \text{ kips/in}$$

Fillet weld size required $= \dfrac{6.30}{1.392} = 4.5/16$

Because of flexibility of end plate and truss vertical flange, there is no need to size the weld to provide ductility. Therefore, use a $5/16$ fillet weld.

b. Bolts and End Plate. The bolts are A325SC-B-X, 1 in ϕ in standard holes. The end plate is 9 in wide and the gage of the bolts is $5\frac{1}{2}$ in. Thus, using the prying action formulation notation of the AISC 2d ed. LRFD Manual (1994), vol. II,

$$b = \frac{5.5 - 0.75}{2} = 2.375$$

$$a = \frac{9 - 5.5}{2} = 1.75 < 1.25 \times 2.375 \qquad \text{OK}$$

$$b' = 2.375 - 0.5 = 1.875$$

$$a' = 1.75 + 0.5 = 2.25$$

$$\rho = \frac{1.875}{2.25} = 0.833$$

$$\delta = 1 - \frac{1.0625}{3} = 0.646$$

Shear per bolt $= V = \dfrac{250}{14} = 17.9$ kips < 28.8 kips \qquad OK

Tension per bolt $= T = \dfrac{87}{14} = 6.21$ kips

$$\phi r_n' = 1.13 \times 51 \times \left(1 - \frac{17.9}{28.8}\right) = 21.8 \text{ kips} > 6.21 \text{ kips} \qquad \text{OK}$$

Try $1/2$ plate

$$t_c = \sqrt{\frac{4.44 \times 21.8 \times 1.875}{3 \times 36}} = 1.30$$

$$a' = \frac{1}{0.646 \times 1.833}\left[\left(\frac{1.30}{0.5}\right)^2 - 1\right] = 4.86$$

Use $a' = 1$

$$T_d = 21.8\left(\frac{0.5}{1.30}\right)^2 \times 1.833 = 5.91 \text{ kips} < 6.21 \text{ kips} \qquad \text{NG}$$

Try ⅝ plate

$$a' = \frac{1}{0.646 \times 1.833}\left[\left(\frac{1.30}{0.625}\right)^2 - 1\right] = 2.81$$

$$T_d = 21.8\left(\frac{0.625}{1.30}\right)^2 \times 1.833 = 9.24 \text{ kips} > 6.21 \text{ kips} \qquad \text{OK}$$

Use ⅝ plate for the end plate

c. Truss vertical flange. The flange thickness of the W 14×61 is 0.645 in which exceeds the end plate thickness as well as being grade 50 steel. The truss vertical flange is therefore OK by inspection, but a calculation will be performed to demonstrate how the flange can be checked. A formula (Mann and Morris, 1979) for an effective bolt pitch can be derived from yield-line analysis as

$$p_{\text{eff}} = \frac{p(n-1) + \pi \bar{b} + 2\bar{a}}{n}$$

where the terms are as previously defined in Fig. 7.26. For the present case

$$\bar{b} = \frac{5.5 - 0.375}{2} = 2.5625$$

$$\bar{a} = \frac{10 - 5.5}{2} = 2.25$$

$$n = 7$$

$$p = 3$$

$$p_{\text{eff}} = \frac{3 \times (7-1) + \pi \times 2.5625 + 2 \times 2.25}{7} = 4.36$$

Once p_{eff} is determined, the prying action theory of the AISC Manual is applied.

$b = \bar{b} = 2.5625$

$b' = 2.5625 - 0.5 = 2.0625$

$a =$ smaller of $\bar{a}$ and a for the end plate $= 1.75 < 1.25 \times 2.5625$ \qquad OK

$a' = 1.75 + 0.5 = 2.25$

$\rho = \dfrac{b'}{a'} = 0.917$

$\delta = 1 - \dfrac{1.0625}{4.36} = 0.756$

$t_c = \sqrt{\dfrac{4.44 \times 21.8 \times 2.0625}{4.36 \times 50}} = 0.957 \text{ in}$

$\alpha' = \dfrac{1}{0.756 \times 1.917}\left[\left(\dfrac{0.957}{0.645}\right)^2 - 1\right] = 0.829$

$T_d = 21.8\left(\dfrac{0.645}{0.957}\right)^2 \times (1 + 0.829 \times 0.756) = 16.1 \text{ kips} > 6.18 \text{ kips}$ OK

3. Truss Vertical to Top Chord Connection. The forces on this connection, from Figs. 7.29 and 7.33, are

Vertical = Q = 298 – 920 × cos(36.7) × tan(17.7) = 63 kips

Horizontal = 87 kips

Converting these into normal and tangential components

$T_{BC} = 87 \cos \gamma - 63 \sin \gamma = 64$ kips

$N_{BC} = 87 \sin \gamma + 63 \cos \gamma = 86$ kips (compression)

a. Bolts. Since the normal force is always compression, the bolts see only the tangential or shear force; thus the number of bolts required is

$\dfrac{64}{28.8} = 2.2$ use 4 bolts

b. Weld. Use a profile fillet weld of the cap plate to the truss vertical, but only the weld to the web of the vertical is effective because there are no stiffeners between the flanges of the top chord. Thus the effective length of weld is

$\dfrac{13.89 - 2 \times 0.645}{\cos \gamma} = 13.23 \text{ in}$

$f_v = \dfrac{64}{2 \times 13.23} = 2.42 \text{ kips/in}$

$f_a = \dfrac{86}{2 \times 13.23} = 3.25 \text{ kips/in}$

$f_r = \sqrt{2.42^2 + 3.25^2} = 4.05 \text{ kips/in}$

The weld size required is $4.05/1.392 = 2.91$. Use $\frac{1}{4}$ FW (AISC minimum size) Check the W14×61 web to support required 2.91/16 FW. For welds of size W on both sides of a web of thickness t_w

$$0.9 \times 0.6 \times F_y x t_w \geq 0.75 \times 0.60 \times 70 \times 0.7071 \times W \times 2$$

or

$$t_w \geq 1.65 W \text{ for grade 50 steel}$$

Thus for $W = 2.91/16 = 0.182$

$$t_{w\,min} = 1.65 \times 0.182 = 0.300 \text{ in}$$

Since the web thickness of a W14 × 61 is 0.375, the web can support the welds.

 c. Cap plate. The cap plate thickness will be governed by bearing. The bearing design strength per bolt is

$$\phi r_p = 0.75 \times 2.4 \times 58 \times t_p \times 1$$

The load per bolt is 64/4 = 16.0 kips. The required cap plate thickness is thus

$$t_p = \frac{16.0}{0.75 \times 2.4 \times 58 \times 1} = 0.153$$

Use a $\frac{1}{2}$-in cap plate. This completes the calculations required to produce the connection of Fig. 7.33.

Hanger Connections ■ The most interesting of the genre is the type that involves prying action, sometimes of both the connection fitting and the supporting member. Figure 7.34 shows a typical example. The calculations to determine the capacity of this connection are as follows: The connection can be broken into three main parts, i.e., the angles, the piece W16×57, and the supporting member, the W18×50. The three main parts are joined by two additional parts, the bolts of the angles to the piece W16 and the bolts from the piece W16 to the W18. The load path in this connection is unique. The load P passes from the angles through the bolts into the piece W16, thence through bolts again into the supporting W18. The latter bolt group is arranged to straddle the brace line of action. These bolts then see only direct tension and shear, and no additional tension due to moment. Statics is sufficient to establish this. Consider now the determination of the capacity of this connection.

 1. Angles. The limit states for the angles are gross tension, net tension, block shear rupture, and bearing. The load can be compression as well as tension in this example. Compression will affect the angle design, but tension will control the above limit states.

 a. Gross tension. The gross area A_{gt} is $1.94 \times 2 = 3.88$ in². The capacity (design strength) is

$$\phi R_{gt} = 0.9 \times 36 \times 3.88 = 126 \text{ kips}$$

 b. Net tension. The net tension area is $A_{nt} = 3.88 - 0.25 \times 1.0 \times 2 = 3.38$ in². The effective net tension area A_e is less than the net area because of shear lag since only

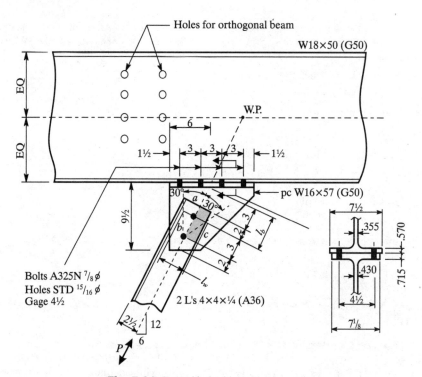

Fig. 7.34 Typical bolted hanger connection.

one of the two angle legs is connected. From the AISC LRFD (1994) commentary on Section B3,

$A_e = UA_{nt} = 0.75 \times 3.38 = 2.54$ in^2

The net tension capacity is

$\phi R_{nt} = 0.75 \times 58 \times 2.54 = 110$ kips

c. Block shear rupture. This failure mode involves the tearing out of the shaded block in Fig. 7.34. The failure is by yield or fracture on the longitudinal line through the bolts (line ab) and a simultaneous yield or fracture failure on the perpendicular line from the bolt's longitudinal line to the angle toe (line bc). Since this is a rupture or fracture failure mode, the mode with the larger fracture component is the controlling mode.

For line ab, the gross shear area is

$A_{gv} = 5 \times 0.25 \times 2 = 2.5$ in^2

and the net shear area is

$A_{nv} = 2.5 - (1.5 \times 0.25 \times 1.0) \times 2 = 1.75$ in^2

For line bc, the gross tension area is

$A_{gt} = 1.5 \times 0.25 \times 2 = 0.75 \text{ in}^2$

and the net tension area is

$A_{nt} = 0.75 - 0.5 \times 1.0 \times 0.25 \times 2 = 0.5 \text{ in}^2$

The shear fracture strength of line ab is $0.6 \times F_u \times A_{nv} = 0.6 \times 58 \times 1.75 = 60.9$ kips, and the tensile fracture strength of line bc is $F_u \times A_{nt} = 58 \times 0.5 = 29.0$ kips. Since $60.9 > 29.0$, the block shear limit state involves shear fracture on line ab and tension yield on line bc. Thus the block shear design strength (capacity) is

$\phi R_{bs} = 0.75 \times (60.9 + 36 \times 0.75) = 65.9$ kips

d. Bearing. The end distance of 2 in and the spacing of 3 in satisfy the criteria of $1\frac{1}{2}d$ and $3d$, respectively, where d is the bolt diameter. Thus the bearing design strength is

$\phi R_p = 0.75 \times 2.4 \times 0.875 \times 0.25 \times 58 \times 2 \times 2 = 91.4$ kips

2. Bolts—Angles to Piece W16. The limit state for the bolts is shear. The shear capacity of one bolt is

$\phi r_v = 0.75 \times 48 \times \pi/4 \times 0.875^2 = 21.6$ kips.

This value can also be obtained from AISC LRFD Manual, vol. II, Table 8-11, p. 8-24. The design strength of two bolts in double shear is

$\phi R_v = 21.6 \times 2 \times 2 = 86.4$ kips

3. Piece W16×57. The limit states for this part of the connection are Whitmore section yield and buckling, bearing, and prying action in conjunction with the W16 flange to W18 flange bolts. Because there is only one line of bolts, tearout is not a limit state.

a. Whitmore section. This is the section denoted by l_w in Fig. 7.34. It is formed by 30° lines from the bolt farthest away from the end of the brace to the intersection of these lines with a line through and perpendicular to the bolt nearest the end of the brace. Whitmore (1952) determined that this 30° spread gave an accurate estimate of the stress in gusset plates at the end of the brace. The length of the Whitmore section $l_w = 3(\tan 30°)2 = 3.46$ in.

i. Whitmore yield:

$\phi R_{wy} = 0.9 \times 50 \times 3.46 \times 0.430 = 67.0$ kips

where 0.430 is the web thickness of a W16×57.

ii. Whitmore buckling: Tests (Gross, 1990) have shown that the Whitmore section can be used as a conservative estimate for gusset buckling. In the present case, the web of the W16×57 is a gusset. If the load P is a compression, it is possible for the gusset to buckle laterally in a sidesway mode. For this mode of buckling, the K factor is 1.2. The buckling length is $l_b = 5$ in in Fig. 7.34. Thus the slenderness ratio is

$$\frac{Kl}{r} = \frac{1.2 \times 5 \times \sqrt{12}}{0.430} = 48.3$$

From AISC LRFD Manual, vol. I, Table 3-50, p.6-148, $\phi F_{cv} = 35.8$ ksi, and the Whitmore buckling capacity is

$$\phi R_{wb} = 35.8 \times 3.46 \times 0.430 = 53.3 \text{ kips}$$

b. Bearing. Since the end and spacing distance requirements are satisfied

$$\phi R_p = 0.75 \times 2.4 \times 0.875 \times 0.430 \times 65 \times 2 = 88.0 \text{ kips}$$

c. Prying action. Prying action explicitly refers to the extra tensile force in bolts which connect flexible plates or flanges subjected to loads normal to the flanges. For this reason, prying action involves not only the bolts but the flange thickness, bolt pitch and gage, and in general, the geometry of the entire connection.

The AISC LRFD manual, vol. II, p.11-6, presents a method to calculate the effects of prying. This method was originally developed by Struik (1969) and presented in the book (Kulak, Fisher, and Struik, 1987). The form used in the AISC LRFD Manual was developed by Thornton (1985), for ease of calculation and to provide optimum results, i.e., maximum capacity for a given connection (analysis) and minimum required thickness for a given load (design). Thornton (1992) has shown that this method gives a very conservative estimate of ultimate load and shows that very close estimates of ultimate load can be obtained by using the flange ultimate strength F_u in place of the yield strength F_y in the AISC LRFD Manual formulas.

From the foregoing calculations, the capacity (design strength) of this connection is 53.3 kips. Let us take this as the design load (required strength) and proceed to the prying calculations. The vertical component of 53.3 is 47.7 kips and the horizontal component is 23.8 kips. Thus the shear per bolt is $V = 23.8/8 = 2.98$ kips and the tension per bolt is $T = 47.7/8 = 5.96$ kips. Since $2.98 < 21.6$, the bolts are OK for shear. The interaction equation for A325 N bolts is

$$\phi F'_t = 0.75(117 - 1.9 f_v) \leq 0.75 \times 90 = 67.5 \text{ ksi}$$

With $V = 2.98$, $f_v = 2.98/0.6013 = 4.96$ ksi and

$$\phi F'_t = 0.75(117 - 1.9 \times 4.96) = 80.7 \text{ ksi} > 67.5 \text{ ksi}$$

so

$$\phi F'_t = 67.5 \text{ ksi} \quad \text{and} \quad \phi r'_t = 67.5 \times 0.6013 = 40.6 \text{ kips}$$

Since the design strength per bolt $\phi r'_t = 40.6$ kips is greater than the required strength (or load) per bolt $T = 5.96$ kips, the bolts are OK. Now, to check prying of the W16 piece, following the notation of the AISC LRFD Manual,

$$b = \frac{4.5 - 0.430}{2} = 2.035$$

$$a = \frac{7.125 - 4.5}{2} = 1.3125$$

Check that $a < 1.25b = 1.25 \times 2.035 = 2.544$. Since $a = 1.3125 < 2.544$, use $a = 1.3125$. If $a > 1.25b$, $a = 1.25b$ would be used.

$$b' = 2.035 - \frac{0.875}{2} = 1.598$$

$$a' = 1.3125 + \frac{0.875}{2} = 1.75$$

$$\rho = \frac{b'}{a'} = 0.91$$

$$p = 3$$

$$\delta = 1 - \frac{d'}{p} = 1 - \frac{0.9375}{3} = 0.6875$$

$$\alpha' = \frac{1}{\delta(1+\rho)}\left[\left(\frac{t_c}{t}\right)^2 - 1\right]$$

$$t_c = \sqrt{\frac{4.44(\phi r_t')b'}{pF_y}} = \sqrt{\frac{4.44 \times 40.6 \times 1.598}{3 \times 50}} = 1.386$$

$$\alpha' = \frac{1}{0.6875 \times 1.91}\left[\left(\frac{1.386}{0.715}\right)^2 - 1\right] = 2.10$$

Since $\alpha' > 1$, use $\alpha' = 1$ in subsequent calculations. $\alpha' = 2.10$ means that the bending of the W16×57 flange will be the controlling limit state. The bolts will not be critical. The design tensile strength T_d per bolt including the flange strength is

$$T_d = \phi r_t'\left(\frac{t}{t_c}\right)^2 (1+\delta) = 40.6 \times \left(\frac{0.715}{1.386}\right)^2 \times 1.6875 = 18.2 \text{ kips} > 5.96 \text{ kips} \qquad \text{OK}$$

The subscript d denotes "design" strength. In addition to the prying check on the piece W16×57, a check should also be made on the flange of the W18×50 beam. A method for doing this was presented in Fig. 7.26. Thus,

$$\bar{b} = \frac{4.5 - 0.355}{2} = 2.073$$

$$\bar{a} = \frac{7.5 - 4.5}{2} = 1.50$$

$$n = 4$$

$$p = 3$$

$$p_{eff} = \frac{3(4-1) + \pi \times 2.073 + 2 \times 1.50}{4} = 4.63$$

now, using the prying formulation from the AISC LRFD Manual,

$b = \bar{b} = 2.073$

$a = 1.3125$

Note that the prying lever arm is controlled by the narrower of the two flanges.

$b' = 2.073 - \dfrac{0.875}{2} = 1.636$

$a' = 1.3125 + \dfrac{0.875}{2} = 1.75$

$\rho = 0.93$

$p = p_{eff} = 4.63$

$\delta = 1 - \dfrac{0.9375}{4.63} = 0.798$

$t_c = \sqrt{\dfrac{4.44 \times 40.6 \times 1.636}{4.63 \times 50}} = 1.13$

$\alpha' = \dfrac{1}{0.798 \times 1.93}\left[\left(\dfrac{1.13}{0.570}\right)^2 - 1\right] = 1.90$

Use $\alpha' = 1$

$T_d = 40.6 \times \left(\dfrac{0.570}{1.13}\right)^2 \times 1.798 = 18.6 \text{ kips} > 5.96 \text{ kips}$ \hfill OK

The prying checks in this example were not critical. If they were and a better estimate of the true failure limit state is desired, the ultimate strength rather than the yield strength can be used in the prying formulas (Thornton, 1992).

The only formula that needs to be modified is

$t_c = \sqrt{\dfrac{4.44 \phi r'_t b'}{p F_u}}$

For the piece W16

$t_c = \sqrt{\dfrac{4.44 \times 40.6 \times 1.598}{3 \times 65}} = 1.215$

$\alpha' = \dfrac{1}{0.6875 \times 1.91}\left[\left(\dfrac{1.215}{0.715}\right)^2 - 1\right] = 1.44$

Use $\alpha' = 1$.

$$T_d = 40.6 \times \left(\frac{0.715}{1.215}\right)^2 \times 1.6875 = 23.7 \text{ kips}$$

Thus T_d is increased to 23.7 kips from 18.2 kips.

Additional checks on the W18×50 beam

Web yielding: Since $5k = 5 \times 1.25 = 6.25 > p = 3$, the web tributary to each bolt at the k distance exceeds the bolt spacing and thus $N = 9$.

$\phi R_{wy} = 1.0 \times (9 + 5 \times 1.25) \times 50 \times 0.355 = 271$ kips > 47.7 kips   OK

Web crippling: Web crippling occurs when the load is compression; thus $N = 12$, the length of the piece W16.

$$\phi R_{wcp} = 0.75 \times 135 \times 0.355^2 \left[1 + 3\left(\frac{12}{17.99}\right)\left(\frac{0.355}{0.570}\right)^{1.5}\right] \sqrt{\frac{50 \times 0.570}{0.355}}$$

$= 227$ kips > 47.7 kips   OK

This completes the design calculations for this connection. A load path has been provided through every element of the connection. For this type of connection, the beam designer should make sure that the bottom flange is stabilized if P can be compressive. A transverse beam framing nearby as shown in Fig. 7.34 by the W18×50 web hole pattern, or a bottom flange stay, will provide stability.

Column Base Plates ■ The geometry of a column base plate is shown in Fig. 7.35. The area of the base plate is $A_1 = B \times N$. The area of the pier which is concentric with A_1 is A_2. If the pier is not concentric with the base plate, only that portion which is concentric can be used for A_2. The design strength of the concrete in bearing is

$$\phi_c F_p = 0.6 \times 0.85 f'_c \sqrt{\frac{A_2}{A_1}}$$

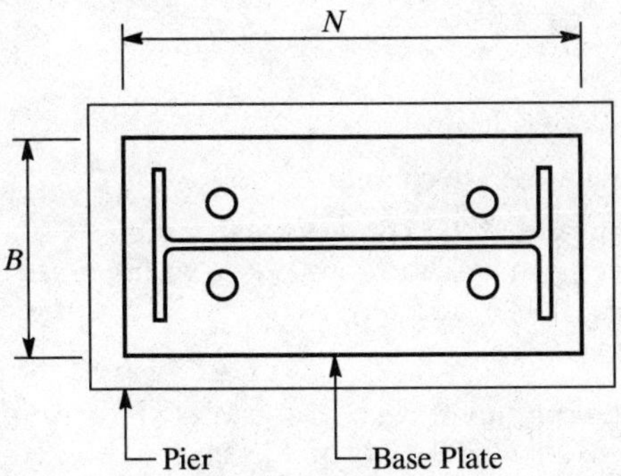

Fig. 7.35 Column base plate.

where f'_c is the concrete compressive strength in ksi and

$$1 \le \sqrt{\frac{A_2}{A_1}} \le 2$$

The required bearing strength is

$$f_p = \frac{P}{A_1}$$

where P is the column load (factored) in kips. In terms of these variables, the required base plate thickness is

$$t_p = l\sqrt{\frac{2f_p}{\phi F_y}}$$

where l = max $\{m, n, 2n'\}$

ϕF_y = base plate design strength = $0.9 F_y$

$m = \dfrac{N - 0.95d}{2}$

$n = \dfrac{B - 0.8b_f}{2}$

$n' = \dfrac{\sqrt{db_f}}{4}$

$\lambda = \dfrac{2\sqrt{x}}{1 + \sqrt{1-x}} \le 1$

$x = \dfrac{4db_f}{(d+b_f)^2} \dfrac{f_p}{\phi_c F_p}$

d = depth of column

b_f = flange width of column

For simplicity λ can always be conservatively taken as unity. The formulation given here was developed by Thornton (1990A, 1990B) based on previous work by Murray (1983), Fling (1970), and Stockwell (1975). It is the method given in the AISC LRFD Manual (1994).

Example ■ The column of Fig. 7.35 is a W24×84 carrying 600 kips. The concrete has f'_c = 4.0 ksi. Try a base plate of A36 steel, 4 in bigger than the column in both directions.

Since $d = 24\frac{1}{8}$ and $b_f = 9$, $N = 24\frac{1}{8} + 4 = 28\frac{1}{8}$, $B = 9 + 4 = 13$. Try a plate 28×13. Assume that 2 in of grout will be used so the minimum pier size is 32×17. Thus $A_1 = 28 \times 13 = 364$ in^2, $A_2 = 32 \times 17 = 544$ in^2, $\sqrt{A_2/A_1} = 1.22 < 2$ (OK), and

$$\phi_c F_p = 0.6 \times 0.85 \times 4 \times 1.22 = 2.49 \text{ ksi}$$

$$f_p = \frac{600}{364} = 1.65 \text{ ksi} < 2.49 \text{ ksi} \qquad \text{OK}$$

$$m = \frac{28 - 0.95 \times 24.125}{2} = 2.54$$

$$n = \frac{13 - 0.8 \times 9}{2} = 2.90$$

$$n' = \frac{\sqrt{24.125 \times 9}}{4} = 3.68$$

$$x = \frac{4 \times 24.125 \times 9.0}{(24.125 + 9.0)^2} \frac{1.65}{2.49} = 0.52$$

$$\lambda = \frac{2\sqrt{0.52}}{1 + \sqrt{1 - 0.52}} = 0.85$$

$l = \max\{2.54, 2.90, 0.85 \times 3.68\} = 3.13$

$$t_p = 3.13 \sqrt{\frac{2 \times 1.65}{0.9 \times 36}} = 0.99 \text{ in}$$

Use a plate $1 \times 13 \times 28$ of A36 steel. If the conservative assumption of $\lambda = 1$ were used, $t_p = 1.17$ in, which indicates a $1\frac{1}{4}$-in-thick base plate.

Erection considerations. In addition to designing a base plate for the column compression load, loads on base plates and anchor bolts during erection should be considered. A common design load for erection is a 1-kip working load, applied at the top of the column in any horizontal direction. If the column is, say, 40 ft high, this 1-kip force at a lever arm of 40 ft will cause a significant couple at the base plate and anchor bolts. The base plate, anchor bolts, and column to base plate weld should be checked for this construction load condition. The paper by Murray (1983) gives some yield-line methods that can be used for doing this.

Splices—Columns and Truss Chords ■ Section J1.4 of the AISC LRFD specification says that finished to bear compression splices in columns need be designed only to hold the parts "securely in place." For this reason, AISC provides a series of "standard" column splices in the AISC LRFD Manual, vol. II, pp. 11-72 through p. 11-91. These splices are nominal in the sense that they are designed for no particular loads. Section J1.4 also requires that splices in trusses be designed for at least 50 percent of the design load (required strength). The difference between columns and "other compression members," such as compression chords of trusses, is that for columns, splices are usually near lateral support points, such as floors, whereas trusses can have their splices at midpanel point where there is no lateral support.

Column Splices. Figure 7.36 shows a standard AISC column splice for a W14×99 to a W14×109. If the column load remains compression, the strong-axis column shear can be carried by friction. The coefficient of static friction of steel to steel is on the order of 0.5 to 0.7, so quite high shears can be carried by friction. Suppose the compression load on this column is 700 kips. How much major-axis bending moment can this splice carry? Even though these splices are nominal, they can carry quite significant bending moment. The flange area of the W14×99 is $A_f = 0.780 \times 14.565 = 11.4$ in^2. Thus, the compression load per flange is 700 × 11.4/29.1 = 274 kips. In order for a bending moment to cause a tension in the column flange, this load of 274 kips must first be unloaded. Assuming that the flange force acts at the flange centroid, the moment in the column can be represented as

$$M = T(d - t_f) = T(14.16 - 0.780) = 13.38T$$

If $T = 274$ kips, one flange will be unloaded, and $M = 13.38 \times 274 = 3666$ kip-in = 306 kip-ft. The design strength in bending for this column (assuming sufficient lateral support) is $\phi M_p = 647$ kip-ft. Thus, because of the compression load, the nominal AISC splice, while still seeing no load, can carry almost 50 percent of the column's bending capacity.

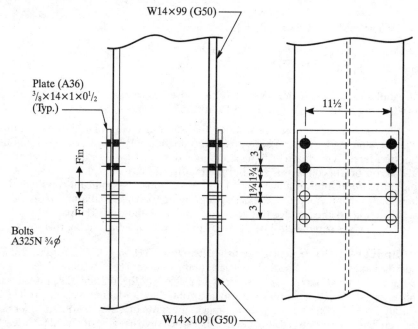

Fig. 7.36 An AISC standard column splice.

The splice plates and bolts will allow additional moment to be carried. It can be shown that the controlling limit state for the splice material is bolt shear. For one bolt $\phi r_v = 15.9$ kips. Thus for 4 bolts $\phi R_v = 15.9 \times 4 = 63.6$ kips. The splice forces are assumed to act at the faying surface of the deeper member. Thus the moment capacity of the splice plates and bolts is $M_s = 63.6 \times 14.32 = 911$ kip-in $= 75.9$ kip-ft. The total moment capacity of this splice with zero compression is thus 75.9 kip-ft, and with 700 kips compression, it is $306 + 75.9 = 382$ kip-ft. The role of compression in providing moment capability is often overlooked in column splice design.

Erection Stability. As discussed earlier for base plates, the stability of columns during erection must be a consideration for splice design also. The usual nominal erection load for columns is a 1-kip horizontal force at the column top in any direction. In LRFD format, the 1-kip working load is converted to a factored load by multiplying by a load factor of 1.5. This load of $1 \times 1.5 = 1.5$ kips will require connections which will be similar to those obtained in allowable stress design (ASD) with a working load of 1 kip. It has been established that for major-axis bending, the splice is good for 75.9 kip-ft. This means that the 1.5-kip load can be applied at the top of a column $75.9/1.5 = 50.6$ ft tall. Most columns will be shorter than 50.6 ft, but if not, a more robust splice should be considered.

Minor-Axis Stability. If the 1.5-kip erection load is applied in the minor or weak-axis direction, the forces at the splice will be as shown in Fig. 7.37. The upper shaft will tend to pivot about point O. Taking moments about point O,

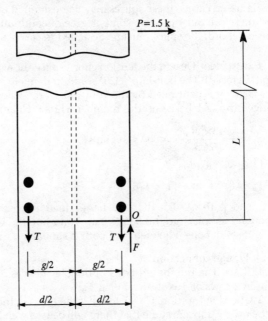

Fig. 7.37 Weak-axis stability forces for column splice.

7-82 ■ Chapter Seven

$$PL = T\left(\frac{d}{2}+\frac{g}{2}\right) + T\left(\frac{d}{2}-\frac{g}{2}\right) = Td$$

Thus the erection load P that can be carried by the splice is

$$P = \frac{Td}{L}$$

Note that this erection load capacity (design strength) is independent of the gage g. This is why the AISC splices carry the note, "Gages shown can be modified if necessary to accommodate fittings elsewhere on the column." The standard column gages are $5\frac{1}{2}$ and $7\frac{1}{2}$ in for beams framing to column flanges. Errors can be avoided by making all column gages the same. The gages used for the column splice can also be $5\frac{1}{2}$ or $7\frac{1}{2}$ in without affecting erection stability.

If the upper column of Fig. 7.36 is 40 ft long and T is the shear strength of four (two per splice plate) bolts,

$$P = \frac{4 \times 15.9 \times 14.565}{40 \times 12} = 1.93 \text{ kips}$$

Since $1.93 > 1.5$, this splice is satisfactory for a 40-ft-long column. If it were not, larger or stronger bolts could be used.

Splices in Truss Chords. These splices must be designed for 50 percent of the chord load, even if the load is compression and the members are finished to bear. As discussed earlier, these splices may be positioned in the center of a truss panel, and therefore must provide some degree of continuity to resist bending. For the tension chord, the splice must be designed to carry the full tensile load.

Example ■ Design the tension chord splice shown in Fig. 7.38. The load is 800 kips (factored). The bolts are A325X, $\frac{7}{8}$ in diameter, $\phi r_v = 27.1$ kips. The load at this location is controlled by the W14×90, so the loads should be apportioned to flanges and web based on this member. Thus the flange load is

$$P_f = \frac{0.710 \times 14.520}{26.5} \times 800 = 311 \text{ kips}$$

and the web load is

$$P_w = 800 - 2 \times 311 = 178 \text{ kips}$$

The load path is such that the flange load P_f passes from the W14×90 (say) through the bolts into the flange plates and into the W14×120 flanges through a second set of bolts. The web load path is similar.

A. Flange connection

1. Bolts. The number of bolts in double shear is $311/(2 \times 27.1) = 5.74$. Use 6 bolts in 2 rows of 3 as shown in Fig. 7.38.

2. Chord net section. Check to see if the holes in the W14×90 reduce its capacity below 800 kips. Assume that there will be two web holes in alignment with the flange holes.

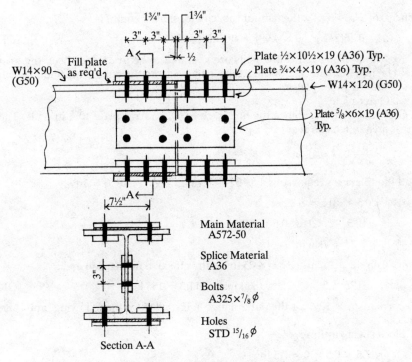

Fig. 7.38 Truss chord tension splice.

$A_{net} = 26.5 - 4 \times 1 \times 0.710 - 2 \times 1 \times 0.440 = 22.8$ in^2

$\phi R_{net} = 22.8 \times 0.75 \times 65 = 1111$ kips > 800 kips OK

3. Bearing

$\phi R_p = 0.75 \times 2.4 \times 65 \times 0.710 \times 0.875 \times 6 = 436$ kips > 311 kips OK

4. Block shear rupture (tearout)

$A_{nv} = (7.75 - 2.5 \times 1.0) \times 0.710 \times 2 = 7.46$ in^2

$A_{nt} = \left(\dfrac{14.520 - 7.5}{2} - 0.5 \times 1\right) \times 0.710 \times 2 = 4.27$ in^2

$A_{gv} = 7.75 \times 0.710 \times 2 = 11.0$ in^2

$A_{gt} = 3.51 \times 0.710 \times 2 = 4.98$ in^2

$F_u A_{nt} = 65 \times 4.27 = 278$ kips

$0.6 F_u A_{nv} = 0.6 \times 65 \times 7.46 = 291$ kips

7-84 ■ Chapter Seven

Since $0.6F_u A_{nv} > F_u A_{nt}$, the tearout capacity (design strength) is

$\phi R_{to} = 0.75(291 + 50 \times 4.98) = 405$ kips > 311 kips OK

5. Flange plates. Since the bolts are assumed to be in double shear, the load path is such that one half of the flange load goes into the outer plate and one half goes into the inner plates.

a. Outer plate

i. Gross and net area: Since the bolt gage is 7½ in, try a plate 10½ in wide. The gross area in tension required is

$$A_{gt} = \frac{311/2}{0.9 \times 36} = 4.8 \text{ in}^2$$

and the thickness required is $4.8/10.5 = 0.46$ in. Try a plate ½ × 10½.

$A_{gt} = 0.5 \times 10.5 = 5.25$ in²

$A_{nt} = (10.5 - 2 \times 1) \times 0.5 = 4.25$ in²

$0.85 \times 5.25 = 4.46$ in²

Since $0.85 A_{gt} > A_{nt}$, use $A_{nt} = 4.25$ in² as the effective net tension area.

$\phi R_{nt} = 0.75 \times 58 \times 4.25 = 185$ kips $> 311/2 = 156$ kips OK

Use a plate ½ × 10½ for the outer flange splice plate for the following limit state checks.

ii. Block shear rupture:

$A_{gv} = 7.5 \times 0.5 \times 2 = 7.5$ in²

$A_{gt} = 1.5 \times 0.5 \times 2 = 1.5$ in²

$A_{nv} = (7.5 - 2.5 \times 1) \times 0.5 \times 2 = 5.0$ in²

$A_{nt} = (1.5 - 0.5 \times 1) \times 0.5 \times 2 = 1.0$ in²

$F_u A_{nt} = 58 \times 1.0 = 58.0$ kips

$0.6 F_u A_{nv} = 0.6 \times 58 \times 5.0 = 174$ kips

Since $0.6 F_u A_{nv} > F_u A_{nt}$,

$\phi R_{to} = 0.75 (174 + 36 \times 1.5) = 171$ kips > 156 kips OK

iii. Bearing:

$\phi R_p = 0.75 \times 2.4 \times 58 \times 0.50 \times 0.875 \times 6 = 274$ kips >156 kips OK

Thus the plate ½ × 10½ (A36) for the outer splice plate is OK.

b. Inner plates

i. Gross and net area: The load to each plate is $156/2 = 78$ kips. The gross area in tension required is

$$A_{gt} = \frac{78}{0.9 \times 36} = 2.41 \text{ in}^2$$

Try a plate 4 in wide. Then the required thickness is 2.41/4 = 0.6 in. Try a plate ³⁄₄ × 4 (A36).

$A_{gt} = 0.75 \times 4 = 3 \text{ in}^2$

$A_{nt} = (4 - 1.0) \times 0.75 = 2.25 \text{ in}^2$

$0.85 \times 3 = 2.55 \text{ in}^2$

$\phi R_{nt} = 0.75 \times 58 \times 2.25 = 97.9 \text{ kips} > 78 \text{ kips}$ OK

ii. Block shear rupture: Since there is only one line of bolts, this limit state is not possible.

iii. Bearing:

$\phi R_p = 0.75 \times 2.4 \times 58 \times 0.75 \times 0.875 \times 3 = 206 \text{ kips} > 78 \text{ kips}$ OK

Use the ³⁄₄ × 4 (A36) inner splice plates.

B. Web Connection. The calculations for the web connection involve the same limit states as the flange connection.

1. Bolts

Number required $= \dfrac{178}{2 \times 27.1} = 3.28$

Use 4 bolts.

2. Web limit states
a. Bearing

$\phi R_p = 0.75 \times 2.4 \times 65 \times 0.440 \times 0.875 \times 4 = 180 \text{ kips} > 178 \text{ kips}$ OK

b. Block shear rupture (tearout). Assume the bolts have a 3-in pitch longitudinally.

$A_{nv} = (4.75 - 1.5 \times 1) \times 0.440 \times 2 = 2.86 \text{ in}^2$

$A_{nt} = (3 - 1 \times 1) \times 0.440 = 0.88$

$A_{gv} = 4.75 \times 0.440 \times 2 = 4.18 \text{ in}^2$

$A_{gt} = 3 \times 0.440 = 1.32 \text{ in}^2$

$F_u A_{nt} = 65 \times 0.88 = 57.2 \text{ kips}$

$0.6 F_u A_{nv} = 0.6 \times 65 \times 2.86 = 112 \text{ kips}$

Since $0.6 F_u A_{nv} > F_u A_{nt}$,

$\phi R_{to} = 0.75 (112 + 50 \times 1.32) = 134 \text{ kips} < 178 \text{ kips}$ NG

Since the tearout limit state fails, the bolts can be spaced out to increase the capacity. Increase the bolt pitch from the 3 in assumed above, to 6 in. Then

$A_{nv} = (7.75 - 1.5 \times 1) \times 0.440 \times 2 = 5.50 \text{ in}^2$

$A_{nt} = 0.88 \text{ in}^2$

$A_{gv} = 7.75 \times 0.440 \times 2 = 6.82 \text{ in}^2$

$A_{gt} = 1.32 \text{ in}^2$

As before, $0.6F_uA_{nv} > F_uA_{nt}$, so

$\phi R_{to} = 0.75(\,0.6 \times 65 \times 5.50 + 50 \times 1.32) = 210$ kips > 178 kips   OK

Use the web bolt pattern shown in Fig. 7.38.

    3. Web plates. Try 2 plates, one each side of the web, 6 in wide and ½ in thick.

    a. Gross area

$\phi R_{gt} = 0.9 \times 36 \times 0.5 \times 6 \times 2 = 194$ kips > 178 kips   OK

    b. Net area

$A_{nt} = (6 - 2 \times 1) \times 0.5 \times 2 = 4.0$ in^2

$0.85A_{gt} = 0.85 \times 0.5 \times 6 \times 2 = 5.1$ in^2

$\phi R_{nt} = 0.75 \times 58 \times 4.0 = 174$ kips < 178 kips   NG

Increase web plates to 5/8 in thick. Net area will be OK by inspection.

    c. Block shear rupture. This is checked as shown in previous calculations. It is not critical here.

    d. Bearing

$\phi R_p = 0.75 \times 2.4 \times 58 \times 0.625 \times 0.875 \times 2 \times 4 = 457$ kips > 178 kips   OK

The completed splice is shown in Fig. 7.38.

If this were a nonbearing compression splice, the splice plates would be checked for buckling. The following paragraph shows the method, which is not required for a tension splice.

    e. Buckling. Because of the 6 in spacing, check web plate buckling between the rows. The load is $178/(2 \times 2) = 44.5$ kips. The slenderness ratio is $0.65 \times 6 \times \sqrt{12}/0.625 = 22$. From AISC LRFD Table 3.36, $\phi F_{cv} = 29.83$ ksi, and $29.83 \times 0.625 \times 6 = 112$ kips > 44.5 kips OK. The plates at the splice line of length 4.0 in can be checked in the same way against a load of $178/2 = 89$ kips/plate. This limit state is OK by inspection, since at a slenderness ratio of 22, each plate is good for 112 kips. This limit state is checked for the flange plates also.

7.4.5 Moment Connections

Figure 7.39 shows the most common type of moment connection—the field-welded connection. Currently (Englehardt et al, 1995) these and other types of moment connections are undergoing heavy scrutiny because of fractures which took place in the Northridge earthquake of January, 1994. In regions of high seismicity, special configurations and welding quality control are being studied. In other regions, the field welded moment connection as shown in Fig. 7.39a is in common use, so it is worthwhile presenting the calculations required for design.

The moment connection of Fig. 7.39a is a three-way moment connection. Additional views are shown in Figs. 7.39b and 7.39c. If the strong axis connection requires stiffeners, there will be an interaction between the flange forces of the strong and weak axis beams. If the primary function of these moment connections

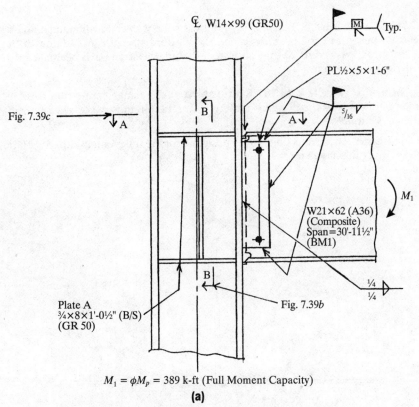

$M_1 = \phi M_p = 389$ k-ft (Full Moment Capacity)

(a)

Fig. 7.39a Field-welded moment connection.

is to resist lateral maximum load from wind or seismic sources, the interaction can generally be ignored because the maximum lateral loads will act in only one direction at any one time. If the moment connections are primarily used to carry gravity loads, such as would be the case when stiff floors with small deflections and high natural frequencies are desired, there will be interaction between the weak and strong beam flange forces. This latter case will be assumed here.

The load path through this connection that is usually assumed is that the moment is carried entirely by the flanges, and the shear entirely by the web. This will be the approach used here. The approach is justified by the lower bound theorem, provided, of course, that the connection is ductile. Generally, the connection will be sufficiently ductile if the moment capacity of the flanges alone is at least 70 percent of the total moment capacity [AISC (1994) seismic specification]. For the W21×62, $\phi M_p = 389$ kip-ft and the flange moment capacity is $\phi M_f = \phi b_f t_f (d - t_f) F_y = 0.9 \times 8.240 \times 0.615 (20.99 - 0.615) \, 36 = 3345$ kip-in $= 279$ kip-ft. Since $279/386 = 0.72 > 0.70$, the assumed load path is satisfactory. Checking the W 21×44, $\phi M_p = 358$ kip-ft, and $\phi M_F = 0.9 \times 6.500 \times 0.450(20.66 - 0.450) \times 50/12 =$

7-88 ■ Chapter Seven

222 kip-ft, so 222/358 = 0.62 < 0.70. In this case, the web connection must take at least 20 percent of the beam web nominal flexural strength in addition to the beam shear. Proceeding to the connection design, the strong-axis beam, beam 1, will be designed first.

Beam 1 W21×62 (A36) Composite ■ The flange connection is a full-penetration weld so no design is required. The column must be checked for stiffeners and doublers.

Stiffeners. The connection is to be designed for the full moment capacity of the beam. Thus the flange force F_f is

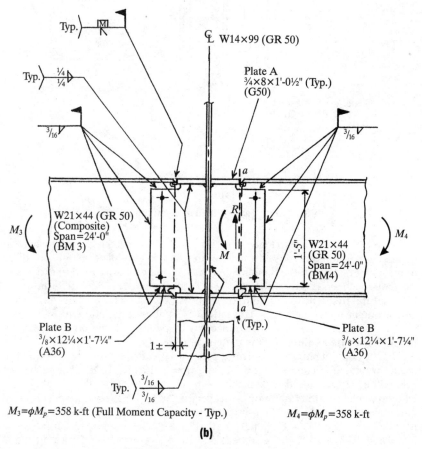

$M_3 = \phi M_p = 358$ k-ft (Full Moment Capacity - Typ.) $\qquad M_4 = \phi M_p = 358$ k-ft

(b)

Fig. 7.39b Section B-B of Fig. 7.39a.

$$F_f = \frac{\phi M_p}{d - t_f} = \frac{389 \times 12}{20.99 - 0.615} = 229 \text{ kips}$$

From the column load tables of the AISC LRFD Manual, vol. I, p. 3-20, web yielding:

$$P_{wy} = P_{wo} + t_b P_{wi} = 174 + 0.615 \times 24.3 = 189 \text{ kips} < 229 \text{ kips}$$

thus stiffeners are required at both flanges.
Web buckling:

$$P_{wb} = 261 \text{ kips} > 229 \text{ kips}$$

no stiffener required at compression flange.
Flange bending:

$$P_{fb} = 171 \text{ kips} < 229 \text{ kips}$$

stiffener required at tension flange.

From the above three checks (limit states), a stiffener is required at both flanges. For the tension flange, the total stiffener force is 229 − 171 = 58 kips and for the compression flange, the stiffener force is 229 − 189 = 40 kips. In this example, because the loads are primarily gravity, the moment will not reverse and the tension (top) and compression (bottom) flanges are known. For design, however, it is simpler to ignore this and use the larger of 58 and 40 as the stiffener force for both flanges. Then the force in *each* stiffener is 58/2 = 29 kips, both top and bottom.

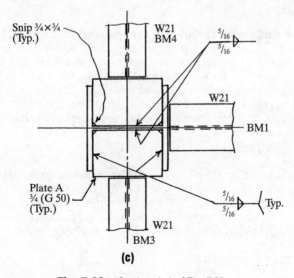

Fig. 7.39c Section A-A of Fig. 7.39a.

Determination of Stiffener Size
■ The minimum stiffener width w_s is

$$\frac{b_{fb}}{3} - \frac{t_{wc}}{2} = \frac{8.24}{3} - \frac{0.485}{2} = 2.5 \text{ in}$$

Use a stiffener 6½ in wide to match column.
The minimum stiffener thickness t_s is

$$\frac{t_{fb}}{2} = \frac{0.615}{2} = 0.31 \text{ in}$$

Use a stiffener at least ³⁄₈ in thick.
The minimum stiffener length l_s is

$$\frac{d_c}{2} - t_{fc} = \frac{14.16}{2} - 0.78 = 6.3 \text{ in}$$

The minimum length is for a "half depth" stiffener which is not possible in this example because of the weak-axis connections. Therefore, use a full-depth stiffener of 12½ in length.

A final stiffener size check is a plate buckling check which requires that

$$t_s \geq w_s \frac{\sqrt{F_y}}{95} = 6.5 \frac{\sqrt{36}}{95} = 0.41 \text{ in}$$

Therefore, the minimum stiffener thickness is ½ in. The final stiffener size for the strong axis beam is ½ × 6½ × 12½. The contact area of this stiffener against the inside of the column flange is 6.5 − 0.75 = 5.75 due to the snip to clear the column web to flange fillet. The stiffener design strength is thus 0.9 × 36 × 5.75 × 0.5 = 93.2 kips > 29 kips, OK.

Welds of Stiffeners to Column Flange and Web
■ Putting aside for the moment that the weak axis moment connections still need to be considered and will affect both the strong axis connection stiffeners and welds, the welds for the ½ × 6½ × 12½ strong-axis stiffener are designed as follows. For the weld to the inside of the flange, the *connected* portion of the stiffener must be developed. Thus the 5¾ contact, which is the connected portion, is designed for 93.2 kips rather than 29 kips, which is what the load the stiffener actually "sees." The weld to the flange is thus

$$D_f = \frac{93.2}{2 \times 5.75 \times 1.392 \times 1.5} = 3.9$$

A ¼ fillet weld is indicated, or use the AISC minimum if larger. The factor 1.5 in the denominator above comes from the AISC LRFD specification Appendix J, Section J2.4, for transversely loaded fillets. The weld to the web has a length 12.5 − 0.75 − 0.75 = 11.0. and is designed to transfer the unbalanced force in the stiffener to the web. The unbalanced force in the stiffener is 29 kips in this case. Thus,

$$D_w = \frac{29}{2 \times 11.0 \times 1.392} = 0.95$$

An AISC minimum fillet is indicated.

Connections ■ 7-91

Doublers. The beam flange force (required strength) delivered to the column is $F_f = 229$ kips. The design shear strength of the column $\phi V_v = 0.9 \times 0.6 \times 50 \times 0.485 \times 14.16 = 185$ kips < 229 kips, so a doubler appears to be required. However, if the moment that is causing doublers is $\phi M_p = 389$ kip-ft, then from Fig. 7.40, the column story shear is

$$V_s = \frac{\phi M_p}{H}$$

where H is the story height. If $H = 13$ ft,

$$V_s = \frac{389}{13} = 30 \text{ kips}$$

and the shear delivered to the column web is $F_f - V_s = 229 - 30 = 199$ kips. Since 199 kips > 185 kips, a doubler (or doublers) is still indicated. Remember that the moment connections in this building, which is a low-rise building, are not primarily for resistance to lateral load. Therefore, some panel zone deformation is acceptable, and the AISC LRFD specification Section K1.7, Formula K1-11 or K1-12, contains an extra term which increases the panel zone strength. The term is

$$\frac{3b_{fc}t_{fc}^2}{d_b d_c t_{wc}} = \frac{3 \times 14.565 \times 0.780^2}{20.66 \times 14.16 \times 0.485} = 0.187$$

and if the column load is less than $0.75 P_y = 0.75 \times A_c F_{yc} = 0.75 \times 29.1 \times 50 = 1091$ kips, which is the usual case,

$$\phi V_v = 185 \times 1.189 = 220 \text{ kips}$$

Since 220 kips > 199 kips, no doubler is required. In a high-rise building where the moment connections are used for drift control, the extra term can still be used, but an analysis which includes inelastic joint shear deformation should be considered.

Shear Connections

Shear Connection for Beam 1. The specified shear for the web connection is $R = 163$ kips, which is the shear capacity of the W21×62 (A36) beam. The connection is a shear plate with two erection holes for erection bolts. The shear plate is shop-welded to the column flange and field-welded to the beam web. The limit states are plate gross shear, weld strength, and beam web strength.

Plate Gross Shear ■ Try a plate $1/2 \times 18$

$\phi R_{gv} = 0.5 \times 18 \times 0.9 \times 0.6 \times 36 = 175$ kips > 163 kips   OK

Plate net shear need not be checked here because it is not a valid limit state.

Weld to Column Flange ■ This weld sees shear only. Thus

$$D = \frac{163}{2 \times 18 \times 1.392} = 3.25 \quad \text{use 1/4 FW}$$

7-92 ■ Chapter Seven

Weld to Beam Web ■ This weld sees the shear plus a small couple. Using AISC Manual, Table 8-42, $l = 18$, $kl = 4.25$, $k = 0.24$, $x = 0.04$, $xl = 0.72$, $al = 4.28$, $a = 0.24$, $c = 2.04$, and

$$D = \frac{163}{2.04 \times 18} = 4.44$$

Thus a 5/16 fillet weld is satisfactory.

Beam Web ■ To support a 5/16 fillet weld on both sides of a plate, AISC LRFD Manual, Table 9-3, shows that a 0.72-in web is required. For a 5/16 fillet on one side, a 0.36-in web is required. Since the W21×62 web is 0.400 in thick, it is OK.

Beams 3 and 4 W21×44 (G50) Composite. The flange connection is a full-penetration weld, so again, no design is required. Section A-A of Fig. 7.39a shows the arrangement in plan. See Fig. 7.39c. The connection plates A are made 1/4 in thicker than the W21×44 beam flange to accommodate under- and overrolling and other minor misfits. Also, the plates are extended beyond the toes of the column flanges by 3/4 to 1 in to improve ductility. See Fig. 7.39c. The plates A should also be welded to the column web, even if not required to carry load, to provide improved ductility. A good discussion of this is contained in the AISC LRFD Manual, vol. II, pp. 10-61 through 65.

The flange force for the W21×44 is based on the full moment capacity as required in this example, so $\phi M_p = 358$ kip-ft and

$$F_f = \frac{358 \times 12}{20.66 - 0.45} = 213 \text{ kips}$$

Figure 7.41 shows the distribution of forces on the plates A, including the forces from the strong-axis connection. The weak-axis force of 213 kips is distributed 1/4 to each flange and 1/2 to the web. This is done to cover the case when the beams may not be reacting against each other. In this case, all of the 213 kips must be passed to the flanges. To see this, imagine that beam 4 is removed and plate A for beam 4 remains as a backup stiffener. One half of the 213 kips from beam 3 passes into the beam 3 near-side column flanges, while the other half is passed

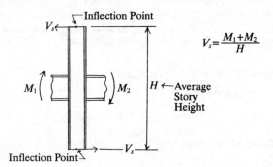

Fig. 7.40 Relationship between column story shear and the moments which induce it.

through the column web to the backup stiffener, and thence into the far-side flanges, so that all of the load is passed to the flanges. This is the load path usually assumed, although others are possible.

Merging of Stiffeners from Strong- and Weak-Axis Beams ■ The strong-axis beam, beam 1, required stiffeners $1/2 \times 6 1/2 \times 12 1/2$. The weak-axis beams 3 and 4 require plates $A \ 3/4 \times 8 \times 12 1/2$. These plates occupy the same space because the beams are all of the same depth. Therefore, the larger of the two plates is used, as shown in Fig. 7.39a.

Since the stiffeners are merged, the welds that were earlier determined for the strong-axis beam must be revisited.

Weld to Flanges ■ From Fig. 7.41, the worst-case combined flange loads are 53 kips shear and 29 kips axial. The length of weld is $6 1/4$ in. Thus,

$$D_f = \frac{\sqrt{29^2 + 53^2}}{2 \times 6.25 \times 1.392} = 3.5$$

which indicates a $1/4$ fillet weld, which is also the AISC minimum. However, remember that for axial load, the contact strength of the stiffener must be developed. The contact strength in this case is $0.9 \times 36 \times 6.25 \times 0.75 = 152$ kips because the stiffener has increased in size to accommodate the weak-axis beams. But the delivered load to this stiffener cannot be more than that which can be supplied by the beam, which is $229/2 = 114.5$ kips. Thus

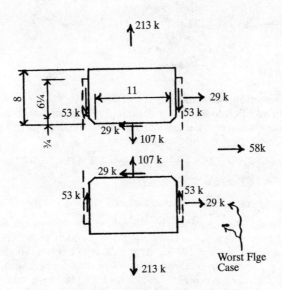

Fig. 7.41 Distribution of forces on plates A.

7-94 ■ Chapter Seven

$$D_f = \frac{114.5}{2 \times 6.25 \times 1.392 \times 1.5} = 4.39$$

which indicates that a 5/16 fillet weld is required. This is the fillet weld that should be used and is shown in Fig. 7.39c.

Weld to Web ■ From Fig. 7.41,

$$D_w = \frac{\sqrt{29^2 + 107^2}}{2 \times 11.0 \times 1.392} = 3.62$$

Use a $1/4$ fillet weld.

Stresses in Stiffeners (Plate A). The weak-axis beams are G50 steel and are butt-welded to plates A. Therefore, plates A should also be G50 steel. Previous calculations involving this plate assumed it was A36, but changing to G50 will not change the final results in this case because the stiffener contact force is limited by beam 1 delivered force rather than the stiffener strength.

The stiffener stresses for the flange welds are, from Fig. 7.41,

$$f_v = \frac{53}{0.75 \times 6.25} = 11.3 \text{ ksi} < 0.9 \times 0.6 \times 50 = 27 \text{ ksi} \qquad \text{OK}$$

$$f_a = \frac{29}{0.75 \times 6.25} = 6.19 \text{ ksi} > 0.9 \times 50 = 45 \text{ ksi} \qquad \text{OK}$$

and for the web welds

$$f_v = \frac{29}{0.75 \times 11} = 3.5 \text{ ksi} < 27 \text{ ksi} \qquad \text{OK}$$

$$f_a = \frac{107}{0.75 \times 11} = 13.0 \text{ ksi} < 45 \text{ ksi} \qquad \text{OK}$$

Shear Connection for Beams 3 and 4. The specified shear for these beams is $R = 107$ kips. In addition to this shear, for ductility reasons, as discussed earlier, the web connections must carry at least 20 percent of the beam web's nominal flexural strength. The section modulus of the beam web is

$$Z = 1/4 t_w (d - 2t_f)^2 = 0.25 \times 0.350 \times (20.66 - 2 \times 0.450)^2 = 34.2 \text{ in}^2$$

and $\phi M_{pw} = 34.2 \times 50 = 1710$ kip-in

The design moment for the web connection is thus $M = 0.2 \times 1710 = 342$ kip-in.

Weld to Beam Web. As with the strong-axis beam web connection, this is a field-welded connection with bolts used for erection only. The design loads (required strengths) are $R = 107$ kip-in and $M = 342$ kip-in. The beam web shear R is essentially constant in the area of the connection and is assumed to act at the edge of plate A (section a-a of Fig. 7.39b). This being the case, there will be a small eccentricity on the C-shaped field weld. Following AISC LRFD Manual, Table 8-

42, $l = 17$, $kl = 4$, $k = 0.24$, $x = 0.04$, $xl = 0.68$, $al = 4.25 - 0.68 = 3.57$. But the moment M will decrease the eccentricity, as can be seen from Fig. 7.39b, where the forces R and M shown on section a-a are acting on a free body of the beam web including part of plate B on the beam side of a-a. Because of the assumed position of the load R at the end of the plates A, the eccentricity can conservatively be taken as either $al = 3.57$ or $342/107 = 3.20$, whichever is larger. Thus $al = 3.57$, $a = 0.21$, and from Table 8-42 by interpolation, $c = 2.10$, and the weld size required is

$$D = \frac{107}{2.10 \times 17} = 2.99$$

which indicates that a 3/16 fillet weld is required.

Plate B (Shear Plate) Gross Shear. Try a 3/8 plate of A36 steel. Then

$\phi R_v = 0.9 \times 0.6 \times 36 \times 0.375 \times 17 = 124$ kips > 107 kips   OK

Plate B Bending. Because of the couple $M = 342$ kip-in, the capability of the plate B to transmit this moment must be checked. The section modulus is $Z = 0.25 \times 0.375 \times (17)^2 = 27.1$ in^2, so

$\phi M_n = 0.9 \times 36 \times 27.1 = 878$ kip-in > 342 kip-in   OK

Weld of Plate B to Column Web. This weld carries all of the beam shear $R = 107$ kips. The length of this weld is 17.75 in. Thus

$$D = \frac{107}{2 \times 17.75 \times 1.392} = 2.17$$

A 3/16 fillet weld is indicated. Because this weld occurs on both sides of the column web, the column web thickness should satisfy the relationship $0.9 \times 0.6 \times 50 \times t_w \geq 1.392 \times D \times 2$ or $t_w \geq 0.103 D = 0.103 \times 2.17 = 0.23$ in. Since the column web thickness is 0.485 in, the web can support the 3/16 fillets. The same result can be achieved using AISC LRFD Manual, Table 9-3.

Weld of Plate B to Plates A. These welds carry the ductility couple $M = 342$ kip-in, which is statically equivalent to equal and opposite horizontal forces at the plates A to plate B interface of $H_m = 342/19.25 = 17.8$ kips. In addition to this force, there is a shear flow $q = VQ/I$ acting on this interface, where $V = R = 107$ kips and Q is the statical moment of plate A with respect to the neutral axis of the I section formed by plates A as flanges and plate B as web. Thus

$$I = \frac{1}{12} \times 0.375 \times 19.25^3 + 0.75 \times 12.5 \times \left(\frac{19.25 + 0.75}{2}\right)^2 \times 2 = 2100 \text{ in}^4$$

$Q = 0.75 \times 12.5 \times 10 = 93.8$ in^3

and

$$q = \frac{107 \times 93.8}{2100} = 4.78 \text{ kips/in}$$

The contact length between plates A and B is $8 - 1 - 0.75 = 6.25$ in, so the total shear flow is $4.78 + 17.8/6.25 = 7.63$ kips/in and the weld size required is

$$D = \frac{7.63}{2 \times 1.392} = 2.74$$

Since plate A is $3/4$ in thick, the AISC minimum fillet weld is $1/4$ in.

Reconsideration of Plate A Welds. It will be remembered that in the course of this analysis these welds have already been looked at twice. The first time was as stiffener welds for the strong-axis beam. The second time was for the combination of forces from the weak- and strong-axis beam flange connections. Now, additional forces are added to plates A from the weak-axis beam web connections. The additional force is $17.8 + 4.78 \times 6.25 = 48$ kips. Figure 7.42 shows this force and its distribution to the plate edges. Rechecking the plate A welds, for the flange weld,

$$D_f = \frac{\sqrt{29^2 + (53+12)^2}}{2 \times 6.25 \times 1.392} = 4.09$$

which indicates that a 5/16 fillet is required. This is the size already determined. For the web weld,

$$D_w = \frac{\sqrt{(107+24)^2 + 29^2}}{2 \times 11.0 \times 1.392} = 4.38$$

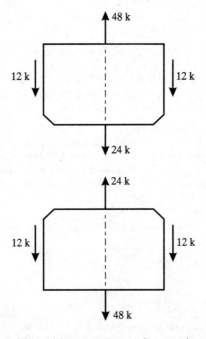

Fig. 7.42 Additional forces in plates A due to web connections.

Connections ■ 7-97

This weld, which was previously determined to be a $1/4$ fillet weld, must now be increased to a $5/16$ fillet weld.

This completes the calculations required to design the moment connections of Fig. 7.39A. Figure 7.39A, B, and C shows the completed design. Load paths of sufficient strength to carry all loads from the beams into the column have been provided.

A Further Consideration. It sometimes happens in the design of this type of connection that the beam is much stronger in bending than the column. In the example just completed, this is not the case. For the strong-axis W21×62 beam, $\phi M_p = 389$ kip-ft, while for the column, $\phi M_p = 647$ kip-ft. If the ϕM_p of the column were less than half the ϕM_p of the beam, then the connection should be designed for $2(\phi M_p)$ column because this is the maximum moment that can be developed between the beam and column. Similar conclusions can be arrived at for other arrangements.

7.4.6 Simple Shear Connections

These are the most common type of connection on every job. They are generally considered to be "simple" connections in that the beams supported by them are "simple" beams, i.e., no bending moment at the beam ends. There are two basic types of shear connections, framed and seated.

Framed Connections ■ These are the familiar double-angle, single-angle, shear plate, and shear end plate connections. They are called framed connections because they connect beams, web to web directly. Figure 7.43 shows a typical double-angle connection and Fig. 7.44 shows a shear end plate connection. These and other types of framed connections can be easily designed using the design aids (charts, tables) contained in the AISC LRFD Manual, vol. II, Part 9. A shear end plate connection will be designed in detail in the next example. The other types are designed in a similar manner.

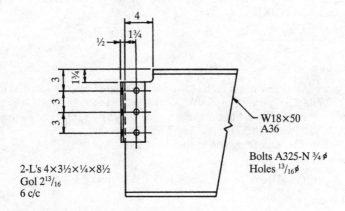

Fig. 7.43 Double-angle framed connection.

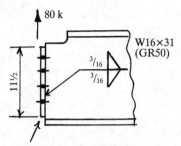

Plate: $^3/_8 \times 8 \times 11^1/_2$ (A36)
Bolts: A325SC-A-N $^7/_8 \phi$ at $5^1/_2$ gage
Holes: HSS

Fig. 7.44 Shear end plate connection.

Example—Shear End Plate Design ▪ The final design is shown in Fig. 7.44. The calculations to achieve it are as follows. The bolts are required to be slip-critical, class A, A325 bolts with threads included in the shear planes. In shorthand notation, the bolts are A325SC-A-N. The bolt size is $^7/_8$ in diameter. Horizontal short slots are used in the end plate to assist in erection.

Bolts. The slip-critical design strength per bolt is $\phi r_v = 0.85 \times 1.13 \times 0.33 \times 39 = 12.4$ kips. For bearing $\phi r_v = 0.75 \times 48 \times 0.6013 = 21.6$ kips. Since $12.4 < 21.6$, use $\phi r_v = 12.4$ kips/bolt. The number of bolts required is $80/12.4 = 6.5$. Use 8 bolts, 2 rows of 4.

Weld. The length of the shear plate for 4 rows of bolts is $11^1/_2$ in with 3-in spacing and $1^1/_4$-in edge distance. Because it is difficult to weld the full length of the plate, a distance equal to the weld size top and bottom is discounted. Thus the weld length is $11.5 - 2 \times 0.25 = 11.0$ in, where a $^1/_4$ fillet has been assumed. The required weld size is

$$D = \frac{80}{2 \times 11 \times 1.392} = 2.6$$

A $^3/_{16}$ fillet weld is satisfactory.

Beam Web. The beam web to support $^3/_{16}$ fillets, both sides, is 0.31 in thick from AISC Table 9-3. The web of the W16×31 is 0.275 in thick, which appears to be too thin. However, the $^3/_{16}$ fillet is the nominal size. The actual size required is 2.6/16. So, $(2.6/3) \times 0.31 = 0.269 < 0.275$ and the web is OK.
End plate: The end plate is A36 steel $^3/_8$ in thick.
Gross shear:

$$\phi R_{gv} = 0.9 \times 0.6 \times 36 \times 0.375 \times 11.5 \times 2 = 168 \text{ kips} > 80 \text{ kips} \qquad \text{OK}$$

Net shear:

$$\phi R_{nv} = 0.75 \times 0.6 \times 58 \times (11.5 - 4 \times 1.0) \times 0.375 \times 2 = 147 \text{ kips} > 80 \text{ kips} \qquad \text{OK}$$

Bearing: Because the edge distance of $1.25 < 1.5 \times 0.875 = 1.3125$, the top bolt has a nominal bearing capacity $r_p = 1.25 \times 0.375 \times 58 = 27.2$ kips. For the remaining bolts, the spacing of $3.0 > 3 \times 0.875 = 2.625$, so $r_p = 2.4 \times 0.875 \times 0.375 \times 58 = 45.7$ kips. Thus

$$\phi R_p = 0.75 (27.2 \times 2 + 45.7 \times 6) = 246 \text{ kips} > 80 \text{ kips} \qquad \text{OK}$$

This completes the design calculations for the connection of Fig. 7.44.

One of the principal uses of shear end plate connections is for skewed connections. Suppose the W16 beam is skewed $9\frac{1}{2}°$ (a 2 on 12 bevel) from the supporting beam or column as shown in Fig. 7.45. The nominal weld size is that determined from the analysis with the plate perpendicular to the beam web (Fig. 7.45a). This is denoted by W', where $W' = 2.6/16 = 0.1625$. The effective throat for this weld is $t_e = 0.7071 W' = 0.7071 \times 0.1625 = 0.115$ in. If the beam web is cut square, the gap on the obtuse side is $0.275 \sin 9.5 = 0.0454 < \frac{1}{16}$, so it can be ignored.

The weld size W for a skew weld is

$$W = t_e \left(2 \sin \frac{\Phi}{2} \right) + g$$

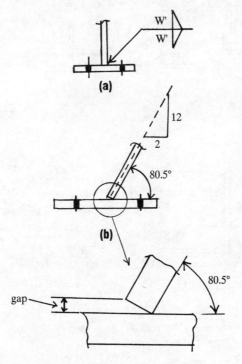

Fig. 7.45 Geometry of skewed joint.

For the obtuse side, $\Phi = 90 + 9.5 = 99.5$.

$$W = 0.115\left(2\sin\frac{99.5}{2}\right) + 0 = 0.1755 \to \frac{3}{16}Fw$$

For the acute side $\Phi = 90 - 9.5 = 80.5$.

$$W = 0.115\left(2\sin\frac{80.5}{2}\right) + 0 = 0.1486 \to \frac{3}{16}Fw$$

In this case, the fillet sizes remain the same as the orthogonal case. In general the obtuse side weld will increase and the acute side weld will decrease.

Seated Connections ■ The second type of shear connection is the seated connection, either unstiffened or stiffened (Fig. 7.46). As with the framed connections, there are tables in the AISC LRFD Manual, Part 9, which aid in the design of these connections.

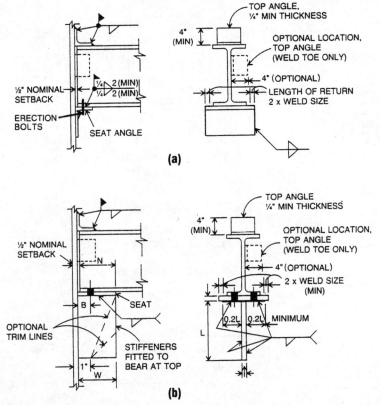

Fig. 7.46 Standardized weld seat connections. (*a*) Unstiffened seat. (*b*) Stiffened seat.

The primary use for this connection is for beams framing to column webs. In this case, the seat is inside the column flange toes or nearly so, and is not an architectural problem. Its use also avoids the erection safety problem associated with most framed connections where the same bolts support beams on both sides of the column web.

When a seat is attached to one side of the column web, the column web is subjected to a local bending pattern because the load from the beam is applied to the seat at some distance e from the face of the web. For stiffened seats, this problem was addressed by Sputo and Ellifrit (1991). The stiffened seat design table (Table 9.9) in the AISC LRFD Manual reflects the results of their research. For unstiffened seats, column web bending also occurs, but no research has been done to determine its effect. This is the case because the loads and eccentricities for unstiffened seats are much smaller than for stiffened seats. Figure 7.47 presents a yield-line analysis which can be used to assess the strength of the column web. The nominal capacity of the column web is

$$R_w = \frac{2m_p L}{e_f}\left(2\sqrt{\frac{T}{b}} + \frac{T}{L} + \frac{L}{2b}\right)$$

where the terms are defined in Fig. 7.47, and

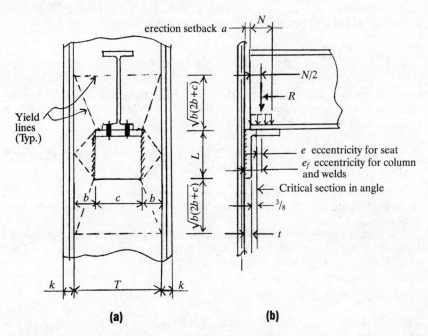

Fig. 7.47 Column web yield lines and design parameters for unstiffened seated connection.

7-102 ■ Chapter Seven

$$m_p = \frac{1}{4}t_w^2 F_y$$

$$b = \frac{T-c}{2}$$

Since this is a yield limit state, $\phi = 0.9$.

Example ■ A W14×22 beam of grade 50 steel is to be supported on an unstiffened seat to a W14×90 (G50) column. The given reaction (required strength) is 33 kips. Design the unstiffened seat.

The nominal erection setback $a = \frac{1}{2}$ in. For calculations, to account for underrun, use $a = \frac{3}{4}$. Try a seat 6 in long ($c = 6$). From AISC LRFD Manual, Table 9.7, with beam web of approximately $\frac{1}{4}$ in, a $\frac{5}{8}$-in angle gives a capacity 37.7 kips > 33 kips, OK. Choosing an angle 6 × 4 × $\frac{5}{8}$, Table 9-7 indicates that a $\frac{5}{16}$ fillet weld of the seat vertical leg (the 6-in leg) to the column web is satisfactory (41.0 kips). Consider this to be a preliminary design, which needs to be checked.

The first step in the checking of this design is to determine the required bearing length N. Note that N is not the horizontal angle leg length less a, but rather cannot exceed this value. The bearing length for an unstiffened seat starts at the end of the beam and spreads from this point, because the toe of the angle leg tends to deflect away from the bottom flange of the beam. The bearing length cannot be less than k and can be written in a general way as

$$N = \max\left\{\frac{R-\phi R_1}{\phi R_2}, \frac{R-\phi R_3}{\phi R_4} \text{ or } \frac{R-\phi R_5}{\phi R_6}, k\right\}$$

where R_1 through R_6 are defined in the AISC LRFD Manual, p. 4-33, and are tabulated in the factored uniform load tables. For the W14×22, $\phi R_1 = 25.2$, $\phi R_2 = 11.5$, $\phi R_3 = 23.0$, $\phi R_4 = 2.86$, $\phi R_5 = 20.4$, and $\phi R_6 = 3.81$. Thus

$$N = \max\left\{\frac{33-25.2}{11.5}, \frac{33-23.0}{2.86} \text{ or } \frac{33-20.4}{3.81}, 0.875\right\} = \max\{0.67, 3.50 \text{ or } 3.31, 0.875\}$$

Therefore, N is either 3.50 or 3.31, depending on whether $N/d \le 0.2$ or $N/d > 0.2$, respectively. With $d = 13.74$, $3.50/13.74 = 0.255$, and $3.31/13.74 = 0.241$. Since clearly $N/d > 0.2$, $N = 3.31$ in.

It was stated earlier that $(N + a)$ cannot exceed the horizontal angle leg. Using $a = \frac{1}{2} + \frac{1}{4} = \frac{3}{4}$, $N + a = 3.31 + 0.75 = 4.06$, which is close enough to 4 in to be OK.

The design strength of the seat angle critical section is

$$\phi R_b = \frac{1}{4}\frac{ct^2}{e}\phi F_y$$

where the terms are defined in Fig. 7.47. From Fig. 7.47, $e_f = N/2 + a = 3.31/2 + 0.75 = 2.41$ and $e = e_f - t - 0.375 = 2.41 - 0.625 - 0.375 = 1.41$. Then

$$\phi R_b = \frac{0.25 \times 6 \times 0.625^2 \times 0.9 \times 36}{1.41} = 13.5 \text{ kips}$$

Since 13.5 kips < 33 kips, the seat is unsatisfactory. The required seat thickness can be determined from

$$t_{req'd} = \sqrt{\frac{4Re}{\phi F_y c}} = \sqrt{\frac{4 \times 33 \times 1.41}{0.9 \times 36 \times 6}} = 0.98 \text{ in}$$

Therefore, an angle 1 in thick can be used, although a thinner angle between 5/8 and 1 may work since e depends on t. There is no L 6 × 4 × 1 available, so use a 6 × 6 × 1. The extra length of the horizontal leg is irrelevant.

It can be clearly seen from the above result that the AISC Manual, Table 9-7, should not be relied upon for final design. The seats and capacities given in Table 9-7 are correct for the seats themselves. It is the beam that is failing. Tables 9-6 and 9-7 were originally derived based on the bearing length required for web yielding (i.e., ϕR_1 and ϕR_2). The new requirements of web crippling ϕR_3 and ϕR_4, or ϕR_5 and ϕR_6) must be considered in addition to the capacities given in Tables 9-6 and 9-7. If web yielding is more critical than web crippling, the tables will give satisfactory capacities.

Next, the weld of the seat vertical leg to the column web is checked. Table 9-7 indicated a 5/16 fillet was required. This can be checked using AISC Manual, Table 8-38. With $e_x = e_f = 2.41$, $l = 6$, $a = 2.41/6 = 0.40$, $c = 2.00$, and

$$\phi R_{weld} = 2.00 \times 5 \times 6 = 60.0 \text{ kips} > 33 \text{ kips} \qquad \text{OK}$$

The weld sizes given in Tables 9-6 and 9-7 will always be found to be conservative because they are based on using the full horizontal angle leg minus a as the bearing length N. Finally, checking the column web,

$m_p = 0.25 \times 0.440^2 \times 50 = 2.42$ kip-in/in

$T = 11.25$

$c = 6$

$L = 6$

$b = \dfrac{11.25 - 6}{2} = 2.625$

$$\phi R_{web} = \frac{0.9 \times 2 \times 2.42 \times 6}{2.41}\left(2\sqrt{\frac{11.25}{2.625}} + \frac{11.25}{6} + \frac{6}{2 \times 2.625}\right) = 77.6 \text{ kips} > 33 \text{ kips OK}$$

This completes the calculations for this example. The final design is shown in Fig. 7.48.

Beam Shear Splices ■ If a beam splice takes moment as well as shear, it is designed with flange plates in a manner similar to the truss chord splice treated in Sec. 7.4.4. The flange force is simply the moment divided by the center-to-center

7-104 ■ Chapter Seven

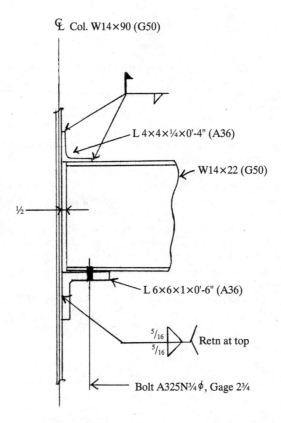

Fig. 7.48 Unstiffened seat design.

flange distance for inside and outside plate connections, or the moment divided by the beam depth for outside plate connections. The web connection takes any shear. Two typical shear splices are shown in Fig. 7.49. These are common in cantilever roof construction. Figure 7.49a shows a four clip angle splice. The angles are shop bolted (as shown) or shop welded to the beam webs. The design of this splice is exactly the same as that of a double-angle framing connection. The shear acts at the faying surface of the field connection and each side is designed as a double-angle framing connection. If shop bolted, all the bolts are in shear only; there is no eccentricity on the shop bolts. If shop welded, the shop welds see an eccentricity from the location of the shear at the field faying surface to the centroids of the weld groups on each side. This anomaly is historical. The bolted connections derive from riveted connections which were developed before it was considered necessary to satisfy "the niceties of structural mechanics," according to McGuire (1968).

A second type of shear splice uses one or two plates in place of the four angles. This type, shown in Fig. 7.49b, has moment capacity but has been used for many years with no reported problems. It is generally less expensive than the angle type. Because it has moment capability, eccentricity on the bolts or welds cannot be neglected. It has been shown by Kulak and Green (1990) that if the stiffness on both sides of the splice is the same, the eccentricity is one half the distance between the group centroids on each side of the splice. This will be the case for a shop-bolted–field-bolted splice as shown in Fig. 7.49b. A good discussion on various shear splice configurations and the resulting eccentricities is given in the AISC LRFD Manual, vol. II (1994), pp. 9-179 and 9-180.

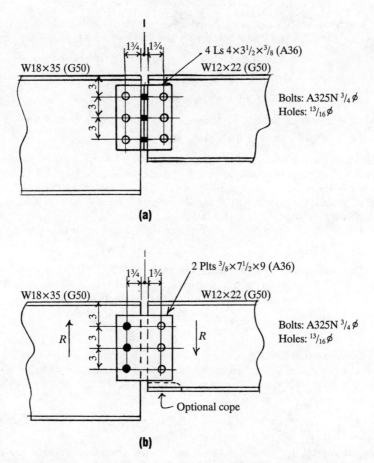

Fig. 7.49 Typical shear splices. (a) Shear splice with 4 angles. (b) Shear splice with 1 or 2 plates.

7-106 ■ Chapter Seven

Example ■ As an example of the design routine for the Fig. 7.49b splice, its capacity (design strength) will be calculated.

Bolts. The design strength per bolt in single shear is $\phi r_v = 15.9$ kips. The eccentricity is $e_x = 2.25$. From AISC LRFD Manual, Table 8-18, for $e_x = 2.25$ and $n = 3$, $c = 2.11$, and

$$\phi R_v = 2.11 \times 15.9 \times 2 = 67.1 \text{ kips}$$

Bearing. Bearing will be critical on the W12×22 web, which has a thickness $t_w = 0.260$ in. Since the edge distances and spacings satisfy the $1\frac{1}{2}d$ and $3d$ criteria, respectively,

$$\phi R_p = 0.75 \times 2.4 \times 65 \times 0.260 \times 0.75 \times 2.11 = 48.1 \text{ kips}$$

Net Shear on Splice Plates. This will be more critical than block shear rupture because the edge and end distances are the same and $0.6 \times 58 = 34.8 < 36$.

$$\phi R_{nv} = (9 - 3 \times 0.875) \times 0.375 \times 2 \times 0.75 \times 0.6 \times 58 = 125 \text{ kips}$$

Gross Shear on Splice Plates

$$\phi R_{gv} = 9 \times 0.375 \times 2 \times 0.9 \times 0.6 \times 36 = 131 \text{ kips}$$

Gross Shear on Beam Web (Uncoped). The W12×22 will control.

$$\phi R_{gv} = 0.260 \times 12.31 \times 0.9 \times 0.6 \times 50 = 86.4 \text{ kips}$$

Note that the net shear and block shear are not limit states for the web of either beam unless there is a cope toward which the shear acts. In cantilever construction, the shear R usually acts as shown in Fig. 7.49b; i.e., the W12×22 is pushed down against the bolts so the shear acts away from the cope. Since the shear acts away from the cope, block shear is not a limit state.

The cope will affect the gross shear strength, however. If the cope is 1 in deep

$$\phi R_{gv} = 0.260 \times (12.31 - 1) \times 0.9 \times 0.6 \times 50 = 79.4 \text{ kips}$$

Net Section Bending on Splice Plates. From AISC LRFD Manual, Table 12.1, the net section modulus of a splice plate is $S_{net} = 3.75$ in², $\phi M_n = 0.75 \times 58 \times 3.75 = 163$ kip-in. The design strength for net bending is thus

$$\phi R_{nb} = \frac{163 \times 2}{2.25} = 145 \text{ kips}$$

Gross Section Bending on Splice Plates. For convenience, this is usually considered at the bolt line unless there is a greater moment elsewhere. The section modulus is $Z = 0.25 \times 0.375 \times 9^2 = 7.59$ in, so $\phi M_n = 0.9 \times 36 \times 7.59 = 246$ kip-in, and

$$\phi R_{gb} = \frac{246 \times 2}{2.25} = 219 \text{ kips}$$

Considering all of the limit states, and taking the least capacity, the capacity (design strength) of the splice in Fig. 7.49b is 48.1 kips, as determined from bearing on the W12×22 web.

7.4.7 Miscellaneous Connections

Simple Beam Connections under Shear and Axial Load ■ As its name implies, a simple shear connection is intended to transfer shear load out of a beam while allowing the beam to act as a simply supported beam. The most common simple shear connection is the double-angle connection with angles shop-bolted or welded to the web of the carried beam and field-bolted to the carrying beam or column. This section, which is from Thornton (1995A), will deal with this connection.

Under shear load, the double-angle connection is flexible regarding the simple beam end rotation, because of the angle leg thickness and the gage of the field bolts in the angle legs. The AISC LRFD Manual, 2d ed., vol. II, p. 9-12, recommends angle thicknesses not exceeding ⅝ in with the usual gages. Angle leg thicknesses of ¼ in to ½ in are generally used, with ½-in angles usually being sufficient for the heaviest shear load. When this connection is subjected to axial load in addition to the shear, the important limit states are angle leg bending and prying action. These tend to require that the angle thickness increase or the gage decrease, or both, and these requirements compromise the connection's ability to remain flexible to simple beam end rotation. This lack of connection flexibility causes a tensile load on the upper field bolts which could lead to bolt fracture and a progressive failure of the connection and the resulting collapse of the beam. It is thought that there has never been a reported failure of this type, but it is perceived to be possible.

Even without the axial load, some shear connections are perceived to have this problem under shear alone. These are the single plate shear connections (shear tabs) and the tee framing connections. Recent research on the tee framing connections (Thornton, 1996) has led to a formula (AISC Manual, LRFD 2d ed., vol. II, p. 9-170) which can be used to assess the resistance to fracture (ductility) of double-angle shear connections. The formula is

$$d_{b\,min} = 0.163t \sqrt{\frac{F_y}{\tilde{b}}\left(\frac{\tilde{b}^2}{L^2}+2\right)}$$

where $d_{b\,min}$ = minimum bolt diameter (A325 bolts) to preclude bolt fracture under a simple beam end rotation of 0.03 radian
  t = angle leg thickness
  $\tilde{b}$ = distance from bolt line to k distance of the angle (Fig. 7.50)
  L = length of connection angles

Note that this formula can be used for ASD and LRFD designs in the form given above. It can be used to develop a table (Table 7.12) of angle thicknesses and gages for various bolt diameters which can be used as a guide for the design of double-angle connections subjected to shear and axial tension. Note that Table 7.12 validates AISC's long-standing (AISC, 1970) recommendation (noted above) of a maximum ⅝-in angle thickness for the "usual" gages. The usual gages would be 4½ to 6½ in. Thus, for a carried beam web thickness of, say, ½ in, GOL will range from 2 to 3 in. Table 7.12 gives a GOL of 2½ in for ¾-in bolts (the most critical as well as

7-108 ■ Chapter Seven

the most common bolt size). Note also that Table 7.12 assumes a significant simple beam end rotation of 0.03 radian, which is approximately the end rotation that occurs when a plastic hinge forms at the center of the beam. For short beams, beams loaded near their ends, beams with bracing gussets at their end connec-

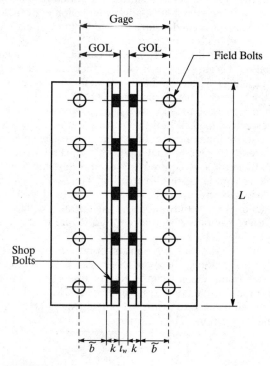

Fig. 7.50 Geometry of double angles (shop bolted shown).

Table 7.12 Estimated Minimum Angle Gages (GOL) for A36 Angles and A325 Bolts for Rotational Flexibility.

Angle thickness, in	Minimum Gage of Angle (GOL)*		
	$3/4$-in-dia bolt, in	$7/8$-in-dia bolt, in	1-in-dia bolt, in
$3/8$	$1 3/8$	$1 1/4$	$1 1/8$
$1/2$	$1 7/8$	$1 5/8$	$1 1/2$
$5/8$	$2 1/2$	$2 1/8$	$1 7/8$
$3/4$	$3 1/4$	$2 11/16$	$2 5/16$
1	6	$4 5/16$	$3 1/2$

* Driving clearances may control minimum GOL.

tions, and beams with light shear loads, the beam end rotation will be small and Table 7.12 does not apply.

As an example of a double-clip angle connection, consider the connection of Fig. 7.51. This connection is subjected to a shear load of 33 kips and an axial tensile load of 39 kips.

Shop Bolts. The shop bolts "see" the resultant load $R = \sqrt{33^2+39^2} = 51.1$ kips. The design shear strength of one bolt is $\phi r_v = 15.9$ kips, so these bolts in double shear have a design capacity

$\phi R_v = 15.9 \times 3 \times 2 = 95.4$ kips > 51.1 kips OK

Beam Web. The limit states for the beam web are bearing and block shear rupture (tearout).

Bearing. The edge distances are $1^3/_4$ and $1^1/_4$ in. The minimum edge distance for the bearing stress to be $2.4F_u$ is $1.5 \times 0.75 = 1.125 < 1.25$, which satisfies the $1^1/_2d$ criterion. The spacing is 3 in and $3 \times 0.75 = 2.25 < 3$, which satisfies the $3d$ criterion. Thus

$\phi R_p = 0.75 \times 2.4 \times 58 \times 0.75 \times 0.355 \times 3 = 83.4$ kips > 51.1 kips OK

If the loads of 33 kips shear and 39 kips axial always remain proportional, i.e., maintain the bevel of $10^1/_8$ to 12 as shown in Fig. 7.52, the spacing requirement is irrelevant because there is only one bolt in line of force and the true edge distance is 1.94 in or 2.29 in rather than 1.25 in as used above. When there is only one bolt in the line of force

$R_n = L_e t F_u \le 2.4 \, dt F_u$

and thus if $L_e \le 2.4d$, the bearing strength will be reduced from the 83.4 kips thus far determined. Since $d = 0.75$, $2.4 \times 0.75 = 1.8$ in, which is less than the provid-

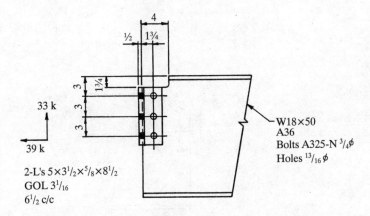

Fig. 7.51 Framed connection subjected to axial and shear loads.

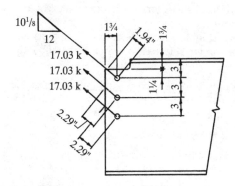

Fig. 7.52 Edge distances along the line of action.

ed edge distances of 1.94 and 2.29 in, as shown in Fig. 7.52, the bearing strength is not reduced from 83.4 kips, but situations can develop where some reduction occurs. The designer should keep an eye on this.

Block Shear Rupture (Tearout). A simple conservative way to treat block shear when shear and tension are present is to treat the resultant as a shear. Then, from Figs. 7.51 and 7.53,

$A_{gv} = 7.25 \times 0.355 = 2.57 \text{ in}^2$

$A_{nv} = (7.25 - 2.5 \times 0.875) \times 0.355 = 1.80 \text{ in}^2$

$A_{gt} = 1.75 \times 0.355 = 0.621 \text{ in}^2$

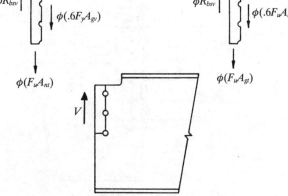

Fig. 7.53 Block shear rupture under shear V.

$A_{nt} = (1.75 - 0.5 \times 0.875) \times 0.355 = 0.466 \text{ in}^2$

$0.6F_uA_{nv} = 0.6 \times 58 \times 1.80 = 62.6$

$F_uA_{nt} = 58 \times 0.466 = 27.0$

Since $0.6F_uA_{nv} > F_uA_{nt}$,

$\phi R_{bsv} = 0.75 (62.6 + 36 \times 0.621) = 63.7 \text{ kips} > 51.1 \text{ kips}$ OK

An alternate approach is to calculate a block shear rupture design strength under tensile axial load. From Fig. 7.54,

$A_{gv} = 0.621 \text{ in}^2$

$A_{nv} = 0.466 \text{ in}^2$

$A_{gt} = 2.57 \text{ in}^2$

$A_{nt} = 1.80 \text{ in}^2$

$0.6F_uA_{nv} = 0.6 \times 58 \times 0.466 = 16.2$

$F_uA_{nt} = 58 \times 1.80 = 104$

Since $F_uA_{nt} > 0.6 F_uA_{nv}$,

$\phi R_{bst} = 0.75 (104 + 0.6 \times 36 \times 0.621) = 88.1 \text{ kips}$

Using an elliptical interaction equation, which is analogous to the vonMises (distortion energy) yield criterion,

$$\left(\frac{V}{\phi R_{bsv}}\right)^2 + \left(\frac{T}{\phi R_{bst}}\right)^2 \leq 1 \quad \text{OK}$$

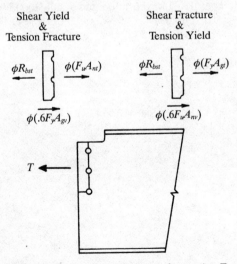

Fig. 7.54 Block shear rupture under tension T.

where V is the factored shear and T the factored tension. Then

$$\left(\frac{33}{63.7}\right)^2 + \left(\frac{39}{88.1}\right)^2 = 0.464 < 1 \qquad \text{OK}$$

This interaction approach is always less conservative than the approach using the resultant $R = \sqrt{V^2 + T^2}$ as a shear because $\phi R_{bst} > \phi R_{bsv}$ for the geometries of the usual bolt positioning in double-angle connections with two or more bolts in a single vertical column. The resultant approach, being much simpler as well as conservative, is the method most commonly used.

Connection Angles. Figure 7.51 shows angles $5 \times 3\frac{1}{2} \times \frac{5}{8}$, but assume for the moment that 3/8 angles are to be checked. The shop legs are checked for the limit states of bearing, gross shear and gross tension, and net shear and net tension. Net shear rupture and net tension rupture will control over block shear rupture with the usual connection geometries, i.e., $1\frac{1}{4}$ edge and $1\frac{1}{4}$ end distances. Since the sum of the clip angle thicknesses = 0.375 + 0.375 = 0.75 >> 0.355, the beam web and not the shop legs of the clip angles will control.

Connection Angles—Field Legs and Field Bolts. The field legs of the angles can be checked for gross and net shear using the resultant $R = 51.1$ kips

$\phi R_{gv} = 0.9 \times 36 \times 8.5 \times 0.375 \times 2 = 207$ kips > 51.1 kips OK

$\phi R_{nv} = 0.75 \times 58 \, (8.5 - 3 \times 0.875) \times 0.375 \times 2 = 192$ kips > 51.1 kips OK

These limit states seldom control, especially when there is an axial force which, because of angle leg bending, tends to increase the required angle leg thickness far beyond anything that might be required for shear. Thus the critical limit state for axially loaded clip angles is leg bending (prying action).

Prying Action. The AISC LRFD Manual has a table to aid in the selection of a clip angle thickness. The preliminary selection table (Table 11.1) indicates that a $\frac{5}{8}$ angle will be necessary. Trying L's $5 \times 3\frac{1}{2} \times \frac{5}{8}$, and following the procedure of the AISC Manual:

$$b = \frac{6.5 - 0.355 - 2 \times 0.625}{2} = 2.45$$

$$a = \frac{10.355 - 6.5}{2} = 1.93 \, (< 1.25 \times 2.45 = 3.06 \text{ OK})$$

$$b' = 2.45 - \frac{0.75}{2} = 2.08$$

$$a' = 1.93 + \frac{0.75}{2} = 2.31$$

$$\rho = \frac{2.08}{2.31} = 0.90$$

$$p = \frac{8.5}{3} = 2.83$$

$$\delta = 1 - \frac{0.8125}{2.83} = 0.71$$

The shear per bolt $V = 33/6 = 5.5$ kips < 15.9 kips OK. The tension per bolt $T = 39/6 = 6.5$ kips. Because of interaction,

$$\phi F'_t = 0.75\,(117 - 1.9 f_v) \leq 0.75 \times 90 = 67.5 \text{ ksi}$$

With $f_v = 5.5/0.4418 = 12.5$ ksi,

$$\phi F'_t = 0.75\,(117 - 1.9 \times 12.5) = 70.0 \text{ ksi} > 67.5 \text{ ksi}$$

Use $\phi F'_t = 67.5$ ksi, and $\phi r'_t = 67.5 \times 0.4418 = 29.8$ kips/bolt. Since $T = 6.5$ kips < 29.8 kips, the bolts are satisfactory independent of prying action. Returning to the prying action calculation,

$$t_c = \sqrt{\frac{4.44 \times 29.8 \times 2.08}{2.83 \times 36}} = 1.64 \text{ in}$$

$$\alpha' = \frac{1}{0.71 \times 1.90}\left[\left(\frac{1.64}{0.625}\right)^2 - 1\right] = 4.36$$

Since $\alpha' = 4.36$, use $\alpha' = 1$. This means that the strength of the clip angle legs in bending is the controlling limit state. The design strength is

$$T_d = 29.8 \left(\frac{0.625}{1.64}\right)^2 (1 + 0.71) = 7.4 \text{ kips} > 6.5 \text{ kips} \qquad \text{OK}$$

The L's $5 \times 3\tfrac{1}{2} \times \tfrac{5}{8}$ are satisfactory.

Ductility Considerations. The $\tfrac{5}{8}$-in angles are the maximum thickness recommended by AISC for flexible shear connections. Using the formula introduced at the beginning of this section,

$$d_{b\min} = 0.163 t \sqrt{\frac{F_y}{\tilde{b}}\left(\frac{\tilde{b}^2}{L^2} + 2\right)}$$

with $t = 0.625$, $F_y = 36$, $\tilde{b} = 3.0625 - 1.125 = 1.94$, $L = 8.5$,

$$d_{b\min} = 0.163 \times 0.625 \sqrt{\frac{36}{1.94}\left(\frac{1.94^2}{8.5^2} + 2\right)} = 0.63 \text{ in}$$

Since the actual bolt diameter is 0.75 in, the connection is satisfactory for ductility.

As noted before, it may not be necessary to make this check for ductility. If the beam is short, is loaded near its ends, or for other reasons is not likely to experience very much simple beam end rotation, this ductility check can be omitted.

This completes the calculations for the design shown in Fig. 7.51.

Reinforcement of Axial Force Connections ■ It sometimes happens that a simple beam connection, designed for shear only, must after fabrication and erection be strengthened to carry some axial force as well as the shear. In this case, washer plates can sometimes be used to provide a sufficient increase in the axial capacity. Figure 7.55 shows a double-angle connection with washer plates which extend from the toe of the angle to the k distance of the angle. These can be made for each bolt, so only one bolt at a time need be removed, or if the existing load is small, they can be made to encompass two or more bolts on each side of the connection. With the washer plate, the bending strength at the "stem" line, section a-a of Fig. 7.55, is

$$M_n = \tfrac{1}{4}F_y p t^2$$

while that at the bolt line, section b-b, is

$$M'_n = \alpha\delta \tfrac{1}{4} F_y p(t^2 + t_p^2) = \alpha\delta \tfrac{1}{4} F_y p t^2 \left(1 + \frac{t_p^2}{t^2}\right) = \alpha\delta\eta M_n$$

where $\eta = 1 + (t_p/t)^2$ and the remaining quantities are in the notation of the AISC LRFD Manual (1994). With the introduction of η, the prying action formulation of the AISC LRFD Manual can be generalized for washer plates by replacing δ wherever it appears by the term $\delta\eta$. Thus

$$\alpha' = \frac{1}{\delta\eta(1+\rho)}\left[\left(\frac{t_c}{t}\right)^2 - 1\right]$$

and

$$T_d = \phi r_t \left(\frac{t}{t_c}\right)^2 (1 + \alpha'\delta\eta)$$

All other equations remain the same.

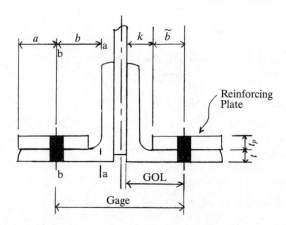

Fig. 7.55 Prying action with reinforcing (washer) plates.

As an example of the application of this method, consider the connection of Fig. 7.56. This was designed for a shear of 60 kips but now must carry an axial force of 39 kips when the shear is at 33 kips. Let us check the axial capacity of this connection. The most critical limit state is prying action because of the thin angle leg thickness. From Fig. 7.56,

$$b = \frac{5.5 - 0.355 - 2 \times 0.375}{2} = 2.20$$

$$a = \frac{8 + 0.355 - 5.5}{2} = 1.43$$

$$1.25 \times 2.20 = 2.75 > 1.43$$

Use $a = 1.43$. Then $b' = 1.82$, $a' = 1.81$, $\rho = 1.01$, $\delta = 0.72$, $V = 33/8 = 4.125$ kips/bolt. The holes are HSSL (horizontal short slots), so $\phi r_v = 8.88$ kips/bolt. Since $4.125 < 8.88$, the bolts are OK for shear (as they obviously must be since the connection was originally designed for 60 kips shear). Because this is a shear connection, the shear capacity is reduced by the tension load by the factor $1 - T/(1.13T_b)$, where T is the applied load per bolt and T_b is the specified pretension. Thus the reduced shear design strength is

$$\phi r'_v = \phi r_v \left(1 - \frac{T}{1.13T_b}\right)$$

This expression can be inverted to a form usable in the prying action equations as

$$\phi r'_t = 1.13T_b \left(1 - \frac{V}{\phi r_v}\right) \leq \phi r_t$$

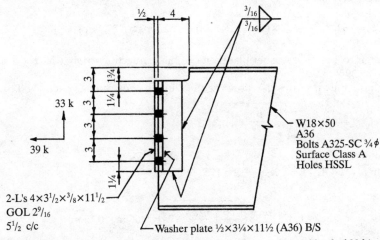

Fig. 7.56 A shear connection needing reinforcement to carry axial load of 39 kips.

For the present problem

$$\phi r'_t = 1.13 \times 28\left(1 - \frac{4.125}{8.8}\right) = 16.8 \text{ kips} < 29.8 \text{ kips}$$

Use $\phi r'_t = 16.8$ kips. Since $T = 39/8 = 4.875$ kips < 16.8 kips, the bolts are OK for tension/shear interaction exclusive of prying action. Now, checking prying action, which includes the bending of the angle legs,

$$t_c = \sqrt{\frac{4.44 \times 16.8 \times 1.82}{2.875 \times 36}} = 1.15$$

$$\alpha' = \frac{1}{0.72 \times 2.01}\left[\left(\frac{1.15}{0.375}\right)^2 - 1\right] = 5.81$$

Since $\alpha' > 1$, use $\alpha' = 1$, and

$$T_d = 16.8 \times \left(\frac{0.375}{1.15}\right)^2 \times (1.72) = 3.07 \text{ kips} < 4.875 \text{ kips} \qquad \text{NG}$$

Thus, the 3/8 angle legs fail. Try a 1/2 washer plate. Then

$$\eta = 1 + \left(\frac{0.5}{0.375}\right)^2 = 2.78$$

$$\alpha' = \frac{1}{0.72 \times 2.78 \times 2.01}\left[\left(\frac{1.15}{0.375}\right)^2 - 1\right] = 2.08$$

Since $\alpha' > 1$ use $\alpha' = 1$

$$T_d = 16.8 \times \left(\frac{0.375}{1.15}\right)^2 \times (1 + 0.72 \times 2.78) = 5.32 \text{ kips} > 4.875 \text{ kips} \qquad \text{OK}$$

Therefore, the 1/2-in washer plates enable the connection to carry $5.32 \times 8 = 42.6$ kips > 39 kips, OK.

If ductility is a consideration, the ductility formula can be generalized to

$$d_{b\min} = 0.163 t \sqrt{\eta} \sqrt{\frac{F_y}{\tilde{b}}\left(\frac{\tilde{b}^2}{L^2} + 2\right)}$$

With $\tilde{b} = \text{GOL} - k = 2\frac{9}{16} - \frac{13}{16} = 1.75$,

$$d_{b\min} = 0.163 \times 0.375 \sqrt{2.78} \sqrt{\frac{36}{1.75}\left(\frac{1.75^2}{11.5^2} + 2\right)} = 0.66 \text{ in} < 0.75 \text{ in} \qquad \text{OK}$$

Use of Ultimate Tensile Strength to Increase Axial Capacity in the Presence of Prying Action ■ In the example of the previous section, washer plates were used to increase the axial capacity of the connection. Research (Thornton, 1992) has shown that the prying action formulation of the AISC Manual is very conservative with respect to strength, and that if F_y, the yield strength, is replaced by F_u, the tensile strength, a much better prediction of strength is achieved. This idea of replacing F_y with F_u in prying calculations was originally suggested by Kato and McGuire (1973), who also showed that it yielded more accurate predictions of strength. The prying action equations are unchanged when F_u replaces F_y, except for

$$t_c = \sqrt{\frac{4.44\phi r_t' b'}{pF_u}}$$

Following the calculations of the preceding section for the example of Fig. 7.56,

$$t_c = \sqrt{\frac{4.44 \times 16.8 \times 1.82}{2.875 \times 58}} = 0.902$$

$$\alpha' = \frac{1}{0.72 \times 2.01}\left[\left(\frac{0.902}{0.375}\right)^2 - 1\right] = 3.31$$

Since $\alpha' > 1$, use $\alpha' = 1$.

$$T_d = 16.8\left(\frac{0.375}{0.902}\right)^2 (1 + 0.72 \times 1) = 4.99 \text{ kips} > 4.875 \text{ kips} \qquad \text{OK}$$

Thus, if the tensile strength approach is used, the $4 \times 3\frac{1}{2} \times \frac{3}{8}$ angles are capable of carrying $4.99 \times 8 = 40.0$ kips > 39 kips and are satisfactory without the reinforcing washer plates.

References

1. American Institute of Steel Construction, 1970, *Manual of Steel Construction*, 7th ed., AISC, Chicago.

2. American Institute of Steel Construction, 1992, *Manual of Steel Construction*, vol. II, AISC, Chicago.

3. American Institute of Steel Construction, 1994, *Manual of Steel Construction*, (two volumes), LRFD, 2d ed., AISC, Chicago.

4. Blodgett, Omer W., 1966, *Design of Welded Structures*, The James F. Lincoln Arc Welding Foundation, Cleveland.

5. Englehardt, M. D., T. A. Sabol, R. S. Aboutaha, and K. H. Frank, 1995, "An Overview of the AISC Northridge Moment Connection Test Program," Proceedings of the AISC National Steel Construction Conference, San Antonio.

6. Fling, R. S., 1970, "Design of Steel Bearing Plates," *Engineering Journal*, vol. 7, no. 2, 2d quarter, AISC, Chicago.

7. Gaylord, E. H., and C. N. Gaylord, 1972, *Design of Steel Structures*, McGraw-Hill, New York, pp. 139–141.

8. Gross, J. L., 1990, "Experimental Study of Gusseted Connections," *Engineering Journal*, vol. 27, no. 3, 3rd quarter, pp. 89–97, AISC, Chicago.

9. Hardash, S., and R. Bjorhovde, 1985, "New Design Criteria for Gusset Plates in Tension," *Engineering Journal*, vol. 23, no. 2, 2d quarter, pp. 77–94, AISC, Chicago.

10. Kato, B. and W. McGuire, 1973, "Analysis of T-Stub Flange to Column Connections," *Journal of the Structural Division*, ASCE, vol. 99, no. ST5, May, pp. 865–888, New York.

11. Kulak, G. L., J. W. Fisher, and J. H. A. Struik, 1987, *Guide to Design Criteria for Bolted and Riveted Joints*, Wiley—Interscience, New York.

12. Kulak, G.L., and D. L. Green, 1990, "Design of Connections in Web Flange Beam or Girder Splices," *Engineering Journal*, vol. 27, no. 2, 2d quarter, pp. 41–48, AISC, Chicago.

13. Mann, A.P., and L. J. Morris, 1979, "Limit Design of Extended End Plate Connections," *Journal of the Structural Division*, ASCE, vol. 105, no. ST3, March, pp. 511–526, New York.

14. McGuire, W., 1968, *Steel Structures*, Prentice Hall, Englewood Cliffs.

15. Murray, T., 1983, "Design of Lightly Loaded Column Base Plates," *Engineering Journal*, vol. 23, No. 4, 4th quarter, pp. 143–152, AISC, Chicago.

16. Richard, R.M., et al., 1983, "Analytical Models for Steel Connections," *Behavior of Metal Structures*, Proceedings of the W. H. Munse Symposium, Edited by W. J. Hall and M. P. Gaus, May 17.

17. Richard, R.M., 1986, "Analysis of Large Bracing Connection Designs for Heavy Construction, "National Steel Construction Conference Proceedings, pp. 31.1–31.24, AISC, Chicago.

18. Sputo, T., and D. S. Ellifrit, 1991, "Proposed Design Criteria for Stiffened Seated Connections to Column Webs," Proceedings of the AISC National Steel Construction Conference, Washington, D.C., pp. 8.1–8.26, AISC, Chicago.

19. Stockwell, F. W., Jr., 1975, "Preliminary Base Plate Selection," *Engineering Journal*, vol. 21, no. 3, 3rd quarter, AISC, Chicago.

20. Struik, J. H. A., and J. deBack, 1969, "Tests on T-Stubs With Respect To A Bolted Beam To Column Connections," Report 6-69-13, Stevin Laboratory, Delft University of Technology, Delft, the Netherlands. [As referenced in Kulak, Fisher, and Struik (1987)].

21. Thornton, W. A., 1985, "Prying Action—A General Treatment," *Engineering Journal*, vol. 22, no. 2, 2d quarter, pp. 67–75, AISC, Chicago.

22. Thornton, W. A., 1990A, "Design of Small Base Plates for Wide Flange Columns," *Engineering Journal*, vol. 27, no. 3, 3rd quarter, pp. 108–110, AISC, Chicago.

23. Thornton, W. A., 1990B, "Design of Small Base Plates for Wide Flange Columns—A Concatenation of Methods," *Engineering Journal*, vol. 27, no. 4, 4th quarter, pp. 173–174, AISC, Chicago.

24. Thornton, W. A.,1991, "On the Analysis and Design of Bracing Connections," National Steel Construction Conference Proceedings, pp. 26.1–26.33, AISC, Chicago.

25. Thornton, W. A., 1992, "Strength and Serviceability of Hangar Connections," *Engineering Journal*, vol. 29, no. 4, 4th quarter, pp. 145–149, AISC, Chicago. (Errata (1996), Engineering Journal, Vol. 33, No. 1, 1st quarter, pp. 39–40, AISC, Chicago).

26. Thornton, W. A., 1995A, "Treatment of Simple Shear Connections Subject to Combined Shear and Axial Forces," *Modern Steel Construction*, vol. 35, no. 9, September, pp. 9–10, AISC, Chicago.

27. Thornton, W. A., 1995B, "Connections—Art, Science and Information in the Quest for Economy and Safety," *Engineering Journal*, vol. 32, no. 4, 4th quarter, pp. 132–144, AISC, Chicago.

28. Thornton, W. A., 1996, "Rational Design of Tee Shear Connections," *Engineering Journal*, vol. 33, no. 1, 1st quarter, pp. 34–37, AISC, Chicago.

29. Timoshenko, S. P., and J. N. Goodier, 1970, *Theory of Elasticity*, Third Edition, McGraw-Hill, New York, pp. 57–58.

30. Vasarhelyi, D. D., 1971, "Tests of Gusset Plate Models," *Journal of the Structural Division*, vol. 97, ST2, February, ASCE, New York.

31. Whitmore, R. E., 1952, *Experimental Investigation of Stresses in Gusset Plates*, University of Tennessee Engineering Experiment Station Bulletin 16, May.

8

Robert Dexter
John Fisher

FATIGUE AND FRACTURE

8.1 Introduction

Laboratory tests show that structural steel assemblages which are properly proportioned and detailed can consistently exceed their yield strength, deform to several times the displacement at the yield point, and achieve the calculated plastic limit load.[1] Because of this experience, design specifications have evolved which are based on the plastic limit load rather than an allowable stress. One example is the Load and Resistance Factor Design (LRFD) Specification for Structural Steel Buildings from the American Institute of Steel Construction (AISC).[2]

In order to achieve plastic limit states, ductility is required to allow yielding and redistribution of the stresses. However, details such as welds, holes, or copes can concentrate strain on the net section, prevent the development of a plastic hinge in the gross section, and consequently limit overall structural ductility. Notches, fatigue cracks, and other cracklike defects concentrate stress and strain to an extreme degree. In the event of overload due to extreme or accidental load, cracklike defects can limit tensile ductility or even maximum load by inducing premature fracture. Therefore, it is essential that fatigue and fracture limit states be considered in design of buildings and other structures.

Fatigue cracks can form and propagate from weld discontinuities and/or stress concentrations if a member is subjected to significant cyclic live loads, even if the maximum stresses are well below the yield strength. In highly redundant structures, fatigue cracking is usually just a serviceability problem; i.e., the repair of the cracks is a nuisance rather than a threat to structural integrity. For example, if a fatigue crack forms in one element of a riveted built-up structural member, the crack cannot propagate directly into neighboring elements. Usually, a riveted member will not fail until a second crack forms in another element. Therefore, riveted built-up structural members are inherently redundant.

Welding began to be used for some steel structures in the 1930s. Once a fatigue crack forms, it can propagate directly into all elements of a continuous welded member and cause failure at service loads. The lack of inherent redundancy in welded members is one reason that fatigue and fracture changed from a nuisance to a significant structural integrity problem as welding became widespread. Welded structures are not inferior to bolted or riveted structures; they just require more attention to design, detailing, and quality.

Just before the Second World War, several welded Vierendeel truss bridges in Europe failed suddenly with relatively light load shortly after being put into service.[3] These types of fractures, at nominal stresses below the yield strength with little plastic deformation, are referred to as brittle fractures. More than 20 percent of the 4694 merchant ships built in the US during the war suffered brittle fractures, and 145 broke in two.[4]

The need to understand the failures of the merchant ships led to significant research after the war at the Naval Research Laboratory. George R. Irwin[5] developed the engineering discipline of linear elastic fracture mechanics (LEFM) to predict brittle fracture. The investigations showed the failures occurred because the welded construction contained notches and the material had poor fracture toughness. Fracture toughness is a measure of the material's resistance to fracture in the presence of a notch. It can be directly measured in terms of a quantifiable fracture parameter that can be used directly in design (these parameters are explained in Sec. 8.3) or it can be inferred from the results of Charpy V-notch (CVN) tests.

The collapse of the Point Pleasant Bridge in West Virginia in 1967 brought about increased awareness of the possibility of brittle fracture among structural engineers in the United States. This collapse, as well as many other cases of cracking in bridges, are examined using LEFM in Ref. 6.

In bridges, fracture is typically preceded by fatigue due to the cyclic live load. Therefore, the emphasis in bridge design is on fatigue limit states and the use of fatigue-resistant details. In addition to controlling the stress range and the notch size through design, detailing, and inspection, the bridge fracture control plan also strives to screen out brittle materials. For example, the AASHTO specifications[7] require a minimum CVN "notch toughness" at a specified temperature for the base metal and the weld metal of members loaded in tension or tension due to bending. As long as large defects do not exist, the notch toughness requirement assures that the fracture will not be brittle. Since 1978, there has been an AASHTO Guide Specifications for Fracture Critical Non-Redundant Steel Bridge Members[8] with more stringent detailing, fabrication, and notch-toughness requirements for fracture-critical members (FCM). FCM are defined as those members which, if fractured, would result in collapse of the bridge. Almost two decades of experience with these bridge specifications have proved that they are successful in significantly reducing the number of fatigue cracks and brittle fractures.

Fatigue cracking may occur in industrial buildings subjected to loads from cranes or other equipment or machinery. Although it has not been a problem in the past, fatigue cracking could occur in high-rise buildings frequently subjected to large wind loads. Because of the relatively large number of cycles before fatigue failure, these types of loading are referred to as high-cycle fatigue. Design for high-cycle fatigue is covered in Appendix K3 in the LRFD specification.

Since fatigue is uncommon in buildings, fracture is also less of a possibility than in bridges or ships. If a fracture occurs in a building, it typically must occur directly from a notch and/or fabrication defect without being preceded by the formation and growth of a fatigue crack. If a fracture occurs in a building, it will usually occur during construction as the members are loaded for the first time and may also be exposed to low temperatures.

For example, fractures occurred in the 1980s as jumbo shapes, i.e., shapes in groups 4 and 5, began to be used for tension chords of trusses in large buildings.[9] These jumbo shapes are normally used for columns, where they are not subjected to tensile stress. These sections often have low fracture toughness, particularly in the region of the web and flange junction. The low toughness has been attributed to the relatively low rolling deformation and slow cooling in these thick shapes. The low toughness is of little consequence if the columns remain in compression.

The fractures of jumbo tension chords occurred during construction at welded splices at groove welds (Fig. 8.1) or at flame-cut edges of cope holes (Figure 8.2).[9] In both cases the cracks formed at cope holes in the hard layer formed from ther-

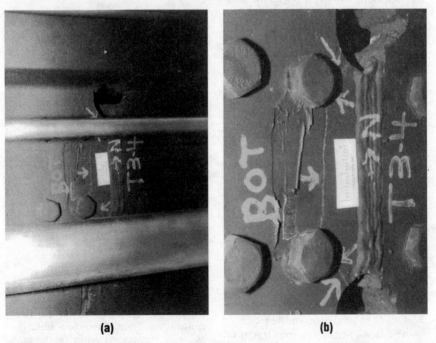

Fig. 8.1 View from the side (*a*) of splice in jumbo sections used as tension chord in a roof truss and close-up view (*b*) of fracture in web originating from weld access holes.

mal cutting. These cracks propagated in the core region of these jumbo sections which has very low toughness. Figure 8.2 showed the cracks curved down into the flange such that they were perpendicular to the primary tensile stress. The cross section of the brittle fracture surface on the flange shows the dark spot which constitutes the original fabrication defect. As a consequence of these brittle fractures, AISC specifications now have a supplemental CVN notch-toughness requirement for shapes in groups 4 and 5 and (for the same reasons) plates greater than 51 mm thick, when these are welded and subject to primary tensile stress from axial load or bending (see Section A3.1c). Poorly prepared cope holes have resulted in cracks and fractures in lighter shapes as well.

Other building fractures have occurred in the tension flange of welded girders. In January 1985, as the temperature decreased to record lows, an exposed steel box girder that supports part of the roof of the new Filene Center at Wolf Trap Farm Park in Vienna, Va., was discovered to be fractured. Figure 8.3 shows the fracture surface of the box girder, which exhibits the typical "herringbone" patterns that point to the origin of brittle failure. In this case, the fracture originated from lack of fusion in the splice welds in the backing bar used for the corner longitudinal welds of the box. Typically, the welds splicing the backing bar are thought to be unimportant and are not given the same attention as the welds in the member itself. However, the backing bar is fused to the corner longitudinal weld, and therefore this defect became a defect in the corner longitudinal weld. Such defects in backing bar splices have led to fatigue cracking in bridges in several cases in the past.[6]

The lack-of-fusion defects fractured the corner longitudinal groove welds because of the unusually low toughness of these welds. These low-tough-

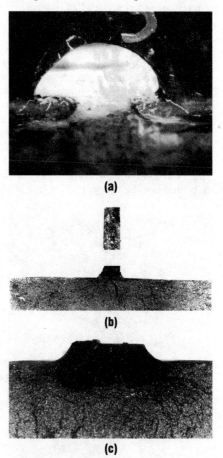

Fig. 8.2 Fabrication crack originating in a weld access hole (*a*) curves down into flange [darkened area in flange cross section (*b*) and close-up of web-flange junction (*c*)] and then causes brittle fracture during construction. Note macroscopic features of cleavage fracture, chevron markings, or river patterns, which point to the origin of the brittle fracture at the location of the initial fabrication crack.

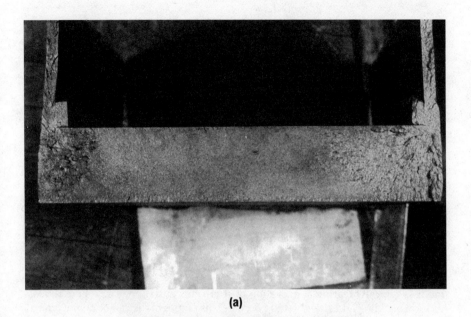

Fig. 8.3 The fractured flange of a box girder (*a*) and close-up of web-flange junction (*b*) showing chevron markings which point to the origin of the crack in the corner weld.

ness welds resulted from mixing two weld processes. Tack welds were used to place the backing bar. These tack welds were made with a self-shielded flux-core arc weld (FCAW-S) which contained a high aluminum content. These tack welds were partially remelted in the longitudinal submerged-arc welds (SAW), giving these welds very low toughness which was comparable to the low toughness observed in the core region of jumbo sections. As a consequence of the fracture at the Filene Center, there is now a warning in the AISC LRFD code (Section J.2.7 and commentary) regarding the mixing of weld metals.

Fracture in a building may also occur in service from unusually large ductility demand such as from support settlement, explosions, or earthquakes. For example, although no steel-framed structures collapsed in the earthquake which occurred on Jan. 17, 1994, in Northridge, Calif., many suffered significant fractures. Among the steel-framed structures, damage included fractured column base plates, fractured brace members, and brittle fracture of beam-column connections in welded special moment frames (WSMF). The latter type of damage to WSMF, as shown in Fig. 8.4, was widespread and alarming. At the time of this writing, over one hundred buildings with WSMF are known to have beam-column connection failures.

Steel frames that may be seismically loaded are expected to withstand cyclic plastic deformation. If brittle fracture of connections is suppressed, the failure mode is typically tearing at a location of strain concentration, e.g., where repeated local buckling and straightening occurs. This failure mode can be characterized as low-cycle fatigue. Low-cycle fatigue has been studied for pressure vessels and some other types of mechanical engineering structures. Since low-cycle fatigue is an inelastic phenomenon, the strain range is the key parameter rather than the stress range. However, at this time very little is understood about low-cycle fatigue in buildings and the tearing failure mode that occurs after repeated cyclic plastic deformation. For example, it is not known whether low-cycle fatigue from low-intensity earthquakes can induce undetected cracks that influence the fracture behavior in subsequent large earthquakes.

The detailing rules that are used to prevent fatigue are intended to avoid notches and other stress concentrations. These detailing rules are useful for the avoidance of brittle fracture as well as fatigue. For example, the detailing rules in AASHTO bridge design specifications[7] would not permit a backing bar to be left in place because of the unfused notch perpendicular to the tensile stress in the flange. This type of backing bar notch was a significant factor in the brittle fracture of WSMF connections in the Northridge earthquake. Detailing rules similar to the AASHTO detailing rules are included in American Welding Society (AWS) D1.1 Structural Welding Code—Steel[8] for dynamically-loaded structures. "Dynamically loaded" has been interpreted to mean fatigue loaded. Unfortunately, most seismically loaded building frames have not been required to be detailed in accordance with these rules. Even though it is not required, it might be prudent in seismic design to follow the AWS D1.1 detailing rules for dynamically loaded structures.

This section on fatigue and fracture is intended for practicing civil and structural engineers engaged in regulation, design, inspection, repair, and retrofit of steel-framed buildings. The section is intended to summarize:

Fatigue and Fracture   8-7

1. Aspects of fatigue and fracture which are relevant to structural steel buildings
2. The ability to predict and avoid fatigue and fracture
3. The guidance in the LRFD Manual on fatigue and fracture

The AISC LRFD design procedures for fatigue in Appendix K3 are well developed and sufficient for most high-cycle fatigue applications. Section 8.2 provides some discussion of these fatigue design procedures and adds some additional design guidance for special situations which are not covered in the AISC specifi-

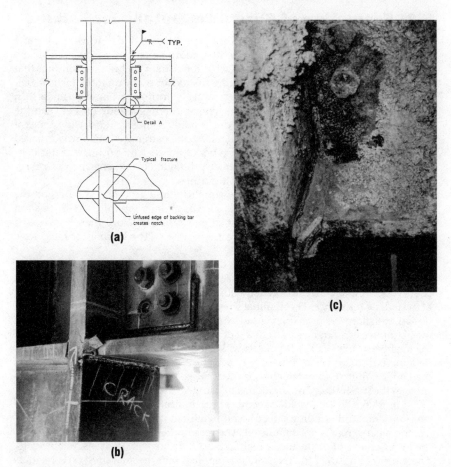

Fig. 8.4 Welded steel moment frame (WSMF) connection (*a*) showing location of typical fractures, a fractured WSMF in a building under construction (*b*) in which the crack propagated into the column flange, and another fractured WSMF with fire protection partially removed (*c*) in which the crack propagated vertically along the fusion line in the weld metal.

cations. The current LRFD guidance on avoiding fracture is relatively vague and is only qualitative. Therefore, some additional fracture evaluation and design methods are presented in Sec. 8.3. These relatively simple methods are based on LEFM. Similar fracture mechanics methods can be found in the Eurocode 3 (Annex C: Design against Brittle Fracture)[18] and have been used for decades to design and reevaluate the service life of aircraft, pressure vessels, as well as pipelines and steel structures in the oil and gas industry.

8.2 Evaluation of Structural Details for Fatigue

In most cases of strength checking in design, a calculated load is compared to a measured material property and a liberal amount of empiricism is used in the development of the design rules. The calculation usually includes only things that are easily quantifiable whereas everything that is not easily quantifiable is usually lumped in with the material property and determined from experiments. In predicting the fatigue life of welded structures, there is a trade-off between the complexity of the analysis (which introduces many sources of error) and the cost to acquire adequate fatigue resistance data. At one extreme is an approach based on the results of tests on full-scale prototypes of the structure using realistic loading. Such testing is usually considered prohibitively expensive, except for extremely expensive structures such as airframes.

The next best approach (in terms of simplicity and accuracy) is based on the results of tests on structural components such as girders. Typically, small-scale test specimens will result in longer apparent fatigue lives. Therefore, there seems to be no alternative other than to test large-scale specimens. The reasons for these scale effects are discussed in Sec. 8.2.2.

When structural members are tested, the loading is characterized in terms of the nominal stress in the structural member remote from the weld detail. The local stress concentration effect associated with the shape of the weld is considered part of the fatigue resistance. The nominal stress is conveniently obtained from standard design equations such as those used elsewhere in the LRFD specification using member forces and moments from a global analysis. Usually, the nominal stress in the members can be easily calculated without excessive error. However, the proper definition of the nominal stresses may become a problem in regions of high stress gradients. If finite-element analysis is used, the computed stresses will be very sensitive to the mesh size. One solution is to use elements of a standardized size.

Testing on full-scale welded members has indicated that the primary effect of constant-amplitude loading can be accounted for in the live-load stress range;[12–14,17] i.e., the mean stress is not significant. The reason the dead load has little effect on the lower bound of the results is that, locally, there are very high residual stresses. Therefore, the mean of the total stresses (applied plus residual stresses) is relatively high regardless of the dead load.[28,29] However, without residual stress, the life for a given stress range would be expected to be greater if the minimum stress were in compression (reversal loading) than would be the case if the minimum stress were in tension (pulsating loading). The reason for the longer life in reversal loading is the crack is closed during the compression part of the stress range and therefore only the tensile part of the stress range is effective in causing crack growth.[29] In details that are not welded, such as anchor bolts, there is a strong mean stress effect.[30] A worst-case conservative assumption, i.e., a high tensile mean stress, is made in the testing and in the design of these nonwelded details.

Multiaxial live-load stresses and stresses from second-order distortion effects can also complicate the determination of the nominal stress range. Guidance (beyond the scope of the AISC LRFD specification) for design in case of multiaxial or distortion-induced stresses is presented in Sec. 8.2.3. Otherwise, the following discussion assumes details are subjected to essentially uniaxial live-load stress ranges.

The strength and type of steel have only a negligible effect on the fatigue resistance expected for a particular detail.[12-15,17] The welding process also does not typically have an effect on the fatigue resistance.[15,16] This material independence was confirmed in a recent study by a significant number of large-scale tests on welded HSLA-80 ship details weldments.[15] The fatigue resistance of these details was consistent with the existing data base, which was generated mainly on carbon-manganese steels. The independence of the fatigue resistance from the type of steel greatly simplifies the development of design rules for fatigue since it eliminates the need to generate data for every type of steel.

Fatigue test data generally consist of the number of cycles to failure for a particular detail subjected to a particular constant-amplitude stress range. The results are in general highly variable; therefore, a statistically significant number of replicate tests must be performed. The large variance in the number of cycles to failure is primarily due to variance in both the weld geometry and weld discontinuities. This large variance makes it difficult to distinguish the secondary effects of many variables such as type of steel and filler metal, rate of loading, mean stress, and the environment. An exception is exposure to seawater or moist sea air, which may significantly reduce the fatigue life in some cases. Recommendations in the British Standard for Fatigue Design BS 7608[38] are that the fatigue life be reduced by a factor of 2 for all detail categories for exposure to seawater.

Fatigue tests are performed at a number of different stress ranges, and the data are generally plotted with the logarithm of the nominal stress range on the ordinate and the logarithm of the number of cycles to failure on the abscissa (even though the number of cycles is the dependent variable). The relationship used to represent the lower bound to these data is referred to as an *S-N* curve (see Fig. 8.5). An *S-N* curve is an exponential equation of the form

$$N = CS^{-m} \tag{8.1}$$

or

$$\log N = \log C - m \log S$$

where N = number of cycles to failure
  C = constant dependent on detail category
  S = applied constant amplitude stress range
  m = inverse of the slope of the *S-N* curve

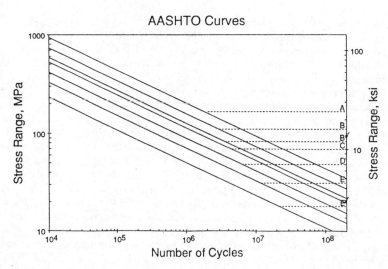

Fig. 8.5 S-N curves represented by the fatigue strength values given in Table A-K 3.3 of the AISC LRFD specifications. Dotted lines are the constant-amplitude fatigue limits (CAFL) and indicate the detail category.

The term "slope" has become synonymous with the exponent m. Note that the exponent m is actually the negative of the slope of the function $\log N = f(\log S)$. But S-N curves are traditionally plotted with $\log N$ (the dependent variable) on the abscissa, whereas most functions are plotted with the independent variable on the abscissa. Therefore, m is actually the inverse of the apparent slope of the plotted S-N curve. Greater values of m, referred to as greater "slope," will actually result in S-N curves which appear flatter.

It has been observed that the logarithm of N is approximately normally distributed at a particular stress range.[15] The mean S-N curve is found by performing a linear regression analysis, minimizing the error in $\log N$ using the method of least squares with $\log S$ as the independent variable. The runouts, or tests that did not fail, are excluded from this regression analysis.

The slope of the mean regression line in most cases is in the range 2.9 to 3.1. [12–15,17] Therefore, in the AISC and AASHTO codes as well as in Eurocode 3,[18] the slopes have been standardized at 3.0. The lower-bound S-N curves (Figure 8.5) are based on the lower 95 percent confidence limit (using $\log N$), which is approximately two standard deviations of $\log N$ below the mean S-N curve. Considering that the slope will be fixed to 3.0, Eq. 8.1 is used to pool the results at various stress ranges to determine the statistics of the quantity $\log C$. The lower-bound S-N curve is defined by the lower-bound value (mean minus two standard deviations) of $\log C$.[15]

There are many complexities in fatigue that are difficult if not impossible to model accurately. The fatigue life is governed by random phenomena such as the size and shape of the welds and their inherent discontinuities. The distribution of applied and residual stresses in the local vicinity of the weld is also very complex. These modeling difficulties and associated uncertainty are avoided in the nominal-stress-range approach since these complexities are properly represented in the full-scale tests.

The nominal-stress-range approach is used in the AISC LRFD specification and most other specifications for welded structures. Some details in offshore structures and ships are designed in terms of a "hot spot-stress range," which includes the effect of stress concentration factor (SCF).[24,26,27] The hot-spot stress range approach will not be discussed but is explained in AWS D1.1[10] and BS 7608.[38]

In the AISC LRFD specification, the fatigue design provisions are given in Appendix K 3. The design procedures are based on control of the nominal stress range and knowledge of the fatigue strength of the details. (Fatigue strength is the nominal stress range corresponding to the lower-bound S-N curve at a particular number of cycles, usually 2 million cycles.) Since fatigue is typically only a serviceability problem, fatigue design is carried out using service loads. Consequently, the LRFD fatigue design provisions are essentially the same as the allowable stress design (ASD) fatigue design procedures.[11] In the AISC fatigue design provisions, the loading is assumed to be simple constant-amplitude loading. Four different numbers of cycles are used for design as specified in Table A-K 3.1, i.e., 100,000, 500,000, 2,000,000, and over 2,000,000. Special procedures for variable-amplitude loading are discussed in Sec. 8.2.4.

8.2.1 Classification of Structural Details According to Fatigue Strength

It is standard practice in fatigue design of welded structures to separate the weld details into categories having similar fatigue resistance in terms of the nominal stress.[12–14,17] Table A-K 3.3 in the AISC LRFD specification has eight discrete categories with a fatigue strength specified at discrete numbers of cycles. Most common details can be idealized as analogous to one of the drawings in Figure A-K 3.1. Table A-K 3.2 associates these details in Figure A-K 3.1 with the categories in Table A-K 3.3.

The fatigue detail category actually accounts for several other variables which are highly variable and difficult to quantify in practice, such as weld discontinuities and local stress concentration due to the geometry of the weld detail. In most cases, there is some link between the cracking mode and the stress concentration of the various details in a category. For example, transverse stiffeners and transverse butt welds with reinforcement are both category C details. The cracking mode for these details is expected to be similar; i.e., cracks form and grow in a similar manner from the weld toe. The stress concentration at the toe of the weld is also similar for these two details. However, in some cases details are grouped together because they happened to have similar fatigue strength, even though the cracking mode and stress concentration are completely different. For example, properly tensioned slip-critical bolted joints and longitudinal welds are both category B details.

The points in Table A-K 3.3 correspond to the lower-bound S-N curves for each category. Figure 8.5 showed seven of the S-N curves which are the basis for the stress ranges in Table A-K 3.3. The way the LRFD rules are to be applied, the function in

Table A-K 3.3 is like a step function which is below the S-N curve and is only equal to the lower-bound S-N curve at the limiting number of cycles in each column. Therefore, for numbers of cycles between the limits, the rules are very conservative because the number of cycles is essentially always rounded up to the next limit.

The fatigue strength for "more than 2 million cycles" in Table A-K 3.3 is referred to as the constant-amplitude fatigue limit (CAFL). The CAFL is a nominal stress range below which test specimens apparently do not fail. As shown in Fig. 8.5, the CAFL for the AISC/AASHTO curves was defined based on experimental observations and actually occurs at a greater number of cycles than 2 million for category B and below. When a detail is designed for "more than 2 million cycles," it is expected that the fatigue life will be at least 100 million cycles, which is about the longest tests are typically run before terminating. A structure loaded continuously at an average rate of three times per minute (0.05 Hz) would last 60 years before exceeding 100 million cycles. Most structures are not subjected to significant live load this frequently, and therefore a life of 100 million cycles is considered essentially infinite. Some additional reduction on the allowable stress range might be considered if there was machinery or something that subjected the structure continuously to significant loading at greater than 0.05 Hz.

The categories in Fig. 8.5 range from A to E' in order of decreasing fatigue strength. There is an eighth category F in Table A-K 3.3 for fillet welds loaded in shear. However, there have been very few if any failures related to shear, and the stress ranges are typically very low such that fatigue rarely would control the design. Therefore, the shear stress category F will not be discussed further.

In fact there have been few if any failures which have been attributed to details which have a fatigue strength greater than category C. An expedient approach to fatigue design is first determine if there will be category C or worse details near where the maximum stress ranges occur. If this is the case, the preliminary fatigue design can be based on limiting the live-load stress range at all locations to the value in Table A-K 3.3 for category C. (This approach could be easily implemented as an initial screening procedure in CAD programs.) Then particular attention should be focused on identifying and checking the stress range for only those details which are in category D, E, or E'. Most structures have many of these more severe details, and these will generally govern the fatigue design. Therefore, unless all connections in highly stressed elements of the structure are high-strength bolted connections rather than welded, it is usually a waste of time to check category C and better details.

In addition to being used by AISC and AASHTO specifications, the S-N curves in Fig. 8.5 and detail categories are essentially the same as those adopted by the American Welding Society (AWS) Structural Welding Code D1.1.[10] The AISC S-N curves are also the same as seven of the eleven S-N curves in the Eurocode 3.[18] The British Standard 7608[38] has slightly different S-N curves, but these can be correlated to the nearest AISC S-N curve for comparison.

The following is a brief simplified overview of the categorization of fatigue details. In all cases, the exact specifications of Appendix K 3 should be checked. Several reports have been published which show a large number of illustrations of details and their categories in addition to those in Figure A-K 3.1.[19,20] Also, the

Eurocode 3[18] and the British Standard 7608[38] have more detailed illustrations for their categorization than do the AISC or AASHTO specifications. A book by Maddox[41] discusses categorization of many details in accordance with BS 7608, from which roughly equivalent AISC categories can be inferred.

In most cases, the fatigue strength recommended in these European standards is similar to the fatigue strength in the AISC and AASHTO specifications. However, there are several cases where the fatigue strength is significantly different; usually the European specifications are more conservative. Some of these cases are discussed in the following, as well as the fatigue strength for details that are not found in the AISC specifications.

Detail Categories for Base Metal and Thermal-Cut Edges. ■ The base metal must conform to Section 3.2 of AWS D1.1,[8] which has requirements for the plate or shape condition, thermal-cut edges, repairs, reentrant corners, beam copes, and weld-access holes. Category A represents base metal or rolled shapes, including thermal-cut edges within a roughness no greater than 1000 μin (0.025 mm). Very few engineers can actually discern the smoothness, however. Experiments were recently conducted at the ATLSS Center at Lehigh University[21] on welded girders with thermal-cut edges that were not within this tolerance because of nicks and gouges. The results of these experiments indicate that thermal-cut edges that are not strictly controlled to the smoothness tolerance of 1000 should be considered category C. Fortunately, this difference rarely matters because most welded structures have details such as stiffeners, butt welds, or worse details which will control the fatigue life of the member. Therefore, except at holes or other cutouts such as the coped and blocked flanges at end connections, the base metal and thermal-cut edges rarely control the fatigue design.

However, there is a potential of fabrication cracks emanating from a hard surface layer (martensite) which forms on the thermal-cut edges, particularly at a hole or reentrant corner. As discussed in the introduction, such cracks have led to fracture and collapse of trusses. The fatigue strength will be reduced by the presence of these fabrication cracks. The propensity for such cracks increases when there are nearby welds, especially under highly restrained conditions which develop multiaxial residual stress. Great care is required in designing, executing, and inspecting holes or other cutouts, especially with thick members which may have low-toughness regions such as the core region of some group 4 and 5 shapes.

Weld access holes (cope holes or ratholes) have been the source of numerous fabrication cracks and fatigue cracks but are not explicitly treated as a fatigue category in the AISC specifications. For example, Fig. 8.6 shows a fabrication crack which has formed at an inadequate cope hole and the longitudinal web-flange weld. Figures 8.1 and 8.2 showed fabrication cracks at cope holes in jumbo sections.

Section 3.2.5 of AWS D1.1 provides detailing requirements, some of which are illustrated in Figure C-J1.2 of the commentary to Chapter J of the AISC LRFD specification. The height of the cope hole must exceed the thickness of the web material, and the width should extend at least 1.5 times from the edge of the weld and the thickness. The angles where the cope hole intersects the flange must be right angles or greater, with very flat angles for welded built-up sections. Usually,

8-14 ■ Chapter Eight

Fig. 8.6 View from the side of a fabrication crack originating from web-flange fillet weld termination at a particularly poor weld access hole in a welded truss tension chord which fractured at another weld access hole.

increasing the size of the access hole decreases the problems with fabrication; however, the total length should not exceed 100 mm. Of course, the net section should not be reduced significantly. A cope hole with edges conforming to the ANSI smoothness of 1000 which is less than 100 mm across may be considered a category D detail. Some jagged flame cutting may be done at the cope hole or the welds are not terminated well. If these problems may occur, the cope hole must be treated as a category E detail.

At a location of copes, blocks, or cuts, there is a potentially large stress concentration at the reentrant corner. The radius of the corner must be 25 mm or greater. Some detailing recommendations are provided in Part 8, particularly in Figures 8-59 and 8-60, of Volume II of the LRFD Manual. A detail which is especially trouble-free is to drill or saw a hole of diameter 50 mm or greater at the location of the corner, and then cut lines almost tangent to the hole. If there is fatigue loading, these details can be evaluated by using a stress range equal to the nominal stress range times the stress-concentration factor (SCF), and comparing this concentrated stress range to category A or, if edges are not within tolerance, to category C.

In any cutouts, a generous radius of at least 25 percent of the width of the opening should be used. Drilling holes for the corners will improve the fatigue resistance of thermal-cut cutouts. Grinding the edges of the cutout may be helpful. These cutouts may have to be reinforced for strength considerations; however, excessive reinforcement can often increase the likelihood of fatigue cracking. Such openings in the decks were a major cause of fractures in the Second World War merchant ships. Large openings must be specially evaluated to determine the SCF. The concentrated stress range should be compared to the allowable stress range for the thermal-cut edges, i.e., category A or, if edges are not within tolerance, to category C.

Severely corroded members should be evaluated to see what the stress range is with respect to the reduced thickness and loss of section. Notches and pits should be specially evaluated,[43] but otherwise severely corroded members may be treated as category E^{22} when the element thickness has been reduced by more than 50 percent.

Detail Categories for Mechanically-Fastened Joints. ■ Small holes are considered category D details. Therefore, riveted and mechanically fastened joints (other than high-strength bolted joints) loaded in shear are evaluated as category D in terms of the net-section nominal stress. Pin plates and eyebars are designed as category E details in terms of the stress on the net section.

The fatigue strength of small empty holes can be improved by installing a highly torqued high-strength bolt. The appropriate fatigue category for fully tightened high-strength bolted joints in slip-critical connections loaded in shear is category B in terms of the gross-section nominal stress. Similar high-strength bolted joints in bearing are evaluated as category B in terms of the net-section nominal stress.

Bolted joints loaded in direct tension are evaluated in terms of the maximum unfactored tensile load, including any prying load. Typically, these provisions are applied to hanger-type or bolted flange connections where the bolts are tensioned against the plies. If the number of cycles exceeds 20,000, the allowable load is reduced relative to the allowable load for static loading only as shown in Table A-K 3.4. Prying and calculation of prying forces is discussed in Part 11 (Volume II) of the LRFD Manual. Prying is very detrimental to fatigue, so if the number of cycles exceeds 20,000, it is advisable to minimize prying forces. Equation 8.2 is given in Part 11 to determine the minimum thickness of the flange or angle t_{req} to minimize prying:

$$t_{req} = \sqrt{\frac{4.44 r_{ut} b'}{p F_y}} \qquad (8.2)$$

where r_{ut} = factored tensile force per bolt
  b' = distance between edge of bolt hole and face of web (see Figure 11-2 in Volume II of the LRFD Manual)
  p = tributary length of the flange or angle per bolt
  F_y = yield strength of the flange or angle

When bolts are tensioned against the plies, the total fluctuating load is resisted by the whole area of the precompressed plies, so that the bolts are subjected to only a fraction of the total load.[39] The analysis to determine this fraction is difficult,[39] and this is one reason that the bolts are designed in terms of the maximum load rather than a stress range in the AISC LRFD specifications.

8-16 ■ Chapter Eight

Note that the bolt strength given in Section J3 is based on the nominal unthreaded body (shank) area of the bolt A_b. The allowable total service load in fatigue, given in Table A-K3.4, is also given in terms of A_b; e.g., for more than 500,000 cycles the allowable service load is 25 percent of the product of A_b and the ultimate tensile strength F_u. F_u for the A490 bolts is 1030 MPa (150 ksi), while F_u for A325 bolts (less than 28 mm in diameter) is 830 MPa (120 ksi).

In some codes, the fatigue strength of bolts is given in terms of the stress range on the "tensile stress area," which is analogous to what is defined as the "net tensile area" in Table 8-7 of Part 8 (Volume II) of the LRFD Manual. The tensile stress area A_t is given by

$$A_t = \frac{\pi}{4}\left(d_b - \frac{0.9743}{n}\right)^2 \tag{8.3}$$

where d_b = nominal diameter (body or shank diameter)

n = threads per inch, given in Table 8-7. (Note that the constant would be different if SI units are used.)

Equation 8.3 gives a tensile stress area which is on the order of 0.75 times the body area A_b. For example, for 25-mm (1-in) bolts with n of 8, A_t is equal to 0.77 times A_b. Thus the allowable total service stress for greater than 500,000 cycles, derived for the tensile stress area from the values in Table A-K3.4, is actually about 0.33 times F_u.

In BS7608[38] a slightly different approach is used for bolts in tension which achieves approximately the same result as the AISC LRFD specification for high-strength bolts. The stress range, on the tensile stress area of the bolt, is taken as 20 percent of the total applied load, regardless of the fluctuating part of the total load. The S-N curve for bolts is proportional to F_u, so that for high-strength bolts the result is an S-N curve between category E and E' for cycles less than 2 million. (In BS7608, the F_u values are taken as no greater than 785 MPa, even for higher-strength bolts.) The CAFL for bolts is taken at 2 million cycles and is equal to 6 percent of the F_u. (There is a reduction in the fatigue strength for bolts larger than 25 mm in diameter.) For the maximum F_u of 785 MPa (about the same as A325 bolts), BS7608 would give a CAFL of 47 MPa, which is about the same as the AISC LRFD CAFL for category D.

Since the stress range is only 20 percent of the peak load, BS7608 essentially limits total load to five times the stress range, or $0.3F_u$, which, for A325 bolts, is about the same as the AISC provisions for greater than 500,000 cycles. The AISC provisions would allow an even greater allowable total load for A490 bolts, but the BS7608 would not because of the strength limit.

In the Eurocode 3[18] the fatigue strength in the bolts is given in terms of the actual stress range in the bolts, although it is not clear how to calculate this for pre-tensioned connections. The recommended fatigue strength is given in terms of the tensile stress area of the bolt and does not depend on tensile strength. The design S-N curve from Eurocode 3 is about the same as category E', which is consistent with BS7608 for high-strength bolts. However, the AISC and BS7608 provisions would give more conservative S-N curves for lower-strength bolts.

The CAFL in the Eurocode is 23 MPa for all types of bolts, which is much less than that implied by the AISC provisions or the BS7608 provisions for high-strength bolts. Since the BS7608 provisions give the CAFL as a function of F_u, they would give a CAFL comparable to the Eurocode CAFL for an F_u of about 380 MPa, which is comparable to the lowest grade (e.g., grade 36) bolts.

Anchor bolts in concrete cannot be adequately pretensioned and therefore do not behave like hanger-type or bolted flange connections. At best they are pretensioned between nuts on either side of the column base plate, and the part below the bottom nut is still exposed to the full load range. Some additional test data were recently generated at the ATLSS Center at Lehigh University[30] for grade 55 and grade 105 anchor bolts. When combined with the existing data from Karl Frank at the University of Texas,[40] the data show that the fatigue strength for anchor bolts is slightly greater than category E' in terms of the stress range on the tensile stress area of the bolt.[30] Some of the bolts were tested with an intentional misalignment of 1:40, and these had only slightly lower fatigue strength than the aligned bolts, bringing the lower bound to the data closer to the category E' S-N curve.

The ATLSS data indicate that proper tightening between the double nuts changes the failure mode from the first thread that engages the top nut to some location below the bottom nut. This change occurs because proper tensioning reduces the actual stress range between the nuts (in a manner analogous to the properly tightened tension connections discussed above) and also probably because some bending and stress concentration is eliminated. Tightening of one-third turn past snug-tight is more than enough to induce this change. The fatigue strength for the properly tightened double-nut anchor bolts is slightly greater than for the untightened bolts; however, the improvement in the fatigue strength is not enough to exceed one category.[30]

The S-N curve derived from the ATLSS data is consistent with the S-N curve for all types of bolts given in Eurocode 3 as discussed above. The ATLSS data show that the grade 105 bolts had slightly greater fatigue strength than grade 55 bolts when tested at the same minimum stresses, but not enough to exceed one category. For grade 55 anchor bolts, F_u is 520 MPa (75 ksi), and the BS7608 code would give an S-N curve well below the Category E' S-N curve. However, the AISC provisions and BS7608 for the grade 105 strength level, which is comparable to A325 bolts, would give an S-N curve which was in agreement with the ATLSS data for both types of bolts. Hence the AISC LRFD and BS7608 provisions would appear to be excessively conservative for lower-strength bolts such as the grade 55 bolts.

The ATLSS data show the CAFL for all anchor bolts is slightly greater than the category D CAFL (48 MPa). As in the case of the S-N curve in the finite-life regime, this is consistent with the AISC design provisions and the BS7608 provisions, but only for high-strength bolts. The ATLSS data and Karl Frank's data show that proper tightening between the double nuts had a slight beneficial effect on the CAFL, but not enough to increase it by one category.

When considering the effect of ultimate strength on anchor bolts, it is important to note that in the tests recently done at ATLSS, the higher-strength bolts were tested at a higher maximum stress than the lower-strength bolts. For the lower-strength anchor bolts, the maximum stress in the tests represented a worst-case value of 0.6 times the yield strength. It is likely that, in service, the higher-strength bolts would be loaded at even higher levels of maximum stress and could possibly exhibit even lower fatigue strength. High maximum stress has a deleterious effect on fatigue. Thus there are two competing factors that affect the fatigue strength of the higher-strength bolts that tend to cancel each other out.

Also, depending on the application, lower-strength anchor bolts are typically governed by strength considerations rather than fatigue, whereas higher-strength anchor bolts are more likely to be fatigue-critical (because the stress ranges will be larger). Higher-strength bolts are more likely to have lower fracture toughness or to be susceptible to stress-corrosion cracking. Therefore, for numerous reasons, the use of lower-strength anchor bolts, i.e., grade 55 (F_u of 520 MPa) or lower, is preferred.

In summary, for all types of anchor bolts, it is recommended that (1) for finite life, the category E' S-N curve be used; and (2) for infinite life, the CAFL equivalent to that for category D be used (48 MPa).

Detail Categories for Welded Joints. ■ Welded joints are considered longitudinal if the axis of the weld is parallel to the primary stress range. Continuous longitudinal welds are category B details. A special category B' was added to account for the lower fatigue strength of complete-penetration longitudinal welds with permanent backing bars as well as partial-penetration longitudinal welds. The termination of longitudinal fillet welds is more severe (category E). (The termination of full-penetration groove longitudinal welds requires a ground transition radius but gives a greater fatigue strength, depending on the radius.) If longitudinal welds must be terminated, it is better to terminate at a location where the stress ranges are less severe.

Category C includes transverse full-penetration groove welds (butt joints) subjected to nondestructive evaluation (NDE). Experiments conducted at Lehigh University show that groove welds that are not subject to NDE may contain large internal discontinuities, and the fatigue strength is reduced to category E. The British Standard 7608[38] and the British Standards Institute published document PD 6493[25] have reduced fatigue strength curves for groove welds with defects that are generally in agreement with these experimental data. Transverse groove welds with a permanent backing bar are reduced to category D.[17,18,38,41]

Cope holes for weld access and to avoid intersecting welds were discussed in Sec. 8.2.1. A cope hole with edges conforming to the ANSI smoothness of 1000 which is less than 100 mm across may be considered a category D detail. Poorly executed cope holes must be treated as a category E detail. In some cases small cracks have occurred from the thermal-cut edges if martensite is developed. In those cases, crack extension will occur at lower stress ranges. Testing performed at ATLSS[15] as well as at TNO in the Netherlands[23] has shown that the cope hole has lower fatigue strength than overlapping welds, which is less expensive but has traditionally been avoided because of the discontinuity at the overlap.

There have been many fatigue-cracking problems in structures at miscellaneous and seemingly unimportant attachments to the structure for such things as racks and hand rails. Attachments are a "hard spot" on the strength member which create a stress concentration at the weld. Often, it is not realized that such secondary members become part of the girder, i.e., that these secondary members stretch with the girder and therefore are subject to large stress ranges. Consequently, problems have occurred with fatigue of such secondary members.

Attachments normal to flanges or plates that do not carry significant load are rated category C if less than 51 mm long in the direction of the primary stress range, D if between 51 and 101 mm long, and E if greater than 101 mm long. (The 101-mm limit may be smaller for plate thinner than 9 mm.) If there is not at least 10-mm edge distance, then category E applies for an attachment of any length. The category E', slightly worse than category E, applies if the attachment plates or the flanges exceed 25 mm in thickness. Transverse stiffeners are treated as short attachments (category C).

The cruciform joint where the load-carrying member is discontinuous is considered a category C detail because it is assumed that the plate transverse to the load-carrying member does not have any stress range.[44] A special reduction factor for the fatigue strength is provided when the load-carrying plate exceeds 13 mm in thickness. This factor accounts for the possible crack initiation from the unfused area at the root of the fillet welds (as opposed to the typical crack initiation at the weld toe for thinner plates).[44]

Transverse stiffeners that are used for cross bracing or diaphragms are also treated as category C details with respect to the stress in the main member. In most cases, the stress range in the stiffener from the diaphragm loads is not considered, because these loads are typically small and unpredictable. In any case, the stiffener must be attached to the flanges so even if the transverse loads were significant, most of the load would be transferred in shear to the flanges. (The web has very little out-of-plane stiffness.) Theoretically, the shear stress range in the fillet welds to the flanges should be checked, but shear stress ranges rarely govern design.

In most other types of load-carrying attachments, there is interaction between the stress range in the transverse load-carrying attachment and the stress range in the main member. In practice, each of these stress ranges is checked separately. The attachment is evaluated with respect to the stress range in the main member and then it is separately evaluated with respect to the transverse stress range. The combined multiaxial effect of the two stress ranges is taken into account by a decrease in the fatigue strength; i.e., most load-carrying attachments are considered category E details. Multiaxial effects are discussed in greater detail in Sec. 8.2.3.

If the fillet or groove weld ends of a longitudinal attachment (load-bearing or not) are ground smooth to a transition radius greater than 50 mm, the attachment can be considered category D (load-bearing or not). If the transition radius of a groove-welded longitudinal attachment is increased to greater than 152 mm (with the groove-weld ends ground smooth), the detail (load-bearing or not) can be considered category C.

The AWS Structural Welding Code D1.1[10] has separate rules for fatigue design of circular tubular sections. The details represented by the category A, B, D, and E curves for circular sections are analogous to the details represented by these categories in the AISC specifications, i.e., A is for plain steel pipe, B is for longitudinal welds in pipe, etc. The S-N curves for category A, B, D, and E for circular sections are slightly lower than the corresponding curves in the AISC specifications. The

AWS rules for tubular joints also contain several S-N curves to be used for particular types of T, Y, and K joints based on the nominal stresses in each member. The AWS curves C1 and C2 for tubular joints are used with the hot-spot stress approach, which is discussed below. The British Standard 7608[38] and the Eurocode 3[18] have extensive guidance on rectangular and circular tubular members and connections.

Misalignment is a primary factor in susceptibility to cracking. The misalignment causes eccentric loading, local bending, and stress concentration. If the ends of a member with a misaligned connection are essentially fixed, the stress concentration factor associated with misalignment is:

$$\text{SCF} = 1.0 + \frac{3e}{t} \tag{8.4}$$

where e is the eccentricity and t is the smaller of the thicknesses of two opposing loaded members. The nominal stress times the SCF should then be compared to the appropriate category. Generally, such misalignment should be avoided at fatigue-critical locations. Equation 8.2 can also be used where e is the distance that the weld is displaced out of plane due to angular distortion. In either case, if the ends are pinned, the SCF is twice as large. A thorough guide to the SCF for various types of misalignment and distortion, including plates of unequal thickness, can be found in British Standards Institute published document PD 6493.[25]

8.2.2 Scale Effects in Fatigue

As previously mentioned, fatigue tests on small-scale specimens will give higher apparent fatigue strength and are therefore unconservative.[45–48] There are several possible reasons for the observed scale effects. First, there is a well-known thickness effect in fatigue.[45] To explain this thickness effect, it is useful to consider a plate with a welded attachment on the surface, and a scale model of this detail. The plate thicknesses are different but the stress gradients are proportional. Fatigue cracks grow at exponentially increasing growth rates such that the majority of the fatigue life is associated with the growth of the crack up to a depth of about 10 mm. In the thicker full-scale plate, this zone where the majority of the fatigue life is spent is a smaller percentage of the total thickness. Near the top surface of the plate, the residual stress is greater and the applied stress range is greater. Thus, in the thicker plate, most of the life is spent in this higher-stress zone, whereas in the thinner plate, the crack may grow out of the high-stress zone and still have a significant remaining life. Therefore, thinner plates will generally have longer fatigue lives. This thickness effect is reflected in many places in Table A-K3.2 (in the LRFD specifications) where the fatigue strength is reduced for details with plate thickness greater than 20 or 25 mm in certain cases.

In BS 7608[38] the fatigue strength of many details is keyed to plates with thickness 16 mm and less. For plates exceeding 16 mm, an equation is given which reduces the fatigue strength for thicker plates. A similar equation is used in Eurocode 3 for plates greater than 25 mm thick. These equations produce reductions in fatigue strength proportional to the $1/4$ power of the ratio of the thickness to the base thickness (i.e., 16 or 25 mm).

Another effect is that the applied stress range may be different in small-scale specimens. For example, the stress concentration associated with welded attachments varies with the length of the attachment in the direction of the stresses. This effect is reflected in Table A-K3.2 for attachments where the fatigue strength is decreased for increasing attachment length. Also, in large-scale specimens, even though the nominal stress state is uniaxial or bending, unique local multiaxial stress states may develop naturally in complex details from random stress concentrations (e.g., poor workmanship and weld shape) and eccentricities (e.g., asymmetry of the design, tolerances, misalignment, distortion from welding). These complex natural stress states may be difficult to simulate in small-scale specimens, and are difficult if not impossible to simulate analytically.

The state of residual stress from welding may be significantly different for small specimens due to the lack of constraint.[48] Even if the specimens are cut from large-scale members, the residual stress will be altered.[28] Finally, the volume of weld metal in full-scale members is sufficient to contain a structurally relevant representative sample of discontinuities, e.g., microcracks, pores, slag inclusions, hydrogen cracks, tack welds, and other notches.[46,47]

8.2.3 Distortion and Multiaxial-Loading Effects in Fatigue

In the AISC fatigue design provisions, the loading is assumed to be simple uniaxial loading. However, the loading and response of buildings is known to be more complex than is commonly assumed in design. For example, fatigue design is based on the primary tension and bending stress ranges. Torsion, racking, transverse bending, and membrane action in plating are considered secondary loads and are typically not considered in fatigue analysis.

However, it is clear from the type of cracks which occur in bridges that a significant proportion of the cracking is due to distortion which results from such secondary loading.[6,24,31,34] The situation in buildings is similar in many ways. Stresses due to distortion have been measured in bridge girders. It is possible to analyze the fatigue life of these details with distortion in terms of the local hot-spot stress ranges as discussed in Ref. 24.

However, it has been observed that the magnitude and even the sign of these stresses are difficult to predict in the case of web gap in girders because of sensitivity to as-built dimensions.[31] Therefore, the solution to the problem of fatigue cracking due to secondary loading usually relies on the qualitative art of good detailing. Unfortunately, detailing is often anathema to engineers with sophisticated finite-element analysis tools. Some of the good detailing rules of thumb that have been effective for bridges are expected to be applicable to buildings as well.

A related problem is the reduction in cross section at coped flanges which results in a large reduction in cross section and increases stresses at the cope.

Often, the best solution to distortion cracking problems may be to stiffen the structure. Fortunately, distortion problems are often localized and can be reduced just by modification of the connections or details.[32] Typically, the better connections are more rigid. For example, transverse stiffeners or gusset plates on welded girders should be welded directly to the flanges as well as the web. There has been a tendency to avoid welding to the tension flange due to an unfounded concern about brittle fracture. Numerous fatigue cracks have occurred due to distortion in the narrow gap between the stiffener and the flange.

In some cases a better solution is to allow the distortion to take place over a greater area so that lower stresses are created; i.e., the detail should be made more flexible. For example, if the transverse stiffener is not welded to a flange, it is important to assure that the gap between the flange and the end of the stiffener is sufficiently large, between 4 to 6 times the thickness of the web.[32] Another example where the best details are more flexible is connection angles for "simply supported" beams. Despite our assumptions, such simple connections transmit up to 40 percent of the theoretical fixed-end moment, even though they are designed to transmit only shear forces. For a given load, the moment in the connection decreases significantly as the rotational stiffness of the connection decreases. The increased flexibility of connection angles allows the limited amount of end rotation to take place with reduced bending stresses. A criterion has been developed for the design of these angles to provide sufficient flexibility.[33,34] The criterion states that the angle thickness t must be

$$t < 12 \frac{g^2}{L} \qquad (8.5)$$

where g is the gauge in inches and L is the span length in inches. For example, for connection angles with a gauge of 76 mm and a beam span of 7 m, the angle thickness should be just less than 10 mm. To solve a connection-angle cracking problem in service, the topmost rivet or bolt may be removed and replaced with a loose bolt to ensure the shear capacity. For loose bolts, steps are required to ensure that the nuts do not back off.

Significant stresses from secondary loading are often in a different direction than the primary stresses. Many questions arise regarding multiaxial loading, e.g.:

1. How do you combine stress cycles in the different directions that are random and may be out of phase or at different frequencies?

2. Should principal stresses be examined or stresses in the longitudinal and transverse directions?

3. What type of cumulative damage model should be used to predict fatigue cracking?

Fortunately, experience with multiaxial loading experiments on large-scale welded structural details indicates the loading perpendicular to the local notch or the weld toe dominates the fatigue life. The cyclic stress in the other direction has no effect if the stress range is below 83 MPa and only a small influence above 83 MPa.[24,31]

This finding is contrary to some results of multiaxial fatigue experiments on small specimens.[35,36] However, these small specimen experiments do not contain a large enough sample of weld metal to contain "worst-case" discontinuities. Therefore, crack initiation was a significant portion of the cycles to failure. The effect of multiaxial loading on the initiation of a crack is very significant. For fatigue problems associated with welds in large-scale structures, the crack initiation life is negligible due to the existence of "worst-case" weld discontinuities. The crack propagation life is governed by the stress range perpendicular to the plane of the crack and is relatively insensitive to the stress ranges in the orthogonal directions. Ironically, the unfortunate existence of weld discontinuities at least has the benefit of reducing the complexity of a multiaxial fatigue analysis, since the combination of multiaxial loading does not have to be considered. The recommended approach for multiaxial loads is[24,31]

1. Decide which loading (primary or secondary) dominates the fatigue cracking problem (typically the loading perpendicular to the weld axis or perpendicular to where cracks have previously occurred in similar details).
2. Perform the fatigue analysis using the stress range in this direction (i.e., ignore the stresses in the orthogonal directions).

8.2.4 The Effective Stress Range for Variable-Amplitude Loading

An actual service load history is likely to consist of cycles with a variety of different load ranges, i.e., variable-amplitude loading.[42,49] However, the LRFD fatigue design provisions are based on constant-amplitude loading and do not give any guidance for variable-amplitude loading. A procedure is shown below to convert variable stress ranges to an equivalent constant-amplitude stress range with the same number of cycles.[42,49] This procedure is based on the damage summation rule jointly credited to Palmgren and Miner (referred to as Miner's rule).[37]

To use Miner's rule, a histogram of the stress ranges in the load history is developed by counting and grouping the stress ranges into discrete intervals of stress range. The contribution of damage D for each interval i is then equal to the ratio of the number of stress ranges (cycles) within that interval of stress range n_i to the expected number of cycles to failure at that stress range from a constant-amplitude test N_i. A fatigue crack is expected to develop when the sum of the contributions of damage from all the stress range intervals equals 1, i.e.,

$$D = \sum_i \frac{n_i}{N_i} < 1 \qquad (8.6)$$

Miner's rule assumes that there is no significant sequence effect of the loads and no significant interaction of the loads. While there is considerable evidence of load interaction effects in fatigue, these effects must cancel out in a realistic service load history because Miner's rule seems to work adequately.

As discussed above, constant-amplitude fatigue test results exhibit a fatigue limit (CAFL) or a stress range below which the fatigue life appears to be infinite. The fatigue limit only occurs for constant-amplitude testing. When occasional large load cycles are mixed with cycles that are below the fatigue limit, the test data appear to continue to follow the linear S-N curve. There are at least two reasons why this might occur. The first has to do with the effect on crack initiation. It is hypothesized that the large load cycles induce plasticity on a microscopic scale around discontinuities. Cyclic slip then persists along these slip bands that were established during the large load cycle. Soon these slip bands become microscopic fatigue cracks. The second reason is based on crack propagation modeled using fracture mechanics. For the purposes of this argument, it is only necessary to understand that the threshold stress intensity factor for crack growth depends on the crack length as well as the stress range. Each time the threshold is exceeded, the crack is extended slightly and the threshold is lowered slightly. This process accelerates with time as more stress ranges exceed the threshold as time goes on.

The variable-amplitude data from full-scale tests[42] are plotted in Figure 8.7 in terms of the effective stress range. The effective stress range is derived using Miner's rule.[37] If the slope of the S-N curve is equal to 3, then the relative damage of stress ranges is proportional to the cube of the stress range. Therefore, the effective stress range is equal to the cube root of the mean cube of the stress ranges, i.e.,

$$S_{\text{effective}} = \left(\sum_i \frac{n_i}{N_{\text{total}}} S_i^3 \right)^{1/3} \tag{8.7}$$

The LRFD version of the AASHTO specification implies such an effective stress range using the straight-line extension of the constant-amplitude curve. This is essentially the approach for variable amplitude loading in BS 7608.[38] In the AASH-

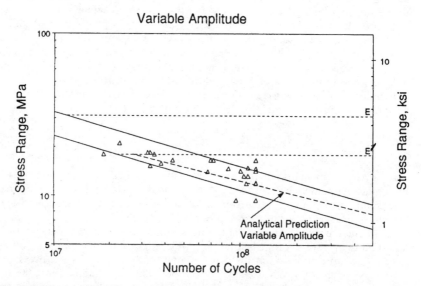

Fig. 8.7 Results of variable-amplitude fatigue tests on thick web attachment details (category E' detail) expressed as effective stress range showing that the results continue to follow the category E' S-N curve with a slope of 3 below the CAFL provided more than 0.01 percent of the stress ranges exceed the CAFL.

TO LRFD specifications, if the effective stress range is less than half the CAFL, the fatigue life is considered infinite. This choice of half the CAFL is based on the observed shape of stress range histograms for bridges. Eurocode 3[18] also uses the effective stress range concept. However, the S-N curves in Eurocode 3 have a change in slope at 5 million cycles to a flatter curve. More complex equations for the effective stress range are given in Ref. 49 for such bilinear S-N curves.

For offshore structures, the use of the straight-line extension of the constant-amplitude curves together with Miner's rule is thought to be overconservative and gives predictions of fatigue life that are much smaller than observed from service experience. Therefore, there are currently different recommendations in various codes regarding the S-N curve and changes in slope in the long-life regime. In practice, use of the straight-line provision might result in a substantial weight penalty as member size is increased to ensure adequate fatigue strength. However, this problem of overconservatism is due to an overconservative estimate of the loads. For example, Faulkner[50] estimated that loads on semisubmersibles were overestimated by at least 30 percent. This is also certainly the case with bridges, where experience with bridge field measurements of stress has shown that stress ranges are much less than predicted due to composite action, higher continuity, and better distribution of load among members than was considered in conservative design calculations. This lack of accuracy should be dealt with directly rather than giving the fatigue curves an offsetting but unwarranted reduction in conservatism.[42]

8.3 Evaluation of Structural Details for Fracture

Unlike fatigue, fracture behavior depends strongly on the type and strength level of the steel or filler metal. In general, fracture toughness has been found to decrease with increasing yield strength of a material, suggesting an inverse relationship between the two properties. In practice, however, fracture toughness is more complex than implied by this simple relationship since steels with similar strength levels can have widely varying levels of fracture toughness. That similar materials with the same strength level can also possess widely different fracture properties indicates that the metallurgical condition of the material has a significant influence on its fracture toughness.

Steel exhibits a transition from brittle to ductile fracture behavior as the temperature increases. For example, Fig. 8.8 shows a plot of the energy required to fracture Charpy V-notch (CVN) impact test specimens of A588 structural steel at various temperatures. These results are typical for ordinary hot-rolled structural steel. The transition phenomena shown in Fig. 8.8 is a result of changes in the underlying microstructural fracture mode. There are really at least three distinct types of fracture with distinctly different behavior.

1. Brittle fracture is associated with cleavage, which is transgranular fracture on select crystallographic planes on a microscopic scale. This type of fracture occurs at the lower end of the temperature range, although the brittle behavior can persist up to the boiling point of water in some low-toughness materials. This part

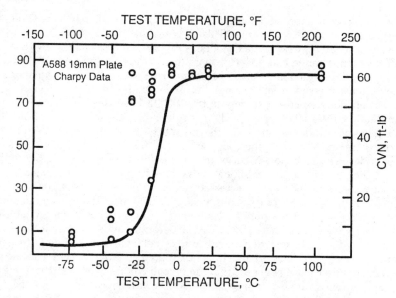

Fig. 8.8 Typical Charpy impact toughness vs. temperature curve for structural steel.

of the temperature range is called the lower shelf because the minimum toughness is fairly constant up to the transition temperature. Brittle fracture is sometimes called elastic fracture because the plasticity that occurs is negligible and consequently the energy absorbed in the fracture process is also negligible.

2. Transition-range fracture occurs at temperatures between the lower shelf and the upper shelf and is associated with a mixture of cleavage and fibrous fracture on a microstructural scale. Because of the mixture of micromechanisms, transition-range fracture is characterized by extremely large variability. Fracture in the transition region is sometimes referred to as elastic-plastic fracture because the plasticity is limited in extent but has a significant impact on the toughness.

3. Ductile fracture is associated with a process of void initiation, growth, and coalescence on a microstructural scale, a process requiring substantial energy, and occurs at the higher end of the temperature range. This part of the temperature range is referred to as the upper shelf because the toughness levels off and is essentially constant for higher temperatures. Ductile fracture is sometimes called fully plastic fracture because there is substantial plasticity across most of the remaining cross section ahead of a crack. Ductile fracture is also called fibrous fracture due to the fibrous appearance of the fracture surface, or shear fracture due to the usually large slanted shear lips on the fracture surface.

Unfortunately, these terms are often used ambiguously. For example, fracture in the transition region is often called brittle or ductile, depending on the relative toughness. Some materials go through a transition but still exhibit relatively low toughness on the upper shelf, despite a fibrous fracture appearance. Here, brittle fracture will imply fracture which is substantially cleavage.

In structural steels and weld filler metals with exceptionally high toughness, and in some modern "high-performance" steels, even test specimens and structural members with large notches or cracks will exhibit ductile fully plastic behavior in tension. In this case, a fracture will slowly propagate or "tear" from the notch or crack. The stability of ductile tearing depends on the type of loading. For example, in displacement control, the tearing may be stable but in load control the tearing will be unstable and catastrophic. Fortunately, most structures have a high degree of redundancy and are close to displacement control; therefore, tearing is typically stable. This fully ductile type of response is necessary for some welded structures. Gas-transmission pipelines, for example, must be able to arrest propagating fractures subject to extremely high-energy driving force, i.e., the pressurized gas.

While completely ductile behavior is obviously the most desirable fracture mode, the type of steel with very low transition temperature which is required is very expensive. Ordinary structural steel such as A36 or A572 is typically only hot-rolled, while to achieve high toughness steels must be controlled-rolled, i.e., rolled at lower temperatures, or must receive some auxiliary heat treatment such as normalization. In contrast to the weld metal, the cost of the steel is a major part of total costs. The expense of the high-toughness steels has not been found to be warranted for most buildings and bridges, whereas the cost of high-toughness filler metal is easily justifiable. For bridges, it is more important to control the fabrication quality and prevent large welding defects to minimize the possibility of fatigue cracking. Hot-rolled steels, which fracture in the transition region at the lowest service temperatures, have sufficient toughness for the required performance of most welded buildings and bridges.

Brittle fracture can be analyzed using the principles of linear-elastic fracture mechanics which are explained below. Fracture behavior in the elastic-plastic regime is similar to brittle fracture except that some ductility and additional energy is absorbed in fracture. Therefore, elastic-plastic fracture is also typically analyzed using linear-elastic fracture mechanics, with some modifications. On the other hand, fully ductile fracture is a completely different phenomenon from brittle fracture or fracture in the transition range. The effect of certain variables such as strain rate and size may not be the same as the effect of these variables on brittle fracture and elastic-plastic fracture. Therefore, the models for analyzing fully ductile fracture are different from the linear-elastic fracture mechanics models for brittle and transition-range fracture. Knowledge of these ductile-fracture models may be important for the special applications such as line pipe, offshore structures, and ships, where it is necessary to demonstrate an extremely large fracture resistance because of fatigue cracking or the necessity to arrest a propagating crack. However, due to the limited length of this chapter, the ductile-fracture models will not be discussed further.

For the purpose of the design of buildings and most other welded steel structures, it is sufficient to assure that brittle fracture does not occur.[63] Even if ductile fracture occurs before local buckling or other failure mode, ductile fracture is considered to give acceptable ductility.

8-28 ■ Chapter Eight

8.3.1 Specification of Steel and Filler Metal

Current AISC specifications refer to ASTM specifications for structural steel such as A36, A572, and A588. Without supplemental specifications, these steel specifications do not require the Charpy test to be performed. Past experience suggested that these steels provided minimum levels of fracture toughness for buildings, and the supplemental specifications were seldom invoked.

If there is concern about brittle fracture and either (1) high-ductility demand, (2) concern with low-temperature exposed structures, or (3) dynamic loading, then the Charpy V-notch (CVN) impact test should be specified by the purchaser of steel as a supplemental requirement. The results of the CVN test, impact energies, are often referred to as "notch-toughness" values. In the LRFD Manual, the abbreviation CVN is used to represent the impact energy in foot-pounds. Thus a steel with "CVN 20" at a certain temperature is specified to have a minimum impact energy of 27 J (20 ft-lb).

The Charpy test ensures a certain level of toughness, which enhances ductility. The term "fracture toughness" is usually reserved for quantities such as K_c which were determined from a standard fracture mechanics test method other than the Charpy impact test. Because the Charpy test is relatively easy to perform, it will likely continue to be the measure of toughness used in steel specifications. The general term "toughness" can be used to refer to any measure of fracture resistance. This ambiguity is usually not a problem because all of these toughness values are related. For example, at least in the range of temperatures called the brittle region, the Charpy notch toughness is approximately correlated with K_c from fracture mechanics tests.

The transition behavior of steel is exploited as a means to screen out brittle materials. For example, the fractures in bridges from fabrication defects and/or fatigue cracks ultimately led to mandatory requirements for a minimum CVN energy (notch toughness) at some temperature for the base metal and the weld metal to assure that the transition temperature is below the lowest service temperature. As long as large defects do not exist, the notch-toughness requirement ensures that brittle fracture will not occur unless large fatigue cracks develop.

Because the Charpy test is relatively easy to perform, it will likely continue to be the measure of toughness used in steel specifications. Often 34 J (25 ft-lb), 27 J (20 ft-lb), or 20 J (15 ft-lb) are specified at a particular temperature. The intent of specifying any of these numbers is the same, i.e, to make sure that the transition starts below this temperature.

Some Charpy toughness requirements for steel and weld metal for bridges and buildings are compared in Table 8.1. This table is simplified and does not include all the requirements. Note that the bridge steel specifications require a CVN at a temperature which is 38°C *greater* than the minimum service temperature. This "temperature shift" accounts for the effect of strain rates, which are lower in the service loading of bridges (on the order of 10^{-3}) than in the Charpy test (greater than 10^1). It is possible to measure the toughness using a Charpy specimen loaded at a strain rate characteristic of bridges, called an intermediate strain rate, although the test is more difficult and the results are more variable. When the

Fatigue and Fracture 8-29

CVN energies from an intermediate strain rate are plotted as a function of temperature, the transition occurs at a temperature about 38°C lower for materials with yield strength up to 450 MPa.

The temperature shift is shown schematically in Fig. 8.9. For high-strength steels the temperature shift diminishes with increasing yield strength. For example, the temperature shift for 690 MPa yield strength steel is only 16°C. For brittle

Table 8.1 Minimum Charpy Impact Test Requirements for Bridges and Buildings

	Minimum service temperature		
	−18°C	−34°C	−51°C
Material	Joules at °C	Joules at °C	Joules at °C
Steel: nonfracture critical members*,†	20 at 21	20 at 4	20 at 12
Steel: fracture critical members*,†	34 at 21	34 at 4	34 at 12
Weld metal for nonfracture critical*	27 at 18	27 at 18	27 at 29
Weld metal for fracture critical*,†	34 at 29°C for all service temperatures		
AISC: jumbo sections and plates thicker than 50 mm†	27 at 21°C for all service temperatures		

*These requirements are for welded steel with minimum specified yield strength up to 350 MPa up to 38 mm thick. Fracture critical members are defined as those which if fractured would result in collapse of the bridge.

†The requirements pertain only to members subjected to tension or tension due to bending.

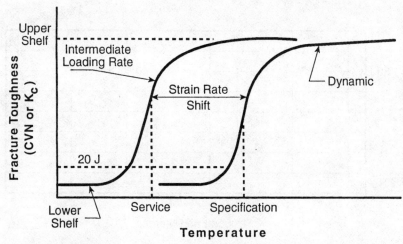

Fig. 8.9 Schematic of the temperature shift which occurs in the CVN and K_c fracture toughness vs. temperature curves due to the difference in strain rates between dynamic impact-test conditions and intermediate service-loading conditions.

materials like some low-toughness weld metal and the core region of some jumbo rolled shapes, there is no significant temperature shift since the toughness is always on the lower shelf. The term "fracture toughness" is usually reserved for quantities such as K_c which, as explained later, are determined from a standard fracture mechanics test other than the Charpy impact test. The temperature shift as shown in Fig. 8.9 would also occur for the fracture toughness K_c.

Figure 8.10 shows some data from fracture tests on full-scale welded girders with fatigue cracks which support the temperature-shift concept.[59,60] In these tests, the specimen was cooled to a target temperature and then the crack was grown by fatigue at 4.5 Hz until instability occurred. Figure 8.10 shows that for temperatures above the 38°C temperature shift, the critical surface crack length along the surface of the girder flange exceeded 75 mm. The depth of these larger cracks exceeded 13 mm. This size crack is considered to be detectable in inspection with some reliability. At higher temperatures the critical crack size increases rapidly in a manner analogous to the CVN vs. temperature curve.

As shown in Table 8.1, the AWS D1.5 Bridge Welding Code specifications for weld metal toughness are more demanding than the specifications for base metal. This is reasonable because the weld metal is always the location of discontinuities and high tensile residual stresses. The commentary in the AISC LRFD Manual, Section C-A3, says weld metal is usually not critical. Similarly, there is no requirement for weld metal toughness in AWS D1.1. This lack of requirements was rationalized because typically the weld deposits are higher toughness than the base metal. However, this is not always the case; e.g., the self-shielded flux-cored arc welds (FCAW-S) used in many of the WSMF that fractured in the Northridge earthquake were reported to be very low toughness. The commentary in the AISC

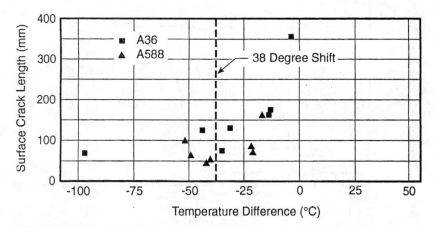

Fig. 8.10 Results from Lehigh University full-scale tests on structural-steel girders with fatigue surface cracks at welded details on the flange showing the critical crack size at fracture as a function of the difference between the test temperature minus the temperature associated with 20 J CVN for the steel.

Manual does warn that for "dynamic loading, the engineer may require the filler metals used to deliver notch-tough weld deposits."

ASTM A673 has specifications for the frequency of Charpy testing. The H frequency requires a set of three CVN specimens to be tested from one location for each heat or about 50 tons. These tests can be taken from a plate with thickness up to 9 mm different from the product thickness if it is rolled from the same heat. The P frequency requires a set of three specimens to be tested from one end of every plate, or from one shape in every 15 tons of that shape. For bridge steel, the AASHTO code requires CVN tests at the H frequency as minimum. For fracture critical members, the guide specifications require CVN testing at the P frequency. In the AISC code, CVN tests are required at the P frequency for thick plates and jumbo sections. A special test location in the core of the jumbo section is specified, as well as a requirement that the section tested be produced from the top of the ingot.

Even the P testing frequency may be insufficient for as-rolled structural steel. In a recent report for NCHRP,[60] CVN data were obtained from various locations on bridge steel plates. The data show that because of extreme variability in CVN across as-rolled plates, it would be possible to miss potentially brittle areas of plates if only one location per plate is sampled. For plates which were given a normalizing heat treatment, the excessive variability was eliminated.

The toughness of steel plate or shapes is primarily determined by composition, processing, and resulting grain size. Alloy elements are added for strength, but most alloy elements are detrimental to toughness, especially carbon. Consequently, A572 steel specifies only small amounts of alloying elements. Steel with carbon less than 0.15 percent by weight typically has good toughness and also excellent weldability. Nickel increases both strength and toughness. Weld metal with 0.5 to 1.0 percent nickel typically has high toughness.

Unintentional elements from scrap metal such as copper, lead, and tin, and residual elements such as sulfur, phosphorus, and nitrogen may decrease toughness and should be minimized. These elements may also adversely affect weldability, particularly at thermal-cut edges at reentrant corners or at highly constrained joints such as butt splices in columns. If steel is to be loaded in tension through the thickness of the plate, such as in column flanges in fully welded beam-to-column connections, sulfur should be limited to levels less than 0.01 percent by weight in order to minimize the potential for lamellar tearing. Weathering steel, because of the relatively higher alloy, may have increased susceptibility to lamellar tearing.

Typically, structural steel for buildings such as A572 are hot-rolled, although normalization can be specified to refine grain size. Fine grain size increases both strength and fracture toughness. The grain size is controlled by the cooling rate. Therefore, thicker sections will have larger grain size and lower toughness.

Steel plate is made in Europe and Asia by thermomechanical controlled processing (TMCP) to produce fine-grained low-carbon microstructure with high fracture toughness. TMCP steel is used widely for line pipe, ships, and offshore structures. A913 jumbo shapes are produced by a type of TMCP, the quenching and self-tempering (QST) process, and have high fracture toughness. Other processing and chemistry is used in the US to produce high-fracture-toughness plate such as A710.

Because of variability in the cooling rate and resultant microstructure and grain size, weld metal toughness can vary widely from manufacturers' certification, to weld procedure qualification test, to the fabrication of the structure.[61] The toughness is dependent on proper shielding gas and flux to reduce impurities. Therefore, welding procedures must be monitored to control toughness as well as to avoid defects. Typically, higher heat input decreases cooling rate and toughness. Electroslag welding and electrogas welding are particularly high heat input processes which may produce low-toughness welds. Qualification tests are often carried out on plates 25 mm thick. The procedure may then be applied to thinner plate, where cooling rates will decrease and the toughness may be lower than qualification tests indicate.

Usually weld metal has low carbon and toughness greater than the steel plate. The self-shielded flux-cored arc welding (FCAW-S) process produces welds with high aluminum (used in the flux for deoxidation). The FCAW-S process can produce welds with adequate toughness with good filler metal composition and carefully controlled weld procedures. However, prior to the 1994 Northridge earthquake, the E70T-4 electrode was used in California for field welds in WSMF connections. The welders typically used high heat input to increase productivity, and the resulting welds have very low toughness.

8.3.2 Fracture Mechanics Analysis

Fracture mechanics is based on the mathematical analysis of solids with notches or cracks. Relationships between the material toughness, the crack size, and the stress or displacement will be derived below using fracture mechanics. The objective of a fracture mechanics analysis (as outlined here) is to assure that brittle fracture does not occur. Even if ductile fracture occurs before local buckling or another failure mode, ductile fracture is considered to give acceptable ductility. Brittle fracture occurs with nominal net-section stresses below or just slightly above the yield point. Therefore, the relatively simple principles of linear-elastic fracture mechanics (LEFM) can be used to conservatively assess whether a welded joint is likely to fail by brittle fracture rather than fail in a ductile manner.

It significantly simplifies the presentation and practical use of fracture mechanics if the discussion is confined to brittle fracture only. Worst-case assumptions are made regarding numerous factors that can enhance fracture toughness, e.g., temperature, strain rate, constraint, and notch acuity or sharpness. These assumptions eliminate the need for extensive discussion of these effects.

If necessary, these effects can be considered and more advanced principles of fracture mechanics can be used to estimate the maximum monotonic or cyclic rotation before ductile tearing failure. Fracture mechanics can also be used to predict the "subcritical" propagation of cracks due to fatigue and/or stress-corrosion cracking that may precede fracture. In order to provide a thorough discussion of the brittle fracture problem in a limited number of pages, these and many other interesting topics in fracture mechanics cannot be presented in detail. Several excellent books on fracture mechanics cover these topics in detail.[3,6,51,52] A recent review of fatigue and fracture of fabricated steel structures is particularly relevant.[53]

Fatigue and Fracture   8-33

Although cracks can be loaded by shear, experience shows that only the tensile stress normal to the crack is important in causing fatigue or fracture in steel structures. This tensile loading is referred to as "Mode I." When the plane of the crack is not normal to the maximum principal stress, a crack which propagates subcritically or in stable manner will generally turn as it extends such that it becomes normal to the principal tensile stress. Therefore, it is typically recommended that a welding defect or cracklike notch which is not oriented normal to the primary stresses can be idealized as an equivalent crack with a size equal to the projection of the actual crack area on a plane which is normal to the primary stresses (see BSI PD6493, for example[25]).

Brittle fracture occurs with nominal net-section stresses below or just slightly above the yield point. Therefore, the relatively simple principles of linear-elastic fracture mechanics (LEFM) can be used to conservatively assess whether a welded joint is likely to fail by brittle fracture rather than fail in a ductile manner.

It significantly simplifies the presentation and practical use of fracture mechanics if the discussion is confined to brittle fracture only. Worst-case assumptions are made regarding numerous factors that can enhance fracture toughness, e.g., temperature, strain rate, constraint, and notch acuity. These assumptions eliminate the need for extensive discussion of these effects. If necessary, these effects can be considered and more advanced principles of fracture mechanics can be used to estimate the maximum monotonic or cyclic rotation before ductile tearing failure. Fracture mechanics can also be used to predict the "subcritical" propagation of cracks due to fatigue and/or stress corrosion that may precede fracture.

LEFM gives a relatively straightforward method for predicting fracture, based on a parameter called the stress intensity factor K which characterizes the stresses at notches or cracks.[62] The applied K is determined by the size of the crack (or cracklike notch) and the nominal gross-section stress remote from the crack. Cracklike notches and weld defects are idealized as cracks, and the term crack includes cracklike notches and weld defects as well. In the case of linear elasticity, the stress-intensity factor can be considered as a measure of the magnitude of the crack tip stress and strain fields. Solutions for the applied stress-intensity factor K for a variety of geometries can be found in handbooks.[55-58] Most of the solutions are variations on standard test specimens which have been studied extensively. The following discussion presents a few useful solutions and examples of their application to welded joints.

In general, the applied stress-intensity factor is given as

$$K = F_c * F_s * F_w * F_g * \sigma \sqrt{\pi a} \tag{8.8}$$

where the F terms are modifiers on the order of 1.0, specifically:

F_c is the factor for the effect of crack shape.

F_s is the factor, equal to 1.12, that is used if a crack originates at a free surface.

F_w is a correction for finite width which is necessary because the basic solutions were generally derived for infinite or semi-infinite bodies.

F_g is a factor for the effect of nonuniform stresses, such as bending stress gradient.

A stress-concentration factor (SCF) is defined as the ratio of the peak stress near the stress raiser to the nominal gross-section stress remote from the stress raiser. SCF are often used in fracture assessments when the crack is located near a stress raiser. For example, a crack may be located at a plate edge which is badly corroded. Any SCF would also be included in F_g.

The stress-intensity factor has the unusual units of MPa-m$^{1/2}$ or ksi-in$^{1/2}$. The material fracture toughness is characterized in terms of the applied K at the onset of fracture in simplified small test specimens, called K "critical" or K_c. The fracture toughness K_c is considered a transferable material property; i.e., fracture of structural details is predicted if the value of the applied K in the detail exceeds K_c. Equation 8.8 relates the important factors that influence fracture: K_c represents the material, σ represents the design, and a represents the fabrication and inspection.

In this section, K_c is used as any type of critical K associated with a quasistatic strain rate, derived from any one of a variety of test methods. One measure of K_c is the plane-strain fracture toughness, which is given the special subscript I for plane strain K_{Ic}. K_{Ic} must be measured in specimens which are very thick and approximate plane strain. If the fracture toughness is measured in an impact test, the special designation K_d is used, where the subscript d is for dynamic. In practice, K_c is often estimated from correlations with the result from a CVN test, because the CVN is much cheaper to perform and requires less material than a fracture-mechanics test, and all test laboratories are equipped for the CVN test. A widely accepted correlation for the lower shelf and lower transition region between K_d and CVN is credited to Rolfe and Barsom[3]

$$K_d = 11.5*\sqrt{CVN} \qquad (8.9)$$

where CVN is given in J and K_d is given in MPa-m$^{1/2}$. A different constant is used for English units.[3] This correlation is used to construct the lower part of the curve for dynamic fracture toughness K_d as a function of temperature directly from the curve of CVN vs. temperature. There is a temperature shift between the intermediate load rate values of K_c and the impact load rate values of K_d which is approximately equal to the temperature shift that occurs for CVN data as shown in Fig. 8.9. Therefore, K_c values for structural steel are obtained by shifting the K_d curve to a temperature which is 38°C lower. However, for brittle materials there is essentially no temperature shift and therefore K_c is approximately equal to K_d.

Figure 8.11 shows some typical data for a W14×605 "jumbo" section which shows the gradient in toughness from the web-flange core to the plate surface. The lower-bound CVN energy for the core region ranges from 4 to 14 J. Using the correlation of Eq. 8.9, the scatterband for the predicted K_c ranges from 23 to 43 MPa-m$^{1/2}$. The scatterband for K_c based on the Charpy data is shown in Fig. 8.12 with some measured K_{Ic} data. These data show that the correlation of Eq. 8.9 is conservative, since the actual K_{Ic} ranges between 45 and 50 MPa-m$^{1/2}$.

Prior to the 1994 Northridge earthquake, the welds in the WSMF connections were commonly made with the self-shielded flux-cored arc welding (FCAW-S) process using an E7XT-4 weld wire. For the connections which fractured in the Northridge earthquake which have been investigated so far, the weld metal CVN is plotted in Fig. 8.13. The lower-bound impact energy is between 4 and 14 J for temperatures up to 50°C. If recommended weld procedures are followed, the fracture toughness increases slightly but remains inadequate.

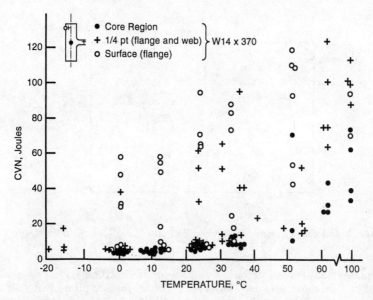

Fig. 8.11 Charpy-impact-energy data for as-rolled jumbo shapes without mandatory notch toughness.

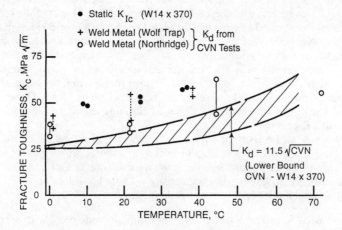

Fig. 8.12 Results from correlation of the lower-bound CVN data for the W14×370 jumbo shape (from Fig. 8.11) to K_c compared to K_{Ic} from fracture-mechanics tests on this shape as well as K_d values correlated from CVN data from the weld metal in the box girder in Fig. 8.3 and from the weld metal in several Northridge WSMF connections. These low-toughness materials have a similar lower-bound fracture toughness.

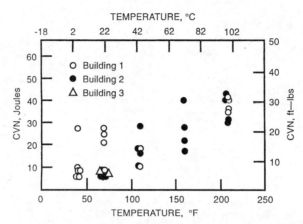

Fig. 8.13 Typical Charpy impact energy from E7XT-4 FCAWS-S weld from Northridge WSMF connections.

It happens that the value of the CVN and K_c for this Northridge weld metal also falls within the scatterband for "lower-shelf" brittle-fracture-prone materials as shown in Fig. 8.12. The lower bounds of the CVN and K_c data are not sensitive to temperature or strain rate because these materials do not undergo a transition at temperatures of interest. This similarity in the data suggests that there may be a "lower-bound" value of the fracture toughness that can be assumed for brittle ferritic weld metal, structural steel, and the heat-affected zone (HAZ). The lower-bound fracture toughness reflects the worst effects of temperature and strain rate. For these materials, the lower-bound fracture toughness was between 45 and 50 MPa-m$^{1/2}$. This concept of a lower-bound fracture toughness is very useful for fracture assessment.

As a consequence of the brittle fractures in jumbo sections, AISC specifications now have a supplemental Charpy requirement for shapes in groups 4 and 5 and (for the same reasons) plates greater than 51 mm thick, when these are welded and subject to primary tensile stress from axial load or bending. These jumbo shapes and thick plates must exhibit 27 J at 21°C. Using the correlation of Eq. 8-9, 27 J will give a K_c of 60 MPa-m$^{1/2}$.

There are size and constraint effects and other complications which make the LEFM fracture toughness K_c less than perfect as a material property. This is especially true when K_c is only estimated based on a correlation to CVN. Nevertheless, as illustrated in the following, the conservative lower-bound value of K_c can be used by structural engineers to avoid brittle fracture.

Center Crack. ■ The K solution for an infinitely wide plate with a through crack subject to uniform tensile membrane stress is

$$K = \sigma\sqrt{\pi a} \qquad (8.10)$$

where σ = nominal gross-section stress remote from the crack
$2a$ = total overall crack length

If the total width of the panel is given as 2W, F_w for this crack geometry can be approximated by the Fedderson or secant formula:

$$F_w \sqrt{\sec \frac{\pi a}{2W}} \qquad (8.11)$$

This formula gives a value that is close to 1.0 and can be ignored for a/W less than a third. For a/W of about 0.5, the secant formula gives F_w of about 1.2. However, the values from the secant equation go to infinity as a approaches W. The secant formula is reasonably accurate for a/W up to 0.85. The F_w may be used for other crack geometries as well.

Many common defects and notches in welded joints can be idealized as a center crack in tension. For example, Fig. 8.14 shows several defects which were revealed from cores of a WSMF flange connection which had indications from ultrasonic testing. The defect shown as in Fig. 8.14a may be analyzed as center cracks on a vertical plane. The length of the crack is determined from the projection of the defect on the vertical plane. In this case, the dimension $2a$ is the maximum width of the pore, which is about 6 mm and the dimension $2w$ is the thickness of the plate or 25 mm. The a/W ratio is less than 0.25 and therefore F_w can be taken as 1.0. The stress-intensity factor can therefore be calculated using Eq. 8.10. As explained above, a good lower bound to the fracture toughness of steel is 45 MPa-m$^{1/2}$. With this toughness, Eq. 8.10 gives an allowable total (applied plus residual) stress of 460 MPa for a of 3 mm. For a grade 50 steel, minimum specified yield strength (MSYS) of 350 MPa,

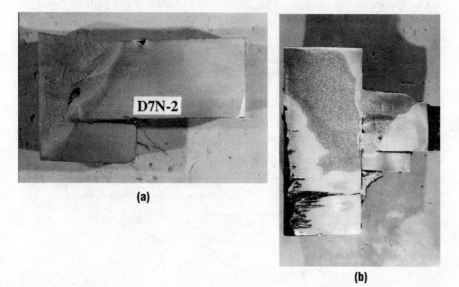

Fig. 8.14 Cross sections from cores where ultrasonic tests of WSMF revealed defects showing lack-of-fusion defect on beam flange side of weld (*a*) and incomplete-penetration defect and lamellar tear in column plate near backing bar notch (*b*).

yielding should occur before fracture for this small defect. Therefore, unless the defect could grow by fatigue, this type of welding defect is inconsequential.

The large pore in Fig. 8.14a gave a much larger indication than the lack of fusion defects in Fig. 8.14b, although it is clear that the latter defects are more critical. The defects in Fig. 8.14b could also be idealized as a center crack. Because the defects are close to each other, the overall length 2a would be taken as the sum of the length of these defects, about 22 mm, even though they may not be linked together. Therefore, these defects would be predicted to fracture if the lower-bound toughness is used.

Another type of notch that can be idealized as a center-cracked panel is a backing bar with a fillet as shown in Fig. 8.15. In this case, the unfused area of the backing bar creates a cracklike notch with one tip in the root of the fillet weld and one tip at the root of the groove weld. The crack is asymmetrical, but since the connection is subjected to uniform tension, the crack can be analyzed as if it were in a symmetric center-cracked panel. Of course, the applied K is higher on the crack tip, which is in the root of the fillet weld because there is a high F_w for this side. Assuming the weld metal of the groove weld and the fillet weld comparable toughness, the fillet weld side of the backing bar will govern the fracture limit state. Therefore, the panel is idealized as being symmetric with respect to the center of the backing bar, i.e., having a width (2W) of 50 mm.

Assuming negligible weld root penetration, the crack size 2a is taken as being equal to the backing bar thickness, or 13 mm. Therefore, for a/W of 0.5, Eq. 8.11 gives F_w equal to 1.2. Although this idealization seems like a gross approximation at best, the validity of the K solution for this particular weld joint was verified based on observed fatigue crack propagation rates.

If this is a grade 50 steel, the yield strength could be up to 450 MPa. The notch tip could be subjected to full tensile residual stress. Therefore, Eq. 8.8 is solved with the gross-section stress equal to 450 MPa, with the F_w factor of 1.2, and a of 6 mm, giving 74 MPa-m$^{1/2}$. It can be seen that this configuration could cause a brit-

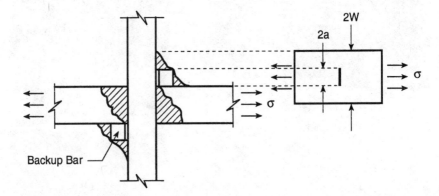

Fig. 8.15 Cross section of one-sided groove-welded cruciform-type connection with loaded plate discontinuous and idealization of the notches from a backing bar with a fillet weld as a center-cracked tension panel.

tle fracture for very brittle materials. However, weld metal and base steel with modest toughness could easily withstand this defect.

Edge Crack ▪ The stress-intensity factor for an edge crack in an infinitely wide plate is

$$K = 1.12\sigma\sqrt{\pi a} \tag{8.12}$$

where σ is the remote gross-section nominal stress and a is the depth of the edge crack or cracklike notch. It can be seen that the edge crack equation is treated like half of a center crack, i.e., Eq. 8.10, where a and W are the total length and width, respectively, for the edge crack. The F_s of 1.12 is applied to account for the free edge, which is not restrained as it is in the center-cracked geometry. Equation 8.12 is also modified with the F_w as in Eq. 8.11.

As was shown in Figs. 8.1 and 8.2, brittle fractures have occurred in group 4 and 5 shapes, i.e., jumbo shapes. These fractures initiated from fabrication cracks which occurred in the hard layer of the thermal-cut edge of the cope holes. Figure 8.2 showed the crack originated from the cope holes, but they curved down into the flange such that they were perpendicular to the primary tensile stress. The cross section of the brittle fracture surface on the flange shows the dark spot which constitutes the original fabrication defect. The length of the defect is about 25 mm deep into the flange. Since the a/W ratio is small, F_w can be taken as 1.0. Setting K equal to 45 MPa-m$^{1/2}$, Eq. 8.12 would predict the total stress, at fracture of 143 MPa, which is below the allowable service stress.

In January 1986 after the temperature decreased to less than –20°C, an exposed steel box girder that supports part of the roof of the new Filene Center at Wolf Trap Farm Park in Vienna, Va., was discovered to be fractured. Figure 8.3 showed the fracture surface of the box girder, which exhibits the typical "herringbone" patterns that point to the origin of failure. In this case, the fracture originated from lack of fusion in the splice welds in the backing bar used for the corner longitudinal welds of the box. Often the backing bar is not even spliced; this led to several fractures in the 1995 Kobe, Japan, earthquake. If the splices are welded, the welds splicing the backing bar are mistakenly thought to be unimportant and are not given the same attention as the welds in the member itself. However, the backing bar is fused to the corner longitudinal weld, and therefore any unfused area is like an edge crack in the corner longitudinal weld. The size of the edge crack can be taken as the width of the backing bar.

In the case of the Filene Center box girder, the lack of fusion defects fractured the corner longitudinal groove welds because of the unusually low toughness of these welds. These low-toughness welds resulted from mixing two weld processes. Tack welds were used to place the backing bar. These tack welds were made with a self-shielded flux-core arc weld (FCAW-S) which contained a high aluminum content. These tack welds were partially remelted in the longitudinal submerged-arc welds (SAW), giving these welds very low toughness which was comparable to the low toughness observed in core region of jumbo sections. In fact, the data are also shown in Figure 8.12, and it is seen that these data fall in the same region as the data from the brittle jumbo core region. As a consequence of the fracture at the Filene Center, there is now a warning in the AISC LRFD code (Section J.2.7 and commentary) regarding the mixing of weld metals.

8-40 ■ Chapter Eight

The base metal had good static toughness and could have tolerated a rather large crack statically. However, the base metal dynamic toughness was insufficient to arrest these cracks. Cleavage cracks propagate at speeds on the order of 1000 m/s, and the strain rates ahead of the cracks are extremely large. The Navy and the gas-transmission pipeline industry have studied the crack-arrest capability of plates. If a steel is required to arrest dynamic cracks, it must have a very high dynamic toughness. Steel with crack-arrest capability is too expensive for buildings. Rather, the goal should be to avoid cleavage initiation.

Another problem that can be idealized as an edge crack is the unfused ends of runout tabs. As shown in Fig. 8.16, the backing bar is fused to the flange and therefore becomes part of the flange. The unfused edge between the runout tab and the transverse member acts just like a crack in the edge of the flange. This type of detail has caused fracture in several buildings.

Figure 8.17 shows a typical connection between a brace and a gusset. This detail has given rise to fatigue cracks in crane trusses and is also commonly used for seismic braced frames. Fatigue cracks initiate easily at the unfused notch at the top of the cut of the brace and the edge of the gusset. It is also known that simply by adding a cope hole at the top of the cut, as shown in Fig. 8.17, the notch and the associated high-cycle fatigue cracking problem are eliminated. This small change in the detailing would be expected to increase the resistance of this connection to low-cycle seismic loading as well. However, reducing the net section at the cope hole location could also have a deleterious effect by concentrating strain at this location and preventing the gross-section yielding of the brace.

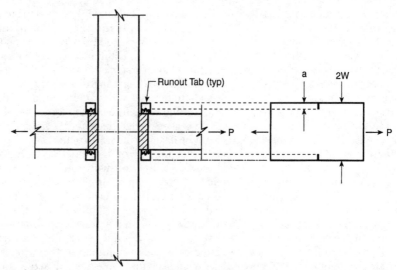

Fig. 8.16 Plan view of one-sided groove-welded connection of beam flanges to transverse member flanges and idealization of the notches from a runout tab as a double-edge-notched panel.

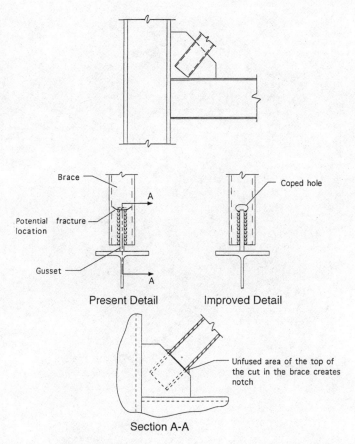

Fig. 8.17 Tubular-brace-to-gusset connection showing potential fatigue cracking or fracture from notch created by unfused area at the top of the cut in the brace and improved detail with cope hole to eliminate notch.

Figure 8.18 shows a cross section near the crack origin from each of the WSMF connections shown in Fig. 8.4 which fractured in the Northridge earthquake and a fracture surface from the fracture that stayed in the weld. The fracture surfaces indicate that the fractures originate in the root of the weld, typically at a lack-of-fusion defect. This lack-of-fusion defect is difficult to avoid when the weld must be stopped on one side of the web and started on the other side. The weld fracture surface in Fig. 8.18 shows the cracklike notch formed by the combination of a lack-of-fusion defect and the unfused edge of the backing bar. On a cross section at the deepest point of the lack-of-fusion defect, the total depth of the notch, including the unfused edge of the backing bar, is between 13 and 19 mm.

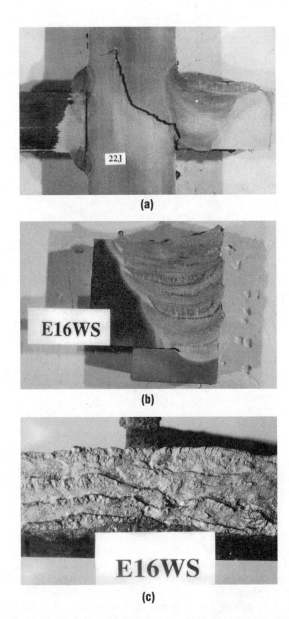

Fig. 8.18 Polished cross section of the bottom beam flange connection: (*a*) the WSMF (with backing bar removed) shown in Fig. 8.4*b*; (*b*) the WSMF shown in Fig. 8.4*c*; and (*c*) the fracture surface of the beam flange end for the WSMF shown in Fig. 8.4*c* showing the lack-of-fusion defect in the center of the weld under the web and the brittle fracture in the weld metal.

The value of 45 MPa-m$^{1/2}$ can be used as a lower bound to the fracture toughness of structural steel or weld metal. Equation 8.9 may be used to predict brittle fracture for the WSMF connection welds when K exceeds 45 MPa-m$^{1/2}$. For a notch depth of 13 to 19 mm, Eq. 8.12 would predict that brittle fracture is likely to occur for gross-section stress between 160 and 200 MPa, well below the yield point. These types of LEFM calculations, had they been performed prior to the earthquake, would have predicted that brittle fracture would occur in the WSMF connections before yielding.

As explained previously in the discussion of the fractures at Filene Center at Wolf Trap, the strain rates are extremely high ahead of a propagating cleavage crack tip, making the arrest of a cleavage crack very difficult. Therefore, even if the toughness of the WSMF column section was very good, it was typically not sufficient to arrest the dynamic crack emanating from the weld into the steel of the column flange and web as shown in Figs. 8.4 and 8.18. The scenario where a crack is initiated in a brittle material and then fractures a reasonably tough structural member is common in many fractures which have been investigated, e.g., the Filene Center. Often, seemingly insignificant but brittle tack welds are responsible. Another example which recently was investigated is a fracture due to seismic loading where a tack weld crossed the unfused abutting ends of a longitudinal backing bar for the corner welds inside a box section. This well-known scenario is exploited in one of the original fracture tests, ASTM E 208 "Standard Test Method for Conducting Drop-Weight Tests to Determine Nil-Ductility Transition Temperature of Ferritic Steel." In this test, a crack is initiated from a brittle weld bead with a notch cut in it.

The propagation path of the unstable dynamic crack is seemingly chaotic, as it is influenced by dynamic stress waves and complex residual stress fields. The critical event was the initiation of the unstable crack in the brittle weld. There is little significance to whether the crack propagated in the weld or turned and entered the column.

Fracture mechanics can be used to establish required levels of fracture toughness as well as limits on the size of defects. One obvious component of the solution to the WSMF cracking problem is elimination of the possibility of such large defects (i.e., $a \approx 19$ mm) that result from the backing bar and any lack of fusion defect. The backing bar removal and back gouging to minimize the lack of fusion will eliminate most large cracklike conditions. Appropriate weld notch toughness requirements will avoid the use of low-toughness weld metal.

Fracture-mechanics calculations can be used to examine the expected stress-intensity factor using the maximum yield strength of the steel beam and defect sizes considered acceptable by AWS D1.1. The steel can be expected to have a yield strength up to 450 MPa. AWS D1.1 Section 8.15 indicates that embedded defects up to 19 mm and edge defects up to 1.5 mm are tolerable in such a thick plate. With the embedded defect of 19 mm, a is equal to 9.5 mm and Eq. 8.10 indicates that the maximum applied K is 78 MPa-m$^{1/2}$, which, without a temperature shift, corresponds to a Charpy energy of about 46 J. Although CO_2 shielded FCAW will typically produce even better results, this level of fracture toughness plus a comfortable margin can even be provided with the self-shielded FCAW-S process, for example with E70TG-K2 (Lincoln NR 311Ni) wire. This wire has 1 weight percent Ni, which can improve toughness, as discussed previously. The tolerable edge crack of 1.5 mm gives an applied K of 35 MPa-m$^{1/2}$.

There are certainly many other equally important design issues that influenced these fractures. The overall lack of redundancy, i.e., the reliance on only one or two massive WSMF to resist resist Seismic lateral loads in each direction, contribute to large forces, increase the thickness of the members, and the high constraint of the connections. Even if brittle fracture is avoided, welds will typically fail at a lower level of plastic strain than base metal. Therefore, it can also be argued that it is imprudent to rely upon welds for extensive plastic deformation. Several improved WSMF connections have been proposed, most of which are designed such that the plastic hinge develops out in the span away from the connection. Nevertheless, in the event of unexpected loading, it is still desirable that these weld joints have a ductile failure mode. Therefore, while these improved connection designs may be worthwhile, the low-toughness weld metal and joint design with a built-in notch should still not be used under most circumstances.

Note that having a fillet on the open side of the backing bar improves the edge-crack condition (as in the WSMF connection welds). The crack size becomes equal to half of the backing bar width when the fillet is added because the notch is now treated as a center crack of length $2a$ as in Fig. 8.15 rather than an edge crack of length a. The calculations for the WSMF connection weld can be reevaluated to see what effect this would have had on the potential for brittle fracture. These calculations showed that without the fillet, a 19-mm-deep notch would fail at 160 MPa stress. With the fillet, the tolerable stress could be increased to 220 MPa, which at least exceeds the design allowable stress but will not achieve yielding. Thus by simply adding a fillet to the backing bar, the allowable stress can be increased by 37 percent. With or without the fillet, however, a higher-toughness weld metal would be required.

The fracture scenario of these WSMF connections is similar to the fracture scenario of the box sections at the Filene Center at Wolf Trap. Both examples illustrate the importance of weld metal toughness as the first line of defense against fracture. Building connections, like any other welded structures, are susceptible to brittle fracture, especially if the connections are likely to be loaded dynamically above the yield strength to deformations several times the yield deformations. Low temperature can also contribute to fracture when the steel is exposed.

Buried Penny-Shaped Crack. ■ Many internal weld defects are idealized as an ellipse or a circle which is circumscribed around the projection of the weld defect on a plane perpendicular to the stresses. Often, the increased accuracy accrued by using the relatively complex elliptical formula is not worth the effort, and the circumscribed penny-shaped or circular crack is always conservative. The stress-intensity factor for the penny-shaped crack in an infinite body is given as

$$K = \frac{2}{\pi}\sigma\sqrt{\pi a} \qquad (8.13)$$

where a is the radius of the circular crack. As for the other types of cracks, the F_w can be calculated using Eq. 8.11. In terms of Eq. 8.8, the crack shape factor F_c in this case is $2/\pi$ or 0.64. Using (1) the lower-bound fracture toughness of 45 MPa-m$^{1/2}$; and, (2) an upper-bound residual stress plus applied stress equal to the upper-bound yield strength for grade 50 steel (450 MPa), Eq. 8.13 shows that a penny-shaped crack would have to have a radius exceeding 8 mm to be critical; i.e., the diameter of the allowable welding defect would be 15 mm (provided that fatigue is not a potential problem).

The crack shape factor F_c is more favorable (0.64) for buried cracks as opposed to F_s of 1.12 for edge cracks, and the defect size is equal to $2a$ for the buried crack and only a for the edge crack. These factors explain why edge cracks of a given size are much more dangerous than buried cracks of the same size.

8.3.3 Fracture-Mechanics Test Methods

Fracture tests can be divided according to the objective or use of the data. Screening tests, like the CVN test, can rank materials and give a relative indication of toughness but the result cannot be directly used in a quantitative analysis. On the other hand, fracture mechanics tests are intended to get a quantitative value of fracture toughness that can be used directly to predict fracture in structural members. As explained previously, it is sometimes possible to indirectly infer a quantitative value of K_c from a correlation to a screening tests result like CVN.

One of the first fracture-mechanics tests was ASTM E399, "Standard Test Method for Plane-Strain Fracture Toughness of Metallic Materials." The K_c value determined from this test is given the special subscript I for plane strain K_{Ic}. K_{Ic} is commonly measured on the compact-tension (CT) specimen such as that shown in Fig. 8.19, although single-edge-notched bend (SENB) bars may also be used (Fig. 8.20). In all fracture-mechanics tests, the specimen must be fatigue pre-

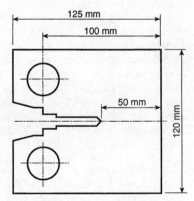

Fig. 8.19 Typical compact-tension (CT) specimen planar dimensions (other sizes which are proportional can also be used). Specimen should be full thickness of plate or shape.

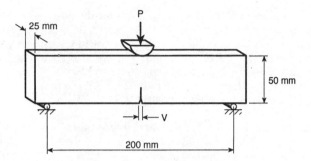

Fig. 8.20 Typical single-edge-notch bend (SENB) specimen dimensions (other sizes which are proportional can also be used). Specimen should be full thickness of plate of shape.

cracked. The load and crack mouth displacement are monitored in the test, and K is computed from the load either at the point of instability or at some small offset from the elastic slope. In order for the test to be considered valid, the specimens must have large planar dimensions and be very thick, approximating plane strain. Specifically, the remaining ligament b and the thickness B must be

$$b, B > 2.5 \left(\frac{K_{Ic}}{\sigma_y} \right)^2 \tag{8.14}$$

This requirement is intended to assure that the specimen size dimensions are on the order of 50 times bigger than the plastic zone at the crack tip.

Consider the very low toughness materials with K_{Ic} of 45 MPa-m$^{1/2}$ and a yield strength of 450 MPa. Even for these brittle materials, a specimen thickness greater than 25 mm would be required. If the plate or flange thickness were less thick, valid K_{Ic} could not be obtained. Materials with adequate toughness, greater than 100 MPa-m$^{1/2}$, for example, require specimens thicker than 120 mm. Clearly, it can be seen that this is a test which is impractical for all but the most brittle materials. Typically, invalid K_c values obtained from this test with specimens that are too small will be larger than the valid K_{Ic}. However, if the test specimen and the structural member have the same thickness, invalid data are often used with caution.

The J integral is a parameter for elastic-plastic fracture much as K is a parameter for elastic fracture. The J-integral tests were developed for elastic-plastic fracture where the fracture mode was ductile tearing rather than cleavage. The most simple of these is ASTM E813, "Standard Test Method for J_{Ic}, A Measure of Fracture Toughness" gives a value of J at the initiation of ductile tearing. This test is typically performed on CT specimens, such as that shown in Fig. 8.19, although SENB specimens may also be used. In these J tests, the load and crack mouth displacement are monitored and J is computed from the work done on the specimen, i.e., from the area under the load displacement curve. In order to identify the initiation of ductile tearing, changes in compliance are monitored by performing periodic partial unloading of the specimen. The crack extension is determined from these compliance measurements.

J is often converted to an "equivalent" K by the following relation:

$$K = \sqrt{J*E} \qquad (8.15)$$

where E is the modulus of elasticity. K_c values that are obtained from this J test are commonly given the subscript J, i.e., K_{Jc}. The specimen size requirements for ASTM E813 are much less stringent than E399, i.e.,

$$b, B > 20 \frac{J_{Ic}}{\sigma_y} \qquad (8.16)$$

For a given value of fracture toughness the specimen may be about 50 times thinner than for ASTM E399 (K_{Ic}). For moderate toughness of about 100 MPa-m$^{1/2}$ ($J=48$ kJ/m^2), the specimen ligament and thickness are required to be greater than 3 mm, which can be easily met.

ASTM E1290, "Standard Test Method for Crack-Tip Opening Displacement (CTOD) Fracture Toughness Measurement," gives a slightly different test which is easier to perform but gives results which are more variable. This test is typically performed on SENB specimens, although the CT specimen can also be used. The specimens are the full thickness of the plate or shape and there are no validity requirements. The load and crack-mouth displacement are monitored during the test, and the CTOD is inferred from the crack-mouth displacement. A variety of outcomes are possible including short propagation or pop-in of the crack without instability. The critical CTOD is either at the point of "pop-in" or at the maximum load for more ductile behavior.

The CTOD concept and test method were developed at the Welding Institute (TWI) in the United Kingdom. An empirically based conservative fracture assessment procedure called the "CTOD design curve" was developed around the CTOD test results and was verified through extensive wide-plate tension tests of weldments. This procedure has been extensively used in the oil and gas industry for pipelines and offshore structures. In 1980, the British Standards Institute published a document called PD6493, "Guidance on Some Methods for the Derivation of Acceptance Levels for Defects in Fusion Welded Joints."[25] This document is based on fracture-mechanics fitness-for-purpose concepts, and is meant to resolve disputes about minor fabrication defects. Originally, the PD6493 procedure was based on fracture toughness measured using the CTOD test and the empirical CTOD design curve. While this approach is still embodied in PD6493, the procedure was generalized in a new version in 1991 to permit any measure of toughness and analysis method. The PD6493 document has an easy-to-follow codified procedure that can be agreed to in advance by the owner, the engineer, the regulator, and the fabricator. More economical fabrication can result when such an understanding exists beforehand. This document has been widely accepted by industry in the United Kingdom, and by the oil and gas industry around the world. Further full-scale wide-plate testing and now more than 15 years of experience with this document establishes the reliability of this approach.

As CTOD testing procedures became more sophisticated in the 1980s, tests were performed where the notch was placed precisely in the coarse-grained region of the heat-affected zone (HAZ) of weldments. It was found that low values of toughness can be obtained, called local brittle zones (LBZ). Most welded structures have such LBZ. Yet to date there have been no significant failures that can be attributed to such LBZ. Laboratory fatigue tests of full-scale welded details have had cracks which propagated through such zones, yet never experienced any distress or crack instability. Such small brittle regions are of little consequence because even if these LBZ are assumed to be cracks, nearby tough material in the weld and in the base metal do not allow instability.

The CTOD can be related to J and also to K, and therefore CTOD fracture toughness values are often converted to equivalent K values using:

$$J = 1.6 * \sigma_y * \text{CTOD} \tag{8.17}$$

Substituting Eq. 8.17 into Eq. 8.15 gives

$$K = \sqrt{1.6 * \sigma_y * E * \text{CTOD}} \tag{8.18}$$

Recognizing that all of these tests are performed on similar specimens and that all of the various fracture toughness measures can be related, BSI has recently developed a unified testing procedure BS 7448, "Fracture Mechanics Toughness Tests." Using this method, a test is performed and then, based on the results, it is decided how the test should be interpreted. ASTM is currently working on a similar unified test method.

8.4 Summary

1. If there is a significantly fluctuating live load, building members and connections are potentially susceptible to fatigue. The fatigue design procedures in the AISC LRFD Manual are based on control of the stress range and knowledge of the fatigue strength of the various details.

2. Welded building connections and thermal cut holes, copes, blocks, or cuts are susceptible to brittle fracture. Many interrelated design variables can increase the potential for brittle fracture including lack of redundancy, large forces and moments with dynamic loading rates, thick members, geometrical discontinuities, and high constraint of the connections. Low temperature can be a factor for exposed structures. There is only qualitative information in the AISC LRFD Manual for the control of fracture.

3. Many factors in fabrication can increase the potential for fatigue and brittle fracture including notches, misalignment and other geometrical discontinuities, cold work, thermal cutting, flame straightening, weld heat input, weld joint design, particularly backing bars, weld sequence and residual stress, nondestructive eval-

uation and weld defects, intersecting welds, and inadequate weld access holes. In general, these factors are much more difficult to control in field welding.

4. Simple, linear-elastic fracture-mechanics concepts can be used to predict the potential for brittle fracture in buildings. These procedures could eventually be considered as a quantitative design check to preclude brittle fracture.

5. Several examples of fracture in buildings illustrate the importance of weld metal toughness as the first line of defense against fracture. Very brittle base metal, such as in jumbo sections without mandatory toughness, must not be welded or thermally cut and subjected to tension. In most other cases, unless there is fatigue loading to propagate the crack out of a weld into the base metal, the toughness of the base metal is less important.

References

1. ASCE-WRC (1971), *Plastic Design in Steel, A Guide and Commentary*, 2d ed., Joint Committee of the Welding Research Council and the American Society of Civil Engineers, ASCE, New York, 1971.

2. *Load and Resistance Factor Design Specification for Structural Steel Buildings*, 2d ed., American Institute of Steel Construction (AISC), Chicago, 1994.

3. Barsom, J. M., and S. T. Rolfe, *Fracture and Fatigue Control in Structures*, 2d ed., Prentice-Hall, Englewood Cliffs, N.J., 1987.

4. Irving, Bob, "Way Back When," *Welding Journal*, vol. 73, no. 9, p. 22, Sept. 1994.

5. Irwin, G. R., and J. A. Kies, "Critical Energy Rate Analysis of Fracture Strength of Large Welded Structures," *Welding Journal*, vol. 33, Research Supplement, pp. 193s–198s, April 1954.

6. Fisher, J. W., *Fatigue and Fracture in Steel Bridges*, Wiley, New York, 1984.

7. AASHTO LRFD Bridge Design Specifications, American Association of State Highway Transportation Officials, Washington, D.C., 1994.

8. Guide Specifications for Fracture Critical Non-Redundant Steel Bridge Members, American Association of State Highway and Transportation Officials, Washington, D.C., 1978 (with interims).

9. Fisher, J. W., and A. W. Pense, "Experience with Use of Heavy W Shapes in Tension," *Engineering Journal*, American Institute of Steel Construction, vol. 24, no. 2, 1987.

10. ANSI/AWS D1.1, "Structural Welding Code—Steel," American Welding Society, Miami, 1992.

11. Specification for Structural Steel Buildings—Allowable Stress Design and Plastic Design, 9th ed., American Institute of Steel Construction (AISC), Chicago, 1989.

12. Fisher, J. W., K. H. Frank, M. A. Hirt, and B. M. McNamee, *Effect of Weldments on the Fatigue Strength of Steel Beams*, National Cooperative Highway Research Program (NCHRP) Report 102, Highway Research Board, Washington, D.C., 1970.

13. Fisher, J. W., P.A. Albrecht, B. T. Yen, D. J. Klingerman, and B. M. McNamee, *Fatigue Strength of Steel Beams with Welded Stiffeners and Attachments*, National Cooperative Highway Research Program (NCHRP) Report 147, Transportation Research Board, Washington, D.C.,1974.

14. Gurney, T. R., and Maddox, S. J., "A Re-Analysis of Fatigue Data for Welded Joints in Steel," *Welding Research International*, vol. 3, no. 4, pp. 1–54, 1973; or Report R/RB/E44/71, The Welding Institute, Abington Hall, Abington, Cambridge, CB1 6AL, UK, Sept. 1971.

15. Dexter, R. J., J. W. Fisher, and J.E. Beach, "Fatigue Behavior of Welded HSLA-80 Members," *Proceedings, 12th International Conference on Offshore Mechanics and Arctic Engineering*, vol. III, part A, Materials Engineering, pp. 493–502, ASME, New York, 1993.

16. Petershagen, H., and W. Zwick, *Fatigue Strength of Butt Welds Made by Different Welding Processes*, IIW-Document XIII-1048-82, 1982.

17. Keating, P. B., and J. W. Fisher, *Evaluation of Fatigue Tests and Design Criteria on Welded Details*, NCHRP Report 286, National Cooperative Highway Research Program, September 1986.

18. ENV 1993-1-1, "Eurocode 3: Design of Steel Structures—Part 1.1: General Rules and Rules for Buildings," European Committee for Standardization (CEN), Brussels, April 1992.

19. Demers, C., and J. W. Fisher, *Fatigue Cracking of Steel Bridge Structures*, vol. I: A Survey of Localized Cracking in Steel Bridges—1981 to 1988, *Report* FHWA-RD-89-166, also vol. II, A Commentary and Guide for Design, Evaluation, and Investigating Cracking, *Report* FHWA-RD-89-167, FHWA, McLean, Va., March 1990.

20. Yen, B. T., T. Huang, L.-Y. Lai, and J. W. Fisher, *Manual for Inspecting Bridges for Fatigue Damage Conditions, Report* FHWA-PA-89-022+85-02, Fritz Engineering Laboratory *Report* 511.1, Pennsylvania Department of Transportation, Harrisburg, Pa., January 1990.

21. Dexter, R. J., and M. R. Kaczinski, "Large-Scale Fatigue Tests of Advanced Double Hull Weld Joints," Advanced (Unidirectional) Double Hull Technical Symposium, *Proceedings*, Oct. 25–26, 1994, Gaithersburg, Md., Paper 8, pp. 1–20, 1994.

22. Out, J. M. M., J. W. Fisher, and B. T. Yen, *Fatigue Strength of Weathered and Deteriorated Riveted Members, Report* DOT/OST/P-34/85/016, Department of Transportation, Federal Highway Administration, Washington, D.C., October 1984.

23. Djikstra, O. D., J. Wardenier, and A. A. Hertogs, "The Fatigue Behavior of Welded Splices with and without Mouse Holes in IPE 400 and HEM 320 Beams", *Proceedings of the International Conference on Weld Failures—Weldtech '88*, The Welding Institute, Abington Hall, Abington, Cambridge, CB1 6AL, UK, November 1988.

24. Dexter, R. J., J. E. Tarquinio, and, J. W. Fisher, "Application of the Hot-Spot Stress Fatigue Analysis to Attachments on Flexible Plate," *Proceedings, 13th International Conference on Offshore Mechanics and Arctic Engineering*, ASME, vol. III, Materials Engineering, 85–92.

25. BSI PD 6493, "Guidance on Some Methods for the Derivation of Acceptance Levels for Defects in Fusion Welded Joints," British Standards Institute, London, 1991.

26. Marshall, P.W., "Tubular Joint Design," contributing chapter in *Planning and Design of Fixed Offshore Platforms*, B. McClelland and M. Reifel (eds.), Van Nostrand Reinhold, New York, 1986.

27. Hugill, P. N., and J. D. G. Sumpter, "Fatigue Life Prediction at a Ship Deck/Superstructure Intersection," *Strain*, vol. 26, no. 3, pp. 107–112, August 1990.

28. Maddox, S. J., *Influence of Tensile Residual Stress on the Fatigue Behaviour of Welded Joints in Steel*, ASTM STP 776, American Society for Testing and Materials, Philadelphia, 1982.

29. Maddox, S. J., "The Influence of Mean Stress on Fatigue Crack Propagation—A Literature Review," *International Journal of Fracture*, vol. 11, no. 3, 1975.

30. VanDien, J. P., M. R. Kaczinski, and R. J. Dexter, "Fatigue Testing of Anchor Bolts," *Structures Congress XIV*, Chicago, April 1996.

31. Fisher, J. W., J. Jian, D. C. Wagner, and B. T. Yen, "Distortion-Induced Fatigue Cracking in Steel Bridges," National Cooperative Highway Research Program (NCHRP) *Report* 336, Transportation Research Board, Washington, D.C., 1990.

32. Fisher, J. W., and P. B. Keating, "Distortion-Induced Fatigue Cracking of Bridge Details with Web Gaps," *Journal of Constructional Steel Research*, no. 12, pp. 215–228, 1989.

33. Fisher, J. W., et al, *Fatigue and Fracture Evaluation for Rating Riveted Bridges*, National Cooperative Highway Research Program (NCHRP) *Report* 302, Transportation Research Board, Washington, D.C., 1987.

34. Yen, B. T., et. al., "Fatigue Behavior of Stringer Floorbeam Connections," *Proceedings of the Eighth International Bridge Conference*, paper IBC-91-19, pp. 149–155, Engineers' Society of Western Pennsylvania, 1991.

35. Stambaugh, K. A., P. R. Van Mater, Jr., and W. H. Munse, *Fatigue Performance Under Multiaxial Loading in Marine Structures*, SSC-356, Ship Structures Committee, Washington, D.C., 1990.

36. Siljander, A., P. Kurath, and F. V. Lawrence, Jr., "Nonproportional Fatigue of Welded Structures", *Advances in Fatigue Lifetime Predictive Techniques*, ASTM STP 1122, M. R. Mitchell and R.W. Landgraf (eds.), American Society for Testing and Materials, Philadelphia, pp. 319–338, 1992.

37. Miner, M. A., "Cumulative Damage in Fatigue," *Journal of Applied Mechanics*, vol. 12, p. A-159, 1945.

38. BS 7608, "Code of Practice for Fatigue Design and Assessment of Steel Structures," British Standards Institute, London, 1994.

39. Kulak, G. L., J. W. Fisher, and J. H. Struick, *Guide to Design Criteria for Bolted and Riveted Joints*, 2d ed., Prentice-Hall, Englewood Cliffs, N.J., 1987.

40. Frank, K. H., "Fatigue Strength of Anchor Bolts," *Journal of the Structural Division*, ASCE, vol. 106, no. ST, June 1980.

41. Maddox, S. J., *Fatigue Strength of Welded Structures*. 2d ed., Abington Publishing, Cambridge, UK, 1991.

42. Fisher, J. W., et. al., "Resistance of Welded Details under Variable Amplitude Long-Life Fatigue Loading," *National Cooperative Highway Research Program Report* 354, Transportation Research Board, Washington, D.C., 1993.

43. Kulicki, et. al., "Guidelines for Evaluating Corrosion Effects in Existing Steel Bridges," *National Cooperative Highway Research Program Report* 333, Transportation Research Board, Washington, D.C., December 1990.

44. Frank, K. H., and, J. W. Fisher "Fatigue Strength of Fillet Welded Cruciform Joints," *Journal of the Structural Division, ASCE*, vol. 105, no. ST9, pp. 1727–1740, September 1979.

45. Marsh, K. J. (ed.), *Full-Scale Fatigue Testing of Components and Structures*, Butterworths, London, 1988.

46. Nishimura, T., and C. Miki, "Fatigue Strength of Longitudinally Welded Members of 80 kg/mm^2 Steel," *Technical Report* 22, Tokyo Institute of Technology, January 1978.

47. Miki, C., T. Nishimura, J. Tajima, and A., Okukawa, "Fatigue Strength of Steel Members Having Longitudinal Single-Bevel Groove Welds," *Transactions of the Japan Welding Society*, vol. 11, no. 1, pp. 43–56, April 1980.

48. Petershagen, H., "Fatigue Problems in Ship Structures," *Advances in Marine Structures*, Elsevier Applied Science, London, pp. 281–304, 1986.

49. Smith, I. F. C., C. A. Castiglioni, and P. B. Keating, "Analysis of Fatigue Recommendations Considering New Data," *IABSE Periodica* 3/1989, IABSE proceedings P-137/89, pp. 97–109, 1989.

50. Faulkner, D., "Unified Design Codes for Floating Systems," *Integrity of Offshore Structures-4*, D. Faulkner, M.J. Cowling, and A. Incecik (eds.), pp. 517–532, Elsevier Applied Science, London, 1991.

51. Broek, D., *Elementary Fracture Mechanics*, 4th Ed., Martinis Nijhoff Publishers, Dordrecht, Netherlands, 1987.

52. Anderson, T. L., *Fracture Mechanics—Fundamentals and Applications*, 2d ed., CRC Press, Boca Raton, Fla, 1995.

53. Kanninen, M. F., and, C. H. Popelar, *Advanced Fracture Mechanics*, Oxford University Press, New York, 1985.

54. Kulak, G. L., and I. F .C. Smith, "Analysis and Design of Fabricated Steel Structures for Fatigue: A Primer for Civil Engineers," University of Alberta Department of Civil Engineering, *Structural Engineering Report* 190, July 1993.

55. Murakami, Y., et. al. (eds.), *Stress Intensity Factors Handbook* (vols. 1 and 2), Pergamon Press, Oxford, UK, 1987.

56. Tada, H., *The Stress Analysis of Cracks Handbook*, Paris Productions, Inc., Saint Louis, 1985.

57. Rooke, D. P., and D. J. Cartwright, *Compendium of Stress Intensity Factors*, Her Majesty's Stationery Office, London, 1974.

58. Rooke, D. P., *Compounding Stress Intensity Factors*, Research Reports in Materials Science (Series One), The Parthenon Press, Lancashire, UK, 1986.

59. Roberts, R., J. W. Fisher, G. R. Irwin, K. D. Boyer, H. Hausammann, G. V. Krishna, V. Morf, and R. E. Slockbower, "Determination of Tolerable Flaw Sizes in Full Size Welded Bridge Details," *Report* FHWA-RD-77-170, Federal Highway Administration, Washington, D.C., December 1977.

60. Frank, K. H., et al., "Notch Toughness Variability in Bridge Steel Plates," *NCHRP Report* 355, National Cooperative Highway Research Program, 1993.

61. Easterling, K., *Introduction to the Physical Metallurgy of Welding*, Butterworths Monographs in Materials, London, 1983.

62. Irwin, G. R., "Analysis of Stresses and Strains Near the End of Crack Traversing a Plate," *Transactions ASME, Journal of Applied Mechanics*, vol. 24, 1957, also reprinted in ASTM volume on classic papers.

63 Shank, M.E. (ed.), *Control of Steel Construction to Avoid Brittle Fracture*, Welding Research Council, New York, 1957.

9

Linda M. Hanagan
Thomas M. Murray

SERVICEABILITY CONSIDERATIONS FOR FLOOR AND ROOF SYSTEMS

9.1 Introduction

The LRFD specification[1] recognizes two categories for designing steel structures: strength limit states and serviceability limit states. The strength limit states are related to the safety of the structure with respect to the load-carrying capacity. The serviceability limit states seek to preserve the function of the building. Satisfying the serviceability limit states for floor and roof systems is the subject of this chapter.

Designing for the serviceability limit states can be a complex task because of the vagueness of the criteria involved. Serviceability limit states, as they relate to floor and roof systems, primarily address excessive displacements and vibrations. The descriptive term "excessive" is the source of the vagueness and depends on many variables. These variables include the perceptions and expectations of building owners and users as well as their relationship with cost. Because of the subjective nature of the requirements and the fact that life safety is not an issue, the serviceability limit states are not as rigid as those related to strength. Design criteria related to serviceability are considered guidelines which require careful judgment on the part of the engineer. The discussion that follows seeks to provide insight as well as guidelines for assessing serviceability conditions in roof and floor systems.

Exceeding deflection limit states can result in many adverse phenomena. Excessive deflections can result in unsightly cracking of nonstructural components, an unnerving sag of floor and roof systems, and leakage of the building envelope, to name a few. In Sec. 9.2, some of the more widely accepted recommendations for deflection limits to prevent some common serviceability problems are discussed. Exceeding vibration limits can result primarily in complaints from the occupants or malfunctioning of equipment. The widely accepted guidelines for limiting excessive vibrations in a floor system are much less concise in the literature and are therefore discussed in several sections of this chapter.

9.2 Serviceability Requirements for Static Deflections

Excessive deflections of roof and floor systems can result in many problems that are not strength-related but are of major concern nonetheless. Common problems associated with excessive deflections of floor and roof members were presented by an ad hoc committee[2] of ASCE and are reviewed below.

Visually sagging members are aesthetically undesirable and can also cause unnecessary concern in the occupants for structural safety. Sagging floor or roof members can also be functionally objectionable. They can cause damage to nonstructural elements and dysfunction of equipment. Sagging roof members can cause drainage problems possibly leading to ponding. Ponding must be considered with respect to strength limits also.

The ad hoc committee[2] also presented deflection and span ratios associated with the onset of certain serviceability problems. The nonvisible cracking of brickwork begins at ratios greater than 1/1000; cracking of partition walls ensues at ratios greater than 1/500. At ratios greater than 1/300 many visible serviceability problems begin to occur: general architectural damage, cracking in reinforced walls, damage to ceilings and flooring, facade damage, cladding leakage and visual annoyance. For ratios between 1/300 and 1/200, the list of problems is expanded to include improper drainage, damage to lightweight partitions, windows, and finishes. Finally, in the range of 1/200 to 1/100, the list includes impaired operation of movable components such as doors, windows, and sliding partitions.

For reasons noted earlier, U.S. codes and specifications do not quantify serviceability requirements. Because of this, one needs to turn to other sources for guidance. One particular source is the *Supplement to the National Building Code of Canada*.[3] The recommendations of two sources presented in the supplement, NBC 1990[4] and CAN/CSA-S16.1-M89,[5] for limiting deflections in steel floor systems are particularly relevant. These recommendations are reviewed below. Where the recommendations of the code and standard differ, both guidelines are presented.

Maximum Live Load Deflections and Span Ratios

a. Roof or floor members supporting plastered ceilings, partitions, etc.   1/360

b. Floor members not supporting plastered ceilings or partitions:

   CAN/CSA-S16.1-M89   1/300

   NBC 1990   1/360

   NBC 1990 (for bedrooms only)   1/240

Serviceability Considerations for Floor and Roof Systems ■ 9-3

c. Roof members not supporting plastered ceilings or partitions:

CAN/CSA-S16.1-M89 (for asphaltic built-up roofs)	1/240
CAN/CSA-S16.1-M89 (for sheet metal or elastic membrane covering)	1/180
NBC 1990	1/240
NBC 1990 (for roofs with no ceilings)	1/180

9.3 Dynamic Behavior of Floor Systems

To understand the criteria presented to assess floor vibration, it is first necessary to address the generalities and theory related to the dynamic behavior of floor systems.

9.3.1 The Nature of Floor Vibrations

A floor system can be a very complex dynamic system, often possessing several closely spaced natural frequencies which contribute significantly to the vibration response. In a case study of an elementary school,[6] the floor was shown to possess six different natural frequencies and corresponding mode shapes which contributed significantly to the response at the location of the measurement. These natural frequencies existed over a range of 5 to 15 Hz. This floor is considered typical for steel-framed floors where the first natural frequency is commonly found to be in the range of 5 to 8 Hz.[7] This is unfortunate because automobile and aircraft comfort studies have found that humans are most acutely sensitive to vibrations in this range,[8] a phenomenon that is explained by the fact that many of the major organs in the human body resonate in this frequency range.[7]

As noted in the introduction, excessive vibration can be characterized as too large for sensitive equipment or too large for occupant comfort. Determining these permissible levels is an entire research area in itself; however, some of the more widely accepted levels are discussed in the following sections. These levels are expressed by researchers in terms of either acceleration, velocity, or displacement amplitudes and are often frequency-dependent. There is no consensus as to the most relevant measure for describing acceptable levels.

Comfort of the occupants is a function of human perception. This perception is affected by factors including the task or activity of the perceiver, the remoteness of the source, and the movement of other objects in the surroundings. A person is distracted by acceleration levels as small as 0.5 percent g in an office or residential environment. People involved in an activity such as aerobics may be comfortable with acceleration levels up to 5 percent g.[9] Multiple-use occupancies must therefore be carefully considered.

Webster and Vaicaitis[10] describe a facility that has both dining and dancing in a large open area. The floor was noted to have a first natural frequency of 2.4 Hz, which is in resonance with the beat of many popular dance songs. This resonance response produced maximum acceleration and displacement levels of 7 percent g and 0.13 in, respectively. Such levels actually caused sloshing waves in drinks and noticeable bouncing of the chandeliers. The occupants found these levels to be quite objectionable.

9-4 ■ Chapter Nine

Many different scales and criteria are available which address the subjective evaluation of floor vibration. Factors included in these subjective evaluations include the natural frequency of the floor system, the maximum dynamic amplitude (acceleration, velocity, or displacement) due to certain excitations, and the amount of damping present in the floor system. At the present time, most of the design criteria utilize either a single impact function to assess vibrations which are transient in nature or a sinusoidal function to assess steady-state vibrations from rhythmic activities.

9.3.2 Basics of Structural Dynamics

The basic element in structural dynamics is the single-degree-of-freedom system. Many of the available vibration criteria utilize a strategy to simplify a complex floor system into this basic element. The single-degree-of-freedom system is represented by a single mass m, spring k, and damper c, as shown in Fig. 9.1. The governing differential equation of motion for this system follows.

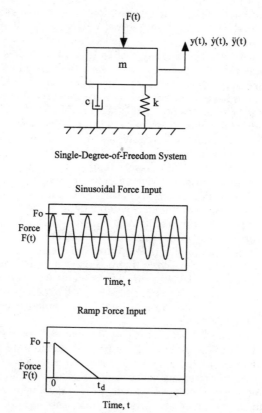

Fig. 9.1 Single-degree-of freedom system and two common input forces.

Equation of motion for single degree of freedom:

$$m\ddot{y}(t) + c\dot{y}(t) + ky(t) = F(t)$$

When the mass m is subjected to a time-dependent input force $F(t)$, the result is a vibration response which can be described by the displacement $y(t)$, the velocity, $\dot{y}(t)$, and the acceleration, $\ddot{y}(t)$. The equation of motion for a single-degree-of-freedom system can also be formulated in terms of the natural frequency of the free vibration and ratio of critical damping.

$$\ddot{y}(t) + 2\zeta\omega_0\dot{y}(t) + \omega_0^2 y(t) = \frac{F(t)}{m}$$

where ω_0 = circular natural frequency, radians/s
    = $\sqrt{k/m} = 2\pi f_0$
 f_0 = natural frequency, Hz
 ζ = ratio of critical damping
    = c/c_{cr}
 c_{cr} = critical damping, the value of damping for which the roots of the characteristic equation are equal
 c_{cr} = $2\sqrt{k\,m}$

Also shown in Fig. 9.1 are two input forces commonly used to represent different sources of floor excitations. A sinusoidal force input function is often used to predict floor response due to rhythmic excitations. The ramp force input function is often used to assess the floor system response to transient excitations such as walking. The closed-form solutions for the response of a single-degree-of-freedom system subjected to these input forces can be found in most structural dynamics textbooks and will not be presented here.

Continuous systems, such as beams or platelike structures, contain an infinite number of free vibration modes. Each of these modes can be characterized by its mode shape and its associated natural frequency. Figure 9.2 illustrates the first three modes of vibration for a simply supported beam with a uniform mass distribution. The vibration response at any point on a beam can be approximated by the sum of the individual modal contributions, truncated at some finite mode, at that point in space and time.

The fundamental natural frequency f_1 of a simply supported beam with a uniform mass distribution, as shown in Fig. 9.2, can also be conveniently expressed in terms of the static deflection due to distributed weight. The derivation of this expression is as follows:

$$f_1 = \frac{\pi}{2}\sqrt{\frac{EI}{\overline{m}L^4}} = \frac{\pi}{2}\sqrt{\frac{5g}{384\Delta}} = 0.18\sqrt{\frac{g}{\Delta}}$$

$$\Delta = \frac{5wL^4}{384EI} = \frac{5g}{384}\frac{\overline{m}L^4}{EI}$$

$$\frac{EI}{\overline{m}L^4} = \frac{5g}{384\Delta}$$

9-6 ■ Chapter Nine

where w = uniformly distributed load on a beam = $\overline{m} \cdot g$
$\overline{m}$ = uniformly distributed mass on beam
g = acceleration of gravity = 386.4 in/s² or 9800 mm/s²
L = beam length
E = modulus of elasticity
I = moment of inertia for the beam cross section

The expression for the fundamental natural frequency, in terms of static deflection, is often misused in determining the natural frequency for other beam con-

Simple Beam

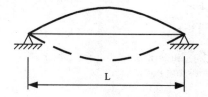

Mode 1: $f_1 = \dfrac{\pi}{2}\sqrt{\dfrac{EI}{mL^4}}$

Mode 2: $f_2 = 2\pi\sqrt{\dfrac{EI}{mL^4}}$

Mode 3: $f_3 = \dfrac{9\pi}{2}\sqrt{\dfrac{EI}{mL^4}}$

Continuous Beam

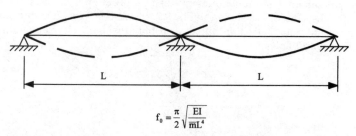

$$f_0 = \dfrac{\pi}{2}\sqrt{\dfrac{EI}{mL^4}}$$

Fig. 9.2 Modes of vibration for beams with uniformly distributed mass.

Serviceability Considerations for Floor and Roof Systems ■ 9-7

figurations. In particular, the expression f_1 above cannot be used for continuous beams. There is a common misconception that providing continuity of beams over a support will raise the fundamental frequency of the system. While it is true that continuity reduces the maximum static deflection, the fundamental natural frequency remains the same. This concept is illustrated in Fig. 9.2.

Platelike structures, such as beam and girder systems, also possess an infinite number of natural frequencies and mode shapes. Figure 9.3 illustrates the natural frequencies and mode shapes, for the first four modes, for a one-bay floor system comprised of a slab, joists, and girders. In addition to the mass distribution, the frequencies and mode shapes are affected by the slab, joist (or beam), and girder

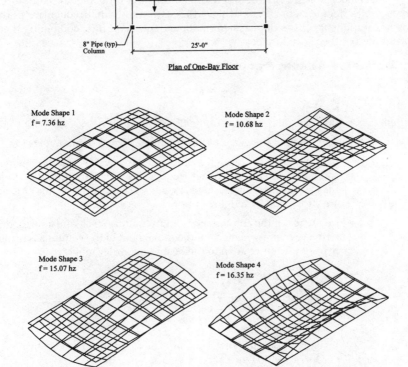

Fig. 9.3 Mode shapes and natural frequencies from a computer analysis.

properties. This concept is explored in the following subsection. Close inspection of Fig. 9.3 and some intuition reveals that an activity like jumping at the center of the floor would cause dynamic amplitudes consisting of the superposition of modes 1, 4 and higher-order modes with a modal amplitude at that point.

One particular phenomenon to carefully consider and, if possible, avoid is that of resonance. Resonance occurs when a component of a harmonic excitation corresponds to one of the natural frequencies of the structure. Vibration amplitudes are greatly amplified in lightly damped structures such as steel floor systems. This concept is expanded in Sec. 9.12.

9.3.3 Evaluation of Fundamental Natural Frequency for a Floor System

As illustrated in Fig. 9.3, the dynamic behavior of a floor system is very complex. There are, however, commonly accepted procedures to determine dynamic characteristics of floor systems. The following discussion provides necessary information and a procedure to estimate the frequency of the first mode of free vibration for a steel floor system. A close approximation of the fundamental natural frequency of a floor system can be achieved by considering the frequencies of the major components of the floor system independently and then combining them as outlined in the procedure below.

Estimated System Frequency

$$\frac{1}{f_s^2} = \frac{1}{f_b^2} + \frac{1}{f_g^2} + \frac{1}{f_c^2}$$

where f_s = first natural frequency of the floor system, Hz

f_b = frequency of the beam or joist member, Hz; see equations below.

f_g = frequency of the girder members; the lowest girder frequency should be used if the girder frequencies differ; the girder term in the system expression above can be neglected if the beams or joists are supported by a rigid support such as a wall

f_c = frequency of the column, Hz; except in unusual circumstances, this term is generally neglected; the movement of the columns is usually insignificant relative to the beam and girder motion

Beam or Joist Frequency

$$f_b = K\sqrt{\frac{gEI_t}{wL^4}}$$

where K = 1.57 for simply supported beams; 0.56 for cantilevered beams; refer to Murray and Hendrick[11] for overhanging beams

g = acceleration of gravity; 386.4 in/s^2 or 9800 mm/s^2

E = modulus of elasticity for transformed section, 29,000 ksi for steel

Serviceability Considerations for Floor and Roof Systems   9-9

I_t = transformed moment of inertia; when the steel deck supporting the concrete rests directly on the beam or joist (connected by welds, screws, mechanical shear connectors, etc.), assume composite action between the steel member and the concrete slab; see Sec. 9.3.4 for more information on the computation of composite member properties

w = floor weight per unit length of beam; value should be the actual expected service load on the beam; overestimating this value can result in a nonconservative prediction of acceptability; 10 percent to 25 percent of the live load used in strength calculations is suggested for design[7]

L = beam or joist span

Girder Frequency

$$f_g = K\sqrt{\frac{gEI_t}{wL^4}}$$

where I_t = transformed moment of inertia; see Sec. 9.3.4 for more information on the computation of the girder properties

w = floor weight per unit length of girder; value should be the actual expected service load on the girder; loads from the beams or joists framing into the girder can usually be treated as continuous regardless of the spacing

Note: All other variables are as defined for the beam or joist frequency above.

Column Frequency

$$f_c = \frac{1}{2\pi}\sqrt{\frac{gAE}{PL}}$$

where A = area of the column section

P = load on the column; value should be the actual expected service load

L = length of column

Note: All other variables are as defined previously.

9.3.4 Evaluation of Transformed Section Properties

As noted previously, when the deck supporting the concrete rests directly on the beam, joist, or girder (connected by welds, screws, mechanical shear connectors, etc.), composite action should be assumed between the steel member and the concrete slab.[7,12] Particular information for computing transformed section properties for beams, joists, and girders is presented below.

Transformed Beam Properties ■ To calculate the transformed moment of inertia for a beam section with concrete on metal deck, the concrete within the profile of the deck is ineffective and should be neglected as illustrated in Fig. 9.4. The parameters in this figure are defined as follows:

b = effective width of the concrete; should be taken as the tributary width of the beam

n = modular ratio

$$n = \frac{E_s}{E_c} = \frac{\text{modulus of elasticity of steel}}{\text{modulus of elasticity of concrete}}$$

Transformed Beam Section Properties

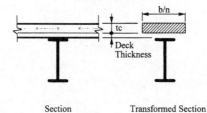

Section   Transformed Section

Transformed Joist Section Properties

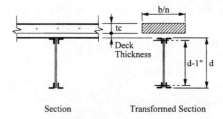

Section   Transformed Section

Transformed Girder Section Properties

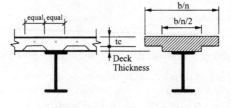

Section   Transformed Section

Fig. 9.4 Transformed section properties.

Serviceability Considerations for Floor and Roof Systems ■ 9-11

Transformed Joist Properties ■ Because of the nature of the data provided for a joist section, calculation of the necessary joist properties requires several assumptions. A procedure for calculating joist properties from joist table data is presented below. This procedure should be limited to standard K-series joists.

1. Determine the allowable moment for the joist section.

$$M_{all} = \frac{wL_n^2}{8}$$

where w = allowable total load per foot from joist table
L_n = clear span of joist
L_n = total span (ft) - 0.33 ft

2. Determine the effective area of the joist.

A_j = effective area of joist
= 0.85 A_{chord}

A_{chord} = $A_{top} + A_{bot}$

A_{bot} = area of the bottom chord of joist, in^2

$$A_{bot} = \frac{M_{all}}{(d-1 \text{ in})f_{all}}$$

where d = total depth of joist (see Fig. 9.4), in
f_{all} = allowable stress of steel, 30 ksi typical
A_{top} = area of the top chord of joist, in^2
= 1.25 A_{bot}

3. Determine the location of the neutral axis of the joist.

The neutral axis of the joist $\bar{y}$ is computed with the assumption that the centroid of the top chord of the joist is 0.5 in from the top of the joist and the centroid of the bottom chord of the joist is 0.5 in above the bottom of the joist as illustrated in Fig. 9.4.

$\bar{y}$ = neutral axis of joist section, measured from the top of the joist

$$\bar{y} = 0.5 \text{ in} + \frac{A_{bot}(d-1 \text{ in})}{A_{chord}}$$

4. Determine the moment of inertia of the joist section.

I_j = effective moment of inertia of joist, includes the effect of shear deformation

$I_j = 0.85 I_{chord}$

I_{chord} = moment of inertia of chords

$$I_{chord} = A_{top}(\bar{y} - 0.5 \text{ in})^2 + A_{bot}(d - \bar{y} - 0.5 \text{ in})^2$$

The properties, as determined above, can be used in computing the transformed moment of inertia for a joist section as defined for the beam in Fig. 9.4.

Transformed Girder Properties ■ Because the deck runs parallel to a girder, the portion of the concrete inside of the deck profile is effective as illustrated in Fig. 9.4. Dividing the total effective width of concrete by 2, as shown in the figure, is accurate for deck profiles with deck ribs of equal dimension. The effective width of the concrete b for the girder is the tributary width for the girder or 40 percent of the girder span, whichever is smaller.

When girders support joist members, creating a gap between the concrete and the girder, composite behavior is somewhat uncertain. Allen and Murray[13] suggest an assumption of partially composite behavior pending future research. The recommendations are as follows:

For joists with seat heights less than or equal to 3 in:

$$I_g = I_{nc} + \frac{I_c - I_{nc}}{2}$$

where I_g = transformed moment of inertia of girders supporting joists
 I_{nc} = noncomposite girder moment of inertia
 I_c = fully composite girder moment of inertia.

For joists with seat heights greater than 4 in:

$$I_g = I_{nc} + \frac{I_c - I_{nc}}{4}$$

Preliminary experimental research[14] indicates noncomposite behavior for girders supporting joists.

9.4 Reiher-Meister Scale for Steady-State Vibrations

This and the following three sections discuss four scales useful in assessing human perception to vibration levels. They are presented to provide insight with respect to the vibration levels which annoy occupants as well as a historical perspective on the development of floor vibration criteria. Reiher and Meister[15] published a frequently referenced scale concerning human perception to steady-state vibration. While this scale was not derived specifically for the evaluation of floor systems, it has been extrapolated by other researchers for such purposes.[12,16] The scale represented by the right-hand axis of the graph in Fig. 9.5 was derived from the subjective evaluations of 10 persons standing on a vibrating platform. The subjects were exposed to vertical steady-state vibration episodes, each lasting approximately 5 minutes, and were

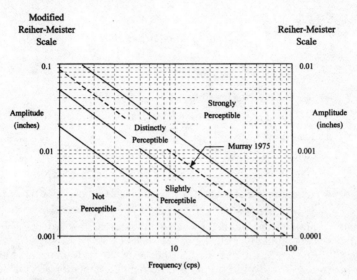

Fig. 9.5 Modified Reiher-Meister and Reiher-Meister scales.[12, 15, 16]

asked to classify the vibration as (1) slightly perceptible, (2) distinctly perceptible, (3) strongly perceptible, (4) disturbing, and (5) very disturbing. The frequency and displacement ranges of the episodes were 5 to 70 Hz and 0.001 to 0.40 in, respectively.

9.5 Modified Reiher-Meister Scale for Transient Vibrations

Many years after Reiher and Meister developed their vibration scale, research began to develop a scale pertaining to vibration in floor systems. After studying two laboratory floors and 46 in-place floors around the Kansas City area, Lenzen[12] presented several concepts useful in evaluating steel joist–concrete slab floor systems. Lenzen noted that full composite action, as noted in Sec. 9.3.4, between the slab and the steel joists was found to exist in low-level occupant-induced floor vibrations. Another concept introduced by Lenzen[12] is that occupant-induced floor vibration is transient in nature. Human perception to transient vibration is dependent on three factors (frequency, amplitude, and damping) of which damping is the most important. Lenzen's observations pertaining to human perception of the transient vibration led to an adjustment of the vertical axis of the original Reiher-Meister scale. The modified Reiher-Meister scale, represented by the left-hand axis of Fig. 9.5, shifts the original vertical axis of the scale by a factor of 10 to evaluate human perception to transient vibration.

Lenzen suggested that, in the absence of partitions, the modified Reiher-Meister scale be used to evaluate the floor system using the following formulas to estimate the parameters.

Lenzen's formulas:

$$f_s = \text{same as defined in Sec. 9.3.3}$$

$$A_o = \frac{PL^3}{48EI_t N_{\text{eff}}}$$

where P = 300 lb for spans less than 24 ft
- = L^2/d for spans greater than 24 ft, L and d expressed in ft and P in lb
- L = joist span
- E = modulus of elasticity for transformed section
- I_t = transformed moment of inertia of composite section
- N_{eff} = total number of joists ≤ 10, based on a 24-in joist spacing and a 2 ½-in concrete slab.

A note of caution was extended by Lenzen that research was necessary before the above concepts could be included in the specification.

In his 1975 paper, Murray[16] added further refinements to Lenzen's Modified Reiher-Meister scale and procedure. First, he defined limits for acceptable floor parameters. These limits are defined by the line noted "Murray 1975" in Fig. 9.5. Floors with open areas (free of partitions) and damping between 4 and 10 percent which plot above this line will result in complaints from the occupants. Second, Murray refined the calculation for the displacement amplitude A_o. These refinements are as follows:

Murray's formulas:

- f_s = same as defined in Section 9.3.3.

- A_o = maximum dynamic displacement computed from a theoretical heel-drop impact.

Murray continued to develop this approach to floor vibration assessment. The Murray criterion is presented in Sec. 9.10.

9.6 Wiss and Parmelee Rating Factor for Transient Vibrations

Wiss and Parmelee[17] also conducted research to refine the findings of Lenzen's research. In particular, they attempted to quantify, in a more scientifically rigorous manner, human perception to transient floor motion. They subjected 40 persons, standing on a vibrating platform, to transient vibration episodes with different combinations of frequency (2.5 to 25 Hz), peak displacements (0.0001 to 0.10 in), and damping (0.1 to 0.16, expressed as a ratio of critical). After each episode, the subject was asked to rate the vibration on a scale of 1 to 5 with the following definitions: (1) imperceptible, (2) barely perceptible, (3) distinctly perceptible, (4) strongly perceptible, and (5) severe. Using regression analysis, an equation was developed which related the three variables of the vibration episode to the subjective perception ratings. This equation is presented below.

Serviceability Considerations for Floor and Roof Systems 9-15

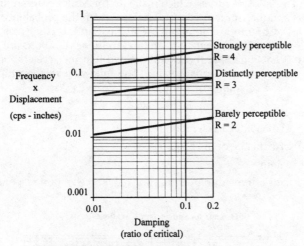

Fig. 9.6 Wiss and Parmelee rating factor scale.[17]

Wiss and Parmelee rating factor:

$$R = 5.08\left(\frac{FA}{D^{0.217}}\right)^{0.265}$$

where R = response rating; 1=imperceptible; 2=barely perceptible; 3=distinctly perceptible; 4=strongly perceptible; 5=severe.

F = frequency of the vibration episode, Hz

A = maximum displacement amplitude, in

D = damping ratio, expressed as a ratio of critical

A graph of this subjective rating system is shown in Fig. 9.6. It should be noted that the lines represent a mean for that particular rating. The authors suggest that the boundaries for each rating lie halfway between the mean lines. The boundaries defining $R=1$ and $R=5$ are not identified by the authors. These ratings are unbounded; therefore, a mean line cannot be computed.

9.7 ISO Scale for Human Response to Building Vibrations

The International Organization for Standardization publishes a document which is particularly relevant to the serviceability of floor systems with respect to vibration. The title of this document is "Basis for the Design of Structures—Serviceability of Buildings against Vibration."[18] Contained in this document is a scale which can be used to assess human perception to building vibration. This scale is presented in Fig. 9.7. The curves which make up this scale represent magnitudes of continuous and intermittent vibration below which the probability of adverse comment is low. Continuous vibrations are defined, in this context as those with a duration exceeding 30 min per 24 h, and intermittent vibrations are those with more than 10 events per 24 h. It should be noted that the acceleration limits are expressed as the root-mean-square value. The following relationship is convenient in assessing acceleration magnitudes:

$$a(t) = a_0 \sin 2\pi f t$$

where $a(t)$ = sinusoidal acceleration response
 f = frequency of the response, Hz.
 t = time
 a_0 = maximum amplitude of sinusoidal acceleration response

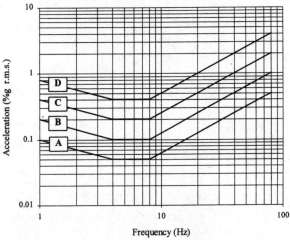

A: Baseline Curve, acceleration limit for critical working areas (some hospital operating rooms, some precision laboratories, etc.)

B: 2 x Baseline Curve, low end of acceleration limit for daytime in residential occupancies.

C: 4 X Baseline Curve, high end of acceleration limit for daytime in residential occupancies, acceleration limit for offices, schools, etc.

D: 8 X Baseline Curve, acceleration limit for workshops, etc.

Fig. 9.7 ISO scale for satisfactory vibration limits.[18]

$$a_{rms} = \left\{ \frac{1}{T} \int_0^T [a(t)]^2 dt \right\}^{0.5}$$

$a_{rms} = 0.707 a_o$ for a sinusoidal acceleration response

The root-mean-quadrature method is suggested for evaluating the effects of a vibration signal containing two or more discrete frequency components.

9.8 Vibration Criteria for Sensitive Equipment and Facilities

As noted in the introduction, occupant comfort is not the only consideration with respect to levels of floor vibration. Many high-technology facilities contain equipment which is particularly sensitive to vibration. In some cases, the equipment may be so sensitive as to prohibit its use on any kind of framed floor system. To keep vibrations to a minimum, even the site must be carefully selected for very sensitive facilities. Ungar, Sturz, and Amick[19] have developed some general vibration guidelines for different types of sensitive facilities. These guidelines are presented in Table 9.1. The velocity limits stated are considered applicable in the range of 8 to 80 Hz. This information is presented to provide the reader with a frame of reference regarding sensitive facilities. The design of these facilities is very complicated and should not be undertaken without careful research.

9.9 CSA Criterion for Walking Vibrations

Although scales can be very useful in assessing the severity of an existing floor vibration problem, they are not particularly useful for design purposes unless a procedure for predicting vibration behavior is determined. Such a prediction is dependent on both floor system properties and the source of the excitation. It is the combination of a rating scale and a procedure for predicting vibration behavior that makes up a useful criterion for design purposes. Several criteria for assessing a floor design, with respect to levels of floor vibration, are presented in this and the following three sections.

The Canadian Standards Association[20] recommends the criterion described below for assessing floors subjected to walking excitations. The use of this criterion is limited to floor systems with first natural frequencies below 10 Hz and member spans greater than 7 m.

9.9.1 Summary of the Criterion

An acceptable floor design is predicted if the computed peak acceleration a_o is less than the limits defined in the CSA scale shown in Fig. 9.8. The criterion is applicable to floors subjected to walking excitations. The sensitivity level of the scale is targeted at residential, school, and office occupancies. For more sensitive occupancies, such as operating rooms and special laboratories, acceptable levels are less than those represented in Fig. 9.8.

Table 9.1 Vibration Criteria[19]

Facility equipment or use	Velocity (μin/sec)
Ordinary workshops	32,000
Offices	16,000
Residences; Computer systems	8,000
Operating rooms; Surgery; Bench microscopes at up to 100x magnification; Laboratory robots	4,000
Bench microscopes at up to 400x magnification; Optical and other precision balances; Coordinate measuring machines; Metrology laboratories; Optical comparators; Microelectronics manufacturing equipment - Class A	2,000
Micro-surgery, eye surgery, neuro-surgery; Bench microscopes at magnification greater than 400x; Optical equipment on isolation tables; Microelectronics manufacturing equipment - Class B	1,000
Electron microscopes at up to 30,000x magnification; Microtomes; Magnetic resonance imagers; Microelectronics manufacturing equipment - Class C	500
Electron microscopes at greater than 30,000x magnification; Mass spectrometers; Cell implant equipment; Microelectronics manufacturing equipment - Class D	250
Microelectronics manufacturing equipment - Class E; Unisolated laser and optical research systems	130

Class A: Inspection, probe test, and other manufacturing support equipment.
Class B: Aligners, steppers, and other critical equipment for photolithography with line widths of 3 microns or more.
Class C: Aligners, steppers, and other critical equipment for photolithography with line widths of 1 micron.
Class D: Aligners, steppers, and other critical equipment for photolithography with line widths of $1/2$ micron; includes electron-beam systems.
Class E: Aligners, steppers, and other critical equipment for photolithography with line widths of $1/4$ micron; includes electron-beam systems.

9.9.2 Procedure

To check a floor design, an estimate of the floor system damping and the peak acceleration due to an impulse must be made. Estimates for floor system damping are provided with descriptions as follows: (1) bare floor, 3 percent of critical; (2) finished floor—ceiling, ducts, flooring, and furniture, 6 percent of critical; (3) finished floor with partitions, 12 percent of critical. These estimates are to be used to identify the appropriate acceleration limit on the CSA scale. The equation below is specified in the standard for estimating the peak acceleration. The coefficients in the equation have been derived for an impulse function of 70 N·s. The reader is cautioned that the equations below, unlike those presented in Sec. 9.3, are unit-dependent. When using this criterion, the SI units specified should be used.

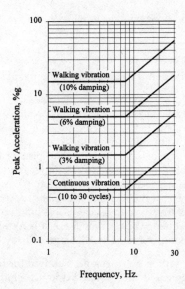

Fig. 9.8 CSA annoyance criteria for floor vibrations.[20]

CSA equation for peak acceleration:

$$a_o = \frac{60f}{WBL}$$

where f = frequency, Hz

For a simply supported one-way system,

$$f_1 = 156\sqrt{\frac{EI_t}{wL^4}}$$

where E = modulus of elasticity for steel, MPa
 I_t = moment of inertia for transformed tee-beam, assume composite action, even for noncomposite construction, mm^4
 w = dead load of the tee beam, N/mm of span
 L = span, mm

For one-way systems supported by steel girders,

$$\frac{1}{f^2} = \frac{1}{f_1^2} + \frac{1}{f_2^2}$$

where f_2 = frequency of the floor supported on steel girders (specifics for computing this frequency are not specified in the standard; the recommendations for computing f_g in Sec. 9.3 can be followed)

W = weight of the floor plus contents, kPa
L = span, m
B = width of equivalent beam, m

The following is quoted from the standard for the purposes of estimating the parameters above.

For steel joist or beam and concrete deck systems on stiff supports, L is the joist span and B can be approximated as $40t_c$ ($20t_c$ for edge panel adjacent to interior edge openings) where t_c is the thickness of the concrete deck as determined from the average mass of concrete, including ribs. For joists or beams and concrete deck supported on flexible girders, where the girder frequency is much lower than the joist and therefore girder vibration predominates, L is the girder span and B can be approximated as the width of the floor supported by the girder. For cases where the larger of the two frequencies is less than 1.5 times the smaller, or where the combined frequency is less than 5 Hz, it is recommended to use the design criterion for walking vibration in Allen and Murray (1993).[20]

After estimating the damping and peak acceleration for the floor system, the acceptability of floor design can be determined from the chart in Fig. 9.8.

9.10 Murray Criterion for Walking Vibrations

9.10.1 Summary of the Criterion

In the criterion presented by Murray,[7] an acceptable steel floor system is predicted, with respect to vibration levels due to walking excitation, if the dynamic criterion below is met. This criterion is applicable to offices and residences with fundamental natural frequencies below 10 Hz.

Murray criterion:

$$D > 35A_o f + 2.5$$

See the procedure that follows for design estimates of the parameters below.

D = damping in floor system, expressed as a percent of critical

A_o = maximum initial amplitude of the floor system due to a heel-drop excitation, in

f = first natural frequency of the floor system, Hz

This criterion is only applicable for the units specified. The reader is cautioned against using other units.

9.10.2 Design Procedure

The following steps outline a procedure for determining D, A_o, and f for a specific floor design.

Serviceability Considerations for Floor and Roof Systems ■ 9-21

1. Estimate the floor system damping.
2. Determine f and A_o for beam or joist.
3. Determine f and A_o for girder.
4. Determine f and A_o for system.
5. Evaluate the criterion.

A review of these five steps follows.

Step 1: Estimate Damping ■ The author recommends ranges for estimating the damping in a floor system. These estimates are outlined in Table 9.2.

Step 2: Determine f and A_o for the Beam or Joist ■ f_b=frequency of the beam or joist, Hz; see Sec. 9.3.3 for details.

$$A_{oj} = \frac{A_{ot}}{N_{eff}}$$

where A_{ot} = initial amplitude of a single tee beam due to a heel-drop impact; a heel-drop impact is estimated as a linearly decreasing ramp function having a maximum amplitude of 600 lb and a duration of 50 ms

$$A_{ot} = \text{DLF} * d_s$$

where DLF = dynamic load factor for a unit ramp function with a 50 ms duration, determined from the chart presented in Ref. 7 or by the equations below (the equations below can be derived from structural dynamics)

$$= 19.68 * 0.10 - \frac{\tan^{-1}(0.1\pi f)}{\pi f} \quad \text{for } f \geq 7.42 \text{ Hz}$$

Table 9.2 Estimates of Floor System Damping[7]

Component	Damping (% of critical)	Description
Bare floor	1–3%	Lower limit for thin slab of lightweight concrete; upper limit for thick slab of normal weight concrete
Ceiling	1–3%	Lower limit for hung ceiling; upper limit for sheetrock on furring attached to beams of joists
Ductwork and mechanical	1–10%	Depends on amount and attachment
Partition	10–20%	If attached to the floor system and not spaced more than every five floor beams of the effective joist floor width

$$= \frac{19.68}{2\pi f}\sqrt{2[1-(0.1\pi f)\sin(0.1\pi f)-\cos(0.1\pi f)]+(0.1\pi f)^2} \quad \text{for } f < 7.42 \text{ Hz}.$$

d_s = static displacement due to a 600-lb load at the center of the beam span.

$$= \frac{PL^3}{48EI_t} \quad \text{where } P=600 \text{ lb and } L, E, \text{ and } I_t \text{ are as defined in Sec. 9.3.3}$$

N_{eff} = number of effective tee-beams.

For beams or joists with spacing < 30 in,

$$N_{eff} = 1+2\sum\left(\cos\frac{\pi x}{2x_0}\right) \quad \text{for } x \leq x_0$$

where x = distance from the center joist to the joist under consideration, in
 x_0 = distance from the center joist to the edge of the effective floor, in
  = $1.06\varepsilon L$
 L = joist span, in
 ε = $(D_x/D_y)^{0.25}$
 D_x = flexural stiffness perpendicular to the joists
  = $E_c t_c^3/12$, where E_c is the modulus of elasticity for concrete and t_c is the concrete thickness (use the average thickness for ribbed decks)
 D_y = flexural stiffness parallel to the joists
  = EI_t/S where E and I_t are as defined previously in this section and S is the beam or joist spacing

For beams or joists with spacing > 30 in,

$$N_{eff} = 2.97 - \frac{S}{17.3t_c} + \frac{L^4}{1.35EI_t}$$

where the variables are as defined previously in this section.

Limitations:

$$15 \leq \frac{S}{t_c} < 40 \quad \text{and} \quad 1\times10^6 \leq \frac{L^4}{I_t} \leq 50\times10^6$$

Step 3: Determine f and A₀ for the Girders

f_g = frequency of the girder, Hz (see Sec. 9.3.3 for details)

$$A_{og} = \frac{A_{ot}}{N_{eff}}$$

Serviceability Considerations for Floor and Roof Systems ■ 9-23

where A_o = initial amplitude of girder due to a heel-drop impact (see step 2 of this procedure for details)

N_{eff} = 1.0 for girders

Step 4: Determine f and A_0 for the System

$$\frac{1}{f_s^2} = \frac{1}{f_b^2} + \frac{1}{f_g^2}$$

where f_s = estimated system frequency, Hz
f_j = joist or beam frequency, Hz
f_g = girder frequency, Hz

$$A_{os} = A_{oj} + \frac{A_{og}}{2}$$

where A_{os} = system amplitude, in
A_{oj} = joist or beam initial amplitude, in
A_{og} = girder initial amplitude, in

Step 5: Evaluate the Criterion. ■ A satisfactory floor system is predicted, with respect to walking excitation, if the following inequality is satisfied:

$$D > 35 A_o f_s + 2.5$$

where D is estimated in step 1 and A_o and f_s are estimated in step 4.

9.11 Allen and Murray Criterion for Walking Vibrations

Allen and Murray[13] present a design criterion which extends the restrictions of the previous criteria. This criterion is described below.

9.11.1 Summary of the Criterion

An acceptable steel footbridge or floor system is predicted with respect to vibration levels due to walking excitation if the dynamic criterion below is met. These two equations express the same criterion in two different formats. This criterion is applicable to offices, residences, churches, shopping malls, and footbridges with fundamental natural frequencies below 9 Hz. For floor systems with fundamental frequencies between 9 and 18 Hz, the stiffness criterion below, when applied in addition to the criterion below, will ensure adequate floor performance for walking vibrations. Above 18 Hz, the authors note that the stiffness criterion will govern the design and the dynamic criterion below need not be checked.

9-24 ■ Chapter Nine

Dynamic criterion:

$$\beta W \geq K \exp(-0.35 f_0) \quad \text{or} \quad f_0 \geq 2.86 \ln\left(\frac{K}{\beta W}\right)$$

Stiffness criterion:

$$\Delta_{conc} \leq 1 \text{ mm } (0.04 \text{ in})$$

Definition of variables:

K = occupancy constant, dependent on acceleration limits for the prescribed occupancy (see Table 9.3)

β = damping ratio, primarily dependent on nonstructural components and furnishings (see Table 9.3)

f_0 = fundamental natural frequency of the floor system (see procedure below)

W = equivalent mass weight of the floor system (see procedure below)

9.11.2 Design Procedure

The following steps outline a procedure for determining f_o and W for joist (or beam) and girder systems.

1. Determine joist (or beam) properties: joist and beam panel mode properties or joist and beam edge panel mode properties
2. Determine girder properties: girder mode properties or girder edge mode properties
3. Determine floor system properties: combined mode properties
4. Evaluate the performance of the floor

A review of these four steps follows

Step 1a: Joist and Beam Panel Mode Properties

f_j = frequency of the beam or joist, Hz (the expression below is derived in Sec. 9.3.2)

Table 9.3 Values of K and β for use in Allen and Murray Criterion[13]

Occupancy classification	K kN (kips)	β
Offices, residences, churches	58 (13.0)	0.03*
Shopping Malls	20 (4.5)	0.02
Footbridges	8 (1.8)	0.01

*0.05 for full-height partitions, 0.02 for floors with few nonstructural components (ceilings, ducts, partitions, etc.) as can occur in churches.

Serviceability Considerations for Floor and Roof Systems ■ 9-25

$$f_j = 0.18\sqrt{\frac{g}{\Delta_j}}$$

where g = acceleration of gravity
   = 386.4 in/s² = 9800 mm/s²

Δ_j = maximum static deflection of a beam or joist, relative to its support, due to the weight supported by the individual beam or joist.

$$\Delta_j = \frac{5wL^4}{384EI} \text{ for simply supported beams or joists with a uniformly distributed load}$$

where w = floor weight per unit length of beam; value should be the actual expected service load on the beam; overestimating this value can result in a nonconservative prediction of acceptability; a live load of 11 psf is suggested by the authors for office occupancies

I_t = moment of inertia; when the steel deck supporting the concrete rests directly on the beam or joist (connected by welds, screws, mechanical shear connectors, etc.), assume composite action between the steel member and the concrete slab; see Sec. 9.3.4 for more information on the computation of composite member properties

$$W_j = wB_jL_j$$

where w = floor weight per unit area (note that this w is different from that noted above); again, this value should be the actual expected service load on the beam; overestimating this value can result in a nonconservative prediction of acceptability; a live load of 11 psf is suggested by the authors for office occupancies

L_j = joist or beam span

B_j = effective joist panel width

$$B_j = 2\left(\frac{D_s}{D_j}\right)^{1/4} L_j \leq \tfrac{2}{3} B_{max}$$

where D_s = flexural rigidity per unit width in the slab direction
   = $\dfrac{E_c t_c^3}{12}$

E_c = estimated dynamic modulus of elasticity for concrete (the modulus found by current structural standards should be increased by a factor of 1.35)

t_c = concrete thickness (use average thickness for ribbed decks)

D_j = flexural rigidity per unit width in the joist direction

$$D_j = \frac{EI_t}{S}$$

where E = modulus of elasticity for steel
         I_t = moment of inertia for joist (same as that computed for the joist frequency above)
         S = joist or beam spacing
         B_{max} = total width of the floor perpendicular to the joists or beams

Note: Where beams or joists are continuous over their supports, and an adjacent span (on the other side of the girder) is $0.7L_j$ or greater, the effective joist panel weight W_j can be increased by 50 percent. This increase is justified because the continuity over the supports engages the participation of adjacent floor panels. Wyatt[21] suggests further increases as follows: 70 percent increase where the adjacent span is $0.8\,L_j$ and 100 percent increase when the span is $1.0\,L_j$.

Step 1b: Joist and Beam Edge Panel Mode Properties. ■ Unsupported edges of floors need special consideration regarding the computation of the equivalent mass W. A distinction is made between exterior floor edges and interior floor edges. Exterior floor edges are rarely a problem for two reasons. First, they are generally stiffened by the exterior cladding and, second, walkways are not generally located near exterior walls. Interior edges (e.g., around an atrium, etc.) are often problematic with respect to excessive floor vibration due to pedestrian movement. When an interior edge is supported by a beam or joist, the effective weight W_j of the joist panel is computed by the expression below.

$$W_j = wB_jL_j$$

where B_j = effective joist panel width
            = $1\left(\dfrac{D_s}{D_j}\right)^{1/4} L_j \le \tfrac{1}{3}B_{max}$

Note: All undefined variables for the edge panel effective weight are the same as defined in step 1a.

Step 2a: Girder Panel Mode Properties

$$f_g = 0.18\sqrt{\frac{g}{\Delta_g}} \quad \text{approximated as a uniformly distributed load}$$

where Δ_g = maximum static deflection of girder, relative to its support, due to the weight supported by the girder

$$\Delta_g = \frac{5wL^4}{384EI_t} \quad \text{for simply supported girders}$$

Serviceability Considerations for Floor and Roof Systems ■ 9-27

where w = uniformly distributed floor weight per unit length of girder; value should be the actual expected service load on the floor; overestimating this value can result in a nonconservative prediction of acceptability; a live load of 11 psf is suggested by the authors for office occupancies

I_t = moment of inertia; when the steel deck supporting the concrete rests directly on the girder (connected by welds, screws, mechanical shear connectors, etc.), assume composite action between the steel member and the concrete slab; see Sec. 9.3.4 for more information on the computation of composite member properties

$$W_g = w B_g L_g$$

where w = floor weight per unit area (note that this w is different from that noted above); again, this value should be the actual expected service load on the beam; overestimating this value can result in a non-conservative prediction of acceptability; a live load of 11 psf is suggested by the authors for office occupancies

L_g = girder span

B_g = effective girder panel width as calculated below with a lower limit equal to the tributary panel width supported by the girder

$$B_{min} \leq B_g = 1.6 \left(\frac{D_j}{D_g} \right)^{1/4} L_g \leq \tfrac{2}{3} B_{max}$$

where D_j = flexural rigidity per unit width in the joist direction

$$= \frac{E I_{tj}}{S}$$

E = modulus of elasticity for steel

I_{tj} = moment of inertia for joist (same as computed for joist frequency above)

S_b = joist or beam spacing

D_g = flexural rigidity per unit width in the joist direction

$$= \frac{E I_{tg}}{S}$$

E = modulus of elasticity for steel

I_{tg} = moment of inertia for girder (same as computed for girder frequency above)

S_g = girder spacing

B_{min} = tributary panel width supported by the girder

B_{max} = total width of the floor perpendicular to the girder

Notes: 1. Where girders are continuous over their supports, and an adjacent span is $0.7L_j$ or greater, the effective girder panel weight W_g can be increased by 50 percent.

2. The 1.6 factor in the equation for B_g takes into account the discontinuity of joist systems over their supports. For a system that consists of rolled beams the factor 1.6 can be increased to 1.8.

Step 2b: Girder Edge Panel Mode Properties ▪ The general concept of an unsupported edge was discussed under step 1b. When an interior edge is supported by a girder, the effective weight W_g of the girder panel mode is equal to the tributary weight supported by the girder. Other variables are as defined in step 2a.

Step 3: Combined Mode Properties

f_0 = fundamental natural frequency of the floor system (estimated by the Dunkerly relationship)

$$f_0 = 0.18 \sqrt{\frac{g}{\Delta_j + \Delta_g}}$$

W = equivalent mass

$$W = \frac{\Delta_j}{\Delta_j + \Delta_g} W_j + \frac{\Delta_g}{\Delta_j + \Delta_g} W_g$$

Note: g, Δ_j, Δ_g, W_j, and W_g are as defined in steps 1 and 2 with the following exception: If $L_g < B_j$, Δ_g must be modified as follows:

$$\Delta_g = \frac{L_g}{B_j} \Delta_g \quad \text{where } 0.5 \leq \frac{L_g}{B_j} \leq 1.0$$

Step 4: Evaluation of Floor Performance ▪ $f_0 \leq 9$ Hz: For floors with a first natural frequency less than or equal to 9 Hz, the floor is predicted to be satisfactory with respect to walking excitations if the following inequality is satisfied:

$$\beta W \geq K \exp(-0.35 f_0) \quad \text{or} \quad f_0 \geq 2.86 \ln\left(\frac{K}{\beta W}\right)$$

9 Hz $< f_0 \leq 18$ Hz: For floors with a first natural frequency greater than 9 Hz and less than or equal to 18 Hz, the floor is predicted to be satisfactory with respect to walking excitations if the following inequalities are satisfied.
Dynamic criterion:

$$\beta W \geq K \exp(-0.35 f_0) \quad \text{or} \quad f_0 \geq 2.86 \ln\left(\frac{K}{\beta W}\right)$$

Stiffness Criterion: $\Delta_{\text{conc}} \leq 1$ mm(0.04 in). Δ_{conc} is the maximum static deflection due to a concentrated force of 1 kN (225 lb); Δ_{conc} is calculated using the same properties (i.e., composite properties where appropriate) as the dynamic criterion.

Serviceability Considerations for Floor and Roof Systems ■ 9-29

$f_0 > 18$ Hz: For floors with a first natural frequency greater than 18 Hz, the floor is predicted to be satisfactory with respect to walking excitations if the stiffness criterion ($\Delta_{conc} \leq 1$ mm) defined above is satisfied.

9.12 Vibration Considerations for Rhythmic Excitations

Activities such as dancing, foot stomping, jumping exercises, aerobics, and marching can be particularly problematic with respect to floor serviceability requirements. Many researchers have been involved in quantifying the effects of rhythmic activities on floor systems. A summary of the most widely accepted analytical procedure and design criterion for this type of load is presented in the following sections.

9.12.1 Dynamic Analysis for Rhythmic Activities

The rhythmic activities described above subject floor systems to periodic forces with fundamental frequencies in the 1- to 4-Hz range. These forces can be approximated by a sinusoidal dynamic load or, in the case of jumping and aerobics, by a series of sinusoidal dynamic loads as shown below.

Dynamic Forces Due to Rhythmic Activities[22]

$$F(t) = \sum_{i=1}^{n}\left[\alpha_i w_p \sin(2\pi i f_f t + \varphi_n)\right]$$

where α = dynamic load factor for specified activity (see Table 9.4 for recommended values)

w_p = weight of the participants (see Table 9.4 for recommended values)

f_f = forcing frequency of the dynamic event (see Table 9.4 for typical ranges)

Table 9.4 Parameters for Assessing Rhythmic Excitations[3]

	Estimated loading during rhythmic events		
Activity	Forcing frequency, f_f, Hz	Weighting of participants W_p, psf	Dynamic Load Factor, α
Dancing	1.5 – 3	12.5 (27 sq. ft./couple)	0.5
Lively concert or sports event	1.5 – 3	31.3 (5 sq. ft./person)	0.25
Jumping exercises:		4.2 (42 sq. ft./person)	
First harmonic	2 – 2.75		1.5
Second harmonic	4 – 5.5		0.6
Third harmonic	6 – 8.25		0.1

i = harmonic multiple of the forcing frequency
n = number of terms in the series to approximate the forcing function
  = 1 for dancing, lively concert, or sporting event
  = 3 for jumping exercises or aerobics
φ_n = phase angle

The maximum dynamic amplitude of the floor system can be determined for each harmonic multiple of the dynamic force from the steady state solution as expressed below.

Steady-State Response Due to Sinusoidal Loading[22]

$$\frac{a_i}{g} = \frac{1.3 \alpha_i w_p / w_t}{\sqrt{\left[\left(\frac{f_0}{i \cdot f_f}\right)^2 - 1\right]^2 + \left(2\beta \frac{f_0}{i \cdot f_f}\right)^2}}$$

where a_i = peak acceleration due to the ith harmonic of the loading function
 w_t = total floor weight
 f_0 = fundamental natural frequency of the floor system

In the case of jumping and aerobics, the maximum dynamic amplitude can be determined from the expression below.

Combining Harmonic Contributions [9]

$$a_{max} = \left[\Sigma(a_i^{1.5})\right]^{\frac{1}{1.5}}$$

Finally, the maximum acceleration, as determined from the dynamic analysis presented above, can be evaluated using the recommended acceleration limits presented in Table 9.5.

Table 9.5 Acceleration Limits Recommended by NBC[3]

Recommended acceleration limits for vibrations due to rhythmic activities	
Occupancies affected by the vibration	Acceleration limit, % gravity
Office and residential	0.4 – 0.7
Dining and weightlifting	1.5 – 2.5
Rhythmic activity only	4 – 7

Serviceability Considerations for Floor and Roof Systems ■ 9-31

Care should be taken when evaluating the maximum acceleration expression for design purposes. In particular, the possibility of resonance must be considered in selecting if_f. Additionally, if $f_0 \leq if_f$, higher-order vibration modes of the floor system must be considered because the possibility of resonance with these modes also exists. These considerations and cautions are implicit in the design criterion presented in the following section.

9.12.2 Design Criterion for Rhythmic Excitations

An acceptable floor design is predicted, with respect to annoying levels of floor vibration, if the inequality below is satisfied.

Design Criterion for Rhythmic Excitations[3]

$$f_0 \geq if_f \sqrt{1 + \frac{K}{a_o/g} \frac{a_i w_p}{w_t}}$$

f_0, I, f_f, α_i, w_p, w_t are as defined in Sec. 9.12.1

a_0/g = recommended acceleration limit for the given occupancy (see Table 9.5; the lower value is recommended for design)

K = 1.3, except for jumping and aerobics; 2.0 for jumping and aerobics; the increased K factor accounts for the increased amplitude due to multiple harmonics contributing to the response.

One should note that aerobic and jumping excitations require that three separate cases of the inequality above need to be satisfied (i.e., $i=1$, $i=2$, $i=3$).

In the case of very heavy, highly damped floor systems, resonance may not produce unacceptable levels of floor vibration, rendering the criterion above unnecessarily conservative. The analytical procedure of the previous section should be used to address this possibility.

9.13 Design Examples

The calculations in the examples that follow were performed using FLOORVIB,[23] software by StructuralEngineers, Inc.

9.13.1 Murray Criterion for Walking Vibrations

This example illustrates the use of the Murray criterion[7] presented in Sec. 9.10. The floor system in Fig. 9.9 is evaluated for an office or residential occupancy subjected to walking excitations.

Step 1: Estimate Damping ■ From Table 9.2,

Bare floor	2%
Ceiling	2%
Mechanical, etc.	1%
Total	5% = D

9-32 ■ Chapter Nine

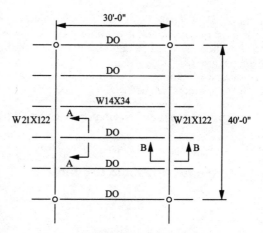

PLAN OF INTERIOR BAY
Adjacent bays similar

Concrete Properties:	Beam Properties:	Girder Properties:
Concrete Weight = 145 pcf	W14X34	W21X122
Concrete Strength = 4 ksi	$A = 10 \text{ in}^2$	$A = 35.9 \text{ in}^2$
Ec = 3492 ksi	d = 13.98 in	d = 21.68 in
Modular Ratio = n = 8.3	$Ix = 340 \text{ in}^4$	$Ix = 2960 \text{ in}^4$

Fig. 9.9 Floor system A.

Step 2: Determine f and A_0 for the Beam or Joist

$$f_b = K\sqrt{\frac{gEI_t}{wL^4}} = 1.57\sqrt{\frac{(386.4)(29{,}000)(1263)}{(0.5732)(30)^4(1728)}} = 6.59 \text{ Hz}$$

where K = 1.57 for simply supported beams

w = floor weight per unit length

Loading data:

Slab + 1 psf deck = 52.4 psf

Superimposed dead load = 4.0 psf

Live load = 11.0 psf

Beam weight = 34.0 plf

$w = (52.4+4.0+11.0)(8 \text{ ft})+34.0 = 573.2 \text{ plf} = 0.5732 \text{ klf}$

$E = 29{,}000$ ksi for steel

$I_t = 1263 \text{ in}^4$ (see Fig. 9.10 for calculations of transformed section properties)

Serviceability Considerations for Floor and Roof Systems ■ 9-33

$$A_{0j} = \frac{A_{0t}}{N_{eff}} = \frac{0.0144}{2.0} = 0.0072 \text{ in}$$

$$A_{0t} = \text{DLF} * d_s = (0.903)(0.0159) = 0.0144 \text{ in}$$

For $f < 7.42$ Hz,

$$\text{DLF} = \frac{19.68}{2\pi f} \sqrt{2\left[1 - (0.1\pi f)\sin(0.1\pi f) - \cos(0.1\pi f)\right] + (0.1\pi f)^2}$$

Transformed Beam Section Properties:

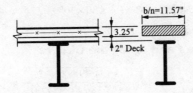

Section A-A   Transformed Section

$$\bar{y} = \frac{(11.57)(3.25)(3.25/2) + (10)(12.24)}{(11.57)(3.25) + 10} = 3.855 \text{ in}^4$$

$$I_x = \frac{(11.57)(3.25)^3}{12} + (11.57)(3.25)(2.23)^2 + 340 + (10)(8.385)^2 = 1263 \text{ in}^4$$

Transformed Girder Section Properties:

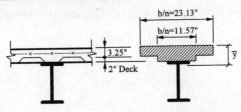

Section B-B   Transformed Section

$$b = 192.0 \text{ in} = \min \begin{cases} 0.4 L_g = 0.4(40 \text{ ft})(12 \text{ in/ft}) = 192.0 \text{ in} \\ \text{tributary width} = (30 \text{ ft})(12 \text{ in/ft}) = 360.0 \text{ in} \end{cases}$$

$$\bar{y} = \frac{(23.13)(3.25)(3.25/2) + (11.57)(2.0)(4.25) + (35.9)(16.09)}{(23.13)(3.25) + (11.57)(2.0) + 35.9} = 5.947 \text{ in}^4$$

$$I_x = \frac{(23.13)(3.25)^3}{12} + (23.13)(3.25)(4.322)^2 + \frac{(11.57)(2.0)^3}{12} + (11.57)(2.0)(1.697)^2$$
$$+ 2960 + (35.9)(10.14)^2 = 8196 \text{ in}^4$$

Fig. 9.10 Transformed section properties for floor A.

$$= \frac{19.68}{2\pi \cdot 6.59} \sqrt{2[1-(0.1\pi \cdot 6.59)\sin(0.1\pi \cdot 6.59) - \cos(0.1\pi \cdot 6.59)] + (0.1\pi \cdot 6.59)^2}$$
$$= 0.903$$
$$d_s = \frac{PL^3}{48EI_t} = \frac{600(30)^3(1728)}{48(29,000,000)(1263)} = 0.0159 \text{ in}$$

For beams or joists with spacing > 30 in

$$N_{eff} = 2.97 - \frac{S}{17.3t_c} + \frac{L^4}{1.35EI_t} = 2.97 - \frac{96.0}{(17.3)(4.25)} + \frac{(30 \cdot 12)^4}{1.35(29,000,000)(1263)} = 2.00$$

Limitations: $15 \leq \frac{S}{t_c} < 40$; $1 \times 10^6 \leq \frac{L^4}{I_t} \leq 50 \times 10^6$

Step 3: Determine f and A_o for the Girders

$$f_b = K\sqrt{\frac{gEI_t}{wL^4}} = 1.57\sqrt{\frac{(386.4)(29,000)(8197)}{(2.056)(40)^4(1728)}} = 4.99 \text{ Hz}$$

where K = 1.57 for simply supported beams
 w = floor weight per unit length
 w = $\frac{(4.5)(573.2)(30)}{40.0} + 122 = 2056 \text{ plf} = 2.056 \text{ klf}$
 E = 29,000 ksi for steel
 I_t = 8197.4 in^4 (see Fig. 9.10 for calculations of transformed section prop

$$A_{0g} = \frac{A_{0t}}{N_{eff}} = \frac{0.0042}{1.0} = 0.0042 \text{ in}$$

$A_{0t} = \text{DLF} * d_s = (0.720)(0.00581) = 0.0042 \text{ in}$

For $f < 7.42$ Hz,

$$\text{DLF} = \frac{19.68}{2\pi f}\sqrt{2[1-(0.1\pi f)\sin(0.1\pi f) - \cos(0.1\pi f)] + (0.1\pi f)^2}$$
$$= \frac{19.68}{2\pi \cdot 4.99}\sqrt{2[1-(0.1\pi \cdot 4.99)\sin(0.1\pi \cdot 4.99) - \cos(0.1\pi \cdot 4.99)] + (0.1\pi \cdot 4.99)^2}$$
$$= 0.720$$
$$d_s = \frac{PL^3}{48EI_t} = \frac{600(40)^3(1728)}{48(29,000,000)(8197)} = 0.00581 \text{ in}$$

$N_{eff} = 1.0$ for girders

Serviceability Considerations for Floor and Roof Systems ■ 9-35

Step 4: Determine f and A_0 for the System

$$\frac{1}{f_s^2} = \frac{1}{f_b^2} + \frac{1}{f_g^2} = \frac{1}{6.59^2} + \frac{1}{4.99^2} = \frac{1}{3.98^2}$$

$f_s = 3.98$ Hz

$$A_{0s} = A_{0j} + \frac{A_{0g}}{2} = 0.0072 + \frac{0.0042}{2} = 0.0093 \text{ in}$$

Step 5: Evaluate the Criterion

■ A satisfactory floor system is predicted, with respect to walking excitation, if the following inequality is satisfied for the beam, the girder, and the system.

For the beam: $35A_0f_s + 2.5 = 35(0.0072)(6.59) + 2.5 = 4.2 < 5$ percent as estimated in step 1.

For the system: $35A_0f_s + 2.5 = 35(0.0042)(4.99) + 2.5 = 3.2 < 5$ percent.

For the system: $35A_0f_s + 2.5 = 35(0.0093)(3.98) + 2.5 = 3.8 < 5$ percent.

The floor satisfies the criterion.

9.13.2 Modified Reiher-Meister Scale for Transient Vibrations

An evaluation of the floor system in Fig. 9.9 can be made using the modified Reiher-Meister scale with the amplitude and frequency calculated in Sec. 9.13.1.

From Sec. 9.13.1,

$$f_s = 3.98 \text{ Hz}$$

$$A_0 = 0.0093 \text{ in}$$

The point defined by the values above plots below the line labeled "Murray 1975" in Fig. 9.5. The floor is therefore predicted to be acceptable for an office or residential occupancy.

9.13.3 Allen and Murray Criterion for Walking Vibrations

This example illustrates the use of the Allen and Murray criterion[13] presented in Sec. 9.11. The floor system in Fig. 9.9 is evaluated for an office or residential occupancy subjected to walking excitations.

Step 1a: Joist and Beam Panel Mode Properties

$$\Delta_b = \frac{5wL^4}{384EI_t} = \frac{5(0.5732)(30)^4(1728)}{384(29,000)(1263)} = 0.285 \text{ in}$$

where w = floor weight per unit length

Loading data:

> Slab+1 psf deck=52.4 psf
> Superimposed dead load=4.0 psf
> Live load=11.0 psf
> Beam weight=34.0 plf
> $w = (52.4+4.0+11.0)(8 \text{ ft}) + 34.0 = 573.2 \text{ plf} = 0.5732 \text{ klf}$
>
> $E = 29{,}000$ ksi for steel
> $I_t = 1263$ in^4 (see Fig. 9.10 for calculations of transformed section properties)
> $L = 30$ ft, beam span
>
> $$W_b = 1.5 w B_b L_b = 1.5(0.0716)(31.8)(30) = 102.5 \text{ kips}$$
>
> 1.5 = 50 percent increase due to beam continuity over supports
>
> $$w = \frac{0.5732}{8 \text{ ft}} = 0.0716 \text{ ksf}$$
>
> $$B_b = 2\left(\frac{D_s}{D_b}\right)^{0.25} L_b \leq \frac{2}{3} B_{max} = 80 \text{ ft}$$
>
> $$= 2(30{,}157 / 381{,}531)^{0.25}(30) = 31.8 \text{ ft}$$
>
> $$D_s = \frac{E_c t_c^3}{12} = 4714(4.25^3)/12 = 30{,}157$$
>
> $E_c = 1.35(3492) = 4714$ ksi
>
> $$D_b = \frac{EI_t}{S} = 29{,}000(1263)/96 \text{ in} = 381{,}531$$
>
> $I_t = 1263$ in^4 from Fig. 9.10
> $B_{max} = 3(40 \text{ ft}) = 120 \text{ ft}$

Step 2a: Girder Panel Mode Properties (Δ_g and W_g)

$$\Delta_g = \frac{5wL^4}{384EI_t} = \frac{5(2.056)(40)^4(1728)}{384(29{,}000)(8197)} = 0.498 \text{ in}$$

where w = floor weight per unit length

$$w = \frac{(4.5)(573.2)(30)}{40.0} + 122 = 2056 \text{ plf} = 2.056 \text{ klf}$$

$E = 29{,}000$ ksi for steel
$I_t = 8197.4$ in^4 (see Fig. 9.10 for calculations of transformed section properties)

Serviceability Considerations for Floor and Roof Systems ■ 9-37

$L = 40$ ft

$W_g = 1.5 w B_g L_g = 15(0.0685)(64.8)(40) = 258.0$ kips
where $1.5 = 50$ percent increase due to girder continuity over supports

$w = \dfrac{2.056}{30 \text{ ft}} = 0.0685$ ksf

$B_{\min} \leq B_g = 1.8 \left(\dfrac{D_b}{D_g} \right)^{0.25} \quad L_g \leq \dfrac{2}{3} B_{\max} = 60$ ft

$B_g = 1.8 \left(\dfrac{381{,}531}{660{,}314} \right)^{0.25} (40) = 62.8$ ft

$B_{\min} = 30$ ft

$B_{\max} = 3(30) = 90$ ft

$D_g = \dfrac{EI_t}{S} = \dfrac{29{,}000(8197)}{360 \text{ in}} = 660314$

$I_t = 8197$ in^4 from Fig. 9.10

$B_{\max} = 3(40 \text{ ft}) = 120$ ft

$L_g = 40$ ft

Step 3: System Parameters

$f_0 = 0.18 \sqrt{\dfrac{g}{\Delta_b + \Delta_g}} = 0.18 \sqrt{\dfrac{386.4}{0.285 + 0.498}} = 3.99$ Hz

Since $L_g = 40$ ft > 31.8 ft, no modification to Δ_g is necessary.

$W = \dfrac{\Delta_b}{\Delta_b + \Delta_g} W_b + \dfrac{\Delta_g}{\Delta_b + \Delta_g} W_g$

$= \dfrac{0.285}{0.285 + 0.498}(102.5) + \dfrac{0.498}{0.285 + 0.498}(258.0) = 201.4$ kips

Step 4: Evaluation of Floor Performance ■
For $f_0 < 9$ Hz, an acceptable floor system is predicted, with respect to walking excitation, if the following inequality is satisfied.

$$f_0 \geq 2.86 \ln\left(\dfrac{K}{\beta W} \right) = f_{\text{req'd}}$$

$f_0 = 3.99$ Hz as computed in step 3

$f_{\text{req'd}} = 2.86 \ln\left[\dfrac{13.0}{(0.03)(201.4)} \right] = 2.19$ Hz where K = 13.0, occupancy constant for office or residential uses (see Table 9.3)

β = 0.03, recommended damping value from Table 9.3

The criterion is satisfied; therefore, the floor system is predicted to be acceptable for the intended occupancy.

9.13.4 Design Evaluation for Rhythmic Excitations

This example illustrates the use of the design criterion presented in Sec. 9.12.2 to assess a floor subjected to rhythmic excitations. The floor system defined in Fig. 9.9 is evaluated for a mixed occupancy of aerobics and weightlifting.

For aerobics, three separate cases must be evaluated for the following criterion:

$$f_0 \le i \cdot f_f \sqrt{1 + \frac{K}{a_o/g} \frac{a_i w_p}{w_t}} = f_{\text{req'd}} \quad \text{for } i = 1, 2, \text{ and } 3$$

Step 1: Evaluation of Floor System Frequencies

Calculate the fundamental natural frequency of the beam as follows:

$$f_b = K\sqrt{\frac{gEI_t}{wL^4}} = 1.57\sqrt{\frac{(386.4)(29{,}000)(1263)}{(0.519)(30)^4(1728)}} = 6.93 \text{ Hz}$$

where K = 1.57 for simply supported beams

w = floor weight per unit length

Loading data:

Slab+1 psf deck=52.4 psf

Superimposed dead load=4.0 psf

Live load=4.2 psf (see Table 9.4)

Beam weight=34.0 plf

w=(52.4+4.0+4.2)(8 ft)+34.0=519 plf=0.519 klf

E=29,000 ksi for steel

I_t=1263 in^4 (see Fig. 9.10 for calculations of transformed section properties)

Calculate the fundamental natural frequency of the girder as follows:

$$f_g = K\sqrt{\frac{gEI_t}{wL^4}} = 1.57\sqrt{\frac{(386.4)(29{,}000)(8197)}{(1.873)(40)^4(1728)}} = 5.23 \text{ Hz}$$

where K = 1.57 for simply supported beams

w = floor weight per unit length

Serviceability Considerations for Floor and Roof Systems ■ 9-39

$$w = \frac{(4.5)(518.8)(30)}{40.0} + 122 = 1873 \text{ plf} = 1.873 \text{ klf}$$

$E = 29{,}000$ ksi for steel
$I_t = 8197.4$ in^4 (see Fig. 9.10 for calculations of transformed section properties)

Calculate the fundamental natural frequency of the floor system as follows:

$$\frac{1}{f_o^2} = \frac{1}{f_b^2} + \frac{1}{f_g^2} = \frac{1}{6.93^2} + \frac{1}{5.23^2} = \frac{1}{4.17^2}$$

$f_0 = 4.17$ Hz

Step 2: Determine of Remaining Parameters

$f_f = 2.5$ Hz estimated forcing frequency for first harmonic of jumping exercises (see Table 9.4)

$\dfrac{a_0}{g} = 0.02$ recommended acceleration limit for weightlifting occupancy from Table 9.5

$K = 2.0$ for jumping and aerobics
$w_p = 4.2$ psf weight of participants for jumping exercises (see Table 9.4)
$w_t = 518.8$ plf / 8 ft = 64.8 psf; floor weight per square foot
$\alpha_i = 1.5, 0.6,$ and 0.1 for $i = 1, 2,$ and 3, respectively

Step 3: Evaluation of the Criterion ■
A satisfactory floor system is predicted, for a floor subjected to jumping activities, if the following inequality is satisfied.

$f_0 \geq f_{\text{req'd}}$ for $i = 1, 2,$ and 3

For $i = 1$,

$$f_{\text{req'd}} = (1)(2.5)\sqrt{1 + \frac{2.0}{0.02}\frac{1.5(4.2)}{64.8}} = 8.19 \text{ Hz}$$

For $i = 2$,

$$f_{\text{req'd}} = (2)(2.5)\sqrt{1 + \frac{2.0}{0.02}\frac{0.6(4.2)}{64.8}} = 11.06 \text{ Hz}$$

For $i = 3$,

$$f_{\text{req'd}} = (3)(2.5)\sqrt{1 + \frac{2.0}{0.02}\frac{0.1(4.2)}{64.8}} = 9.63 \text{ Hz}$$

$$f_{\text{req'd}}(\max) = 11.06 \text{ Hz} > f_0 = 4.17 \text{ Hz}$$

The floor is unacceptable for a weightlifting occupancy combined with jumping activities.

9-40 ■ Chapter Nine

References

1. AISC, *Manual of Steel Construction, Load and Resistance Factor Design*, American Institute of Steel Construction, Chicago, Ill, 1994.

2. Ad Hoc Committee on Serviceability Research, "Structural Serviceability: A Critical Appraisal of Research Needs," *Journal of Structural Engineering, ASCE*, vol. 112, no. 12, pp. 2646–2664, 1986.

3. National Research Council of Canada, *Supplement to the National Building Code of Canada*, Ottawa, Canada, 1990.

4. NBC, *Supplement to the National Building Code of Canada*, Chapter 4, Commentary A, National Research Council of Canada, Ottawa, Ont., 133–140, 1990.

5. CSA, *Steel Structures for Buildings—Limit States Design*, CAN/CSA-S16.1-M89, Canadian Standards Association, Rexdale, Ont., Canada, 1989.

6. Pernica, G., "Dynamic Load Factors for Pedestrian Movements and Rhythmic Exercises," *Canadian Acoustics*, vol. 18, no. 2, pp. 3–18, 1990.

7. Murray, T. M., "Building Floor Vibrations," *Engineering Journal, AISC*, vol. 29, no. 3, pp. 102–109, 1991.

8. Hanes, R. M., *Human Sensitivity to Whole Body Vibration: A Literature Review*, Silver Spring, Md., Applied Physics Laboratory, The Johns Hopkins University, 1970.

9. Allen, D. E., "Building Vibrations from Human Activities," *Concrete International Design and Construction*, vol. 12, no. 6, pp. 66–73, 1990.

10. Webster, A. C., and R. Vaicaitis, "Application of Tuned Mass Dampers to Control Vibrations of Composite Floor Systems," *Engineering Journal, AISC*, vol.29, no. 3, pp. 116–124, 1992.

11. Murray, T. M. and W. E. Hendrick, "Floor Vibrations and Cantilevered Construction," *Engineering Journal, AISC*, vol. 14, no. 3, 1977.

12. Lenzen, Kenneth H., "Vibration of Steel Joist-Concrete Slab Floors," *Engineering Journal, AISC*, vol. 3, no. 3, pp. 133–136, 1966.

13. Allen, D. E., and T. M. Murray, "Design Criterion for Walking Vibrations," *Engineering Journal, AISC*, vol. 31, no. 3, pp. 117–129, 1993.

14. Hanagan, L. M., *Active Control of Floor Vibrations*, Ph.D. Dissertation, Technical Report CE/VPI-ST 94/13, Charles E. Via Department of Civil Engineering, Virginia Polytechnic Institute and State University, Blacksburg, Va., 1994.

15. Reiher, H., and F. J. Meister, "The Effect of Vibration on People" (in German). Forschung auf dem Gebiete des Ingenieurwesens, vol. 2, no. II, p. 381. (Translation: Report F-TS-616-RE H.Q. Air Material Command, Wright Field, Ohio), 1931.

16. Murray, T. M., "Design to Prevent Floor Vibrations," *Engineering Journal, AISC*, vol. 12, no. 3, pp. 82–87, 1975.

17. Wiss, J. F., and R. A. Parmelee, "Human Perception of Transient Vibrations," *Journal of the Structural Division, ASCE*, vol. 100(ST4), pp. 773–787, 1974.

18. ISO, International Standards ISO 10137, "Basis for the Design of Structures—Serviceability of Buildings Against Vibration," International Standards Organization, pp. 41-43, 1992.

19. Ungar, E. E., D. H. Sturz, and C. H. Amick, "Vibration Control Design of High Technology Facilities," *Sound and Vibration*, July 1990, pp. 20–27, 1990.

20. Canadian Standards Association, *Steel Structures for Buildings—Limit States Design, Appendix G: Guide for Floor Vibrations*, Canadian Standards Association, CAN/CSA-S16.1-94, Rexdale, Ont., pp. 116–123, 1994.

21. Wyatt, T. A., *Design Guide on the Vibration of Floors*, The Steel Construction Institute Silwood Park, Ascot, Berkshire, 1989.

22. Allen, D. E., J. H. Rainer, and G. Pernica, "Vibration Criteria for Assembly Occupancies" *Canadian Journal of Civil Engineering*, vol. 12, no. 3, pp. 617–623, 1985.

23. Murray, T. M., "FLOORVIB User's Manual," Structural Engineers, Inc., Radford, VA 1993.

10

Ali A. K. Haris

Composite Design

10.1 Introduction

To the structural engineer, the term "composite construction" is typically applied to systems in which hot-rolled-steel or precast-concrete structural shapes are mechanically interlocked with cast-in-place concrete. Composite construction is comprised of a variety of specific forms including composite beams, composite trusses, composite columns, and composite walls. This chapter is limited to a discussion of the most common forms of composite construction, composite beams and columns fabricated using rolled-steel shapes. Further, the discussion is limited to applications in building construction.

In composite beams, a concrete floor slab is mechanically connected to a steel shape. For beams in positive bending, the floor slab acts as a compression flange for the steel beam, thereby allowing the composite beam to have both greater strength and stiffness than the steel beam alone. Compared with noncomposite construction, therefore, the structural-steel section is usually lighter when using a composite design. In most applications, the slab concrete is cast on a ribbed metal deck which is placed on the top flange of the steel shape. The metal deck has the dual functions of providing a permanent formwork and acting as a positive moment reinforcement for the slab. The use of the metal deck allows rapid, unshored construction of the building slabs and minimizes the amount of reinforcing steel required in the slab.

For construction in the United States, the shear connection between the slab and the steel shape is typically provided by headed studs welded through the metal deck. Although the installation of the studs requires the use of specialized welding equipment and material, the installation costs are relatively low when the savings gained by reduced structural-steel weights are considered. For construction in Europe and Canada, other types of shear connections are used, including connectors installed with powder-actuated pins and welded connectors that interlock with the slab reinforcement. Practical considerations, however, limit the discussion and sample problems in this chapter to this most common form of construction in the United States.

Prior to the introduction of the composite deck and weld-through studs in the 1960s, composite floor systems were constructed using removable formwork, similar in construction to that used for reinforced-concrete floors. For fire protection, concrete was used to encase the steel shape. The metal-deck system now used does not allow the encasement of the steel section in concrete. Therefore, fire protection of the section is provided by using a rated ceiling or, more commonly, by spraying the steel section with a cementlike, expanding material with no structural properties. Because of the expense of spraying the underside of the metal deck, designers normally select a minimum concrete deck thickness that allows a fire-rated slab system not requiring spraying a fireproofing material on the underside of the metal deck. The structural-steel beams, however, have to be sprayed for fireproofing. Figure 10.1 is a 35-story building under construction, one of the first major buildings built in the late 1980s utilizing the LRFD method for all composite beams and steel columns.

For the design of composite beams, the designer is concerned with the service limit states as well as ultimate limit states. The service limit states for a floor system include tolerable levels of vibration, deflection, and slab cracking. The ultimate limit states include the stability and strength of the noncomposite steel section (before the concrete reaches reasonable strength) as well as the stability and strength of the composite system.

Many assumptions used in the design of composite beams are analogous to those used in reinforced-concrete design. The service properties are calculated by using an effective moment of inertia that will account for slip, or partial composite action, between the steel section and the concrete slab. These properties can be calculated using modified procedures of basic mechanics, which have been shown to provide good agreement with experimental data. The strength of the composite system is calculated using a rigid, perfectly plastic material model for the steel and the uniform compression stress block for the concrete.

Fig. 10.1 312 Walnut, Cincinnati, Ohio: Composite structure designed using the LRFD method. (Courtesy of Duke Realty.)

The encasement of a wide-flange steel column with concrete provides several advantages, such as fireproofing the steel, providing an architecturally finished surface, and adding structural value to the column's strength and stiffness. In most cases, it is less costly to spray the column with a cementlike, expanding material than to encase the steel section in concrete to provide fireproofing.

In tall buildings, composite columns can be an economical solution for both strength and lateral stiffness requirements. The steel section is designed to carry only the weight of a few floors, while the total concrete-steel composite section carries the final total loads. This method enables the structural-steel construction crew to continue the erection of the steel without interruption from the concreting crew.

A concrete-filled steel tube has been used in multistory buildings to achieve the additional strength and stiffness of the composite column. This type of composite column may be economical in proper applications by eliminating costly concrete form work.

10.2 Composite Beams

A number of parameters must be considered in the design of composite beams. The slab thickness, concrete type and strength, gauge of the steel deck, beam spacing, slab reinforcement, steel section size and grade, and quantity of studs must be set by the engineer. Figure 10.2 shows a typical composite beam with metal deck ribs parallel to the steel beams.

The slab thickness and concrete type are often dictated by fire-rating requirements. These requirements are set by the governing model building code and reflected in tabulated information provided by deck manufacturers. For typical designs, the engineer sets these parameters so as to preclude the need for sprayed-on fireproofing of the slab system.

The gauge and depth of the formed steel deck is normally controlled by construction loading (the weight of the plastic concrete plus a minimum construction live load), the beam spacing, and the number of continuous spans anticipated for the deck. The beam spacing itself is a function of the selected deck gauge as well as the practical limitation of using equal spacing within the building grid. The most commonly used depths for the steel deck are $1\frac{1}{2}$, 2, and 3 in. Figure 10.3 shows a typical composite deck section.

The degree of slab reinforcement requires a consideration of the intended building function. In most building applications, the interior composite slabs will be covered by flexible flooring. Slab cracking due to shrinkage, temperature change, and negative moment will not, therefore, be explicitly considered in the design. In these cases, the slab reinforcement is minimal, consisting of

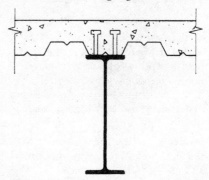

Fig. 10.2 Typical composite beam construction.

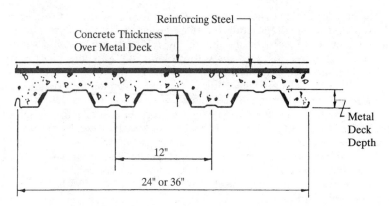

Fig. 10.3 Typical composite deck construction.

small-gauge welded-wire mesh. In building applications where slab cracking is of concern (parking garages, public areas with exposed concrete, and floors supporting brittle flooring materials), the designer must consider the use of additional reinforcement, particularly negative moment reinforcement above the beams and girders.

The section size and grade, as well as the quantity of studs, will be found after a trial-and-error process of design for stiffness and strength. The three parameters are varied and costs are estimated, using approximate values for the unit costs of steel and installed studs. The optimum selection will provide a minimum-cost design satisfying all design requirements. Because of the complexity of the calculations and the iterative nature of the selection, this process can be effectively performed with a computer program.

10.2.1 Design Procedure for Composite Beams

The design of composite floor beams utilizing an unshored metal deck and steel beams is a two-phase process, as follows:

1. During construction, the steel beam, as noncomposite, is designed to support the wet weight of the concrete, the metal deck, and the steel as well as additional temporary loads such as construction workers and equipment. The additional loads may be considered as live loads. The weight normally assumed is 20 lb/ft^2 for the additional temporary construction loads.

2. After the concrete is hardened, the composite beam is designed to support the live loads and the total dead loads, which are the weight of the concrete deck, the steel weight, and the superimposed dead loads such as partitions, ceilings, and mechanical equipment.

The load factors for both phases 1 and 2 to generate the ultimate loads are 1.4 × dead loads or 1.2 × dead loads + 1.6 × live loads, whichever governs the design.

The ultimate applied moment must be equal to or less than the resistance moment capacity:

$$M_u < \phi \times M_n$$

where ϕ = resistance factor (equal to 0.90 for noncomposite steel section design and 0.85 for composite design)

M_n = nominal moment capacity of the flexural member

Chapter F of the American Institute of Steel Construction (AISC) *Manual of Steel Construction*[1] is followed for the design of the noncomposite steel section. The top flange of the steel section is considered to be braced laterally by the metal deck. For a beam with end moments, such as those due to cantilever or end fixity, the beam section is designed for negative moment with the consideration of the unbraced length of the bottom flange. For the construction-phase design, the maximum positive moment is determined by assuming that the end moment due to cantilever is zero. This assumption is based upon the possibility that during construction the concrete may be placed at certain stages between the supports only, but not over the cantilever portion. For the final-phase design, the maximum positive moment is determined by assuming that only the floor weight is the applied load at the cantilever and all dead and live loads are over the beam span between the supports.

For a typical composite section, as in a concrete slab over a ribbed metal deck, a portion or all of the concrete slab may contribute to the flexural capacity of the beam depending on the beam spacing and span length.

The effective width of the concrete deck as a flange of the composite beam is specified by the AISC as follows: The effective width of the concrete slab on each side of the beam centerline should not exceed:

- 1/8 × beam span
- 1/2 × distance from beam centerline to centerline of adjacent beam
- Distance from beam centerline to edge of slab

With reference to Fig. 10.4, the sum of the compression forces must equal the sum of the tension forces in the composite cross section. The compression force C in the concrete in kips is the smallest of the following three values:

(a) $C = 0.85 f_c A_c$

where f_c = concrete compressive strength, ksi

A_c = area of the concrete within the effective slab width, in^2. If the metal-deck ribs are perpendicular to the beam, only the concrete above the metal deck is considered. If, however, the ribs are parallel to the beam, the total concrete area, including the concrete in the ribs, is considered

(b) $C = A_s F_y$

where A_s = area of the steel section, in^2

F_y = yield strength of the steel, ksi

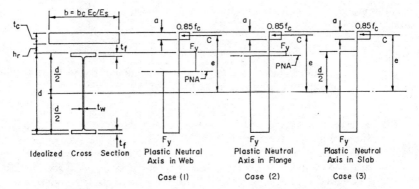

Fig. 10.4 Plastic stress distribution in a composite beam section.

(c) $C = \Sigma Q_n$

where ΣQ_n = Sum of the nominal strength in kips of the shear studs between the point of maximum positive moment and zero moment on either side

The three possible locations of the plastic neutral axis and the corresponding nominal moment capacities (see Fig. 10.4) are:

Case 1 ▪ The plastic neutral axis is located in the web of the steel section. This case occurs when the concrete compressive force is less than the web force ($C \leq P_{yw}$). Thus, the equation for the nominal moment capacity is

$$M_n = M_p - \left(\frac{C}{P_{yw}}\right)^2 M_{pw} + Ce$$

where e = $0.5d + h_r + t_c - 0.5a$
a = $C/0.85 f_c b$
M_{pw} = $0.25 P_{yw}(d - 2t_f)$
P_{yw} = $(d - 2t_f)t_w F_y$
M_p = ZF_y

Case 2 ▪ The plastic neutral axis is located within the thickness of the top flange of the steel section. This case occurs when $P_{yw} < C < P_y$. The equation for the nominal moment capacity becomes

$$M_n = 0.5(P_y - C)\left(d - \frac{P_y - C}{2P_{yf}}t_f\right) + Ce$$

where $P_y = A_s F_y$
$P_{yf} = 0.5(A_s F_y - P_{yw})$

Case 3 ■ The plastic neutral axis is located in the concrete slab. This case occurs when $C = P_y$. (When the plastic axis occurs in the concrete slab, the tension in the concrete is neglected.) In this case, the equation for the nominal moment capacity is

$$M_n = P_y e$$

10.2.2 Stud Shear Connectors

The transfer of horizontal shear forces from the steel section to the concrete slab can be achieved by welding shear studs to the top flange of the steel beam after placing the metal deck. The nominal strength of each stud Q_n in kips embedded in a concrete slab is

$$Q_n = 0.5 A_{sc} \sqrt{f_c E_c} \leq A_{sc} F_u$$

where A_{sc} = cross-sectional area of the stud, in²
f_c = specified concrete compressive strength, ksi
F_u = minimum tensile strength of the stud, ksi
E_c = modulus of elasticity of the concrete, ksi

The type of metal deck commonly used for composite floor construction is 1½-, 2-, or 3-in-deep ribbed deck. A reduction in the nominal strength of the studs may be necessary to account for stud spacing, stud length, or the metal deck configuration. The expressions for the reduction of the nominal strength of the shear studs are listed below.

1. When the metal deck ribs are perpendicular to the steel beam, the reduction factor is

$$\frac{0.85}{\sqrt{N_r}} \frac{w_r}{h_r} \left(\frac{H_s}{h_r} - 1.0 \right) \leq 1.0$$

where h_r = nominal rib height, in
H_s = length of stud after welding ($H_s < h_r + 3$), in
N_r = number of stud connectors in one rib ($N_r < 3$)
w_r = average width of concrete rib, in

2. When the metal deck ribs are parallel to the steel beam, the reduction factor, if $w_r/h_r < 1.5$, is:

$$0.6 \frac{w_r}{h_r} \left(\frac{H_s}{h_r} - 1.0 \right) \leq 1.0$$

The minimum spacing of studs is 6 diameters along the longitudinal axis of the beam (4 in for $^3/_4$-in studs) and 4 diameters transverse to the beam (3 in for $^3/_4$-in studs).

The maximum number of studs placed in one rib of the metal deck perpendicular to the axis of the beam is 3. When the metal-deck ribs are parallel to the axis of the beam, the number of rows of studs will depend upon the flange width of the beam.

The number of shear studs required for the composite beam design is

$$N_s = \frac{C}{Q_n}$$

where N_s = number of shear studs between maximum positive moment and zero moment on each side

For a beam with asymmetric loading, the distances between the maximum positive moment and zero moment on either side of the point of maximum moment will not be equal. Or, if one end of the beam has negative moment, then the zero moment will not be located at the support. Therefore, the total number of shear connectors ($2 \times N_s$) distributed equally throughout the beam length may not be adequate. Instead, the stud spacing should be based upon the shortest distance between maximum positive moment and zero moment, as follows:

$$\text{Stud spacing} = \frac{X}{N_s}$$

where X = shortest distance calculated to place N_s studs

Thus, the total number of studs to be designated for the beam should be

$$N = \frac{\text{span length}}{\text{stud spacing}}$$

Note: N does not necessarily equal $2 \times N_s$.

When concentrated loads occur on the beam, the number of shear connectors between the concentrated loads and the zero moment should be adequate to develop the positive moment at each concentrated load.

10.2.3 Partial Composite Design

When the full composite beam moment capacity is much greater than the applied moment, a partial composite beam design may be utilized requiring fewer shear connectors to reduce construction costs. In addition, a partial composite design may be used when the number of shear connectors required for a full composite beam section cannot be provided owing to limited flange width and length.

Figure 10.5 shows eight possible locations of the plastic neutral axis (PNA) in the steel section. The horizontal shear between the steel section and the concrete slab, which is equal to the compressive force in the concrete C, can be determined as follows:

Location of PNA	Horizontal Shear C
1. Above steel flange	$C_1 = A_s F_y$
2. Within top flange thickness	$C_2 = A_s F_y - 2 \times A_f F_y/4$
3. " " "	$C_3 = A_s F_y - 4 \times A_f F_y/4$
4. " " "	$C_4 = A_s F_y - 6 \times A_f F_y/4$
5. Bottom surface of top flange	$C_5 = A_s F_y - 8 \times A_f F_y/4$
6. Within web	$C_6 = [C(5) + C(7)]/2$
7. " "	$C_7 = 0.25\, A_s F_y$
8. Centroid of steel section	$C_8 = 0.0$

where A_f = area of top flange of structural-steel section

The concrete compressive force should be equal to C above, provided that the value of C is equal to or less than $0.85\, f_c A_c$ and the number of shear studs between the maximum moment and zero moment is adequate for a horizontal shear equal to the value of C.

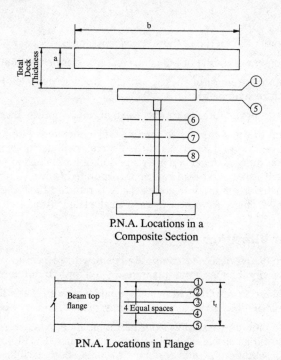

P.N.A. Locations in a Composite Section

P.N.A. Locations in Flange

Fig. 10.5 Plastic neutral axis (PNA) locations in a composite section.

10.2.4 Deflection of Composite Beams

Composite steel-beam properties are used for calculating beam deflections under superimposed dead loads and live loads. The composite section moment of inertia is determined by transforming the area of the concrete portion to an equivalent steel area. Therefore,

$$A_e = \frac{A_c E_c}{E_s}$$

or

$$b_e = \frac{b E_c}{E_s}$$

Therefore, the moment of inertia of the section based on the elastic material behavior can be computed for an asymmetric, all-steel section. In this computation of the elastic moment of inertia, the section is assumed to be fully composite.

For partial composite steel beams, the effective moment of inertia is reduced owing to slippage and shear deformations of the shear studs. The expression I_{eff} is

$$I_{\text{eff}} = I_s + \sqrt{\frac{\Sigma Q_n}{C_f}} (I_{tr} - I_s)$$

where I_s = moment of inertia of the steel section alone, in^4
  I_{tr} = moment of inertia of full composite transformed section, in^4
  ΣQ_n = sum of strength of shear connectors between points of maximum and zero moments, kips
  C_f = compressive force in concrete slab of full composite beam, kips

Since a portion of the compressive forces in the concrete is due to sustained loads, such as superimposed dead loads, the concrete will be subject to a creep effect producing additional deflection after a period of time. The long-term effect on deflection due to sustained load can be incorporated in the computation by assuming that the concrete modulus of elasticity is reduced to 50 percent when calculating b_e above. Therefore, a new I_{eff} can be calculated with a reduced b_e.

10.2.5 Floor Vibration

The performance of the designed composite beam with respect to human perceptibility of floor vibration is commonly investigated for two different criteria:

1. Random vibration resulting from individuals walking on the floor, which occurs, for example, in offices, residential buildings, schools, and shopping centers.

2. Continuous vibration resulting from harmonic motion, such as motors or individuals performing aerobic exercises or similar activities. The buildings affected by this type of vibration include equipment rooms, sports facilities, and exercise facilities.

Based upon tests conducted by Wiss and Parmelee to evaluate floor framing for random vibration by the "drop-of-the-heel" method, the following empirical equation has been generated:

$$R = 5.08 \left(\frac{fA_o}{D^{0.217}} \right)^{0.265}$$

where R = mean response rating
 = 1, imperceptible vibration
 = 2, barely perceptible vibration
 = 3, distinctly perceptible vibration
 = 4, strongly perceptible vibration
 = 5, severe vibration
 f = frequency of the composite beam
 A_o = maximum amplitude
 D = damping ratio

The frequency f of a composite simple-span beam is given by the following equation:

$$f = 1.57 \left(\frac{gEI_t}{WL^3} \right)^{0.5}$$

where g = gravitational acceleration, 386.4 in/sec^2
 E = steel modulus of elasticity, ksi
 I_t = transformed moment of inertia of the composite section, in^4
 W = total weight on the beam, kips
 L = span length, in

The amplitude of a single beam is calculated by dividing the total floor amplitude by the number of effective beams N_{eff}:

$$A_0 = \frac{A_{0t}}{N_{eff}}$$

The total amplitude of the floor A_{0t} is

$$a = 0.1\, \pi f$$

When the constant $t_0 = \dfrac{1}{\pi f} \tan^{-1} a \leq 0.05$, then

$$A_{0t} = 246 L^3 \frac{0.10 - t_0}{EI_t}$$

Or, when the constant $t_0 > 0.05$,

10-12 ■ Chapter Ten

$$A_{0t} = \frac{246L^3}{EI_t}\left[\frac{1}{2\pi f}\sqrt{2(1-a\sin a - \cos a) + a^2}\right]$$

$$N_{eff} = 2.967 - 0.05776\frac{S}{d_e} + 2.556 \times 10^{-8}\frac{L^4}{I_t} + 0.0001\left(\frac{L}{S}\right)^3 \geq 1.0$$

where S = spacing of beams in the floor, in
 d_e = effective depth of the slab, in. The value for d_e is equal to the average slab thickness when the metal-deck ribs are perpendicular to the beam and equal to the concrete thickness above the metal deck when the deck ribs are parallel to the beam.

Another empirical equation, suggested by Thomas Murray[3], evaluates the random vibration in a floor system by calculating D:

$$D = 35A_0f + 2.5$$

If the available floor damping is equal to or greater than D given in the above equation, the floor vibration is acceptable. Otherwise the floor system is not acceptable, as it will likely have perceptible vibration problems.

10.2.6 Sample Problem 1 — Composite Beam Design

The following problem illustrates a typical composite beam design supporting the composite deck directly. The beam design starts by assuming grade 50 steel. The saving in material weight exceeds the relatively small additional cost of using grade 50 instead of grade 36 steel. If, however, the deflection criteria require larger beam sizes, then switching to grade 36 for the material may be appropriate. The experienced designer may start with grade 36, knowing that the deflection will control the design. Therefore, in this problem, the steel material is assumed to be grade 36. The general information of the building construction type, specifically the composite floor construction, is given below:

- Metal-deck ribs are perpendicular to steel beam.
- Concrete compressive strength is 3.0 ksi.
- Concrete unit weight is 115 pcf.
- Depth of metal deck is 2 in.
- Concrete thickness over metal deck is 3.25 in.
- An additional load of 0.3 times the dead loads is used during construction.
- Unshored construction is assumed.
- Shear connectors are ¾ in in diameter and 3.5 in long.
- Critical damping is 5 percent.

The beam's loads and span are shown in Fig. 10.6. The spacing to the adjacent beams on either side is 10 ft. The values of the uniform loads W are:

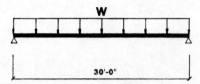

Fig. 10.6 Beam data in Example 1.

Due to construction dead loads = 0.50 kip/ft

Due to superimposed dead loads = 0.25 kip/ft

Due to live loads (before live load reduction) = 0.50 kip/ft

Solution

a. It is easier to start with a steel-beam size that is adequate to support the live loads, normally construction loads. The construction live load is normally assumed to be 20 psf (0.2 lb/ft) to account for the weight of equipment, assuming the weight of the beam is 26 lb/ft:

Ultimate load = 0.526 × 1.4 = 0.736 kip/ft or
= 0.526 × 1.2 + 0.2 × 1.6 = 0.951 kip/ft

Ultimate moment = 0.951 × 30²/ 8 = 107 kip/ft

$$Z_{req} = \frac{M_u}{\phi \times F_y} = \frac{107 \times 12}{0.9 \times 36} = 39.63 \text{ in}^3$$

Use W16×26 (Z = 44.2 in³).

To provide the flat surface of the finished concrete with the steel beam deflecting under the weight of the concrete slab, the designer must choose between varying the thickness of the concrete and cambering the steel beam to compensate for the deflection. The camber then is a negative value of the calculated beam deflection. Beams deflecting less than 0.75 in may not be cambered according to the steel mill practice. Many steel fabricators, however, may have in their shop facilities the equipment that produces camber. Therefore, any value of camber may be specified.

$$\text{Camber} = \frac{5 \times 0.526 \times 30^4 \times 12^3}{384 \times 29{,}000 \times 301} = 1.1 \text{ in}$$

The camber can be specified on the drawings as 1 in, using ¼-in increments.

b. The composite steel section is designed next to support the total loads. Based on the Uniform Building Code, the live loads may be reduced according to the area supported by the beam. The live load reduction is

$$R = \frac{0.8}{100}(A_{sup} - 150)$$

with $A_{sup} = 10 \times 30 = 300$ ft²

$$R = \frac{0.8}{100}(300 - 150) = 0.12$$

Reduced live load = $0.5 \times (1-0.12) = 0.44$ kip/ft

Ultimate load = $1.2 \times (0.5+0.25+0.026) + 1.6 \times 0.44 = 1.635$ kip/ft or
= $1.4 \times (0.5+0.25+0.026) = 1.086$ kip/ft

The governing ultimate load value is 1.635 kip/ ft.

$$\text{Ultimate moment} = \frac{1.635 \times 30^2}{8} = 183.94 \text{ kip-ft}$$

Next, the effective width of the concrete slab must be determined based on 10-ft spacing and 30-ft span length:

Concrete flange width $b = 10 \times 12 = 120$ in

or

$$b = \left(\frac{30 \times 12}{8}\right) \times 2 = 90 \text{ in (govern)}$$

For a determination of the effective area of concrete, the slab thickness above the deck is considered. No concrete within the ribs of the metal deck should be considered when the metal deck ribs are perpendicular to the steel beam. Therefore, the maximum concrete compression force for full composite action is

$C = 0.85 \times f_c \times A_c = 0.85 \times 3 \times 90 \times 3.25 = 745.9$ kips

or

$C = F_y \times A_s = 36 \times 7.68 = 276.48$ kips (govern)

Since the maximum compressive force C is less than the maximum concrete strength, the LRFD method assumes that the plastic neutral axis occurs within the concrete slab thickness. Any tensile stresses in the concrete below the plastic neutral axis are neglected. The effective concrete thickness is

$$a = \frac{C}{0.85 \times f_c \times b} = \frac{276.48}{0.85 \times 3 \times 90} = 1.205 \text{ in}$$

And the distance of the compressive force to the neutral axis of the steel section is:

$e = 0.5d + 5.25 - 0.5a$
$= 0.5 \times 15.69 + 5.25 - 0.5 \times 1.205 = 12.493$ in

Therefore, the resisting capacity of the composite section is

$$\phi M_n = 0.85 P_y e = 0.85 \times 276.48 \times \frac{12.493}{12} = 244.67 \text{ kip-ft}$$

Since the capacity of the full composite section is more than required, the partial composite section may be considered to reduce the number of studs required.

Composite Design ■ 10-15

The following seven values of the composite section are based on specific plastic-neutral-axis locations:

Value 1: Plastic neutral axis is above the top flange (full composite):

$\Sigma Q_n = A_s F_y = 276.48$ kips

$\phi M_n = 244.67$ kip-ft

Value 2: Plastic neutral axis is at a distance equal to one-fourth of the flange thickness from its top surface:

$\Sigma Q_n = A_s F_y - \dfrac{2 A_f F_y}{4}$

$A_f = 5.5 \times 0.345 = 1.898$ in^2

$\Sigma Q_n = 276.48 - 2 \times 1.898 \times \dfrac{36}{4} = 242.3$ kips

$a = \dfrac{242.3}{0.85 \times 3 \times 90} = 1.0558$ in

$e = \dfrac{15.69}{2} + 5.25 - \dfrac{1.0558}{2} = 12.567$ in

$M_n = 0.5(276.48 - 242.3)\left(15.69 - \dfrac{276.48 - 242.3}{2 \times 1.9 \times 36} \times 0.345\right)$
$\qquad + 242.3 \times 12.567 = 3311.68$ kip-in

$\phi M_n = 0.85 \dfrac{3311.68}{12} = 234.6$ kip-ft

Value 3: Plastic neutral axis is at a distance equal to one-half of the flange thickness from its top:

$\Sigma Q_n = 208.18$ kips

$\phi M_n = 223.97$ kip-ft

Value 4: Plastic neutral axis is at a distance equal to three-fourths of the flange thickness from its top:

$\Sigma Q_n = 174.03$ kips

$\phi M_n = 212.78$ kip-ft

Value 5: Plastic neutral axis is at the bottom of the top flange:

$\Sigma Q_n = 139.9$ kips

$\phi M_n = 201.02$ kip-ft

Value 6: Plastic neutral axis is within the web. This value of ΣQ_n is the average of 5 and 7:

10-16 ■ Chapter Ten

$$\Sigma Q_n = \frac{139.9 + 69.12}{2} = 104.5 \text{ kips}$$

$\phi M_n = 186.47$ kips

Value 7: Plastic neutral axis is at the web, producing a concrete compressive force equal to 25 percent of the full-composite compressive force:

$\Sigma Q_n = 0.25 \times 276.48 = 69.12$ kips

$\phi M_n = 166.68$ kip-ft

From the above partial composite values 2 to 7, value 6 is just greater than the applied ultimate moment. Therefore, the number of shear studs should be based on (C = 104.5 kips).

The AISC *Manual of Steel Construction* contains tables for composite beams with the standard wide-flange sections. With these tables, the procedure is much easier than using the calculations above for determining the ultimate capacity of the composite section in flexure. The tables show the seven values of the composite action for each beam, starting with full composite. The assumption for ΣQ_n can be used. The other variable required in using the tables is Y_2, which is

$$Y_2 = Y_{conc} - \frac{a}{2}$$

where Y_{conc} = distance from the top of the concrete to the top flange of the steel section, in

a = concrete compression block depth, in

To illustrate using the tables for this example, the value for *a* must be found first:

$$a = \frac{\Sigma Q_n}{0.85 \times f_c \times b}$$

$\Sigma Q_n = 276$ kips for full composite

$$a = \frac{276}{0.85 \times 3 \times 90} = 1.2 \text{ in}$$

$Y_{conc} = 5.25$ in

$$Y_2 = \frac{5.25 - 1.2}{2} = 4.65 \text{ in}$$

From the AISC tables, the moment capacity of the section is:

$\phi M_n = 242$ kip-ft for $Y_2 = 4.5$ in

and

$\phi M_n = 252$ kip-ft for $Y_2 = 5$ in

By extrapolation, the moment capacity corresponding to $Y_2 = 4.65$ in is

$$\phi M_n = \frac{252 - 242}{5 - 4.5} \times (4.65 - 5) = 245 \text{ kip-ft}$$

For partial composite, if the conservative values of ϕM_n under $Y_2 = 4.5$ are followed, perhaps the fifth value of the plastic neutral axis is adequate when $\phi M_n = 197$ kips. Since the ΣQ_n values are much smaller than the full composite value, the Y_2 is greater than the calculations above for full composite. Therefore, the sixth value of plastic neutral axis may produce adequate results. For $\Sigma Q_n = 104$ kips,

$$a = \frac{104}{0.85 \times 3 \times 90} = 0.453 \text{ in}$$

$$Y_2 = 5.25 - \frac{0.453}{2} = 5.02 \text{ in}$$

From the AISC tables, $\phi M_n = 186$ kip-ft > 183.94, which is acceptable.

c. The required number of shear studs is then determined. For a $\frac{3}{4}\phi$ shear stud, the shear capacity is

$$Q_n = 0.5 A_{sc}\sqrt{f'_c E_c} \leq A_{sc} F_u$$
$$= 0.5 \times 0.442\sqrt{3 \times 2136} = 17.68 \text{ kips}$$

The capacity of $\frac{3}{4}$-in shear studs for lightweight concrete with $f'_c = 3$ ksi is 17.7 kips. The number of shear studs is $2 \times (104.5/17.7) = 11.82$. The number of shear connectors used is 12. The total number of metal-deck ribs supported on the steel section is 30. Therefore, only one row of shear studs is required, and no reduction factor is needed. The designer must be careful in specifying the length of the shear studs. If the studs are too short, then the stud's capacity may be reduced.

d. The final deflection is calculated. Deflections due to live loads and superimposed dead loads occur when composite action is in place. The procedure is to transfer the concrete area to an equivalent steel material area, thereby using the steel properties for the modulus of elasticity. The concrete material modulus of elasticity E_c is

$$E_c = 115^{1.5}\sqrt{3} = 2136 \text{ ksi}$$

With the steel modulus of elasticity $E_s = 29{,}000$ ksi, the ratio of the steel modulus of elasticity and concrete modulus of elasticity n is

$$n = \frac{29{,}000}{2136} = 13.6$$

The effective width of the concrete slab transferred to the equivalent steel area is:

$$b_e = \frac{90}{13.6} = 6.62 \text{ in}$$

The transferred slab area is

$$A_1 = 6.62 \times 3.25 = 21.52 \text{ in}^2$$

and the steel section area is

$$A_2 = 7.68 \text{ in}^2$$

The location of the elastic neutral axis is

$$X = \frac{(21.52 \times 3.25/2) + [7.68 \times (0.5 \times 15.69 + 5.25)]}{21.52 + 7.68} = 4.64 \text{ in}$$

The elastic transformed moment of inertia with full composite action is

$$I_{tr} = \left(\frac{6.62 \times 3.25^3}{12}\right) + \left[21.52 \times \left(4.64 - \frac{3.25}{2}\right)^2\right]$$
$$+ \left[7.68 \times \left(\frac{15.69}{2} + 5.25 - 4.64\right)^2\right] + 301 = 1065 \text{ in}^4$$

Since partial composite is used, the effective moment of inertia is

$$I_{eff} = I_s + \sqrt{\sum \frac{Q_n}{C_f}} (I_{tr} - I_s)$$

$$I_{eff} = 301 + \sqrt{\frac{104.5}{276.48}} \times (1065 - 301) = 770.7 \text{ in}^4$$

The above effective moment of inertia is used to calculate the immediate deflection without long-term effect. For long-term effect due to creep of the concrete, the moment of inertia is reduced to correspond to a 50 percent reduction in E_c value. Accordingly, the new transformed moment of inertia with full composite action is

I_t(reduced) = 900.3 in^4

and the effective moment of inertia for a partial-composite section with long-term effect is

$$I_{eff}(\text{reduced}) = 301 + \sqrt{\frac{104.5}{276.48}} \times (900.3 - 301) = 669.4 \text{ in}^4$$

Since unshored construction is specified, the deflection under the wet weight of the concrete and the steel weight is compensated for by the camber specified. The long-term effect due to these weights should not be considered because the concrete is not stressed by these weights.

Deflection due to long-term superimposed dead loads:

$$D1 = \frac{5 \times 0.25 \times 30^4 \times 12^3}{384 \times 29,000 \times 669.4} = 0.235 \text{ in}$$

Deflection due to short-term live loads:

$$D2 = \frac{5 \times 0.44 \times 30^4 \times 12^3}{384 \times 29,000 \times 770.7} = 0.358 \text{ in}$$

Final deflection:

$D = D1 + D2 = 0.235 + 0.358 = 0.593$ in

e. The vibration study. The total dead load considered in the vibration equations is the actual self-weight of the concrete and steel beam plus a realistic percentage of the superimposed dead load. The percentage of superimposed dead load is 30 percent in this example. For the drop-of-the-heel-vibration, the dead loads help to reduce the impact of human movement.

Total dead loads = $0.500 \times 30 + 0.026 \times 30 + 0.3 \times 0.25 \times 30 = 18.03$ kips

$$\text{Frequency } f = 1.57 \left[\frac{386.4 \times 29{,}000 \times 770.7}{18.03 \times (30 \times 12)^3} \right]^{0.5} = 5.03 \text{ s}^{-1}$$

$$N_{eff} = 2.967 - 0.05776 \frac{120}{4.25} + 2.556 \times 10^{-8} \times \frac{360^4}{120} + 10^{-4} \times \left(\frac{360}{770.7} \right)^3 = 1.90$$

$a = 0.1 \times \pi \times 5.03 = 1.58$

$$A_{0t} = \frac{246 \times 360^3}{29 \times 10_6 \times 770.7} \left[\frac{1}{2 \times \pi \times 5.04} \sqrt{2(1 - 1.58 \sin 1.58 - \cos 1.58) + 1.58^2} \right] = 0.0188 \text{ in}$$

$$A_0 = \frac{A_{0t}}{N_{eff}} = \frac{0.0188}{1.9} = 0.0099 \text{ in}$$

$$R = 5.08 \left(\frac{5.03 \times 0.0099}{0.05^{0.217}} \right)^{0.2653} = 2.7 > 2.5 \text{ and } < 3.5 \text{ (distinctly perceptible)}$$

According to Thomas Murray's[3] equation, the minimum acceptable damping is

$D = 35 \times 0.0099 \times 5.03 + 2.5 = 4.2 < 5.0$ (acceptable)

10.2.7 Sample Problem 2 — Composite Girder Design

This sample building construction type is the same as presented in Sample Problem 1. The general information in this sample of a construction floor system is the same with the following exception:

- The metal-deck ribs are parallel to the girder.

The girder's loads and span are shown in Fig. 10.7. The spacing to the left adjacent girder is 30 ft and to the right girder is 20 ft.

The values of the concentrated loads (P_1 and P_2) are

Due to construction dead loads = 14.85 kips

Due to superimposed dead loads = 7.5 kips

Due to live loads (50 psf) = 15.0 kips

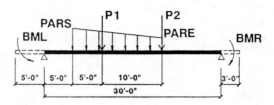

Fig. 10.7 Girder data in Example 2.

Left end negative moments (BML) are:
   Due to dead load = 22.5 kip-ft
   Due to superimposed dead loads = 7.5 kip-ft
   Due to live loads = 20 kip-ft
Right end negative moments (BMR) are
   Due to dead load = 7.5 kip-ft
   Due to superimposed dead loads = 2.5 kip-ft
   Due to live loads = 7 kip-ft
The values of the start of partial load (PARS) are
   Due to construction dead loads = 0.50 kip/ft
   Due to superimposed dead loads = 0.75 kip/ft
   Due to live loads = 0.50 kip/ft
The values of the end of partial load (PARE) are
   Due to construction dead loads = 0.20 kip/ft
   Due to superimposed dead loads = 0.30 kip/ft
   Due to live loads = 0.20 kip/ft
The requirement is to design an economical composite girder.

Solution

a. **Additional Dead Loads.** The steel girder is designed to support the construction dead loads as unshored construction using 20 psf live load in addition to construction dead loads. To avoid additional bookkeeping of live load during construction, the additional dead load can be assumed to be 30 percent of the construction dead loads. For computation of the maximum positive moments during construction, the negative end moments are neglected because of the possibility that the concrete may be placed over the span between the supports but not over the cantilever portion.

Composite Design ■ 10-21

The ultimate dead loads (in this case, the additional 30 percent loads during construction are assumed as dead loads) are

$P_u = 14.85 \times 1.3 \times 1.4 = 27.027$ kips

$PARS_u = 0.5 \times 1.3 \times 1.4 = 0.91$ kip/ft

$PARE_u = 0.2 \times 1.3 \times 1.4 = 0.364$ kip/ft

$W_u = 0.044 \times 1.4 = 0.0616$ kip/ft (using W21×44)

The maximum positive ultimate moment without any end moment is

$M_u = 328.0$ kip-ft ($\phi M_u = 358$ kip-ft for W21×44, grade 50)

The maximum negative moment during construction is

$22.5 \times 1.3 \times 1.4 = 40.95$ kips/ft

W21×44 with a 5-ft unsupported length is adequate.

b. Maximum Deflection Due to Construction Dead Loads. The camber computations correspond to the maximum deflection due to actual service construction dead loads. For this computation, only the ends' dead load moments were considered. The loads are

$P = 14.85$ kips

$PARS = 0.50$ kip/ft

$PARE = 0.20$ kip/ft

$W = 0.044$ kip/ft

$BML = 22.5$ kip-ft

$BMR = 7.5$ kip-ft

The corresponding deflection = 1.093 in (camber = 1 in)

c. Design for Maximum Negative Moment Due to Total Dead and Live Loads. For the maximum possible unbraced length of the bottom flange of the steel section, only the dead loads are used between supports. Therefore, the unbraced length for the bottom flange is the distance from the support to the point of zero moment.

The ultimate minimum loads over the span between the supports are

$P_u = 1.2 \times 14.85 = 17.82$ kips

$PARS_u = 1.2 \times 0.5 = 0.60$ kip/ft

$PARE_u = 1.2 \times 0.2 = 0.24$ kip/ft

$W_u = 1.2 \times 0.044 = 0.053$ kip/ft

The maximum ultimate end moments are

(left) $BML_u = 1.2 \times (22.5 + 7.5) + 1.6 \times 20 = 68$ kip-ft

(right) $BMR_u = 1.2 \times (7.5 + 2.50) + 1.6 \times 7 = 23.2$ kip-ft

Unbraced length of the bottom flanges between left support and zero moment=2.9 ft

Cantilever length=5.0 ft (govern)

$\phi M_n = 350$ kip-ft with $L_b = 5.0$ ft for W21×44, grade 50.

d. Design for Maximum Positive Moment Due to Final Total Loads Utilizing Composite Action. The end moments must be minimized to produce the maximum positive moment. For these computations, the load factor used for the negative dead load end moments is 1.2.

According to building codes, the live loads on beams and girders can be reduced, depending on the area supported. The maximum live load reduction is 60 percent. Therefore, the live loads are

For the concentrated loads P_1 and P_2,

Live load = 15 × 0.6 = 9 kips

For the partial distributed load,

PARS (the starting value) = 0.5 × 0.6 = 0.3 kip/ft

For the partial distributed load,

PARE (the end value) = 0.2 × 0.6 = 0.12 kip/ft

Therefore, the ultimate loads producing the maximum positive moment are

$P_u = 1.2 \times (14.85 + 7.5) + 1.6 \times 9 = 41.22$ kips

$PARS_u = 1.2 \times (0.5 + 0.75) + 1.6 \times 0.3 = 1.98$ kips/ft

$PARE_u = 1.2 \times (0.2 + 0.3) + 1.6 \times 0.12 = 0.792$ kip/ft

The self-weight of the beam is

$W_u = 1.2 \times 0.044 = 0.053$ kip/ft

The minimum ultimate end moments are

$BML_u = 1.2 \times 22.5 = 27$ kip-ft

$BMR_u = 1.2 \times 7.5 = 9$ kip-ft

From an analysis of the single-span beam width above the ultimate loads and ultimate end moments, the maximum positive moment is

$M_u = 509.6$ kip-ft

The composite capacity of the section W21×44 in flexure must be equal to or greater than the applied ultimate positive moment. For a determination of the capacity of the composite section, the effective concrete flange width is

$$b = (30 + 20) \times \frac{12}{2} = 300 \text{ in}$$

or

$$b = 30 \times \frac{12}{4} = 90 \text{ in (govern)}$$

The maximum concrete compression force (based on an average concrete thickness of 4.25 in utilizing the concrete in the ribs when the ribs are parallel to the beams) is

$C = 0.85 \times 3 \times 90 \times 4.25 = 975.4$ kips

or

$C = 50 \times 13 = 650$ kips (govern)

From the AISC tables for composite beams corresponding to W21×44, grade 50, the value of $\phi M_n > 509.6$ can be easily obtained for $\Sigma Q_n = 160$ kips.

Try $\Sigma Q_n = 260$ kips.

$a = \dfrac{260}{0.85 \times 3 \times 90} = 1.133$ in

$Y_2 = 4.68$ in

$\phi M_n = 543 + \dfrac{552-543}{5.0-4.5} \times (4.68-4.5) = 546$ kip-ft $> M_u$

From an analysis of the one-span beam with the above ultimate loads and moments, the location of M_u from the left support is

$X_m = 13.25$ ft

The location of the left zero moment from the left support is

$XL_0 = 0.49$ ft

and the location of the right zero moment from the right support is

$XR_0 = 0.19$ ft

e. Shear Studs. The total number of shear studs is determined to develop the maximum positive moment within the beam length between zero moment and the maximum positive moment location.

The number of studs required on each side of the maximum moment location is

$\dfrac{260}{17.7} = 14.69$ studs

Therefore, the minimum required left spacing is

$(13.25 - 0.49) \times \dfrac{12}{14.69} = 10.4$ in

and the minimum required right spacing is

$(16.75 - 0.19) \times \dfrac{12}{14.69} = 13.5$ in

The required studs at the location of the first left P:

$M_{1u} = 502.1$ kip-ft

$\Sigma Q_n = 260$ kip-ft

$\phi M_n = 546$ kip-ft

The required number of studs = 14.69 studs
The required minimum spacing of studs (must be ≤ 10.4 in):
$$(10 - 0.49) \times \frac{12}{14.69} = 7.77 \text{ in}$$
The number of studs for the first 10-ft segment:
$$\frac{120}{7.77} = 15.44 \text{ (use 16 studs)}$$
The required studs at location of the next P:
$M_{2u} = 481.2$ kip-ft
$\Sigma Q_n = 163$ kips
$\phi M_n = 486$ kips

The required number of studs:
$$\frac{163}{17.7} = 9.21 \text{ studs}$$
Since this load location is to the right of the maximum moment, this stud layout should be between the location of the load and the right zero moment. The required minimum spacing of studs (must be ≤ 13.5 in):
$$(10 - 0.19) \times \frac{12}{9.21} = 12.78$$
The number of studs for the last 10-ft segment:
$$\frac{120}{12.78} = 9.4 \text{ (use 10 studs)}$$
The required studs between the two concentrated loads:
Left, the maximum M_u:
$$\text{Number of studs} = \frac{13.25 \times 12}{10.4} - 16 = -0.71 \text{ (use maximum stud spacing of 32 in)}$$
Right, the maximum M_u:
$$\text{Number of studs} = \frac{16.75 \times 12}{13.5} - 10 = 4.89 \text{ (use 5 studs)}$$
Stud spacing:
$$(16.75 - 10) \times \frac{12}{5} = 16.2 \text{ in}$$

The number of studs between the two concentrated loads should be based on the smallest spacing on either side of the maximum moment.
$$\text{Number of studs} = 10 \times \frac{12}{16.2} = 7.4 \text{ (use 8 studs)}$$
The engineer may indicate the stud layout on the drawings; however, the possibility of field error may result in an unsafe design. Therefore, the installation is

easier if the spacing of the studs is uniform for the full span length of the beam. In this case, the minimum stud spacing shown in the above calculations should be used. The minimum stud spacing required to satisfy the maximum moment on either side of the moment or the moments at each concentrated load is 7.77 in. The total number of studs based on 7.77-in spacing is:

$$30 \times \frac{12}{7.77} = 46.3 \text{ (use 47 studs)}$$

f. **Final Deflection.** The elastic properties of the composite W21×44 steel section with a concrete-slab deck that is 5.25 in deep (an average of 4.25 in thick) and 90 in wide are

$$E_c = 115^{1.5} \times \sqrt{3} = 2136 \text{ ksi}$$

$$n = \frac{29{,}000}{2136} = 13.57$$

$$\frac{b}{n} = \frac{90}{13.57} = 6.632$$

$$I_{tr} = 2496 \text{ in}^4$$

Since partial composite is used for determining the number of studs, the effective moment of inertia at the location of the maximum moment is reduced. The empirical method for the reduced moment of inertia is:

$$I_{eff} = I_s - \sqrt{\frac{\Sigma Q_n}{C_f}}(I_{tr} - I_s)$$

$$\Sigma Q_n = 2604 \text{ kips}$$

$$C_f = 0.85 \times 3 \times 4.25 \times 90 = 975.4 \text{ kips}$$

$$= 50 \times 13 = 650 \text{ kips (govern)}$$

$$I_{eff} = 843 + \sqrt{\frac{260}{650}} \times (2496 - 843) = 1888.4 \text{ in}^4$$

An additional reduction factor in the moment of inertia is due to creep of the concrete when the loading is a permanent dead load.

The reduced moment of inertia R-I_{eff} due to long-time effect is determined by reducing the E_c value by 50 percent.

$$E_c = 0.5 \times 2136 = 1068 \text{ ksi}$$

$$n = \frac{29{,}000}{1068} = 27.154$$

$$\frac{b}{n} = \frac{90}{27.154} = 3.315$$

$$R\text{-}I_{tr} = 2088.3 \text{ in}^4$$

The effective moment of inertia due to partial composite is

$$R\text{-}I_{\text{eff}} = 843 + \sqrt{\frac{260}{650}} \times (2088.3 - 843) = 1630.6 \text{ in}^4$$

The deflection computations for unshored construction exclude the self-weight of the concrete and steel. Whether the steel beam is adequately cambered or not, the concrete will be finished as a level surface. More concrete thickness may be expected at the midspan of the beams and deck when the beam is not cambered. Only the superimposed dead-load moments are assumed for the negative end moments. The working superimposed dead loads are

$P_{sid} = 7.5$ kips

$\text{PARS}_{sid} = 0.75$ kip/ft

$\text{PARE}_{sid} = 0.30$ kip/ft

$\text{BML}_d = 7.5$ kip-ft

$\text{BMR}_d = 2.5$ kip-ft

From the beam analysis, the maximum deflection due to the superimposed dead loads using the reduced I_{eff} due to long-term and partial-composite effects is

$\text{DEF}_d = 0.38$ in

which occurred at 14.34 ft from the left support.

For generation of the maximum deflection due to live loads, the end negative moments due to live loads are assumed to be zero. The working live loads are

$P_1 = 9.0$ kips

$\text{PARS}_1 = 0.3$ kip/ft

$\text{PARE}_1 = 0.12$ kip/ft

and, from the beam analysis, the maximum deflection is

$\text{DEF}_1 = 0.32$ in

The total deflection due to superimposed dead loads and live loads is

$0.38 + 0.32 = 0.70$ in

10.3 Reinforcing Existing Composite Beams

One of the advantages of a steel building is the flexibility it allows in modifying the floor load-carrying capacity by welding cover plates to the flanges, split-tee steel section, or any other type of steel section. In most office buildings, changes in the floor layout are common because of the tenants' changing needs. Spaces in buildings may be converted to accommodate unusual live loads, such as equipment, library books, storage, or high-density files. The common method used to reinforce an existing composite steel beam is to weld a plate below the bottom flange. The plate is normally wider than the flange to avoid overhead welding. A composite beam with concrete slabs on both sides of the beam may provide more concrete area and strength than required for the compression resistance force. Therefore, for a small composite-steel section subject to excessive loads, reinforcing the top flange is normally not required. For a beam with a large steel section and a relatively small concrete slab, reinforcing the top flange may be required when it is subject to excessive loads. Figure 10.8 is a generalized composite-steel beam section with added plates at the top and bottom flanges. The design formulas and examples illustrated in this section are limited to a plate at the bottom flange of a beam and two plates at the top flange, as shown in Fig. 10.8.

10.3.1 Design Procedure

The ultimate loads and load factors are described in Sec. 10.2. The procedure to evaluate the flexural resisting capacity of the composite section is also similar to that described previously.

For a composite beam section subject to positive bending, the sum of compression forces above the plastic neutral axis must equal the sum of tension forces below the plastic neutral axis. The compression force C in the concrete is the smallest of the following three values:

$C = 0.85 f_c A_c$

$C = A_s F_y$

$C = \Sigma Q_n$

Five locations of the plastic neutral axis and the corresponding nominal moment capacity are possible, depending on the size of the wide-flange shape, the size of the added plates, and the applied moment (see Figs. 10.9 to 10.13). The following is a list of these cases of plastic neutral axis locations and the corresponding expressions for M_n.

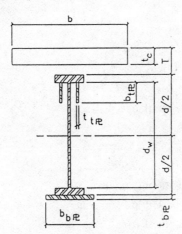

Fig. 10.8 Typical composite beam with top and bottom reinforcing plates.

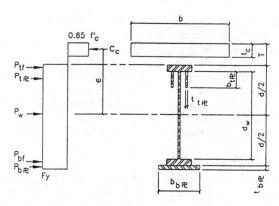

Fig. 10.9 Stress distribution when PNA is above top flange.

Case 1. The plastic neutral axis is located above the top flange of the steel section (see Fig. 10.9). This case occurs when the concrete compressive force is $C = P_{st}$, where P_{st} is the total force acting on the steel section, including the added plates:

$$P_{st} = A_{wf} \times F_{ywf} + 2A_{tp} \times F_{yep} + A_{bp} \times F_{ybp}$$

Thus, the equation for the nominal moment capacity is

$$M_n = Ce - [P_{tf} \times 0.5 \times (T_{tf} + d_w)] + [P_{bf} \times 0.5 \times (T_{bf} + d_w)] - [P_{tp} \times 0.5 \times (d_w - b_{tp})] + [P_{bp1} \times 0.5 \times (d_w + t_{bf} + t_{bp})]$$

where e = $0.5d + T - 0.5a$
a = $C / 0.85 f_c b$
P_{tf} = $b_{tf} \times t_{tf} \times F_y$
P_{bf} = $b_{bf} \times t_{bf} \times F_y$
P_{tp} = $2 \times b_{tp} \times t_{tp} \times F_{ytp}$
P_{bp} = $b_{bp} \times t_{bp} \times F_{ybp}$
d = depth of wide-flange shape, in
T = total thickness of deck, in
b_{tf} = width of top flange, in
t_{tf} = thickness of top flange, in
b_{bf} = width of bottom flange, in
t_{bf} = thickness of bottom flange, in
F_y = yield strength of wide-flange steel shape, ksi
b_{tp} = width of top plates, in

t_{tp} = thickness of top plates, in
F_{ytp} = yield strength of top plates, in
b_{bp} = width of bottom plate, in
t_{bp} = thickness of bottom plate, in
F_{ybp} = yield strength of bottom plate, ksi

Case 2. The plastic neutral axis is located within the thickness of the top flange of the steel section (see Fig. 10.10). This case occurs when $P_{st} > C > (P_{st} - 2 \times P_{tf})$. On the assumption that the location of the plastic neutral axis from the top of the steel beam is $X \times t_{tf}$, then the value for X is:

$$X = \frac{P_{st} - C}{2 \times P_{tf}}$$

Therefore, the expression for the nominal moment capacity is

M_n = $Ce - [P_{tf} \times 0.5 \times (T_{tf} + d_w)] + [P_{bf} \times 0.5 \times (T_{bf} + d_w)] - [P_{tp} \times 0.5 \times (d_w - b_{tp})] + [P_{bp1} \times 0.5 \times (d_w + t_{bf} + t_{bp})] + [2 \times X \times P_{tf} \times 0.5 \times (d_w - X \times t_{tf})]$

Case 3. The plastic neutral axis is located below the top flange and within the width of the top reinforcing plate (see Fig. 10.11). This case occurs when

$(P_{st} - 2 \times P_{tf}) > C > [P_{st} - 2 \times P_{tf} - 2 \times (P_{tp} + b_{tp} \times t_w \times F_y)]$

Assuming the location of the plastic neutral axis is at $X \times b_{tp}$, then

$$X = \frac{P_{st} - C - 2 \times P_{tf}}{2 \times P_{tp} + 2 \times b_{tp} \times t_w \times F_y}$$

In this case, the equation for the nominal moment capacity is

M_n = $C + [P_{tf} \times 0.5 \times (d_w + t_{tf})] - [P_{tp} \times 0.5 \times (d_w - b_{tp})] + [P_{bf} \times 0.5 \times (d_w + t_{bf})] + [P_{bp} \times 0.5 \times (d_w + 2 \times t_{bf} + t_{bp})] + [2 \times X \times (P_{tp} + b_{tp} \times t_w \times F_y) \times 0.5 \times (d_w - X \times b_{tp})]$

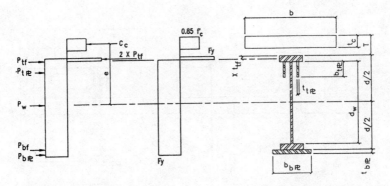

Fig. 10.10 Stress distribution when PNA is within top flange thickness.

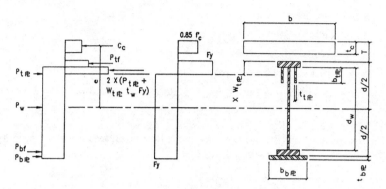

Fig. 10.11 Stress distribution when PNA is within top plate width.

Case 4. The plastic neutral axis is located within the web and above the $d_w/2$ axis (see Fig. 10.12). This case occurs when

$$C < [P_{st} - 2 \times P_{tf} - 2 \times (P_{tpl} + b_{tp} \times t_w \times F_y)]$$

and

$$C > (P_{st} - 2 \times P_{tf} - 2 \times P_{tp} - 2 \times P_w \times 0.5)$$

The location of the plastic neutral axis from the wide-flange shape mid-depth is

$$y = \frac{C + P_{tf} + P_{tp} - P_{bf} - P_{bp}}{2 \times t_w \times F_y}$$

In this case, the equation for the nominal moment capacity is

$$M_n = (C \times e) + [P_{tf} \times 0.5 \times (d_w + t_{tf})] + [P_{tp} \times 0.5 \times (d_w - b_{tp})]$$
$$+ [P_{bf} \times 0.5 \times (d_w + t_{bf})] + [P_{bp} \times 0.5 \times (d_w + 2 \times t_{bf} - t_{bp})]$$
$$+ \left(P_w \times \frac{d_w}{4}\right) - \left(2 \times y \times t_w \times F_y \times \frac{y}{2}\right)$$

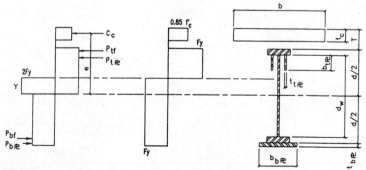

Fig. 10.12 Stress distribution when PNA is within web and above mid-depth of steel section.

Case 5. The plastic neutral axis is located within the web and below the $d_w/2$ axis (see Fig. 10.13). This case occurs when

$$C < (P_{st} - 2 \times P_{tf} - 2 \times P_{tp} - 2 \times P_w \times 0.5)$$

and

$$C > (P_{st} - 2 \times P_{tf} - 2 \times P_{tp} - 2 \times P_w)$$

The location of the plastic neutral axis from the wide-flange shape mid-depth is:

$$Y = \frac{-C - P_{tf} - P_{tp} + P_{bf} + P_{bp}}{2 \times t_w \times F_y}$$

$$\begin{aligned} M_n &= (C \times e) + [P_{tf} \times 0.5 \times (d_w + t_{tf})] + [P_{tp} \times 0.5 \times (d_w - b_{tp})] \\ &+ [P_{bf} \times 0.5 \times (d_w + t_{bf})] + [P_{bp} \times 0.5 \times (d_w + 2 \times t_{bf} - t_{bp})] \\ &+ \left(P_w \times \frac{d_w}{4}\right) - \left(2 \times y \times t_w \times F_y \times \frac{y}{2}\right) \end{aligned}$$

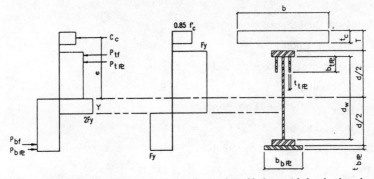

Fig. 10.13 Stress distribution when PNA is within web and below mid-depth of steel section.

10.3.2 Sample Problems of Investigation and Reinforcement of Existing Composite Beams

An existing girder in an office building was originally designed for the 50 psf live load. The tenant wanted to utilize the space at the location of this girder for storage, which increased the live loads significantly. The girder size is W21×44, grade 50. The stud layout on the first left 10 ft is 17 studs, on the next 10 ft is 8 studs, and on the last 10 ft is 11 studs. The spacing to the left adjacent girder is 30 ft and to the right girder is 20 ft. This problem is to investigate the beam-carrying capacity and compare it with the applied loads. A plate added to the bottom flange may be designed if reinforcement is necessary to increase the beam load-carrying capacity. The general information regarding a composite floor construction of the office building is given below.

10-32 ■ Chapter Ten

- Metal deck ribs are parallel to the steel beam.
- Shored construction is assumed.
- Concrete compressive strength is 3.0 ksi.
- Concrete unit weight is 115 pcf.
- Depth of metal deck is 2 in.
- Concrete thickness over metal deck is 3.25 in.
- Minimum acceptable span-to-deflection ratio is 360 and the maximum acceptable deflection value is 0.75 in.
- Shear connectors are ¾ in in diameter and 3.5 in long.

The girders' new loads and span are the same as shown in Fig. 10.7. The values of the concentrated loads (P_1) are

Due to construction dead loads	= 14.85 kips
Due to superimposed dead loads	= 2.5 kips
Due to live loads	= 31.25 kips

The values of the left end moments are

Due to dead loads	= 45 kip-ft
Due to superimposed dead loads	= 15 kips
Due to live loads	= 40 kip-ft

The values of the right end moments are

Due to dead loads	= 15 kip-ft
Due to superimposed dead loads	= 5 kip-ft
Due to live loads	= 14 kip-ft

The values of the start of partial load (PARS) are

Due to construction dead loads	= 0.50 kip/ft
Due to superimposed dead loads	= 0.75 kip/ft
Due to live loads	= 1.25 kips/ft

The values of the end of partial load (PARE) are

Due to construction dead loads	= 0.20 kip/ft
Due to superimposed dead loads	= 0.30 kip/ft
Due to live loads	= 1.25 kips/ft

Solution ■ The following calculations illustrate the steps in the investigation of an existing composite-steel girder and the design of additional reinforcement using the LRFD method.

a. Investigation of the steel section W21×44, grade 50, for its resistance to the end negative moments. The ultimate left end moment:

$$-M_{u1} = (60 \times 1.2) + (40 \times 1.6) = 136 \text{ kip-ft}$$

and the ultimate right end moment:

$$-M_{ur} = (20 \times 1.2) + (14 \times 1.6) = 46.4 \text{ kip-ft}$$

The applied loads between the supports must be considered as the minimum possible loads (dead loads only) to investigate the maximum possible unbraced length of the bottom flange within the back span between the supports. Therefore, the minimum ultimate loads are:
The uniform loads:

$$W_u = 0.044 \times 1.2 = 0.0616 \text{ kip/ft}$$

The concentrated loads are

$$P_u = 14.85 \times 1.2 = 17.82 \text{ kips}$$

and the partial loads

$$\text{PARS}_u = 0.5 \times 1.2 = 0.6 \text{ kip/ft}$$

$$\text{PARE}_u = 0.2 \times 1.2 = 0.24 \text{ kip/ft}$$

A simple beam analysis with the above loads results in the left zero moment point being 5.6 ft from the left support, which is farther than the left cantilever span of 5.0 ft. The right zero moment is calculated at 2.3 ft from the right support, which is shorter than the right cantilever span of 3.0 ft. Therefore:

Moment capacity with unsupported length of 5.6 ft = 340.8 kip-ft > 136

Moment capacity with unsupported length of 3.0 ft = 357.8 kips/ft > 46.4

b. Investigation of the composite-steel section capacity at maximum ultimate positive moment due to total loads. The ultimate loads' combined dead loads and live loads are

$$W_u = 0.044 \times 1.2 = 0.0616 \text{ kip/ft}$$

$$P_u = [(14.85 + 2.5) \times 1.2] + (31.25 \times 1.6) = 70.82 \text{ kips}$$

$$\text{PARS}_u = [(0.5 + 0.75) \times 1.2] + (1.25 \times 1.6) = 3.5 \text{ kips/ft}$$

$$\text{PARE}_u = [(0.2 + 0.3) \times 1.2] + (1.25 \times 1.6) = 2.6 \text{ kips/ft}$$

The ultimate end moments are based on the dead loads only, excluding the superimposed dead loads:

$$-M_{ul} = 45 \times 1.2 = 54 \text{ kip-ft}$$

$$-M_{ur} = 15 \times 1.2 = 18 \text{ kip-ft}$$

A simple beam analysis reveals the points of zero moments to be

Left point at 0.58 ft from left support.
Right point at 0.23 ft from right support.

Therefore, from the analysis of the beam span with the above loads and end moments, the maximum positive moment is

$+M_u = 840.65$ kip-ft at 11.67 ft from left support

To determine the moment capacity of the beam, the number of studs and the total horizontal shear capacity ΣQ_n between the concrete and steel are determined at the left and right side of the maximum positive moment. The number of studs between maximum $+M$ and left zero moment point is

$$\text{Number of studs} = \left(17 - \frac{17}{10} \times 0.58\right) + \left[\frac{8}{10} \times (11.67 - 10)\right] = 17.35 \text{ (say 17)}$$

and its total shear capacity is:

$\Sigma Q_n = 17 \times 17.68 = 306.75$ kips

The number of studs between maximum $+M$ and right zero moment point is:

$$\text{Number of studs} = \left(11 - \frac{11}{10} \times 0.23\right) + \left[\frac{8}{10} \times (10 - 1.67)\right] = 17.39 \text{ (say 17)}$$

and its total shear capacity is

$\Sigma Q_n = 17 \times 17.68 = 306.55$ kips

$$\text{Effective concrete flange width} = \frac{30 \times 12}{4} = 90 \text{ in}$$

For a determination of the horizontal shear value C, the smaller of

$0.85 f'_c A_c = 0.85 \times 3 \times 90 \times 4.25 = 975.4$ kips

$A_s f_y = 13 \times 50 = 650$ kips

$\Sigma Q_n = 306.55$ kips (govern)

$$a = \frac{306.55}{0.85 \times 3 \times 90} = 1.336 \text{ in}$$

$$e = \frac{20.66}{2} + 5.25 - \frac{1.336}{2} = 14.9 \text{ in}$$

$P_y = 13 \times 50 = 650$ kips

The actual beam depth is 20.66 in, the flange thickness is 0.45 in, and the web thickness is 0.35 in. Therefore, the web force P_{yw} is

$P_{yw} = (20.66 - 2 \times 0.45) \times 0.35 \times 50 = 345.8$ kips

Since $C < P_{yw}$, the plastic neutral axis is in the web.

$$M_n = M_p - \left(\frac{C}{P_{yw}} 2 \times M_{pw}\right) + (C \times e)$$

$M_{pw} = 0.25 P_{yw} \times d_w = 0.25 \times 345.8 \times [20.66 - (2 \times 0.45)] = 1708.25$ kip-in

$M_p = Z \times F_y = 95.4 \times 50 = 4770$ kip-in

$M_n = 4770 - \left[\left(\dfrac{306.55}{345.8}\right)^2 \times 1708.25\right] + (306.55 \times 14.9) = 8002$ kip-in

$\phi M_n = 0.85 \times \dfrac{8002}{12} = 566.8$ kip-in < applied moment = 840.65

Therefore, the existing steel section must be reinforced.

c. The design of an additional plate at the bottom flange of the existing steel section. There is no direct solution to arrive at the required plate size. A trial-and-error approach is presented in this problem.

Try 12-in wide × 7/8-in thick plate, grade 36.

P_{st} = 650 + 0.875 × 12 × 36 = 1028 kips

P_{tf} = 6.5 × 0.45 × 50 = 146.25 kips

P_w = 345.8 kips

P_{bp} = 12 × 0.875 × 36 = 378 kips

By adding a plate at the bottom flange, the plastic neutral axis location is moved down. For determining which case is applicable for the plastic neutral axis location within the web of the beam, the values of R_3, which is the total steel force p_{yst} less twice the force generated by the top flange and the upper half of the web are used. In addition, R_4 needs to be determined, which is the total force in the steel p_{ys} less twice the force generated by the top flange and the full web area. Therefore,

$R_3 = P_{ys} - 2 \times (P_{tf} + P_w \times 0.5)$

$R_3 = 1028 - 2 \times (146.25 + 345.8 \times 0.5) = 389.7$ kips

$R_4 = P_{ys} - 2 \times (P_{tf} + P_w)$

$R_4 = 1028 - 2 \times (146.25 + 345.8) = 43.9$ kips

Therefore,

(R_3 = 389.7) > (C = 306.55) > (R_4 = 43.9)

Case 5 is applicable.

$y = \dfrac{-306.55 + 378}{2 \times 0.35 \times 50} = 2.03$ in

$d_w = 20.66 - (2 \times 0.45) = 19.76$ in

Using equation:

$M_n = (C \times e) + [2 P_{tf} \times 0.5 \times (d_w + t_{tf})]$

$+ \left(P_w \times \dfrac{d_w}{4}\right) - \left(2 \times Y \times t_w \times F_y \times \dfrac{y}{2}\right) + [P_{bf} \times 0.5 \times (d_w + 2 \times t_{bf} + t_{bp})]$

10-36 ■ Chapter Ten

$$M_n = (306.55 \times 14.9) + [2 \times 146.25 \times 0.5 \times (19.76 + 0.45)]$$
$$+ \left(345.8 \times \frac{19.76}{4}\right) - (0.35 \times 50 \times 2.0232)$$
$$+ [378 \times 0.5 \times (19.76 + 2 \times 0.45 + 0.875)]$$
$$= 13{,}241.4 \text{ kip-in}$$

$$\phi M_n = 0.85 \times \frac{13{,}241.4}{12} = 937.9 \text{ kip-ft} > M_u = 840.65$$

Since there are applied concentrated loads, the capacity of the beam at each concentrated load location must be determined and compared with the applied ultimate moment at the same location. The beam capacity may be less than its capacity at midspan or at the maximum positive moment, depending on the number of shear studs on the beam between the concentrated load and zero moment.

To verify the capacity at the first left concentrated load:

$$\text{Number of studs} = 17 - \left(\frac{17}{10} \times 0.58\right) = 16.014 \quad \text{(say 16)}$$

$$\Sigma Q_n = 16 \times 17.68 = 282.9 \text{ kips}$$

$$a = \frac{282.9}{0.85 \times 3 \times 90} = 1.23 \text{ in}$$

$$e = \frac{20.66}{2} + 5.25 - \frac{1.23}{2} = 14.963 \text{ in}$$

$$y = \frac{-282.9 + 378}{2 \times 0.35 \times 50} = 2.7 \text{ in}$$

The plastic neutral axis is located below the centroid of the wide-flange section; therefore, the equation corresponding to case 5 is applicable for the moment capacity.

$$\phi M_n = 909.7 \text{ kip-ft} > M_u = 835.44$$

To verify the capacity at the next concentrated load:

$$\text{Number of studs} = 11 - \left(\frac{11}{10} \times 0.23\right) = 10.75 \quad \text{(say 11)}$$

$$\Sigma Q_n = 11 \times 17.68 = 194.5 \text{ kips}$$

$$a = \frac{194.5}{0.85 \times 3 \times 90} = 0.85 \text{ in}$$

$$e = \frac{20.66}{2} + 5.25 - \frac{0.85}{2} = 15.2 \text{ in}$$

$$y = \frac{-194.5 + 378}{2 \times 0.35 \times 50} = 5.24 \text{ in}$$

$$\phi M_n = 794.0 \text{ kip-ft} > M_u = 782.5$$

Composite Design ■ 10-37

10.4 Computer Program (HCOMPL)

The design of composite beams, if done manually, requires greater effort by the engineer than the design of noncomposite steel members. The LRFD method has made engineers spend more time in the design process as compared with the ASD method. The strength criteria are based on ultimate design, and the serviceability criteria are based on the working loads. The engineer must determine the elastic properties of the section to determine the deflections. Therefore, the utilization of computers for automated design with a comprehensive design computation and serviceability computations is almost essential when using the LRFD method. One of the first computer programs was made available in 1987 to the engineering community, that of HCOMPL by Haris Engineering Software Company (HESCO). This program was developed by Haris to design the first major multistory building utilizing the LRFD method—The 312 Walnut, a 35-story office building in Cincinnati, Ohio. (Fig. 10.1)

This computer program has been enhanced significantly in the way data are entered since the time it was developed. Today's engineers are accustomed to user-friendly design and analysis programs that require little effort to produce data for any problem. Therefore, HCOMPL was developed with interactive data screens that make each data entry self-explanatory. The data entry program is a separate file from the design program written in BASIC language for DOS applications. The program recalls a default data file which can be edited at the start of a project and tailored for that project's specific construction type. The name of the project, the engineer's name, type of metal deck, thickness and properties of the concrete over the metal deck, size and length of studs, and acceptable deflection criteria all can be changed in the default data file. Therefore, such data do not have to be entered for each beam design. Once the input data files have been saved, they can be recalled and changes in the input data can he made. This process of data-interactive entry enables the engineer to run the program without review of the manual. The manual becomes a reference to cross-check the results or understand how the program performs the computations. The first screen of the program offers the option of either design or investigation mode.

A metric version of the program is also available as an option. A metric default data file is called up when the program is run in a metric environment. Since engineers in the United States are accustomed to design results in English units rather than metric units, the program offers the options of entering data in metric units and obtaining results in either metric or English units. Therefore, the engineer can first review the results in English units. If the results of the design are reasonable, the engineer can rerun the program selecting metric units for the design results.

The design program is a separate file written in Microsoft FORTRAN. It is loaded and run after the interactive data entry program execution has been completed. The design program recalls the AISC standard section data base for the wide-flange sections. After reading all data including loads at service values, the program produces the ultimate load combinations and performs the section selection starting with the least-weight section with the nominal depth entered as the minimum section depth. The size selection is based on the following criteria:

1. If negative end moments exist, the program will recall the size that satisfies the end moment as a noncomposite steel section with unbraced length of the bottom flange.

2. For unshored construction, the program will select the least-weight steel section as a noncomposite section to support the construction loads consisting of the self-weight, the wet weight of the concrete, and an additional construction live load.

3. Utilizing the size found in 2 above, the program generates the composite beam capacity with the assumption that the full composite action is achievable. The flexural nominal capacity of the composite section is compared with the ultimate maximum positive moment. If the flexural capacity of the composite section is less than the applied ultimate moment, the program selects a larger section and repeats all computations until the strength requirements are satisfied.

4. A partial composite action is generated and the same seven values as the AISC for composite actions are generated. The composite action value just greater than the applied moment is selected.

5. Deflection due to superimposed dead loads and live loads is generated, utilizing the elastic properties of the composite section with consideration of partial composite action. If the deflection is greater than the allowable deflection specified in the input data, then the program moves to greater composite action until it reaches full composite. If full composite still produces deflection greater than allowable, then a larger member size is selected and the composite properties are generated.

6. If concentrated loads occur, the program checks the adequacy of the section based on the studs available between the concentrated load location and the zero moment (or support).

7. Based on the final member size selected, the program will determine the camber values based on the construction dead load values.

The investigation program is similar to the design program in the sequence of computations except that the size of the beam and number of studs are given. The variables that may change based on the requirements to meet strength and deflection criteria are the bottom and top additional plates. The bottom plate is first added without any top plates. Top plates are added only when the maximum bottom plate size is reached before meeting the strength or deflection criteria.

10.5 Composite Columns

The AISC introduced composite column design in 1986 in the first edition of *Manual of Steel Construction Load and Resistance Factor Design*.[1] With the American Concrete Institute (ACI) design procedure for composite columns, the engineer today has the option to design composite columns using either standard. The results can be significantly different. In this chapter is presented an exact procedure used by the author. According to this exact procedure using computer programs, the basic composite column types are (Fig. 10.14)

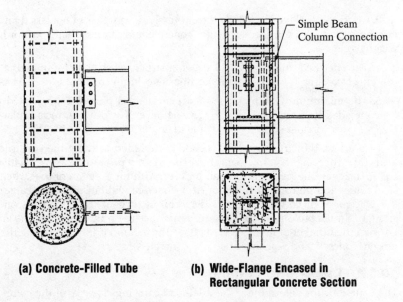

Fig. 10.14 Composite column sections. (*a*) Concrete-filled tube. (*b*) Wide-flanged section encased in rectangular concrete section.

1. Steel wide-flange section encased in a rectangular-shaped concrete section with vertical reinforcing-steel bars and ties.
2. Concrete-filled steel pipe with or without vertical reinforcement.

The AISC procedure is applicable for columns when the area of the structural steel shape is at least 4 percent of the gross section area (including the concrete). If the structural-steel section is less than 4 percent, the column is considered as a reinforced-concrete column, making the ACI procedure applicable for the design.

The limitations for the composite column to qualify for the AISC design procedure, as listed in the *Manual of Steel Construction*, are:

1. The cross-sectional area of the steel shape, pipe, or tubing shall comprise at least 4 percent of the total composite cross section.

2. Concrete encasement of a steel core shall be reinforced with longitudinal load-carrying bars, longitudinal bars to restrain concrete, and lateral ties. Longitudinal load-carrying bars shall be continuous at framed levels; longitudinal restraining bars may be interrupted at framed levels. The spacing of ties shall not be greater than two-thirds of the least dimension of the composite cross section. The cross-sectional area of the transverse and longitudinal reinforcement shall be at least 0.007 in^2 per inch of bar spacing. The encasement shall provide at least $1\frac{1}{2}$ in of clear cover outside of both transverse and longitudinal reinforcements.

3. Concrete shall have a specified compressive strength f'_c of not less than 3 ksi nor more than 8 ksi for normal-weight concrete and not less than 4 ksi for lightweight concrete.

4. The specified minimum yield stress of structural steel and reinforcing bars used in calculating the strength of a composite column shall not exceed 55 ksi.

5. The minimum wall thickness of structural-steel pipe or tubing filled with concrete shall be equal to $b\sqrt{F_y/3E}$ for each face of width b in rectangular and $D\sqrt{F_y/8E}$ for circular sections of outside diameter D.

In practice, the structural-steel section is designed to carry the construction loads from the weight of the floors above the column prior to encasing it with concrete. This way, the steel erectors can proceed with the erection of the structural-steel frame and pouring of the floor deck concrete without the delay caused by the placement of concrete around the column to gain the adequate concrete strength. In tall buildings, the designer must specify on the drawings the limitations on the number of floors erected above the column prior to encasing the column in concrete and reaching adequate concrete strength.

10.5.1 Composite Columns

The equations for determining the axial-load carrying capacity of the composite sections are the same as those for steel sections:

$$\phi P_n = 0.85 \times A_s F_{cr}$$

where

$$F_{cr} = (0.658 \lambda_c^2) F_{my} \quad \text{for } \lambda_c \leq 1.5$$

or

$$F_{cr} = \frac{0.877}{\lambda_c^2} F_{my} \quad \text{for } \lambda_c > 1.5$$

$$\lambda_c = \frac{KL}{r_m \pi} \sqrt{\frac{F_y}{E_m}}$$

The value of the yield strength of F_{my} is

$$F_{my} = F_y + C_1 F_{yr} \frac{A_r}{A_s} + C_2 f'_c \frac{A_c}{A_s}$$

and the value of the modulus of elasticity E_m is

$$E_m = E + C_3 E_c \frac{A_c}{A_s}$$

where r_m = radius of gyration of steel shape but not less than 0.3 times the overall dimension of the composite section in the plane of buckling, in

A_c = area of concrete, in²

A_r = area of longitudinal reinforcing bars, in^2
A_s = area of structural-steel shape, in^2
E = modulus of elasticity of steel, ksi
E_c = modulus of elasticity of concrete, ksi
F_y = yield stress of steel shape, ksi
F_{yr} = yield stress of reinforcing steel bars, ksi
f'_c = concrete compressive strength, ksi
C_1, C_2, C_3 = numerical coefficients, as follows:

For concrete-filled pipe and tubing,

$C_1 = 1.0$
$C_2 = 0.85$
$C_3 = 0.4$

For concrete-encased shapes,

$C_1 = 0.7$
$C_2 = 0.6$
$C_3 = 0.2$

The combined compression and flexure follow the basic interaction equations according to the AISC, as follows: For $P_u / \phi P_n \leq 0.3$,

$$\frac{P_u}{\phi P_n} + \frac{8}{9}\left(\frac{M_{ux}}{\phi_b M_{nx}} + \frac{M_{uy}}{\phi_y M_{ny}}\right) \leq 1.0$$

The nominal axial strength of the composite section ϕP_n is determined as discussed earlier in this section.

The nominal flexural strengths M_{nx} and M_{ny} are determined using an approximate equation. The following approximate equation is limited for relatively large axial compression loads (for $P_u/\phi P_n > 0.3$):

$$M_n = M_p = ZF_y + 1/3(h_2 - 2C_r)A_r F_{yr} + \left(\frac{h_2}{2} - \frac{A_w F_y}{1.7 f'_c h_1}\right) A_w F_y$$

where A_w = web area of encased steel shape; for concrete-filled pipes and tubes, use $A_w = 0$, in^2

Z = plastic section modulus of structural steel section, in^3

C_r = average of distance from compression face to vertical reinforcing bars located near the compression face and distance of tension face from vertical bars located near the tension face

h_1 = width of composite cross section perpendicular to the plane of bending, in

h_2 = width of composite cross section parallel to the plane of bending, in

10-42 ■ Chapter Ten

The AISC *Manual of Steel Construction Load and Resistance Factor Design* provides design aid tables of a variety of wide-flange sections encased in a rectangular reinforced-concrete section. In addition, the tables include concrete-filled steel pipes and tubes without reinforcing steel.

10.5.2 Numerical Method Using Strain Compatibility Approach

The near exact solution of the composite cross section for a generalized interaction of axial load (compression or tension) and the biaxial bending moment is introduced in the computer program HCONCOL by Haris Engineering Software Co. (HESCO) as a numerical solution of the integrations of forces acting on the cross-section's elements. The calculations divide the cross section of the rectangular column into a mesh of elements of equal width and height for the concrete. For columns subject to compression load and biaxial moments, the maximum concrete compressive strain $\varepsilon = 0.003$ occurs at one corner of the rectangular section. The minimum compression strain (or tensile strain) occurs at the opposite corner from the maximum compression corner. Figure 10.15 shows the basic concept of the strain-stress analysis of the column cross section.

The strength elements of a column section are: (1) concrete material, (2) reinforcing steel, (3) structural-steel section in composite columns.

The contributions of each of the above elements to the axial loads and moments are described below.

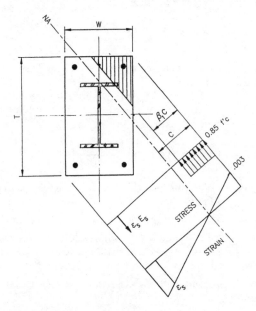

Fig. 10.15 Strain and stress in rectangular column section subject to biaxial moments.

Concrete Material ■ A generalized strain distribution along both axes of the rectangular section produces a stress block, as shown in Fig. 10.14. The tension in the concrete is neglected. With the assumption that c is the distance between the neutral axis and the corner or section edge where maximum compression strain occurs, the following equation determines the equivalent stress block size:

$a = \beta_1 c$

According to ACI, the value of β_1 is, for $f'_c \leq 4000$ psi,

$\beta_1 = 0.85$

and for $f'_c > 4000$ psi,

$b_1 = 1.05 - 0.05 \times \dfrac{f'_c}{1000}$

The axial load and moments capacity contributed by the concrete material can be expressed as follows:

$P_c = 0.85 \times f'_c \times A_{cb}$

$M_{xc} = P_c \times Y_c$

$M_{yc} = P_c \times X_c$

where A_{cb} is the area of the concrete compression block and Y_c and X_c are the distances of the centroid of the compression block from the centroid of the column section.

Reinforcing Steel Bars ■ The strain value of any bar location in the column section can he determined given the maximum strain of 0.003 at a corner of the section and a linear strain distribution in both directions. Therefore, the stress at any bar is

$f_{si} = e_i \times E_s$

$E_s = 29,000$ psi

provided that f_{si} is less than f_y of the steel material. The reinforcing bars' stress can be compression or tension. If reinforcing bars are placed at the concrete compression zone, then the concrete stresses produced by the area displaced by the bars must be subtracted from the steel material capacity. The same, however, can be achieved by producing the following effective bars' stresses when they are in compression:

$f_{esi} = f_{si} - f_{ci}$

where f_{ci} is the concrete compressive stress at the location of the steel bar. The contributions of the reinforcing bars to the axial and moments' capacity of the column section are as follows:

$P_r = \Sigma f_{esi} \times A_{ri}$

$M_{xr} = \Sigma f_{esi} \times A_{ri} \times Y_{ri}$

$M_{yr} = \Sigma f_{esi} \times A_{ri} \times X_{ri}$

where A_{ri} = area of the reinforcing bars, in^2
Y_{ri} = distance of the bar from the major axis of the section, in
X_{ri} = distance of the bar from the minor axis of the section, in

Structural-Steel Wide-Flange Section in Composite Columns ■ The performance of the structural-steel section in the column section is similar to the reinforcing steel assuming full compatibility in strain between the concrete and steel section. The steel wide-flange section is divided into 15 elements. The flanges are divided into five elements each, and the web is divided into five elements. Strain is calculated at the centroid of each element. Therefore, stresses are calculated as follows:

$$f_{sti} = \varepsilon_i \times E_s$$

When the stresses are compressive, then

$$f_{sti} = \varepsilon_i \times E_s - f_{ci}$$

The contribution of the structural-steel section to the axial and bending moment capacity is as follows:

$$P_{st} = \Sigma f_{sti} \times A_{sti}$$
$$M_{xst} = \Sigma f_{sti} \times A_{sti} \times Y_{sti}$$
$$M_{yst} = \Sigma f_{sti} \times A_{sti} \times X_{sti}$$

where A_{sti} = area of the flange or web steel element, in^2
Y_{sti} = distance of the steel element's centroid from the major axis of the column section, in
X_{sti} = distance of the steel element's centroid from the minor axis of the column section, in

The total capacity of the column section is the sum of the forces and moments in the three sections above.

The basic principle of ultimate strength in a round reinforced-concrete section is the same as for a rectangular section. A round section, however, does not need to be analyzed for biaxial bending. If moments are applied around two assumed principal axes, then the resultant moment can be found and compared with the biaxial capacity of the section (see Fig. 10.16). The strength elements of the round column section are the same as in the rectangular composite sections:

1. The concrete material
2. The reinforcing steel
3. The structural-steel pipe section in a composite column.

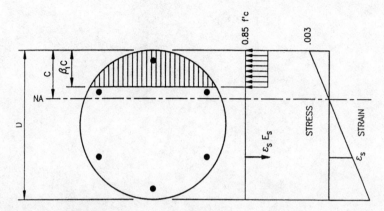

Fig. 10.16 Strain and stress in circular column section subject to moments.

References

1. *Manual of Steel Construction—Load and Resistance Factor Design*, AISC, 1986.

2. Murray, Thomas M., "Design to Prevent Floor Vibrations," *Engineering Journal*, AISC, Third Quarter, 1975.

3. Murray, Thomas M., "Acceptability Criterion for Occupant-Induced Floor Vibrations," *Engineering Journal*, AISC, Second Quarter, 1989.

11

Demetrios E. Tonias

LIMIT DESIGN OF STEEL HIGHWAY BRIDGES

The design of highway bridges in general, and steel highway bridges in particular, has undergone considerable change since the advent of modern bridge construction in the first half of the twentieth century. The origins of the modern highway bridge, as we know it today, can be traced to the nascent development of the U. S. Interstate. The construction of over 44,000 miles (70,800 km) of highway and associated bridge structures led to the creation of design and construction techniques which, for the most part, remained unchanged for several decades after their introduction.

Recently, however, the changing nature of design, along with the development of new design methodologies, has led to a departure from the design specifications with which the majority of our highway bridges were designed. This new approach, known as *limit design*, is rapidly changing not only the way engineers design bridges, but also the way they think about them.

11.1 Overview of Limit States Design of Highway Bridges

While the use of limit states design for steel highway bridges is not necessarily *new* (in terms of its development and use throughout the world), its widespread and accepted use in the United States has occurred only recently. Indeed, it was not until the latter part of 1994 that the American Association of State Highway and Transportation Officials (AASHTO) issued their first, stand-alone specifications which utilized a limit states approach. Known as the AASHTO LRFD Bridge Design Specifications, these design specifications represent a landmark departure from the more traditional working stress design (WSD) approach used previously by transportation departments in the United States and throughout the world.

Prior to discussing the nature and impact of these new specifications, however, it is important to understand the differing philosophies of working stress and limit states design as they apply to the design of steel highway bridges.

11.1.1 Working Stress and Limit States

In bridge engineering, two principal methods of design are in use today. The names used to define these design methods vary depending on the structural material being used, the design code being referenced, or even the era of a publication. For the purposes of this section, we classify the two design methods as

1. Working stress design
2. Limit states design

For most of the century, the working stress design approach was the standard by which bridges and other structural engineering projects were designed. By the 1970s, however, limit states design began to gain acceptance by the general engineering community. What are these two approaches to design and how do they differ? Is one better than the other? To answer these questions, it is first necessary to understand the concepts behind each approach. The following offers both a background and overview of these two design methods and how they apply to the design of structures in general and steel highway bridges in particular.

Working Stress Design ▪ Working stress design is an approach in which structural members are designed so that unit stresses do not exceed a predefined allowable stress. The allowable stress is defined by a limiting stress divided by a factor of safety, so that, in general, working stress is expressed in the form of

$$f_{actual} \leq f_{allowable} \tag{11.1}$$

For a beam this actual stress would be defined by

$$f_{actual} = \frac{Mc}{I}$$

where M = maximum moment
c = distance to the neutral axis from the extreme fiber
I = moment of inertia of the beam cross section

and the allowable stress could be given by

$$f_{allowable} = \frac{f_y}{FS}$$

where f_y = minimum yield stress
FS = factor of safety

Limit Design of Steel Highway Bridges ■ 11-3

The allowable stress could also be defined by some other controlling criterion such as the buckling stress for steel, compressive strength of concrete, etc. Thus the allowable stress can be thought of as a fraction of some failure stress for a given material like steel or concrete.

Under the working stress approach, the actual stresses are representative of stresses due to the service or working loads that a structure is supposed to carry. The entire structure is designed to fall well within the elastic range of the material the element or component is constructed with. When the strain, or deformation, of a material is proportional to the applied stress, the material is said to behave elastically. Figures 11.1 and 11.2 show stress-strain diagrams for steel and concrete, respectively. The point where a material ceases to behave elastically is defined as the proportional limit (i.e., stress and strain are no longer proportional and the stress-strain curve is no longer linear). Once stress and strain are no longer proportional, the material enters the plastic range.

For elastic materials (i.e., materials that behave elastically up to their yield stress) the working stress approach seemed to make a great deal of sense. Since, if a material is loaded past the yield point, a permanent or plastic deformation will occur, the elastic range offers a known, safe region within which an engineer could confidently design structures. In addition to this, the load a member can carry prior to failure is easily measured.

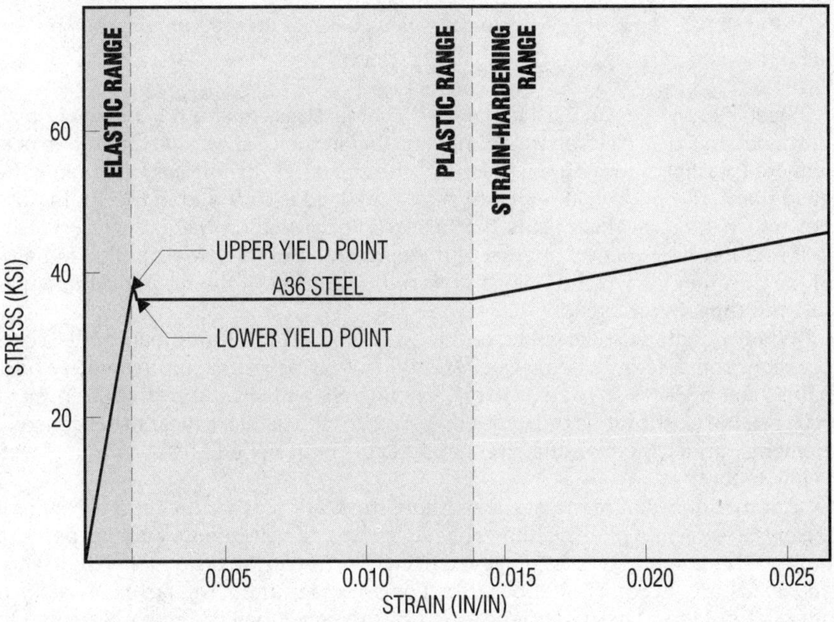

Fig. 11.1 Stress-strain diagram for A36 steel in tension.

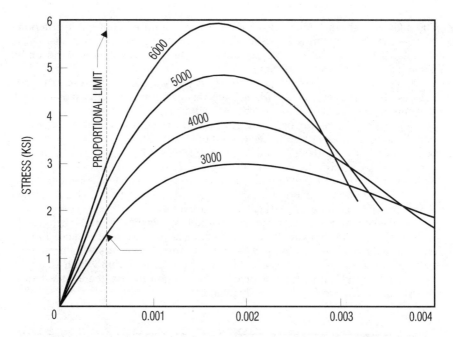

Fig. 11.2 Stress-strain diagram for various concrete strengths in compression.

Steel is known as an elastic material. That is, stress to strain is relatively proportional up to the yield point. This means that in the elastic range there is no permanent (plastic) deformation. As long as the stress (i.e., loading) is kept below the yield point, the strain will return to zero if the load is removed. If the strain does not return to zero, this means that a plastic deformation has occurred. Plastic deformation is a function not just of the magnitude of the stress applied but also of the duration for which the load is placed. The effect of the former is known as slip and the latter as creep.

What of materials like concrete, though? Figure 11.2 shows that, under compression, concrete only behaves elastically to a stress that is approximately one-half its compressive strength. That is, for concrete with a compressive strength of 3000 psi, the elastic range only goes up to about 1500 psi. This means that concrete elements, under the working stress approach, are designed at a level that is well below failure.

Another question mark for the working stress approach is the incorporation of factors of safety. While the allowable stress has, for all intents and purposes, a built-in factor of safety, it is, however, fixed. This means that no matter how variable loads are, in terms of either frequency or magnitude, the factor of safety is always the same. These deficiencies led to the development of an alternative to the working stress design method based on the limit states of a material.

Limit States Design ▪ The limit states design method was, in part, developed to address the drawbacks to the working stress approach mentioned above. This approach makes use of the plastic range for the design of structural members and incorporates load factors to take into account the inherent variability of loading configurations.

The AISC *Manual of Steel Construction* defines a limit state as a condition representing "structural usefulness." As mentioned previously, working stress suffered from the inability of the factor of safety to adequately address the variable nature of loading conditions. One of the advantages of the limit states approach is that it takes into account this variance by defining limit states which address strength and serviceability. The bridge designer can think of these terms in the following way:

Strength is the limit state which defines the safe operation and adequacy of the structure. The criteria which are used to define this are yielding, plastic strength, fatigue, buckling, overturning, etc.

Serviceability is the limit state which defines the performance and behavior of the structure. Some serviceability criteria are deflection, vibration, drift, etc.

From the above, it is easy to see why limit states design codes, like those published by AISC and AASHTO, place a great deal of importance on the strength limit state, since this is the one that is concerned with "public safety for the life, limb and property of human beings." This is why the strength limit state is also often referred to as the safety limit state. Obviously, the limit states for strength will vary depending on the type of member being designed, its material properties, and the given loading condition.

Therefore, like working stress design, limit states design methods vary depending on the material being used and its related design specification. In general, though, we can define the limit states equation as

$$\text{Strength provided} \geq \begin{pmatrix} \text{strength required:} \\ \text{axial force, shear,} \\ \text{or bending caused} \\ \text{by factored loads} \end{pmatrix} \tag{11.2}$$

The strength provided is defined by the specification applicable to the design of the member (e.g., ACI, AISC, AASHTO). The strength required is computed using conventional analysis methods and multiplying computed values by appropriate load factors. This can be translated symbolically into an equation whose form is

$$\phi S_n \geq \sum \psi_i L_i \tag{11.3}$$

where ϕ = strength reduction factor
S_n = nominal strength
L_i = a service load acting on the member
ψ_i = a load factor pertaining to uncertainty of L_i

Thus the right half of Eq. 11.3 represents the sum of individual loads, each multiplied by its specific load factor. Therefore, if we were simply considering dead load and live load, the strength required would be dead load times some factor plus live load times another factor. Specific values for load factors are provided by the applicable design codes.

Development and Application ▪ The development, application, and acceptance of a design methodology is not a trivial concern. Civil engineers spend 4 years at a university being educated in the general profession and specific disciplines. Another 4 years are spent gaining experience and becoming licensed as a professional. Another decade or so could be invested afterward in the formulation of a specific expertise. All of this educational effort is often built around a core design approach. One does not come by such an accumulation of expertise without a significant degree of plain, hard work.

Then, all of a sudden, a designer can be faced with a radically new way of doing things. This was the case in 1971 for a great many engineers designing concrete structures. The working stress method was an accepted and proven way of building structures. No science, however, is static. Research and development initiated as early as the 1930s began investigating the ultimate strength of concrete beams. In 1963, the new release of the ACI Building Code (ACI 318-63) had a limit states approach published along with the traditional working stress approach. By 1971 the transformation was complete. For concrete structures, working stress was now the alternative rather than the norm. Today, the latest release of the ACI Building Code (ACI 318-89, revised 1992) relegates the working stress method to an appendix.

Still, the progression toward a new approach is not an easy affair. Although concrete design has moved away from working stress, the design of steel structures still maintains a dual standards approach. The road that AISC and AASHTO took toward limit states for the design of steel structures, however, was somewhat different from that of ACI for those constructed with concrete.

It would be easy, from a cursory review, to say that ACI was much more progressive than AISC and AASHTO in moving to limit states. The elastic properties of steel, however, are quite different from those of concrete. Unlike concrete, which reaches the proportional limit at only half its compressive strength, steel performs pretty much linearly right up to its yield point. This makes working stress better suited for steel than for concrete. A comparison of Figs. 11.1 and 11.2 illustrates why concrete professionals made a big push to refine the elastic design approach. Conversely, it also shows why steel professionals were in less of a hurry.

It should also be noted that, even though it was not until 1986 when AISC first issued its comprehensive limit states design specification, the working stress methodology was revised in 1978 to take into account the performance of steel at the limit states. The important advantage steel designers realize with the limit states approach, now codified in the AISC and AASHTO specifications, is the ability to "give proper weight to the degree of accuracy with which the various loads and resistances can be determined."

11.1.2 Development of the AASHTO LRFD Code

Traditionally, the design of engineering structures follows industry-formulated guidelines and standards. It has been said that standards are a wonderful thing, which is why we have so many of them. Where organizations like the American Institute of Steel Construction (AISC) and the American Concrete Institute (ACI) formulate design standards based on a specific material (i.e., steel and concrete), the AASHTO design specifications address the implementation and application of material design specifications for the design of a specific type of structure: the highway bridge.

As mentioned earlier, these bridge design specifications originated in the first half of the twentieth century. At one point in time, only large states with a major transportation infrastructure, such as New York and California, could afford the luxury of maintaining such specifications.

In 1931, however, AASHTO introduced the first national standard for the design and construction of highway bridges. This specification, which came to be known as the Standard Specifications for Highway Bridges, has been continuously updated, with major and interim revisions, and is currently in its 15th edition (1992).

These design standards were based solely on the working stress approach until the early 1970s when adjustments were made to take into account the variable nature of load types (e.g., wind loads, live loads). These "adjustments" took the form of adjusting design factors and were contained within a design methodology known as load factor design (LFD).

While LFD can be considered a form of limit states design, it differs from a more comprehensive load and resistance factor design (LRFD) methodology which takes into account the variability in the behavior of structural elements, in an explicit manner, through the use of statistical methods.

It should be noted that the AASHTO design specifications parallel those of related material design specifications. For example, the AASHTO specifications covering the design of concrete elements draw heavily on the ACI Building Design Code. The same is true for steel. There is not, however, a one-to-one correspondence between AASHTO and other material design codes. The allowable stress for steel in bending, for example, is defined differently by AASHTO than by AISC.

Since there is a natural relationship between the AASHTO and related material design specifications, it was only a matter of time before AASHTO would follow AISC's suit and introduce its own LRFD specification (the latter of which originated in 1986 and has since seen a major revision).

Like AISC, which issues both working stress and LRFD specifications, AASHTO now maintains a dual standard approach with its classic Standard Specifications for Highway Bridges now existing alongside the newly adopted LRFD Bridge Design Specifications.

11.1.3 The Canadian Experience—OHBDC

The AASHTO design specifications are indeed the standard on which the United States and many other countries base the design of their highway bridges. However, the advanced use of LRFD in the design of steel highway bridges has been ongoing for many years now. Indeed, Americans need look no farther than their northern border to see one of the first and most widespread uses of LRFD in the analysis and design of bridges.

The Ontario Ministry of Transportation and Communication, like many agencies, relied on the AASHTO specifications for many years. In 1979, however, the Ministry switched to its own LRFD-based standards. A major reason for making this switch was the ability of the LRFD code to take into account reserve capacity in existing bridges which the more "conservative" working stress approach was unable to address.

The Canadian LRFD specifications, known as the Ontario Highway Bridge Design Code (OHBDC), allowed the province to raise the permissible vehicle loading from roughly 100,000 to 140,000 lb. An example of how the OHBDC recognizes the reserve capacity of bridges can be found in the part of the code covering reinforced-concrete deck slabs. The OHBDC recognizes the arching or compressive membrane effect in slabs, calling for half the conventional amount of reinforcing steel. The code, however, can be more conservative in other areas, such as in its requiring significantly more shear reinforcing in concrete members.

The Canadian experience with LRFD serves as an excellent illustration of the forces driving the switch to LRFD south of the border in the United States. With the maturation of the bridges designed and constructed during the boom days of interstate development, the twin issues of maintenance and rehabilitation have rapidly come to the forefront. Today's engineers are now faced with maintaining and fixing what their predecessors built. Recent estimates for bringing deficient bridges up to current specifications call for staggering expenditures in the billions of dollars. Therefore, one of the most attractive features of LRFD is its ability to offer a more "refined" analysis of the capacity of a structure. The "refinement" of structural capacity values, however, also comes with heightened responsibilities with regard to the specific load factors introduced.

With increased emphasis being placed on the proper allocation of funds in maintaining and rehabilitating highway bridges, LRFD will play an increasingly important role in more accurately assessing the condition of a transportation agency's bridges and thereby provide for a more reasonable maintenance and rehabilitation plan.

11.1.4 Related Design Codes

As mentioned above, the AASHTO design specification draws upon related material design codes. While this section pertains specifically to steel highway bridges, the predominant use of composite concrete slab-steel stringer bridges makes the ACI design specifications also important.

Also, it is important to recognize that a highway bridge is not a discrete structure but rather a component in an overall highway network. The bridge engineer, therefore, will also have to be familiar with related nonstructural design specifications such as AASHTO's "A Policy on Geometric Design of Highways and Streets."

11.2 Steel Highway Bridges

Prior to discussing the actual LRFD design of a steel highway bridge, it is important to understand the various components of this type of structure. Typical steel highway bridges consist of a concrete deck resting on a set of steel girders which are in turn supported at their ends by concrete structures. All elements above these supporting structures are collectively known as the *superstructure*. The supporting structural systems are classified as the *substructure*. This section is principally concerned with the design of steel superstructure elements.

Described below are the principal components and forms of steel superstructures and their associated functionality in various design settings.

11.2.1 Constituent Components

As mentioned above, the superstructure comprises all the components of a bridge above the supports. Figure 11.3 shows a typical superstructure. The basic superstructure components consist of the following:

Wearing Surface ▪ The wearing surface (course) is that portion of the deck cross section which resists traffic wear. In most instances this is a separate layer made of bituminous material. The wearing course usually varies in thickness from 2 to 4 in (51 to 102 mm); however, this thickness can sometimes be larger owing to resurfacing of the overpass roadway, which occurs throughout the life cycle of a bridge.

Deck ▪ The deck is the physical extension of the roadway across the obstruction to be bridged. In this example, the deck is a reinforced-concrete slab. In an orthotropic bridge, the deck is a stiffened steel plate. The main function of the deck is to distribute loads along the bridge cross section or transversely. The deck either rests on or is integrated with a frame or other structural system designed to distribute loads along the length of the bridge or longitudinally.

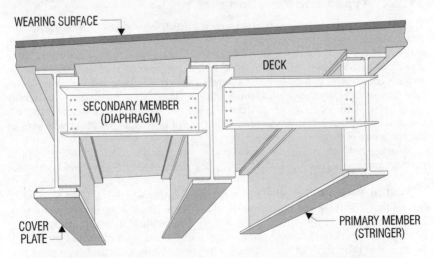

Fig. 11.3 Constituent components of a slab-on-stringer superstructure.

Primary Members ■ Primary members distribute loads longitudinally and are usually designed principally to resist flexure. In Fig. 11.3, the primary members consist of rolled, wide-flange beams. In some instances, the outside or fascia primary members possess a larger depth and may have a cover plate welded to the bottom of them to carry an additional parapet or curb load as well as provide for a more aesthetically pleasing structure.

Beam-type primary members such as this are also called stringers or girders. These stringers could be steel plate girders (i.e., steel plates welded together to form an I section), prestressed concrete, glued laminated timber, or some other type of beam. Rather than have the slab rest directly on the primary member, a small fillet or haunch can be placed between the deck slab and the top flange of the stringer. It is also possible for the bridge superstructure to be formed in the shape of a box (either rectangular or trapezoidal). Box girder bridges can be constructed out of steel or prestressed concrete and are used in situations where large span lengths are required.

Secondary Members ■ Secondary members are bracing between primary members designed to resist cross-sectional deformation of the superstructure frame and help distribute part of the vertical load between stringers. In Fig. 11.3 a detailed view of a bridge superstructure shows channel-type diaphragms used between rolled section stringers. The channels are bolted to steel connection plates which are in turn welded to the wide-flange stringers shown. Other types of diaphragms are short-depth, wide-flange beams or crossed, steel angles. Secondary members, composed of crossed frames at the top or bottom flange of a stringer, are used to resist lateral deformation. This type of secondary member is called lateral bracing.

11.2.2 Types and Functionality

The two principal materials utilized in superstructure construction are steel and concrete. Materials such as timber and aluminum are also utilized to a lesser extent. When compared to concrete, steel has the advantage of lighter weight and more rapid construction. Steel also lends itself well to prefabrication at the factory, which reduces the amount of field labor for operations such as bolting and welding. Recent advancements in fabrication and construction methods have made steel much more competitive with concrete. Automation of element assembly and welding and the use of stay-in-place forms are just two examples of how steel has improved to meet the challenge of the concrete bridge.

A superstructure frame is composed of primary and secondary members. Steel primary members come in a variety of types, some of which are listed below.

Rolled Beam ■ The rolled beam is a steel girder which has been formed by hot rolling. The most common type of rolled beam used as a primary member in highway bridges is the wide-flange variety. The wide flange differs from its I-beam cousin in that its flanges are parallel rather than tapered. When the term I-beam is used throughout this section, it implies a beam with an I-type cross section, not the American Standard I-beam cross section.

Rolled Beam with Cover Plate ■ To maintain an economy of material, rolled beams are sometimes equipped with a rectangular plate, or cover plate, at the bottom flange. The cover plate increases the ability of the stringer to resist flexure without having to use a larger-size rolled beam or plate girder. The cover plate, however, also increases the potential for fatigue damage by introducing stress concentrations at the ends of the plate.

Plate Girder ■ A plate girder, like a rolled beam, has an I-type cross section. Rather than being hot-rolled, however, the girder is constructed from steel plate elements which are connected together with welds, bolts, or rivets. For modern highway bridges, shop welding is the predominant method. Since the designer is specifying the section properties of the stringer (i.e., flange width and thickness, web depth, etc.), a greater economy of materials results. To further reduce the amount of steel used, plate girders can be varied in depth, or haunched (see Fig. 11.4), to accommodate regions of low and high moment and/or shear. Plate girders gain an advantage over rolled beams as span lengths become great.

Box Girder ■ A box girder is a form of plate girder which utilizes four rather than three plates. Since box girders possess excellent torsional stiffness, they usually do not require secondary members to provide bracing. Although the box girder provides an aesthetically pleasing structure, the amount of steel required can sometimes exceed that for a standard I cross-section plate girder.

Prestressed Steel ■ A recent advancement in steel bridges utilizes prestressing of I cross-section stringers (such as a wide-flange rolled beam). Beams can be prestressed through use of one of the following methods:

1. High-strength wires or bars are anchored to the ends of the girder. This can be accomplished with either a rod and turnbuckle or a draped high-strength cable.

2. A part of the I-beam cross section (e.g., top flange) is stressed and, while in the stressed state, welded to the lower T cross section. A similar technique involves welding high-strength cover plates to the top and bottom of a deflected I-beam.

3. Another method of prestressing steel is to deflect the beam and then cast in place the concrete deck slab underneath it. In other words, the bridge is

Fig. 11.4 A haunched plate girder economizes the amount of steel used for long spans.

upside down. The beam is then prestressed by loading the entire system at the bottom flange (which is facing up in this configuration). After the deck has cured, the loads are released and the slab and girder inverted for placement.

Prestressed steel offers the advantages of increased ultimate strength, better resistance to fatigue, and a reduction in the amount of steel required. Also, since the concrete slab has already been compressed, it is better suited to handle thermal forces resulting from seasonal changes. A drawback to prestressed steel bridges is a result of one of its principal advantages: reduced amount of steel. While the economy of material is good from a cost standpoint, it does impact the overall stiffness of the superstructure, which can lead to excessive deflections.

Steel Arch ■ The arch is most often used for major crossings like the Hell Gate and Sydney Harbor bridges. Figure 11.5 shows a picture of the twin Thaddeus Kosciuszko bridges crossing the Mohawk River in upstate New York. In this particular site, the steel arches provide for an attractive-looking structure while also eliminating the need for a pier in the river. When the deck, as is the case with the structures in Fig. 11.5, is suspended from the steel arch, the structure is called a through arch. When the deck is supported on top of the arch, this is called a deck arch. An arch bridge generates large forces at its end supports and therefore requires excellent subsurface conditions.

Fig. 11.5 Twin steel through arches cross the Mohawk River in upstate New York.

Steel Rigid Strut Frame ■ A steel bridge with integral steel supporting legs is another form of structure which utilizes steel as its principal component. In such a configuration, not only is the superstructure made of steel, but the substructure as well.

Large Structures ■ Steel is also an excellent material for large structures requiring spans of significant length. Suspension, arch, truss, and cable-stayed structures all provide solutions for this class of bridge.

11.2.3 Limit States Construction-Related Benefits

In terms of construction-related benefits, a limit states approach can offer several advantages over a working stress methodology. We have already mentioned that limit states allows designers to take into account reserve capacity in existing structures. Similar features of the limit states design approach apply to the design and construction of new structures as well.

Prior to the release of the new AASHTO LRFD code, an *enhanced* limit states criterion, known as the *Autostress* method, has been used for the purpose of economizing bridge members. Autostress is an extension of AASHTO's load factor design procedure mentioned earlier. Autostress expands on AASHTO's load factor design by utilizing enhanced limit states which allow for inelastic load redistribution for continuous structures. Under the Autostress approach, a bridge is overloaded by an initial live loading of the structure. This overload has a pre-stressing effect on the bridge by inducing stresses over the yield point in the negative moment region and relieving some residual stresses. The name Autostress is derived from the automatic load redistribution which occurs.

The Autostress method allows designers to use prismatic members in continuous spans along the entire length of a structure. For rolled beams this means that cover plates can be eliminated and for plate girders, fewer flange splices and changes in thickness are required.

Therefore, it can be seen that the use of limit states in the design of a highway bridge can reduce fabrication costs while at the same time providing enhanced performance characteristics by eliminating fatigue-critical details such as cover plates.

11.3 AASHTO LRFD Code

The AASHTO LRFD code builds on the basic limit states concepts detailed in Eq. 11.3 to create a complete, stand-alone specification for the design of highway bridges. The following details the general principles behind this specification.

11.3.1 Using Load and Resistance Factors

AASHTO defines the limit states criteria for a highway bridge using the following expression:

$$\eta \sum y_i Q_i \leq \phi R_n = R_r \tag{11.4}$$

where η = a factor relating to ductility, redundancy, and operational importance
y_i = load factor: a statistically based multiplier applied to force effects (ψ_i in Eq. 11.3)
Q_i = force effect (L_i in Eq. 11.3)
ϕ = resistance factor: a statistically based multiplier applied to nominal resistance (ϕ in Eq. 11.3)
R_n = nominal resistance (S_n in Eq. 11.3)
R_r = factored resistance = ϕR_n

The factor η is defined such that

$$\eta = \eta_D \eta_R \eta_I > 0.95 \tag{11.5}$$

where η_D = a factor relating to ductility
η_R = a factor relating to redundancy
η_I = a factor relating to importance

Ductility and redundancy factors relate to the physical strength of a bridge while importance refers to the potential effects of a bridge being out of service. These factors illustrate the dynamic nature of limit states criteria. Limit states criteria are continually refined to more accurately reflect the capabilities of a structure to withstand loads. Given the recent development of the AASHTO standard, it is quite natural that these and other factors will be refined as the code is applied on a more widespread basis. Specific values for the η factors are listed below:

η_D = 1.05 for nonductile components and connections
= 0.95 for ductile components and connections
= 1.00 for all other limit states

η_R = 1.05 for nonredundant members
= 0.95 for redundant members
= 1.00 for all other limit states

η_I ≥ 1.05 if bridge is of operational importance
≥ 0.95 if bridge is not of operational importance

In many instances, the force effects of Eq. 11.4 are determined through the use of an elastic analysis. Since the LRFD approach makes use of the inelastic range of a material in determining the ability of components and connections to resist loads, this will naturally lead to some inconsistencies. This is a problem inherent in limit states methodologies, however, because of our incomplete knowledge of inelastic structural behavior.

Strength and Serviceability ▪ It was discussed earlier that any limit states methodology is based on a structure fulfilling strength and serviceability limit states. The AASHTO LRFD code quantifies five strength and three service limit states. In addition to these are two *extreme event* and one *fatigue* limit state. Table 11.1 describes the general nature of these limit states.

Load combinations consist of groupings of loads acting on a structure. These "groupings" represent probable occurring combinations of loads acting on a structure. All structural members should be designed to withstand all load groups. In the design of a highway bridge, the load combinations are selected to "obtain realistic extreme effects." Naturally, not all of the loads in a given load combination are going to apply for each structure (e.g., stream loads do not apply to a bridge crossing another highway).

Table 11.2 shows the various load combinations and load factors present in the AASHTO LRFD code. The concept behind this table is similar to that of the group loading coefficients and load factors present in the original AASHTO specification.

Table 11.1 AASHTO LRFD Strength and Serviceability Limit States

Limit State	Description
Strength I	Basic load combination relating to the normal vehicular use of the bridge without wind.
Strength II	Load combination relating to the use of the bridge by owner specified special design vehicles and/or evaluation permit vehicles, without wind.
Strength III	Load combination relating to the bridge exposed to wind velocity exceeding 55 mph.
Strength IV	Load combination relating to very high dead load to live load force effect ratios.
Strength V	Load combination relating to normal vehicular use of the bridge with wind of 55 mph velocity.
Extreme event I	Load combination including earthquake.
Extreme event II	Load combination relating to ice load, collision by vessels and vehicles, and to certain hydraulic events with reduced live load, other than that which is part of the vehicular collision load, CT.
Service I	Load combination relating to the normal operational use of the bridge with 55 mph wind, and with all loads taken at their nominal values. Also related to deflection control in buried metal structures, tunnel liner plate and thermoplastic pipe, and to control crack width in reinforced concrete structures.
Service II	Load combination intended to control yielding of steel structures and slip of slip-critical connections due to vehicular live load.
Service III	Load combination relating only to tension in prestressed concrete structures with objective of crack control.
Fatigue	Fatigue and fracture load combination relating to repetitive gravitational vehicular live load and dynamic responses under a single design truck having a constant axle spacing of 30 ft between 32 k axles (AASHTO LRFD 3.6.1.4.1).

The AASHTO LRFD code also presents load factors for construction and jacking loads. These load factors take into account situations such as the loading of a structure with special construction equipment or jacking of a structure to maintain bearings with traffic passing over the bridge.

As can be seen from Table 11.2, the AASHTO LRFD code provides for different limit states which are defined by various combinations of loads. AASHTO defines these loads using a two-letter designation which pertains to an individual type of *permanent* or *transient* load. The designation of these loads is as follows:

Permanent Loads

DD = downdrag
DC = dead load of structural components and nonstructural attachments
DW = dead load of wearing surfaces and utilities
EH = horizontal earth pressure load
ES = earth surcharge load
EV = vertical pressure from dead load of earth fill

Transient Loads

BR = vehicular braking force
CE = vehicular centrifugal force
CR = creep
CT = vehicular collision force
CV = vessel collision force
EQ = earthquake
FR = friction
IC = ice load
IM = vehicular dynamic load allowance
LL = vehicular live load
LS = live load surcharge
PL = pedestrian live load
SE = settlement
SH = shrinkage
TG = temperature gradient
TU = uniform temperature
WA = water load and stream pressure
WL = wind on live load
WS = wind load on structure

Limit Design of Steel Highway Bridges ■ 11-17

Table 11.2 AASHTO LRFD Load Combinations and Load Factors

Limit state	DC DD DW EH EV ES	LL IM CE BR PL LS	WA	WS	WL	FR	TU CR SH	TG	SE	EQ	IC	CT	CV
										Use one of these at a time			
Strength I	γ_p	1.75	1.00	–	–	1.00	0.50/1.20	γ_{TG}	γ_{SE}	–	–	–	–
Strength II	γ_p	1.35	1.00	–	–	1.00	0.50/1.20	γ_{TG}	γ_{SE}	–	–	–	–
Strength III	γ_p	–	1.00	1.40	–	1.00	0.50/1.20	γ_{TG}	γ_{SE}	–	–	–	–
Strength IV EH, EV, ES, DW	γ_p	–	1.00	–	–	1.00	0.50/1.20	–	–	–	–	–	–
DC only	1.5	–	1.00	–	–	1.00	0.50/1.20	–	–	–	–	–	–
Strength V	γ_p	1.35	1.00	0.40	0.40	1.00	0.50/1.20	γ_{TG}	γ_{SE}	–	–	–	–
Extreme event I	γ_p	γ_{EQ}	1.00	–	–	1.00	–	–	–	1.00	–	–	–
Extreme event II	γ_p	0.50	1.00	–	–	1.00	–	–	–	–	1.00	1.00	1.00
Service I	1.00	1.00	1.00	0.30	0.30	1.00	1.00/1.20	γ_{TG}	γ_{SE}	–	–	–	–
Service II	1.00	1.30	1.00	–	–	1.00	1.00/1.20	–	–	–	–	–	–
Service III	1.00	0.80	1.00	–	–	1.00	1.00/1.20	γ_{TG}	γ_{SE}	–	–	–	–
Fatigue - LL, IM, & CE only	–	0.75	–	–	–	–	–	–	–	–	–	–	–

For various combinations of loads, the individual load types are multiplied by the appropriate load factor listed in Table 11.2. In addition to the factors provided directly in the table, load factors for permanent loading (γ_p), settlement (γ_{SE}), and temperature gradient (γ_{TG}) are taken from Table 11.3 and related design criteria (see below).

The goal when selecting the appropriate load combinations is to create a loading configuration which produces extreme yet realistic results. Each load in the load combination is multiplied by the appropriate *load factor* and *multiple presence factor* (see below). These products are then summed using Eq. 11.4 described earlier.

In the *TU, CR, SH* column of Table 11.2, two values are given (e.g., 0.50/1.20). The larger value is used to account for deformations and the smaller one for all other effects. The reader is referred to the AASHTO LRFD code for a more detailed description concerning the selection of load factors and load combinations.

11.3.2 Loading the Structure

In Sec. 11.3.1 we discussed how load factors are applied to permanent and transient loads. These loads are classified by AASHTO in the following fashion.

11-18 ■ Chapter Eleven

Permanent Loads ■ These loads are comprised of dead and earth loads. The *dead load* on a structure is the aggregate weight of all structure elements. This would include, but not be limited to, the deck, wearing surface, stay-in-place forms, sidewalks and railings, parapets, primary members, secondary members (including all bracing, connection plates, etc.), stiffeners, signing, and utilities. One of the first steps in any design of a highway bridge is to compile a list of all the elements which contribute to dead load. Table 11.4 provides a list of some dead load unit weights that are used in computing dead loads. The designations used by AASHTO for dead load are *DC, DW,* and *EV* (see Sec. 11.3.1).

Earth loads are those loads induced by earth pressure, earth surcharge, and downdrag loads and are designated as *EH, ES,* and *DD*. While these loads primarily affect substructure elements, they have the potential of impacting superstructure elements as well at points where these two components interface (e.g., at the abutment backwall).

Transient Loads ■ Transient, or temporary loads are those loads which are placed on a bridge for only a short period of time. Just as dead loads are the prin-

Table 11.3 Permanent Load Factors, γ_p

Type of Load	Load Factor Maximum	Minimum
DC: Component and Attachments	1.25	0.90
DD: Downdrag	1.80	0.45
DW: Wearing Surfaces and Utilities	1.50	0.65
EH: Horizontal Earth Pressure		
Active	1.50	0.90
At-Rest	1.35	0.90
EV: Vertical Earth Pressure		
Overall Stability	1.35	N/A
Retaining Structure	1.35	1.00
Rigid Buried Structure	1.30	0.90
Rigid Frames	1.35	0.90
Flexible Buried Structures other than Metal Box Culverts	1.95	0.90
Flexible Metal Box Culverts	1.95	0.90
ES: Earth Surcharge:	1.50	0.75

cipal permanent loading condition, *live loads* (LL) represent the major temporary loading condition. There are, however, several other classes of temporary loads which the designer must consider. Discussed below are the major forms of temporary loading.

Vehicle Live Load. The term live load means a load that moves along the length of a span. Therefore, a person walking along the bridge can be considered live load. Obviously, however, a highway bridge has to be designed to withstand more than pedestrian loading. To give designers the ability to accurately model the live load on a structure, hypothetical design vehicles based on truck loading were developed.

To describe the effects of live load, either a *design truck* or a *design tandem* can be used to simulate the effects of a single design vehicle on a bridge. To simulate the effect of a train of trucks on a bridge, a so-called *lane loading* can be used which consists of a uniformly distributed load applied longitudinally along the length of a span.

Table 11.4 Dead Load Unit Weights

Material	Unit Weight
Aluminum Alloys	0.175 k/ft^3
Bituminous Wearing Surfaces	0.140 k/ft^3
Cast Iron	0.450 k/ft^3
Cinder Filling	0.060 k/ft^3
Compacted Sand, Silt, or Clay	0.120 k/ft^3
Concrete	
Lightweight	0.110 k/ft^3
Sand-Lightweight	0.120 k/ft^3
Normal	0.150 k/ft^3
Loose Sand, Silt, or Gravel	0.100 k/ft^3
Soft Clay	0.100 k/ft3
Rolled Gravel, Macadam, or Ballast	0.140 k/ft^3
Steel	0.490 k/ft^3
Stone Masonry	0.170 k/ft^3
Wood	
Hard	0.060 k/ft^3
Soft	0.050 k/ft^3
Transit Rails, Ties, and Fastening	
per Track	0.200 k/ft^3
Water	
Fresh	0.0624 k/ft^3
Salt	0.0640 k/ft^3

Design Truck. The AASHTO LRFD code utilizes a design truck which is analogous to the HS20-44 design vehicle found in the standard specifications (see Fig. 11.6). From Fig. 11.6 we see that the design truck has a variable spacing between the two rear axles. This distance between axles, varying from 14 to 30 ft (4.27 to 9.14 m), is used to create a live loading situation which will induce maximum moment in a span. For simply supported bridges, this value will be the 14 ft minimum. In continuous spans, however, the distance between axles is varied to position the axles at adjacent supports in such a fashion as to create the maximum negative moment.

Design Tandem. An alternative to the design truck is the design tandem. The tandem is comprised of two 25-kip axles that are spaced 4 ft apart. Transversely the wheels are spaced at 6 ft.

Design Lane loading. As mentioned earlier, a lane load is a uniformly distributed load which is applied along the length of a span (i.e., longitudinally) to simulate the effects of a train of trucks moving along the bridge. In the AASHTO LRFD code, design lane loading is defined by a 0.640 kip/ft uniform load distributed transversely over a 10-ft width of the structure.

Design Lanes. The number of design lanes is determined by taking the integer portion of the ratio defined as

$$\frac{w}{12} \quad (11.6)$$

where w = clear roadway width between curbs and/or barriers, ft

It is also important for the designer to anticipate the possibility of any future expansion of the structure which would result in widening of the bridge and possibly increasing the number of design lanes.

Multiple Presence of Live Load. To account for the effects of multiple vehicles on a structure at a given time, a multiple presence factor is applied as outlined in Table 11.5. These factors are based on an average daily truck traffic ($ADTT$) of 5000 trucks going in one direction. AASHTO allows for a reduction in the multiple presence factor if a bridge at a particular site has an

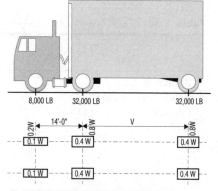

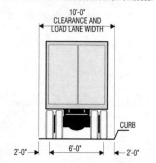

Fig. 11.6 AASHTO LRFD design truck.

ADTT lower than 5000. The extreme live load force effect is found by taking into account each possible combination of the number of loaded lanes multiplied by the appropriate multiple presence factor.

Table 11.5 Multiple Presence Factors m

Number of Loaded Lanes	Multiple Presence Factors m
1	1.20
2	1.00
3	0.85
> 3	0.65

Application of Live Load. The live load configuration selected will be that which creates maximum moment in the span. The criterion used for determining the appropriate live loading scheme is to take the larger of the following:

1. The effect of the design tandem combined with the effect of the design lane load.
2. The effect of one design truck with the variable axle spacing set to create maximum moment (see above), combined with the effect of the design lane load.
3. For both negative moment between points of dead load contraflexure (i.e., inflection points), and reaction at interior piers only, 90 percent of the effect of two design trucks spaced a minimum of 50.0 ft between the lead axle of one truck and the rear axle of the other truck, combined with 90 percent of the effect of the design lane load; the distance between the 32.0-kip axles of each truck should be taken as 14 ft.

Transversely, the design lanes and the position of the 10-ft-wide lane loading should be located so as to create maximum force effects. Axles which are found not to contribute to the maximum force effect are to be neglected.

Other Structure Loads ▪ As can be seen from the discussion on load factors, a highway bridge can be subjected to a wide variety of loads depending on the location of the structure. Many of these loads are a product of *natural forces* which are dependent on the geographic location of the bridge. In general, there are four major natural forces with which the bridge engineer must be concerned:

1. Seismic forces
2. Wind forces
3. Channel forces
4. Thermal forces

Like the vehicle live loads discussed above, seismic, wind, and channel forces are temporary loads on a structure which act for a short duration.

Other types of loads which affect a bridge are *cyclical* in nature. These forces tend to "work" on a bridge over a period of time, causing the gradual propagation of small, localized distresses into larger more severe areas of deterioration. Fatigue loading represents the principal form of cyclical loading acting on a highway bridge structure.

While it is beyond the scope of this text to describe in detail the LRFD criteria pertaining to the many different forms of natural and cyclic loading, it is important to examine fatigue loading in particular, since its effects on a highway bridge can be profound.

Fatigue. Fatigue is the propagation of localized cracks through the repetitive loading and unloading of a structure. As bridges began to evidence fatigue in the late 1960s and early 1970s, AASHTO moved to introduce fatigue criteria which would address this situation. The first fatigue criteria appeared in the 1965 specifications. Subsequently in 1971 and 1974, revisions and enhancements based on experimental data were issued. The issue of fatigue is very broad and pertains to a wide variety of elements. Shear connectors and cover plates are examples of two steel details which can be adversely affected by fatigue loading.

Since fatigue is a result of the cyclical loading and unloading of a structure, the AASHTO LRFD code bases its fatigue loading on the single-lane average daily truck traffic ADTT$_{SL}$. That is, the fatigue limit state cannot be expressed in terms of the load itself but must also take into account the accumulated stress-range cycles as well.

When more accurate data are not available, the single-lane average daily truck traffic is taken as

$$\text{ADTT}_{SL} = p \times \text{ADTT} \tag{11.7}$$

where ADTT = number of trucks per day in one direction averaged over the design life

ADTT$_{SL}$ = number of trucks per day in a single lane averaged over the design life

p = see Table 11.6

For live-load-induced fatigue, each steel detail must satisfy the following condition:

$$\gamma(\Delta f) \leq (\Delta F)_n \tag{11.8}$$

where γ = load factor listed in Table 11.2 for the fatigue load combination

Δf = force effect with a live load stress range resulting from the passage of truck traffic as detailed above, ksi

$(\Delta F)_n$ = nominal fatigue resistance, ksi

In general, the nominal fatigue resistance is given by the following expression:

$$(\Delta F)_n = \left(\frac{A}{N}\right)^{1/3} \geq \frac{1}{2}(\Delta F)_{TH} \tag{11.9}$$

where N = $(365)(75)(n)(\text{ADTT}_{SL})$

A = constant; see Table 11.7

n = number of stress cycles per truck passage; see Table 11.8

ADTT$_{SL}$ = number of trucks per day in a single lane averaged over the design life, Eq. 11.7

$(\Delta F)_{TH}$ = constant-amplitude fatigue threshold; see Table 11.7

In computing N the quantity 75 represents a design life of 75 years. If the owner desires a design life other than 75 years, the desired number of years may be substituted into the expression for N in lieu of 75.

Table 11.7 makes reference to specific components and details identified with the AASHTO LRFD code that are susceptible to the adverse effects of fatigue loading. The category selected depends on the specific detail in question (i.e., a fillet-welded connection vs. a groove-welded connection). The reader is referred to the AASHTO code for a complete description of the various detail categories.

11.3.3 Composite Slab-on-Girder Bridges

The use of two dissimilar materials to form a single structural member is not a new technique. The advent of composite construction in bridge design, however, did not take effect until the mid- to late 1940s. From a historical perspective, the time of the acceptance of composite construction is critical since it came almost concurrently with the nascent development of the U.S. Interstate. The new-found economy of composite construction in the 1940s and 1950s led to the construction of thousands of composite steel beam bridges in the United States alone.

What is composite construction, though, and how does it offer such economy of materials? From basic strength of materials, we know that the ability of a beam cross section to resist loads is a function of the size of its cross-sectional properties. Therefore, it is in the best interest of the designer to increase the section properties of the

Table 11.6 Fraction of Truck Traffic in a Single Lane, p

Number of lanes available to trucks	p
1	1.00
2	0.85
3 or more	0.80

Table 11.7 Detail Category Constant, A, and Constant Amplitude Fatigue Thresholds, $(\Delta F)_{TH}$

Detail category	Constant, A $\times 10^8$	Threshold $(\Delta F)_{TH}$, ksi
A	250.0	24.0
B	120.0	16.0
B'	61.0	12.0
C	44.0	10.0
C'	44.0	12.0
D	22.0	7.0
E	11.0	4.5
E'	3.9	2.6
M164 (A325) bolts in axial tension	17.1	31.0
M253 (A490) bolts in axial tension	31.5	38.0

Table 11.8 Cycles per Truck Passage, n

Longitudinal members	Span length	
	> 40.0 ft	≤ 40.0 ft
Simple-span girders	1.0	2.0
Continuous girders		
near interior support	1.5	2.0
elsewhere	1.0	2.0
Cantilever girders	5.0	
Trusses	1.0	
Transverse members		
spacing > 20.0 ft	1.0	
spacing ≤ 20.0 ft	2.0	

11-24 ■ Chapter Eleven

resisting cross section as much as possible. This is where the principal advantage of composite action comes into play.

If a concrete slab simply rests on top of a steel beam, a phenomenon known as slippage occurs. As loads are placed on top of the slab, the tops of both the slab and beam are in compression and the bottoms of the slab and beam are in tension. In essence, both elements deflect like a beam, albeit independently. Since the bottom of the slab is in tension (i.e., pushing outward toward the ends of the beam) and the top of the beam is in compression (i.e., pushing inward toward the center of the beam), the resulting effect is one of the slab extending out over the ends of the beam (see Fig. 11.7). In analyzing such a configuration, the slab and beam are treated independently, with the geometry of each element defining the neutral axis and moment of inertia of the slab and beam.

If the slab and beam, however, were somehow integrated, they could resist loads as a single unit. In this arrangement, the neutral axis would be located somewhere in the middle of the section defined by the top of the slab and the bottom of the beam. With proper integration, the slab-beam would act as a unit with the top of the slab in compression and the bottom of the beam in tension and no slippage in between. This integration is accomplished through the incorporation of a shear connector between the slab and beam.

The shear connector is generally a metal element which extends vertically from the top flange of the supporting beam and is embedded into the slab. Several of these connectors are placed along the length of the beam to prevent slippage which is caused by the compression in the top of the beam at the slab-beam interface. Shear studs are the most common form of shear connector used today in com-

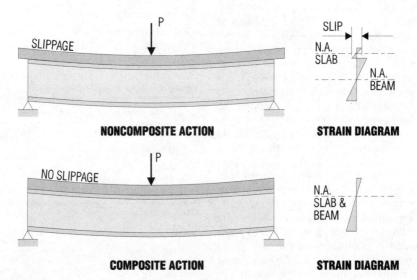

Figure 11.7 Slippage occurs at the slab-beam interface in noncomposite beams.

posite concrete-steel beam bridges. The shear stud's installation is facilitated by the use of an automatic welding gun. This ease of installation and the relatively low cost of stud connectors have led to their popularity in composite construction.

Advantages of Composite Action ▪ With shear connectors in place, the slab and beam can now be analyzed as a single unit. Since the size of the section modulus has been increased, with the incorporation of the concrete slab into the resisting section, the composite beam can resist heavier loads. In essence, the I-shaped beam is replaced by a T-shaped cross section composed of the slab and stringer, the advantages being:

1. A decrease in the size (and weight) of stringer required
2. Longer possible span lengths
3. A stiffer cross section
4. A reduction in live load deflections
5. An increase in overload capacity
6. Enhanced resistance to lateral loads

The economy of material realized by composite construction quickly led to its growth as a staple in highway bridge design. Standard rolled wide-flange beams could now be used for much longer span lengths. The decrease in size of stringers also meant that beams could now be less deep than before. Another added benefit is that the connection between slab and stringer acts like a diaphragm between adjacent stringers to resist lateral deformation.

The reader should be aware, however, that while we use the expression composite beam to imply a concrete slab resting on a steel girder, there are other forms of composite beams. In addition to the steel beam-concrete slab-type bridge described in this text, there also exist composite timber-concrete bridges which utilize a timber, rather than steel, beam, as well as composite concrete-concrete type structures with a concrete slab and concrete girder. For the purposes of the discussion in this section, though, we use the expressions composite beam, composite action, and composite construction to describe a concrete slab-on-steel stringer-type bridge.

So it can be seen that composite construction offers numerous advantages to the bridge designer. The AASHTO LRFD code recommends the use of composite rather than noncomposite construction because of the advantages detailed above.

Shored and Unshored Construction ▪ The dead loads on a slab-on-stringer bridge consist principally of the slab and beam itself. Since concrete has to cure for 28 days before it reaches full strength, the slab and beam cannot be considered to work compositely in resisting dead loads. This means that, until the concrete has hardened to a sufficient point, the steel stringer itself must resist all dead loads.

Shored construction minimizes the amount of dead load the beam has to carry by providing support at intermediate locations along the length of the span. Once the concrete has reached sufficient strength, the shores are removed and the loads are carried by the composite section. Therefore, with shoring, smaller beams are required. Conversely, unshored construction provides no support during curing and expects the beam to resist all dead loads. For most spans, the costs associated with erecting shoring to support the beam are greater than the savings in steel realized. In other instances, the provision of shoring may not be possible.

Loading of the Composite Section ▪ In discussing shored vs. unshored construction, it can be seen that *when* loads are applied to a composite beam is very important. Specifically, the AASHTO LRFD code identifies three major points in the loading sequence for a composite section. These points are defined by loads applied to the following resisting sections

1. Steel only
2. Short-term composite section
3. Long-term composite section

A load is not assumed to be permanent unless the concrete slab has attained 75 percent of its compressive strength f_c'.

Effective Flange Width ▪ Even though the deck runs continuously across the supporting stringers, only a portion of the slab is taken to work in a composite fashion with the stringer. This portion of the slab acts as the top flange of a T-shaped girder (see Fig. 11.8). This section is termed the effective flange width. The definition of the effective flange width varies depending on whether the slab forms a T-shaped top flange (interior stringer) or is present on only one side of the stringer (exterior girder). For T-shaped cross sections, the effective flange width is defined by the AASHTO LRFD code as the minimum of:

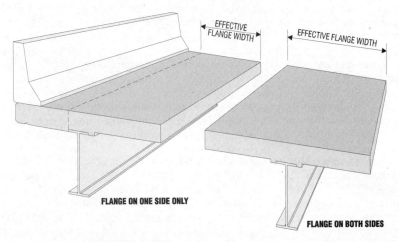

Figure 11.8 Effective flange width for a slab acting as a stringer on one side (exterior stringer) and on both sides (interior stringer).

1. One-fourth the effective span length
2. Twelve times the average thickness of the slab, plus the greater of web thickness or one-half the width of the top flange of the girder
3. The average spacing of adjacent beams

From inspection it can be seen that the first criterion will govern more often for short spans and the second for thin slabs. Otherwise the stringer spacing will most likely hold true. If the slab is present on only one side of the stringer (exterior beam), the effective flange width should not exceed

1. One-eighth of the effective span length
2. Six times the average thickness of the slab, plus the greater of half the web thickness or one-quarter the width of the top flange of the basic girder
3. The width of the overhang

Note that the exterior girder shown in Fig. 11.8 shows a conservative interpretation of the AASHTO effective flange width criteria. For overhangs where the deck is integrally reinforced and monolithically poured with the rest of the slab, the exterior slab-stringer configuration can be analyzed as a symmetrical T-beam section.

The effective flange width is used to compute the section properties of the composite section and represent the portion of the deck which, together with the stringer, resists loads. It is also used in computing the section of the slab acting as dead load on an unshored stringer.

Transformed Section ▪ In order to analyze a composite section, the section of concrete slab must be *transformed* to an equivalent section of steel. This transformation is accomplished through use of the modular ratio n. The modular ratio is defined as

$$n = \frac{E_s}{E_c} \qquad (11.10)$$

where E_s = modulus of elasticity for steel
E_c = modulus of elasticity for concrete

Figure 11.9 shows a visualization of what a transformed section could look like. The accepted value for the modulus of elasticity for steel is 29,000,000 psi. If the actual weight of concrete used is known, the modulus of elasticity can be computed using

$$E_c = 33,000 w_c^{1.5} \sqrt{f_c'} \qquad (11.11)$$

where w = weight of concrete, kips/ft^3
f_c' = compressive strength of concrete, ksi

For composite members, however, the AASHTO LRFD code offers a list of set modular ratio values which vary depending on the ultimate cylinder strength of concrete used based on the use of normal-weight concrete. These modular ratios are as follows:

$2.4 \leq f_c' < 2.9 \qquad n = 10$

$2.9 \leq f_c' < 3.6 \qquad n = 9$

$3.6 \leq f_c' < 4.6 \qquad n = 8$

$4.6 \leq f_c' < 6.0 \qquad n = 7$

$6.0 \leq f_c' \qquad n = 6$

Once the modular ratio has been determined, the transformed width of concrete slab can be calculated by simply dividing the effective width of the slab by the modular ratio:

$$b_{tr} = \frac{b_{eff}}{k \cdot n} \qquad (11.12)$$

where b_{eff} = effective width of concrete slab

b_{tr} = width of transformed concrete slab

k = multiplier accounting for creep (see Effects of Creep, below)

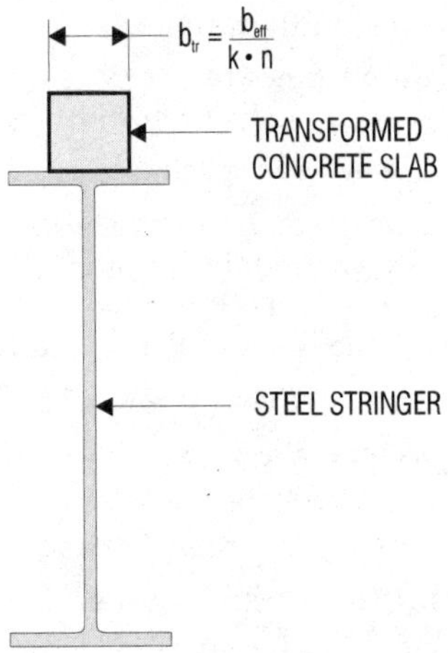

Figure 11.9 Transformed composite section.

This is the width used when computing the resisting moment of inertia of the composite section. The reader should keep in mind, however, that the full effective width is used when computing dead loads on the stringer.

Effects of Creep ▪ Creep is the deformation of concrete caused by loads sustained over a period of time. When considering a composite member, this has an impact on superimposed dead loads such as curbs, railings, and parapets, which are placed after the deck has cured and are sustained for the life of the structure (i.e., dead loads acting on the composite section).

To account for the effects of creep, the AASHTO LRFD code specifies a multiplier to be applied to the modular ratio when computing the width of the transformed slab. A multiplier of $k = 1$ is used for live loads and dead loads on the stringer only. For superimposed dead loads acting on the composite section, a multiplier of $k = 3$ is used.

Therefore, if a modular ratio of $n = 10$ were being used for the computation of stresses on the composite section due to live loads, a value of $n = 30$ would be used for superimposed dead loads. This has the effect of reducing the width of the transformed slab and thereby reducing the size of the composite section modulus.

Yield and Plastic Moment ▪ In the AASHTO LRFD code, both yield and plastic moments M_y and M_p, respectively, must be calculated.

The *yield moment* M_y is required for the strength limit state analysis only. This moment consists of the sum of the moments applied separately to the steel, short-term, and long-term composite sections which causes yielding in either the top or bottom flange (web yielding in hybrid sections is disregarded).

The *plastic moment* M_p is the first moment of plastic forces about the plastic neutral axis. These plastic forces act at the following locations:

1. Midthickness for the flanges and slab
2. Middepth of the web
3. Center of the reinforcement (top and bottom of concrete slab)

The AASHTO LRFD code presents equations for locating the plastic neutral axis and calculating the plastic moment for commonly occurring conditions in composite sections.

Depth of Web in Compression ▪ The depth of web in compression must then be calculated for both the elastic and plastic moments. For the elastic moment, this depth D_c is simply the length of web over which the sum of compressive stresses resulting from dead and live load plus impact acting on the steel and in the short-term composite sections.

At the plastic moment, the depth D_{cp} is determined for either positive or negative flexure. For portions of the span in positive bending where the plastic neutral axis is in the web, the following expression is used:

$$D_{cp} = \frac{D}{2} \frac{F_{yt}A_t - F_{yc}A_c - 0.85 f'_c A_s - F_{yr}A_r}{F_{yw}A_w} + 1 \qquad (11.13)$$

where D_c = depth of web in compression at the plastic moment, in
D = web depth, in
A_s = area of the slab, in^2
A_t = area of the tension flange, in^2
A_c = area of the compression flange, in^2
A_w = area of the web, in^2
A_r = area of the longitudinal reinforcement included in the section, in^2
F_{yt} = specified minimum yield strength of the tension flange, ksi
F_{yc} = specified minimum yield strength of the compression flange, ksi
F_{yw} = specified minimum yield strength of the web, ksi
F_{yr} = specified minimum yield strength of the longitudinal reinforcement included in the section, ksi
f'_c = specified minimum 28-day compressive strength of the concrete, ksi

For any other condition in positive bending, the depth of web in compression should be taken as zero.

For sections of the span in negative bending, where the plastic neutral axis is located in the web, the following expression is used:

$$D_{cp} = \frac{D}{2A_w F_{yw}} \left(F_{yt}A_t + F_{yw}A_w + F_{yr}A_r - F_{yc}A_c \right) \qquad (11.14)$$

For any other condition in negative bending, the depth of web in compression shall be taken as equal to D.

Compact Sections ▪ Whether a section is *compact* or *noncompact* depends on its ability to satisfy a limiting web slenderness condition. This web slenderness criterion is based on the plastic depth of web in compression (see Eqs. 11.13 and 11.14), the thickness of the web, the yield strength of the compression flange, and the modulus of elasticity of the steel used. Specifically, this web slenderness constraint is defined as

$$\frac{2D_{cp}}{t_w} \leq 3.76 \sqrt{\frac{E}{F_{yc}}} \tag{11.15}$$

If this condition is not meant, the section is said to be noncompact. The condition described in Eq. 11.15 is the same as that adopted by AISC in their 1986 LRFD code.

A compact section of a simply supported span in positive flexure has a nominal flexural resistance M_n equal to the plastic moment M_p. This condition also applies to continuous spans with compact interior supports.

Noncompact Sections ▪ Any section which does not satisfy the limiting web slenderness criteria described in Eq. 11.15 is considered to be noncompact. When this condition occurs, the nominal flexural resistance of each flange in terms of stress is taken as

$$F_n = R_b R_h F_{yf} \tag{11.16}$$

where R_b and R_h = flange stress-reduction factors

F_{yf} = specified minimum yield strength of the flange, ksi

The reader is referred to the AASHTO LRFD code for more on the design criteria governing the calculation of flange stress-reduction factors.

11.3.4 Design Example

Overview ▪ The following is a basic design example following the general design outline given in the AASHTO LRFD code. The general LRFD design procedure will be applied to the design of an interior girder of a composite concrete-steel I-beam highway bridge superstructure. All code referenced in parentheses are to the 1st edition of the AASHTO LRFD code.

Problem Definition ▪ Design the interior stringer for flexure and the Service-I Limit State. The bridge cross section and elevation are shown below. Neglect the effects of wind loading.

Given:

1. Bridge to carry two traffic lanes. Travelway width = 44 ft. No-skew.
2. Bridge loading specified to be HS20-44.
3. Concrete strength f'_c = 4,500 psi.

4. Account for 25 psf future wearing surface.
5. Span length of 45 ft centerline to centerline of bearings.
6. Average haunch depth of 2 in.
7. Unshored construction.
8. Account for 25 psf future wearing surface.
9. Span is simply supported.
10. Overpass is a major highway with *ADTT* of 3,000.

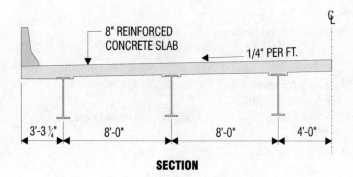

SECTION

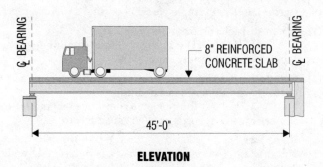

ELEVATION

The design of the interior stringer will be performed using the AASHTO LRFD approach. As shown in the illustration, the beam spacing is 8.0 ft and the deck thickness is 8 in. For simplicity's sake, we will take the entire 8 in as resisting loads. An average haunch depth of 2 in is specified. The haunch is a small layer of concrete between the stringer and concrete slab. The span is simply supported (i.e., not continuous), so it may be analyzed as a simple beam. Unshored construction is specified so the beam must accommodate the dead load of the concrete slab.

General Considerations

A. Design Philosophy (1.3.1)
AASHTO specifies that a bridge is to be designed for specified limit states to achieve the objectives of constructibility, safety, and serviceability, with due regard to issues of inspectability, economy, and aesthetics.

B. Limit States (1.3.2)
Each component and connection shall satisfy Eq. 11.4 (see Section 11.3.1).

C. Design and Location Features (2.3) (2.5)
N/A. These are already addressed in our problem definition.

Superstructure Design

A. Develop General Section

  1. **Roadway Width (Highway Specified)**
     44 ft (Given)

  2. **Span Arrangements (2.3.2) (2.5.4) (2.5.5) (2.6)**
     N/A for our example.

  3. **Select Bridge Type - assumed to be I or Box Girder**
     I-girder

B. Develop Typical Section and Design Basis

  1. **I Girder**

   a. **General (6.10.1)**
   These provisions apply to I-sections which are rolled or fabricated (i.e., plate girder). In this example we will be using a rolled girder.

   (1) **Elastic or Inelastic Analysis (6.10.2.2)**
   N/A. AASHTO LRFD 6.10.2.2 applies to continuous spans and the problem span is simply supported.

   (2) **Composite (6.10.5)**
   (6.10.5.1.1a - Sequence of Loading)
   Elastic stress at any location on a composite section is the sum of the stresses caused by loads applied separately to:

   - The Steel   (Dead Load - DL)
   - Short-Term Composite Section (Superimposed Dead Load - SDL)
   - Long-Term Composite Section   (Live Load - LL)

   (6.10.5.1.1b - Positive Flexure)
   For positive flexure, the flexural stresses are computed by transforming an effective width of concrete slab into an equivalent section of steel. This is accomplished using the so-called modular ratio, n. The value for the modular ratio used is obtained from AASHTO LRFD 6.10.5.1.1b. The size of the transformed section varies depending on whether the loads are placed for a long duration after the slab and stringer have begun to work in a composite fashion. Live loads

do not fall into this category because they are short duration loads. The dead load of the slab and stringer do not because they are placed prior to composite action taking effect. Superimposed dead loads, however, are in place after the concrete has cured, and therefore must use the multiplier of $k= 3$ (AASHTO LRFD 6.10.5.1.1b).

For $f'_c = 4500$ psi, use: $n = 8$

(6.10.5.1.1c - Negative Flexure)
N/A. Since we have one, simply supported span under truck loading, negative flexure is not a consideration.

(6.10.5.1.1d - Effective Width of Slab)
The provisions for the effective width of slab are found in AASHTO LRFD 4.6.2.6 and were described in Section 11.3.3. With the current thickness of concrete slabs in use, the average spacing between stringers generally wins out over the other two criteria. We will use the effective flange width to compute the amount of concrete slab acting as dead load on the stringer.

The effective flange width is defined as:

Minimum of	
1/4 × Effective Span Length = (0.25)(45.00)	= 11.25 ft
12 × Average Slab Thickness = (12)(8 in)(1 ft/12 in) = (see note below)	8.00 ft
Average Spacing of Adjacent Beams	= 8.00 ft

so: $b_{eff} = 8.00$ ft

Note: According to the AASHTO LRFD code, we should add the greater of the web thickness or one-half the width of the top flange; either of which will make this value greater than the average spacing of adjacent beams, which is clearly the minimum.

(6.10.5.1.2 - Yield Moment)
N/A. The yield moment is required only for the strength limit state investigation of compact positive bending sections in continuous spans, negative bending sections, and hybrid negative bending sections. Since we have no negative bending and our span is simply supported, we do not need to calculate a yield moment.

(6.10.5.1.3 - Plastic Moment)
The plastic moment of a composite section in positive bending is calculated using the following steps described by AASHTO in AASHTO LRFD 6.10.5.1.3 and Appendix A of that section:

- Calculate the element forces and use them to determine whether the plastic neutral axis is in the web, top flange, or slab.

- Calculate the location of the plastic neutral axis within the element determined in the step above.
- Calculate M_p.

In order to calculate the element forces, we need to have an initial guess at a trial section. To find a trial section we must compute the applied moment due to dead load, superimposed dead load, and live load. Then, based on this section we can determine the location of the plastic neutral axis and the plastic moment.

Step 1: Compute the Dead Load on Stringer
The dead load on the stringer is all weight placed prior to the concrete deck hardening and reaching full strength. In this case, the dead load on the stringer consists of the stringer itself, the slab and the haunch (the wearing course, parapet, etc., will be accounted for below). Since we do not yet know the exact size of the stringer, we will make an initial guess of 150 lb/ft for the stringer. We will assume 5 percent of the stringer weight to account for miscellaneous connection plates, diaphragms, etc.

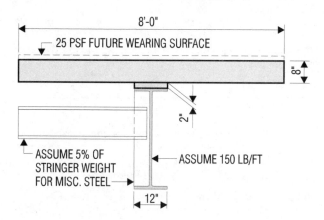

The dead load is composed of the following items:

DL_{slab} = (b_{eff})(slab thickness)(w_{conc})
= (8.0 ft)(8.0 in)(1 ft/12 in)(0.150 k/ft³) = 0.800 k/ft

DL_{haunch} = (haunch width)(haunch thickness)(w_{conc})
= (1.0 ft)(2.0 in)(1 ft/12 in)(0.150 k/ft³) = 0.025 k/ft

DL_{steel} = (assumed stringer weight) + (misc. steel)
= (0.150 k/ft) + (5%)(0.150 k/ft)(1/100) = 0.157 k/ft

DL = Total Dead Load in Stringer = 0.982 k/ft

Limit Design of Steel Highway Bridges ■ 11-35

Step 2: Compute the Superimposed Dead Load on Stringer
Superimposed dead loads are those loads placed on the bridge after the deck has cured. These loads are resisted by the composite section. In this case, the superimposed dead load consists of the future wearing surface and parapet. These loads are to be distributed equally among all stringers. The future wearing surface is applied over the travelway width of roadway. We are given a total travelway width 44 ft.

Account for the wearing surface:

$$SDL_{ws} = \frac{(\text{width of roadway})(\text{future wearing surface})}{\text{number of stringers}}$$

$$= \frac{(44 \text{ ft})(0.025 \text{ k}/\text{ft}^2)}{6 \text{ stringers}} = 0.183 \text{ k}/\text{ft}$$

Calculate weight of parapet by first computing the area of its cross section:

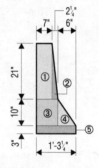

$A_1 = (7 \text{ in})(21 \text{ in}) = 147.00 \text{ in}^2$

$A_2 = 1/2(2.25 \text{ in})(21 \text{ in}) = 23.62 \text{ in}^2$

$A_3 = (9.25 \text{ in})(10 \text{ in}) = 92.50 \text{ in}^2$

$A_4 = 1/2(6 \text{ in})(10 \text{ in}) = 30.00 \text{ in}^2$

$A_5 = (15.25 \text{ in})(3 \text{ in}) = 45.75 \text{ in}^2$

$\qquad\qquad\qquad A_p = 338.87 \text{ in}^2$

$w_p = (338.87 \text{ in}^2)(1 \text{ ft}^2/144 \text{ in}^2)(0.150 \text{ k}/\text{ft}^3) = 0.35 \text{ k}/\text{ft}$

There are two parapets on the bridge which are distributed over all six stringers:

$$SDL_p = 2 \text{ Parapets} \cdot \frac{w_p}{\text{no. stringers}} = 2 \cdot \frac{0.35 \text{ k}/\text{ft}}{6 \text{ stringers}} = 0.117 \text{ k}/\text{ft}$$

$SDL = SDL_{ws} + SDL_p = 0.183 \text{ k}/\text{ft} + 0.117 \text{ k}/\text{ft}$

$SDL = \text{Total Superimposed Dead Load} = 0.300 \text{ k}/\text{ft}$

Step 3: Compute the Transformed Width of Slab
The value for the modular ratio used is obtained from AASHTO LRFD 6.10.5.1.1b (see B.1.a(2) above and Fig. 11.9). The width of the transformed section varies depending on whether the loads are placed for a long duration after the slab and stringer have begun to work in a composite fashion. Live loads do not fall into this category because they are short duration loads. The dead load of the slab and stringer do not because they are placed prior to composite action taking effect. Superimposed dead loads, however, are in place after the concrete has cured, and therefore must use the multiplier of $k = 3$ (AASHTO LRFD

6.10.5.1.1b). The two values are converted to inches to facilitate computation of the moments of inertia.

For $f'_c = 4500$ psi, use: $\quad n = 8$

For live loads and dead loads acting on the stringer only ($k = 1$):

$$b_{tr1} = \frac{b_{eff}}{k \cdot n} = \frac{8.0 \text{ ft}}{(1)(8)} = 1.0 \text{ ft} \quad \Rightarrow b_{tr1} = 12 \text{ in}$$

For superimposed dead loads ($k = 3$):

$$b_{tr3} = \frac{b_{eff}}{k \cdot n} = \frac{8.0 \text{ ft}}{(3)(8)} = 0.33 \text{ ft} \quad \Rightarrow b_{tr3} = 4 \text{ in}$$

Step 4: Compute Dead and Superimposed Deal Load Moments:

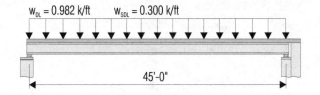

Dead Load Moment is given by:

$$M_{DL} = \frac{wL^2}{8} = \frac{(0.982 \text{ k / ft})(45.00 \text{ ft})^2}{8} \quad \Rightarrow M_{DL} = 248.57 \text{ ft-k}$$

Superimposed Dead Load Moment is given by:

$$M_{SDL} = \frac{wL^2}{8} = \frac{(0.300 \text{ k / ft})(45.00 \text{ ft})^2}{8} \quad \Rightarrow M_{SDL} = 75.94 \text{ ft-k}$$

Step 5: Compute Distribution of Live Load per Lane:
We must first compute the distribution of live load per lane. Referring to AASHTO LRFD 4.6.2.2.2b (Interior Beams with Concrete Decks) and Table 4.6.2.2.2b-1, for a bridge with:

1. Concrete Deck
2. On Steel Beams
3. Two or More Design Lanes

we should use:

$$DF = 0.075 + \left(\frac{S}{9.5}\right)^{0.6} \left(\frac{S}{L}\right)^{0.2} \left(\frac{K_g}{12.0Lt_s^3}\right)^{0.1}$$

where the term $K_g/(12.0Lt_s^3)$ may be taken as 1.0 for preliminary design purposes

Limit Design of Steel Highway Bridges ■ 11-37

The load distribution factor is for the entire live load per lane, so we apply this to an axle load not a wheel load. In the Standard Specifications (i.e., non-LRFD AASHTO code) the distribution factor given was for wheel loads. Under the older specifications a wheel load distribution factor of S/5.5 would have been used or 1.45 per wheel or 0.725 for an entire axle. For our purposes, we will use a wheel load distribution factor, so we will multiply 0.69 by 2:

Use Wheel Load $DF = 1.38$

Step 6: Compute Dynamic Load Allowance (Impact):
In the AASHTO LRFD code, the concept of impact is now called Dynamic Load Allowance (AASHTO LRFD 3.6.2.1). For an interior girder we use a factor of:

$I = (1 + IM/100)$ where: $IM = 33\%$ (Table 3.6.2.1-1)   Use $I = 1.33$

Step 7: Compute Design Truck Live Load Plus Impact Moment
As per AASHTO LRFD 3.6.1.3.1, live loading shall consist of both a design truck and a lane load. The truck live load is placed as described in AASHTO LRFD 3.6.1.2.2. First we calculate the reactions by summing moments about the left support (Point A). This yields the reaction at the right support. Both reactions combined must equal the total downward load of 36k. Once we know the reactions, we can solve for the maximum live load moment which occurs at Point C (the center 16k wheel load). We can solve for this moment by taking a section from the point in question to one support or the other. The free body diagram at the bottom of the calculation sheet shows that we will use the left support, but we just as easily could have used the right one. The resulting live load moment is 269.35 ft-kips.

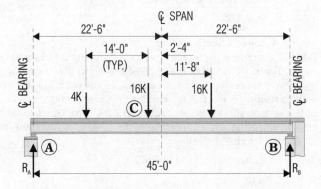

First, solve for the reactions by summing moments about Point A:

$\curvearrowright_+ \sum M_A = 0$: $(4 \text{ k} \cdot 6.167 \text{ ft}) + (16 \text{ k} \cdot 20.167 \text{ ft})$
   $+ (16 \text{ k} \cdot 34.167 \text{ ft}) - (R_B \cdot 45 \text{ ft}) = 0$

$$R_B = \frac{894 \text{ ft} \cdot \text{k}}{45 \text{ ft}} = 19.867 \text{ k} \quad \text{so, } R_A = 36 \text{ k} - 19.867 \text{ k} = 16.133 \text{ k}$$

Now, compute the maximum live load moment:

$$M_{LL} = M_{MAX} = (R_A \cdot 20.167 \text{ ft}) - (4 \text{ k} \cdot 14 \text{ ft}) = 269.35 \text{ ft} \cdot \text{k}$$

Apply the impact and wheel load distribution factors

$$M_{LL+I} = M_{LL} \cdot DF \cdot I$$

$$M_{LL+I} = (269.35 \text{ ft} \cdot \text{k})(1.38)(1.33)$$

$$M_{LL+I} = 494.36 \text{ ft} \cdot \text{k}$$

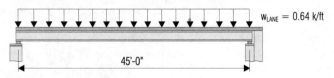

Step 8: Compute Lane Load Live Load Moment
We must now compute the moment for a lane load of 0.64 k/ft (AASHTO LRFD 3.6.1.2.4). This moment will not have a dynamic load allowance (i.e., impact) factor applied as per AASHTO LRFD 3.6.2.1.

Lane Load Moment is given by:

$$M_{LANE} = \frac{wL^2}{8} = \frac{(0.640 \text{ k}/\text{ft})(45.00 \text{ ft})^2}{8} \Rightarrow M_{LANE} = 162 \text{ ft} \cdot \text{k}$$

Step 9: Choose a Section
We will guess at a preliminary section by using an old working stress rule of thumb. First all of the applied moments are added up and then the aggregate is divided by an allowable stress of 20 ksi. This yields a section modulus for the resisting composite section. Using the computed section modulus, we enter the AISC properties table for rolled girders and choose a section three sizes smaller (to account for the slab working with the girder).

Required composite section modulus:

$$S_{tr} = \frac{M_{DL} + M_{SDL} + M_{LL+I} + M_{LANE}}{F_b}$$

$$= \frac{(248.57 + 75.94 + 494.36 + 162) \text{ft} \cdot \text{k}(12 \text{ in}/\text{ft})}{20 \text{ k}/\text{in}^2} = 588.522 \text{ in}^3$$

Limit Design of Steel Highway Bridges ■ 11-39

Entering the tables with a section modulus of approximately 580 in³ we obtain an initial section of W36×170, we then go down three sizes, so that our preliminary section we will use is:

⇒ W 36×135

Step 10: Calculate Element Forces
The element forces acting on the composite section are illustrated below: AASHTO C6.10.5.1.3 allows the forces in the longitudinal reinforcement may be conservatively neglected. So set P_{rb} and P_{rt} equal to 0.

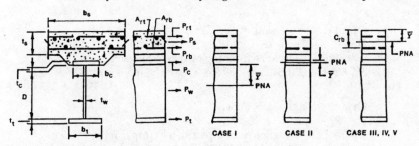

$P_{rt} = F_{yrt}A_{rt} = 0$ (assume conservatively)

$P_s = 0.85f'_cbt_s = (0.85)(4.5 \text{ k/in}^2)(8.0 \text{ ft})(12 \text{ in/ft})(8 \text{ in}) = 2937.6 \text{ k}$

$P_{rb} = F_{yrb}A_{rb} = 0$ (assume conservatively)

$P_c = F_{yc}b_ct_c = (36 \text{ ksi})(11.95 \text{ in})(0.79 \text{ in}) = 339.858 \text{ k}$

$P_w = F_{yw}Dt_w = (36 \text{ ksi})(35.55 - (2)(11.95 \text{ in})(0.60 \text{ in}) = 251.640 \text{ k}$

$P_t = F_{yt}b_tt_t = (36 \text{ ksi})(11.95 \text{ in})(0.79 \text{ in}) = 339.858 \text{ k}$

where: F_{yc} = specified minimum yield strength in the compression flange (ksi) (Table 6.4.1-1) = 36 ksi

F_{yt} = specified minimum yield strength in the tension flange (ksi) (Table 6.4.1-1) = 36 ksi

F_{yw} = specified minimum yield strength of the web (ksi) (Table 6.4.1-1) = 36 ksi

AASHTO Table A6.1-1 gives various criteria for determining whether the plastic neutral axis falls:

- Case I: In the web
- Case II: In the top flange
- Case III: In the slab, below P_{rb}
- Case IV: In the slab, at P_{rb}
- Case V: In the slab, above P_{rb}

Case I: In the web the condition is:
$P_t + P_w \geq P_c + P_s + P_{rb} + P_{rt}$
339.858 k + 251.640 k ≥ 339.858 k + 2937.6 k + 0 + 0
591.498 k ≱ 3277.458 k ✗

Case II: In the top flange the condition is:
$P_t + P_w + P_c \geq P_s + P_{rb} + P_{rt}$
339.858 k + 251.640 k + 339.858 k > 2937.6 k + 0 + 0
931.356 k ≱ 2937.6 k ✗

Case III: In the slab, below P_{rb}:
$P_t + P_w + P_c \geq (C_{rb}/t_s)P_s + P_{rb} + P_{rt}$
where: C_{rb} = Distance from top of slab to lower rebar = 7 in
339.858 k + 251.640 k + 339.858 k ≥ (7 in / 8 in)(2937.6 k) + 0 + 0
931.356 k ≱ 2570.4 k ✗

Case IV: In the slab, at P_{rb}:
$P_t + P_w + P_c + P_{rb} \geq (C_{rb}/t_s)P_s + P_{rt}$
where: C_{rb} = Distance from top of slab to lower rebar = 7 in
339.858 k + 251.640 k + 339.858 k + 0 ≥ (7 in / 8 in)(2937.6 k) + 0
931.356 k ≱ 2570.4 k ✗

Case V: In the slab, above P_{rb}:
$P_t + P_w + P_c + P_{rb} \geq (C_{rt}/t_s)P_s + P_{rt}$
where: C_{rt} = Distance from top of slab to upper rebar = 2.5 in
339.858 k + 251.640 k + 339.858 k + 0 ≥ (2.5 in / 8 in)(2937.6 k) + 0
931.356 k ≥ 917.969 k ✓

Step 11: Calculate Location of Plastic Neutral Axis and Plastic Moment
Based upon the five cases examined in Step 9 above, the location of the plastic neutral axis (and corresponding plastic moment) will vary. For our condition, where the plastic neutral axis falls above the lower reinforcement:

$$\overline{Y} = (t_s)\left[\frac{P_{rb} + P_c + P_w + P_t - P_{rt}}{P_s}\right]$$

$$= (8 \text{ in})\left[\frac{0 + 339.858 \text{ k} + 251.640 \text{ k} + 339.858 \text{ k} - 0}{2937.6 \text{ k}}\right] = 2.536 \text{ in}$$

From the figure above, for Case V, the distance to the plastic neutral axis is measured from the top of slab, therefore the plastic neutral axis is 2.536 in from the top of slab. In the equation for the plastic moment below, all "d" distances are from the plastic neutral axis to the element force in question (e.g., d_c is the distance from the plastic neutral axis to the center of the compression flange).

$$M_p = \left(\frac{\overline{y}^2 P_s}{2t_s}\right) + \left[P_{rt}d_{rt} + P_{rb}d_{rb} + P_c d_c + P_w d_w + P_t d_t\right]$$

d_c = [(8 in slab + 2 in haunch + (1/2)(0.790 flange thickness)] - 2.536 in
    = 7.859 in

$d_w = [(8 \text{ in slab} + 2 \text{ in haunch} + (1/2)(35.55 \text{ girder depth})] - 2.536 \text{ in}$
$= 25.239 \text{ in}$

$d_t = [(8 \text{ in slab} + 2 \text{ in haunch} + 35.55 \text{ girder depth} - (1/2)(0.790 \text{ flange thickness})] - 2.536 \text{ in} = 42.619 \text{ in}$

$M_p = \left(\dfrac{(2.536 \text{ in})^2 (2937.6 \text{ k})}{(2)(8 \text{ in})}\right) + [0 + 0 + (339.858 \text{ k})(7.859 \text{ in})$
$+ (251.640 \text{ k})(25.239 \text{ in}) + (339.858 \text{ k})(42.619 \text{ in})]$
$= 24,687.28 \text{ in} \cdot \text{k} = 2057.27 \text{ ft} \cdot \text{k}$

(6.10.5.1.4 - Depth of Web in Compression)
(6.10.5.1.4a - At Elastic Moment)

The depth of the web in compression, D_c, at the elastic design moment is the depth over which the stresses applied to the short-term composite section are compressive. To determine this distance, we need to locate the neutral axis for the steel girder and the concrete slab with a transformed width calculated using b_{eff}/kn, where $k = 1$ for short-term loads and n = the modular ratio. This was calculated earlier as $b_{tr1} = 12$ in.

Moment of inertia resisting live load plus impact:

The area and moment of inertia of the transformed slab are computed using the $k = 1$ value and a modular ratio of $n = 8$ (see above).

$A_{ctr1} = b_{tr1} t = (12 \text{ in})(8 \text{ in}) = 96 \text{ in}^2$

$I_{tr1} = \dfrac{bh^3}{12} = \dfrac{(12 \text{ in})(8 \text{ in})^3}{12} = 512 \text{ in}^4$

Element	A (in^2)	Y (in)	AY (in^3)	AY^2 (in^4)	I_0 (in^4)
W36×135	39.7	17.775	705.667	12,543.240	7800.00
Slab ($k = 1$)	96.0	41.550	3988.800	165,734.640	512.00
Totals	174.0		4694.467	178,277.880	8312.00

$I_Z = \Sigma I_0 + \Sigma AY^2 = 8312.0 \text{ in}^4 + 178,277.880 \text{ in}^4 = 186,589.880$

$\overline{Y} = \dfrac{\Sigma AY}{\Sigma A} = \dfrac{4694.467 \text{ in}^3}{174.0 \text{ in}^2} = 26.980 \text{ in}$

$I_{LL+I} = I_Z - (\Sigma A)(Y')^2 = 186,589.880 \text{ in}^4 - (174.0 \text{ in}^2)(26.98 \text{ in})^2$

$= 59,931.730 \text{ in}^4$

$\Rightarrow D_c = 35.55 \text{ in} - 26.980 \text{ in} = 8.57 \text{ in}$

(6.10.5.1.4 - Depth of Web in Compression - Continued)
(6.10.5.1.4a - At Plastic Moment)
For positive flexure where the plastic neutral axis is not in the web, the depth of web in compression at the plastic moment, D_{cp}, can be taken as:

$D_{cp} = 0$

(6.10.5.2 - Compact Sections)
Our W36×135 stringer may be considered compact.

(6.10.5.2.2 - Positive Flexure)
(6.10.5.2.2a - Nominal Flexural Resistance)
For simple spans the nominal flexural resistance is taken as:

$M_n = M_p = 2057.27$ ft·k

(3) Non-Composite (6.10.6)
N/A. We have a composite section.

(4) Homogeneous or Hybrid (6.10.5.4)
N/A. This section applies to hybrid girders, and we do not have a hybrid girder.

2. Box Girder
N/A. We are using rolled girders, not a box girder.

C. Design Conventionally Reinforced Concrete Deck
This is not within the scope of our problem.

D. Select Resistance Factors

1. Strength Limit State (6.5.4.2)
For flexure, the two basic strength limit states we shall take are:

$\phi_f = 1.00$ (for flexure)

$\phi_v = 1.00$ (for shear)

E. Select Load Modifiers

1. Ductility (1.3.3)
For the strength limit state we use:

$\eta_D = 0.95$ (for ductile components)

2. Redundancy (1.3.4)
For strength limit state we use:

$\eta_R = 0.95$ (for redundant members)

3. Operational Importance (1.3.5)
For a bridge of no operational importance we use:

$\eta_I = 0.95$

F. Select Load Combinations and Load Factors (3.4.1)
Based on Table 11.2, for the Service-I Limit States, we will use load combinations of:

SERVICE-I: $1.00DC + 1.00DW + 1.00(LL+I)$

where: DC = Dead Load (Components)
DW = Superimposed Dead Load (Wearing surface)
$LL+I$ = Live Load Plus Impact

1. **Strength Limit State (6.10.2)**
   The factored flexural resistances shall be taken as:
   $M_r = \phi_f M_n = 1.00 M_n = M_n = 2057.27$ ft·k

2. **Service Limit State (6.10.3)**
   Not within the scope of our problem.

3. **Fatigue and Fracture Limit State (6.10.4)**
   Webs without longitudinal stiffeners shall satisfy the following requirement.

   If $\dfrac{2D_c}{t_w} \leq 5.67\sqrt{\dfrac{E}{F_{yc}}}$ then $f_{cf} \leq R_h F_{yc}$

   $\dfrac{(2)(18.57 \text{ in})}{0.600 \text{ in}} \leq 5.67\sqrt{\dfrac{29{,}000 \text{ ksi}}{36 \text{ ksi}}}$ so: $61.9 \leq 163.5$ O.K.

   so our fatigue and fracture limit state will be:

   $f_{cf} \leq R_h F_{yc}$

   where: f_{cf} = maximum compressive elastic flexural stress in the compression flange due to unfactored permanent load and the fatigue loading.
   R_h = flange-stress reduction factor
   F_{yc} = yield strength of the compression flange, ksi

G. Calculate Live Load Force Effects

1. **Select Live Loads (3.6.1) and Number of Lanes (3.6.1.1.1)**
   The number of lanes is taken as the integer portion of the ratio $w/12$ where w is the clear roadway width in feet between barriers and/or curbs. This would produce 44 ft/12 = 3.67 or 3 lanes. Our problem stipulates only 2 design lanes.

   2 Design Lanes (Given)

2. **Multiple Presence (3.6.1.1.2)**
   Based in AASHTO Table 3.6.1.1.2-1, for 2 design lanes we use a multiple presence factor of:
   $m = 1.00$

3. **Dynamic Load Allowance (3.6.2)**
   As mentioned above, in the AASHTO LRFD code, the concept of impact is now called Dynamic Load Allowance (AASHTO LRFD 3.6.2.1). For an interior girder we use a factor of:
   $I = (1+IM/100)$ where: $IM = 33\%$ (Table 3.6.2.1-1)   Use $I = 1.33$

11-44 ■ Chapter Eleven

4. **Distribution Factor for Moment (4.6.2.2.2)**
   a. **Interior Beams with Concrete Decks (4.6.2.2.2b)**
      As shown earlier when we were choosing a preliminary section, we will use a wheel load distribution factor of: Use Wheel Load $DF = 1.38$
   b. **Exterior Beams (4.6.2.2.2d)**
      N/A.
   c. **Skewed Bridges (4.6.2.2.2e)**
      N/A.
5. **Distribution Factor for Shear (4.6.2.2.3)**
   Beyond the scope of our problem.
6. **Wind Effects (4.6.2.7) (6.10.5.7)**
   Neglect as per problem definition.
7. **Reactions to Substructure (3.6)**
   Beyond the scope of our problem.

H. **Calculate Force Effects From Other Loads Identified in Step B6.3.F**
Recall that:

SERVICE-I: $1.00DC + 1.00DW + 1.00(LL+I)$

$M_{S1} = M_{DL} + M_{SDL} + M_{LL+I} + M_{LANE}$

$M_{S1} = (248.57 + 75.94 + 494.36 + 162)$ ft·k $= 980.87$ ft·k

I. **Design Required Sections - Illustrated for Analysis of I-Girder**
   1. **Check D_c/t_w for Fatigue Induced by Web Flexure or Shear (6.10.4)**
      Recall that our fatigue and fracture limit state is:

      $f_{cf} < R_h F_{yc}$

      where: f_{cf} = maximum compressive elastic flexural stress in the compression flange due to unfactored permanent load and the fatigue loading (see below)

      R_h = flange-stress reduction factor = 1.0 [for homogeneous (i.e., non-hybrid) section AASHTO LRFD 6.10.5.4.1a]

      F_{yc} = yield strength of the compression flange, ksi = 36 ksi

      Moment Due to Permanent Loads
      $M_{DL} = 248.57$ $M_{SDL} = 75.94$ ft·k

      Moment Due to Fatigue Loads
      AASHTO 6.10.4.2 instructs us to take twice the fatigue load calculated using the fatigue load combination in Table 3.4.1-1 Fatigue Loading = TWICE [0.75(Fatigue load as specified in 3.6.1.4.1)] = 1.5(Fatigue Load)

      In AASHTO 3.6.1.4.1 a design truck is specified with a constant spacing of 30 ft between the two 32k axles (see Figure 11.6). This moment is calculated in a similar fashion to that shown earlier for the 14 ft spacing and produces a fatigue load moment:

$M_{FL} = 164.73$ ft·k

We must take into account the Dynamic Load Allowance for fatigue, which is specified in 3.6.2 and Table 3.6.2.1-1 as:

$I = (1+IM/100)$ where: $IM = 15\%$ (Table 3.6.2.1-1) Use $I = 1.15$

Since we are using an approximate method (AASHTO 3.6.1.4.3b), use the previously calculated wheel load distribution factor: Use DF = 1.38

So we get a moment due to fatigue load of:

$M_{FL+I} = (164.73$ ft·k$)(1.15)(1.38) = 261.43$ ft·k

Take into account twice the load factor as discussed above:

$M_{FL+I} = (1.5)(261.43$ ft·k$) = 392.14$ ft·k

Elastic Stress in Compression Flange

Dead load acts on the steel only:

$M_{DL} = 248.57$ ft·k

$c =$ Distance from Neutral Axis to compression flange $= 35.55$ in $/ 2 = 17.775$ in

$I =$ Moment of Inertia $= 7800$ in^4

$f_{DL} = Mc / I = (248.57$ ft·k$)(12$ in /ft$)(17.775$ in$) / 7800$ in$^4 = 6.79$ ksi

Superimposed dead loads act on the composite section with $k=3$ (similar to calculation shown earlier):

$M_{SDL} = 75.94$ ft·k

$c =$ Distance from Neutral Axis to compression flange with ($k=3$)
$= 35.55$ in $- 28.38$ in $= 7.17$ in

$I =$ Moment of Inertia ($k = 3$) $= 17,985.96$ in^4

$f_{SDL} = Mc / I = (75.94$ ft·k$)(12$ in /ft$)(7.17$ in$) / 17,985.96$ in$^4 = 0.36$ ksi

Fatigue loads act on the composite section:

$M_{FL+I} = 392.14$ ft·k

$c =$ Distance from Neutral Axis to compression flange $= D_c = 8.57$ in

$I =$ Moment of Inertia $= 59,931.730$ in^4

$f_{FL+I} = Mc / I = (392.14$ ft·k$)(12$ in /ft$)(8.57$ in$) / 59931.73$ in$^4 = 0.67$ ksi

$f_{cf} =$ Elastic stress in compression flange $= f_{DL} + f_{FL+I} + f_{SDL}$
$= 6.79$ ksi $+ 0.36$ ksi $+ 0.67$ ksi $= 7.82$ ksi

So that, checking we find:

$f_{cf} \leq R_h F_{yc}$
7.82 ksi $\leq (1.0)(36$ ksi$)$
7.82 ksi ≤ 36 ksi ✔

2. **For Composite Sections:**
   a. **Consider Sequence of Loading and Pouring Sequence (6.10.5.1.1)**
      See above.
   b. **Determine Effective Flange Width (4.6.2.6)**
      See above. $b_{eff} = 8.0$ ft
   c. **Determine if Section is Compact**
      Our rolled section is compact.
      (1) **Flexural Resistance for Compact Sections:**
         (a) **Web Slenderness (6.10.5.2.2c) (6.10.S.2.3b)**
            Web slenderness of sections shall satisfy:
            $$\frac{2D_{cp}}{t_w} \leq 3.76\sqrt{\frac{E}{F_{yc}}} \quad \text{since } D_{cp} \text{ is 0, we are O.K.}$$
            where: D_{cp} = depth of web in compression at plastic moment, in
            F_{yc} = minimum yield strength of the compression flange, ksi
         (b) **Flange Slenderness (6.10.5.2.3c)**
            Compression flange slenderness of sections shall satisfy:
            $$\frac{b_f}{2t_f} \leq 0.382\sqrt{\frac{E}{F_{yc}}} \quad \text{or:} \quad \frac{11.95 \text{ in}}{(2)(0.790)} \leq (0.382)\sqrt{\frac{29{,}000 \text{ ksi}}{36 \text{ ksi}}}$$
            or: $7.56 \leq 10.84$ O.K.
         where: b_f = width of the compression flange, in
         t_f = flange thickness, in
         (c) **Flange Bracing (6.10.5.2.3d)**
            N/A.
         (d) **Calculate Flexural Resistance (6.10.5.2.2a) (6.10.5.2.3a)**
            For simple spans the nominal flexural resistance is taken as:
            $M_n = M_p = 2057.27$ ft·k
         (e) **Check Positive Flexure Ductility (6.10.5.2.2b)**
            To protect the concrete deck from premature crushing and spalling when the composite section approaches the plastic moment, the following requirement should be met:
            $$D_p \leq \frac{d + t_s + t_h}{7.5} \quad \text{so: } 2.536 \text{ in} \leq \frac{35.55 \text{ in} + 8 \text{ in} + 2 \text{ in}}{7.5} \leq 6.07 \text{ in,}$$
            so we are O.K.
            where: D_p = distance from the top of the slab to the neutral axis at the plastic moment
            d = depth on the steel section (in)

Limit Design of Steel Highway Bridges ■ 11-47

t_s = thickness of the concrete slab (in)

t_h = thickness of the concrete haunch above the top flange (in)

   (2) **Flexural Stress Limits for Non-Compact Sections:**
   N/A.

3. **For Non-Composite Sections:**
   N/A.

   The following are additional checks, where appropriate, to be performed for a complete girder design.

4. **Shear Design**
   a. Unstiffened Web (6.10.7.2)
   b. Stiffened Web (6.10.7.3)
      (1) Minimum Requirements for Handling (6.10.7.3.2)
      (2) Homogeneous Section (6.10.7.3.3)
      (3) Hybrid Sections (6.10.7.3.4)
   c. Stiffener Design (6.10.8)
      (1) Transverse (6.10.8.1)
      (2) Bearing (6.10.8.2)
      (3) Longitudinal (6.10.8.3)
   d. Shear Connectors (6.10.7.4)
      (1) Fatigue Resistance (6.10.7.4.2)
      (2) Strength Limit State (6.10.7.4.4)
      (3) Details (6.10.7.4.1 a, b, & c) (6.10.7.4.3)

5. **Constructability**
   a. General Proportions (6.10.1.1)
   b. Flexure (6.10.10.2)
   c. Shear (6.10.10.3)

J. **Dimension and Detail Requirements**
   1. Material Thickness (6.7.3)
   2. Optional Deflection Control (2.5.2.6.2)
   3. Bolted Connections (6.13.2)
      a. Minimum Design Capacity (6.13.1)
      b. Net Sections (6.8.3)
      c. Bolt Spacing Limits (6.13.2.6)
      d. Slip-Critical Bolt Resistance (6.13.2.2) (6.13.2.8)
      e. Shear Resistance (6.13.2.7)
      f. Bearing Resistance (6.13.2.9)
      g. Tensile Resistance (6.13.2.10)
   4. Welded Details (6.13.3)
   5. Diaphragms and Cross-Frames (6.7.4)
   6. Lateral Bracing (6.7.5)
   7. Splices (6.13.6)
   8. Fatigue and Fracture Compliance (6.5.3)

11.4 Load Rating

With rapid aging of our infrastructure in general, and highway bridges in particular, greater emphasis is being placed on the analysis on existing bridge structures to determine their ability to sustain current vehicular loading configurations. Such analyses, known as *load ratings*, are nothing new, to be sure. The recent increase in the use of limit states design methods, coupled with the need to accurately assess capacity (and more importantly, *reserve capacity*) of highway bridge structures, has put a new spin on this important aspect of bridge engineering.

The foundation of any sound load rating analysis is reliable record and inspection data. One of the main functions of these data is to provide an agency with feedback as to exactly how their bridges are performing in the field. In addition to sound inspection data, however, a maintenance department requires more quantitative information regarding how an individual bridge can withstand loads.

While the load rating is based upon the precepts set forth in the general AASHTO specifications, and now the AASHTO LRFD code, the guidelines for developing a highway bridge load rating value are set forth in the AASHTO publication *Manual for Maintenance Inspection of Bridges*. It is important for the reader to recognize that the calculated load rating value of a bridge is not a magical number. The load rating, like inspection data, is only a gauge of bridge performance and one component in the overall profile of a structure.

The use of load rating values varies depending on the specific bridge in question. At one end of the spectrum a load rating may be used to determine the type and scope of rehabilitation a structure receives in order to bring the bridge up to current standards. Another use of load rating values would be in the posting of a bridge, limiting the type and/or weight of vehicles passing over the structure.

The following discussion is meant only to offer the novice designer an overview of the general concepts behind the load rating of a highway bridge. For more information on the technical aspects of load rating, the reader is referred to *Manual for Maintenance Inspection of Bridges* by AASHTO.

11.4.1 Inventory and Operating Ratings

AASHTO differentiates between lower and upper ranges of bridge performance. The lower range of performance implies safe use of a highway bridge on a day-to-day basis. Naturally, since this day-to-day use of a bridge will imply the largest amount of traffic passing over a structure, the factor of safety built into the rating will be relatively large. This lower rating is called the inventory rating. This rating represents "a load level which can safely utilize an existing structure for an indefinite period of time."

There are instances, however, when vehicles (generally trucks) have to carry abnormally large loads over a structure. While a structure can withstand these loads on special occasions, it is not desirable to have them repeatedly pass over a structure. This upper range of bridge load capacity uses a smaller factor of safety to maximize the capabilities of the bridge to withstand loads and is known as the operating rating. This rating represents "the absolute maximum permissible load level to which the structure may be subjected."

Limit Design of Steel Highway Bridges ■ 11-49

Unlike the bridge inspector's rating assigned to a bridge, the inventory and operating ratings are calculated using analytical rather than subjective methods. The live loadings used are the standard AASHTO design vehicles discussed earlier. The final load rating values can be developed using either the working stress or load factor methods, with the former approach generally yielding more conservative results.

When an existing structure is being load rated, the presence of as-built plans greatly aids in calculation of a load rating value. Should as-built plans be unavailable, then a thorough site inspection must be conducted, complete with detailed field measurements to fill in the missing gaps.

11.4.2 Field Measurements and Inspection

We have already touched upon the importance of field measurements. Regardless of whether detailed as-built plans are available or not, field measurements which describe the present-day condition of a structure are absolutely essential. With regard to steel superstructures, measurements of primary members (e.g., flange and web thickness) aid in creating an accurate picture of what loads a bridge can withstand.

If a bridge has lost 20 percent of its cross-sectional properties, this can result in a greatly reduced load rating (depending on the location of the deterioration). This same principle applies to concrete and timber superstructures as well. In addition to loss of section, any other defects such as misalignment, bends, eccentricities, or kinks in compression members should also be noted. In general, the inspecting engineer should be on the lookout for any physical conditions which he or she feels can impact the load-carrying capacity of the structure.

11.4.3 Loading the Structure

The load rating of a bridge is based on the structure's ability to withstand both dead loads and live loads. Dead loads are computed using the minimum unit weights of materials and is a relatively straightforward affair. Live loads are usually calculated using the standard AASHTO design vehicles. Special truck loading configurations, however, are used by different agencies. The *Manual for Maintenance Inspection of Bridges* itself has its own set of design vehicles. Designers should check with the owner prior to beginning a load rating calculation to ascertain the loading configuration desired by the agency.

Placement of loads on a bridge is made in accordance with the AASHTO specifications. The AASHTO *Manual for Maintenance Inspection of Bridges*, however, adds the following requirements:

1. Roadways whose widths are 18 to 20 ft (5.5 to 6.1 m) are given two design lanes equal to half the roadway width each. Live loads are centered in the design lanes.
2. Roadways whose widths are less than 18 ft are given only one traffic lane. Spans greater than 200 ft (61 m) in length are considered with more than one maximum load.

AASHTO permits a reduction in traffic lanes if the designer believes that the present volume of traffic and movement over the structure warrants such a reduction.

As would be expected, impact is also applied to all live loads. Again, AASHTO allows a reduction in impact should alignment conditions, enforced speed limit postings, etc., warrant such a reduction. Such conditions, if utilized in reducing the impact factor, should be deemed to significantly reduce the speed of vehicles passing over the structure.

Distribution of loads across the superstructure is also performed in accordance with the general AASHTO specifications. If special loading conditions are being utilized, then the applicable distribution factors should be applied. The design vehicles contained within *Manual for Maintenance Inspection of Bridges* possess their own distribution criteria which are documented in that publication.

As mentioned above, either the working stress or load factor methods can be used in calculating the load rating of a structure. The following discussion, however, will focus on the working stress approach for calculating a highway bridge's load rating.

11.4.4 Working Stress Method

As we have already seen, the working stress method is built around the concept of the stress in members falling under a specified allowable stress. This also leads to the method being known as the allowable stress method. The AASHTO *Manual for Maintenance Inspection of Bridges* provides guidelines for applying the working stress (and load factor) method in computing the load rating for structures constructed out of a variety of materials such as:

1. Steel and wrought iron
2. Conventionally reinforced and prestressed concrete
3. Timber

The following provides an overview of the basic criteria governing the rating of steel bridges.

Rating Steel Bridges ▪ In rating an existing steel structure, these allowable stresses will vary depending on the following:

1. Whether an inventory or operating rating is calculated
2. The type of steel present

With regard to the latter item, it is quite possible that, for older structures, the type of steel used is unknown. Such a situation can arise from the absence of as-built plans. However, even if as-built or other plans are available, it is quite possible that the type of steel used was not labeled on the plans in the first place.

For this reason, AASHTO provides minimum tensile strength and yield strength values for steel constructed during various years. These values are based on the probable type of steel used in construction during certain eras (e.g., A7 steel was quite popular during the 1950s).

Limit Design of Steel Highway Bridges 11-51

In addition to this, AASHTO also provides various allowable stresses based on the loading condition the member is subjected to. Some examples of these loading conditions are:

1. Axial tension
2. Compression in extreme fiber of beams
3. Compression in concentrically loaded columns
4. Shear in girder webs

Criteria are also provided for various types of bolts and rivets. For many loading conditions, the principal difference between inventory and operating ratings is the reduction factor applied to the minimum yield point and tensile strength. For inventory ratings the following reduction factors are applied:

$0.55F_y$ where F_y = minimum yield point

$0.50F_u$ where F_u = minimum tensile strength

and for operating ratings the reduction factors applied are:

$0.75F_y$ where F_y = minimum yield point

$0.67F_u$ where F_u = minimum tensile strength

There are, however, exceptions to these criteria depending on the loading and physical conditions (e.g., bearing on milled stiffeners). When a girder with transverse intermediate stiffeners is being investigated, the spacing of the stiffeners does not require checking unless the web has severely deteriorated and/or the spacing of the stiffeners exceeds the depth of the web.

Allowable stresses for various grades of reinforcing steel in tension are also presented. Inventory and operating allowables are given for structures constructed prior to 1954 (18 and 25 ksi), grade 40 billet (20 and 28 ksi), grade 50 rail (20 and 32.5 ksi), and grade 60 (24 and 36 ksi). These values are used without reduction.

11.4.5 Strength Evaluation

So far we have covered some of the basic guidelines for rating a bridge as set forth in the AASHTO *Manual for Maintenance Inspection of Bridges*. Because the interpretation of these guidelines varied so much from state to state (even to the point where the line between inventory and operating rating was becoming blurred), AASHTO decided to develop an additional method for rating bridges. This method is based on a limit states approach and is detailed in the *Guide Specifications for Strength Evaluation of Existing Steel and Concrete Bridges*.

This new method for rating bridges was primarily developed to address the following concerns:

1. Reduce the amount of variation in ratings between structures as mentioned above
2. Provide more realistic ratings through use of a load and resistance factor methodology

In addition to these two goals, the new method was also designed to account for the influence of parameters which were not covered in the AASHTO *Manual for Maintenance Inspection of Bridges*. These parameters are specifically:

1. Deck condition
2. Structural condition (i.e., deterioration)
3. Traffic condition

A belief that is fairly widespread in the transportation community is that the working stress approach in general, and the previous load rating methodology in particular, is overly conservative in the analysis of bridges. A major focus of the new strength evaluation criteria is to offer a more realistic image of a structure's actual capacity. Specifically, the new method states that the older approach does not recognize "the large safety margins present." The strength evaluation methodology is based upon practices in the United States, and while building a bridge in one location is very much the same as building it in another, local practices in bridge construction could influence some load limit values.

It is beyond the scope of this book to offer a thorough examination of the strength evaluation method for rating bridge structures. However, the following is an overview of the basic concepts behind the method and some of the major points of concern designers will have in utilizing this approach.

Overview ▪ When rating any structure, the basic requirement an engineer seeks to check is whether the capacity of the structure is greater than the loads it is subjected to. AASHTO's strength evaluation method defines this question in terms of three fundamental parameters:

1. Available capacity
2. Applied loading
3. Response to the applied loading

These are the key items which the method seeks to define and quantify in order to develop a rating for the structure.

Where the allowable stress method compensates for uncertainty with a factor of safety, the limit states approach (used in the strength evaluation method) takes a "probabilistic approach" to safety through the use of load and resistance factors. The resulting load rating is developed by comparing factored loads with the factored resistance of key sections.

An initial screening rating is developed which, if found to be unsatisfactory, can be further refined should the engineer so desire. The AASHTO Guide Specification believes that the strength evaluation rating is more advantageous than previous solutions because it provides for:

1. A more uniform and consistent evaluation of bridges
2. Flexibility in making evaluations
3. Uniform levels of reliability based on performance histories

4. An incorporation of traffic and response data
5. The use of site-specific data in the evaluation

The Concept of Safe Evaluation ▪ The strength evaluation method is based on the concept of safe evaluation. By safe evaluation, we imply an approach which takes into account the function and site specifics of a structure in order to render a rating. This is made in contrast to the methodology previously discussed, which is, for all intents and purposes, a straight analytical approach (although it may be argued that a similar safe evaluation has been built into the allowable stresses through their refinement over the years).

At first blush, the reader may get the impression that the strength evaluation approach will consistently produce ratings which are higher than the traditional, working stress approach. This, however, is not always the case. A bridge which possesses a redundant load path, is well maintained, and has well enforced traffic regulations will most likely produce an operating rating larger than those previously computed for it. A structure, however, which has significant deterioration, nonredundant components, and heavy truck traffic may produce an inventory rating which is below its current rating.

Like any limit states method, the strength evaluation approach for rating bridges lives and dies by the quality of the load and resistance factors used. Therefore, factors describing the wide variety of variables present at a bridge site such as heavy volume of traffic, maintenance work, dead loads, live loads, rough approaches, heavy deterioration, etc., all have to be taken into account.

A sound limit states approach, however, is more than the simple presence of these load factors. Their magnitude must also be accurate and reflect the practices and desires of the owner. This accuracy must be in terms of both structural and economic performance. Quite naturally, it would take any design engineer or transportation department a period of time before confidence in both the method and the ratings it produces is achieved.

Conclusions ▪ From the above, it is obvious that the strength evaluation approach to rating bridges places a great deal of responsibility, not only on the engineer as an individual, but on the transportation department as a whole. It is one thing to say, "this road doesn't have that much truck traffic" and quite another to incorporate that into a final rating. It is quite accurate to say that a limit states approach offers flexibility, but with flexibility comes responsibility.

This subjectivity could also lead to some conflict between those who design and those who maintain. No engineer can say that he or she has never encountered disagreement with fellow engineers examining the same situation. For example, it is not uncommon to have two separate traffic analyses performed, with each producing a different conclusion. Like anything new, however, a certain comfort factor has to be built into the process. The more limit states approaches, like the one described above, are used, the more engineers will begin to trust both the method and their own judgment.

11.5 LRFD and Working Stress in Bridge Engineering

The preceding discussion has offered a general overview of the LRFD approach and its application in conventional highway bridge design. It is very easy to put one's self in the position of contrasting these LRFD with working stress in an attempt to assess "which one is better." In the end, however, such a contrast is difficult, if not impossible.

The working stress design approach has, for the most part, produced structures with an excellent track record. While working stress may be criticized for being too conservative, it should be emphasized that what is conservative to one engineer may be well designed and durable to another.

In order to fully appreciate the move to a limit states approach, one has to examine the reasons motivating the change. At first glance, it is easy to point to optimization of the design as a clear and principal reason, but this is only part of the story. As we saw in the discussion on load rating and strength evaluation, the need to assess the reserve capacity of structures in order to more efficiently maintain and rehabilitate them is of critical importance. Also, while LRFD, on the whole, will produce designs which are "less conservative," we have seen that this is not always the case. Perhaps the best argument that can be made for the use of LRFD in general, and for highway bridges in particular, is that it is not more or less conservative, but that it is *more realistic*.

The AASHTO LRFD code discussed here is presently in its first edition. There is little doubt that this relatively young specification has plenty of room for improvement. As the specification gains wider use and acceptance, however, there will undoubtedly be enhancement and refinement which will allow the next generation of bridge engineers to design and analyze bridges in a much more *realistic* fashion than their predecessors ever imagined.

12

Robert E. Shaw, Jr.

Fabrication, Erection, and Quality Control

12.1 Shop Detailing Process

12.1.1 Material Takeoff, Mill Order

Upon receipt of award of the contract for structural steel, the fabricator begins the process of material takeoff. The purpose of this takeoff is to allow for early ordering of the structural steel so that it arrives in a timely manner when required for fabrication. Working from the design drawings, the size, grade, and approximate length of each piece required are listed. For main members that are expected to be ordered cut to length from either a steel mill or a warehouse, the exact length is determined. For common sections to be cut from the fabricator's inventory, lengths are estimated to the nearest inch or two. Connection material sizes and lengths are generally unknown until designed and detailed, and are typically not included in the material takeoff at this point unless special requirements are noted.

The material takeoff, also called an advance bill or material list, is used to place orders with the suppliers of the structural steel. Material can be ordered directly from a steel mill, from the steel mill's warehouse or depot, or from an independent steel service center. These centers purchase steel directly from the mills to resell to fabricators and other users, often cutting to specific lengths requested by the customer.

Steel mills commonly produce steel shapes in standard stock lengths of 40, 50, and 60 ft but will also furnish lengths of 45, 55, and 65 ft without extra charge. No metric stock lengths have been established. They will also roll longer sections and cut shapes to specific lengths when ordered by the customer. Mill cutting tolerances are such that the fabricator generally orders material longer than the finished piece. Lengths may be limited by total weight for heavy sections, or by shipping restrictions between the mill and fabricator.

Bar stock is produced to common thicknesses and widths and cut to stock lengths. Flat steel products over ¼ in (6 mm) thick and up to 8 in (200 mm), inclusive, in width are considered bars. Wider products are considered plates. Fabricators purchase bar stock for inventory for use as connection, stiffeners, and other detail material. For large projects that require a large quantity of material, mill orders may be placed.

Plates are available in virtually any thickness and in widths generally up to 150 in (3800 mm). For plate girder flanges and other built-up sections, the plates are ordered to their exact width and thickness. Plate girder webs are typically ordered wider than the final depth of the web to allow for camber and trimming to the final dimension. In some cases where plate girder flanges are spliced along their length and there is sufficient duplication, the girder flange material may be ordered so that several splices can be made in a single wide plate, then cut longitudinally to provide the required width of flange.

Mill orders are placed stating section size, grade, and length, plus any special requirements. The mill fills the order from depot stock or places the order on a rolling schedule. The rolling schedule states the week when that particular grouping of structural shapes is scheduled to be run by the mill.

12.1.2 Erection Plans and Shop Detail Drawings

Using the architectural plans and the engineer's structural drawings, the detailer prepares two different types of drawings—erection plans and shop detail drawings.

Shop detail drawings are for the fabrication of the individual pieces of structural steel. Shop drawings provide exact dimensions for piece sizes, lengths, cuts, bevels, hole locations, hole diameters, weld sizes and locations, and any other pertinent information necessary to make the completed piece. Attachments such as clip angles and stiffeners are itemized and provided a "mark," and the completed piece is assigned a "piece mark."

The piece mark is placed at one end of the member, normally the left end of the piece as drawn on the shop detail drawing. The piece mark is also placed on the erection plan at the same end of the piece. The most common piece mark system uses a combination of letter and number. Beam D3 would be found on shop detail drawing number 3. Some shops also use a letter to denote beams B, columns C, girders G, and so forth. In tier buildings, more complex marking systems may be used to designate tier or floor numbers.

In some cases, shop drawings use "opposite hand" details. These details are used to reduce the total number of sketches and drawings required and to take advantage of duplication. An opposite hand piece is identical to the standard piece as drawn but is reversed in dimension from right to left. Pieces that are opposite hand receive a different piece mark than the piece drawn. Shop drawings also commonly use "near side" and "far side" to indicate attachment to only one side of the piece.

Shop detail drawings are prepared by the fabricator or the fabricator's detailer based upon information supplied by the engineer. This information should

Fabrication, Erection, and Quality Control ■ 12-3

include the service load and factored load for each connection. Shears, moments, and axial forces, if any, are required to design and detail connections. If various load combinations change the connection requirements, these combinations should be noted. Transfer forces between bracing elements must also be provided because the bracing forces themselves do not always indicate the flow of force from one brace or strut to the next.

Critical connections, especially moment connections, are designed by the structural engineer. Information on the type, size, number, and location of bolts, as well as the size, type, and location of welds is provided to the detailer for final dimensioning on the shop detail drawings.

12.1.3 Erection Plans and Field Bolt Lists

The erection plans are used to identify locations, sizes, and piece marks of individual members. The detailed information of the shop detail drawings is not required for field erection because a copy of the shop drawings is made available to the field for reference.

The erection plans are commonly simplified reproductions of the engineer's design drawings for the structural steel. Anchor bolt locations are presented on an anchor bolt plan for use by the foundation contractor.

The piece marks are placed on the erection plan at the same end as the piece marks placed on the beams, girders, and other members. This enables the erection crew to place the members in the right location with the proper orientation.

Field connections requiring special attention, particularly field welding, are noted and detailed on the erection plans as required. If field fabrication is required, such as drilling holes in existing members, these details are provided in the same manner as a shop drawing.

The detailer also prepares a field bolt list. Rather than list bolt diameters and lengths on the erection plan, a list is prepared of each connection, by piece mark. The number, type, diameter, and length of bolt required, along with the number of washers required, is given on the list.

12.1.4 CAD/CAM Interface

The use of computer-aided drafting packages, coordinated with computer software that performs structural analysis, has led to new opportunities to improve and speed the transfer of information from the structural engineer to the fabricator and detailer.

The traditional sequence of drawing preparation has been to use the engineer's drawings for the creation of the advance bill of materials. The erection plans almost copy the structural drawings, deleting information not necessary for the steelwork and expanding information necessary for field erection. Shop detail drawings are prepared from the information on the erection plans and the original structural and architectural drawings. Connection design forces shown on the engineer's drawings are used in designing and detailing the connections, and assumptions are made for those forces not provided.

Using a CAD interface, the erection plans can be easily and accurately prepared using the engineer's original information, without the need to redraw. The information provided as a part of the CAD transfer includes member sizes, locations, elevations, and forces. CAD systems can take this information directly from the creation plan to produce finished shop detail drawings.

Many CAD packages for structural-steel detailing also have an interface for computer-aided manufacturing (CAM). Information is provided to cutting lines, shears, drill, and punch lines to automate these operations when so equipped. Most commonly, beams and columns are drilled by computer numerically controlled (CNC) equipment. Also, clip angles and similar common detail parts are fabricated using CNC equipment.

The CAD/CAM interface from engineering design drawings to fabricated structural steel continues to develop, reducing the time of delivery and the cost of the completed structure.

12.2 Shop Fabrication

12.2.1 Materials Control

Structural-steel fabrication requires a variety of steel shapes, plates, bars, bolts, washers, and nuts, plus welding materials and coating materials.

Most fabricators maintain an inventory of commonly used structural shapes, common bar sizes, and common plate thicknesses. This allows for expedient deliveries of small jobs and reduces detailed ordering for connection and detail materials. These materials are ordered to meet applicable ASTM specifications. Upon receipt of these materials from the supplier, the mill test certificates are reviewed and the materials are placed into storage. The fabricator typically does not maintain traceability of these materials to a particular mill test report.

Upon receipt of an order, the fabricator prepares an advance bill of material and orders steel directly from a mill or steel service center. Commonly, mill test reports are requested to be supplied for the steel. Upon delivery, the fabricator may place the steel into inventory, only verifying that the supplied material meets ASTM specifications. Again, traceability of a steel piece to a particular test report is not maintained.

The fabricator may also, if requested by the customer, maintain traceability of the steel back to a particular mill test report. This traceability is typically maintained only until the piece is fabricated on the shop floor but can be retained further if necessary. As an example, there may be 50 beams of mark B8 required on the project. Twenty of these beams may be fabricated from steel supplied from a heat with one mill test report, the other thirty from another heat. Once assembled, the fabricator would be unable to determine from which heat the completed piece was fabricated, only that the steel met ASTM specifications. On critical structural elements, the customer may require such traceability and the fabricator may provide it through additional shop controls and by placing the heat number adjacent to the piece mark. Such traceability is again lost when the piece is erected and receives a field-applied coating, unless field records are maintained showing such information.

When a customer requests material test reports, the fabricator supplies the reports for steel ordered specifically for the project. For steels drawn from inventory, the fabricator may supply reports representative of materials drawn from that inventory. The AISC Code of Standard Practice, Section 5, further explains this situation.

Bolts, nuts, and washers are typically purchased as commodity items and are placed into inventory. Because the shop bolt list is not completed until the shop detail drawing is done, and the field bolt list is not done until the erection plans and shop details are done, bolts are ordered in advance using estimates of quantities and lengths.

Bolts, nuts, and washers should be maintained in protected storage with the manufacturer's certification available. Bolts and nuts should not be mixed with others of the same grade and length because the RCSC bolting specifications require preinstallation testing for most fastener assemblies by production lot. If torque control methods are used for installation, inventory control by lot is especially important.

Certificates of compliance with ASTM specifications are required for bolts, nuts, and washers only when requested by the customer. However, ASTM specifications require that such certificates be prepared by the manufacturer, whether or not they are requested by the customer.

Only a few fastener manufacturers place their lot number on the fastener itself. All others place their lot identification on the keg or box only. Once removed from the container, lot identification can be maintained only through established shop control procedures.

Welding materials such as electrodes, fluxes, and shielding gases also have manufacturer's certificates of compliance that they meet applicable American Welding Society and ANSI standards. These certificates of compliance may be requested by the owner. Lot numbers may be placed on electrodes, but otherwise identification is limited to the containers. Once welding has been completed, traceability back to a particular lot is impossible. Only strict shop controls can ensure that the lots for which certification is provided have been used for that customer's project.

Proper storage of welding materials, especially low-hydrogen SMAW electrodes, is extremely important to achieve the quality of welding desired. Electrode fluxes may absorb atmospheric moisture, reducing the desired low-hydrogen properties. The same holds true for SAW fluxes.

Coatings systems such as primers and paints require proper storage, many in a controlled environment. Manufacturer lot numbers and certificates of compliance to SSPC and other standards may be required.

12.2.2 Cutting

Wide-flange structural shapes are typically cut to length by cold-saw or hacksaw. Another option is flame cutting. The cold-saw equipment is also capable of producing a square, smooth cut that also qualifies as a finished surface for many bearing conditions. Hacksaws are used for smaller sections. Fabricators may also purchase their steel cut to length by the mill or service center.

Angles are typically sheared in a mechanical shear, integrated with the hole-punching operation. Plates are usually sheared in a plate shear but they may also be cut using plasma-arc or flame-cutting equipment. Tracing tables with optical guidance to follow a drawing may be used for cutting unusual configurations or multiple copies of the same piece. Many cutting tables of this type have multiple cutting heads.

For bearing joints such as column splices, where steel-to-steel contact is assumed to transfer part of the load, a cold saw or milling equipment is used to provide a smooth surface. In column splices, a small gap may remain between the column shafts under the provisions of LRFD specification, Section M4.4.

The tops of base plates may also need to be milled under the provisions of LRFD specification, Section M2.8. If the base plate thickness is 2 in (50 mm) or less, no milling is required as long as satisfactory bearing is obtained. If the plate thickness exceeds 2 in (50 mm), but not over 4 in (100 mm), it can be flattened with a press or milled at the contact area. Base plates over 4 in (100 mm) in thickness must have the contact area milled. An exception to these provisions is made if the column is to be welded to the base plate with complete-penetration groove welds.

Flame-cut edges must meet the surface roughness provisions found in LRFD specification, Section M2.2, and/or AWS D1.1, Section 5.15.4. AISC modifies the provisions of AWS 5.15.4.4 when the cut edge is free and subjected to static tensile stress, requiring repair of sharp V notches.

12.2.3 Layout and Assembly

The assembly of the various components of fabricated structural steel is done by hand. The vast majority of structural steel is custom designed and built for a specific project, so automation and robotics are not a part of current practice except for specific functions such as hole drilling and welding.

Detail parts such as stiffeners, clip angles, gusset plates, and such are fabricated individually in specific sections of the fabricating shop. There may be considerable duplication of these detail parts, as they are used on many members on the project. Some fabricators use standardized connection details that are applicable to many projects.

The main member is brought from the cutting area to the assembly floor and placed on workhorses. The detail parts are brought from the detail fabrication area. Other necessary items such as fasteners are also brought to the assembly area.

The fitter checks the main member length and the location of any holes that may have been punched or drilled in previous operations. The fitter then measures and marks the piece showing the locations of all attachments. If welded, the attachments are clamped or tack-welded in place. If bolted, the attachments are positioned and the bolts are snugged. After checking to verify that the piece has been properly configured, the welding is completed and the bolts are pretensioned, if necessary. Bolt holes that were not made prior to assembly are drilled using magnetic drills or similar equipment. The finished piece is inspected and sent on for cleaning and painting or shipping.

Special pieces like plate girders require extra fixtures to aid in cost-effective and accurate assembly. An adjustable frame is often used to hold the flange and web in the proper position, tightly clamped, so that automated welding can be performed in the flat or horizontal position.

Trusses also require custom fixturing to aid in assembly. Because most trusses are cambered, the truss jig allows for adjustments in camber for each truss. The truss chord members, often WT sections, are clamped to the jig to follow the required camber. The web members for one side of the piece are added and welded and the end connections are made. The piece is then carefully removed from the jig and flipped onto the opposite side. The remaining web members and end connection parts are added and welded into place.

In most cases, each fabricated item is shipped to the jobsite without checking to see if it properly fits with other pieces. Good detailing and good workmanship enable on-site assembly to progress quickly. Should errors occur or if tolerances are exceeded, repairs and adjustments are made in the field.

Certain structures may require preassembly at the fabricator's plant prior to shipping. Typically, these are continuous bridge members that are cambered or curved for the roadway alignment. Such preassembly verifies fit and eliminates additional field work and repairs. Often for bridges, holes that splice members together are made by drilling the parts in place as they are preassembled in the fabricator's yard.

12.2.4 Cambering and Curving

Long spans in floor and roof framing often require camber to be applied to compensate for dead-load, and sometimes live-load, deflections. The fabricator's standard practice is to fabricate with the beam's natural camber up, but often the natural camber is inadequate for long spans. Camber is added to structural shapes using either heat or mechanical means.

The straightness tolerance provided by ASTM A6, Table 24, for a wide-flange section in the strong-axis direction is $1/8$ in per 10 ft of length (1 mm per m of length). This can be reduced somewhat if the section is ordered as a column. Camber can be ordered from the steel mill within limits, as shown in Table 12.1. This table ensures

Table 12.1 AISC Table 1-10, Mill Cambering Limits for Rolled Beams

Section nominal depth, in	Specified length of beam, ft				
	Over 30 to 42, incl.	Over 42 to 52, incl.	Over 52 to 65, incl.	Over 65 to 85, incl.	Over 85 to 100, incl.
	Maximum and minimum camber acceptable, in				
W shapes, 24 and over	1 to 2, incl.	1 to 3, incl.	2 to 4, incl.	3 to 5, incl.	3 to 6, incl.
W shapes, 14 to 21, incl., and S shapes, 12 in and over	$3/4$ to $2 1/2$, incl.	1 to 3, incl.			

that there is enough camber specified that the mill is able to induce sufficient plastic strain through mechanical means to induce and hold the camber, but not so much camber as to risk inducing a kink in the beam web or buckling the member.

The camber as supplied by the mill or fabricator has a plus tolerance of $\frac{1}{2}$ in (50 mm) and a minus tolerance of zero for beams up to 50 ft (15,000 mm) in length. For beams over 50 ft (15,000 mm) in length, an additional plus tolerance of $\frac{1}{8}$ in (3 mm) for each 10 ft (3000 mm) in length, or fraction thereof, is allowed, but the minus tolerance remains zero.

An induced camber may be applied by the fabricator using localized heating. If positive (upward) camber is desired, the lower flange is heated. It is suggested to begin the heat in the web and use a V pattern downward, finishing with passes across the flange. AISC Section M2.1 limits the temperature of the steel to 1200°F (650°C) for standard structural steels and 1100°F (590°C) for quenched and tempered steel such as A514 and A852. AWS D1.1, Section 5.26.2, places similar limits on the application of heat.

If mechanical presses are used, the beam is taken beyond the yield point in selected areas of the beam, inducing a permanent set. Care must be taken not to buckle the beam web when using such methods.

Camber measurements are taken in the unloaded condition at the time and location of cambering. Since cambering involves the use of residual stresses, camber may be lost during shipment, handling, and storage. Generally, an average of about 25 percent of camber is lost before erection but will vary considerably from no camber loss to 100 percent.

12.3 Bolted Connections

12.3.1 Hole Punching and Drilling

The AISC specification addresses the use of punching and drilling for making bolt holes. The size of hole used for a particular size bolt may vary with the type of joint and hole selected by the engineer. Table 12.2 states the given hole sizes for each diameter of bolt, as required by AISC Table J3.3. Metric holes sizes have not yet been established by adopted specifications, but anticipated values are given.

Oversized holes may be used only in slip-critical joints. Slotted holes may be used in shear-bearing joints only when the load is transverse to the direction of the slot. Otherwise they may be used only in slip-critical joints.

Under AISC specification, Section M2.5, holes may be punched in the material provided the thickness of the material does not exceed the diameter of the hole plus $\frac{1}{8}$ in (3 mm), except for holes to be punched in quenched and tempered steels such as A514. For these steels, the limit for punching is $\frac{1}{2}$ in (13 mm) thickness.

AASHTO Section 11.4.8.1 limits the punching thickness to $\frac{3}{4}$ in (19 mm) for structural steel, $\frac{5}{8}$ in (16 mm) for high-strength steel (50 ksi 0.350 MPa, and over), and $\frac{1}{2}$ in (13 mm) for quenched and tempered steel. If the thickness of the steel exceeds these limits, the holes must be drilled or subpunched or subdrilled and reamed to the proper diameter. AASHTO also requires that holes penetrating through five or more layers of steel be either subdrilled and reamed to the proper diameter or drilled full size while preassembled.

Table 12.2 Nominal Hole Dimensions - AISC Table J3.3

Table J3.3 Nominal Hole Dimensions

Bolt Diameter	Hole dimensions, in			
	Standard	Oversize	Short slot	Long slot
$1/2$	$9/16$	$5/8$	$9/16 \times 11/16$	$9/16 \times 1\,1/4$
$5/8$	$11/16$	$13/16$	$11/16 \times 7/8$	$11/16 \times 1\,9/16$
$3/4$	$13/16$	$15/16$	$13/16 \times 1$	$13/16 \times 1\,7/8$
$7/8$	$15/16$	$1\,1/16$	$15/16 \times 1\,1/8$	$15/16 \times 2\,3/16$
1	$1\,1/16$	$1\,1/4$	$1\,1/16 \times 1\,5/16$	$1\,1/16 \times 2\,1/2$
$\geq 1\,1/8$	$d + 1/16$	$d + 5/16$	$(d + 1/16) \times (d + 3/8)$	$(d + 1/16) \times (2.5\,d)$

Metric Table J3.3 Nominal Hole Dimensions

Bolt Diameter	Hole dimensions, mm			
	Standard	Oversized	Short slot	Long slot
M16	18	20	18×22	18×40
M20	22	24	22×26	22×50
M22	24	28	24×30	24×55
M24	27	30	27×32	27×60
M27	30	35	30×37	30×67
M30	33	38	33×40	33×75
$\geq$ M36	$d + 3$	$d + 8$	$(d + 3) \times (d + 10)$	$(d + 3) \times (2.5\,d)$

The size of the completed hole may exceed the nominal diameter of the hole by a maximum of $1/32$ inch (1 mm). If the hole size is larger, then it must be considered oversized, which may change the design assumption, the design strength for the bolt, and the bearing strength of the steel.

Holes may also be flame-cut, although this method is not addressed by specification except for the slotting of holes. Research[1] indicates that flame-cut holes in A36 steel to $1/2$ in (13 mm) thickness are suitable provided they meet the hole size requirements, a difficult task. Piercing with the torch and reaming to the proper diameter is suggested.

12.3.2 Bolt Installation Methods

The RCSC specification allows two conditions of bolt installation, "tensioned" and "snug-tightened." The AISC specification requires that bolts in buildings be "tensioned" to the amount given in Table 4 of the RCSC specification if they are used

12-10 ■ Chapter Twelve

in oversized holes, slip-critical connections, tension connections, or connections of structures meeting the descriptions given in J1.9.1 of the AISC specification or otherwise specified by the engineer. Bolts need only be "snug-tight" in all other connections. AASHTO requires that all bolts in bridges be "tensioned."

The definition of "snug-tight" is stated in the LRFD specification in Section J3.1: "The snug-tight condition is defined as the tightness attained by either a few impacts of an impact wrench or the full effort of a worker with an ordinary spud wrench that brings the connected plies into firm contact."

The first step in installing bolts to be tensioned is the proper snugging of the joint. If the joint is not properly snugged, no tensioning method will work correctly. Bringing all bolts in a connection to a snug-tight condition will put the plies of the joint in solid contact. If the joint is not in solid contact, the tensioning method employed may fail to achieve the proper tension of the bolts in the joint. Tensioning the first bolt in the group will only serve to further draw down the gap between the steel elements. The installer erroneously assumes the first bolt is tight. The next bolt tightened further draws down any remaining gap and the initial bolt becomes looser still. This can become a compounding series in some joints. Only after all the bolts in the joint are snug tight and the plies of the joint in solid contact should the bolts be tensioned.

Turn-of-Nut Installation Method ■ The principle of the turn-of-nut method is controlled elongation of the bolt. Turning the nut a prescribed rotation, after proper snugging, elongates the bolt a certain amount. The elongation has a direct correlation to bolt tension. Table 12.3 provides the required turns for given bolt length-to-diameter ratios, as provided in Table 5 of the RCSC specification. No such table is available for metric bolts. The same degree of turn will provide at least the required amount of tension for a given length-diameter ratio because as bolts become larger in diameter, the number of threads per inch decreases.

Table 12.3 RCSC Table 5 - Turn-of-Nut Rotation

	Disposition of outer face of bolted parts		
Bolt length (underside of head to end of bolt)	Both faces normal to bolt axis	One face normal to bolt axis and other sloped not more than 1:20 (beveled washer not used)	Both faces sloped not more than 1:20 from normal to bolt axis (beveled washers not used)
Up to and including 4 diameters	1/3 turn	1/2 turn	2/3 turn
Over 4 diameters but not exceeding 8 diameters	1/2 turn	2/3 turn	5/6 turn
Over 8 diameters but not exceeding 12 diameters	2/3 turn	5/6 turn	1 turn

As an example, with flat surfaces and bolts less than or equal to four diameters in length, say a $^3/_4$- by 3-in bolt, $^1/_3$ turn must be provided. A $^3/_4$-by 5-in bolt would receive $^1/_2$ turn. A $^3/_4$- by 6 $^1/_2$-in bolt would receive $^2/_3$ turn. There is a tolerance to the amount of applied rotation. For turns of $^1/_2$ turn or less, the tolerance is plus or minus 30°. For turns of $^2/_3$ or more, the tolerance is plus or minus 45°.

For bolts over 12 diameters in length, there is too much variation to provide tabular values. It is required that the installer use a bolt tension calibrator, such as manufactured by Skidmore-Wilhelm, to determine the number of turns required to provide the required bolt tension.

The installation sequence should start with snugging the joint. Following inspection for snug, there are two options. In the first, the crew match-marks the end of the bolt and the nut and turns the nut the amount specified in Table 12.3. The joint would then be inspected to verify the applied turns by checking the match-mark rotation. RCSC says, "Bolt tightening to final tension may be accomplished with greatest assurance by match-marking the outer face of the nut with the protruding bolt point after the snug tightening operation." The installation crew may also use the "watch the wrench chuck" method for turn-of-nut, electing not to match-mark. However, the inspector must monitor the crew closely to verify that the proper turns are routinely applied.

Calibrated Wrench Installation Method ■ The calibrated wrench method uses a pneumatic impact wrench which can be adjusted to stop impacting when a certain torque is felt by the wrench. The wrench is adjusted or "calibrated" to stop impacting when the torque applied equals the torque believed adequate to tension the bolt. Functionally, these wrenches depend upon an internal cam unit for control. When the "calibrated" torque is reached, the cam unit shifts and the wrench stalls out.

In this method the wrench must be "calibrated" or set with a given air supply condition. The wrench should be calibrated using the same compressor and pressure settings, air hose and air hose length that will be used at the work. If an additional wrench is to be driven off the compressor, the wrench calibration should be checked with both wrenches in operation simultaneously as well as individually. If a significant length of hose from compressor to wrench is either added or removed, then the wrench should be recalibrated. All of this is necessary because if the air pressure or air volume is inadequate, the control mechanism in the wrench will not function properly and will continue to impact the fastener, although at a slower and weaker level which will not tension the bolt.

Calibration of the wrench is required every day, before installation begins, with three fastener assemblies of each diameter, length, grade, and lot. An assembly, by specification, is comprised of a bolt from a specific production lot and a nut from a specific production lot. Washers representative of those being used in the work must be included in the test, but lot control for washers is not mandated except under AASHTO requirements for bridge work.

If there is a significant difference in the quality of fastener lubrication, then the wrench must be calibrated for the varying lubrication conditions. A well-oiled bolt, nut, and washer assembly will require considerably less torque than one that is nearly dry or one that exhibits some indications of rust. Hence, if the wrench is calibrated using well-oiled bolts and then used on poorly lubricated bolts, the resultant bolt tension will be less than required. The same concerns apply if the bolt, nut, or washer surfaces contain dirt, grit, or sand.

Efficient calibration becomes key to the use of the calibrated wrench method of installation. Some projects have used a separate calibrated wrench for each given diameter, length, and grade being installed, changing wrenches when a different bolt group is being installed. Every wrench is calibrated each morning. It is possible to simplify operations by calibrating a wrench to properly install a wider range of bolts. This would be accomplished by setting the wrench to install at a tension well above the required tension, but yet not high enough to exceed the turns table. This same wrench setting could be tested on another bolt length of the same diameter and grade, and then another. Perhaps one wrench setting could be used throughout the day for all bolts of the same diameter and grade being installed, or perhaps only two wrenches would be needed instead of several.

Snugging the joint can be done with either an impact wrench (releasing the trigger when snug is achieved) or with a hand wrench on smaller bolts in lighter framing. After snugging, the joint should be inspected to verify snug, then approved for tensioning of the bolts. After the wrenches are calibrated, tensioning can begin. The wrench operator should tension the bolts using a systematic pattern, observing the chuck rotation as tensioning proceeds. If the rotation of the nut exceeds the turns table for turn-of-nut, the wrench calibration should be checked. After tensioning all the bolts in the pattern the operator should return to "touch up" each bolt in the pattern. Only the calibrated wrench method calls for such "touch up."

Direct Tension Indicator Installation Method ▪ Direct tension indicators (DTIs) are hardened-steel, washer-shaped devices with protrusions, "bumps," pressed out on one face, manufactured according to the provisions of ASTM F959. When a DTI is installed on a bolt with the bumps placed against the underside of the bolt head or a hardened washer, there are noticeable gaps between the "bumps." As the bolt is tensioned, the "bumps" are compressed. When the gaps have been reduced to the required dimension, the bolt has been properly tensioned. A feeler gauge is used to verify that the gaps have suitably closed.

The preferred installation arrangement is to position the DTI under the head of the bolt with the "bumps" facing the bolt head and the nut turned to tension the bolt. During installation, the bolt head must be held from turning to prevent abrasion of the "bumps," possibly giving a false reading of when the bolt is tensioned. For building applications the criterion for bolt tension is to have a 0.015-in feeler gauge refused in half or more of the gaps. For bridges a 0.005-in feeler gauge is used to assure that the remaining gap can be covered by the bridge's paint coating, thus excluding moisture from the gap area.

DTIs can be placed under the nut with the nut turned, or under the bolt head with the bolt head turned. In either case, a hardened F436 washer must be placed between the bolt head and the DTI. In each of these cases, a 0.005-in feeler gauge is used.

Fabrication, Erection, and Quality Control 12-13

A DTI is not a washer, so the rules for using washers with bolts in oversize holes should be followed or else the DTI will cup into the hole and give an invalid reading.

As with other installation methods, the joint must be snugged first. This is done by turning the nut until the "bumps" show visible signs of deforming. At this time, the bolt will have almost 50 percent of the tension required. The inspector should verify that the appropriate feeler gauge fits into more than half of the gap spaces.

After snugging, the installation crew should tension the bolts until it is apparent that the appropriate feeler gauge will be refused in half or more of the gaps. Compression of the "bumps" to a nil gap should not be a cause for rejection. However, on A490 bolts the installers should try to stop before this condition is reached. If there is a concern about overtightening, remove a sample number of bolts from the work and inspect them for deformation by running the nut down to the thread runout. If the nut runs down, there has been no overtightening.

Alternate Design Fastener (Twist-off Bolt) Installation Method ▪ A twist-off bolt is similar to a high-strength steel bolt, except that it has a splined segment connected by a notched section to the thread end. A special electrically driven tool is used for the installation. The spline end is engaged by an interior socket and the nut is engaged by an exterior socket. As the exterior socket rotates clockwise, turning the nut, the interior socket reacts against the spline end so all the torque forces are contained within the tool. When the nut begins to rotate and the bolt head grabs the steel, the bolt shank itself does not turn and only the nut rotates.

When the torsional reaction against the spline is sufficient to shear through the notched section, the spline end separates from the bolt and is contained in the interior socket until a trigger is pulled, "spitting out" the separated spline end. The manufacturers of twist-off bolts produce these bolts such that if they are installed in the condition in which they were manufactured, the torque necessary to break off the spline end should develop the tensions required. There is no ASTM for twist-off bolts (at the time of writing) and there is diversity among manufacturers as to the amount of torque to break off the spline and the quality of the lubricants on the assemblies.

Because of the interdependence of the bolt, nut, and washer upon the torque used for installation, the twist-off bolt unit is preassembled by the manufacturer. Substitutions of other nuts or washers may adversely affect performance and cause bolt tensions to be too low. Should these assemblies become dry, rusty, or dirty or otherwise change their character so they are not in the condition in which they were produced, they should be returned to the manufacturer for reprocessing.

The joint is first snugged using a systematic method, as with all installation procedures previously discussed. Care must be used to make sure that the spline is not twisted off during the snugging operation. Any bolts that twist off during snugging must be replaced. In some cases, deep sockets are used on conventional impact wrenches to snug the joints, therefore protecting the splines. Upon completion of snugging, the snug condition is verified to make sure no improper gaps exist. Upon acceptance of the snugged joint, the installation crew proceeds to tighten each twist-off bolt with the installation wrench until the spline shears off. A systematic pattern should be used for this step.

Alternate Design Fastener (Lock Pin and Collar) Installation Method ▪ Another form of alternate design fastener is the lock pin and collar fastener. With this type of fastener, a pin with a set of concentric locking grooves and a breakneck is tensioned with a hydraulic tool. As the tool pulls on the pin, the unit also swages a locking collar onto the locking grooves to retain the pin's tension. At a point above the required tension for the fastener, the breakneck portion of the pin fractures in tension, stopping the tensioning process. The pin maintains the required residual tension because of the locking collar.

12.3.3 Bolt Inspection Procedures

Bolt inspection is a multistep operation that begins before the bolt installation starts. The following steps should be included in a bolt inspection program.

1. Check the materials and certifications. Verify that the materials supplied comply with the project specifications. Review the product certification papers for ASTM and project compliance. Check that the certifications match with the supplied product.

2. Check for proper storage conditions. The various fastener components should be kept separate by lot until time for installation. Preassembly of bolts, nuts, and washers prior to use by the installer is satisfactory as long as lot control is maintained. Proper storage also includes protection from the elements, maintaining adequate lubrication and keeping the materials free from dirt, sand, grit, and other foreign material. In addition the RCSC specification states in 8(a) "Only as many fasteners as are anticipated to be installed and tightened during a work shift shall be taken from protected storage. Fasteners not used shall be returned to protected storage at the end of the shift." If fasteners are left in the work for any extended period of time prior to tensioning, they may become dry, rusty, or dirty. This will, in turn, significantly change the amount of torque which will be required to tension them, invalidating any torque/tension installation method (calibrated wrench or twist-off bolts) and significantly increasing the amount of torque required to install bolts by other methods.

3. Perform the assembly tests prescribed in the RCSC or AASHTO specification in a bolt-calibration device. This will assure that the assembly is capable of achieving the required tension without breaking, thread stripping, or excessive installation effort. AASHTO also requires that two assemblies of each bolt lot, nut lot, and washer lot, as supplied and to be assembled and installed in the shop or field, be tested with a special rotational-capacity test. This test verifies material strength and ductility, thread-stripping resistance, and the efficiency of the lubrication.

4. Check the validity of the installation technique for that group of fasteners. Perform the installation technique in a bolt-calibration device or with a calibrated DTI if the bolt is too short to fit into the calibrator. Verify that at least the minimum required tension, plus 5 percent, is achieved using the specified technique. For the calibrated wrench method, observe the calibration of the wrenches before the start of work each day.

5. Verify the knowledge of the installation crew. The previous two steps should be performed by the installation crew. By observing these tests, the crew demonstrates to the inspector their knowledge of the proper technique.

6. Visually check after snugging to verify that the snug condition has been achieved. For DTIs also verify that the installers check for an adequate number of remaining gaps. For twist-off-bolts, check that the splines are intact.

7. Observe the installation crew for proper technique. Observation of the installation crew does not mean that the installation of each individual bolt is observed, but that it is verified that the crew understands and follows the proper techniques on a uniform basis.

Torque testing is applicable only when there is a dispute, per Section 9(b) of the RCSC Bolt Specification. The methodology given is not intended as an inspection method, but as a way to resolve claims that the crew may not have followed the proper techniques for a particular joint. The specification states that this section is to be used only when inspection per 9(a) has been performed—the visual observation of the preinstallation testing, checking for snug, and observation of the installation technique of the crew.

For AASHTO bridge work, torque testing is still required for 10 percent of the bolts in each connection, minimum two per connection, for bolts installed using turn-of-nut or calibrated wrench methods. The torque testing procedures of AASHTO are similar to the RCSC Bolt Specification, Section 9(b), except that only three bolts are used to determine the inspection torque, not five. For DTI, twist-off, and lock pin and collar installation, no torque testing is required.

12.4 Welded Connections

12.4.1 Welding Process

Four welding processes predominate in structural steel fabrication. These are:

Shielded metal arc welding (SMAW)

Flux cored arc welding (FCAW)

Gas metal arc welding (GMAW)

Submerged arc welding (SAW)

In addition, three other processes may be used from time to time. Electroslag welding (ESW) and electrogas welding (EGW) may be used for large, thick weldments in some applications. Gas tungsten arc welding (GTAW), also commonly called tungsten inert gas welding (TIG), may be used for small root passes and for joining specialty steels.

The first four processes mentioned (SMAW, FCAW, GMAW, and SAW) are considered prequalified welding processes under AWS D1.1, Section 3.2.1. However, the short-circuiting transfer mode of GMAW, abbreviated GMAW-S, is not prequalified. The benefit of prequalified welding processes is that no testing of the process is required for a particular joint provided the joint design and welding procedures fall within the limits of the D1.1 specification. Nonprequalified procedures require additional documentation and testing of the Welding Procedure Specification (WPS).

Shielded Metal Arc Welding (SMAW) ▪ The shielded metal arc welding process is a common welding process used for a variety of applications. The electrode is a fixed-length wire or rod of a given steel and diameter, covered by a coating that serves as a shielding and fluxing agent. See Fig. 12.1. The electric current passes through the electrode to the steel, or from steel to electrode, depending upon polarity, forming an arc between the electrode and steel. The heat of the arc melts a portion of the steel and the end of the electrode. The electrode steel and alloying material is transferred to the weld puddle by the electrical forces generated. The electrode material and steel are mixed together by the arc action, then solidify to form the weld metal. The fluxing agent supplied by the electrode coating, as well as impurities generated by the welding itself, solidify on the top surface of the weld in the form of slag.

SMAW electrodes are categorized as low-hydrogen and non-low-hydrogen. Low-hydrogen electrodes have coatings designed for minimum moisture absorption from the atmosphere and provide a weld and heat-affected zone with a minimum amount of hydrogen. Hydrogen is considered a cause of low ductility and underbead cracking. Low-hydrogen electrodes are identified as EXXX5, EXXX6, and EXXX8. Low-hydrogen electrodes also require special care in storage and handling, including baking and drying, to retain their low-hydrogen characteristics.

Electrodes can be from AWS/ANSI A5.1 Specification for Mild Steel Covered Arc Welding Electrodes or AWS/ANSI A5.5 Specification for Low Alloy Steel Covered Arc Welding Electrodes. A5.5 electrodes are identified as EXXXX-X. A5.5 electrodes are optional for steels of groups I and II as listed in AWS Table 3.1. They are required for AWS group III steels and the steels listed in AWS Annex M, unless WPS qualification testing is performed. The nomenclature used to identify SMAW electrodes is shown in Table 12.4.

Flux Cored Arc Welding (FCAW). ▪ Flux cored arc welding (FCAW) is another popular welding process. The process uses a tubular wire containing fluxing and alloying agents inside the tube. The wire is fed into the weld area using a welding "gun." The heat of the arc vaporizes the fluxing agents in the core, releas-

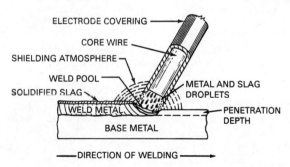

Fig. 12.1 Shielded metal arc welding. (*Adapted from American Welding Society Welding Handbook Vol. 2.*)

Fabrication, Erection, and Quality Control ■ 12-17

Table 12.4 Electrode Classification System for Shielded Metal Arc Welding

EXXXX-X		
E	SMAW electrode	
XX	Minimum tensile strength of undiluted weld metal, ksi, in the as-welded condition	
X	Permitted position 1. All 2. Horizontal (fillets only) and flat 4. Flat, horizontal, overhead, and vertical down	
X	Coating and operating characteristics 5, 6, and 8—low hydrogen	
-X	Major alloying elements (A5.5 electrodes)	

es the alloying elements, melts the electrode wire to enable transfer by the arc, and melts the steel being welded in the area of the arc. See Fig. 12.2.

One option used with FCAW, typically used as self-shielded, is gas shielding. Such a welding process is designated FCAW-G. FCAW-G is also considered a prequalified welding process. Because of the gas shielding, welding must not be done in wind speeds above 5 mi/h (8 km/h) under the provisions of AWS D1.1, Section 5.12.1.

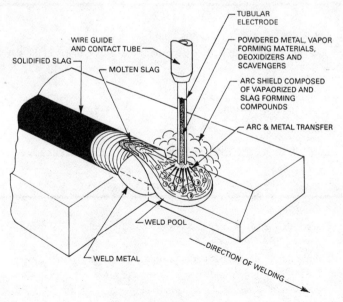

Fig. 12.2 Flux cored arc welding. (*Adapted from American Welding Society Welding Handbook Vol. 2.*)

Flux cored electrodes for structural steel may be of two types, AWS/ANSI A5.20 Specification for Carbon Steel Electrodes for Flux Cored Arc Welding and AWS/ANSI A5.29 Specification for Low Alloy Steel Electrodes for Flux Cored Arc Welding. Flux cored electrodes are designated as shown in Table 12.5.

Flux cored arc welding is known for higher productivity rates because the wire is continuously fed from a coil compared to the fixed-length electrodes of SMAW, and because of higher deposition rates.

Gas Metal Arc Welding (GMAW) ■ Gas metal arc welding (GMAW) is similar to FCAW in that the electrode wire is continuous and welding is performed using a welding "gun." The GMAW wire is a solid wire, rather than tubular like FCAW. In some specialty steels other than carbon and low-alloy structural steels, a metal-cored wire may be used. The shielding of the weld region is supplied by an external shielding gas fed into the weld through a hose. See Fig. 12.3. Gas metal arc welding is also commonly referred to as MIG (metal inert gas) welding.

GMAW electrodes can be of two types, AWS/ANSI A5.18 Specification for Carbon Steel Filler Metals for Gas Shielded Arc Welding or ANSI/AWS A5.28 Specification for Low Alloy Steel Filler Metals for Gas Shielded Arc Welding. A5.28 electrode wires must be used for AWS Table 3.1, Group III steels and AWS Annex M steels, unless otherwise qualified by test. Electrodes for gas metal arc welding are identified as shown in Table 12.6.

Table 12.5 Electrode Classification System for Flux Cored Arc Welding

EXXT-X		
E		Electrode
X		Minimum tensile strength of deposited weld metal, 10 ksi, in the as-welded condition
	X	Permitted position 0. Flat and horizontal 1. All positions
	T	Tubular wire
	X	Alloy information (A5.29 electrodes)

Table 12.6 Electrode Classification System for Gas Metal Arc Welding

ERXXS-X		
ER		Electrode rod (if ER, may also be used as GTAW filler rod; if E, may not be used for GTAW)
	XX	Minimum tensile strength of deposited weld metal, ksi, in the as-welded condition
	S	Solid wire (C designates metal-cored wire)
	-X	Alloy information (A5.28)

Common shielding gases used for structural steel include mixtures of argon and oxygen, argon and carbon dioxide, and straight carbon dioxide. Similar to FCAW-G, GMAW welding cannot be performed in winds greater than 5 mi/h (8 km/h) and therefore is rarely used in field applications.

GMAW has two advantages for welding structural steel. The first is that there is minimal slag to be removed. The second is that there are minimal welding fumes to be exhausted from the work space.

Submerged Arc Welding (SAW)

The submerged arc welding (SAW) process is a unique process that employs both a wire-fed welding gun and a deposit of flux powder. The arc is buried beneath a blanket of granular flux. The flux shields the weld region from atmospheric impurities, as well as providing fluxing and alloying elements. See Fig. 12.4. Of the four principal prequalified welding processes, SAW provides the most penetration. As an automatic or semiautomatic process, it is also the fastest once

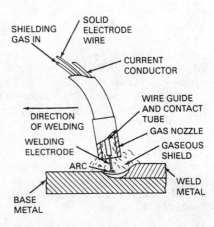

Fig. 12.3 Gas metal arc welding. (*Adapted from American Welding Society Welding Handbook Vol. 2.*)

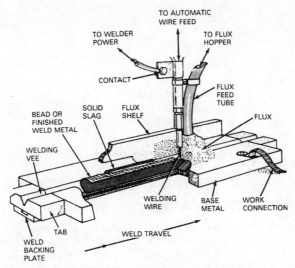

Fig. 12.4 Submerged arc welding. (*Adapted from American Welding Society Welding Handbook Vol. 2.*)

setup is complete. Because the weld puddle is buried beneath the flux, not visible to the welder, establishing proper welding parameters and accurately tracking the root of the weld is critical to the satisfactory completion of the weld.

Submerged arc welding may be performed with multiple welding heads, set either in parallel, with both electrodes controlled by the same feeder and power supply, or in tandem, with separate feeders and controls.

The proper combination of electrode wire and flux is important for quality welding. For this reason, the electrode and flux combination is specified within one document. ANSI/AWS A5.17 covers Specification for Bare Mild Steel Electrodes and Fluxes for Submerged Arc Welding, and ANSI/AWS A5.23 covers Specification for Low Alloy Electrodes and Fluxes for Submerged Arc Welding. The nomenclature system used for SAW electrodes and fluxes under ANSI/AWS A5.23 is shown in Table 12.7.

ANSI/AWS A5.23 specifications use a more complex nomenclature system with an additional -XX at the end of the electrode section to provide additional information on the system.

Table 12.7 Flux and Electrode Classification System for Submerged Metal Arc Welding

FXXX-EXXX

F		Flux
X		Strength level intended for flux, 10 ksi units
X		A or P, A for weld deposit strength determined in the as-welded condition, P for strength determined in postweld heat-treated condition
X		Impact properties temperature Z. No impact properties O. 20 ft-lb at 0°F (27 J at –18°C) 2. 20 ft-lb at –20°F (27 J at –29°C) 4. 20 ft-lb at –60°F (27 J at –51°C) 6. 20 ft-lb at –60°F (27 J at –51°C)
	-E	Electrode (EC for composite electrode)
	X	Manganese content L. Low manganese (0.6% max.) M. Medium manganese (1.40% max.) H. High manganese (2.25% max.)
	X	Nominal carbon content, in 1/1000% Ex: 12=0.12%
	K	Killed steel (if applicable)

12.4.2 Welding Procedures

The use of written established welding procedure specifications (WPS) is mandated by AWS D1.1 in Section 3.1. The WPS may be either prequalified or qualified by test. A prequalified WPS must fall within the limits prescribed in Section 3 of D1.1, notably Table 3.7, must use a prequalified welding process, and must be on a joint deemed prequalified in Section 3. All other WPSs must be qualified by test using the procedures set forth in Section 4 of D1.1. Appendix E of D1.1 provides example WPS forms.

Welding procedure specifications may be furnished by electrode suppliers, welding equipment suppliers, technical organizations, or consultants, or may be developed by the contractor. WPSs are specific to the following parameters:

Welding process

Base metal (steel classification, strength, type)

Base metal thickness (range)

Electrode classification

Flux classification

Shielding gas

Joint type (butt, tee, corner)

Weld type (groove, fillet, plug)

Joint design details (root opening, groove angle, use of backing)

Use of backgouging

Position (flat, horizontal, vertical, overhead, tubular application)

Using these parameters, the following items are established:

Number and position of passes

Electrode diameter

Polarity

Current or wire feed speed

Travel speed

Voltage

Technique

Shielding gas (if used) flow rate

Preheat, interpass, and postheat requirements

Cleaning requirements

Inspection requirements

12.4.3 Welding Inspection

The responsibilities and levels of welding inspection must be established in the contract documents. Neither the AWS nor AISC nor the model building codes provides a complete listing of welding inspection duties. Inspection duties may be assigned to the contractor (fabrication and erection inspection) or to an inspector who reports to the owner or engineer (verification inspection). Under the special inspection requirements of the Uniform Building Code and National Building Code, certain welding inspection must take place by an inspector responsible to the owner or engineer. The level of inspection varies according to the type of project, the structural system employed, and the seismic category.

The welding inspection provisions of AWS do not assign specific inspection tasks to either the contractor or verification inspector. Verification inspection is not mandated. Any forms of nondestructive testing, beyond visual inspection, must be specified in the contract documents, as AWS requires only visual inspection by the welder.

Welding inspection is a start-to-finish task. Inspection can be broken into three timing categories—before welding, during welding, and after welding. Both general and specific welding inspection requirements are provided in AWS D1.1. Section 6 on inspection covers procedural matters and acceptance criteria. Some workmanship requirements are found in Section 5. Various inspector checklists have been compiled and published in numerous sources. The following is a composite list of critical welding inspection tasks.

Before Welding ▪ Check the qualifications of the welding personnel. Welders, tack welders, and welding operators must be qualified by the employer under the requirements of AWS D1.1, Section 4.1.2. The testing of these personnel is specified in Section C of Part 4. Welders are those who manipulate the welding electrode and filler metal by hand. Tack welders may make only small welds necessary for the preassembly of parts, holding them in proper alignment until the final welds are made. Welding operators set up and operate automatic welding equipment, robotic welders, and other mechanized welding operations.

Welder performance qualification (WPQ) documentation must be made available for the inspector's review prior to the start of welding. In many cases, the welder's testing is conducted by an independent testing laboratory. If a previous employer's testing results are to be used, then the engineer must approve the current employer's reliance upon these previous tests.

Welder and welding operator qualification remains effective indefinitely as long as the welder continues to use the welding process for which the welder is qualified. Should the welder not use the welding process for a period exceeding 6 months, then a requalification procedure is required. Also, if the welder produces work of a quality below that required by the code, the inspector can require that the welder be requalified by testing. Tack welder qualification remains effective indefinitely unless the quality of their work falls below required standards, in which case retesting is required.

Welding Procedures ▪ The inspector should check for and review the welding procedure specifications (WPS) to be used for the project. The WPS is specific to the material, welding process, position, type of joint, and configuration of joint. Verify that a proper WPS is available for all welds to be completed, and that the welding personnel follow the WPS as prescribed.

Base Metal Quality ▪ Verify that the quality of the base metal is suitable for welding. The steel to be welded must be clean, smooth, without surface discontinuities such as tears, cracks, fins, and seams. Such surface discontinuities could propagate into the weld after welding. The surface should also be free of excessive rust, mill scale, slag, moisture, grease, oil, and any other material that could cause welding problems. Some materials may be permitted such as thin mill scale (mill scale that withstands a vigorous wire brushing), thin rust-inhibitive coatings, and antispatter compounds made specifically for weld-through applications. AWS Section 5.15 provides additional Information and exceptions to these provisions.

Joint Preparation and Fit-up. ■ Fillet weld fit-up tolerances are given in AWS Section 5.22.1. Gaps between parts of $1/16$ in (1.6 mm) or less are permitted without correction. If the gap exceeds $1/16$ in (1.6 mm) but does not exceed $3/16$ in (5 mm), then the leg size of the fillet weld should be increased to compensate for the gap between the parts. Gaps over $3/16$ in (5 mm) are permitted only in thick materials over 3 in (76 mm). In these cases, the use of a backing material is required as well as compensation in the weld leg size. Such provisions cannot be used for gaps over $5/16$ in (8 mm). Similar provisions are used for partial-penetration groove welds when the welds are parallel to the length of the member.

When groove welds are used, tolerances to the root opening, groove angle, and root face apply. The specific tolerances depend upon the type of groove weld, the presence of backing, and the use of backgouging. AWS Section 5.22.4 and AWS Figure 5.3 provide these values. Groove tolerances are also provided in the prequalified groove weld details in AWS Figures 3.3 and 3.4. For tubular joints, Section 5.22.4.2 governs.

Part alignment on butt joints can be critical, depending upon application. AWS Section 5.22.3 requires alignment within 10 percent of the part thickness, not to exceed $1/8$ in (3 mm), when the parts are restrained from bending from such misalignment. No provisions are given for cases where such restraint does not exist. For girth welds in tubular joints, the alignment tolerances are provided in Section 5.22.3.1.

Welding Equipment and Consumables ■ In order to properly follow the parameters of a welding procedure specification (WPS), the equipment used for welding must be in good repair and properly calibrated. The inspector should check the maintenance and testing records of the equipment to be employed and if necessary use testing equipment to verify that the equipment settings and the welding machine output are accurate.

Welding electrodes, fluxes, and shielding gases should be checked to be in conformance with AWS Section 5.3. Low-hydrogen SMAW electrodes require special controls, including requirements for baking and storage temperatures and exposure time limits. Fluxes for SAW require dry, contamination-free storage, with removal of the top 1 in (25 mm) of material from previously opened bags prior to use. Drying of flux from damaged bags may be required.

Shielding gases must be of welding grade and have a dew point of –40°F (–40°C) or lower. The gas manufacturer's certification of dew point may be required by the engineer.

Welding Conditions. ■ For good welding, the welder and the operating equipment must have conditions suitable for welding. The environmental conditions for welding must also be adequate. Limits are given in AWS D1.1, Section 5.12. The temperature of the area immediately surrounding the welding must be above 0°F (–18°C). The temperature in the general vicinity can be lower, but heating must be provided to raise the temperature immediately around the weld to at least this temperature. The surfaces to be welded must not be wet or exposed to moisture. High winds must be avoided. For GMAW, GTAW, EGW, and gas-shielded FCAW, the wind speed must not exceed 5 mi/h (8 km/h), requiring protective enclosures in most field applications. No maximum wind speed is specified for welding processes requiring no shielding gases, but a practical level is generally around 25 mi/h (40 km/h).

Preheat ■ Preheating of the steel is necessary for thick steels, certain high-strength steels, and steels when their temperature is below 32°F (0°C). The preheating requirements should appear in the welding procedure specification (WPS). AWS preheat requirements may also be found in Table 3.2, Minimum Prequalified Preheat and Interpass Temperature. In this table, the required preheat is given for a specified steel specification group, welding process, and thickness range of part. When the temperature of the steel is below 32°F (0°C), the steel must be heated to at least 70°F (21°C). The thicker the steel, the higher the required preheat temperature. Higher-strength steels require higher preheats. Certain high-strength steels listed in AWS D1.1, Annex M, have preheats limited to a maximum of 400°F (205°C). Preheat requirements may also be modified using the provisions of AWS D1.1, Appendix XI, which evaluates the welding process, restraint, and the weldability (carbon equivalency) of the steel.

During Welding ■ After checking the welder qualifications, WPS, welding consumables, steel materials, welding conditions, equipment, joint fit-up, and preheat, the welding inspection performed during welding is limited to verifying that the welding procedures are properly followed. This includes the maintenance of interpass temperature during welding, usually the same temperature required for preheat. Each pass should be thoroughly cleaned and visually inspected. Control of electrodes, especially low-hydrogen SMAW electrodes, must be maintained. In some cases, nondestructive testing may be performed at various stages during welding.

After Welding ■ After the weld has been completed, the final size and location is checked to verify that it meets the plans and specifications of the contract. Visual inspection is performed, and nondestructive testing of the completed weld may be performed if required by the contract documents. If repairs are required, the inspection should include the repair work and reinspection of the repaired weld. The inspector responsible for the completed weld should place an identifying mark near the weld. Written documentation of the weld's quality, including any noted significant discontinuities, should be prepared and submitted.

Nondestructive Testing of Welded Joints ■ Several methods of nondestructive testing (NDT), also called nondestructive examination (NDE), may be used on a structural-steel project. The frequency and location of NDT must be stated in the contract documents. Little NDT is mandatory under the various codes except for certain types of joints in seismic applications.

The first common form of NDT is *visual* (VT). Most visual inspection is performed without the use of magnifiers. Magnifying glasses can be used to more closely examine areas that are suspected of cracks and other small but potentially significant discontinuities. Adequate light and good visual acuity is necessary. Various weld gauges are used to determine weld size and various other measurements.

An expanded form of visual inspection is *penetrant testing* (PT). The weld surface and surrounding steel is thoroughly cleaned. A penetrating liquid dye is applied to the weld surface and allowed time to penetrate cracks, pores, and other surface discontinuities. After an allotted time (dwell time), the penetrant is removed and a developer is applied to the surface. The developer draws the penetrant back to the surface of the weld. The developer is of a color (usually white) that contrasts with the color of the dye in the penetrant. The inspector observes the dye in the developer, then removes the developer and dye to more closely inspect the weld surface visually. Some penetrant testing uses an ultraviolet solution, rather than a dye, to aid in visibility when an ultraviolet lamp is available. Penetrant testing can detect surface discontinuities only.

Magnetic particle testing (MT) can be used to detect surface and slightly subsurface discontinuities. The general limit to depth of inspection is around $5/16$ in (8 mm). Electricity is induced into the region of the weld through the use of prods or a yoke. The electricity generates a magnetic field on and near the surface of the steel. Magnetic particles such as fine iron particles are then applied to the surface of the steel. These particles may be in the form of a dry powder or may be in a liquid emulsion.

When cracks or other discontinuities are on or near the surface, the flux lines generated by the current are interrupted. In essence, the interruption has created two new magnetic poles in the steel, attracting the particles. The inspector then observes and interprets the position and nature of the accumulated particles, judging them to indicate a crack or other discontinuity on the surface or subsurface. For best performance, the flux lines must flow perpendicular to the discontinuity. Therefore, the MT technician must rotate the yoke, prods, or other source at 90° angles along the length of the weld to inspect it for both longitudinal and transverse discontinuities. Although MT is generally a visual technique, permanent records of discovered defects can be made with the use of adhesive tape placed over attracted magnetic particles during testing.

Ultrasonic testing (UT) is a very popular method of nondestructive testing. It is capable of testing weldments from approximately $5/16$ in (8 mm) to 8 in (200 mm) in thickness. The most common method of testing uses a pulse-echo mode similar to radar or sonar. The control unit sends high-frequency electronic signals into a transducer made of piezoelectric material. The electrical energy is transformed by the transducer into vibration energy. The vibration is transmitted into the weldment through a coupling liquid. The vibration carries through the steel until a discontinuity or other interruption, such as an edge or end of the material, disrupts the vibration. The disruption causes the vibration to return the ultrasound wave back toward the transducer. The return vibration is then converted back into electrical energy by the transducer, sending a signal to the display unit. The return signal's configuration, strength, and time delay are then interpreted by the testing technician.

The interpretation by the technician uses the height of the response signal to indicate size and severity. Using a calibration setup, the distance on the display unit from the initial pulse to the reflection is used to determine the distance from the transducer to the discontinuity. The operator can also manipulate the transducer in various patterns to determine a better understanding of the location, length, depth, orientation, and nature of the discontinuity.

AWS D1.1, Table 6.6, prescribes the testing procedures for butt, tee, and corner joints of various thicknesses. The search angle and faces to be used are given.

Annex K of AWS D1.1 gives alternative techniques for ultrasonic testing and the evaluation of weld discontinuities.

Radiographic testing (RT) is another common method of NDT for welds in structural steel. RT is performed using either x-rays or gamma rays, sending energy into the steel weldment. Film is placed on the side of the weldment opposite the energy source. The steel and weld metal absorbs energy, preventing it from exposing the film, but weld discontinuities allow more energy to get to the film. This exposes the film more, producing a darkened area on the film to be interpreted by the radiographer.

Radiographic testing is effective in steels up to about 9 in (230 mm) in thickness. X-ray machine capabilities depend upon the voltage setting of the machine, with 2000 kV required for an 8-in (200-mm) thickness. Gamma-ray machine capabilities depend upon the isotope used, usually cobalt-60 or iridium-192, but sometimes cesium-137. Exposure times and film selection are varied according to conditions and thicknesses. Image quality indicators (IQIs), either wire-type or hole penetrameter, are used to verify the sharpness and sensitivity of the film image, as well as provide a measurement scale on the exposed film. Because of the radiation-exposure hazards and time and equipment involved, radiographic testing is typically the most expensive of the methods mentioned above.

12.4.4 Weld Acceptance Criteria

The acceptance criteria to be used for the required weld quality are to be established by the engineer. Commonly, the workmanship and inspection criteria found in AWS D1.1, Sections 5 and 6, are adopted. However, the use of alternate criteria is both accepted and encouraged. Section 6.8 of AWS states that "The fundamental premise of the Code is to provide general stipulations applicable to most situations. Acceptance criteria for production welds different from those specified in the Code may be used for a particular application, provided they are suitably documented by the proposer and approved by the engineer." The Commentary to Section 6.8 provides additional insights into the development and use of alternate acceptance criteria.

The visual acceptance criteria for welds are summarized in Table 6.1 of AWS D1.1. This table is broken down into three categories of connections—statically loaded nontubular, cyclically loaded nontubular, and tubular. Generally, cyclically loaded and tubular connections require higher standards of quality. These values are also used when penetrant testing (PT) and magnetic particle testing (MT) is used.

When ultrasonic testing is used, AWS Table 6.2 is used for statically loaded nontubular connections and Table 6.3 for cyclically loaded nontubular connections. For tubular connections, use AWS Section 6.13.3. Alternately, the techniques of AWS Annex K may be employed when approved by the engineer.

When radiographic testing is used, AWS Figure 6.1 is used for statically loaded nontubular connections, Figure 6.4 for cyclically loaded nontubular tension connections, and Figure 6.5 for cyclically loaded nontubular compression connections. For tubular connections, AWS Sections 6.12.3 and 6.18 apply.

Because many of the acceptance criteria found in AWS D1.1 are based upon what a qualified welder can provide, rather than the quality necessary for structural integrity, alternate acceptance criteria can be used to save both time and money.

In addition, repairs to some welds with innocuous discontinuities may result in more damage to the material in the form of additional discontinuities, lower toughness, larger heat-affected zones, more distortion, and higher residual stresses.

Alternate acceptance criteria have been published by several organizations in various forms. In the United States, the Electric Power Research Institute has published *Visual Weld Acceptance Criteria*, document NP-5380, for use in reinspections of welds in existing nuclear power plant facilities. These weld acceptance criteria were accepted for use by the Nuclear Regulatory Commission. The Welding Research Council has published several WRC bulletins providing suggested criteria. The ASME Boiler and Pressure Vessel Code, Section IX, provides acceptance criteria for welds that can be also used for structural welds. The International Institute of Welding has published several documents providing suggested acceptance criteria, with considerable research documentation justifying the criteria. British Standards Institution document PD 6493:1991, "Guidance on Methods for Assessing the Acceptability of Flaws in Fusion Welded Structures," is one of the most thorough documents currently available.

12.5 Materials, Fabrication, and Erection Tolerances

12.5.1 Mill Material

Structural-steel materials are manufactured to meet the dimensional requirements of ASTM A6/A6M. Each product category (W, S, HP, C, MC, L, WT, etc.) has a particular set of rolling tolerances. The tolerances are established to allow the mill some leeway in manufacturing, including roll adjustment and wear, as well as variations from cooling and residual stresses. Steel detailers, fabricators, and erectors must be, in turn, familiar with these tolerances and provide for some adjustment during detailing, fabrication, and erection.

For wide-flange (W) sections, the principal variations are in member depth, flange width, flange tilt, web off center, straightness, and length. The cross-section tolerances are set forth in the ASTM A6/A6M specification in Table 16, length tolerances in Table 22, and straightness tolerances in Table 24.

Tolerances are not set for web thickness or flange thickness, but the member itself must be within 2.5 percent of its theoretical weight. Thus a beam manufactured shallower and narrower than theoretical will have a flange thicker than theoretical.

Angle sections have leg length, thickness, and squareness tolerances provided in Table 17. Angles may be out-of-square by a slope of 3/128, or approximately $1^1/_2°$.

Tee sections cut from other sections such as WT, ST, and MT have the same cross-sectional tolerances as the shapes from which they are cut. However, the depth tolerance of the tee section, as well as the out-of-straightness, is increased to the values provided in Table 25. Table 18 for rolled tees is applicable only to shapes rolled in a tee configuration, sections not typically used in steel construction.

Plates have tolerances for width, thickness, length, flatness, and waviness. Thickness tolerances are in Table 1, width and length in Table 3, flatness in Tables 13 and 14, and waviness in Tables 15. Thicker and wider plates have increased tolerances for thickness, width, and flatness. Flatness is the curvature of the plate when measured across the width. Waviness considers the irregularity at the surface of the plate from rolling. The number of peaks found along the surface of the plate is counted; then the allowable variation from flatness between adjacent peaks is determined. The waviness tolerance is dependent upon the number of peaks in a 12-ft length of plate and the flatness tolerance for the width and thickness of plate. Plates with more waves have reduced allowances for waviness, and thicker plates allow more waviness.

Several of the ASTM A6/A6M tables applicable to structural steel are reproduced for convenience in the AISC *Manual of Steel Construction* in Section 1. Cross-sectional tolerances for shapes are also interpreted in Figure 1 in the Commentary of the AISC Code of Standard Practice.

For pipe sections, manufacturing tolerances are found in ASTM A53. For square and rectangular hollow structural sections, tolerances are in ASTM A500, A501, and A618. Primary tolerances are also reproduced in the AISC *Manual of Steel Construction*, Part 1.

12.5.2 Shop Tolerances

Shop tolerances for the fabrication of structural steel are provided by the AISC Code of Standard Practice in Section 6. Tolerances are provided for member length, straightness for columns, and camber for other members. *No metric values for the Code of Standard Practice have yet been adopted by AISC, so the metric conversions noted are approximate.*

For beams and other members 30 ft (9000 mm) and under in length, the length tolerance is $\pm 1/16$ in (2 mm). For members over 30 ft (9000 mm) in length, the length tolerance is $\pm 1/8$ in (3 mm). Columns and other members that are finished for contact bearing at both ends have a length tolerance of $1/32$ in (1 mm).

Column sections and other compression members should be straight to within $1/1000$ between braced points. For a 13-ft (4000-mm) floor-to-floor height, this would allow column curvature of up to $5/32$ in (4 mm) at midheight. This tolerance is slightly tighter than the typical mill rolling straightness tolerance of a rolled section, which allows $1/8$ in per 10 ft of length, maximum $3/8$ in, for members up to 45 ft in length (1 mm per meter of length, maximum of 10 mm, for members up to 14 m in length), when ordered as columns. The straightness tolerance at the mill is measured from end to end. The straightness in fabrication is measured from brace point to brace point.

When a specific camber for beams, girders, trusses, or other members is required by the design documents, the camber tolerance is plus $1/2$ in (13 mm), minus 0 for members to 50 ft (15,000 mm) in length. For members over 50 ft (15,000 mm) in length, the tolerance is $1/2$ in (13 mm) plus $1/8$ in (3 mm) for each additional 10 ft (3000 mm) of length. Camber is measured in the fabricating shop in the unstressed condition. Camber may be lost during storage, shipment, and erection handling as the residual stresses used to induce the camber are partially relieved by handling stresses. The average loss of camber from fabrication to erection has been estimated at 25 percent.

Fabrication, Erection, and Quality Control ■ 12-29

For built-up sections made by welding, fabrication tolerances are provided in AWS D1.1, Section 5.23. Column, beam, and girder straightness tolerances provide values similar to rolled shapes meeting ASTM A6. Applied camber tolerances are more liberal than those allowed by the AISC Code of Standard Practice.

AWS Section 5.23.6 contains provisions for the flatness of welded girder webs, with values dependent upon the depth-thickness ratio, the presence of stiffeners, and whether the loading is static or cyclical.

Other AWS tolerances for built-up sections include web off-center of $1/4$ in (6 mm), flange warpage and tilt of 1 percent or $1/4$ in (6 mm), whichever is greater, and depth variations which depend upon the member depth. For members up to 36 in (900 mm) in depth, the depth tolerance is $\pm 1/8$ in (3 mm), for members over 36 in (900 mm) to 72 in (1800 mm), $\pm 3/16$ in (5 mm), and members over 72 in (1800 mm), plus $5/16$ in (8 mm), minus $3/16$ in (5 mm). Section 5.23.11 provides specific tolerances for the location and fit of stiffeners.

12.5.3 Field Tolerances

Field tolerances are established in the AISC Code of Standard Practice in Section 7. The erection tolerances must consider several variables, including variations permitted in the location of the anchor bolts, elevation of the base plates, plus mill rolling tolerances and fabrication tolerances. *Caution should be used in working with the metric conversions provided, as they have not been adopted by the AISC.*

The owner sets the building lines for theoretical location of columns and anchor bolts. For anchor bolt locations, an "established column line" is defined in Section 7.5.1(d) as "the actual field line most representative of the as-built anchor bolt groups along a line of columns." An individual column anchor bolt group must not vary from this established column line by more than $1/4$ in (6 mm). From one column to the next, the anchor bolt groups must be within $1/4$ in (6 mm) of the proper distance. For long column lines, the maximum variation of spacing over a 100-ft (30,000-mm) distance must not exceed $1/4$ in (6 mm). For each additional 100 ft (30,000 mm) of length, an additional variation of spacing of $1/4$ in (6 mm) can be added, but not to exceed a total of 1 in (25 mm) variation, applicable at 400 ft (120,000 mm). Individual anchor bolts within an anchor bolt group should be located within $1/8$ in (3 mm) of their proper location relative to the other anchor bolts. These tolerances were established using typical values for the diameter of oversized holes in column base plates. The elevation setting of base plates, leveling plates, or other bearing devices should be within $1/8$ in (3 mm) of the established grade.

Steel frame tolerances are provided in Section 7.11 of the Code of Standard Practice. A critical element is the plumbness of the columns, which must be within 1:500 of plumb, measured at the centerline of the column cross section at the top and bottom of the column shaft. This allows a column to be out of plumb 1 in over a 42-ft (25 mm over a 12,500-mm) height. Column plumbness should not be confused with column straightness as discussed above.

In multistory buildings, the columns surrounding the elevator shafts have a maximum deviation from the "established column line," as defined above, of 1 in (25 mm) through the first 20 stories. Above 20 stories, this deviation is increased by $1/32$ in (1 mm) per story. At the 52nd floor (45th floor) and above, the maximum deviation from the established column line is 2 in (50 mm).

Exterior columns are also limited in their deviation from the established column line. Up to 20 stories, the columns may vary no more than 1 in (25 mm) toward the building line or more than 2 in (50 mm) away from the line. Above 20 stories, an additional $1/16$ in (2 mm) per story is permitted, with a maximum of 2 in (50 mm) toward and 3 in (75 mm) away from the building line at and above the 36th floor (33rd floor).

Exterior columns may also deviate from the established column line in a plane that runs parallel to the building line. This variation cannot exceed 2 in (50 mm) in the first 20 stories. Above 20 stories, an additional $1/16$ in (2 mm) tolerance is provided, with a maximum of 3 in (75 mm) at and above the 36th floor (33rd floor).

Exterior columns, as aligned parallel to the building line, must not fall outside a horizontal envelope $1\frac{1}{2}$ in (38 mm) wide when the building is 300 ft (90,000 mm) in length or less. This envelope may be widened an additional $1/2$ in (12 mm) per 100 ft (30,000 mm) of additional building length, with a maximum envelope width of 3 in (75 mm) for buildings 600 ft (180,000 mm) long or more.

The alignment of beams, girders, and other members is dependent upon the alignment of the columns or other supporting member. If a horizontal piece is shipped in two or more sections, it is considered properly aligned it is does not deviate from a straight line by more than 1:500. The same 1:500 alignment variance is permitted for cantilever members.

The elevation of floor and other horizontal members framing into columns is determined by the elevation of the columns, as measured from the upper column splice. The floor member at the column connection may vary $3/16$ in (5 mm) below or $5/16$ in (8 mm) above the working point for the member. For floor and other horizontal members framing into girders and other horizontal members, the elevation is determined solely by the elevation of the girder at the connection point.

In tier buildings, floor framing elevations can vary because of several factors, including base plate elevation tolerance, column length tolerance, differential column shortening, and the floor framing elevation tolerance noted in the previous paragraph. Additional consideration should be made for camber tolerances, potential loss of camber, end connection rigidity, and concrete ponding in establishing screed locations when pouring the slab on deck.

The accumulation of mill, shop, and erection tolerances can sometimes require adjustment to connections, the realignment of columns, and other adjustments to the steel frame.

12.6 Coating Systems

12.6.1 Surface Preparation

Prior to coating application, the surface of the steel must be cleaned to a required level of cleanliness, depending upon the type of coating to be applied. For some coatings, the profile of the surface must be established within a certain height and roughness range.

Fabrication, Erection, and Quality Control ■ 12-31

Not all structural steel requires a coating. Certain structural applications are classified as "Environmental Zone Zero" by the Steel Structures Painting Council in the *Steel Structures Painting Manual*, Volume 2. Zone Zero is defined as a dry interior where the steel is embedded in concrete, encased in masonry, or protected by membrane or spray on fireproofing. This is typical for many office buildings and other enclosed structures.

Zone Zero applications require the least surface preparation. The AISC's *Quality Criteria and Inspection Standards*, third edition, Section 5.III.H, calls for the removal of heavy deposits of oil and grease if the steel is not to be coated or fireproofed. If the steel receives a contact type fireproofing, then the steel is cleaned of dirt, oil, and grease. The removal of mill scale is not required. However, the Association of the Wall and Ceiling Industry—International *Technical Bulletin 12-A* calls for the removal of loose mill scale when fireproofing is to be applied. In either case, the surface preparation requirements for Zone Zero are a step below SSPC SP1 and SP2 requirements.

Environmental Zone 1A is also an interior application that is normally dry or may require only temporary protection. Industrial buildings that do not involve the use of corrosive chemicals fall into this category when the steel is exposed. Other environmental zones are listed in Table 12.8.

Table 12.8 SSPC Environmental Zones

Environmental zone	Zone conditions
0	Dry interiors where structural steel is embedded in concrete, encased in masonry, or protected by membrane on noncorrosive contact-type fireproofing
1A	Interior, normally dry (or temporary protection). Very mild (oil-base paints now last 10 years or more)
1B	Exteriors, normally dry (includes most areas where oil-base paints now last 6 years or more)
2A	Frequently wet by fresh water. Involves condensation, splash, spray, or frequent immersion. (Oil-base paints now last 5 years or less)
2B	Frequently wet by salt water. Involves condensation, splash, spray, or frequent immersion. (oil-base paints now last 3 years or less)
2C	Fresh water immersion
2D	Salt water immersion
3A	Chemical exposure, acidic (pH 2.0 to 5.0)
3B	Chemical exposure, neutral (pH 5.0 to 10.0)
3C	Chemical exposure, alkaline (pH 10.0 to 12.0)
3D	Chemical exposure, presence of mild solvents. Intermittent contact with aliphatic hydrocarbons (mineral spirits, lower alcohols, glycols, etc.)
3E	Chemical exposure, severe. Includes oxidizing chemicals, strong solvents, extreme pHs, or combinations of these with high temperatures

* SSPC *Steel Structures Painting Manual, Vol. 2*, chap. 1, Table 3.

The Steel Structures Painting Council has nine surface-preparation specifications applicable to structural steel. These specifications are as listed in Table 12.9. In order of increasing cleanliness, they are SP1, SP2, SP3, SP11, SP6, SP7, SP10, and SP5, with SP8 (pickling) used principally prior to galvanizing rather than for spray- or hand-applied coatings.

The resultant cleanliness of the steel after cleaning will depend upon the quality of the steel prior to cleaning. Prior to surface preparation beginning, the initial condition of the steel should be established visually. If SP1, SP2, SP3, or SP11 cleaning is sought, the surface condition should be checked with SSPC-Vis 3, "Visual Standard for Power- and Hand-Tool Cleaned Steel." The steel is classified as one of four conditions if previously unpainted, or one of three if previously painted. These classes are listed in Table 12.10. The completed cleaning is then visually compared to reference photographs for the initial condition and preparation level to be achieved. The photographs, however, are used only as a guide to the language used in the specifications, not as a replacement to the specifications themselves.

If blast cleaning to levels SP6, SP7, SP10, or SP5 is to be performed, the surface condition first should be established using SSPC-Vis 1-89 "Visual Standard for Abrasive Blast Cleaned Steel." As before, unpainted steel is classified into one of

Table 12.9 SSPC Surface Preparation Specifications

SSPC specification		Description
1	Solvent cleaning	Removal of oil, grease, dirt, salts, and contaminants by cleaning with solvent, vapor, alkali, emulsion, or steam
2	Handtool cleaning	Removal of loose rust, loose mill scale, and loose paint to degree specified, by hand chipping, scraping, sanding, and wire brushing
3	Power-tool cleaning	Removal of loose rust, loose mill scale, and loose paint to degree specified, by power-tool chipping, descaling, scraping, sanding, wire brushing, and grinding
5	White blast cleaning	Removal of all visible rust, mill scale, paint, and foreign matter by blast cleaning by wheel or nozzle (dry or wet) using sand, grit, or shot
6	Commercial blast cleaning	Blast cleaning until at least two-thirds of the surface area is free of all visible residues
7	Brush-off blast cleaning	Blast cleaning of all except tightly adhering residues of mill scale, rust, and coatings, exposing numerous evenly distributed flakes of underlying metal
8	Pickling	Complete removal of rust and mill scale by acid pickling, duplex pickling, or electrolytic pickling
10	Near-white blast cleaning	Blast cleaning nearly to white-metal cleanliness, until at least 95% of the surface area is free of all visible residues
11	Power-tool cleaning to bare metal	Complete removal of all rust, scale, and paint by power tools, with resultant surface profile

* SSPC *Steel Structures Painting Manual, Vol.* 2, chap. 1, Table 1.

Fabrication, Erection, and Quality Control ■ 12-33

four prepreparation conditions. Upon completion of blasting, the resultant cleanliness is compared visually to a guide photograph for the predetermined rust classification of steel and blasting specification to be achieved.

In addition to cleanliness, surface profile is critical to the performance of some coating systems. If the surface is too smooth, the coating will have difficulty adhering to the surface, needing an anchor pattern. If the surface is too irregular, with too high a surface profile, the coating system may not adequately cover the peaks of the surface and premature coating failure may develop at the peaks. Surface profile is measured in mils (microns in SI).

12.6.2 Coating Systems

There are a wide variety of coating systems used for structural steel. Coating systems found in specifications may range from a "shop coat paint" to multicoat systems. The "shop coat" is used only as temporary protection of the steel during storage and erection. It is not intended for permanent protection of the structure, although in many cases it would do an adequate job for Environmental Zone 1A applications. Selection of the coating material is left to the fabricator but is typically an economical latex primer applied to approximately 1 mil (25 μm) thickness on steel cleaned to SP1 and SP2.

The Steel Structures Painting Council has approximately 30 generic coating specifications, along with guide specifications for selection of these and other coating systems. SSPC has suggested various generic specifications for the environmental zones to provide minimum protection for the structural steel, but a more comprehensive consideration is warranted before making the selection of a coating system. The SSPC suggested systems are provided in Table 12.11.

Prior to the application of the coating system, several items should be checked in addition to the cleanliness and surface profile of the surface preparation. The environmental conditions of application should be checked to make sure they fall

Table 12.10 SSPC-Vis 3 Initial Conditions of Steel

Condition	Description
A	Steel surface completely covered with adherent mill scale; little or no rust visible (rust grade A)
B	Steel surface covered with both mill scale and rust (rust grade B)
C	Steel surface completely covered with rust; little or no pitting visible (rust grade C)
D	Steel surface completely covered with rust; pitting visible (rust grade D)
E	Previously painted steel surface; light-colored paint applied over blast-cleaned surface, paint mostly intact
F	Previously painted steel surface; zinc-rich paint applied over blast-cleaned steel, paint mostly intact
G	Painting system applied over mill scale bearing steel; system thoroughly weathered, thoroughly blistered, or thoroughly stained

12-34 ■ Chapter Twelve

within the requirements of the manufacturer. Temperature and relative humidity are important concerns, as well as the surface temperature of the steel. Make sure the steel is at least 5°F (3°C) warmer than the atmospheric dew point to prevent condensation on the steel.

The suitability and cleanliness of the blast cleaning and coating application equipment should be checked. Moisture or oils in the system can foul the surface cleanliness or coating itself. If blast cleaning, check the condition, cleanliness, size, and type of abrasive used. Contaminated abrasives could foul the surface preparation, and spent or improperly sized abrasive could provide an inadequate profile or a profile of improper configuration.

Proper mixing of the coating, particularly with two-part systems, is critical. In some cases, continuous agitation is required. The application life of some coating systems is limited once the components are mixed. Thinning is permitted for

Table 12.11 SSPC Painting System Suggestions

Environmental zone	Painting system suggestions (minimal protection; consult individual specifications for details)	
0	Leave unpainted	
1A	PS 18	Latex
	PS 7	One-coat
1B	PS 1	Oil base
	PS 18	Latex
2A	PS 4	Vinyl
	PS 11	Coal-tar epoxy
	PS 13	Epoxy
	PS 15	Chlorinated rubber
2B	PS 12	Zinc-rich
	PS 4	Vinyl
	PS 11	Coal-tar epoxy
	PS 13	Epoxy
2C	PS 4	Vinyl
	PS 11	Coal-tar epoxy
2D	PS 4	Vinyl
	PS 11	Coal-tar epoxy
3A	PS 4	Vinyl
	PS 11	Coal-tar epoxy
	PS 15	Chlorinated rubber
3B	PS 12	Zinc-rich
	PS 15	Chlorinated rubber
	PS 4	Vinyl
3C	PS 11	Coal-tar epoxy
	PS 15	Chlorinated rubber
3D	PS 13	Epoxy
3E	Use specific exposure data	

* SSPC *Steel Structures Painting Manual, Vol. 2*, chap. 1, Table 3.

Fabrication, Erection, and Quality Control ■ 12-35

some coatings and prohibited for others. Check that approved thinners are used. Viscosity of the coating may also be checked.

Once applied, the wet-film thickness of the applied coating should be checked. Later, the dry-film thickness may also be checked. A number of other tests may also be completed on the finished coating system.

12.6.3 Inspection Methods

To check surface preparation beyond the visual checks mentioned above, test instruments for measuring surface profile may be used. The instruments include dial surface profile gauges, film tapes for use with micrometers, and surface profile comparators, which use a magnified comparison disk for a given blasting medium with surface profile heights.

To check the wet-film thickness, tooth-type gauges, also called notch gauges, and circular dial gauges are commonly used. The film thickness is indicated by the first wet indication on the gauge teeth or marking on the circular dial gauge. The proper use of WFT notch gauges is given in ASTM D4414. Most specifications call for a minimum (and sometimes maximum) dry-film thickness (DFT). To determine the appropriate wet-film thickness (WFT) for the required DFT, divide the desired DFT by the percent solids by volume for that particular coating.

To verify dry-film thickness once the coating has dried, a variety of magnetic-based gauges can be used. Some are as simple as a pencil-type pull-off gauge or "banana" gauge, and others make digital readings and record the information in a data base for downloading into a system. The principle behind most of these gauges is a magnet in the tip of the gauge that is attracted to the steel. The thicker the coating, the weaker the magnetic attraction. Such readings can be affected by the surface profile used, so calibration may be necessary. Other gauges use a magnetic-flux principle, also affected by film thickness and surface profile. Testing methods are given in ASTM D1186.

For verifying the thickness of the layers of coating on a multicoat system, a destructive test method must be used. Typically, a small notch is cut in the coating down to the steel, with the sides of the notch cut at an angle. The cut coating is then viewed with a magnifier that has a built-in scale based upon the angle of cut. The thickness of the various layers of coating is then determined using this scale. For this system to work, each coat must be a different color. After testing, the coating system requires touching up and is therefore more expensive than single-coat testing. Testing using this and similar methods is found in ASTM D4138.

Adhesion of the coating to the steel can be tested destructively. One method is to cut the coating into a pattern of squares or diamonds or a simple x. Adhesive test tape is rubbed onto the coating and then quickly pulled off. Such tests are described in ASTM D3359. Another adhesion test involves making a circular cut in the coating, epoxying a disk "dollie" to the coating, then using a device to pull up on the dollie. The removal force is measured using a scale on the pulling device.

The number of dry-film thickness tests to be conducted is given in the Steel Structures Painting Council's Paint Application Specification No. 2, "Measurement of Dry Film Thickness with Magnetic Gages." The specification calls for three readings to be made in a close area, with the average of these three "readings" known as a "spot measurement." Five spot measurements are taken on each 100 ft^2 of surface area, and the frequency of testing is adjusted depending upon size of project and whether measurements are found below the required DFT. The average of the five spot measurements must be at least equal to the required DFT. Any one of the five spot measurements can underrun the required DFT by 20 percent. Any one of the three readings that make up a spot measurement can underrun below this 80 percent value.

For structures with 300 ft^2 or less of surface area, each 100 ft^2 section should be tested. For structures up to 1000 ft^2, test three 100 ft^2 sections. For structures over 1000 ft^2, test three 100 ft^2 sections at the beginning, then one 100 ft^2 area for each 1000 ft^2 (100 m^2) thereafter. If any 100 ft^2 area test produces DFTs less than that required, each 100 ft^2 area in that group should be tested.

No metric values have yet been adopted for the specification. A suggested substitution is to use 10 m^2 for values of 100 ft^2, 30 m^2 for 300 ft^2, and 100 m^2 for 1000 ft^2.

A general specification regarding coating inspection is found in ASTM D3276, Standard Guide for Painting Inspectors (Metal Substrates).

12.7 Erection Procedures

12.7.1 Frame Stability

Structural-steel frames may be classified as self-supporting or non-self-supporting. A self-supporting steel frame is defined in the AISC Code of Standard Practice, Section 7.9.2, as "one that provides the required stability and resistance to gravity loads and design wind and seismic forces without interaction with other elements of the structure." As such, the erector is responsible only for any temporary supports, guys, braces, falsework, and similar items that may be necessary to erect the steel. Once the steel is completely bolted or welded, or both, the structure is capable of resisting the lateral and gravity forces. No additional support from masonry walls, concrete shear walls, concrete cores, wall elements, roof or floor deck diaphragms, or other items is necessary to maintain the integrity of the structure.

In non-self-supporting steel frames, the structural engineer must note such in the contract documents and provide the construction sequence necessary for stability. The structural system providing the stability is constructed before, during, or following the steel erection. The erector erects the steel with the necessary temporary supports and removes these supports when appropriate.

12.7.2 Operations Sequencing

The sequence of erection operations must be determined in cooperation with the general contractor and fabricator. The overall flow of materials must be planned to meet the project schedule yet fall within the fabricator's ability to deliver but not overwhelm the site with stored material. Site access, obstructions, foundation work, and the number, type, and capacity of cranes used for erection must be determined. With this information, the project can be broken down into groups by area, story, or both. In projects with tight sites and little or no storage, the groups will naturally be smaller. On downtown high-rise projects with virtually no lay-down area, expediters are commonly used to load and deliver materials on trucks in the order in which they will be hoisted to the top of the structure, or perhaps even in erection sequence.

The setting of the steel is done by ironworkers commonly referred to as the "raising gang." Their task is to erect the steel and install only as many bolts as are necessary to hold the pieces in place. They place guy cables to maintain stability and plumb the structure as needed. They are followed by a "detail gang" that completes installation and tightening the bolts as needed, performs the field welding, and completes the plumbing of the structure.

This work is followed by metal decking crews to lay floor and roof deck, the shear connector welding operators, then those responsible for concrete floor reinforcing and concrete placement. Spray-on fireproofing is applied, and the other trades follow.

As erection progresses, it is necessary to notify the appropriate inspectors to inspect the work. Proper inspection includes preerection checks of materials, welder qualifications and procedures, as well as in-progress monitoring of bolting and welding operations. Final inspection is to be completed before the work is covered by the work of other trades such as decking crews and fireproofing applicators.

12.8 Certification, Qualification, and Inspection Programs

12.8.1 Welder and Welding Inspector Qualifications

Welders, tack welders, and welding operators must be qualified by the contractor responsible for the welding prior to the welding being performed, as required by AWS D1.1, Section 4.1.2.

Welders are individuals who manipulate the welding electrode, wire, and/or filler metal by hand to make the weld. A tack welder is a fitter who makes small welds as necessary to hold parts together until final welding by a welder or welding operator. A welding operator sets up and adjusts equipment to perform automatic welding.

The fabricator or erector responsible for welding must have each welder, tack welder, and welding operator tested using the methods of D1.1, Section 4, Part C, to prove their capability to make adequate quality welds. These individuals are tested and categorized by

Welding process

Welding position

Electrode classification

Base metal thickness range

CJP groove welds if made without backing

The testing may be performed by the contractor (fabricator or erector) or by an independent testing laboratory.

Welders who perform and pass such testing at an independent testing laboratory accredited by the American Welding Society can have their test records placed on file with the AWS and receive the designation of AWS Certified Welder.

A welder's or welding operator's qualification for a given employer remains in effect indefinitely as long as that individual continues welding in that given process, although not especially with the tested electrode classification or in the tested position. If the welder fails to use that process for a period exceeding 6 months, the welder must complete and pass a welding test. If the welder's quality becomes subject to question, the welder's qualification may be revoked by the employer, forcing a retest. Tack welder's qualifications remain in effect perpetually, unless there is specific reason to question the tack welder's abilities. Although welder qualification is the responsibility of the contractor, under the provisions of AWS D1.1, Section 6.4.2, the inspector may also force requalification testing if the welder's quality is poor.

Welding Inspector Qualifications ■ Welding inspectors must be qualified to perform the work on the basis of the contract documents and the building code. The engineer has the responsibility to require any inspector credentials exceeding those of AWS D1.1, Section 6.1.3.1. Under this provision, a welding inspector may be qualified by virtue of being an AWS Certified Welding Inspector, a welding inspector certified under the Canadian Welding Bureau, or "an engineer or technician who, by training or experience, or both, in metals fabrication, inspection and testing, is competent to perform inspection of the work." Section 6.1.3.4 requires an eye examination, with or without corrective lenses.

The inspector's qualification to perform the inspection remains effective indefinitely unless there is specific reason to question the inspector's abilities. In this case, the engineer or similar individual should reevaluate the inspector's abilities and credentials.

For individuals performing only nondestructive testing work, the inspector need not be generally qualified for welding inspection. However, the individual must be qualified using the provisions of the American Society for Nondestructive Testing Recommended Practice No. SNT-TC-1A. This document provides recommendations for the training, experience level, and testing of NDT technicians. A suitable alternative to the Recommended Practice, although not referenced in D1.1, is the ASNT Standard for Qualification and Certification of Nondestructive Testing Personnel.

These documents provide specific listings applicable to several areas of NDT:

Radiographic testing

Magnetic particle testing

Ultrasonic testing

Liquid penetrant testing

Electromagnetic testing

Neutron radiographic testing

Leak testing

Acoustic emission testing

Visual testing (SNT-TC-1A only)

NDT technicians are placed into four categories. Formal definitions vary between the Recommended Practice and the Standard. Using the definitions of the Standard, the level III technician has the "skills and knowledge to establish techniques; to interpret codes, standards and specifications; designate the particular technique to be used; and to verify the adequacy of procedures." This individual is responsible for the training and testing of other NDT personnel in the individual's area of certification. The level II technician has "the skills and knowledge to set up and calibrate equipment, to conduct tests, and to interpret, evaluate and document results in accordance with procedures approved by an NDT Level III." The level I technician has "the skills and knowledge to properly perform specific calibrations, specific tests, and with prior written approval of the level III, perform specific interpretations and evaluations for acceptance or rejection and document the results." The trainee is a technician who works under the supervision of a level II or III and cannot independently conduct any tests or report any test results.

Many level III technicians have taken and passed a nationally administered ASNT examination in the particular field of NDT. However, it is possible for an individual to be named by the employer to a level III designation, based upon experience and knowledge in the field.

12.8.2 Building Code Special Inspections

The Uniform Building Code of the International Conference of Building Officials has contained provisions requiring special inspection for many years. The Building Officials and Code Administrators, Inc., adopted special inspection provisions in 1987 into the National Building Code. The Southern Building Code Congress International has yet to adopt special inspection provisions for the Standard Building Code, but additions have been proposed for future adoption.

Chapter 17 of both the Uniform Building Code and the National Building Code contains special requirements that certain portions of the work be inspected by qualified individuals. The specific provisions vary between the two codes but generally require the inspection of steel, bolt, and welding materials, plus inspection of certain bolting and welding operations. Buildings in high-seismic-risk regions using particular structural systems require specific nondestructive testing to be performed at certain types of joints, particularly at specific moment connections, column splices, and joints subjected to lamellar tearing.

Generally, all special inspection is performed on a continuous basis, meaning that the inspector observes the work as it is being performed rather than waiting until completion. However, some portions of structural-steel work, such as deck welding, shear connector welding, small fillet welds, welding of gauge metal steels, and welding of miscellaneous iron, may be exempted from continuous special inspection when an approved procedure for periodic special inspection has been established by the engineer and approved by the building official.

Portions of the special inspection may be waived for fabricators who have been approved in advance by the building official. The fabricator undergoes an evaluation of their management and shop operations, as documented in their quality control manual.

12.8.3 AISC Quality Certification

The American Institute of Steel Construction's Quality Certification Program was instituted to develop a system under which owners, engineers, and building officials could evaluate the quality of steel fabricators. A voluntary program, steel fabricators are evaluated on the basis of their overall quality control program, judging the general management, engineering and drafting, purchasing, shop operations, and quality control functions of the individual plant. Multiplant fabricating companies are certified on a plant-by-plant basis.

The AISC Quality Certification Program is used to determine if the plant "has the personnel, organization, experience, procedures, knowledge, equipment, capability and commitment to produce fabricated steel of the required quality for a given category of structural steelwork."

There are four Building categories of quality certification for conventional structural steelwork. Conventional Steel Structures includes:

Small public service and institutional buildings
Shopping centers
Light manufacturing plants
Miscellaneous and ornamental iron
Warehouses
Sign structures
Low-rise, truss beam and column structures
Simple rolled beam bridges

Complex Steel Building Structures includes:
Large public service and institutional buildings
Heavy manufacturing plants
Powerhouses (fossil, nonnuclear)
Metal producing and rolling facilities
Crane bridge girders
Bunkers and bins
Stadia
Auditoriums
High-rise buildings
Chemical processing plants
Petroleum processing plants

Fabricating plants certified as Complex Steel Building structures are automatically qualified for Conventional Steel Building Structures. Plants certified as Major Steel Bridges are automatically qualified for Simple Steel Bridge Structures. Major

Steel Bridges include all bridge structures other than unspliced simple rolled beam bridges.

A fifth classification of AISC Quality Certification, Category MB—Metal Building Systems, applies to preengineered metal building systems. This category of certification includes a review of the design procedures used by the manufacturer of the building system in designing the structure.

Simple Steel Bridge Structures include unspliced rolled beam bridges, highway sign structures, and bridge parts such as cross frames.

Alternative methods of evaluating fabricators include the ICBO Evaluation Services listing service and the ISO 9000 program.

References

1. Iwankiw, Nestor, and Thomas Schlafly, "Effect of Hole-Making on the Strength of Double Lap Joints," AISC *Engineering Journal*, vol. 19, no. 3, third quarter, 1982.

APPENDIX A
Richard B. Heagler

STEEL DECK DESIGN

A.1 General

Steel deck is used as a structural product. In some instances architectural considerations are important, but the vast majority of deck products are chosen for structural reasons. All deck is made from sheet steel and is formed into its shape at room temperature. The process is defined as "cold forming." The design of the products is governed by the rules set down by the American Iron and Steel Institute (AISI) in the Specification for the Design of Cold Formed Steel Structural Products. In general, the AISI specification is used to determine the section properties of the deck (the I and S values), the bearing capacity, and the shear strength. These values are used with the appropriate steel strength characteristics to produce load tables for the products. In most cases, the deck load-carrying capacity is more than enough for the design; and, because the dead load of the material is quite small, it matters very little whether LRFD or allowable stress design (ASD) is used to calculate the load capacity. However, composite decks, i.e., decks that act in conjunction with concrete to carry service loads, do use LRFD methods to determine their strength capacities. In a few years all deck loads will probably be determined by LRFD methods, but at this time the industry is in a transition period.

Deck products fall into three categories: roof deck, form deck, and composite floor deck. Figure A.1 illustrates the terminology that is common to all three types. Roof deck is the load-carrying element that transfers the roofing dead loads and live loads (such as snow) to the underlying structural elements, which are usually bar joists or structural-steel beams. Roof deck can also be used as a horizontal load-resisting system to brace the building against wind and earthquake forces. Roof deck is not assumed to act compositely with insulation boards in resisting

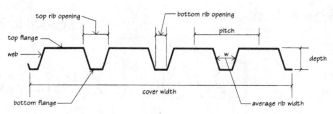

Fig. A.1 Cross-section terminology.

either vertical or horizontal loads. Form deck has a variety of uses, but it is predominantly used as a stay-in-place form for structural concrete. As such, it is designed to carry the concrete and construction loads; the slab is reinforced to carry live loads. Form deck can also be used as a diaphragm to resist horizontal loads. Composite floor deck serves three functions: the first is to provide a working platform for the various trades; the second is to provide the form for the concrete; and the third is to act as the positive bending reinforcement for the slab. When properly connected, it too can act as a diaphragm.

The intent of the Steel Deck Institute (SDI) and of deck manufacturers is to provide design information in an easy-to-use format so that very few calculations are required to properly choose the deck needed for the job. The tables that result from this policy show uniform load capacities and maximum recommended spans for the products. Indeed, for the vast majority of jobs, the entire design can be done by simple table referrals.

A.2 Roof Deck Design

Figure A.2 shows four generic roof deck profiles. The narrow rib (type A) deck is seldom used now, but it does show up on some older existing buildings. The narrow top rib opening of this type of deck was used with thin insulation boards when energy considerations were not too important. The thin, and rather brittle, boards could span the narrow gap without cracking. But the narrow rib made the deck more difficult to install because of the constricted area available for fasteners. The money saved by furnishing the thin insulation may also have been partially used up because the strength of the narrow rib deck is not as great as that of the wide rib (type B). If joist spacings were much more than 5 ft (1.5 m), a heavier-gauge deck could have been needed to carry the loads. The intermediate rib deck (F deck) provides the transition between the narrow and wide rib profiles. However, in today's industry, wide rib (type B) is by far the profile of choice. It is capable of greater spans between supports, and if the insulation board is 2 in (50 mm) or thicker, the insulation can easily span the rib opening. The wide open rib makes attachment an easy job, and a variety of fasteners can be used.

Design for Vertical Load

Figure A.3 shows typical load tables for three types of deck. Figure A.4 shows the data used to produce the tables and also to do a complete design of the deck for vertical loads. The loads in the tables are based on the interaction of shear and bending as required in the AISI specification:

$$\left(\frac{V_{applied}}{V_{allowable}}\right)^2 + \left(\frac{M_{applied}}{M_{allowable}}\right)^2 \leq 1$$

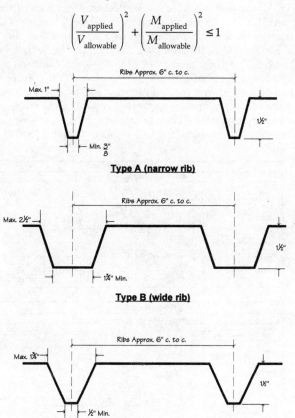

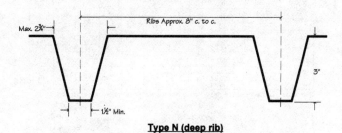

Fig. A.2 Generic roof deck profiles.

A-4 ■ Appendix A

The shear consideration has very little influence on the tabulated load and could safely be neglected unless unusual span and load combinations were being investigated. In Fig. A.4 the maximum recommended spans for each product and gauge are also shown. These spans are based on the Steel Deck Institute (SDI) criteria for construction and maintenance loading. In this loading a 200-lb concentrated load is placed at midspan and distributed over a 1-ft width; the deck stress is limited to 26 ksi and the deflection is limited to 1/240 of the span. Cantilever spans are also determined by the SDI specifications. A construction-phase uniform load of 10 lb/ft² is applied over the cantilever and the adjacent span plus a concentrated load of 200 lb at the end of the cantilever (over a 1-ft width); the stress is limited to 26 ksi. The next loading investigated is a uniform service load of 45 lb/ft² applied over the cantilever and adjacent spans, plus a 100-lb concentrated load at the cantilever

B, BI, BA, BIA Wide Rib Deck
Uniform Total Load (Dead + Live), psf

Span Type	Gage	5'0"	5'6"	6'0"	6'6"	7'0"	7'6"	8'0"	8'6"	9'0"	9'6"
SINGLE	22	99	77	62							
	20	126	97	77	63	52					
	18	173	132	104	84	69	58	50	43		
	16	220	168	132	106	87	72	61	53	46	41
DOUBLE	22	105	87	73	63	54					
	20	136	113	95	81	70	61	54			
	18	188	156	132	112	97	85	74	66	59	53
	16	235	195	164	140	121	106	93	82	74	66
TRIPLE	22	131	108	91	78	67					
	20	169	140	118	101	87	75	63			
	18	234	194	164	140	121	101	85	72	63	55
	16	292	242	204	175	151	127	107	91	78	68

TYPE F, Intermediate Rib Deck
Uniform Total Load (Dead + Live), psf

Span Type	Gage	4'0"	4'6"	5'0"	5'6"	6'0"	6'6"	7'0"	7'6"	8'0"	8'6"
SINGLE	22	108	86	69	57						
	20	133	105	85	71	59					
	18	183	145	117	97	81	67	56			
DOUBLE	22	115	91	74	61	52	44				
	20	140	111	90	74	63	53	46			
	18	189	150	122	101	85	72	62	54	48	42
TRIPLE	22	144	114	92	76	64	55				
	20	174	138	112	93	78	67	58			
	18	236	187	152	126	106	90	78	68	50	53

NS, NI, NSA, NIA, 3" Deep Rib Deck
Uniform Total Load (Dead + Live), psf

Span Type	Gage	10'0"	10'6"	11'0"	11'6"	12'0"	12'6"	13'0"	13'6"	14'0"	14'6"
SINGLE	22	49	45	41	37						
	20	64	57	50	45	41	38	35			
	18	88	77	69	61	55	50	46	42	38	36
	16	116	102	90	80	72	64	58	53	49	45
DOUBLE	22	55	50	46	42	39	36	33	31		
	20	71	65	59	54	50	46	42	39	37	34
	18	98	89	81	74	68	63	58	54	50	47
	16	123	112	102	93	86	79	73	68	63	59
TRIPLE	22	69	63	57	52	48	44	41	38		
	20	89	81	74	67	62	57	53	49	46	43
	18	122	111	101	92	85	78	73	67	63	58
	16	153	139	127	116	107	98	91	85	79	73

Loads shown in **bold italics** are controlled by live load deflection of L/240. Dead load is assumed to be 10psf.

courtesy of:
USD United Steel Deck, Inc.

Fig. A.3 Typical roof deck load tables.

Steel Deck Design ■ A-5

end; the stress is limited to 20 ksi and the deflection to 1/240 of the adjacent span or 1/120 of the cantilever. The smallest span value for the cantilever resulting from the calculations is published. Because the answers are dependent on the adjacent span length, the published values are based on "realistic" adjacent spans. If the adjacent span is less than three times the cantilever span, an individual calculation is required. Figures A.5 and A.6 show the same information as A.3 and A.4 but have had a soft conversion to international units.

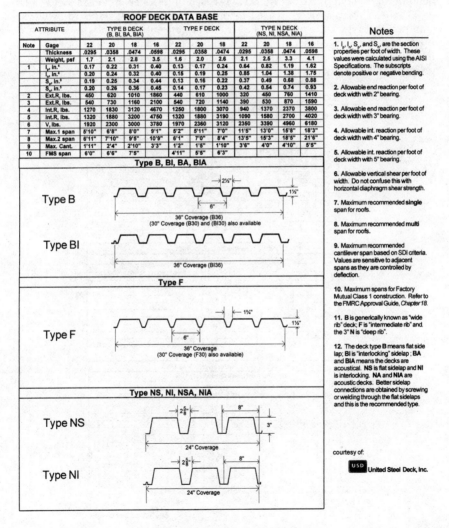

Fig. A.4 Roof deck data base.

ROOF DECK DATA BASE

Note	ATTRIBUTE	TYPE B DECK (B, BI, BA, BIA)				TYPE F DECK			TYPE N DECK (NS, NI, NSA, NIA)				Notes
	Gage	22	20	18	16	22	20	18	22	20	18	16	
	Thickness	.0295	.0358	.0474	.0598	.0295	.0358	.0474	.0295	.0358	.0474	.0598	
	Weight, psf	1.7	2.1	2.8	3.5	1.6	2.0	2.6	2.1	2.5	3.3	4.1	
1	I_p, in.^{4}	0.17	0.22	0.31	0.40	0.13	0.17	0.24	0.64	0.82	1.19	1.62	
	I_n, in.^{4}	0.20	0.24	0.32	0.40	0.15	0.19	0.25	0.85	1.04	1.38	1.75	
	S_p, in.^{3}	0.19	0.25	0.34	0.44	0.13	0.16	0.22	0.37	0.49	0.68	0.88	
	S_n, in.^{3}	0.20	0.26	0.36	0.45	0.14	0.17	0.23	0.42	0.54	0.74	0.93	
2	Ext.R, lbs.	450	620	1010	1860	440	610	1000	320	450	760	1410	
3	Ext.R, lbs.	540	730	1160	2100	540	720	1140	390	530	870	1590	
4	Int.R, lbs.	1270	1830	3120	4670	1250	1800	3070	940	1370	2370	3800	
5	Int.R, lbs.	1320	1880	3200	4750	1320	1880	3190	1090	1580	2700	4020	
6	V, lbs.	1920	2300	3000	3780	1970	2360	3120	2350	3390	4960	6180	
7	Max.1 span	5'10"	6'8"	8'0"	9'1"	5'2"	5'11"	7'0"	11'5"	13'0"	15'8"	18'3"	
8	Max.2 span	6'11"	7'10"	9'5"	10'9"	6'1"	7'0"	8'4"	13'5"	15'3"	18'5"	21'6"	
9	Max. Cant.	1'11"	2'4"	2'10"	3'3"	1'2"	1'5"	1'10"	3'6"	4'0"	4'10"	5'5"	
10	FMS span	6'0"	6'6"	7'5"		4'11"	5'5"	6'3"					

Notes

1. I_p, I_n, S_p, and S_n are the section properties per foot of width. These values were calculated using the AISI Specifications. The subscripts denote positive or negative bending.

2. Allowable end reaction per foot of deck width with 2" bearing.

3. Allowable end reaction per foot of deck width with 3" bearing.

4. Allowable int. reaction per foot of deck width with 4" bearing.

5. Allowable int. reaction per foot of deck width with 5" bearing.

6. Allowable vertical shear per foot of width. Do not confuse this with horizontal diaphragm shear strength.

7. Maximum recommended **single** span for roofs.

8. Maximum recommended **multi** span for roofs.

9. Maximum recommended cantilever span based on SDI criteria. Values are sensitive to adjacent spans as they are controlled by deflection.

10. Maximum spans for Factory Mutual Class 1 construction. Refer to the FMRC Approval Guide, Chapter 18.

11. B is generically known as "wide rib" deck; F is "intermediate rib" and the 3" N is "deep rib".

12. The deck type B means flat side lap; BI is "interlocking" sidelap; BA and BIA means the decks are acoustical. NS is flat sidelap and NI is interlocking. NA and NIA are acoustic decks. Better sidelap connections are obtained by screwing or welding through the flat sidelaps and this is the recommended type.

courtesy of:

USD United Steel Deck, Inc.

Appendix A

The load tables demonstrate that for spans of, or less than, the maximum recommended spans, the load capacity is greater than is required for most roof loading. Even for areas where snow drifting may occur, there is ample load capacity. The most radical action required of a designer would be to move the supporting joists closer together in the area of the high drift, and this would be to aid the joists more than the deck.

The load tables are based on conservative assumptions. The allowable deck bending stress limit of 20 ksi (140 MPa) is based on the steel having a yield stress of 33 ksi (230 MPa). All steel used to produce deck to SDI standards will use stronger steels. Also, the practice of distributing concentrated loads over only a 1-ft width (300 mm) is conservative. Even on bare deck (without insulation boards) the distribution is closer to 18 in (450 mm), and certainly when insulation board is present the distribution will exceed that width.

B, BI, BA, BIA Wide Rib Deck
Uniform Total Load (Dead + Live), kPa

Span Type	Gage	1500	1650	1800	1950	2100	2250	2400	2550	2700	2850
SINGLE	22	5.0	3.8	3.1	2.5						
	20	6.3	4.8	3.8	3.1	2.6	2.2				
	18	8.7	6.6	5.2	4.2	3.5	2.9	2.1	1.9		
	16	11.0	8.4	6.6	5.3	4.3	3.6	3.1	2.6	2.3	2.0
DOUBLE	22	5.2	4.3	3.6	3.1	2.7					
	20	6.7	5.6	4.7	4.0	3.5	3.0	2.7	2.4		
	18	9.3	7.7	6.5	5.5	4.8	4.2	3.7	3.3	2.9	2.6
	16	11.6	9.6	8.1	6.9	6.0	5.2	4.6	4.1	3.6	3.3
TRIPLE	22	6.4	5.3	4.5	3.8	3.3	2.9				
	20	8.3	6.9	5.8	5.0	4.3	3.7	3.1	2.7		
	18	11.5	9.6	8.1	6.9	6.0	5.0	4.2	3.6	3.1	2.7
	16	14.4	12.0	10.1	8.6	7.5	6.4	5.3	4.5	3.9	3.4

TYPE F, Intermediate Rib Deck
Uniform Total Load (Dead + Live), kPa

Span Type	Gage	1200	1350	1500	1650	1800	1950	2100	2250	2400	2550
SINGLE	22	5.4	4.2	3.4	2.8						
	20	6.6	5.2	4.2	3.5	2.9					
	18	9.1	7.2	5.8	4.8	4.0	3.4	2.8			
DOUBLE	22	5.7	4.5	3.7	3.0	2.6	2.2				
	20	6.9	5.5	4.4	3.7	3.1	2.6	2.3			
	18	9.4	7.4	6.0	5.0	4.2	3.6	3.1	2.7	2.4	2.1
TRIPLE	22	7.1	5.6	4.6	3.8	3.2	2.7				
	20	8.6	6.8	5.5	4.6	3.9	3.3	2.8			
	18	11.6	9.2	7.5	6.2	5.2	4.5	3.8	3.4	2.9	2.6

NS, NI, NSA, NIA, 75mm Deep Rib Deck
Uniform Total Load (Dead + Live), kPa

Span Type	Gage	3000	3150	3300	3450	3600	3750	3900	4050	4200	4350
	22	2.4	2.2	2.0	1.8	1.7					
	20	3.2	2.8	2.5	2.3	2.0	1.9	1.7	1.5		
	18	4.4	3.9	3.4	3.1	2.7	2.6	2.3	2.1	1.9	1.8
	16	5.8	5.1	4.5	4.0	3.6	3.2	2.9	2.6	2.4	2.2
	22	2.7	2.5	2.3	2.1	1.9	1.8	1.6	1.5	1.4	1.3
	20	3.5	3.2	2.9	2.7	2.5	2.3	2.1	1.9	1.8	1.7
	18	4.8	4.4	4.0	3.7	3.4	3.1	2.9	2.7	2.5	2.3
	16	6.1	5.5	5.0	4.6	4.2	3.9	3.6	3.3	3.1	2.9
	22	3.4	3.1	2.8	2.6	2.4	2.2	2.0	1.9	1.8	1.6
	20	4.4	4.0	3.6	3.3	3.1	2.8	2.6	2.4	2.3	2.1
	18	6.0	5.5	5.0	4.6	4.2	3.9	3.6	3.3	3.1	2.9
	16	7.6	6.9	6.3	5.7	5.3	4.9	4.5	4.2	3.9	3.6

Loads shown in **bold italics** are controlled by L/240 deflection. Dead load is assumed to be 0.48 kPa.

courtesy of:
USD United Steel Deck, Inc.

Fig. A.5 Typical roof deck load tables.

Steel Deck Design ■ A-7

Design for Horizontal Load

Once the roof deck is attached, it provides in-plane strength capable of bracing the building frame and resisting horizontal loads induced by wind or earthquake. The ability to act as a diaphragm has been the subject of much research, and a considerable effort has gone into the task of providing design tables for the use of structural engineers. The strength and stiffness of a diaphragm are not only dependent on the deck type and gauge used but are also very much related to the

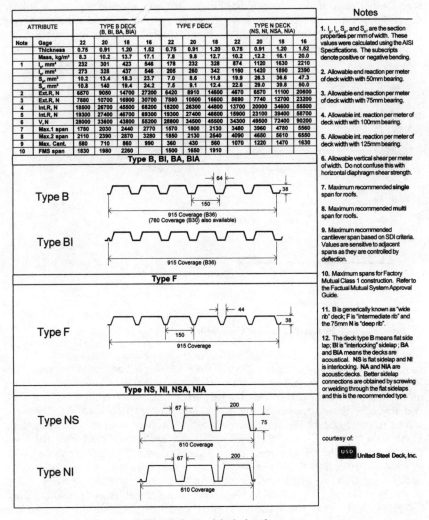

Fig. A.6 Roof deck data base.

A-8 ■ Appendix A

type and spacing of fasteners. Diaphragms are frequently modeled as quasi-plate girders, where the deck acts as the web and the perimeter members of the roof are the flanges. The web (the roof deck) is the shear-resisting element, and indeed the design of the diaphragm is dominated by in-plane shear. Bending deflection can be neglected, but shear deflection is important. Since the "web" formed by the deck is not a solid homogeneous material but is made up of units that are between 2 and 3 ft (600 and 900 mm) wide and are intermittently connected to the frame and to each other at the sides, it is easy to imagine the tendency to slip as the load is applied. This explains why the diaphragm strength and stiffness are so highly dependent on the number and strength of the fasteners. The diaphragm tables published by the SDI reflect this, and tables are published for many combinations of deck types, gauges, spans, and connectors. Figure A.7 shows a plan view of a

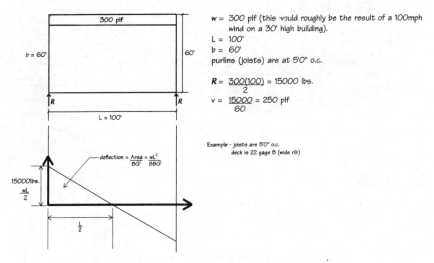

Fig. A.7 Plan view horizontal loading of a building.

simple building with a horizontal load of 300 plf applied along the edge of the roof. The roof is made of 22-gauge deck with joists (purlins) at 5-ft centers. This loading could be the result of a 100 mi/h wind on a building 30 ft high. (Suction on the leeward side has been added to the direct force on the windward side.) The end walls, or braced frames at the ends, carry the reaction loads R, which are solved to be 15,000 lb. The maximum shear in the diaphragm is then seen to be 15,000/60 = 250 plf. The table shown in Fig. A.8 has been copied from the Steel Deck Institute *Diaphragm Design Manual*, second edition. This table shows that 30-in-wide 22-gauge B (WR) deck fastened to each support with $5/8$-in welds on a 30/4 pattern and with one side lap screw will provide a strength of 260 plf. See Fig. A.9. (Note the safety factor is 2.75, which reflects the use of welded connections.) Since 260 plf is greater than the applied 250 plf, the diaphragm is strong enough. The shear deflection of the diaphragm can be calculated by finding the area under the shear diagram and dividing it by the depth of the diaphragm and the stiffness. The stiffness is found by following the directions in Fig. A.9:

$$G' = \frac{K2}{3.78 + 0.3D_{xx}/\text{span} + 3*K1*\text{span}}$$

$$G' = \frac{870}{3.78 + 0.3(1377)/5 + 3(0.536)5} = 9.2 \text{ kips/in}$$

$$\text{Deflections} = \frac{wl^2}{(8bG')} = \frac{0.300(100)^2}{[8(60)9.2]} = 0.68$$

This deflection is approximately $L/1800$, which would probably be quite acceptable. Bending deflection is negligible and does not need to be checked.

Frame Fastening: $\frac{5}{8}''$ Welds on 30/4 Pattern.
Stitch Fastening: #10 screws (buildex)
Safety Factor: 2.75
t = design thickness = .0295"

Stitch Connectors per span	Design Shear, plf Span, ft.									K1
	3.0	3.5	4.0	4.5	5.0	5.5	6.0	6.5	7.0	
0	340	300	270	240	215	195	175	160	150	0.728
1	395	350	315	285	260	235	215	195	180	0.536
2	440	395	355	325	295	275	255	230	215	0.424
3	475	430	395	360	330	305	285	265	245	0.350
4	510	465	425	395	365	335	315	295	275	0.299
5	540	495	460	425	395	365	340	320	300	0.260
6	565	525	485	450	420	395	370	345	325	0.231

$D_{wr} = 1377$
$D_{ir} = 1547$
$D_{nr} = 1608$
$K_2 = 870$

(The subscripts denote wide rib, intermediate rib, or narrow rib deck.)

Substitute these values into the equation for G' as appropriate.

Fig. A.8 Standard 1.5-in deck.

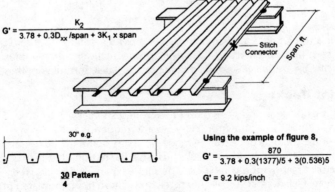

$$G' = \frac{K_2}{3.78 + 0.3D_{xx}/\text{span} + 3K_1 \times \text{span}}$$

30" e.g.

30 Pattern
4

Using the example of figure 8,
$$G' = \frac{870}{3.78 + 0.3(1377)/5 + 3(0.536)5}$$
G' = 9.2 kips/inch

Fig. A.9

Figure A.8 also shows the diaphragm strength can be increased by increasing the side lap fastening. A strength increase could also be obtained by increasing the amount of frame fastening. In general, diaphragm strength and stiffness values are very dependent on the fastening. The general rules for roof deck design are:

1. Choose the deck to carry the design *vertical* loading,
2. Choose the fastening needed to satisfy diaphragm loads,
3. Check the deck to see the spans do not exceed the maximum recommended spans for maintenance and construction,
4. Check the fastening to see that it can resist the anticipated wind uplift loading,
5. Check fire rating compliance or other code or insurance influences.

Roof Deck Fastening

In the diaphragm example problem, the high shear was near the perimeter but abated toward the center of the building. Typically, high shears and high uplifts are at the perimeter, and it makes good sense to zone the deck fastening and put less fastening where the needs are less. In the past, welding was used almost exclusively as the fastening method. During the last few decades other means have developed that do not require the skill of a welder or the power requirements. Self-drilling and -tapping screws can be used to attach deck to the structural framing and, as shown in the diaphragm example problem, to stitch deck to deck at side laps. Screw guns are used to drive the screws because the guns have a clutch and a depth-limiting nosepiece so the screws are not overtorqued. The SDI *Diaphragm Design Manual*, second edition, gives the shear formulas for no. 12 and no. 14 screws attaching deck to structural steel as

$$Q_f = 1.25 F_y t (1 - 0.005 F_y) \text{ kips}$$

where t = deck metal thickness, in

F_y = deck steel yield strength, ksi

Side lap screws (stitch screws) have a shear strength of

$$Q_s = 115\, dt, \text{ kips}$$

where d is the screw diameter. Figure A.10 provides uplift values for screws.

Other frame fasteners are powder or pneumatically driven pins. The SDI *Diaphragm Design Manual*, second edition, provides shear formulas for these fasteners.

Special Decks

Special roof deck sections are made to accomplish specialized tasks. Deep deck sections such as those shown in Fig. A.11 are frequently used in long-single-span applications. Cellular deck can be made from any standard section by welding a plate to the bottom side. These decks furnish a flat bottom that may be a desirable architectural feature and can be used on longer spans, which may be a structural expedience. Cell deck costs roughly three times more than nonplated units and so is used sparingly. Acoustic deck is a popular special adaptation of standard roof

Steel Deck Design ■ A-11

Screw Size	d dia.	d_w nom. head dia.	Average tested tensile strength, kips
#10	0.190	0.415 or 0.400	2.56
#12	0.210	0.430 or 0.400	3.62
1/4	0.250	0.480 or 0.520	4.81

Pull Over Strength, kips = P_{nov} = $1.5 t_1 d_w F_u$; $d_w < 0.50"$.

Washer or head, d_w	16 ga.	18 ga.	20 ga.	22 ga.	24 ga.	26 ga.	28 ga.	Key
0.400	1.61	1.28	0.97	0.80	0.86	0.64	0.54	☐ F_u = 60 ksi
0.415	1.68	1.33	1.00	0.83	0.89	0.67	0.56	☐ F_u = 45 ksi
0.430	1.74	1.38	1.04	0.86	0.92	0.69	0.58	
0.480	1.94	1.54	1.16	0.96	1.03	0.77	0.64	■ F_u = 58 ksi
0.520(0.500)	2.02	1.60	1.20	1.00	1.08	0.81	0.67	

Pull Out Strength, kips = P_{not} = $0.85 t_2 d F_u$; Metal thickness = t_2

Screw	1/4"	3/16"	10 ga. (0.135)	1/8"	12 ga. (0.105)	14 ga. (0.075)	16 ga. (0.060)	18 ga. (0.047)	20 ga. (0.036)	22 ga. (0.030)
#10	2.34	1.78	0.98	1.17	0.76	0.55	0.44	0.35	0.26	0.22
#12	2.66	2.00	1.12	1.33	0.87	0.62	0.50	0.38	0.29	0.24
1/4	3.08	2.31	1.29	1.54	1.00	0.72	0.57	0.45	0.34	0.28

Fig. A.10 Uplift values for screws.

deck. A regular perforation pattern is placed in the webs of deck which is usually the wide-rib (B) profile or the deep-rib (N) type. Sound-absorbing insulating material is then placed in the ribs as shown in Fig. A.12. Because the insulation is trapped between the deck and the insulation board, it is most conveniently installed by the roofing contractor, although it is supplied by the deck supplier. The roof deck is the finished ceiling, which provides good sound-absorbing qualities. Cellular roof decks can also be acoustically treated by perforating the bottom plate and installing sound-absorbing insulation in the cavities.

A.3 Form Deck Design

Form decks are used for a variety of applications, but the most common use is, as the name implies, simply a stay-in-place form for structural concrete. Form decks are also used with lightweight (nonstructural) insulating fills; as exposed roofing and for many temporary construction site purposes. Many profiles have evolved as manufacturers have searched for the most efficient section to do a particular job, but only one generic form deck is produced, which is the pattern shown in Fig. A.13. Other patterns, produced by individual manufacturers, range in depth from 1 in (25 mm) to 3 in (75 mm) and cover the gauge range of 10 (0.135 in, 3.40 mm) to 28 (0.0149 in, 0.38 mm).

Structural Concrete Form

The type and gauge of the form units are most often chosen to support the concrete and construction loads without the use of shoring. Figure A.14 shows the construction loading recommended by the Steel Deck Institute. The 150-lb man

A-12 ■ Appendix A

load is per foot of width of the deck (2.2 kN/m). The distribution of the man load is greater than 1 ft (300 mm), and a one-third stress increase is generally allowed (under allowable stress design) for temporary loading, so the man load allowance of 150 lb/ft of width is conservative.

If the deck is selected so that no shoring is used, and if the deck finish is selected to last the life of the structure (usually galvanized), then the concrete needs only to be reinforced to carry the loads applied after the concrete cures. The weight of the slab will always be carried by the deck. Sometimes, on short deck spans, the deck selected is strong enough to carry the total of the dead and live

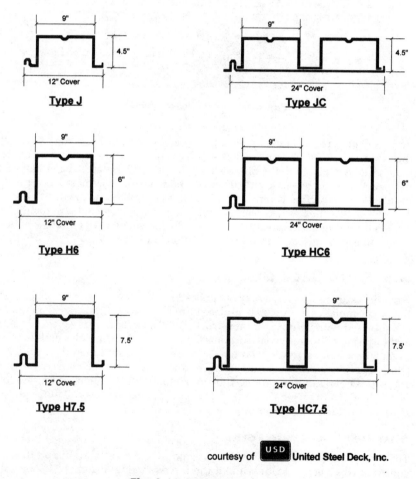

Fig. A.11 Deep deck sections.

loads. In these cases a minimum wire mesh (6 × 6   1.4W × 1.4) is used as a crack-control effort. The slab design is done in a conventional manner with a few idiosyncrasies caused by the deck profile. Although rebars can be used, the most common method is to use "draped mesh." A design example is shown in Fig. A.15. In the example problem 12 in is used as the b width for positive and negative bending. The interference of the deck ribs is compensated for by putting the "bottom" of the slab at the centroid of the steel deck. For deck that is 1.5 in (38 mm) or deeper, the practice is to use the average rib width of the deck as the b width for negative bending (per foot or per meter) and measure the negative bending d from the extreme bottom. See Fig. A.15. When the total slab depth is less than 3 in (75 mm) the mesh is not draped but is centered between the top of the form deck and the top of the slab. Some designers recognize the placement problems when trying to drape mesh on short spans (usually less than 4 ft or 1.3 m) and simply use the mesh at middepth. Diaphragm strength and stiffness for concrete-filled forms are quite substantial and follow design procedures similar to conventional reinforced slabs. Tables are available from individual deck manufacturers.

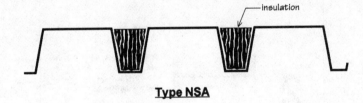

Type NSA

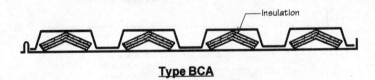

Type BCA

Fig. A.12 Acoustic decks.

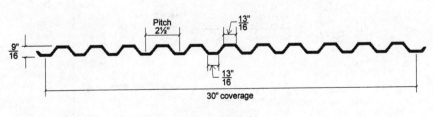

UFS

Fig. A.13 Typical form deck.

A-14 ■ Appendix A

- Clear spans may be used in the formulas.
- For checking web crippling (bearing) the uniform loading case of concrete weight plus 20 psf is used.
- The heavy arrows indicate the formulas that most often control (P = 150lbs.; w_1 = concrete + deck, psf; w_2 = 20 psf construction load; l = clear span.)

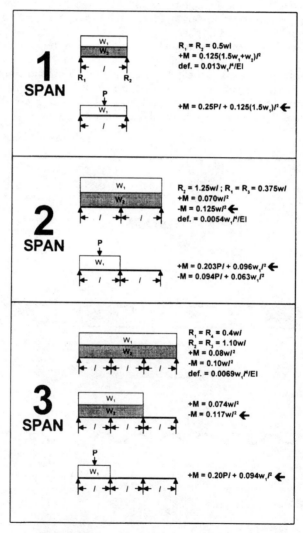

Fig. A.14 SDI formulas for construction loads.

Steel Deck Design ■ A-15

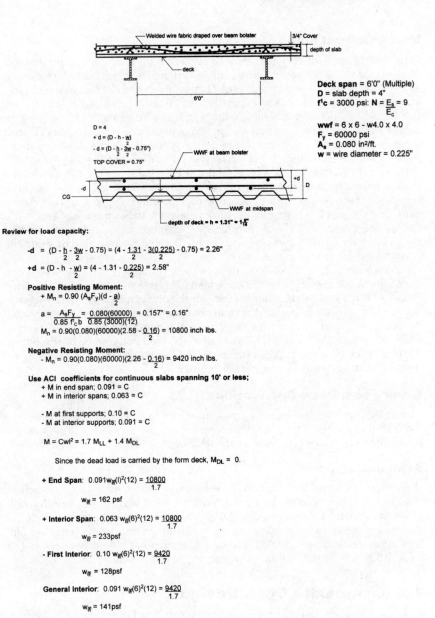

Deck span = 6'0" (Multiple)
D = slab depth = 4"
f'_c = 3000 psi: $N = \frac{E_s}{E_c} = 9$

wwf = 6 x 6 - w4.0 x 4.0
F_y = 60000 psi
A_s = 0.080 in²/ft.
w = wire diameter = 0.225"

Review for load capacity:

$-d = (D - \frac{h}{2} - \frac{3w}{2} - 0.75) = (4 - \frac{1.31}{2} - \frac{3(0.225)}{2} - 0.75) = 2.26"$

$+d = (D - h - \frac{w}{2}) = (4 - 1.31 - \frac{0.225}{2}) = 2.58"$

Positive Resisting Moment:
$+ M_n = 0.90 (A_s F_y)(d - \frac{a}{2})$

$a = \frac{A_s F_y}{0.85 f'_c b} = \frac{0.080(60000)}{0.85 (3000)(12)} = 0.157" = 0.16"$

$M_n = 0.90(0.080)(60000)(2.58 - \frac{0.16}{2}) = 10800$ inch lbs.

Negative Resisting Moment:
$- M_n = 0.90(0.080)(60000)(2.26 - \frac{0.16}{2}) = 9420$ inch lbs.

Use ACI coefficients for continuous slabs spanning 10' or less;
+ M in end span; 0.091 = C
+ M in interior spans; 0.063 = C

- M at first supports; 0.10 = C
- M at interior supports; 0.091 = C

$M = Cwl^2 = 1.7 M_{LL} + 1.4 M_{DL}$

Since the dead load is carried by the form deck, M_{DL} = 0.

+ End Span: $0.091 w_{\ell\ell}(l)^2(12) = \frac{10800}{1.7}$

$w_{\ell\ell}$ = 162 psf

+ Interior Span: $0.063 w_{\ell\ell}(6)^2(12) = \frac{10800}{1.7}$

$w_{\ell\ell}$ = 233 psf

- First Interior: $0.10 w_{\ell\ell}(6)^2(12) = \frac{9420}{1.7}$

$w_{\ell\ell}$ = 128 psf

General Interior: $0.091 w_{\ell\ell}(6)^2(12) = \frac{9420}{1.7}$

$w_{\ell\ell}$ = 141 psf

CONCLUSION:
For end spans the allowable load is 128psf controlled by negative bending. For the general field, the load is also controlled by negative bending and is 141psf.

Fig. A.15 Form deck slab design example.

Nonstructural Fills

Form deck is used with nonstructural cementitious insulating fills in poured roof constructions. The fill material may contain vermiculite or other insulating media as the "aggregate," or even just be very highly air entrained. Polystyrene boards are frequently embedded as layers in the total system. The insulating fill will contain a great deal of water, and indeed the "aggregate" itself may be holding a great deal of water. A roofing membrane is applied over the poured-in-place fill. During the service life of the roof there will be many days when the roof will be directly heated by the sun and high vapor pressures will be developed within the system. If there is no release for the pressure, it can blister the roofing membrane and cause a premature roof failure. To relieve the pressure, top side vents are sometimes used. However, the most frequently used solution is to specify vented form deck. The vents in the bottom of the deck relieve the vapor pressure. Tests run at Granco Steel Products Company in the 1960s showed that if between 1 and 1.5 percent of the projected area was open (vented) the pressure would be sufficiently controlled for most climates.

The strength of the fill material does not add strength to the system for vertical loading but it does increase the horizontal (diaphragm) load capacity. The Steel Deck Institute provides diaphragm tables for the one generic form deck shown in Fig. A.13; for any other profile, the individual manufacturer should be contacted for the tables. An important detail to observe when using these nonstructural filled systems as diaphragms is to hold any embedded polystyrene boards at least 3 ft (1 m) back from the building perimeter or braced frame so that a weak plane does not exist in the area of high shear transfer.

Other Form Deck Applications

Many other applications, including proprietary roofing systems, use form deck as the load-carrying member. The design procedure for the deck follows conventional methods, which generally means selecting from a uniform load table.

Attachments

Screws, powder-actuated fasteners, pneumatically driven nails, or welds may be used to attach form deck to structural supports. Weld washers are required only when the deck thickness is less than 0.028 in (0.7 mm). Insurance needs, uplift, or diaphragm requirements may dictate the type and quantity of the fastening. Side laps can be fastened with welds, screws, and sometimes button punches, but on decks 0.028 in (0.7 mm) or thinner welds are not recommended.

A.4 Composite Deck Design

Composite steel deck acts in the same way as form deck, with the added feature that it interlocks with the concrete to provide the positive bending reinforcement for the slab. Its first duties, then, are to act as a working platform and a stay-in-place form, and for this function it is sized and gauged according to the SDI construction loading shown in Fig. A.14. In most cases the deck is selected so that shoring is not required. The suggested maximum span-to-depth ratio is 32 for continuous slabs and 22 for single spans. The "depth" in this ratio is the total depth of the slab—the distance from the bottom of the deck to the top surface of the slab. The resulting live load capacities of the composite slabs are generally greater than required, so slab depths are most often dictated by fire rating requirements rather than by structural needs. The normal service load deflection limits of either 1/360 or 1/480 of the span will almost never control. The minimum suggested cover of concrete over the top of the composite deck is 2 in (50 mm), which is enough to provide cover for commonly used welded wire fabric. Several 1-, 2-, and 3-h fire ratings are available for both lightweight and normal-weight concrete of 2.5 in (65 mm) cover over the top of the deck. Fireproofing on the bottom surface of the deck may or may not be required, and frequently a design is chosen so that this fireproofing can be eliminated. Figure A.16 shows several common combinations of concrete and fireproofing used on composite deck.

The SDI method for determining the live load capacity of composite deck slabs is shown as an example in Fig. A.17. The methods apply to the types of deck shown in Fig. A.18. The decks shown in Fig. A.18 represent virtually all of the composite deck currently used in the United States. The SDI-sponsored research found that when shear studs are present on beams (perpendicular to the deck) the full ultimate load capacity predicted by conventional underreinforced slab design can be reached. When shear studs are *not* present, the bottom flange of the steel deck can be loaded to its yield point. The present SDI method reduces the ultimate moment based on the number of studs being used. This moment reduction, as illustrated in the exam-

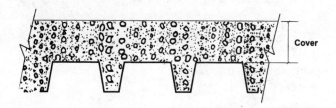

Rating Hours	Concrete cover, LW	Concrete cover, NW	Is fireproofing required?*
1	2.5" (65mm)	3.5"(90mm)	No
1		2.5"(65mm)	Yes
1.5	2.5"(65mm)	2.5"(65mm)	Yes
1.5	3"(75mm)	4"(100mm)	No
2	3.25"(85mm)	4.5"(115mm)	No
2	2.5"(65mm)	2.5"(65mm)	Yes
3	2.5"(65mm)	2.5"(65mm)	Yes
3	4.25"(110mm)	5.25"(135mm)	No

*This column refers to the deck; beams and columns normally need some type of fire protection.

Fig. A.16 Typical fire rated composite slabs.

A-18 ■ Appendix A

ple problem, has been shown to be overly conservative and will no doubt be revised in the near future. Clear spans can be used in either type of system.

The designer does not have to go through the calculation procedures shown in Fig. A.17 to determine maximum unshored spans and uniform load capacities. The SDI and individual manufacturers publish tables for both studded and non-

---

Example calculation for 3" x 12" Composite Floor Deck made from 33 ksi yield steel conforming to ASTM A653-94. For this example a three span form condition is assumed. The slab is 6.25 inches thick consisting of 3 inch deck plus 3.25 inch cover.

The concrete properties are:
 density = 115 pcf; f'c = 3 ksi; f = 3000 psi; n = 14; v = 45 psi(allowable shear stress, ASD)

The 18 gage deck properties are:
 $I = 1.324$; $S_p = .832$; $S_n = .832$; $A_s = .803$; $R_i = 1856$;
 R_i is allowable reaction (4" bearing) at the interior support. Properties are per foot of width.

Deck as a form - use SDI loading as shown.

THREE SPAN CONDITION

$R_1 = R_4 = 0.4wl$
$R_2 = R_3 = 1.10wl$
$+M = 0.08wl^2$
$-M = 0.10wl^2$
def. $= 0.0069w_1l^4/EI$

$+M = 0.074wl^2$
$-M = 0.117wl^2$ ←

$+M = 0.094wl^2$
$+M = 0.20Pl + 0.094w_1l^2$ ←

$P = 150$ lbs.
$w_1 = $ concrete + deck
$w_1 = 48$
$w_2 = 20$ psf
$w = w_1 + w_2$

$+M = [0.20Pl + 0.094w_1l^2]12 = 20000(.832)$
 Solving for l shows $l = 14.49$ ft.
  or
$+M = 0.094(48 + 20)l^2(12) = 20000(.832)$
 Solving for l shows $l = 14.7$ ft.
  or
$-M = 0.117(48 + 20)l^2(12) = 20000(.832)$
 Solving for l shows $l = 13.18$ ft..

Check deflection:
 $y = 0.0069(48)l^4(1728)/(E \times 1.324) = l(12)/180$
 Solving for l shows $l = 16.54$ ft.
  or
 $y = 0.0069(48)l^4(1728)/(E \times 1.324) = 0.75"$
 Solving for l shows $l = 15.02$ ft..
Check web crippling at the interior reaction (exterior reactions can not control).
  $l = 1856/[1.1(48 + 20)] = 24.72$ ft.
The maximum span is controlled by negative bending and is 13.18 ft.

Fig. A.17 Composite floor deck design example.

Calculations for allowable uniform load based on SDI method for no studs on beams. (ASD)
Span length = 13 ft.
The tensile stress caused by casting the concrete on the unshored three span deck is given by:

$f_1 = 0.08 (48) \, l^2(12)/.832 = 9360$ psi

The stress available for live load is then:

$f = 20000 - 9360 = 10640$ psi

The composite deck is assumed to be single span and the uniform design load is:

$W_{ee} = S_e(8)(10640)/(12l^2) \times 1.1$

The 1.1 factor is the 10% increase allowed by the SDI for the use of Welded Wire Mesh.

$S_e = 2.36$; $l = 13$ and $W_{ee} = 110$ psf rounded to neaerest 5psf

Check the allowable load for a deflection of l/360 - single span assumed:

$W_{ee} = EI_{av}(12)/(360 \times 0.013 \times l^3 \times 1728)$
$I_{av} = 13$
$W_{ee} = 259$ psf

Check the allowable load if the concrete stress is 1350 psi:

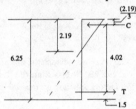

$T = C = 1350 \, (12)(2.19)/2 = 17740$
$M = 17740 \, (4.02)$ inch lbs.
$W = M(8)/(12l^2) = 281$ psf

Check the concrete shear stress.
The area (per foot of width) of concrete available for shear is shown:

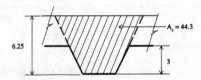

$W_{ee} = 45 \, (44.3)(2)/13 = 306$ psf

The allowable load is based on deck stress and is
$W_{ee} = 110$ psf, rounded to the nearest 5psf.

Fig. A.17 Composite floor deck design example (*Continued*).

A-20 ■ Appendix A

Calculations for the allowable uniform load if studs are present. (LRFD)

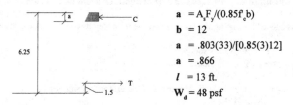

$a = A_s F_y/(0.85 f_c b)$
$b = 12$
$a = .803(33)/[0.85(3)12]$
$a = .866$
$l = 13$ ft.
$W_d = 48$ psf

$M = 0.85(.803)33 (6.25 - 1.5 - .866/2)1000$
$M = 97.24$ inch kips.

The allowable uniform load based on this moment is:
$W_{ee} = (8MR/12l^2) - 1.2W_d)/1.6$ where R is the reduction factor based on the stud spacing.

The current SDI method for calculating the Reduction factor is:
$R = $ stud strength$/K/[F_y(A_s - A_{webs}/2 - A_{bf})]$
where **K** is the number of feet between studs, A_{webs} is the area of the webs per foot and A_{bf} is the area of the bottom flange per foot.

The R values thus calculated cannot exceed one (i.e. increase the moment). **When applying the reduction factor the live load calculated can not be less than the live load calculated using no studs.** For this example, the reduction factors are shown in the table. The live loads are rounded to the nearest 5 psf.

Stud Spacing	R	live load, psf
1'	1	205
2'	.64	120
3'	.43	110

The average stud spacing can be used. Research completed at Virginia Polytechnic Institute has shown the SDI reductions to be very conservative. The R values may be upwardly adjusted in the near future.
The deflection and concrete shear stress calculations would produce the same results as obtained for the "no studs" part of the example.

Fig. A.17 Composite floor deck design example (*Continued*).

studded systems. The SDI tables cover the commonly used profiles on a generic basis in the *Composite Deck Design Handbook*. The Fig. A.19 table (no studs) shows uniform loads based on adding the stresses from concrete placement to the live load stresses on the composite section and allowing the total stress to be $0.66F_y$. If welded wire mesh is not used in the slabs, the allowable stress must be reduced by 10 percent, and therefore the published loads would also be reduced by that amount. The Fig. A.19 table shows the S_c, composite section modulus, and the allowable spans that can carry a 2-kip load over a 30- by 30-in area as required by model building codes for some building types. The Fig. A.20 table (studded beams) shows the same information except it shows the factored resisting moment M in inch-kips instead of the S_c. The uniform loads are based on studs at 12-in average spacings and were calculated using the LRFD procedures shown in Fig. A.17. In both tables an arbitrary upper limit of 400 psf has been applied to the tabulated loads because loads exceeding 400 psf would probably be the result of prorating high concentrated loads. Figures A.21 and A.22 show the same information as Figs. A.19 and A.20 with a soft conversion to international units.

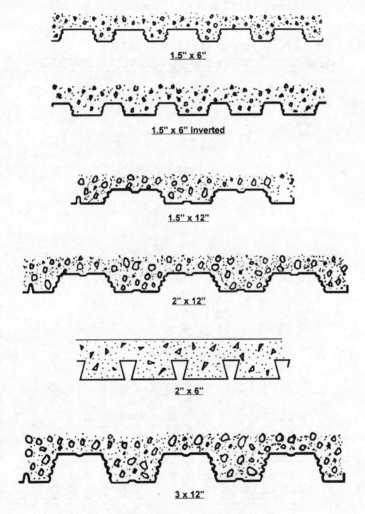

Fig. A.18 Composite deck profiles.

A-22 ■ Appendix A

Connections

Arc puddle welds and shear studs are the usual connections of composite deck to the structural-steel frame. The other deck connectors such as screws or power-driven fasteners are rarely used. The reasons for the preponderance of welding are: Welding was the connection used when composite deck was first introduced, and old habits are hard to change; almost all fire ratings specify welded connections. From a practical standpoint, there is no reason that the other deck connectors could not be used. Shear studs can replace arc puddle welds and are accepted in most fire-rated assemblies. The SDI calls for composite deck to be welded (con-

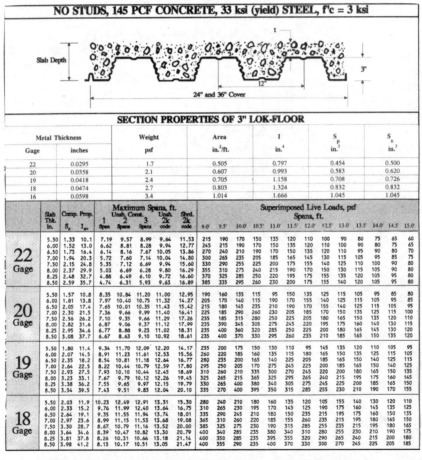

Fig. A.19 Example uniform load table for composite deck—no studs.

Steel Deck Design ■ A-23

nected) to each beam at average spaces not to exceed 12 in (300 mm). Side lap fastenings should not be greater than 36 in (1 m) o.c.

Diaphragms

Diaphragm tables are published for composite deck both with and without structural concrete fill. The unfilled diaphragm values would be useful in reviewing frame bracing during construction. After the concrete hardens, the deck and slab combination provides a very strong and stiff diaphragm. Figure A.23 shows typical strengths for concrete-filled systems.

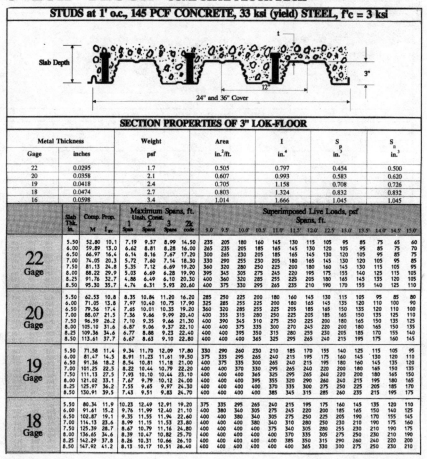

Fig. A.20 Example uniform load table for composite deck—no studs, international units.

A-24 ■ Appendix A

Further Reading

American Iron and Steel Institute, Specification for the Design of Cold Formed Steel Structural Members, AISI, Washington, D.C., 1986.

American Society of Civil Engineers, Standard for the Structural Design of Composite Slabs, ASCE, Washington, D.C., 1992.

Easterling, W. S., and C. S. Young, "Strength of Composite Slabs," *Journal of Structural Engineering*, ASCE, vol. 118, no.9, ASCE, Washington, D.C., 1992.

Factory Mutual System, *Approval Guide*, FMRC, Norwood, Mass., 1995.

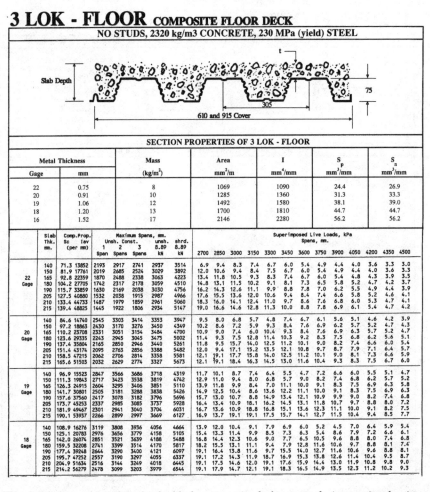

Fig. A.21 Example uniform load table for composite deck—no studs, international units.

Steel Deck Design ■ A-25

Heagler, R. B., L. D. Luttrell, and W.S. Easterling, *Composite Deck Design Handbook*, SDI, Canton, Ohio, 1991.

Luttrell, L. D., *Diaphragm Design Manual*, 2d ed., SDI, Canton, Ohio, 1987.

Steel Deck Institute, *Design Manual (Publication 28)*, SDI, Canton, Ohio, 1992.

Steel Deck Institute, *Manual of Construction with Steel Deck*, SDI, Canton, Ohio, 1992.

Underwriters Laboratory, *Fire Resistance Directory*, UL, Northbrook, Ill., 1995.

United Steel Deck, Inc., *Steel Decks for Floors and Roofs*, USD, Summit, N.J., 1995.

3 LOK - FLOOR COMPOSITE FLOOR DECK
STUDS, 2320 kg/m3 CONCRETE, 230 MPa (yield) STEEL

SECTION PROPERTIES OF 3 LOK - FLOOR

Gage	Metal Thickness mm	Mass (kg/m^2)	Area mm^2/m	I mm^4/mm	S_p mm^3/mm	S_n mm^3/mm
22	0.75	8	1069	1090	24.4	26.9
20	0.91	10	1285	1360	31.3	33.3
19	1.06	12	1492	1580	38.1	39.0
18	1.20	13	1700	1810	44.7	44.7
16	1.52	17	2146	2280	56.2	56.2

Gage	Slab Thk. mm	Comp. Prop. M (per mm)	I av	Maximum Spans, mm. Unsh. 1 Span	Const. 2 Spans	3 Spans	code 8.89 kN	\multicolumn{13}{c}{Superimposed Live Loads, kPa Spans, mm.}
								2700	2850	3000	3150	3300	3450	3600	3750	3900	4050	4200	4350	4500
22 Gage	140	19.57	13850	2195	2915	2740	4405	11.6	10.2	9.0	8.0	7.1	6.4	5.7	5.1	4.6	4.1	3.7	3.3	3.0
	150	22.20	17760	2020	2685	2525	4865	13.2	11.6	10.3	9.1	8.1	7.3	6.5	5.8	5.2	4.7	4.2	3.8	3.4
	165	24.82	22360	1870	2490	2340	5250	14.7	13.0	11.5	10.2	9.1	8.1	7.3	6.5	5.9	5.3	4.8	4.3	3.9
	180	27.44	27705	1740	2315	2180	5580	16.3	14.4	12.7	11.3	10.1	9.0	8.1	7.3	6.5	5.9	5.3	4.8	4.3
	190	30.07	33860	1630	2170	2040	5850	17.9	15.8	14.0	12.4	11.1	9.9	8.9	8.0	7.2	6.5	5.8	5.2	4.7
	205	32.70	40880	1530	2040	1915	6080	19.1	17.2	15.2	13.5	12.1	10.8	9.7	8.7	7.8	7.0	6.3	5.7	5.1
	210	34.01	44735	1485	1980	1860	6180	19.1	17.9	15.8	14.1	12.6	11.2	10.1	9.1	8.1	7.3	6.6	5.9	5.4
	215	35.32	48825	1445	1920	1805	6270	19.1	18.6	16.5	14.6	13.1	11.7	10.5	9.4	8.5	7.6	6.9	6.2	5.6
20 Gage	140	23.17	14740	2545	3305	3415	4950	14.0	12.4	11.0	9.8	8.8	7.9	7.1	6.4	5.8	5.2	4.7	4.3	3.9
	150	26.33	18865	2430	3170	3275	5445	16.0	14.1	12.6	11.2	10.0	9.0	8.1	7.3	6.6	6.0	5.4	4.9	4.4
	165	29.49	23710	2330	3050	3155	5860	17.9	15.9	14.1	12.6	11.2	10.1	9.1	8.2	7.4	6.7	6.1	5.5	5.0
	180	32.64	29335	2245	2945	3045	6205	19.1	17.6	15.6	13.9	12.5	11.2	10.1	9.1	8.2	7.4	6.7	6.1	5.6
	190	35.80	35805	2165	2850	2945	6495	19.1	19.1	17.2	15.3	13.7	12.3	11.1	10.0	9.0	8.2	7.4	6.7	6.1
	205	38.95	43175	2095	2765	2855	6735	19.1	19.1	18.7	16.7	14.9	13.4	12.1	10.9	9.9	8.9	8.1	7.3	6.7
	210	40.53	47215	2060	2705	2815	6840	19.1	19.1	19.1	17.4	15.6	14.0	12.6	11.4	10.3	9.3	8.4	7.7	7.0
	215	42.11	51505	2030	2630	2775	6935	19.1	19.1	19.1	18.1	16.2	14.5	13.1	11.8	10.7	9.7	8.8	8.0	7.2
19 Gage	140	26.53	15525	2845	3565	3685	5430	16.3	14.5	12.9	11.5	10.3	9.3	8.4	7.6	6.9	6.2	5.7	5.1	4.7
	150	30.19	19845	2715	3425	3540	5955	18.6	16.5	14.7	13.1	11.8	10.6	9.6	8.6	7.8	7.1	6.5	5.9	5.4
	165	33.86	24915	2605	3295	3405	6390	19.1	18.5	16.5	14.8	13.2	11.9	10.8	9.7	8.8	8.0	7.3	6.6	6.1
	180	37.53	30800	2505	3180	3290	6755	19.1	19.1	18.3	16.4	14.7	13.2	12.0	10.8	9.8	8.9	8.1	7.4	6.7
	190	41.19	37560	2415	3080	3180	7055	19.1	19.1	19.1	18.0	16.2	14.6	13.1	11.9	10.8	9.8	8.9	8.1	7.4
	205	44.85	45255	2335	2985	3085	7305	19.1	19.1	19.1	19.1	17.6	15.9	14.3	13.0	11.8	10.7	9.8	8.9	8.1
	210	46.69	49465	2300	2940	3040	7415	19.1	19.1	19.1	19.1	18.4	16.5	14.9	13.5	12.3	11.2	10.2	9.3	8.5
	215	48.52	53935	2265	2900	2995	7515	19.1	19.1	19.1	19.1	19.1	17.2	15.5	14.1	12.8	11.6	10.6	9.6	8.8
18 Gage	140	29.78	16275	3120	3810	3935	5865	18.5	16.4	14.7	13.1	11.8	10.6	9.6	8.7	7.9	7.2	6.6	6.0	5.5
	150	33.95	20785	2975	3655	3780	6420	19.1	18.8	16.8	15.0	13.5	12.2	11.0	10.0	9.1	8.2	7.5	6.9	6.3
	165	38.13	26075	2850	3520	3640	6875	19.1	19.1	18.9	16.9	15.2	13.7	12.4	11.2	10.2	9.3	8.5	7.8	7.1
	180	42.30	32210	2740	3400	3515	7255	19.1	19.1	19.1	18.8	16.9	15.2	13.8	12.5	11.4	10.4	9.5	8.6	7.9
	190	46.47	39250	2645	3290	3400	7570	19.1	19.1	19.1	19.1	18.6	16.8	15.2	13.8	12.5	11.4	10.4	9.5	8.7
	205	50.65	47250	2555	3190	3295	7830	19.1	19.1	19.1	19.1	19.1	18.3	16.6	15.0	13.7	12.5	11.4	10.4	9.5
	210	52.74	51635	2515	3145	3250	7940	19.1	19.1	19.1	19.1	19.1	17.3	15.7	14.3	13.0	11.9	10.9	9.9	9.9
	215	54.82	56280	2480	3100	3205	8045	19.1	19.1	19.1	19.1	19.1	18.0	16.3	14.8	13.5	12.4	11.3	10.4	10.4

Fig. A.22 Example uniform load table for composite deck with studs, international units.

A-26 ■ Appendix A

Gage	Span		Design Shear		36/4 weld patterns
	ft.	(mm)	plf	(kN/m)	
22	8	(2450)	1710	(25.0)	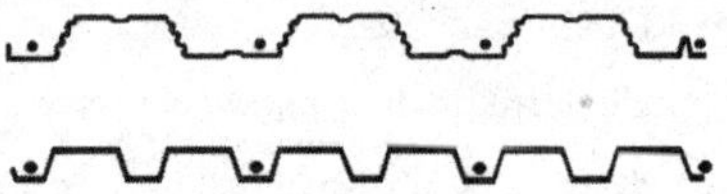
20	8	(2450)	1750	(25.5)	
20	10	(3050)	1750	(25.5)	
18	8	(2450)	1820	(26.6)	
18	10	(3050)	1820	(26.6)	
18	12	(3650)	1770	(25.8)	
16	10	(3050)	1890	(27.6)	
16	12	(3650)	1830	(26.7)	

1. The G' (stiffness) value of 2500 kips/inch (429 kN/mm) can be used for all combinations in the table for either normal weight or light weight concrete.
2. Sidelaps are welded at a maximum of 36" (915mm) on center.
3. Strength values are based on 2.5" (65mm) cover over the ribs of normal weight concrete with f'c = 3000 psi (21 MPa). The table covers 1.5", 2" and 3" (38, 50, 75 mm) composite or form decks. For light weight structural concrete multiply the table shear strength values by 0.7.
4. It may be necessary to increase the number or strength of the perimeter connections to obtain the values shown in the table.
5. The Design Shear values are based on a safety factor of 3.25.
6. The reference basis for the table is the SDI Diaphragm Design Manual, Second Edition.

Fig. A.23 Diaphragm strengths of concrete and steel deck systems.

APPENDIX B

Omer W. Blodgett
Duane K. Miller

SPECIAL WELDING ISSUES FOR SEISMICALLY RESISTANT STRUCTURES

B.2 Introduction and Background

Steel is an inherently forgiving material, and welded construction is a highly efficient, direct method by which several members can be made to function as one. For statically loaded structures, the inherent ductility associated with steel allows the material to compensate for deficiencies in design, materials, and fabrication. For structures containing deficiencies in one or more of these three areas that are subject to cyclical loading, the repeated plastic (or inelastic) deformations can lead to fatigue failure when enough cycles are present. Fabrication of components such as bridges and crane rail supports is therefore more sensitive to deficiencies in design, materials, and fabrication than is fabrication of statically loaded structures. In a similar manner, welded structures subject to seismic loads are sensitive to these three issues as well. This appendix addresses those issues that are necessary to improve the seismic resistance of welded steel structures.

In addition to providing a concise summary of those items that should be considered in the design, fabrication, and erection of steel buildings subject to seismic loading, these items have been separated from the requirements in the body of this handbook in order to avoid the potential introduction of the conservative provisions of this chapter into requirements for structures subject to standard wind and gravity loads. Just as it would be inappropriate, and uneconomical, to impose on all buildings the requirements for bridge fabrication, so it would be undesirable to see these requirements for fabrication in seismic zones automatically transmitted to all structures. The cost increases without any obvious improvement in structure performance would be a disservice to the industry and ultimately the owners of these structures.

The principles presented in Chap. 7 generally apply to seismic construction as well, and the importance of conforming to these principles is even more significant for seismic construction. Many of the details presented in Chap. 8 on fatigue have direct application to seismic loading situations, although at the present time there does not appear to be a cyclic element to the damage that can occur during an earthquake. The very low levels of variable stress typically associated with structures subject to cyclical loading are significantly different from the inelastic loads imposed by an earthquake. Nevertheless, many of the details as they relate to weld backing, weld quality issues, and desirable details are relevant to seismically loaded structures. Chapter 12 provides insight into the available welding processes and the requirement to control quality, in both the shop and the field, when welding any structure. High-quality fabrication is essential for seismically resistant structures. The reader is encouraged to review these other sections, as no attempt has been made to replicate the contents of those chapters as they apply to seismically resistant structures.

Most fabrication work in the United States is performed in accordance with the AWS D1.1 Structural Welding Code—Steel. This code provides general requirements applicable to all welded structures. The D1.1 code does not, however, contain any requirements *unique* to seismic construction. Perhaps this will be done at a future date. The D1.1 code should be taken as the lower bound of acceptable fabrication practice, and the engineer can incorporate additional requirements into contract documents to ensure that the latest requirements for seismic resistance are employed. Section 8 of D1.1 contains the title "Statically Loaded Structures," and Chapter 9, "Cyclically Loaded Structures." This is an improvement over the former title for Section 9 of "Dynamically Loaded Structures." Chapter 9 provisions are specifically geared toward low-stress-range, high-cycle fatigue-type loading, not the high-stress-range, low-cycle stress associated with seismic loading. It is generally recommended that Section 8 criteria be applied for seismic applications, and appropriate modifications be made through contract documents as new information becomes available.

At the time of the writing of this appendix, two significant earthquakes have occurred in the recent past. In both events, significant damage was experienced by welded steel structures, resulting in considerable commitments of resources to research. As a result of these events and the subsequent research, a better understanding of the expected behavior of various structural systems and details has been achieved. While many theories exist regarding various aspects of connection details, some remain unproved, and more testing is required. This appendix represents an accumulation of the best data available to date, and yet recognizes that information will be emerging in the next months and years that may render some of these recommendations incomplete or incorrect. Before adopting these provisions, the reader is cautioned to compare this information to the latest applicable specifications and the state of the art.

Special Welding Issues for Seismically Resistant Structures ■ B-3

One of the best current sources of information is FEMA 267, "Interim Guidelines for Repair and Fabrication of Steel Moment Resisting Frames." This document is the result of the first phase of government-sponsored research performed by the consortium of the Structure of Engineers Association of California (SEAOC), Applied Technology Council (ATC), and California Universities for Research and Earthquake Engineering (CUREe), together known as SAC. The second phase of SAC research, also funded by FEMA, began in 1996 and was expected to continue through 1998. Additional information is expected from these studies.

The principles contained in this appendix are generally based upon well-founded engineering principles. However, not all of the recommendations based upon these principles have been subject to testing. Application of this information to a specific project is of necessity the responsibility of the engineer of record.

B.2 General Review of Welding Engineering Principles

For dynamically loaded structures, attention to detail is critical. This applies equally to high-cycle fatigue loading, short-duration abrupt impact loading, and seismic loading. The following constitutes a review of basic welding engineering principles that apply to all construction, but particularly to seismic applications.

B.2.1 Transfer of Loads

All welds are not evenly loaded. This applies to weld groups that are subject to bending as well as those subject to variable loads along their length. A less obvious situation occurs when steels of different geometries are joined by welding. A rule of thumb is to assume the transfer of force takes place from one member, through the weld, to the member that lies parallel to the force that is applied. Several examples are illustrated in Fig. B.1. For most simple static loading applications, redistribution of stress throughout the member accommodates the variable loading levels. For dynamically loaded members, however, this is an issue that must be carefully assessed in the design. The addition of stiffeners or continuity plates to column webs helps to unify the distribution of stress across the groove weld. Notice that the distribution of stress across the weld joining a beam to an I-shaped column is just opposite that of the same beam joined to a box column.

B.2.2 Minimize Weld Volumes

A good principle of welded design is to always use the least amount of weld metal possible for a given application. Not only does this have sound economic implications, but it reduces the level of residual stress in the connection. All hot expanded metal will shrink as it cools, inducing residual stresses in the connection. By reducing the volume of weld metal, these tendencies can be minimized. Details that will minimize weld volumes for groove welds generally involve minimum root openings, minimum included angles, and the use of double-sided joints. Taken to the extreme, however, these approaches may violate the principles outlined in Sec. B.2.4. By reducing the shrinkage stress, distortion and cracking tendencies can be minimized.

B-4 Appendix B

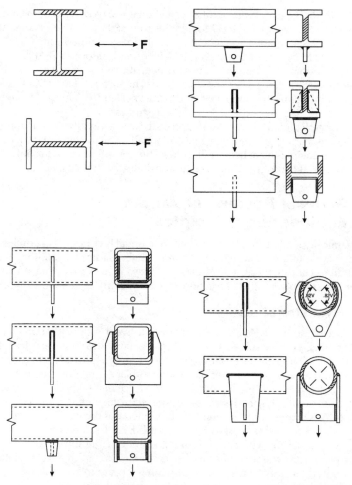

Fig. B.1 Examples of transfer of loads. (*Courtesy of The Lincoln Electric Company.*)

B.2.3 Recognize Steel Properties

Steel is not a perfectly isotropic material. The best mechanical properties usually are obtained in the same orientation in which the steel was originally rolled. This is called the X axis. Perpendicular to the X axis is the width of the steel, or the Y axis. Through the thickness, or the Z axis, the steel will exhibit the least amount of ductility, lowest strength, and lowest toughness properties. When possible, it is always desirable to allow the residual stresses of welding to elongate the steel in the X direction. Of particular concern are large welds placed on either side of the

Special Welding Issues for Seismically Resistant Structures ■ B-5

thickness of the steel where the weld shrinkage stress will act in the Z axis. This can result in lamellar tearing during the time of fabrication, or it can result in subsurface fracture during seismic loading.

B.2.4 Provide Ample Access for Welding

It is essential that the weld joint design as well as the surrounding configuration of material offer adequate access and visibility for the welder and the welding equipment. If the operator cannot adequately observe the joint, weld quality will suffer. As a general rule, if the welder cannot see the joint, neither can the inspector. Weld quality will naturally suffer. It is important that adequate access is provided for the proper placement of the welding electrode with respect to the joint. This is a function of the welding process. Gas-shielded processes, for example, must have ample access for insertion of the shielding gas nozzle into the weld joint. Consideration of these issues has been incorporated into the prequalified groove weld details as listed in AWS D1.1. Overall access to the joint is a function of the configuration of the surrounding material. The designer and detailer should be aware of these general constraints in order to provide adequate access for high-quality fabrication.

B.2.5 No Secondary Members in Welded Design

A fundamental premise of welding design is that there are no secondary members. Anything that is joined by welding can, and will, transfer stress between joined materials. Segmented pieces of steel used for weld backing, for example, can result in a stress-concentration factor at the interface of the backing. Attachments that are merely tack-welded in place may become major load-carrying members, resulting in the initiation of fracture and propagation throughout the structure. These details must be considered in the design stage, and also controlled during fabrication and erection.

B.2.6 Residual Stresses in Welding

As hot expanded weld metal and the surrounding base metal cool to room temperature, they must shrink volumetrically. Under most welding conditions, this contraction is restrained or restricted by the surrounding material, which is relatively rigid and resists the shrinkage. This causes the weld to induce a residual stress pattern where the weld metal is in residual tension, and the surrounding base metal is in residual compression. The residual stress pattern is three-dimensional since the metal shrinks volumetrically. The residual stress distribution becomes more complicated when multiple-pass welding is performed. The final weld pass is always in residual tension, but previous weld beads that were formerly in tension will have compression induced by the subsequent passes.

For relatively flexible assemblages, these residual stresses induce distortion. As assemblages become more rigid, the same residual stresses can cause weld cracking problems, typically occurring near the time of fabrication. If distortion does not occur, or when cracking does not occur, the residual stresses do not relieve themselves but are "locked in." Residual stresses are considered to be at the yield point of the material involved. Because any area that is subject to residual tensile stress is surrounded by a region of residual compressive stress, there is no loss in overall capacity of as-welded structures. This does reduce the fatigue life for low-stress-range, high-cycle applications, which are different from seismic loading conditions.

Small welded assemblies can be thermally stress-relieved where the steel is heated to 1150°F, held for a predetermined length of time (typically 1 h/in of thickness), and allowed to return to room temperature. Residual stresses can be reduced by this method, but they are never totally eliminated. This type of approach is not practical for large assemblies, and care must be exercised to ensure that the components being stress-relieved have adequate support when at the elevated temperature where the yield strength and the modulus of elasticity are greatly reduced as opposed to room-temperature properties. For most structural applications, the residual stresses cause no particular problem to the performance of the system, and owing to the complications of stress-relief activities, welded structures commonly are used in the as-welded condition.

When loads are applied to as-welded structures, there is some redistribution or gradual decrease in the residual stress patterns. Typically called "shake-down," the thermal expansion and contraction experienced by a typical structure as it goes through a climatic season, as well as initial service loads applied to the building, result in a gradual reduction in the residual stresses from welding.

These residual stresses should be considered in any structural application. On a macro level, they will affect the erector's overall sequence of assembly of a building. On a micro level, they will dictate the most appropriate weld bead sequencing in a particular groove-welded joint. For welding applications involving repair, control of residual stresses is particularly important since the degree of restraint associated with weld repair conditions is inevitably very high. Under these conditions, as well as applications involving heavy, highly restrained, very thick steel for new construction, the experience of a competent welding engineer can be helpful in avoiding the creation of unnecessarily high residual stresses, thus alleviating cracking tendencies.

B.2.7 Triaxial Stresses and Ductility

The commonly reported values for ductility of steel generally are obtained from uniaxial tensile coupons. The same degree of ductility cannot be achieved under biaxial or triaxial loading conditions. This is particularly significant since residual stresses are always present in any as-welded structure.

B.2.8 Flat Position Welding

Whenever possible, it is desirable to orient weld details so that the welding can be performed in the flat position, taking advantage of gravity which helps hold the molten weld metal in place. These welds are made with a lower requirement for operator skill, and at the higher deposition rates that correspond to more economical fabrication. This is not to say, however, that overhead welding should be avoided at all costs. An overhead weld may be advantageous if it allows for double-sided welding and a corresponding reduction in the weld volume. High-quality welds can be made in the vertical plane and with the welding consumables available today, can be made at an economical rate.

B.3 Unique Aspects of Seismically Loaded Welded Structures

B.3.1 Demands on Structural Systems

Structures designed for seismic resistance are subject to extreme demands during earthquakes. By definition, any structure designed with an R_w greater than unity will be loaded beyond the yield stress of the material. This is far more demanding than other anticipated types of loading. Because of the inherent ductility of the material, stress concentrations within a steel structure are gradually distributed by plastic deformation. If the materials have a moderate degree of notch toughness, this redistribution eliminates localized areas of high stress, whether due to design, material, or fabrication irregularities. For statically loaded structures, the redistribution of stresses is of little consequence. For cyclically loaded structures, repetition of this redistribution can lead to fatigue failure. In seismic loading, however, it is expected that portions of the structure will be loaded well beyond the elastic limit, resulting in plastic deformation. Localized areas of high stress will not simply be spread out over a larger region by plastic deformation. The resultant design, details, materials, fabrication, and erection must all be carefully controlled in order to resist these extremely demanding loading conditions.

B.3.2 Demand for Ductility

Seismic designs have relied on "ductility" to protect the structural system during earthquakes. Unfortunately, much confusion exists regarding the measured property of ductility and steel, and ductility can be experienced in steel configured in various ways. Because of the versatility of welding, there is the possibility of configuring the materials in ways in which ductility cannot be achieved. It is essential that a fundamental understanding of ductility be achieved in order to ensure ductile behavior in the steel in general, and in the welded connections in particular.

B.3.3 Unique Capabilities of Welded Designs

Welding is the only method by which multiple pieces can be made to act in complete unison as one metallurgical system. Loads can be efficiently transferred through welded connections, and joints can be made to be 100 percent efficient. There are, however, no secondary members in welded designs. Weld backing, for example, participates in conducting forces through members, whether intended or not. The design versatility afforded by the welding process allows for configuring the material in less than optimum orientations. The constraints associated with other systems such as bolting or the use of steel castings provide other constraints that may force compromises for manufacturing but may simultaneously optimize the transfer of stress through the members. It is critical, therefore, that the engineer utilize designs that capitalize upon welding's advantages and minimize potential limitations.

B.3.4 Requirements for Efficient Welded Structures

Five elements are present in any efficient welded structure:

B-8 ■ Appendix B

- Good overall design
- Good details
- Good materials
- Good workmanship
- Good inspection

Each element of the preceding is important, and emphasis on one item will not overcome deficiencies in others. Both the Northridge and Kobe earthquakes have shown that most of the undesirable behavior is traceable to deficiencies in one or more of the preceding areas.

B.4 Design of Seismically Resistant Welded Structures

B.4.1 System Options

Several systems are available to the designer in order to achieve seismic resistance, including the eccentrically braced frame (EBF), concentrically braced frame (CBF), special moment-resisting frames (SMRF), and base isolation. Of the four mentioned, only base isolation is expected to reduce demand on the structure. The other three systems assume that at some point within the structure, plastic deformations will occur in members, thus absorbing seismic energy. In this appendix, no attempt is being made to compare the relative advantages of one system over another. Rather, the focus will be placed on the demands on welded connections and member behavior with respect to the various systems that may be selected by the designer.

In the CBF system, the brace member is the one expected to be subject to inelastic deformations. The welded connections at the termination of a brace are subject to significant tension or compression loads, although rotation demands in the connections are fairly low. Designs of these connections are fairly straightforward, requiring the engineer to develop the capacity of the brace member in compression and tension. Recent experiences with CBF systems have reaffirmed the importance of the brace dimensions (b/t ratio), and the importance of good details in the connection itself. Problems seem to be associated with undersized welds, misplaced welds, missing welds, or welds of insufficient throat due to construction methods. In order to place the brace into the building frame, it is common to utilize a gusset plate welded into the corners of the frame. The brace is slit along its longitudinal axis and rotated into place. In order to maintain adequate dimensions for field assembly, it is necessary to oversize the slot in the tube as compared to the gusset. This results in natural gaps between the tube and the gusset plate. When this dimension increases, as illustrated in Fig. B.2, it is important to consider the effect of the root opening on the strength of the fillet weld. The D1.1 code requires that, for gaps exceeding $1/16$ in, the weld leg size be increased by the amount of the gap. This ensures that a constant actual throat dimension is maintained.

EBFs and SMRFs are significantly different structural systems, but from a welding design point of view, there are principles that apply equally to both systems. It

Special Welding Issues for Seismically Resistant Structures ■ B-9

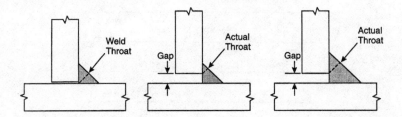

Fig. B.2 Effect of root openings (gaps) on fillet weld throat dimensions. (*Courtesy of The Lincoln Electric Company.*)

is possible to design an EBF so that the "link" consists simply of a rolled steel member. In Fig. B.3, these examples are illustrated by the links designated as C1. In other EBF systems, however, the connection itself can be part of the link, as illustrated by C2. When an EBF system is designed by this method, the welded connections become critical since the expected loading on the connection is in the inelastic region. Much of the discussion under SMRF is applicable to these situations.

A common method applied to low-rise structures is the SMRF system. Advantages of this type of system include desirable architectural elements that leave the structure free of interrupting diagonal members. Extremely high demands for inelastic behavior in the connections are inherent to this system.

When subject to lateral displacements, the structure assumes a shape as shown in Fig. B.4*a*, resulting in the moment diagram shown in Fig. B.4*b*. Notice that the highest moments are applied at the connection. A plot of the section properties is schematically represented in Fig. B.4*c*. Section properties are at their lowest value at the column face, owing to the weld access holes that permit the deposition of the complete joint penetration (CJP) beam flange to column flange welds. These sec-

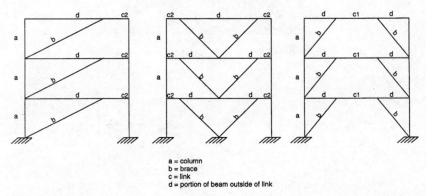

Fig. B.3 Examples of EBF systems. (*From "Seismic Provisions for Structural Steel Building," American Institute of Steel Construction, Inc., 1992.*)

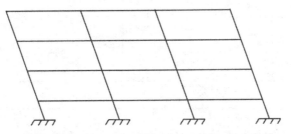
a. SMRF Systems subject to lateral displacements.

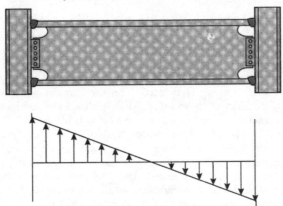
b. Moment diagram of SMRF subject to lateral displacements.

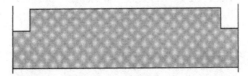
c. Section properties of SMRF subject to lateral displacements.

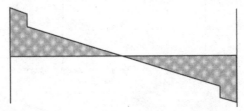
d. Stress distribution of SMRF subject to lateral displacements.

Fig. B.4 Analysis of SMRF behavior. (*Courtesy of The Lincoln Electric Company.*)

Special Welding Issues for Seismically Resistant Structures ■ B-11

tion properties may be further reduced by the deletion of the beam web from the calculation of section properties. This is a reasonable assumption when the beam web to column shear tab is connected by the means of high strength bolts. Greater capacity is achieved when the beam web is directly welded to the column flange with a complete joint-penetration groove weld. The section properties at the end of the beam are least, precisely an area where the moment levels are the greatest. This naturally leads to the highest level of stresses. A plot of stress distribution is shown in Fig. B.4d. The weld is therefore in the area of highest stress, making it critical to the performance of the connection. Details in either SMRF systems or EBF systems that place this type of demand on the weld require careful scrutiny.

B.4.2 Ductile Hinges in Connections

The fundamental premise regarding the special moment-resisting frame (SMRF) is that plastic hinges will form in the beams, absorbing seismically induced energies by inelastically stretching and deforming the steel. The connection is not expected to break. Following the Northridge earthquake, there was little or no evidence of hinge formation. Instead, the connections or portions of the connection experienced brittle fracture, inconsistent with expected and essential behavior. Most of the ductility data are obtained from smooth, slowly loaded, uniaxially loaded tensile specimens that are free to neck down. If a notch is placed in the specimen, perpendicular to the applied load, the specimen will be unable to exhibit its normal ductility, usually measured as elongation. The presence of notchlike conditions in the Northridge connections decreased the ductile behavior.

Initial research on SMRF connections conducted in the summer of 1994 attempted to eliminate the issues of notchlike conditions in the test specimens by removing weld backing and weld tabs, and controlling weld soundness. Even with these changes, "brittle" fractures occurred when the standard details were tested. The testing program then evaluated several modified details with short cover plates, with better success. The reason for these differences can be explained analytically.

Referring to Fig. B.5, the material at point A, whether it be weld metal or base metal, cannot exhibit the ductility of a simple tension test. Ductility can take place

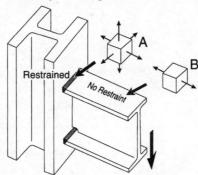

Fig. B.5 Regions to be analyzed relative to potential for ductile behavior. (*Courtesy of The Lincoln Electric Company.*)

B-12 ■ Appendix B

only if the material can slip in shear along numerous slip planes. That is, it must be free to neck down. Four conditions are required for ductility:

1. There must be a shear stress (τ) component resulting from the given load condition.
2. This shear stress must exceed its critical value by a reasonable amount. The more it exceeds this value, the greater will be the resulting ductility.
3. The plastic shear strain resulting from this shear stress must act in the direction which will relieve the particular stress which can cause cracking.
4. There must be sufficient unrestrained length of the member to permit "necking down."

If conditions 1 and 2 are not met, there will be no apparent ductility and no yielding. The stress will simply build up to the ultimate tensile strength with little or no plastic energy absorbed. This condition is called brittle failure.

Figure B.5 shows two regions in question. Point A is at the weld joining the beam flange to the face of the column flange. Here there is restraint against strain (movement) across the width of the beam flange (ε_1) as well as through the thickness of the beam flange (ε_2). Point B is along the length of the beam flange away from the connecting weld. There is no restraint across the width of the flange or through its thickness.

In most strength of materials texts, the following equations can be found:

$$\varepsilon_3 = \frac{1}{E}(\sigma_3 - \mu\sigma_2 - \mu\sigma_1) \tag{1a}$$

$$\varepsilon_2 = \frac{1}{E}(-\mu\sigma_3 + \sigma_2 - \mu\sigma_1) \tag{1b}$$

$$\varepsilon_1 = \frac{1}{E}(-\mu\sigma_3 - \mu\sigma_2 + \sigma_1) \tag{1c}$$

It can be shown that:

$$\sigma_1 = \frac{E[\mu\varepsilon_3 + \mu\varepsilon_2 + (1-\mu)\varepsilon_1]}{(1+\mu)(1-2\mu)} \tag{2a}$$

$$\sigma_2 = \frac{E[\mu\varepsilon_3 + (1-\mu)\varepsilon_2 + \mu\varepsilon_1]}{(1+\mu)(1-2\mu)} \tag{2b}$$

$$\sigma_3 = \frac{E[(1-\mu)\varepsilon_3 + \mu\varepsilon_2 + \mu\varepsilon_1]}{(1+\mu)(1-2\mu)} \tag{2c}$$

Consider the unit cube in Fig. B.6. It is an element of the beam flange from point B in Fig. B.5. The applied force due to the moment is σ_3. There is no restraint against the longitudinal stress of 30 ksi in the flange, so $\sigma_1 = \sigma_2 = 0$. Using Poisson's ratio of $\mu = 0.3$ for steel, Eqs. 1a to 1c yield the following for $\sigma_3 = 30$ ksi:

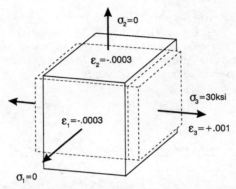

Fig. B.6 Unit cube showing applied stress from Fig. B.5. (*Courtesy of The Lincoln Electric Company.*)

$$\varepsilon_3 = +0.001 \text{ in/in}$$

$$\varepsilon_2 = -0.0003 \text{ in/in}$$

$$\varepsilon_1 = -0.0003 \text{ in/in}$$

From Eqs. 2a to 2c, it is found that

$$\sigma_1 = 0 \text{ ksi}$$

$$\sigma_2 = 0 \text{ ksi}$$

$$\sigma_3 = 30 \text{ ksi}$$

These stresses are plotted in Fig. B.7 as a dotted circle. The larger solid-line circle is for a stress of 70 ksi or ultimate tensile stress. The resulting maximum shear stresses $\tau_{1\text{-}3}$ and $\tau_{2\text{-}3}$ are the radii of these two circles, or 35 ksi. The ratio of shear to

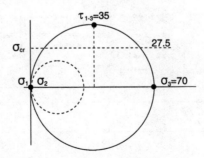

Fig. B.7 A plot of the tensile stress and shear stress from Fig. B.5. (*Courtesy of The Lincoln Electric Company.*)

tensile stress is 0.5. Figure B.8 plots this as line B. Notice at a yield point of 55 ksi, the critical shear value is one-half of this, or 27.5 ksi. When this critical shear stress is reached, plastic straining or movement takes place and ductile behavior will result up to the ultimate tensile strength, here 70 ksi. Figure B.9 shows a predicated stress-strain curve indicating ample ductility.

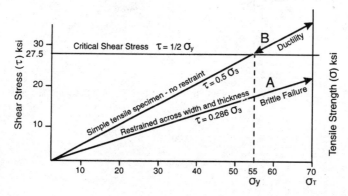

Fig. B.8 The ratio of shear to tensile stress. (*Courtesy of The Lincoln Electric Company.*)

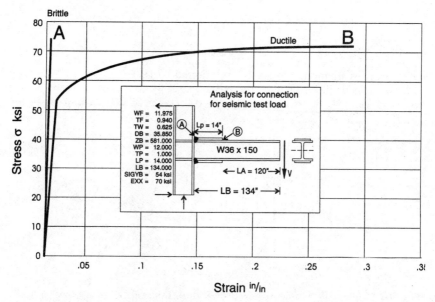

Fig. B.9 Stress-strain curve. (*Courtesy of The Lincoln Electric Company.*)

Special Welding Issues for Seismically Resistant Structures ■ B-15

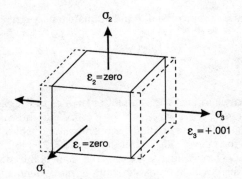

Fig. B.10 The highly restrained region at the junction of the beam and column flange as shown in Fig. B.5. (*Courtesy of The Lincoln Electric Company.*)

Figure B.10 shows an element from point A (Fig. B.5) at the junction of the beam and column flange. Whether weld metal or the material in the column or beam is considered makes little difference. This region is highly restrained. Suppose it is assumed:

$\varepsilon_3 = +0.001$ in/in (as before)

$\varepsilon_2 = 0$ (since it is highly restrained with little strain possible)

$\varepsilon_1 = 0$

Then from the given equations, the following stresses are found:

$\sigma_1 = 17.31$ ksi

$\sigma_2 = 17.31$ ksi

$\sigma_3 = 40.38$ ksi

In Fig. B.11, these stresses are plotted as a dotted circle.

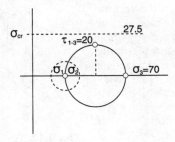

Fig. B.11 A plot of the tensile stress and shear stress from Fig. B.10. (*Courtesy of The Lincoln Electric Company.*)

If these stresses are increased to the ultimate tensile strength, it is found that:

$\sigma_1 = 30.0$ ksi

$\sigma_2 = 30.0$ ksi

$\sigma_3 = 70.0$ ksi

The solid-line circle in Fig. B.11 is a plot of stresses for this condition. The maximum shear stresses are $\tau_{1\text{-}3} = \tau_{2\text{-}3} = 20$ ksi. Notice that these are less than the critical shear stress (27.5 ksi) so no plastic movement, or ductility, would be expected.

In this case, the ratio of shear to tensile stress is 0.286. In Fig. B.8, this condition is plotted as line A. Notice it never exceeds the value of the critical shear stress (27.5 ksi); therefore, there will be no plastic strain or movement, and it will behave as a brittle material. Figure B.9 shows a predicated stress-strain curve going upward as a straight line A (elastic) until the ultimate tensile stress is reached in a brittle manner with no energy absorbed plastically. It would therefore be expected that, at the column face or in the weld where high restraint exists, little ductility would be exhibited. This is where "brittle" fractures have occurred, both in actual Northridge buildings and in laboratory test specimens.

In the SMRF system, the greatest moment (due to lateral forces) will occur at the column face. This moment must be resisted by the beam's section properties, which are lowest at the column face due to weld access holes. Thus the highest stresses occur at this point, the point where analysis shows ductility to be impossible.

In Fig. B.5, material at point B was expected to behave as shown in Fig. B.7, and as line B in Fig. B.8, and curve B in Fig. B.9; that is, with ample ductility. It is essential that plastic hinges be forced to occur in this region.

Several post-Northridge designs have employed details that facilitate use of this potential ductility. Consider the cover-plated design illustrated in Fig. B.12. Notice that this detail accomplishes two important purposes: first, the stress level at point A is reduced as a result of the increased cross section at the weld. This region, incapable of ductility, must be kept below the critical tensile stress, and the increase in area accomplishes this goal. Second, and most important, the most highly stressed region is now at point B, the region of the beam that is capable of exhibiting ductility. Assuming good workmanship with no defects or stress raisers, the real success of this connection will depend upon getting the adjacent beam to bend plastically before this critical section cracks. The way in which a designer selects structural details under particular load conditions greatly influences whether the condition provides enough shear stress component so that the critical shear value may be exceeded first, producing sufficient plastic movement

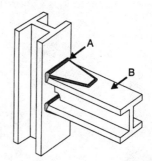

Fig. B.12 Cover-plated detail takes advantage of the region where ductility is possible. (*Courtesy of The Lincoln Electric Company.*)

Special Welding Issues for Seismically Resistant Structures B-17

before the critical normal stress value is exceeded. This will result in a ductile detail and minimize the chances of cracking.

B.4.3 Details of Welded Connections

There are no secondary members in welded construction. Any material connected by a weld participates in the structural system—positively or negatively. Unexpected load paths can be developed by the unintentional metallurgical path that results from the one-component system created by welding. This must be considered in all phases of a welded steel project but is particularly significant in detailing.

Weld Backing ■ Weld backing consists of auxiliary pieces of material used to support liquid weld puddles. The backing can be either temporary or permanent. Permanent backing consists of a steel member of similar composition that is fused in place by the weld. The D1.1 code requires that backing, if used, be continuous for the length of the joint, free of interruptions that would create stress-concentration factors. Segments of backing can be made continuous if complete joint penetration (CJP) groove welds join the various segments of backing. It is essential that these splices be completely fused across the backing cross section.

Weld Tabs ■ Weld tabs are auxiliary pieces of metal on which the welding arc may be initiated or terminated. For statically loaded structures, these are typically left in place. For seismic construction, it is recommended that weld tabs be removed from critical connections that are subject to inelastic loading. It is in the region of these weld tabs that metal of questionable integrity may be deposited. After removal, the end of the weld joint can be visually inspected for integrity.

Weld tab removal is probably most significant on beam-to-column connections where the beam flange width is less than the column flange width. It is reasonable to expect that stress flow would take place through the left-in-place weld tab. In contrast, for butt splices where the same width of material is joined, weld tabs that extend beyond the width of the joint would not be expected to carry significant stress levels, making weld tab removal less important. Tab removal from continuity plate welds is probably not justified.

The presence of weld tabs left-in-place is probably most significant for beam-to-column connections where columns are box shapes. The natural stress distribution under these conditions causes the ends of the groove weld between the beam and column to be loaded to the greatest level, the same region as would contain the weld tab. Just the opposite situation exists when columns are composed of I-shaped members. The center of the weld is loaded most severely, and the areas in which the weld tabs would be located have the lowest stress level. For welds subject to high levels of stress, however, weld tab removal is recommended.

Welds in Combination with Bolts ■ Welds provide a continuous metallurgical path that relies upon the internal metallurgical structure of the fused metal to provide continuity and strength. Mechanical fasteners such as rivets and bolts rely on friction, shear of the fastening element, or bearing of the joint material to provide for transfer of loads between members. When welds are combined with bolts, caution must be exercised in assigning load-carrying capacity to each joining method.

Traditionally it has been assumed that welds that are used in conjunction with bolts should be designed to carry the full load, assuming that the mechanical fasteners have no load-carrying capacity until the weld fails. With the development of high-strength fasteners, it has been assumed that loads can be shared equally between welds and fasteners. This has led to connection details which employ both joining systems. In particular, the welded flange, bolted web detail used for many beam-to-column connections in special moment-resisting frames (SMRF) assumes that the bolted web is equally able to share loads with the welded flanges. While most analysis suggests that vertical loads are transferred through the shear tab connection (bolted) and moments are transferred through the flanges (welded), the web does have some moment capacity. Depending on the particular rolled shape involved, the moment capacity of the web can be significant. Testing of specimens with the welded web detail, as compared to the bolted web detail, generally has yielded improved performance results. This has drawn into question the adequacy of the assumption of high-strength bolts sharing loads with welds when subject to inelastic loading. Research performed after the Northridge earthquake provides further evidence that the previously accepted assumptions may have been inadequate. This is not to suggest that bolted connections cannot be used in conjunction with welded connections. However, previous design rules regarding the capacity of bolted connections need to be reexamined. This may necessitate additional fasteners or larger sizes of shear tabs (both in thickness and in width). The rules regarding the addition of supplemental fillet welds on shear tabs, currently a function of the ratio of Z_f/Z, are very likely also inadequate and will require revision.

Until further research is done, the conservative approach is probably to utilize welded web details. This does not preclude the use of a bolted shear tab for erection purposes but would rely on welds as a singular element connecting the web to the column.

Some of the alternate designs that have been contemplated after the Northridge earthquake (see Sec. B.8.2) increase the moment capacity of the connection, reducing the requirement for the web to transfer moment. These details are probably less sensitive to the degree of interaction between welds and bolts.

Weld Access Holes ■ Weld access holes are openings in the web of a member that permit the welder to gain necessary access for the deposition of quality weld metal in a flange connection. Colloquially known as "rat holes," these openings also limit the interaction of residual stress fields present when a weld is completed. Poorly made weld access holes, as well as improperly sized holes, can limit the performance of a connection during seismic loading.

Consider the beam-to-column connection illustrated in Fig. B.13. A welded web connection has been assumed. As the flange groove weld shrinks volumetrically, a residual stress field will develop perpendicular to the longitudinal axis of the weld, as illustrated in direction X in the figure. Simultaneously, as the groove weld shrinks longitudinally, a residual stress pattern is established along the length of the weld, designated as direction Y. When the web weld is made, the longitudinal shrinkage of this weld results in a stress pattern in the Z direction. At the intersection of the

Special Welding Issues for Seismically Resistant Structures ■ B-19

web and flange of the beam with the face of the column, these three residual stress patterns meet at a point. When steel is loaded in all three orthogonal directions simultaneously, it is impossible for the most ductile steel to exhibit ductility. At the intersection of these three welds, cracking tendencies would be significant. By providing a generous weld access hole, however, the interaction of the Z-axis stress field and the biaxial (X and Y) stress field is physically interrupted. This increases the resistance to cracking during fabrication.

Residual compression in one axis can combine with residual tension in one or more axes, resulting in an increase in the ductility that will be observed. Ideally, weld access holes should be placed in areas where at least one residual compressive stress is present. Consider the longitudinal

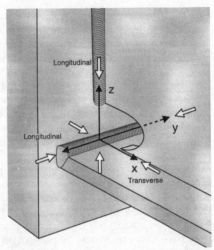

Fig. B.13 A generous weld access hole in this beam-to-column connection provides resistance to cracking. (*Courtesy of The Lincoln Electric Company.*)

residual stress of the groove weld in the Y axis. While residual tension stresses are present in the weld itself, the weld is surrounded by a region of residual compressive stress that counteracts the tensile component. When weld access holes terminate in this area of residual compression, the connection will exhibit enhanced ductility.

For these reasons, AISC and AWS both have minimum weld access hole sizes that must be achieved. These minimums fit into two categories: minimums that must be obtained under all conditions, and minimum dimensions that are a function of the web thickness of the materials being joined. The absolute minimums are a function of the welding processes and good workmanship requirements. The web thickness-dependent dimensions reflect upon the geometric influence of member size upon the level of residual stresses that will be present. With larger member sizes, and correspondingly increased residual stresses, larger weld access holes are somewhat self-compensating.

Initially, the purpose of weld access holes was to provide adequate access for the welder to deposit quality weld metal. Failure to supply the welder with ample visibility of, and access to, the joint will naturally diminish quality. For beam-to-column connections, welding through the access hole when making the bottom beam flange weld has been demonstrated to be a particularly challenging task. Adequate sizing is partially dependent on the weld process being used. For SMAW, the relatively small diameter of the electrodes, coupled with their significant length, allows for easy access into this region. FCAW-ss may employ longer electrode extensions which somewhat duplicate the access flexibility of SMAW. Since FCAW-ss does not utilize a shielding gas, this process is unencumbered with gas nozzles that further restrict visibility. If FCAW-gs is used, however, allowance must be made for the operator to be able to place the welding gun, complete with a gas delivery nozzle, into the joint. For these reasons, weld access holes may have to be larger than the minimum dimensions prescribed in the applicable codes. It is usually best to let the fabricator's detailer select appropriately sized access holes for the particular welding processes to be used.

The quality of weld access holes is an important variable that affects both resistance to fabrication-related cracking as well as resistance to cracking that may result from seismic events. Access holes usually are cut into the steel by the use of a thermal cutting process, either oxy-fuel cutting (frequently called "burning") or plasma arc cutting. Both processes rely on heating the steel to a high temperature and removing the heated material by pressurized gases. In the case of oxy-fuel cutting, oxidation of the steel is a key ingredient in this process. In either process, the steel on either side of the cut (called the "kerf") has been heated to an elevated temperature and rapidly cooled. In the case of oxy-fuel cutting, the surface may be enriched with carbon. For plasma cut surfaces, metallic compounds of oxygen and nitrogen may be present on this surface. The resultant surface may be hard and crack-sensitive, depending on the combinations of the cutting procedure, base metal chemistry, and thickness of the materials involved. Under some conditions, the surface may contain small cracks. These cracks can be the points of stress amplification that cause further cracking during fabrication or during seismic events.

Nicks or gouges may be introduced during the cutting process, particularly when the cutting torch is manually propelled during the formation of the access hole. These nicks may act as stress-amplification points, increasing the possibility of cracking.

To decrease the likelihood of notches and/or microcracks on thermally cut surfaces, AISC has specific provisions that are required for making access holes in heavy group 4 and 5 rolled shapes. These provisions include the need for a preheat before cutting (to minimize the possibility of the formation of hard, crack-sensitive microstructures), requirements for grinding of these surfaces (to provide for a smooth contour, and to eliminate cracks and gouges as well as hard material that may be present), and inspection of these surfaces with magnetic-particle (MT) or dye-penetrant (PT) inspection (to verify a crack-free surface). Whether these requirements are necessary for all connections that may be subject to seismic energies is unknown at this time. However, for connection details that impose high levels of stress on the connection, and specifically those that demand inelastic performance, it is apparent that every detail in this region, including weld access hole geometry and quality, is a critical variable. Some cracking initiated from weld access holes in the Northridge earthquake. Whether this was the result of preexistent cracks that occurred during flange cutting, or the result of strains that were induced by the shrinkage of the welds during fabrication and/or erection, or simply the concentration of seismically induced forces that were amplified in these regions, is not known at this time. It does highlight the importance, however, of paying attention to all construction details, including weld access holes.

B.5 Materials

B.5.1 Base Metal

Base metal properties are significant in any type of steel construction but particularly in structures subject to seismic loading. While most static designs do not require loading beyond the yield strength of the material, seismically resistant structures depend on acceptable material behavior beyond the elastic limit. Although most static designs attempt to avoid yielding, the basic premise of seismic design is to absorb seismic energies through yielding of the material. For static design, additional yield strength capacity in the steel may be desirable. For applications where yielding is the desired method for achieving energy absorption, higher than expected yield strengths have a dramatic negative effect on some designs. This is particularly important as it relates to connections, both bolted and welded.

Figure B.14 illustrates five material zones that occur near the groove weld in a beam-to-column connection. This is the standard connection detail used for SMRF systems, at least prior to the Northridge earthquake. If it is assumed that the web is incapable of transferring any moment (a simple assumption, but probably justified by the lack of confidence in the ability of welds to share loads with bolts), it is critical that the plastic section modulus of the flanges Z_f times the tensile strength be greater than the entire plastic section property Z times the yield strength in the beam. All five material properties must be considered in order for the connection to behave satisfactorily.

Existing ASTM specifications for most structural steels do not place an upper limit on the yield strength but specify only a minimum acceptable value. For example, for ASTM A36 steel, the minimum acceptable yield strength is 36 ksi.

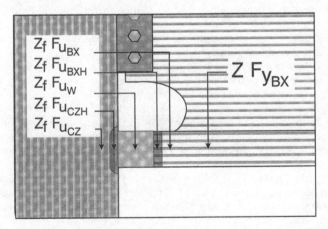

Fig. B.14 Five material zones that occur near the groove weld in a beam-to-column connection. (*Courtesy of The Lincoln Electric Company.*)

This requirement precludes a steel that has a yield strength of 35.5 ksi as being acceptable but does nothing to prohibit the delivery of a 60-ksi steel. The tensile strength range is specified as 58 to 80 ksi. While A36 is commonly specified for beams, columns are typically specified to be of ASTM A572 grade 50. With a 50 ksi minimum yield strength, and a minimum tensile strength of 65 ksi, many designers were left with the false impression that the yield strength of the beam could naturally be less than that of the column. Owing to the specification requirements, it is possible to produce steel that meets the requirements of both A36 and A572 grade 50. This material has been commercially promoted as "dual certified" material. In actual fact, regardless of what the material is called, it is critical for the connection illustrated in Fig. B.14 to have controls on material properties that are more rigorous than the current ASTM standards impose.

Much of the focus has related to the beam yield-to-tensile ratio, commonly denoted as F_y/F_u. This is often compared to the ratio of Z_f/Z, with the desired relationship F_y/F_u being

$$\frac{Z_f}{Z} > \frac{F_y}{F_u}$$

This suggests that not only is F_y (yield strength) important, but the ratio is important as well. For rolled W-shapes, Z_f/Z ranges from 0.6 to 0.9. Based on ASTM *minimum* specified properties, F_y/F_u is as follows:

        A36   0.62

        A572Gr50   0.77

However, when actual properties of the steel are used, this ratio may increase. In the case of one Northridge damaged building, mill test reports indicated the ratio to be 0.83.

ASTM steel specifications need additional controls that will limit the upper value of acceptable yield strengths for materials as well as the ratio of F_y/F_u. A new ASTM specification has been proposed that will address these issues, although its approval will probably not be achieved before 1997.

In Fig. B.14, five zones have been identified in the area of the connection, with the sixth material property being located in the beam. Thus far, only two have been discussed, namely, the beam yield strength and the beam ultimate strength. As shown in the figure, these are designated with the subscript X to indicate that these are the properties in the orientation of the longitudinal axis of the beam.

The properties of interest with respect to the column are oriented in the column Z axis. This is the direction which will exhibit the least desirable mechanical properties. Current ASTM specifications do not require measurement of properties in this orientation. While there are ASTM standards for the measurement of through-thickness properties (ASTM 770), these are not normally applied for structural applications. It is this through-thickness strength, however, that is important to the performance of the connection.

The weld metal properties are also very important, as indicated by the designation F_{u_w} in Fig. B.14. These are discussed in Sec. B.5.2.

Special Welding Issues for Seismically Resistant Structures ■ B-23

On either side of the weld are two heat-affected zones (HAZ), one located on the column side of the weld and the other located on the beam side. These properties are also important to the performance of the connection and are discussed in Sec. B.5.3.

Notch toughness is defined as the ability of a material to resist propagation of a preexisting cracklike flaw while under tensile stress. Pre-Northridge specifications did not have required notch toughness for either base materials or weld metals. When high loads are applied, and when notchlike details or imperfections exist, notch toughness is the material property that resists brittle propagation from that discontinuity. Rolled shapes routinely produced today, specifically for lighter-weight shapes in the group 1, 2, and 3 category, generally are able to deliver a minimum notch toughness of 15 ft-lb at 40°F. This is probably adequate toughness, although additional research should be performed in this area.

For heavy columns made of group 4 and 5 shapes, this level of notch toughness may not be routinely achieved in standard production. After Northridge, many engineers began to specify the supplemental requirements for notch toughness that are invoked by AISC specifications for welded tension splices in jumbo sections (group 4 and 5 rolled shapes). This requirement for 20 ft-lb at 70°F is obtained from a Charpy specimen extracted from the web-flange interface, an area expected to have the lowest toughness in the cross section of the shape. Since columns are not designed as tension members under most conditions, this requirement would not automatically be applied for column applications. However, as an interim specification, it seems to be a reasonable approach to ensure minimum levels of notch toughness for heavy columns.

B.5.2 Weld Metal Properties

Four properties of interest typically are applied to weld metal: yield strength, tensile strength, toughness, and elongation. These properties are generally obtained from data on the particular filler metal that will be employed to make the connection. The American Welding Society (AWS) filler metal classification system contains a "coding" that defines the minimum acceptable properties for the weld metal when deposited under very specific conditions. Most 70 series electrodes (e.g., E7018, E70T-1, E70T-6) have a minimum specified yield strength of 58 ksi and a minimum tensile strength of 70 ksi. As in the specifications for steel, there are no upper limits on the yield strength. However, in welded design, it is generally assumed that the weld metal properties will exceed those of the base metal, and any yielding that would occur in the connection should be concentrated in the base metal, not in the weld metal, since the base metal is assumed to be more homogeneous and more likely to be free of discontinuities that may be contained within the weld. Most of the commercially available filler metals today have a 70 classification, which exceeds the minimum specified strength properties of the commonly used A36 and A572 grade 50.

These weld metal properties are obtained under very specific testing conditions that are prescribed by the AWS A5 filler metal specifications. Weld metal properties are a function of a variety of variables, including preheat and interpass temperatures, welding parameters, base metal composition, and joint design.

Deviations in these conditions from those obtained for the test welds may result in differences in mechanical properties. Most of these changes will result in an increase in yield and tensile strength, along with a corresponding decrease in elongation and, in general, a decrease in toughness. When weld metal properties exceed those of the base metal, and when the connection is loaded into the inelastic range, plastic deformations would be expected to occur in the base metal, not in the weld metal itself. The increase in the strength of the weld metal compensates for the loss in ductility. The general trend to strength levels higher than those obtained under the testing conditions is generally of little consequence in actual fabrication.

There are conditions which may result in lower levels of strength being obtained, and the Northridge earthquake experience revealed that this may be more commonplace and more significant than originally thought. The interpass temperature is the temperature of the steel when the arc is initiated for subsequent welding. There are two aspects to the interpass temperature: the minimum level, which should always be the minimum preheat temperature, and the maximum level, beyond which welding should not be performed. Because of the relatively short length of beam-to-column flange welds, it is possible for a welder to continue welding at a pace that will allow the temperature of the steel at the connection to increase to unacceptably high levels. After one or two weld passes, this temperature may approach the 1000°F range. Under these conditions, a marked decrease in the strength of the weld deposit may occur.

Although it would be unexpected to see the strength drop below the minimum specified property for A572 grade 50 steel, it may fall below the typical strength of the weld deposit made under more controlled conditions. The restraint associated with the geometry at the beam-to-column junction does not encourage yielding, so the decrease in uniaxial yield strength may have less significance than the decrease in tensile capacity.

Much emphasis has been placed on elongation of materials, but as discussed under Sec. B3.1 and B3.2, geometric constraints on ductility would generally preclude welds from being able to deform, regardless of their uniaxial elongation properties.

Weld metal toughness is an area of particular interest in the post-Northridge specifications. Previous specifications did not have any requirement for minimum notch toughness levels in the weld deposits, allowing for the use of filler metals that have no minimum specified requirements. For connections that are subject to inelastic loading, it seems apparent that minimum levels of notch toughness must be specified. The actual limits on notch toughness have not been experimentally determined. With the AWS filler metal classifications in effect in 1996, they are either classified as having no minimum specified notch toughness, or with properties of 20 ft-lb at a temperature of 0°F or lower. As an interim specification, 20 ft-lb at 0°F or lower has been recommended. However, this has been based upon availability, not on an analysis of actual requirements. It is expected that actual requirements will be less demanding, and once these requirements are determined, new filler metals will be developed that will meet the appropriate requirements. It should be recognized that the more demanding notch toughness requirements impose several undesirable consequences upon fabrication, including increased cost of materials, lower rates of fabrication (deposition rates), less operator appeal, and greater difficulty in obtaining sound weld deposits. Therefore, ultraconservative requirements imposed "just to be safe" may be practically and economically unacceptable. Research will be conducted to determine actual toughness requirements.

B.5.3 Heat-Affected Zones

As illustrated in Fig. B.14, the base metal heat-affected zones represent material that may affect connection performance as well. The heat-affected zone (HAZ) is defined as that base metal which has been thermally changed due to the energy introduced into it by the welding process. In the small region immediately adjacent to the weld, the base metal has gone through a different thermal history than the rest of the base material. For most hot-rolled steels, the area of concern is a HAZ that is cooled too rapidly, resulting in a hardened heat-affected zone. For quenched-and-tempered steels, the HAZ may be cooled too slowly, resulting in a softening of the area. In columns, the HAZ of interest is the Z direction properties immediately adjacent to the groove weld. For the beam, these are oriented in the X direction.

Heat-affected zone properties are a function of base metal chemistry and specific welding procedures. Steel makers consider HAZ properties when developing a specific steel composition, and for quenched-and-tempered steels, guidelines are available from the steel producer indicating what precautions must be taken during welding to preclude the formation of undesirable heat-affected zones. The primary welding variables that affect HAZ properties are the preheat and interpass temperatures (both minimum and maximum), and the heat input of welding. Excessively high heat input can negatively affect HAZ properties by causing softening in these areas. Excessively low heat input can result in hardening of the heat-affected zones.

Weld metal properties may be negatively affected by extremely high heat input welding procedures, causing a decrease in both the yield strength and tensile strength, as well as the notch toughness of the weld deposit. Excessively low heat input may result in high-strength weld metal and may also decrease the notch toughness of the weld deposit. Optimum mechanical properties are generally obtained from both the weld metal and the HAZ if the heat input is maintained in the 30- to 80-kJ/in range.

Post-Northridge evaluation of fractured connections has revealed that excessively high heat input welding procedures were commonly used, confirmed by the presence of very large weld beads that, in some cases, exceeded the maximum limits prescribed by the D1.1 code. These may have had some corollary effects on weld metal and HAZ properties.

B.5.4 Material Properties and Connection Design

While a welded structure acts as a one-piece unit, the material properties are not isotropic throughout all zones. When high demands are placed upon connections, each series of material properties must behave as expected.

One approach to obtaining acceptable connections is rigorous control of each of the material properties. Evidence exists that this can be done, even under very demanding conditions that simulate earthquake loading. There is an acceptable approach, however, that relies less on rigorous control of the mechanical properties, and compensates for these concerns by geometrically changing the connection. This alternative is discussed in Sec. B.8.2.

B.6 Workmanship

For severely loaded connections, good workmanship is a key contributor to acceptable performance. In welded construction, the performance of the structural system is often dependent on the ability of skilled welders to deposit sound weld metal. As the level of loading increases, dependence upon high-quality fabrication increases.

B.6.1 The Role of Codes

Design and fabrication specifications such as the AISC *Manual of Steel Construction* and the AWS D1.1 Structural Welding Code—Steel are typical vehicles that communicate minimum acceptable practices. It is impossible for any code to be so inclusive as to cover every situation that will ever be contemplated. These codes and specifications address minimum acceptable standards, relying upon the engineer to specify any additional requirements that supersede these minimum levels.

The D1.1 code does not specifically address seismic issues but does establish a minimum level of quality that must be achieved in seismic applications. Additional requirements are probably warranted. These would include requirements for nondestructive testing and notch tough weld deposits, and additional requirements for in-process verification inspection.

B.6.2 Purpose of WPS

Within the welding industry, the term "welding procedure specification," or simply "WPS," is used to signify the combination of variables that are to be used to make a particular weld. The WPS is somewhat analogous to a cook's recipe. It outlines the steps required to make a good-quality weld under specific conditions. It is the primary method to ensure the use of welding variables essential to weld quality. Also, it permits inspectors and supervisors to verify that the actual welding is performed in conformance with the constraints of the WPS.

Prior to the start of welding on a project, WPSs typically are submitted to the engineer for review. Many engineers with limited understanding of welding will delegate this responsibility to other individuals, often the owner's supplied inspection firm, to verify the suitability of the particular parameters involved. For critical projects, the services of welding engineers may be appropriate. Most importantly, WPSs are not simply pieces of documentation to be filed away—they are intended to be communication tools for maintenance of weld quality. It is essential that all parties involved with the fabrication sequence have access to these documents to ensure conformance to their requirements.

Special Welding Issues for Seismically Resistant Structures ■ B-27

B.6.3 Effect of Welding Variables

A variety of welding variables determine the quality of the deposited weld metal. These variables are a function of the particular welding process being used, but the general trends outlined below are applicable to all welding processes.

Amperage is a measure of the amount of current flowing through the electrode and the work. It is a primary variable in determining heat input. An increase in amperage generally means higher deposition rates, deeper penetration, and more melting of base metal. The role of amperage is best understood in the context of heat input and current density, which are described below.

Arc voltage is directly related to arc length. As the voltage increases, the arc length increases. Excessively high voltages may lead to weld metal porosity, while extremely low voltages will result in poor weld bead shapes. In an electric circuit, the voltage is not constant but is composed of a series of voltage drops. For this reason, it is important to monitor voltage near the arc.

Travel speed is the rate at which the electrode is moved relative to the joint. All other variables being equal, travel speed has an inverse effect on the size of weld beads. Travel speed is a key variable used in determining heat input.

Polarity is a definition of the direction of current flow. Positive polarity (or reverse) is achieved when the electrode lead is connected to the positive terminal of the direct-current power supply. The work lead would be connected to the negative terminal. Negative polarity (or straight) occurs when the electrode is connected to the negative terminal. For most welding processes, the required electrode polarity is a function of the design of the electrode. For submerged arc welding, either polarity could be utilized.

Heat input is generally expressed by the equation

$$H = \frac{60EI}{1000S}$$

where E represents voltage, I is current, and S is the travel speed in inches per minute. The resultant computation is measured in kilojoules per inch. The heat input of welding is also directly related to the cross-sectional area of the weld bead. High heat input welding is automatically associated with the deposition of large weld passes. The AWS D1.1 code does not specify heat input limits but does require that weld bead shapes meet prescribed requirements of maximum height and widths. This has an indirect effect of limiting heat input.

Current density is determined by dividing the welding amperage by the cross-sectional area of the electrode. The current density is therefore proportional to I/d^2. As the current density increases, there will be an increase in deposition rates as well as penetration.

Preheat and interpass temperatures are used to control cracking tendencies, typically in the base material. Excessively high preheat and interpass temperatures will reduce the yield and tensile strength of the weld metal as well as the toughness. When base metals receive little or no preheat, the resultant rapid cooling can promote cracking as well as excessively high yield and tensile properties in the weld metal, and a corresponding reduction in toughness and elongation.

All of the preceding variables are defined and controlled by the welding procedure specification. Conformance to these requirements is particularly sensitive for critical fabrication such as seismically loaded structures, because of the high demand placed upon welded connections under these situations.

B.6.4 Fit-Up

Fit-up is the term that defines the orientation of the various pieces prior to welding. The AWS D1.1 code has specific tolerances that are applied to the as-fit dimensions of a joint prior to welding. It is critical that there is ample access to the root of the joint to ensure good, uniform fusion between the members being joined. Excessively small root openings or included angles in groove welds do not permit uniform fusion. Excessively large root openings or included angles result in the need for greater volumes of weld metal, with their corresponding increases in shrinkage stresses. This in turn increases distortion and cracking tendencies. The D1.1 tolerances for fit-up are generally measured in $1/16$-in increments. As compared to the overall project, this is a very tight dimension. Nevertheless, as it affects the root opening condition, it is critical in order to avoid lack of fusion, slag inclusions, and other unacceptable root conditions.

B.6.5 Field vs. Shop Welding

Many individuals believe that the highest-quality welding naturally is obtained under shop welding conditions. While some aspects of field welding are more demanding than shop welding situations, the greatest differences are not technical but rather are related to control. For shop fabrication, the work force is generally more stable. Supervision practices and approaches are well understood. Communication with the various parties involved is generally more efficient. Under field welding conditions, maintaining and controlling a project seems to be more difficult. There are environmental challenges to field conditions, including temperature, wind, and moisture. However, those issues seem to pose less of a problem than do the management-oriented issues.

Generally, gasless welding processes, such as self-shielded flux cored welding, and shielded metal arc welding are preferred for field welding. Gas metal arc, gas tungsten arc, and gas-shielded flux cored arc welding are all limited due to their sensitivity to wind-related gas disturbances. Regarding managerial issues, it is imperative that field welding conditions receive an appropriate increase in the monitoring and control area to ensure consistent quality. The AWS D1.1 code imposes the same requirements on field welding as on shop welding. This includes qualification of welders, the use of welding procedures, and the resultant quality requirements.

B.7 Inspection

To ensure weld quality, a variety of inspection activities is employed. The AWS D1.1 code requires that all welds be inspected, specifically by means of visual inspection. In addition, at the engineer's discretion and as identified in contract

Special Welding Issues for Seismically Resistant Structures ■ B-29

documents, nondestructive testing may be required for finished weldments. This enables the engineer with a knowledge of the complexity of the project to specify additional inspection methodologies commensurate with the degree of confidence required for a particular project. For seismically loaded structures, and connections subject to high stress levels, the need for inspection increases.

B.7.1 In-Process Visual Inspection

The D1.1 code mandates the use of in-process visual inspection. This activity encompasses those operations performed before, during, and after welding that are used to ensure weld quality. Before start-up, the inspector reviews welder qualification records, welding procedure specifications, and the contract documents to confirm that applicable requirements are met. Before welding is performed, the inspector verifies fit-up and joint cleanliness, examines the welding equipment to ensure it is in proper working order, verifies that the materials involved meet the various requirements, and confirms that the required levels of preheat have been properly applied. During welding, the inspector confirms that appropriate WPS parameters are being achieved and that the intermediate weld passes meet the various requirements. After welding is completed, final bead shapes and welding integrity can be visually confirmed. In spite of its apparent simplicity, effective visual inspection is a critical component for ensuring consistent weld quality.

B.7.2 Nondestructive Testing

There are four major nondestructive testing methods that may be employed to verify weld integrity after the welding operations are completed. None is a substitute for effective visual inspection as outlined above. No process is completely capable of detecting all discontinuities in a weld. The advantages and limitations of each method must be clearly understood in order to ensure an appropriate level of confidence is placed in the results obtained.

Dye penetrant inspection (PT) involves the application of a liquid which is drawn into a surface-breaking discontinuity, such as a crack or porosity, by capillary action. When the excess residual dye is removed from the surface, a developer is applied which will absorb the penetrant that is contained within the discontinuity. The result is a stain in the developer that shows that a discontinuity is present. Dye penetrant testing is limited to surface-breaking discontinuities. It has no ability to read subsurface discontinuities, but it is highly effective in accenting the discontinuities that may be overlooked, or may be too small to detect, by visual inspection.

Magnetic particle inspection (MT) utilizes the change in magnetic flux that occurs when a magnetic field is present in the vicinity of a discontinuity. The change in magnetic flux density will show up as a different pattern when magnetic dustlike particles are applied to the surface of the part. The process is highly effective in locating discontinuities that are surface-breaking or slightly subsurface. The magnetic field can be created in the material in one of two ways: the current is either directly passed through the material, or the magnetic field is induced through a coil on a yoke. The process is most sensitive to discontinuities that lie perpendicular to the magnetic flux path, so it is necessary to energize the part being inspected in two directions in order to fully inspect the component.

Ultrasonic inspection (UT) relies on the transmission of high-frequency sound waves through materials. Solid discontinuity-free materials will transmit the sound throughout the part in an uninterrupted manner. A receiver "hears" the sound reflected off of the back surface of the part being inspected. If a discontinuity is contained between the transmitter and the back of the part, an intermediate signal will be sent to the receiver, indicating the presence of a discontinuity. The pulses are read on a CRT screen. The magnitude of the signal received from the discontinuity indicates its size. The relationship of the signal with respect to the back wall is indicative of its location. UT is most sensitive to planar discontinuities, i.e., cracks. UT effectiveness is dependent upon the operator's skill, so UT technician training and certification is essential. With the available technology today, UT is capable of reading a variety of discontinuities that would be acceptable for many applications. It is important that the acceptance criteria be clearly communicated to the inspection technicians so unnecessary repairs are avoided.

Radiographic inspection (RT) uses x-rays or gamma rays that are passed through the weld to expose a photographic film on the opposite side of the joint. X-rays are produced by high-voltage generators while gamma rays are produced by atomic disintegration of radioisotopes. Whenever radiographic inspection is employed, precautions must be taken to protect workers from exposure to excessive radiation. RT relies on the ability of the material to pass some of the radiation through, while absorbing part of this energy within the material. The absorption rate is a function of the material. As the different levels of radiation are passed through the material, portions of the film are exposed to a greater or lesser degree than the rest. When this film is developed, the resulting radiograph will bear the image of the cross section of the part. The radiograph is actually a negative. The darkest regions are those that were most exposed when the material being inspected absorbed the least amount of radiation. Porosity will be revealed as small dark round circles. Slag is generally dark and will look similar to porosity but will have irregular shapes. Cracks appear as dark lines. Excessive reinforcement will result in a light region.

Radiographic inspection is most effective for detecting volumetric discontinuities such as slag or porosity. When cracks are oriented perpendicular to the direction of a radiographic source, they may be missed with the RT method. Radiographic testing has the advantage of generating a permanent record for future reference. Effective interpretation of a radiograph and its implications requires appropriate training. Radiographic inspection is most appropriate for butt joints and is generally not appropriate for inspection of corner or T joints.

B.7.3 Applications for NDT Methods

Visual inspection should be appropriately applied on every welding project. It is the most comprehensive method available to verify conformance with the wide variety of issues that can affect weld quality. In addition to visual inspection, non-destructive testing can be specified to verify the integrity of the deposited weld metal. The selection of the inspection method should reflect the probable discontinuities that would be encountered and the consequences of undetected discontinuities. Consideration must be made to the conditions under which the inspection would be performed such as field vs. shop conditions. The nature of the joint detail (butt, T, corner, etc.) and the weld type (CJP, PJP, fillet weld) will determine the applicability of the inspection process in many situations. Magnetic particle inspection is generally preferred over dye penetrant inspection because of its relative simplicity. Cleanup is easy, and the sensitivity of the process is good. PT is normally reserved for applications where the material is nonmagnetic and MT would not be applicable. While MT is suitable for surface or slightly subsurface discontinuity detection only, it is in these areas that many welding defects can be located. It is very effective in crack detection and can be utilized to ensure complete crack removal before subsequent welding is performed on damaged structures.

Ultrasonic inspection has become the primary nondestructive testing method used for most building applications. It can be utilized to inspect butt, T, and corner joints, is relatively portable, and is free from the radiation concerns associated with RT inspection. UT is particularly sensitive to the identification of cracks, the most significant defect in a structural system. While it may not detect spherical or cylindrical voids such as porosity, the consequences of nondetection of these types of discontinuities are less significant.

B.8 Post-Northridge Details

Prior to the Northridge earthquake, the special moment-resisting frame (SMRF) with the "pre-Northridge" beam-to-column detail was unchallenged with respect to its ability to perform as expected. This confidence existed in spite of a fairly significant failure rate that had been experienced when testing these connections in previous research. For purposes of this appendix, the "pre-Northridge" detail is considered to exhibit the following:

- CJP groove welds of the beam flanges to the column face, with weld backing left in place and with weld tabs left in place

- No specific requirement for minimum notch toughness properties in the weld deposit

- A bolted web connection with or without supplemental fillet welds of the shear tab to the beam web

- Standard ASTM A36 steel for the beam, and ASTM 572 grade 50 for the column, e.g., no specific limits on yield strength or the F_y/F_u ratio.

As a result of the Northridge earthquake and research performed immediately thereafter, confidence in this detail has been severely shaken. Whether this detail, or a variation thereof, will be suitable for use in the future is unknown at the time of writing. More research must be performed, but we can speculate that, with the possible exception of small-sized members, some modification will be required in order to gain the expected performance from structural systems utilizing this detail.

As was previously stated, testing of this configuration had a fairly high failure rate in pre-Northridge tests. Still, many successful results were obtained. Further research will determine which variables are the most significant in predicting performance success. Some changes, however, have taken place in materials and design practice that should be considered. In recent years, recycling of steel has become a more predominant method of manufacture. Not only is this environmentally responsible, it is economical. However, in the process, residual alloys can accumulate in the scrap charge, inadvertently increasing steel strength levels. In the past 20 years, for example, the average yield strength of ASTM A36 steel has increased approximately 15 percent. Testing done with lower-yield-strength steel would be expected to exhibit different behavior than test specimens made of today's higher-strength steels (in spite of the same ASTM designation).

For practical reasons, laboratory specimens tend to be small in size. Success in small-sized specimens was extrapolated to apply to very large connection assemblies in actual structures. The design philosophy that led to a reduction in the number of special moment-resisting frames throughout a structure necessitated that each of the remaining frames be larger in size. This corresponded to heavier and deeper beams, and much heavier columns, with an increase in the size of the weld between the two rolled members. The effect of size on restraint and triaxial stresses was not evaluated in the laboratory, resulting in some new discoveries about the behavior of the large-sized assemblages during the Northridge earthquake.

There is general agreement throughout the engineering community that the pre-Northridge connection (as defined above) is no longer adequate and some modification will be required. Any deviation from the definition above constitutes a modification for the purposes of this discussion.

B.8.1 Minor Modifications to the SMRF Connection

With the benefit of hindsight, several aspects of the pre-Northridge connection detail seem to be obviously deficient. Weld backing left in place in a connection subject to both positive and negative moments where the root of the flange weld can be put into tension is an obvious prescription for high-stress concentrations that may result in cracking. Failure to specify minimum toughness levels for weld metal for heavily loaded connections is another deficiency. The superior performance of the all-welded web vs. the bolted web in past testing draws into question the assumption of load sharing between welds and bolts. Tighter control of the strength properties of the beam steel and the relationship to the column also seem to be obvious requirements.

The amount of testing that controls each of these variables has been limited to date. Some preliminary results suggest that tightly controlling all these variables will result in acceptable performance. At the time of writing, however, the authors know of no test of unmodified beam-to-column connections where the connection zone has remained crack-free when acceptable rotation limits were achieved. It is speculated that for smaller-sized members, this approach may be technically possible, although the degree of control necessary on both the material properties and the welding operations may not be practical.

Special Welding Issues for Seismically Resistant Structures ■ B-33

B.8.2 Cover-Plated Designs

This concept uses short cover plates that are added to the top and bottom flanges of the beam. Fillet welds transfer the cover-plate forces to the beam flanges. The bottom flange cover plate is shop-welded to the column flange, and the bottom beam flange is field-welded to the column flange and to the cover plate. Both the top flange and the top flange cover plate are field-welded to the column flange with a common weld. The web connection may be welded or high-strength bolted. Limited testing of these connections has been done, with generally favorable results.

The cover-plate approach has received significant attention after Northridge because it offered early promise of a viable solution. Other methods may emerge as superior as time progresses. While the cover-plate solution treats the beam the same as other approaches (that is, it moves the plastic hinge into a region where ductility can be demonstrated), it concentrates all the loading to the column into a relatively short distance. Other alternatives may treat the column in a more gentle manner.

B.8.3 Flange Rib Connections

This concept utilizes one or two vertical ribs attached between the beam flanges and column face. In a flange rib connection, the intent of the rib plates is to reduce the demand on the weld at the column flange and to shift the plastic hinge from the column face. In limited testing, flange rib connections have demonstrated acceptable levels of plastic rotation provided that the girder flange welding is correctly done.

Vertical ribs appear to function very similarly to the cover-plated designs but offer the additional advantage of spreading that load over a greater portion of that column. The single rib designs appear to be superior to the twin rib approaches because the stiffening device is in alignment with the column web (for I-shaped columns) and facilitates easy access to either side of the device for welding. It is doubtful that the single rib would be appropriate for box column applications.

B.8.4 Top and Bottom Haunch Connections

In this configuration, haunches are placed on both the top and bottom flanges. In two tests of the top and bottom haunch connection, it has exhibited extremely ductile behavior, achieving plastic rotations as great as 0.07 radian. Tests of single, haunched beam-column connections have not been as conclusive; further tests of such configurations are planned.

Haunches appear to be the most straightforward approach to obtaining the desired behavior out of the connection, albeit at a fairly significant cost. The treatment to the column is particularly desirable, greatly increasing the portion of the column participating in the transfer of moment. Significant experience was gained utilizing the haunches for the repairs of the SAC-sponsored tests.

B.8.5 Reduced Beam Section Connections

In this configuration, the cross section of the beam is intentionally reduced within a segment to produce a deliberate plastic hinge within the beam span, away

B-34 ■ Appendix B

from the column face. One variant of this approach produces the so-called dog-bone profile.

Reduced section details offer the prospect of a low-cost connection and increased performance out of detailing that is very similar to the pre-Northridge connection. Control of material properties of the beam will still be a major variable if this detail is used. Lateral bracing will probably be required in the area of the reduced section to prevent buckling, particularly at the bottom flange when loaded in compression.

B.8.6 Partially Restrained Connections

Several engineers and researchers have suggested that partially restrained connection details will offer a performance advantage over the special moment-resisting frames. The relative merits of a partially restrained frame vs. a rigid frame are beyond the scope of this appendix. However, many engineers immediately think of bolted PR connections when it is possible to utilize welded connections for PR performance as well.

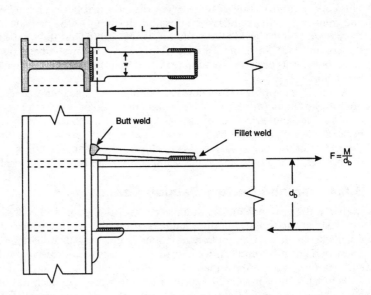

Fig. B.15 Examples of partially restrained connection details. (*Courtesy of The Lincoln Electric Company.*)

Illustrated in Fig. B.15 are a variety of details that can be employed utilizing the PR concept. Detailing rules must be developed, and testing should be performed, before these details are employed. They are supplied to offer welded alternatives to bolted PR connection concepts.

Appendix C
Load Factor Design Selection Table

For shapes used as beams
$\phi_b = 0.90$

	$F_y = 36$ ksi						$F_y = 50$ ksi				
BF, kips	L_r, ft	L_p, ft	$\phi_b M_r$, kip-ft	$\phi_b M_p$, kip-ft	Z_x, in³	Shape	$\phi_b M_p$, kip-ft	$\phi_b M_r$, kip-ft	L_p, ft	L_r, ft	BF, kips
32.5	62.4	16.1	2570	4080	1510	W 36×359	5660	3960	13.7	44.0	56.2
29.4	64.4	15.6	2400	3830	1420	W 33×354	5330	3690	13.2	44.7	51.9
32.2	58.5	16.0	2360	3730	1380	W 36×328	5180	3630	13.6	41.7	54.9
28.9	59.3	15.5	2160	3430	1270	W 33×318	4760	3330	13.1	41.8	49.9
31.6	55.1	16.0	2170	3400	1260	W 36×300	4730	3330	13.5	39.9	52.9
31.0	53.0	15.9	2010	3160	1170	W 36×280	4390	3090	13.5	38.8	51.3
28.2	55.7	15.4	1970	3110	1150	W 33×291	4310	3030	13.0	39.8	48.0
30.3	50.6	15.8	1860	2920	1080	W 36×260	4050	2860	13.4	37.5	49.5
37.0	39.7	11.0	1750	2810	1040	W 36×256	3900	2685	9.4	28.8	62.7
27.7	52.0	15.3	1790	2810	1040	W 33×263	3900	2750	12.9	37.8	46.3
29.6	48.8	15.6	1750	2730	1010	W 36×245	3790	2690	13.3	36.4	47.7

Source: American Institute of Steel Construction.

\multicolumn{5}{c}{$F_y = 36$ ksi}		\multicolumn{5}{c}{$F_y = 50$ ksi}
BF, kips	L_r, ft	L_p, ft	$\phi_b M_r$, kip-ft	$\phi_b M_p$, kip-ft	Z_x, in³	Shape	$\phi_b M_p$, kip-ft	$\phi_b M_r$, kip-ft	L_p, ft	L_r, ft	BF, kips
28.7	47.3	15.5	1630	2550	943	W 36×230	3540	2510	13.2	35.6	45.8
27.0	49.2	15.1	1620	2540	939	W 33×241	3520	2490	12.8	36.2	44.2
36.1	37.2	10.9	1580	2530	936	W 36×232	3510	2430	9.3	27.3	59.9
26.0	46.9	15.0	1480	2310	855	W 33×221	3210	2270	12.7	35.0	41.9
24.4	48.6	14.7	1450	2280	845	W 30×235	3170	2240	12.4	33.9	43.5
34.9	35.0	10.8	1400	2250	833	W 36×210	3120	2160	9.1	26.1	56.8
25.0	44.8	14.8	1330	2080	772	W 33×201	2900	2050	12.6	33.8	39.7
34.0	33.5	10.7	1290	2070	767	W 36×194	2880	1990	9.1	25.3	54.6
21.8	47.9	14.5	1290	2020	749	W 30×211	2810	1990	12.3	35.1	36.0
32.7	32.8	10.6	1210	1940	718	W 36×182	2690	1870	9.0	24.9	52.0
18.2	52.0	13.8	1220	1910	708	W 27×217	2660	1870	11.7	36.8	31.3
21.0	45.4	14.4	1170	1820	673	W 30×191	2520	1790	12.2	33.7	33.9
31.5	31.9	10.5	1130	1800	668	W 36×170	2510	1740	8.9	24.4	49.6
28.3	32.6	10.4	1070	1700	629	W 33×169	2360	1650	8.8	24.5	45.4
17.8	48.0	13.7	1080	1700	628	W 27×194	2360	1670	11.6	34.6	30.0
30.7	30.9	10.4	1060	1680	624	W 36×160	2340	1630	8.8	23.7	48.0
20.2	43.2	14.3	1050	1630	605	W 30×173	2270	1620	12.1	32.5	32.0
29.4	30.2	10.3	983	1570	581	W 36×150	2180	1510	8.7	23.4	45.6
17.5	45.2	13.6	979	1530	567	W 27×128	2130	1510	11.5	33.1	28.8
26.8	31.2	10.3	950	1510	559	W 33×152	2100	1460	8.7	23.7	42.3
25.7	30.1	10.1	874	1390	514	W 33×141	1930	1340	8.6	23.1	40.2
16.9	42.8	13.5	887	1380	512	W 27×161	1920	1370	11.5	31.7	27.4
14.3	47.8	12.7	878	1380	511	W 24×176	1920	1350	10.7	33.8	24.6
27.5	28.8	9.9	856	1370	509	W 36×135	1910	1320	8.4	22.4	42.2
23.7	30.6	9.5	850	1350	500	W 30×148	1880	1310	8.1	22.8	38.6
14.1	45.2	12.7	807	1260	468	W 24×162	1760	1240	10.8	32.4	23.8
24.5	29.1	10.0	792	1260	467	W 33×130	1750	1220	8.5	22.5	37.9
16.2	40.7	13.4	801	1240	461	W 27×146	1730	1230	11.3	30.6	25.8
22.4	29.0	9.4	741	1180	437	W 30×132	1640	1140	8.0	21.9	35.6
10.8	51.7	12.4	741	1170	432	W 21×166	1620	1140	10.5	35.7	19.1
13.8	42.0	12.5	723	1130	418	W 24×146	1570	1110	10.6	30.6	22.8
23.1	27.8	9.7	700	1120	415	W 33×118	1560	1080	8.2	21.7	35.5
21.6	28.2	9.3	692	1100	408	W 30×124	1530	1070	7.9	21.5	34.1
18.9	30.0	9.2	673	1070	395	W 27×129	1480	1040	7.8	22.3	30.9
21.1	27.1	9.1	642	1020	378	W 30×116	1420	987	7.7	20.8	33.0

Load Factor Design Selection Table ■ C-3

$F_y = 36$ ksi						$F_y = 50$ ksi				
BF, kips	L_{rr} ft	L_{p}, ft	$\phi_b M_{rr}$ kip-ft	$\phi_b M_{p}$, kip-ft	Z_x in³	Shape	$\phi_b M_{p}$, kip-ft	$\phi_b M_{rr}$ kip-ft	L_{p}, ft	L_{rr} ft	BF, kips
10.7	46.4	12.3	642	1010	373	W 21×147	1400	987	10.4	32.8	18.4
13.3	39.3	12.4	642	999	370	W 24×131	1390	987	10.5	29.1	21.5
20.2	26.3	9.0	583	934	346	W 30×108	1300	897	7.6	20.3	31.5
18.0	28.2	9.1	583	926	343	W 27×114	1290	897	7.7	21.3	28.7
10.5	43.1	12.2	575	899	333	W 21×132	1250	885	10.4	30.9	17.7
12.7	37.1	12.3	567	883	327	W 24×117	1230	873	10.4	27.9	20.2
7.82	52.2	11.3	550	869	322	W 18×143	1210	846	9.6	35.5	14.0
19.0	25.5	8.8	525	842	312	W 30×99	1170	807	7.4	19.8	29.2
10.3	41.0	12.2	532	829	307	W 21×122	1150	819	10.3	29.8	17.1
17.0	26.8	9.0	521	824	305	W 27×102	1140	801	7.6	20.4	26.7
12.0	35.2	12.1	503	780	289	W 24×104	1080	774	10.3	26.8	18.8
7.78	48.0	11.3	499	786	291	W 18×130	1090	768	9.5	33.0	13.8
14.8	27.1	8.3	478	756	280	W 24×103	1050	735	7.0	20.1	24.1
10.1	38.7	12.1	486	753	279	W 21×111	1050	747	10.3	28.5	16.4
16.2	25.9	8.8	474	751	278	W 27×94	1040	729	7.5	19.9	25.2
7.71	44.1	11.2	450	705	261	W 18×119	979	693	9.5	30.8	13.4
14.3	25.9	8.3	433	686	254	W 24×94	953	666	7.0	19.4	23.0
9.62	37.1	12.0	443	683	253	W 21×101	949	681	10.2	27.6	15.4
15.0	24.9	8.6	415	659	244	W 27×84	915	639	7.3	19.3	23.0
7.61	40.4	11.1	398	621	230	W 18×106	863	612	9.4	28.7	13.0
13.6	24.5	8.1	382	605	224	W 24×84	840	588	6.9	18.6	21.5
11.8	26.6	7.7	374	597	221	W 21×93	829	576	6.5	19.4	19.6
3.86	67.9	15.6	371	572	212	W 14×120	795	570	13.2	46.2	6.82
7.51	38.1	11.0	367	570	211	W 18×97	791	564	9.4	27.4	12.6
12.7	23.4	8.0	343	540	200	W 24×76	750	528	6.8	18.0	19.8
6.08	42.1	10.5	341	535	198	W 16×100	743	525	8.9	29.3	10.7
11.3	24.9	7.6	333	529	196	W 21×83	735	513	6.5	18.5	18.5
3.84	62.7	15.5	337	518	192	W 14×109	720	519	13.2	43.2	6.70
7.28	35.5	11.0	324	502	186	W 18×86	698	498	9.3	26.1	11.9
2.95	75.5	13.0	318	502	186	W 12×120	698	489	11.1	50.0	5.36
12.1	22.4	7.8	300	478	177	W 24×68	664	462	6.6	17.4	18.7
6.05	38.6	10.4	302	473	175	W 16×89	656	465	8.8	27.3	10.3
3.77	58.2	15.5	306	467	173	W 14×99*	647	471	13.4	40.6	6.46
10.7	23.5	7.5	294	464	172	W 21×73	645	453	6.4	17.7	17.0
2.95	67.2	13.0	283	443	164	W 12×106	615	435	11.0	44.9	5.32
6.95	33.3	10.9	285	440	163	W 18×76	611	438	9.2	24.8	11.1

*Indicates noncompact shape; $F_y = 50$ ksi.

C-4 ■ Appendix C

$F_y = 36$ ksi						$F_y = 50$ ksi				
BF, kips	L_r, ft	L_p, ft	$\phi_b M_r$, kip-ft	$\phi_b M_p$, kip-ft	Z_x, in³	Shape	$\phi_b M_p$, kip-ft	$\phi_b M_r$, kip-ft	L_p, ft	L_r, ft	BF, kips
10.4	22.8	7.5	273	432	160	W 21×68	600	420	6.4	17.3	16.5
3.74	54.1	15.4	279	424	157	W 14×90*	577	429	15.0	38.4	6.31
13.8	17.2	5.8	255	413	153	W 24×62	574	393	4.9	13.3	21.4
5.85	34.9	10.3	261	405	150	W 16×77	563	402	8.7	25.2	9.75
2.91	61.4	12.9	255	397	147	W 12×96	551	393	10.9	41.3	5.20
8.29	24.4	7.1	248	392	145	W 18×71	544	381	6.0	17.8	13.8
9.80	21.7	7.4	248	389	144	W 21×62	540	381	6.3	16.6	15.3
4.15	43.0	10.3	240	375	139	W 14×82	521	369	8.8	29.6	7.31
12.7	16.6	5.6	222	362	134	W 24×55	503	342	4.7	12.9	19.6
8.09	23.2	7.0	228	359	133	W 18×65	499	351	6.0	17.1	13.3
2.90	56.4	12.8	230	356	132	W 12×87	495	354	10.9	38.4	5.12
5.57	32.3	10.3	228	351	130	W 16×67	488	351	8.7	23.8	9.02
11.3	17.3	5.6	216	348	129	W 21×57	484	333	4.8	13.1	18.0
4.10	40.0	10.3	218	340	126	W 14×74	472	336	8.8	28.0	7.12
7.90	22.4	7.0	211	332	123	W 18×60	461	324	6.0	16.7	12.8
2.88	51.8	12.7	209	321	119	W 12×79	446	321	10.8	35.7	5.03
4.05	37.3	10.3	201	311	115	W 14×68	431	309	8.7	26.4	6.91
7.63	21.4	7.0	192	302	112	W 18×55	420	295	5.9	16.1	12.2
10.5	16.2	5.4	184	297	110	W 21×50	413	283	4.6	12.5	16.4
2.86	48.2	12.7	190	292	108	W 12×72	405	292	10.7	33.6	4.93
6.42	22.8	6.7	180	284	105	W 16×57	394	277	5.7	16.6	10.7
3.90	34.7	10.2	180	275	102	W 14×61	383	277	8.7	24.9	6.51
7.29	20.5	6.9	173	273	101	W 18×50	379	267	5.8	15.6	11.5
2.80	44.7	12.6	171	261	96.8	W 12×65*	358	264	11.8	31.7	4.72
9.66	15.4	5.3	159	258	95.4	W 21×44	358	245	4.5	12.0	14.9
6.20	21.2	6.6	158	248	92.0	W 16×50	345	243	5.6	15.8	10.1
8.12	16.6	5.4	154	245	90.7	W 18×46	340	236	4.6	12.6	13.0
4.17	28.0	8.0	152	235	87.1	W 14×53	327	233	6.8	20.1	7.02
2.91	38.4	10.5	152	233	86.4	W 12×58	324	234	8.9	27.0	4.96
5.92	20.1	6.5	142	222	82.3	W 16×45	309	218	5.6	15.1	9.43
7.52	15.7	5.3	133	212	78.4	W 18×40	294	205	4.5	12.1	11.7
4.05	26.3	8.0	137	212	78.4	W 14×48	294	211	6.8	19.2	6.70
2.85	35.8	10.3	138	210	77.9	W 12×53	292	212	8.8	25.6	4.77
1.91	48.1	10.7	130	201	74.6	W 10×60	280	200	9.1	32.6	3.38
5.53	19.3	6.5	126	197	72.9	W 16×40	273	194	5.6	14.7	8.67

*Indicate noncompact shape; $F_y = 50$ ksi.

Load Factor Design Selection Table ■ C-5

$F_y = 36$ ksi						$F_y = 50$ ksi				
BF, kips	L_r, ft	L_p, ft	$\phi_b M_r$, kip-ft	$\phi_b M_p$, kip-ft	Z_x, in^3	Shape	$\phi_b M_p$, kip-ft	$\phi_b M_r$, kip-ft	L_p, ft	L_r, ft	BF, kips
3.06	30.8	8.2	126	195	72.4	W 12×50	271	194	6.9	21.7	5.25
3.91	24.7	7.9	122	188	69.6	W 14×43	261	188	6.7	18.2	6.32
1.89	43.9	10.7	117	180	66.6	W 10×54	250	180	9.1	30.2	3.30
6.93	14.8	5.1	112	180	66.5	W 18×35	249	173	4.3	11.5	10.7
3.02	28.5	8.1	113	175	64.7	W 12×45	243	174	6.9	20.3	5.07
5.25	18.3	6.3	110	173	64.0	W 16×36	240	170	5.4	14.1	8.08
4.39	20.0	6.5	106	166	61.5	W 14×38	231	164	5.5	14.9	7.07
1.88	40.7	10.6	106	163	60.4	W 10×49	227	164	9.0	28.3	3.25
2.92	26.5	8.0	101	155	57.5	W 12×40	216	156	6.8	19.3	4.82
4.19	19.0	6.4	94.8	147	54.6	W 14×34	205	146	5.4	14.4	6.58
5.70	14.3	4.9	92.0	146	54.0	W 16×31	203	142	4.1	11.0	8.85
3.46	20.6	6.4	88.9	138	51.2	W 12×35	192	137	5.4	15.2	5.67
3.92	17.9	6.2	81.9	128	47.3	W 14×30	177	126	5.3	13.7	6.06
1.93	31.2	8.3	82.1	126	46.8	W 10×39	176	126	7.0	21.8	3.32
5.12	13.3	4.7	74.9	119	44.2	W 16×26	166	115	4.0	10.4	7.88
3.23	19.1	6.3	75.3	116	43.1	W 12×30	162	116	5.4	14.4	5.10
4.44	13.4	4.5	68.8	109	40.2	W 14×26	151	106	3.8	10.3	6.96
1.89	27.4	8.1	68.3	105	38.8	W 10×33	145	105	6.9	19.7	3.15
3.00	18.1	6.3	65.1	100	37.2	W 12×26	140	100	5.3	13.8	4.64
2.44	20.3	5.7	63.2	98.8	36.6	W 10×30	137	97.2	4.8	14.5	4.13
1.23	35.1	8.5	60.8	93.7	34.7	W 8×35	130	93.6	7.2	24.1	2.16
4.06	12.5	4.3	56.5	89.6	33.2	W 14×22	125	87.0	3.7	9.7	6.26
2.34	18.5	5.7	54.4	84.5	31.3	W 10×26	117	83.7	4.8	13.5	3.85
1.20	32.0	8.4	53.6	82.1	30.4	W 8×31	114	82.5	7.1	22.3	2.07
3.88	11.1	3.5	49.5	79.1	29.3	W 12×22	110	76.2	3.0	8.4	6.24
1.27	27.3	6.8	47.4	73.4	27.2	W 8×28	102	72.9	5.7	18.9	2.22
2.18	16.9	5.5	45.2	70.2	26.0	W 10×22	97.5	69.6	4.7	12.7	3.50
4.51	8.8	3.0	41.1	67.2	24.9	W 14×18	93.4	63.3	2.5	6.9	6.93
3.61	10.4	3.4	41.5	66.7	24.7	W 12×19	92.6	63.9	2.9	7.9	5.70
1.24	24.4	6.7	40.8	62.6	23.2	W 8×24	87.0	62.7	5.7	17.2	2.11
2.60	12.0	3.6	36.7	58.3	21.6	W 10×19	81.0	56.4	3.1	8.9	4.66
1.46	18.6	5.3	35.5	55.1	20.4	W 8×21	76.5	54.6	4.5	13.3	2.47
3.31	9.6	3.2	33.3	54.3	20.1	W 12×16	75.4	51.3	2.7	7.4	5.12
0.74	31.3	6.3	32.6	51.0	18.9	W 6×25	70.9	50.1	5.4	21.0	1.33
2.46	11.2	3.5	31.6	50.5	18.7	W 10×17	70.1	48.6	3.0	8.4	3.97

C-6 ■ Appendix C

\multicolumn{5}{c}{$F_y = 36$ ksi}		\multicolumn{5}{c}{$F_y = 50$ ksi}
BF, kips	L_{r}, ft	L_{p}, ft	$\phi_b M_{r}$, kip-ft	$\phi_b M_{p}$, kip-ft	Z_x, in^3	Shape	$\phi_b M_{p}$, kip-ft	$\phi_b M_{r}$, kip-ft	L_{p}, ft	L_{r}, ft	BF, kips
2.97	9.2	3.1	29.1	47.0	17.4	W 12×14	65.3	44.7	2.7	7.2	4.56
1.40	16.7	5.1	29.6	45.9	17.0	W 8×18	63.8	45.6	4.4	12.2	2.30
2.34	10.3	3.4	26.9	43.2	16.0	W 10×15	60.0	41.4	2.9	7.9	3.69
0.73	25.6	6.3	26.1	40.2	14.9	W 6×20	55.9	40.2	5.3	17.7	1.27
0.65	26.9	5.8	25.3	39.1	14.5	M 6×20	54.4	39.0	5.0	18.2	1.16
3.51	6.6	2.2	23.4	38.6	14.3	M 12×11.8	53.6	36.0	1.9	5.1	5.40
1.53	12.6	3.7	23.0	36.7	13.6	W 8×15	51.0	35.4	3.1	9.2	2.56
1.98	9.5	3.3	21.3	34.0	12.6	W 10×12†	47.0	32.7	2.9	7.4	3.13
0.82	18.3	4.0	19.9	31.6	11.7	W 6×16	43.9	30.6	3.4	12.5	1.46
0.46	30.3	5.3	19.9	31.3	11.6	W 5×19	43.5	30.6	4.5	20.1	0.83
1.43	11.5	3.5	19.3	30.8	11.4	W 8×13	42.8	29.7	3.0	8.5	2.35
0.42	31.1	5.0	18.8	29.7	11.0	M 5×18.9	41.3	28.9	4.2	20.5	0.76
0.69	20.7	6.7	19.0	28.8	10.8	W 6×15*,†	38.6	29.2	6.8	14.9	1.16
0.44	26.3	5.3	16.6	25.9	9.59	W 5×16	36.0	25.5	4.5	17.6	0.80
2.48	5.9	2.0	15.1	24.8	9.19	M 10×9	34.5	23.3	1.7	4.6	3.84
1.30	10.2	3.5	15.2	23.9	8.87	W 8×10†	33.0	23.4	3.1	7.8	2.03
0.78	14.4	3.8	14.3	22.4	8.30	W 6×12	31.1	21.9	3.3	10.2	1.33
0.30	25.5	4.2	10.6	17.0	6.28	W 4×13	23.6	16.4	3.5	16.9	0.54
0.72	12.0	3.8	10.8	16.8	6.23	W 6×9	23.4	16.7	3.2	8.9	1.17
0.26	27.8	3.9	10.2	16.3	6.05	M 4×13	22.7	15.7	3.3	18.3	0.47
1.59	5.3	1.8	9.01	14.6	5.42	M 8×6.5	20.3	13.9	1.5	4.1	2.47
0.94	4.5	1.5	4.68	7.56	2.80	M 6×4.4	10.5	7.20	1.3	3.5	1.49

* Indicates noncompact shape; $F_y = 36$ ksi.
† Indicates noncompact shape; $F_y = 50$ ksi.

Appendix D

SI Metric Conversion Table

Some Conversion Factors, between U.S. Customary and Sl Metric Units, Useful in Structural-Steel Design

	To convert	To	Multiply by
Forces	kip force	kN	4.448
	lb	N	4.448
	kN	kip	0.2248
Stresses	ksi	MPa (i.e., N/mm^2)	6.895
	psi	MPa	0.006895
	MPa	ksi	0.1450
	MPa	psi	145.0
Moments	ft·kip	kN·m	1.356
	kN·m	ft·kip	0.7376
Uniform loading	kip/ft	kN/m	14.59
	kN/m	kip/ft	0.06852
	kip/ft^2	kN/m^2	47.88
	psf	N/m^2	47.88
	kN/m^2	kip/ft^2	0.02089

For proper use of SI, see Standard for Metric Practice (ASTM E380), American Society for Testing and Materials, Philadelphia. Also see Standard Practice for the Use of Metric (SI) Units in Building Design and Construction (Committee E-6 Supplement to E380) (ANSI/ASTM E621), American Society for Testing and Materials, Philadelphia.

Basis of conversions (ASTM E380): 1 in = 25.4 mm: 1 lb force = 4.448 221 615 260 5 newtons.

Basic SI units relating to structural-steel design:

Quantity	Unit	Symbol
length	meter	m
mass	kilogram	kg
time	second	s

Derived SI units relating to structural-steel design:

Quantity	Unit	Symbol	Formula
force	newton	N	kg·m/s²
pressure, stress	pascal	Pa	N/m²
encrgy, or work	joule	J	N·m

Appendix E

Roy Becker

Seismic Design of Steel Buildings Using LRFD

E.1 General Design Information

E.1.1 Specifications and Manuals

The building design is based on the following specifications and standards in order to be applicable to most geographical areas:

1. AISC Seismic Provisions for Structural Steel Buildings—Load and Resistance Factor Design—1992 edition. Referenced as follows in the text: SP Sec. 7.1, SP (6-1), etc.

2. AISC Load and Resistance Factor Design Specifications for Structural Steel Buildings—Sept. 1, 1986. Referenced as follows in the text: LRFD Sec. B5.1, LRFD (E2-1), etc.

3. ASCE 7-95, Minimum Design Loads for Buildings and Other Structures—1995 Edition. Referenced as follows in the text: ASCE Sec. 9.2.3, ASCE (9.2.2.6-1), etc.

4. AISC Load and Resistance Factor Design Manual of Steel Construction, first edition—1986. Referenced as follows in the text: AISC pp. 2–10, etc.

E.1.2 Design Parameters

The building is an Office Occupancy and it requires Type I construction. The building system is a steel frame, and the required fireproofing is achieved by a spray-on type of material to achieve the necessary 2- and 3-h fire ratings.

To provide lateral resistance, the building has a moment frame system in the east-west direction along column lines A and D, and it has a braced frame in the north-south direction along column lines 1 and 5. This is indicated on Fig. E.1, which shows the general building configuration.

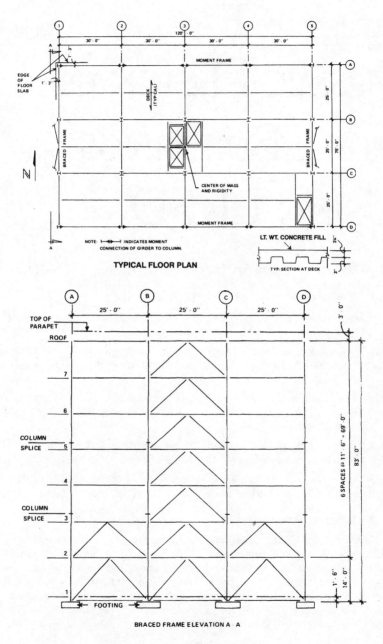

Fig. E.1 Building plan and elevation.

Both floors and roof are constructed using a 3-in-deep metal deck which has a lightweight structural-concrete fill of $3^1/_4$-in thickness over the metal deck. This concrete fill has a unit weight of 110 pcf and a strength of 3000 psi.

The typical story-to-story height is 11 ft, 6 in, which is based on a minimal clear ceiling height of 8 ft. (However, many buildings require story-to-story heights of 13 to 14 ft to provide appropriate clearance for mechanical ducts, lighting fixtures, etc.)

Material specifications are as follows:

- Structural steel, ASTM A36
- High-strength bolts, ASTM A325-SC
- Welding electrodes, AWS E70 series

E.1.3 Seismic Design Parameters

These seismic design parameters are based on ASCE 7-95, Sec. 9:

1. $A_a = A_v = 0.40$ per Map 9-1 and Map 9-2, respectively
2. Seismic hazard exposure group = I per Table 1-1
3. Seismic performance category = D per Table 9.1.4.4
4. Soil profile type midway between type B and C per Table 9.1.4.2
5. Response modification coefficient R per Table 9.2.2.2:

Basic structural system	R
Special moment frame	8
Ordinary moment frame	$4^1/_2$
Concentrically braced frame	5

6. Deflection amplification factor C_d per Table 9.2.2.2:

Basic structural system	C_d
Special moment frame	$5^1/_2$
Ordinary moment frame	4
Concentrically braced frame	$4^1/_2$

E.1.4 Gravity Loads

Roof Loading, psf:

Roofing and insulation		7.0
Metal deck		3.0
Concrete fill		44.0
Ceiling and mechanical		5.0
Steel framing and fireproofing		8.0
Dead load	67.0	
Live load (reducible), ASCE Sec. 4.11		20.0

E-4 ■ Appendix E

Floor loading, psf:

Metal deck	3.0
Concrete fill	44.0
Ceiling and mechanical	5.0
Partitions: ASCE Sec. 9.2.4.3	10.0*
Steel framing, including beams, girders, columns, and spray-on fireproofing	13.0
Dead load	75.0*
Live load (reducible), ASCE Sec. 4.2	50.0
Total load	125.0

Curtain Wall, psf

Average weight including column and spandrel covers	15.0

E.2 Special Moment Frames (SMF)

For seismic performance category D:

E.2.1 East-West Seismic Forces for Stress Analysis

Based on ASCE Sec. 9.2.3, as permitted by ASCE Table 9.2.2.4.3.

$$V = C_s W \qquad \text{ASCE (9.2.3.2-1)}$$

$$C_s = \frac{1.2 C_v}{RT^{2/3}} = \frac{(1.2)(C_v)}{(8)(1.15)^{2/3}} \qquad \text{ASCE (9.2.3.2.1-1)}$$

$C_v = 0.48$ per ASCE Table 9.1.4.2.4B (by interpolation between B and C soil profile types.)

$R = 8$ per ASCE Table 9.2.2.2

$T =$ to be determined per ASCE Sec. 9.2.3.3 and Sec. 9.2.3.7.1

$T_{max} = (C_v)(T_a)$

$C_v = 1.2$ per ASCE Table 9.2.3.3

$T_a = C_T h_n^{3/4}$ per ASCE 9.2.3.3.1-1

$T_a = (0.035)(83.0)^{3/4} = 0.96$ s

$T_{max} = (1.2)(0.96) = 1.15$ s

---

* 10.0 psf partition load used for seismic design computations. However, for vertical load on structural elements, 20.0 psf is utilized for the office occupancy (verify with appropriate code), resulting in 85.0 psf floor dead load.

Based on much prior experience with steel moment frames, the actual period of this 7-story building is usually much larger than 1.15 s, which is the longest period that is permitted for determining *strength* requirements of the frame. The actual period of the building is *at least* 1.80 s, as shown in Sec. E.2.2, which is based on frame drift requirements. Taking:

$T = 1.15$ s

$$C_s = \frac{1.2C_v}{RT^{2/3}} = \frac{(1.2)(0.48)}{(8)(1.15)^{2/3}}$$

$C_s = 0.0655$

or

$V = 0.0655\,W$

$W_{\text{floor}} = (122.5 \times 77.5)(0.075) + (400 \times 11.5)(0.15)$
$= 712 + 70 = 782$ kips

$W_{\text{roof}} = (122.5 \times 77.5)(0.067) + (400 \times 8.75)(0.015)$
$= 636 + 53 = 689$ kips

$W = (6)(782) + 689 = 5380$ kips

$V = (0.0655)(5380) = \underline{352\text{ kips}}$ (total seismic force)

Vertical distribution of seismic forces is as follows:

$$F_x = C_{vx}V \qquad\qquad \text{ASCE (9.2.3.4-1)}$$

$$C_{vx} = \frac{w_x h_x^k}{\sum_{i=1}^{n} w_i h_i^k} \qquad\qquad \text{ASCE (9.2.3.4-2)}$$

$$k = \text{exponent} = 1 + 1\left(\frac{T - 0.50}{2.0}\right) = 1 + 1\left(\frac{1.15 - 0.50}{2.0}\right)$$

$k = 1.32$

Hence seismic load distribution is parabolic over height of building, as shown in Fig. E.2

$$V_x = \sum_{i=x}^{n} F_i \qquad\qquad \text{ASCE (9.2.3.5)}$$

where V_x = story shear at any level, and its magnitude is shown in Table E.1.

E-6 ■ Appendix E

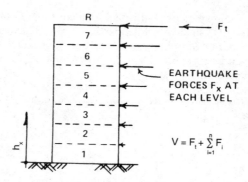

Fig. E.2 Seismic force distribution.

Table E.1 Seismic Forces for Stress Analysis

Floor level	w_x kips	h_x ft	h_x^k	$w_x h_x^k$ × 10⁻³	C_{vx}	F_x kips	V_x kips
R	689	83.0	341	235	0.25	88	
7	782	71.5	280	219	0.23	81	88
6	782	60.0	222	174	0.19	67	169
5	782	48.5	168	131	0.14	49	236
4	782	37.0	118	92	0.10	35	285
3	782	25.5	72	56	0.06	21	320
2	782	14.0	33	26	0.03	11	341
1							352
Σ	5381			933	1.00	352	

1. $k = 1.32$ and $F_x = C_{vx}V = (C_{vx})(352)$.
2. Wind loading is not critical for lateral forces in this design example. However, if wind should control the design of the building frames, then it *may* be necessary to recalculate both the period and the earthquake forces based on stiffness requirements of the frame to resist wind.
3. By comparing the seismic forces for stress (Table E.1) with those seismic forces for drift (Table E.2), the following is evident: At any level it is conservative to obtain the seismic forces for stress by multiplying the values for drift (Table E.2) by a factor of $352/262 = 1.34$.

E.2.2 East-West Seismic Forces for Drift Analysis

In accordance with ASCE Sec. 9.2.3.7.1, it is permissible to use the "computed fundamental period" of the building without the upper-bound limitation of ASCE Sec. 9.2.3.3. To determine this period, this equation, found in many codes, could be utilized:

$$T = 2\pi \sqrt{\left(\sum_{i=1}^{N} w_i \delta_i^2\right) \div g\left(\sum_{i=1}^{N} F_i \delta_i\right)}$$

Seismic Design of Steel Buildings Using LRFD ■ E-7

But this equation, sometimes termed "Rayleigh's," is not practical to use for determining the period of a building which has yet to be designed. However, there is a convenient formula which can be used for a good initial approximation of the period, without having to do any trial-and-error design. This formula, known as "Teal's formula," can be found in the *AISC Engineering Journals*, second and fourth quarters, 1975. It is actually a simplification of Rayleigh's equation, and is as follows:

$$T = 0.25\sqrt{\frac{\Delta}{C_1}}$$

where T = period of building, s
  Δ = lateral deflection at top of building, in
  C_1 = V/W where V produces a deflection Δ

Since drift limitations almost always control the design of a steel moment frame, Teal's formula can be easily used to determine the building period.

ASCE Sec. 9.2.3.7.1 imposes these drift limitations:

$$\delta_x = (C_d)(\delta_{xe}) \qquad \text{ASCE 9.2.3.7.1}$$

where C_d = deflection amplification factor per ASCE Table 9.2.2.2
  δ_{xe} = deflections using the prescribed seismic forces per ASCE Sec. 9.2.3.4
  δ_x = $5.5\,\delta_{xe}$
  δ_x = $0.020\,h$ per ASCE Table 9.2.2.7 for categories I and II
  $0.020h$ = $5.5\,\delta_{xe}$
  δ_{xe} = $0.0036\,h$ (at code seismic force level)

Since maximum drifts are seldom met over the height of a building, let us use *two-thirds* of the allowable drift for determining Δ. (This is a conservative assumption, but not ultraconservative.) Thus,

$$\Delta = \tfrac{2}{3}(\delta_{xe}) = (0.67)(0.0036\,h)$$
$$= (0.67)(0.0036)(12 \times 83.0 \text{ ft}) = 2.42 \text{ in}$$

As related to ASCE Sec. 9.2.3 for seismic forces,

$$V = C_s W = \frac{1.2 C_v W}{RT^{2/3}} = \frac{(1.2)(0.48)(W)}{(8)(T^{2/3})}$$

$$V = \frac{0.072 W}{T^{2/3}} \qquad \text{and} \qquad C_1 = \frac{V}{W} = \frac{0.072}{T^{2/3}}$$

Substituting into Teal's formula for building period,

$$T = 0.25\sqrt{\frac{\Delta}{C_1}} = 0.25\sqrt{\frac{2.42}{0.072/T^{2/3}}}$$

$$T = 0.25\sqrt{33.7 T^{2/3}} = 1.45 T^{1/3}$$

$T^{2/3} = 1.45$ or $T = (1.45)^{3/2}$
$T = 1.74$ s

Since two-thirds of allowable drift is a relatively conservative factor, round up and take $T = 1.80$ s.

Use $T = 1.80$ s.

It is evident that this period is significantly larger than that determined as the maximum period permitted for stress analysis in Sec. E.2.1 ($T = 1.15$ s). And since drift requirements usually control the design of steel moment frames, this "refined" period determination of $T = 1.80$ s is worth the additional effort since it will save substantial amounts of steel framing. Thus,

$$V = \frac{0.072W}{T^{2/3}} = \frac{0.072W}{(1.80)^{2/3}} = 0.0486W$$

$V = 0.0486W = (0.0486)(5380) = 262$ kips (total seismic force)

Vertical distribution of seismic forces as follows:

$F_x = C_{vx}V$ where $V = 262$ kips

$$k = 1 + 1\left(\frac{1.80 - 0.50}{2.0}\right) = 1.65$$

This distribution of seismic forces and shear is shown in Table E.2.

Table E.2 Seismic Forces for Drift Analysis

Floor level	w_x kips	h_x ft	h_x^k	$w_x h_x^k \times 10^{-3}$	C_{VX}	F_x kips	V_x kips
R	689	83.0	1467	1010	0.28	73	
7	782	71.5	1147	897	0.25	66	73
6	782	60.0	859	672	0.19	50	139
5	782	48.5	605	473	0.13	34	189
4	782	37.0	387	303	0.08	21	223
3	782	25.5	209	163	0.05	13	244
2	782	14.0	79	62	0.02	5	257
1							262
Σ				3580	1.00	262	

1. $k = 1.65$ and $F_x = C_{VX}V = (C_{VX})(262)$.
2. When the members of the frame are determined, a verification *must* be made to insure that the actual period of the building is equal to or greater than 1.80 s (the basis for Table E.2). If this is not the case, the seismic forces must be recomputed and the building reanalyzed.

E.2.3 Horizontal Distribution of Seismic Forces

Although the centers of mass and rigidity coincide, ASCE Sec. 9.2.3.5.2 requires designing for a minimum torsional eccentricity e equal to 5 percent of the building dimension perpendicular to the force. However, to be conservative and to simplify the analysis in Sec. E.4, 5 percent of the maximum building dimension is used.

$$e = (0.05)(120) = 6.0 \text{ ft}$$

Both the moment frames and the braced frames will resist this torsion. Because the braced frames are much stiffer than the moment frames, the relative rigidities are assumed as follows:

$$R_A = R_D = 1.00, \qquad R_1 = R_5 = 4.00$$

Shear distribution in the E-W direction:

$$V_{A,x} = R_A \left[\frac{V_x}{\Sigma R_{E-W}} \pm \frac{(V_x e)(d)}{\Sigma R_y (d)^2} \right] = V_{D,x}$$

where d = distance from frame to center of rigidity
R_{E-W} = rigidity of those frames extending in the east-west direction
R_y = rigidity of a braced or moment frame, referred to that frame on column line y
V_x = total earthquake shear on building at story x
$V_{y,x}$ = earthquake shear on a braced or moment frame referred to that frame on column line y at story x

ΣR_{E-W} = $2(1.00) = 2.0$
$\Sigma R_y d^2$ = $2(1.00)(37.5)^2 + 2(4.00)(60.00)^2 = 31{,}600$

$$V_{A,x} = 1.00 \left[\frac{V_x}{2.00} \pm \frac{(V_x \times 6.00)(37.5)}{31{,}600} \right] = 0.5V_x \pm 0.007V_x$$

$$= 0.51V_x = V_{D,x}$$

The second term (torsion) within the bracket being small indicates that the moment frames are resisting little torsion due to eccentricity.

E.2.4 Preliminary Design of Frames, Including Drift Control

Design will be limited to fourth-floor girders and third- to fifth-floor columns (one tier), with the portal method being employed for preliminary design, assuming points of inflection at midlength of members (see Fig. E.3). For irregular framing, it is recommended that the joint coefficient method be used for analysis (PCA method).

$$2F_1 + 3F_2 = V_{A,3} = V_{D,3}$$

Since drift usually controls the design of moment frames, story shear will be that specified for drift calculations.

$V_{A,3} = V_{D,3} = 0.51 V_3 = (0.51)(244) = 124$ kips
$2F_1 + 3F_2 = 177$

E-10 ■ Appendix E

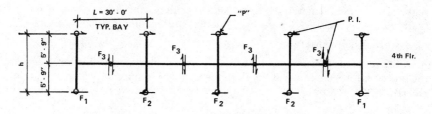

Fig. E.3 Seismic shears acting on frame.

Since $F_1 = F_2/2$, $4F_2 = 124$ and $F_2 = 31.1$ kips. Summation of moments about point P (see Fig. E.3):

$$30.0F_3 = 11.50F_2 = 11.50(31.1) \qquad F_3 = 11.9 \text{ kips}$$

The story drift can be determined by the following relationships (assuming points of inflection at midheight of columns and midspan of girders):

$$\Delta_s = \Delta_c + \Delta_g$$

$$\Delta_c = \frac{Fh^3}{12EI_c} \qquad \Delta_g = \frac{FLh^2}{12EI_g}$$

where F = column shear
 h = story height, in
 I_c = moment of inertia of column
 I_g = moment of inertia of girder
 L = girder length, in
 Δ_s = story drift
 Δ_c = contribution of column to drift
 Δ_g = contribution of girder to drift

This relationship is based on flexure of the column and girder using center line lengths for members. However, it is not conservative in those cases where the panel zone does not possess adequate stiffness or there is excessive axial deformation of the columns (chord drift).

As noted in Sec. E.2.2, the frame drift is limited by ASCE Sec. 9.2.3.7 by

$$\delta_x = (C_d)(\delta_{xe})$$

$$\delta_{xe} = 0.0036 \, h \qquad \text{(at code seismic force level)}$$

$$\Delta_s = \delta_{xe} = (0.0036)(12 \times 11.5) = 0.50 \text{ in}$$

$$0.50 = \frac{Fh^3}{12EI_c} + \frac{FLh^2}{12EI_g}$$

There are many possible column and girder combinations that could satisfy this drift criterion. A preliminary column size can be determined by computing the axial load and moment of an interior column at the third story of the frame. As noted in Table E.1 the reactions for a stress analysis may be obtained by multiplying the drift analysis reactions by a factor of 1.34. For load combinations, utilize SP Sec. 6, where SP (3-5) usually governs:

1.2D + 1.0E + 0.5L

For axial loads on column at third story,

P_D = roof + 4 floors + curtain wall
   = (30.0 × 13.75)(0.067) + 4 (30.0 × 13.75)(0.085) + (30.0 × 60.5)(0.015)

P_D = 28 + 140 + 27 = 195 kips

$P_E = 1.0\, Q_E + 0.5\, C_a D$ \hfill ASCE (9.2.2.6-1)

$Q_E = 0$ (for interior column)

$C_a = 0.40$ per ASCE Table 9.1.4.2.4A (for soil type midway between B and C)

$P_E = 0 + 0.5(0.4)D = 0.2D$

$P_E = (0.2)(195) = 39$ kips

P_L = 4 floors
   = 4(30 × 13.75)(0.022*) = 36 kips

$P_U = 1.2(195) + 1.0(39) + 0.5(36) = 291$ kips

For moment at the third story, about the strong axis,

$M_D = 0$ (interior column)

$$M_E = 1.34\left(\frac{h}{2} \times F_2\right) = 1.34\left(\frac{11.5}{2} \times 31.1\right) = 240 \text{ kip-ft}$$

(see note 3 for Table E.1)

$M_L = 0$

$M_U = 1.2(0) + 1.0(240) + 0.5(0) = 240$ kip-ft

Converting this moment into an equivalent axial load for a W14 column, use AISC, pp. 2–10.

*In accordance with ASCE Sec. 4.8,

$$L = L_o\left(0.25 + \frac{15}{\sqrt{A_1}}\right) \quad \text{and} \quad A_1 = (4)(4)(30.0 \times 13.75)$$

$$L = 50\left(0.25 + \frac{15}{\sqrt{6600}}\right) = 22 \text{ psf}$$

$P_{ueff} = P_u + M_{ux}m$ where $m = 2.0$
$= 291 + (240)(2.0) = 771$ kips

(This also assumes M_{UY} is small). If $KL \approx (1.5)(11.5) = 17$ ft, could use W14×99 (per AISC, pp. 2–20), but

Select: W14×132 column ($\phi_c P_n = 1020$ kips).

As will be shown in the section "Column to Girder and Column Panel Zone Ratio," Sec. E2.8, the W14×132 column was primarily selected in order that its plastic moment capacity would be greater than that of the girders and panel zone. Thus we can now determine the girder size to meet the drift requirements of the moment frame.

$$0.50 = \frac{Fh^3}{12EI_c} + \frac{FLh^2}{12EI_g}$$

$$0.50 = \frac{(31.1)(11.5 \times 12)^3}{(12)(29,000)(1530)} + \frac{(31.1)(30.0 \times 12)(11.5 \times 12)^2}{(12)(29,000)I_g}$$

$$0.50 = 0.15 + \frac{613}{I_g}$$

$I_g = 1750$ in⁴

Select: W24×68 girder ($I_g = 1830$ in⁴).

E.2.5 P-Delta Effects

In accordance with ASCE Sec. 9.2.3.7.2, P-delta effects need *not* be considered if the stability index $\theta \le 0.10$, i.e.;

$$\theta = \frac{P_x \Delta}{V_x h_{sx} C_d} \le 0.10$$

Utilize the first story since the ratio will tend to be the highest at this level (since the ratio of P_x to V_x will be the largest).

$\Delta = \delta_x = C_d \delta_{xe} = 0.020\, h = (0.020)(12 \times 14.0) = 3.36$ in
$V_x = 262$ kips (see Table E.2)
$h_{sx} = 12 \times 14.0 = 168$ in
$C_d = 5.5$
$P_x =$ total unfactored vertical load
   $= P_D + P_L$
   $P_D = 5380$ kips (see Sec. E.2.1)
   $P_L = (6)(122.5 \times 77.5)(0.020) = 1140$ kips
$P_x = 6520$ kips

$$\theta = \frac{(6520)(3.36)}{(262)(168)(5.5)} = 0.09 < 0.10$$

P-delta effects need not be considered for stress or drift analysis.

E.2.6 Strength Check of Frame Members

W24×68 Girder ▪ Use the critical load combination:

$1.2D + 1.0E + 0.5L$ SP (3-5)

$$M_D = \frac{wl^2}{12} \quad \text{(where } w = \text{uniform load)}$$

$w = (0.72 \times 7.5) + 0.10 + (0.015 \times 11.5) = 0.81 \text{ kip/ft}$
     floor   beam   curtain wall

$$M_D = \frac{(0.81)(30.0)^2}{12} = 61 \text{ kip-ft}$$

$M_E = 1.0 Q_E + 0.5 C_a D$
$\quad = Q_E + 0.2(M_D)$

$$M_E = (1.34)(F_3)\frac{L}{2} + 0.2 M_D = (1.34)(11.9)\frac{30.0}{2} + (0.2)(61)$$

$M_E = 239 + 12 = 2531 \text{ kip-ft}$

$$M_L = \frac{wl^2}{12}$$

$w = (0.050 \times 7.5) = 0.38 \text{ kip/ft}$

$$M_L = \frac{(0.38)(30.0)^2}{12} = 29 \text{ kip-ft}$$

$M_u = 1.2(61) + 1.0(251) + 0.5(29)$
$M_u = 73 + 251 + 15 = 339 \text{ kip-ft}$

Per LRFD (F1-3)

$$M_n = C_b \left[M_p - (M_p - M_n)\frac{L_b - L_p}{L_r - L_p} \right] \le M_p$$

$$C_b = 1.75 + 1.05 \frac{M_1}{M_2} + 0.3 \left(\frac{M_1}{M_2} \right)^2 \le 2.3$$

$M_1 = +239 - (73 + 15 + 12) = +139 \text{ kip-ft}$
$M_2 = +239 + (73 + 15 + 12) = +339 \text{ kip-ft}$

$$C_b = 1.75 + 1.05 \frac{139}{339} + 0.3 \left(\frac{139}{339} \right)^2$$

$C_b = 1.75 + 0.43 + 0.05 = 2.23 \le 2.3$ OK

$$L_b = \frac{2500_{ry}}{F_Y} = \frac{2500(1.87)}{36.0} = 130 \text{ in}$$

Place brace at third points, $L_b = 10.0$ ft
Per AISC, p. 3-15, shape is compact ($\lambda \leq \lambda_p$), and

$L_r = 22.4$ ft. $L_p = 7.8$ ft

$\phi_b M_r = 300$ kip-ft, and $\phi_b M_p = 478$ kip-ft

$$\therefore \phi_b M_n = 2.23 \left[478 - (478 - 300) \frac{10.0 - 7.8}{22.4 - 7.8} \right]$$

$\qquad\qquad = 2.23(478 - 27) = 1006 > 478$ max $\qquad\qquad$ NG

Since $\phi_b M_n \leq \phi_b M_p = 478$ kip-ft max

$\phi_b M_n = 478 > 339$ $\qquad\qquad$ OK

Strength stress ratio = 0.71.

In addition, to preclude local buckling, SP Sec. 8.4.6 requires that for flanges a more stringent λ_p be imposed than LRFD Table B5.1.

$$\frac{b}{t} \leq \lambda_p = \frac{52}{\sqrt{F_Y}} = 8.67$$

$$\frac{b}{t} = \frac{b_f}{2t_f} = 7.7 < 8.67 \qquad\qquad \text{OK}$$

Use W24×68 girder.

W14×132 Column ■ Use the critical load combination:

$1.2D + 1.0E + 0.5L$ $\qquad\qquad$ SP (3-5)

As noted in Sec. E.2.4

$P_U = 291$ kips   and $M_U = 240$ kip-ft

P-delta effects need *not* be considered, as noted in Sec. E.2.5. Thus the use of LRFD HI-2 is not required:

$M_{ux} = M_U = 240$ kip-ft

For M_{uy} consider the weak-axis bending due to the shear connection to the girder (Fig. E.4).

$$M_{uy} = \frac{(V_U)(10 \text{ in})}{2*}$$

$V_U = 1.2D + 0.5L$

$$V_D = \left(\frac{30.0 \times 12.5}{2} \right)(0.085) = 16.0 \text{ kips}$$

*Factor of 2 utilized since the bottom and top portions of the column resist equal amounts of the eccentricity.

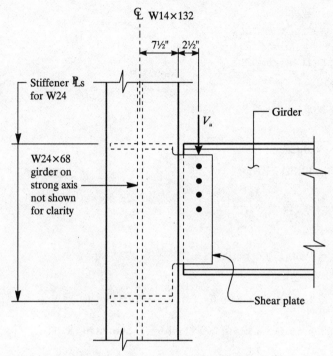

Fig. E.4 Girder shear connection (nonmoment).

$$V_L = \frac{30.0 \times 12.5}{2}(0.050) = 9.4 \text{ kips}$$

$$V_U = (1.2)(16.0) + (0.5)(9.4) = 23.9 \text{ kips}$$

$$M_{uy} = \frac{(23.9)(10)}{2} = 120 \text{ kip-in} = 10 \text{ kip-ft}$$

K_x in accordance with AISC, p. 6-153,

$$G = \frac{\sum\left(I_c/L_c\right)}{\sum\left(I_G/L_G\right)} = \frac{(2)(1530)/11.5}{(2)(1830)/30.0} = 2.2$$

$G_A = G_B = 2.2$

$K_x = 1.65$ per AISC Fig. C-C2.2,

$K_y = 1.00$ since stability is provided by braced frame in other direction

E-16 ■ Appendix E

$$\left(\frac{KL}{r}\right)_x = \frac{(1.65)(11.5 \times 12)}{6.28} = 36$$

$$\left(\frac{KL}{r}\right)_y = \frac{(1.00)(11.5 \times 12)}{3.76} = 37 \qquad \text{(governs)}$$

$\phi_c F_{cr} = 28.5$ ksi per AISC, p. 6-124

$$\frac{P_u}{\phi P_n} = \frac{291}{(38.8)(28.5)} = 0.26 \geq 0.20$$

Thus, utilize LRFD (H1-1a):

$$\frac{P_u}{\phi P_n} + \frac{8}{9}\left(\frac{M_{ux}}{\phi_b M_{nx}} + \frac{M_{uy}}{\phi_b M_{ny}}\right)$$

$$\phi_b M_{ux} = (0.90)(F_Y)(Z_x) = (0.90)(36.0)\left(\frac{234}{12}\right) = 632^* \text{ kip-ft}$$

$$\phi_b M_{ny} = (0.90)(F_Y)(Z_y) = (0.90)(36.0)\left(\frac{113}{12}\right) = 305 \text{ kip-ft}$$

Reducing column moments to face of girder, as shown in Fig. E.5,

$$M_{TF} = M_T \frac{hc}{h} = M_T \frac{9.5}{11.5} = 0.83 M_T$$

$M_{ux_F} = (0.83)(240) = 198$ kip-ft

$M_{uy_F} = (0.83)(10) = 8$ kip-ft

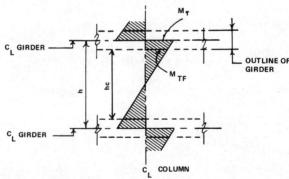

Fig. E.5 Column moment reduction.

*M_p used since $L_{px} = 300r_y/\sqrt{F_{yf}} = (300)(3.76)/\sqrt{36.0}$. $L_{px} = 188$ in $= 15.7$ ft > 11.5 story height. Also, $\lambda \leq \lambda_p$ by inspection of AISC, p. 3-14 (by comparison).

Now substituting into LRFD (HI-1a):

$$\frac{291}{(38.8)(28.5)} + \frac{8}{9}\left(\frac{198}{632} + \frac{8}{305}\right)$$

$$0.26 + \frac{8}{9}(0.31 + 0.03) = 0.56 \leq 1.00 \qquad \text{OK}$$

Use W14×132 column.

E.2.7 Panel Zone

Per SP Sec. 8.3, the shear strength must be capable of resisting the beam bending moments from load combination (3-5) and (3-6):

1.2D + 1.0E + 0.5L and 0.9D ± 1.0E

Since gravity moments will cancel for this interior column, only the effects of the seismic moments need be considered, i.e.:

$M_u = 1.0E = 240$ kip-ft

The shear in the panel zone, shown in Fig. E.6, can be shown to be approximately

$$V_a = \frac{M_A}{0.95d_1} + \frac{M_B}{0.95d_2} - \frac{M_A + M_B}{H}$$

The shear in the column shaft can be obtained by taking moments about the point of inflection of the column, and it is:

$$V_b = \frac{M_A + M_B}{H}$$

where d = girder depth
 H = column height
 M_A = girder moment on left side of column
 M_B = girder moment on right side of column
 V_a = panel zone shear
 V_b = column shaft shear

The column free-body and shear and moment diagrams are as shown in Fig. E.6. As previously determined,

$M_A = 240$ kip-ft $M_B = 240$ kip-ft

where both moments will act either clockwise or counterclockwise, concurrently. Thus, substituting into the equation for V_a

$$V_a = \frac{240}{(0.95)(23.73/12)} + \frac{240}{(0.95)(23.73/12)} - \frac{(2)(240)}{11.5}$$

$$= 128 + 128 - 42$$

$V_a = 214$ kips

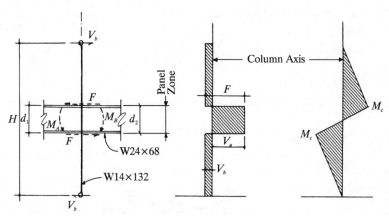

Fig. E.6 Column shear and moment diagram.

Capacity of the panel zone per SP (8-1):

$$\phi_V V_n = 0.55 \phi_V F_y d_c t_p \left(1 + \frac{3 b_{cf} t_{cf}^2}{d_b d_c t_p}\right) \qquad \phi_V = 0.8$$

Setting $V_a = \phi_V V_n$ and solving for t_p using the properties of a W14×132 column and W24×68 girder:

t_p = total thickness of panel zone including doubler plates

$$214 = (0.55)(0.8)(36.0)(14.66)(t_p)\left[1 + \frac{(3)(14.73)(1.03)^2}{(23.7)(14.66)(t_p)}\right]$$

$$214 = 232 t_p \left(1 + \frac{0.139}{t_p}\right)$$

$$214 = 232 t_p + 32$$

$$t_p = \frac{182}{232} = 0.78 \text{ in}$$

$t_{\text{doubler}} = t_p - t_w = 0.78 - 0.65 = 0.13$ in

Use ¼-in web doubler.

Check both column web and web doubler for shear buckling per SP 8-2:

$$t_z \geq \frac{d_z + w_z}{90}$$

$d_z \approx 23.73 - 0.38 = 23.35$ in   (assuming ⅜-in stiffener)

$w_z = d - 2t_f = 14.66 - 2(1.03) = 12.60$ in

Thus,

$$\frac{d_z + w_z}{90} = \frac{23.35 + 12.60}{90} = 0.40 \text{ in}$$

$t_w = 0.65 > 0.40$ OK

$t_{\text{doubler}} = 0.25 < 0.40$ NG

To prevent doubler from buckling, either plug weld to column *or* increase thickness of doubler to at least ³/₈-in. In this case, we will use ¼-in doubler plate, attached as shown in Fig. E.7.

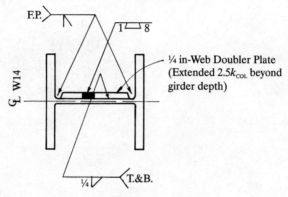

Fig. E.7 Column web doubler.

Alternate Design Using 50 Grade Columns ▪ If 50 grade columns are used ($F_Y = 50$ ksi) such as by using A572 grade 50, then based on SP (8-1):

$$214 = 0.55\phi_V F_Y d_c t_p \left(1 + \frac{3b_{cf} t_{cf}^2}{d_b d_c t_p}\right)$$

$$214 = 322 t_p \left(1 + \frac{0.139}{t_p}\right)$$

$t_p = 0.53 \text{ in} < t_w = 0.65 \text{ in}$

Web doubler not required.

Since the cost of A572 grade 50 material for a W14×132 is only about $3.70 per linear foot more than A36, a savings will result in using 50 grade columns.

E.2.8 Column to Girder and Column to Panel Zone Ratio

Per SP Sec. 8.6, the ratio of column to girder strength or column to panel zone strength should be ≥ 1.0, except for three conditions. In general, most moment frame columns meet the first condition, namely:

$P_{UC} < 0.3\, F_Y A_g$

and per SP Sec. 8.4, local buckling precluded by conforming to SP Table 8-1.

$P_{UC} = 291$ kips (see Sec. E.2.4)

$0.3\, F_Y A_g = (0.3)(36.0)(38.8) = 419$ kips

291 < 419   OK

Now checking ratios per SP Table 8-1 for W14×132:
Flanges:

$$\frac{b}{t} = \frac{b_f}{2t_f} = 7.1$$

$$\lambda_p = \frac{52}{\sqrt{F_Y}} = 8.7 > 7.1 \qquad \text{Flanges OK}$$

Web:

$$\frac{h_c}{t_w} = \frac{14.66 - 2(1.69)}{0.645} = 17.5$$

$$\frac{P_u}{\phi_b P_Y} = \frac{252}{(0.9)(36.0)(38.8)} = 0.20 > 0.125$$

$$\lambda_p = \frac{191}{\sqrt{F_Y}}\left(2.33 - \frac{P_u}{\phi_b P_Y}\right) = \frac{191}{\sqrt{F_Y}}(2.33 - 0.20)$$

$$= \frac{407}{\sqrt{F_Y}} \quad \text{but} \quad \lambda_{p\max} = \frac{253}{\sqrt{F_Y}} = 42$$

$\lambda_p = 42 > 17.5$ Web OK

W14×132 Column does not need to comply with stress ratios.

However, these stress ratios will be checked to determine where the energy will be absorbed inelastically in the event of "the" major earthquake.

$$\frac{\sum Z_c\left[F_{yc} - \left(P_{uc}/A_g\right)\right]}{\sum Z_b F_{yb}} \geq 1.0 \qquad \text{SP (8-3)}$$

$$\frac{(2)(234)\left[36.0 - (291/38.8)\right]}{(2)(160)(36.0)} = 1.16 \geq 1.0 \qquad \text{OK}$$

Column strength exceeds girder by 1.16.

$$\frac{\sum Z_c \left[F_{YC} - \left(P_{uc}/A_g \right) \right]}{V_n d_b \left[H/H - d_b \right]} \geq 1.0 \qquad \text{SP (8-4)}$$

V_n = nominal strength of panel zone

$$V_n = 0.55 F_y d_c t_p \left(1 + \frac{3 b_{cf} t_{cf}^2}{d_b d_c t_p} \right)$$

where t_p = column web plus doubler plate, $t_p = 0.65 + 0.25 = 0.90$ in

$$V_n = (0.55)(36.0)(14.66)(0.90) \left[1 + \frac{(3)(14.73)(1.03)^2}{(23.7)(14.66)(0.90)} \right]$$

$V_n = 300$ kips

Taking ratio of column to panel zone strength,

$$\frac{(2)(234)\left[36.0 - (291/38.8) \right]}{(300)(23.73)\left[11.5/11.5 - 1.98 \right]} = 1.55 \geq 1.0$$

Column strength exceeds panel zone by 1.55.

By observing these ratios, it is evident that the "weakest" link in the system resisting lateral loads is the panel zone where shear yielding will occur at the frame joint under loads causing inelastic behavior (see Fig. E.8).

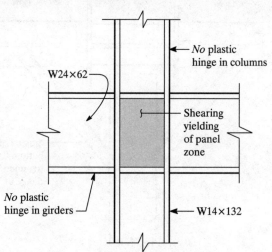

Fig. E.8 Shear yielding of panel zone.

E.2.9 Girder-to-Column Joint Restraint

Per SP Sec. 8.7, the column flanges require lateral support at only the level of the *top* flanges of the girders if *one* of the following is satisfied:

SP (8-3) > 1.25   or   SP (8-4) > 1.25

As shown in Sec. E.2.8, by using SP 8-4, the ratio of column to panel zone strength = 1.55.
Thus,

SP (8-4) = 1.55 > 1.25   OK

Column flanges require bracing only at level of top girder flanges.
This bracing is achieved by attachment of the metal deck floor system to the top flange of the girder. It should be observed in this case that the reason for bracing *not* being required for the column flanges at the level of the *bottom* flanges of the girder is that the column will always remain elastic. Plastic hinges will never occur in the column as shown in Fig. E.8. But if there is a case where

SP (8-3) ≤ 1.25   and   SP (8-4) ≤ 1.25

then the bracing detail per Fig. E.9 might be utilized at the level of the bottom flanges of the girder to *each* column flange.

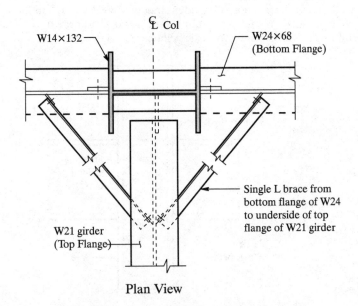

Fig. E.9 Girder bottom flange bracing.

E.2.10 Girder Bracing

Per SP Sec. 8.8, *both* flanges of girders must be laterally supported at intervals of

$$L_{max} = \frac{2500 r_y}{F_Y}$$

and for W24×68

$$L_{max} = \frac{(2500)(1.87)}{36.0} = 130 \text{ in} = 10 \text{ ft - 10 in}$$

With distance between column centerlines = 30 ft and columns providing brace at the girder ends,
 Providing bracing at 10-ft spacing (third points)
 The details in Fig. E.10 could be used. If a beam is framed into the third point of the girder, the detail in Fig. E.11 could be used.

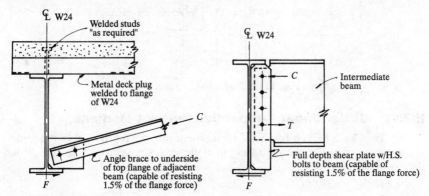

Fig. E.10 Girder bracing detail.   **Fig. E.11** Girder bracing detail.

E.2.11 Girder Moment Connection

Per SP Sec. 8.2.a, the required moment connection is adequate if it will transfer the plastic capacity of the girder. Based on tests at both Lehigh University and University of California at Berkeley University, by utilizing full penetration welds to the flanges, the plastic capacity of the girder is assured (except where the flanges contribute less than 70 percent of the flexural plastic capacity of the girder, as noted in Secs. E.2.12 and E.2.13). Thus, utilize the connection shown in Fig. E.12. This is used 95 percent of the time in "seismic country."

> **Important Note** ▪ Although the connection shown in Fig. E.12 was extensively utilized for about 30 years, it performed poorly in the 1994 Northridge earthquake. It is recommended that this connection be modified in areas subject to major earthquakes in accordance with the FEMA 267 Report Number SAC-95-02 entitled "Interim Guidelines: Evaluation, Repair, Modification and Design of Steel Moment Frames," dated August 1995.

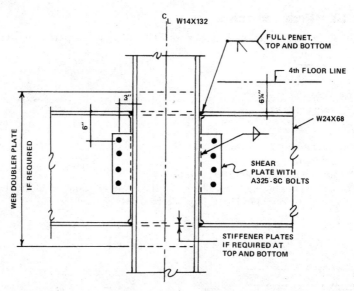

Fig. E.12 Elevation of connection.

E.2.12 Girder Shear Connection without Moment

Per SP Sec. 8.2.c.2, for condition "a," no moment transfer is required to the girder shear connection. Checking the nominal flexural strength of *only* the girder flanges of the W24×68:

$b_f t_f (d - t_f) F_{Yf} \geq 0.7 M_p$

$b_f t_f (d - t_f) F_{Yf} = (8.965)(0.585)(23.73 - 0.585)(36.0) = 4370$ kip-in

$0.7 M_p = (0.7)(177)(36.0) = 4460$ kip-in

$4370 \approx 4460$

(However, if this is not considered as essentially meeting specification requirements, see Sec. E.2.13.)

Thus shear connection design will be based on the transfer of no moment—only shear forces. Per SP Sec. 8.2.a and Sec. 8.2.6, the required shear strength of the connection can conservatively be based on the loading condition in Fig. E.13.

$$V_D = \frac{wL}{2} = \frac{(0.81)(28.8)}{2} = 11.7 \text{ kips}$$

$$V_L = \frac{wL}{2} = \frac{(0.38)(28.8)}{2} = 5.5 \text{ kips}$$

$$V_E = \frac{2M_p}{L} = \frac{(2)(177)(36.0)}{12 \times 28.8} = 36.7 \text{ kips}$$

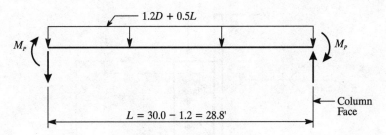

Fig. E.13 Girder free body.

Thus,

$$V_u = 1.5V_D + 1.0V_E + 0.5V_L$$
$$= (1.5)(11.7) + (1.0)(36.7) + (0.5)(5.5) = 57.1 \text{ kips}$$

Bolts ▪ Capacity of bolts taken in accordance with LRFD Specification for Structural Joints Using ASTM A325 or A490 Bolts, dated June 8, 1988. In accordance with this bolt specification, "slip-critical" joints must be used since joints are subject to significant stress reversal. For $^7/_8$-in ϕ, A325 bolt in "slip-critical" joint, single shear, using factored loads, capacity can be computed in accordance with:

$$\phi R_{str} = 1.13_u \, T_m N_b N_s = (1.13)(0.33)(39)(1.0)(1.0) = 14.5 \text{ kips}$$

$$n = \frac{V_u}{\phi R_{str}} = \frac{57.1}{14.5} = 3.9 \text{ bolts}$$

Use four $^7/_8$-in ϕ A325 SC bolts.

Shear Plate ▪ Try $^3/_8$-in by 12-in-long shear plate as shown in Fig. E.14.

$V_U = 57.1$ kips   and $M_U = (57.1)(3.0) = 171$ kip-in

Design shear strength per LRFD Sec. J5.2, using LRFD (J5-3).

$$\phi R_n = (0.80)(0.7 A_g F_y) = (0.80)(0.7)(0.375 \times 12)(36.0)$$
$$\phi R_n = 90.7 > 57.1 \qquad\qquad\qquad\qquad\qquad\qquad\qquad\qquad \text{OK}$$

Design flexural strength per LRFD Sec. F1.3, using LRFD (F1-3) and (F1-5).

$$L_p = \frac{3750 r_y}{M_p} \sqrt{JA}$$

where $r_y = 0.29t = (0.29)(0.375) = 0.11$ in

$$M_p = F_Y Z_x = (36.0) \frac{(0.375)(12.0)^2}{4} = 486 \text{ kip-in}$$

E-26 ■ Appendix E

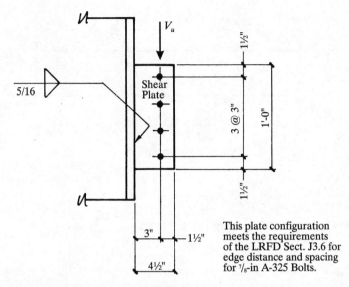

Fig. E.14 Shear plate.

$$A = (0.375)(12.0) = 4.5 \text{ in}^2$$

$$J = \frac{bt^3}{3} = \frac{(12.0)(0.375)^3}{3} = 0.21 \text{ in}^4$$

$$L_p = \frac{(3750)(0.11)}{486}\sqrt{(0.21)(4.5)} = (0.85)(0.97)$$

$$= 0.83 \text{ in}$$

However, $L_b \approx 3 \text{ in} > 0.83$ NG
(This indicates that $M_n < M_p = F_Y Z$)
Now using LRFD (F1-10) and LRFD (F1-11),

$$L_r = \frac{57{,}000 r_y}{M_R}\sqrt{JA}$$

$$M_R = F_Y S = (36.0)\frac{(0.375)(12.0)^2}{6} = 324 \text{ kip-in}$$

$$L_n = \frac{(57{,}000)(0.11)}{324}\sqrt{(0.21)(4.5)} = 18.7 \text{ in}$$

$L_b < 18.7$ OK

Thus, $0.83 < L_b < 18.7$. In lieu of using LRFD (F1-3), simply take $M_n = M_R$ (conservative):

$\phi M_n = (0.90)(324) = 292 > 171$ OK

Use $^3/_8$-in × 1 ft-0 in shear plate

Weldment to Shear Plate

$V_u = 57.1$ kips   and $M_U = (57.1)(3.0) = 171$ kip-in

Computing weld stresses per assumed plastic distribution:

$$f_{uv} = \frac{57.1}{(2)(12)} = 2.4 \text{ kips/in}$$

$$f_{uh} = \frac{171}{(2)(12)^2} = 2.4 \text{ kips/in}$$

$$f_{ur} = \sqrt{(2.4)^2 + (2.4)^2} = 3.4 \text{ kips/in}$$

Weld design strength = ϕF_W, using E70 electrodes

$$\phi F_W A_W = (0.75)(0.60 \times 70.0)\left(\frac{1}{16}\right)(0.707)$$

$\phi F_W A_W = 1.39$ kips/in for $\frac{1}{16}$-in fillet weld

$$n = \frac{3.4}{1.39} = 2.45 \text{ sixteenths}$$

But to meet the criteria of LRFD Table J2.5, a $^5/_{16}$-in fillet weld is required for column flanges being 1.03 in thick.

Use $^5/_{16}$-in fillet weld (both sides).

In lieu of these computations for weld size, LRFD Table XVIII (p. 5-91) may be used which utilizes the concept of the "instantaneous center of rotation."

Special case where $k = 0$:

$P_u = V_u = 57.1$ kips

$aI = 3.0$ in $a = \dfrac{3.0}{12.0} = 0.25$

Per Table XVIII, $C = 2.064$,

$$D = \frac{P_u}{CC_1 I} = \frac{57.1}{(2.064)(1.0)(12)} = 2.31 \text{ sixteenths}$$

(slightly less than using plastic method).

Use $^5/_{16}$-in weld.

E.2.13 Girder Shear Connection with Moment

Per SP Sec. 8.2.c.2, for condition b, moment transfer is required to the girder shear connection (where flanges of girder transfer less than 0.70 M_p). For this case, bolted portion to transfer shear V_u and welded portion to girder to transfer 20 percent of the flexural strength (nominal) of the web.

$V_u = 57.1$ kips

$M_{ut} = (0.20)(Z_{web})(F_Y)$

$$Z_{web} = \frac{(t_w)(d-2t_f)^2}{4} = \frac{(0.415)(23.73 - 2 \times 0.585)^2}{4} = 52.8 \text{ in}^3$$

$M_{ut} = (0.20)(52.8)(36.0) = 380$ kip-in

Bolts ■ Same as Sec. E.2.12 (four $^7/_8$-in bolts).

Shear Plate ■ Try $^1/_2$-in by 1 ft-2 in long shear plate as shown in Fig. E.15. Compute M_u at column face:

$V_u = 57.1$ kips

$M_u = M_{ut} + (V_u)(3.0)$
$\quad = 380 + (57.1)(3.0) = 539$ kip-in

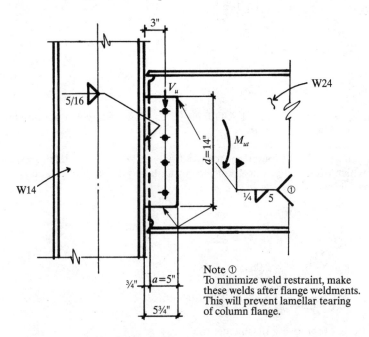

Fig. E.15 Shear plate connection.

Seismic Design of Steel Buildings Using LRFD ■ E-29

Design shear strength OK by inspection.
Design flexural strength per LRFD Sec. F1.3:

$$L_p = \frac{3750 r_y}{M_p}\sqrt{JA}$$

$r_y = (0.29)(0.50) = 0.15$ in

$$M_p = F_Y Z_X = (36.0)\frac{(0.500)(14.0)^2}{4} = 882 \text{ kip-in}$$

$A = (0.500)(14.0) = 7.0$ in^2

$$J = \frac{1}{3}bt^3 = \frac{(14.0)(0.50)^3}{3} = 0.58 \text{ in}^4$$

$$L_p = \frac{(3750)(0.15)}{882}\sqrt{(0.58)(7.0)} = 1.28 \text{ in}$$

$L_b = 3.0$ in > 1.28 NG

$$L_r = \frac{57{,}000 r_y}{M_r}\sqrt{JA}$$

$$M_r = F_Y S = (36.0)\frac{(0.50)(14.0)^2}{6} = 529 \text{ kip-in}$$

$$L_r = \frac{(57{,}000)(0.15)}{529}\sqrt{(0.58)(7.0)} = 32.5 \text{ in}$$

Using LRFD (F1-3), compute M_n. (By observing $L_p = 1.28$ in, $L_r = 32.5$ in and $L_b = 3.0$ in, it can be estimated that $M_n \approx M_p$.)

$$M_n = C_b\left[M_p - (M_p - M_r)\frac{L_b - L_p}{L_r - L_p}\right]$$

$$= 1.0\left[882 - (882 - 529)\frac{3.0 - 1.28}{32.5 - 1.28}\right] = 863 \text{ kip-in}$$

$\phi M_n = (0.90)(863) = 777$ kip-in > 529 OK

Use $1/2$-in × 1 ft-2 in *shear plate.*

Weldment of Shear Plate to Web

$M_u = M_{ut} = 380$ kip-in

Use welds *only* at top and bottom of shear plate.

$$T_u = C_u = \frac{M_u}{d} = \frac{380}{14.0} = 27.1 \text{ kips}$$

If each weld is 5 in long "a" dimension:

$$n = \frac{T_u}{(1.39)(5.0)} = \frac{27.1}{(1.39)(5.0)} = 3.9 \text{ sixteenths}$$

Use $\frac{1}{4}$-in fillet weld top and bottom.

Weldment of Shear Plate to Column

$V_u = 57.1$ kips   and $M_u = M_{ut} + (V_u)(3.0) = 539$ kip-in

$$f_{uv} = \frac{57.1}{(2)(14)} = 2.0 \text{ kips/in}$$

$$f_{uh} = \frac{539}{(2)(14)^2/4} = 5.9 \text{ kips/in}$$

$$f_{ur} = \sqrt{(2.0)^2 + (5.5)^2} = 5.9 \text{ kips/in}$$

$$n = \frac{5.9}{1.39} = 4.2 \text{ sixteenths}$$

Use $\frac{5}{16}$-in fillet weld (both sides).

E.2.14 Column Stiffener Plates

Per LRFD Sec. K1, determine stiffener plate requirements for column. Also, consider the special provisions of SP Sec. 8.5 as related to LRFD (K1-1).

Local Flange Bending of Column

Flange design strength $= \phi R_n$

$\phi R_n = (0.9)(6.25t_f^2)(F_{yf})$ LRFD (K1-1)
$\quad\quad = (0.9)(6.25)(1.03)^2(36.0) = 215$ kips

$T_u = 1.8 F_{yb} b_f t_{bf} =$ girder flange force   SP Sec. 8.5
$\quad\quad = (1.8)(36.0)(8.965)(0.585) = 339$ kips

$\phi R_n = 215 < T_u = 339$ NG

Stiffeners required. Provide stiffeners to resist the difference between T_u and ϕR_n. See Fig. E.16.

$T_u - \phi R_n = 339 - 215 = 124$ kips

Since stiffeners must act in tension,

$\phi_t P_n = (0.90)(F_y A_g) = 124$ kips

$$A_g = \frac{124}{(0.90)(36.0)} = 3.8 \text{ in}^2$$

Try two plates, $\frac{1}{2}$-in × $4\frac{1}{2}$-in wide (clip corners $\frac{1}{2}$ in at column fillets).

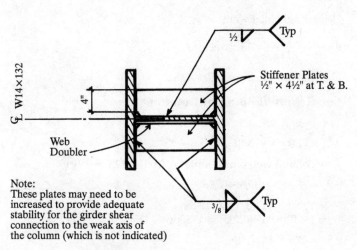

Fig. E.16 Plan of stiffeners.

$A_g = 2(0.50)(4.50 - 0.50) = 4.0 \text{ in} > 3.8$ OK

Use $\frac{1}{2}$-in × $4\frac{1}{2}$-in stiffeners.
Weldments to ends of stiffeners, using fillet welds top and bottom.

Force at each end of stiffener $= \dfrac{T_u - \phi R_n}{2} = 62$ kips

$n = \dfrac{62}{(2)(4)(1.39)} = 5.6$ sixteenths

Use $\frac{3}{8}$-in fillet weld top and bottom.

Weldment to sides of stiffeners assuming both ends resist forces acting in the same direction concurrently (Fig. E.17).

$V_u = (2)(62) = 124$ kips

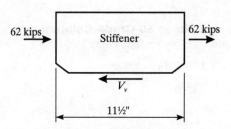

Fig. E.17 Stiffener free body.

Using single fillet weld

$$n = \frac{124}{(1.39)(11.5)} = 7.8 \text{ sixteenths}$$

Use ½-in fillet weld (one side).

Local Web Yielding of Column

Web design strength = ϕR_n

$\phi R_n = (1.0)(5k + N)(F_{yw}t_w)$ LRFD (K1-2)

t_w = column web plus doubler = 0.66 + 0.25 = 0.91 in

$\phi R_n = (1.0)(5 \times 1.69 + 0.585)(36.0)(0.91) = 276$ kips

$P_u = F_{yb} b_f t_{bf}$ = girder flange force
     = (36.0)(8.965)(0.585) = 188 kips

$\phi R_n = 276 > P_u = 188$ OK

Stiffeners not required.

Buckling of Web of Column

Web design strength = ϕR_n

$$\phi R_n = (0.90)\frac{4100 t_w^3 \sqrt{F_{yw}}}{d_c}$$ LRFD (K1-8)

t_w = column web only since doubler is only nominally connected over its width to the column web.

$$\phi R_n = (0.90)\frac{(4100)(0.66)^3 \sqrt{36.0}}{11.25} = 566 \text{ kips}$$

$P_u = F_{yb} b_f t_{bf}$ = girder flange force = 188 kips

$\phi R_n = 566 > P_u = 188$ OK

Stiffeners not required.

Alternate Design Using 50-Grade Columns ■ If 50 grade columns are used (F_y = 50 ksi), and checking for local flange bending of column per LRFD (K1-1),

$\phi R_n = (0.9)(6.25)(1.03)^2 (50.0) = 298$ kips

$T_u = 1.8 F_{yb} b_f t_{bf} = 339$ kips

$\phi R_n = 298 < T_u = 339$ NG (but close)

However, it is evident by increasing to the next larger size of W14, stiffeners can be eliminated by using 50 grade steel.

E.2.15 Check on Building Period

In Sec. E.2.2, seismic forces for drift were determined based on a "conservative" assumption of building deflection. This yielded a building period of 1.8 s. Since $V \propto 1/T^{2/3}$ for determining seismic forces, it should be verified once the frame members have been sized that the assumed period of 1.8 s was conservative; i.e., the actual period of building $\geq$ 1.8 s. This can easily be done by performing a computer analysis of the frame when members are sized and then determining the building period based on Rayleigh's formula or other appropriate methods.

E.3 Ordinary Moment Frames (OMF)

Section E.2 reworked using an ordinary moment frame

E.3.1 East-West Seismic Forces for Stress Analysis

$V = C_s W$ ASCE (9.2.3.2-1)

$$C_s = \frac{1.2C_v}{RT^{2/3}}$$ ASCE (9.2.3.2.1-1)

$C_v = 0.48$ per ASCE Table 9.1.4.2.4.B (by interpolation between B and C soil profile types).

$R = 4.5$ per ASCE Table 9.2.2.2

$T = $ to be determined per ASCE Sec. 9.2.3.3 and Sec. 9.2.3.7.1

Since drift control almost always controls the design of a steel moment frame, utilize Teal's formula and the drift limitations of ASCE Sec. 9.2.3.7.1 to determine building period T.

$\delta_x = (C_d)(\delta_{xe})$ ASCE 9.2.3.7.1 and $C_d = 4.0$ per ASCE Table 9.2.2.2

$\delta_x = 0.020$ h $= 4.0 \, \delta_{xe}$

$\delta_{xe} = 0.0050$ h (at code seismic force level)

To be conservative for building period determination, assume $\Delta = $ two-thirds of the allowable drift at the top of the building.

$\Delta = (2/3)(\delta_{xe}) = (0.67)(0.005 \text{ h})$
$\quad = (0.67)(0.005)(12 \times 83.0) = 3.34$ in

$$T = 0.25 \sqrt{\frac{\Delta}{C_1}} \quad \text{and} \quad C_1 = \frac{V}{W}$$

$$V = \frac{(1.2)(0.4)(1.2)(W)}{(4.5)(T^{2/3})} = \frac{0.128W}{T^{2/3}}$$

$$C_1 = \frac{V}{W} = \frac{0.128}{T^{2/3}}$$

E-34 ■ Appendix E

Thus, using Teal's formula,

$$T = 0.25\sqrt{\frac{3.34}{0.128/T^{2/3}}} = 0.25\sqrt{26.1T^{2/3}} = 1.28T^{1/3}$$

$T^{2/3} = 1.28 \quad$ or $\quad T = (1.28)^{3/2}$

$T = 1.45$ s

Use $T = 1.50$ s.

Hence

$$V = \frac{0.128W}{T^{2/3}}$$

$$V = \frac{(0.128)(W)}{(1.50)^{2/3}} = 0.098W$$

$V = (0.098)(5380) = \underline{526 \text{ kips}}$ (total seismic force)

E.3.2 Comparison of OMF with SMF

Since drift criteria almost always determine the member sizes of moment frames, a comparison of the OMF with the SMF is made on this basis. For SMF drift criteria, see Sec. E.2.2.

SMF:

$V = 262$ kips with $\delta_{xe} = 0.0036$ h

OMF:

$V = 526$ kips with $\delta_{xe} = 0.0050$ h

or comparing equals with equals for a drift index of 0.0036 h

OMF:

$$V = (526)\frac{0.0036}{0.0050} = 378 \text{ kips}$$

Thus, comparing equivalent base shear of the OMF with the SMF, the ratio is

$$\frac{V_{OMF}}{V_{SMF}} = \frac{378}{262} = 1.44$$

OMF 44 *percent greater stiffness than SMF.*

This indicates that the code "discourages" the use of OMF by the imposition of these requirements. It is preferred that a SMF be utilized. If the same depth of framing is used for both the SMF and OMF, the general conclusion is:

OMF 40 *percent heavier than SMF.*

E.3.3 Conclusion

In general, use SMF in lieu of OMF. Exceptions: small buildings, preengineered buildings, etc.

E.4 Braced Frames (CBF)

For seismic performance category D

E.4.1 North-South Seismic Forces (Stress and Drift)

$V = C_S W$ ASCE (9.2.3.2-1)

$C_S = \dfrac{1.2C_v}{RT^{2/3}}$ ASCE (9.2.3.2.1-1)

C_v = 0.48 per ASCE Table 9.1.4.2.4B (by interpolation between B and C soil profile types)

R = 5.0 per ASCE Table 9.2.2.2

T = to be determined per ASCE Sec. 9.2.3.3

Owing to the rigidity of a braced frame, coupled with its tremendous efficiency, it is recommended for most braced frames (not excessively tall or narrow) that a conservative method be used for determining building, namely,

$T_a = C_T h_n^{3/4}$ ASCE (9.2.3.3.1-1)

$T_a = (0.020)(83.0)^{3/4} = 0.55$ s

Use T = 0.55 s.

It should be emphasized that this period should be checked once the building is designed so that the actual period is *at least* 0.55 s. If it is not, the building must be designed for larger seismic forces. Drift limits usually do not control the design of a braced frame, but there are exceptions for tall, narrow braced frames. Taking T = 0.55 s,

$C_S = \dfrac{(1.2)(0.48)}{(5.0)(0.55)^{2/3}} = 0.172$

$V = 0.172W = (0.172)(5380) = 925$ kips (total seismic force)

Vertical distribution of seismic forces as follows:

$F_X = C_{VX} V$ ASCE (9.2.3.4-1)

$C_{VX} = \dfrac{w_x h_x^k}{\sum\limits_{i=1}^{n} w_i h_i^k}$ ASCE (9.2.3.4-2)

$k = 1 + 1\dfrac{0.55 - 0.50}{2.0} = 1.03$

$V_X = \sum\limits_{i=x}^{n} F_i$ ASCE (9.2.3.5)

where V_X = story shear at any level, and its magnitude is shown in Table E.3.

Table E.3 Seismic Forces

Floor level	w_x, kips	h_x, ft	h_x^k	$w_x h_x^k \times 10^{-3}$	C_{VX}	F_x kips	V_X kips
R	689	83.0	95	65.4	0.23	213	
7	782	71.5	82	63.3	0.22	203	213
6	782	60.0	68	53.1	0.18	166	416
5	782	48.5	54	42.2	0.15	139	582
4	782	37.0	41	32.0	0.11	102	721
3	782	25.5	28	21.9	0.07	65	823
2	782	14.0	15	11.7	0.04	37	888
1							925
Σ	5381			290	1.00	925	

1. $k = 1.03$, and $F_x = C_{VX}V = (C_{VX})(925)$.
2. When the members of the braced frame are determined, a verification *must* be made to ensure that the actual period of the building is equal to or greater than 0.55 s (the basis for Table E.3). If this is not the case, the seismic forces must be recomputed and the building reanalyzed.
3. Forces and shears specified are to determine sizes of frame members, frame drift, and overturning at base of building. However, for "V braces" *only*, these forces must be increased by a factor of 1.5, in accordance with SP Sec. 9.4.a.

E.4.2 Horizontal Distribution of Seismic Forces

As determined previously in Sec. E.2.3,

$e = (0.05)(120) = 6.0$ ft
$R_1 = R_5 = 4.00$ (approx.)
$R_A = R_D = 1.00$ (approx.)

Shear distributions in north-south direction:

$$V_{1,x} = (R_1)\left[\frac{V_x}{\Sigma R_{N-S}} \pm \frac{(V_x e)(d)}{\Sigma R_y d^2}\right] = V_{5,x}$$

$\Sigma R_{N-S} = 2(4.00) = 8.00$
$\Sigma R_y d^2 = 2(1.00)(37.5)^2 + 2(4.00)(60.0)^2 = 31{,}600$

$$V_{1,x} = (4.00)\left[\frac{V_x}{8.00} \pm \frac{(V_x \times 6.00)(60.00)}{31{,}600}\right]$$

$$= (4.00)(0.125 V_x + 0.011 V_x) = 0.545 V_x = V_{5,x}$$

E.4.3 Bracing System Configuration

Possible bracing systems that might be utilized are shown in Fig. E.18. An important design consideration in selecting a bracing system is overturning due to earthquake forces. Overturning is relatively easy to resist above the base of the building, but it frequently presents a problem at the foundation since most conventional foundation systems *cannot* resist tension forces.

E.4.4 Foundation Overturning Considerations

Per ASCE Sec. 9.2.3.6, overturning for a building is specified to be

$$M_x = \tau \sum_{i=x}^{n} F_i(h_i - h_x) \qquad \text{ASCE (9.2.3.6)}$$

where $\tau = 1.0$ for the top 10 stories, except $\tau = 0.75$ at the foundation-soil interface for all building heights. The overturning moments are summarized in Table E.4 for the entire building using the relationship indicated.

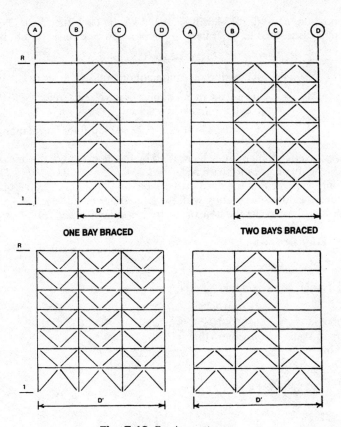

Fig. E.18 Bracing systems.

Table E.4 Seismic Overturning Moment

Floor level	V_x, kips	Story height H, ft	V_xH	$M_X = \Sigma V_X H$	Y	Moment $= \Sigma V_x H$, kip-ft
R						
7	213	11.5	2,450	2,450	1.0	2,450
6	416	11.5	4,780	7,230	1.0	7,230
5	582	11.5	6,690	13,920	1.0	13,920
4	721	11.5	8,290	22,210	1.0	22,210
3	823	11.5	9,460	31,670	1.0	31,670
2	888	11.5	10,200	41,870	1.0	41,870
1	925	14.0	12,950	54,820	1.0	54,820
Σ			54,820			

At foundation, $M = (0.75)(54,820) = 41,120$ kip-ft (soil interface). Overturning moment is distributed to the frames in the same proportion as the shears:

$$M_{1,x} = M_{5,x} = 0.545 M_x$$

where M_x = total earthquake moment on building at story x

$M_{y,x}$ = earthquake moment on a braced frame, referred to that frame on column line y at level x

$M_{1,1} = M_{5,1} = (0.545)(41,120) = 22,400$ kip-ft (at foundation)

This overturning moment must be resisted by the dead load of the braced portion of the frame. In accordance with ASCE (9.2.2.6-2), $E = \pm 1.0\, Q_E - 0.5 C_a D$ and if $C_a = 0.4$, only 80 percent of the dead load can be used (to account for vertical seismic uplift forces acting concurrently with the lateral seismic forces).

Consider the cases at the base of the frames shown in Fig. E.18.

One Bay Braced

$M_{1,1} = M_{5,1} = 22,400$ kip-ft

Dead load of columns on line B and C:

Roof	=	$(407)(0.067)$	= 27 kips
6 Floors	=	$6(407)(0.085)$	= 208
Curtain Wall	=	$(1800)(0.015)$	= 27
Footing	=		= 30
		$P_B = P_C$	= 292 kips

$W_{DL} = 2(292) = 584$ kips

$$M_R = W_{DL} \frac{D'}{2}$$

where D' = width of a braced frame at base
M_R = dead load resisting moment of a frame
W_{DL} = dead load of a braced frame
M_R = (584)(25/2)(0.80)
    = 5840 < 22,400 kip-ft   NG

Overturning exceeds resisting moment. This indicates that the frame is unstable unless the resisting moment is increased by using caissons, piles, or other means which will increase the dead load of the braced portion of the frame.

Two Bays Braced ■ By comparison, with one bay braced, the frame would be unstable unless caissons, etc., are used.

Three Bays Braced

$M_{1,1} = M_{5,1} = 22,400$ kip-ft

Dead load of columns on lines A, B, C, and D:

$W_{DL} = 2(292) + 2(191) = 966$ kips

$$M_R = (966)\frac{75.0}{2}(0.80) = 28{,}980 > 22{,}400 \text{ kip-ft} \qquad \text{OK}$$

This illustrates that "three bays braced" would be required to resist the seismic overturning *without* the use of caissons, piles, or mass concrete. It should be noted that if the "combination of bays braced" is utilized, a dynamic (modal) analysis would be required as required by ASCE Table 9.2.2.4.3 for seismic performance category D. This is due to its vertical structural irregularity. Let us assume for the problem at hand that the frame is a "one bay braced" with appropriate caissons to prevent uplift due to seismic overturning.

E.4.5 Frame Analysis

Design of frame will be limited to fourth floor girders, third to fifth floor columns (one tier), and fourth story braces as shown in Fig. E.19.

Frame Forces Due to Seismic Load (See Fig. E.19a) ■ For the design of both girders and braces, the effect of seismic vertical accelerations will be neglected since they are negligible ($0.5C_aD$ term not used).

Since $V_{1,x} = V_{5,x} = 0.545V_x$, the shears at the third and fourth stories are

$V_{1,4} = V_{5,4} = (0.545)(721) = 393$ kips

$V_{1,3} = V_{5,3} = (0.545)(823) = 449$ kips

Overturning moment at the fourth floor is

$M_{1,4} = M_{5,4} = (0.545)(22{,}210) = 12{,}100$ kip-ft

At Sec. 1-1, taking moments about point f and solving for the axial force in member ad:

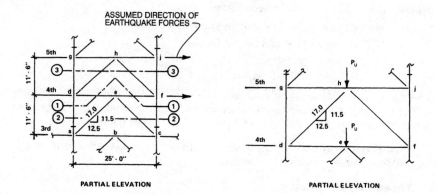

Fig. E.19 Forces acting on frame.

$\Sigma M_f = 0 = 12{,}100 - (25.0)(P_{ad})$

$P_{ad} = \dfrac{12{,}100}{25.0} = 484$ kips

$P_{cf} = 484$ kips

At joint e,

$\Sigma F_y = 0 = -\left(\dfrac{11.5}{17.0}\right)(P_{ae}) - \dfrac{11.5}{17.0}(P_{ce})$

$P_{ae} = P_{ce}$

By taking $\Sigma F_x = 0$, it can be shown that

$P_{de} = -P_{ef}$

At Sec. 2-2,

$\Sigma F_x = 0 = 499 - \dfrac{12.5}{17.0}(P_{ae}) + \dfrac{12.5}{17.0}(P_{ce})$

Since $P_{ae} = -P_{ce}$

$449 = 2\dfrac{12.5}{17.0}(P_{ae})$

$P_{ae} = 305$ kips

$P_{ce} = -305$ kips

At Sec. 3-3,

$\Sigma F_x = 0 = 393 - \dfrac{12.5}{17.0}(P_{dh}) + \dfrac{12.5}{17.0}(P_{fh})$

P_{dh} = 267 kips

P_{fh} = −267 kips

At Sec. 1-1,

$\Sigma F_x = 0 = 449 + P_{de} - P_{ef}$

Since:

$P_{de} = -P_{ef}$

$P_{de} = -225$ kips

$P_{ef} = 225$ kips

Frame Forces Due to Gravity Loads (See Fig. E.19b)

P_u = floor + curtain wall

$P_D = (16.25 \times 12.5)(0.085) + (11.5 \times 12.5)(0.015)$

$P_D = 17.3 + 2.2 = 19.5$ kips

$P_L = (16.25 \times 12.5)(0.050) = 10.2$ kips

$P_u = 1.2P_D + 0.5P_L$ SP (3-5)

$P_u = (1.2)(19.5) + (0.5)(10.2) = 28.6$ kips

At joint h:

$$\Sigma F_y = 0 = -28.6 - \frac{11.5}{17.0}(P_{dh}) - \frac{11.5}{17.0}(P_{fh})$$

From $\Sigma F_X = 0$ along Sec. 1-1 and then joints d and f, it can be shown that $P_{dh} = P_{fh}$. Thus,

$$28.6 = -2\frac{11.5}{17.0}(P_{dh})$$

$P_{dh} = -21.1$ kips $P_{fh} = -21.1$ kips

At joint d:

$$\Sigma F_X = 0 = \frac{12.5}{17.0}(P_{dh}) + P_{de}$$

$$P_{de} = -\frac{12.5}{17.0}(-21.1) = 15.5 \text{ kips}$$

E.4.6 Design of Brace (Fourth Story)

Per SP Sec. 9.4.a, for "V bracing" the forces due to factored loads must be increased by a factor of 1.5. (This factor applies only to the braces, not other members of the braced frame.) Thus, using (SP 3-5):

$P_u = 1.5(1.2\ P_D + 0.5\ P_L + 1.0\ P_E)$

$1.2\ P_D + 0.5\ P_L = 21.1$ kips (C)   (see Sec. E.4.5)

$1.0\ P_E = 267$ kips (C)   (see Sec. E.4.5)

$P_u = 1.5(21 + 267) = 432$ kips (C)

Per SP Sec. 9.2.b, brace must meet the following requirements:

Design strength = $0.8\ \phi_c P_n$

Slenderness = $\dfrac{L}{r} \le \dfrac{720}{\sqrt{F_y}}$

Width to thickness ratio = $\lambda < \lambda_r$

Select W14 brace based on AISC p. 2-21, taking $L = 17.0$ ft and $K = 1.0$ (conservative values).

Try W14×82:

$0.8\phi_c P_u = (0.8)\dfrac{495 + 538}{2} = 413$ kips

$0.8\phi_c P_u = 413 \approx P_u = 432$ kips   OK

$\dfrac{L}{r} \le \dfrac{720}{\sqrt{F_Y}} = \dfrac{720}{6} = 120$

$\left(\dfrac{L}{r}\right)_{actual} = \dfrac{17.0 \times 12}{2.48} = 82 < 120$ OK

$\lambda < \lambda_r$ by inspection of AISC, p. 2-21 OK

Use W14×82 brace.

(In Sec. E.4.9.4, size increased to W14×90 to provide adequate effective net area at bolt holes.)

E.4.7 Design of Column (Third to Fifth Floor)

Use the critical load combination SP (3-5) at the third story:

$P_u = 1.2P_D + 1.0P_E + 0.5P_L$

P_D = roof + 4 floors + curtain wall
  = $(25.0 \times 16.25)(0.067) + 4(25.0 \times 16.25)(0.085) + (25.0 \times 60.5)(0.015)$

$P_D = 27 + 138 + 23 = 188$ kips (C)

P_L = 4 floors
  = $4(25.0 \times 16.25)(L)$

where $L = L_0\left(0.25 + \dfrac{15}{\sqrt{A_1}}\right)$ ASCE (Eq. 1)

$$L = 50\left(0.25 + \frac{15}{\sqrt{4 \times 1625}}\right) = (50)(0.44) = 22 \text{ psf}$$

$P_L = 4(25.0 \times 16.25)(0.022) = 36$ kips (C)

$P_E = Q_E + 0.5C_aD = Q_E + (0.5)(0.4)(D)$ ASCE (9.2.2.6-1)

$P_E = 484 + (0.2)(188) = 522$ kips

$P_u = (1.2)(188) + (1.0)(522) + (0.5)(36)$

$P_u = 766$ kips (C)

Select W14 column based on AISC, p. 2-20, taking $L = 11.5$ ft and $K = 1.0$.
Try W14×90 where $\phi_c P_n = 749 \approx 766$ kips   OK
However, column must also meet the requirements of SP Sec. 6.1:

$$\frac{P_u}{\phi P_n} = \frac{766}{749} \approx 1.00 > 0.5$$

Thus SP (6-1) must be satisfied:

$1.2P_D + 0.5P_L + (0.4R)(P_E) \leq \phi_c P_n$

Since $R = 5$ for a concentrically braced frame, $(0.4R)(P_E) = 2.0P_E$

$P_u = (1.2)(188) + (0.5)(36) + (2.0)(522)$

$P_u = 1288$ kips (C) > 749 N.G.

W14×90 not adequate.
Thus a stronger W14 column must be selected based on AISC, p. 2-19, taking $L = 11.5$ ft and $K = 1.0$.
Try W14×159 where $\phi_c P_n = 1330 > 1288$ kips   OK
Use W14×159 column.

Note: This column size may be excessive since the resistance of the frame may *not* be capable of resisting seismic forces = 2.0 P_E. For limitations, see SP Sec. 6.1.c, especially item 2 regarding capacity of the foundation to resist overturning *uplift*.

E.4.8 Design of Girder (Fourth Floor)

In order to provide a very convenient method of connection of the brace to girder, use a W14 for the girder and orient W14 with the web in a horizontal plane as shown in Fig. E.20. This type of double gusset plate connection can accommodate very large loads, and it is very stable. That is, it can develop the strength of the frame members without a premature or sudden inelastic failure of the connection. Per SP Sec. 9.4.a, item 3, the girder for a "V bracing" system must be capable of supporting gravity loads, assuming the bracing is not effective in supporting these gravity loads. This is indicated in Fig. E.21.

For gravity loads on girder to satisfy item 3, use critical load combination SP 3-2:

$P_u = 1.2P_D + 1.6P_L$

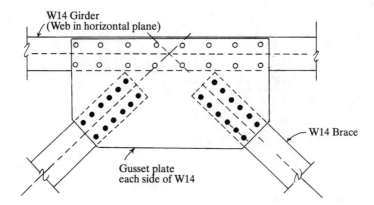

Fig. E.20 Brace to girder connection.

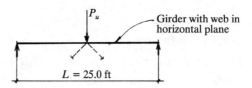

Fig. E.21 Girder free body.

where $P_D = 19.5$ kips and $P_L = 10.2$ kips (see Sec. E.4.5)

$P_u = (1.2)(19.5) + (1.6)(10.2) = 39.7$ kips

Girder moment without braces:

$$M_u = \frac{P_u L}{4} = \frac{(39.7)(25.0)}{4} = 248 \text{ kip-ft}$$

$$\phi M_n = (0.9)(F_y)(Z_y) = M_u = 248; \ Z_y = 92 \text{ in}^3 \text{ required}$$

$$Z_{y,\text{req'd}} = \frac{(248)(12)}{(0.90)(36.0)} = 92 \text{ in}^3$$

Try W14×109 ($Z_y = 92.7 > 92$ OK)
(Do not need to check lateral buckling when members are bent about minor axis per LRFD Sec. F1.7.)

Now check W14×109 for loading condition where seismic forces act on girder with braces intact, using the critical loading condition SP (3-5):

$P_u = 1.2P_D + 0.5P_L + 1.0P_E$

$1.2P_D + 0.5P_L = 0$ (neglect since in girder this causes tension only)

$1.0 P_E = 225$ kips (C) \hfill (see Sec. E.4.5)

$P_u = 0 + 225 = 225$ kips (C)

Concurrent with this compressive force, there exists some transverse loading from floor and curtain wall producing flexural stresses:

$M_u = 1.2 M_D + 0.5 M_L$

$W_D = (3.25)(0.072) + (11.5)(0.015) + 0.130 = 0.53$ kips/ft

$$M_D = \frac{(0.53)(12.5)^2}{8} = 10.4 \text{ kip-ft}$$

$W_L = (3.25)(0.050) = 0.16$ kips/ft

$$M_L = \frac{(0.16)(12.5)^2}{8} = 3.2 \text{ kip-ft}$$

$M_u = (1.2)(10.4) + (0.5)(3.2) = 14.1$ kip-ft

Per LRFD Sec. H1, for combined forces:

$M_{nt} = M_u = 14.1$ kip-ft

and the design

$M_u = B_1 M_{nt} + B_2 M_{lt}$ \hfill LRFD (H1-2)

where $M_{lt} = 0$

$$B_1 = \frac{C_m}{1 - P_u/P_e} \geq 1$$

Since

$$\left(\frac{Kl}{r}\right)_y = \left(\frac{1.0 \times 12.5 \times 12}{3.73}\right) = 40$$

$$\frac{P_e}{A_g} = 179 \hspace{2em} \text{per AISC, p. 6-131 (Table 9)}$$

$P_e = (179)(32.0) = 5728$ kips

$C_m = 1.00$ due to transverse loading

$$B_1 = \frac{1.00}{1 - 225/5728} = \frac{1.00}{0.96} = 1.04$$

$M_{uy} = B_1 M_{nt} = (1.04)(14.1) = 15$ kip-ft

Now checking the interaction of flexure and compression, for $(KL/r)_y = 40$:

$\phi_c F_{cr} = 28.1$ ksi \hfill per AISC, p. 6-124

$\phi P_n = (\phi_c F_{cr})(A_g) = (28.1)(32.0) = 899$ kips

$$\frac{P_u}{\phi P_n} = \frac{225}{899} = 0.25 \geq 0.20$$

Thus, interaction equation LRFD (H1-1a) must be used:

$$\frac{P_u}{\phi P_n} + \frac{8}{9}\frac{M_{uy}}{\phi_b M_{ny}} \leq 1.0$$

where $\phi_b M_{ny} = (0.9)(F_y)(Z_y) = \dfrac{(0.9)(36.0)(92.7)}{12} = 250$ kip-ft

$$\frac{225}{899} + \frac{8}{9}\left(\frac{15}{250}\right) = 0.25 + 0.05 = 0.30 \leq 1.00 \qquad \text{OK}$$

Use W14×109 girder.

E.4.9 Connection of Brace to Girder

Design Criteria for Bolts ■ Design connections using high-strength bolts. Capacity of bolts taken in accordance with LRFD Specification for Structural Joints Using ASTM A325 or A490 Bolts, dated June 8, 1988. In accordance with this bolt specification, "slip-critical" joints must be used since joints are subject to significant stress reversal. For <u>1-in φ, A325 bolt in "slip-critical"</u> joint, single shear, using factored loads, the capacity can be computed in accordance with:

$\phi R_{str} = 1.13_u T_m N_b N_s = (1.13)(0.33)(51)(1.0)(1.0) = 19.0$ kips per bolt

Required Strength of Connection ■ Per SP Sec. 9.3.a, the connections for bracing (*including* beam to column joints that are part of the bracing system) should be the *least* of the following criterion:

(i) Design tension strength of bracing member.

$P_u = \phi_t F_y A_g$

(ii) Force in brace resulting from SP (3-7) or SP (3-8).

$P_u = 1.2 P_D + 0.5 P_L + 2.0 E$

(iii) The maximum force that can be transferred to the brace by the system.

Criterion (iii) may often be the *least* of these three items, especially if overturning about the foundation is governed by uplift capacity. However, in this example let us assume that criterion (iii) is not known at this point in the design.

Connection Configuration ■ Let us assume that the connection detailing shown in Fig. E.22 will be used to join the braces to the girder members. For additional information on this double gusset plate connection, see Sec. E.4.8.

Bolts to Brace ■ The required strength for the W14×82 is:

(i) $P_u = \phi_t F_y A_g = (0.9)(36.0)(24.1) = 781$ kips

(ii) $P_u = 1.2 P_D + 0.5 P_L + 2.0 P_E$

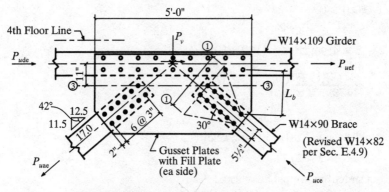

Fig. E.22 Brace to girder connection.

$1.2P_D + 0.5P_L = 21$ kips

$2.0P_E = (2.0)(267) = 534$ kips

$P_u = 21 + 534 = 555$ kips (governs)

Design of brace connection based on 555 kips (smaller force)

$$n = \frac{P_u}{19.0} = \frac{555}{19.0} = 29.2 \text{ bolts}$$

Use 28-1 in A325 SC bolts   (4 percent overstressed)   (OK)

Per SP Sec. 9.3.b, the net section of W14×82 brace must be checked to ensure that in the event of an overload condition, *yielding in lieu of rupture* of the brace would occur:

$$\frac{A_e}{A_g} \geq \frac{1.2 \alpha P_u}{\phi_t P_n}$$ SP (9-1)

where $\alpha = 1.00$

P_u = required strength per SP Sec. 9.3.a = 555 kips

$\phi_t = 0.75$

P_n = nominal tension fracture strength per LRFD Sec. D1

$P_n = F_u A_e$

Per LRFD (B3-1):

$$A_{e\,actual} = UA_n = (0.90)(24.1 - 4 \times 0.855 \times 1.125)$$
$$A_{e\,actual} = (0.90)(20.25) = 18.2 \text{ in}^2$$

$$\frac{A_e}{A_g} \geq \frac{(1.2)(1.0)(555)}{(0.75)(58.0 \times 18.2)} \geq 0.84$$

or

$$A_e \geq 0.84 A_g = (0.84)(24.1) = 20.2 \text{ in}^2$$
$$A_{e\,actual} = 18.2 < 20.2 \qquad\qquad\qquad \text{NG}$$

Thus a larger brace must be selected.
Try W14×90.

$$A_{e\,actual} = UA_n = (0.90)(26.5 - 4 \times 0.710 \times 1.125) = 21.0 \text{ in}^2$$

$$\frac{A_e}{A_g} \geq \frac{(1.2)(1.0)(555)}{(0.75)(58.0 \times 21.0)} \geq 0.73$$

or

$$A_e \geq 0.73 A_g = (0.73)(26.5) = 19.3 \text{ in}^2$$
$$A_{e\,actual} = 21.0 > 19.3 \qquad\qquad\qquad \text{OK}$$

Use W14×90 brace (Revise Sec. E.4.6 brace size).

Bolts to Girder ■ The required strength is based on the condition where the braces are loaded to an ultimate force P_u. See Fig. E.22 for forces acting on connection.

$$P_{u\,ce} = 1.2 P_D + 0.5 P_L + 2.0 E = 21 + 534 = 555 \text{ kips (C)}$$

and

$$P_{u\,ae} = -21 + 534 = 513 \text{ kips (T)}$$

Both $P_{u\,ce}$ and $P_{u\,ae}$ occur concurrently. Thus, to resist the horizontal component of these brace forces (see Fig. E.22)

$$P_{u\,de} + P_{u\,ef} = \frac{12.5}{17.0}(555) + \frac{12.5}{17.0}(513)$$

$$P_{u\,de} + P_{u\,ef} = 785 \text{ kips}$$

Design of girder connection to transfer 785-kip horizontal force (neglecting small vertical force P_v):

$$n = \frac{785}{19.0} = 41.3 \text{ bolts}$$

Use 40-1 in φ A325 SC bolts.
Per SP Sec. 9.3.b, the net section of the W14×109 girder is checked:

$$\frac{A_e}{A_g} \geq \frac{1.2 \alpha P_u}{\phi_t P_n} \qquad \text{SP 9-1}$$

$$P_u = \frac{12.5}{17.0}(555) = 408 \text{ kips}$$

$A_{e\,actual} = (0.90)(32.0 - 4 \times 0.860 \times 1.125) = 25.4 \text{ in}^2$

$$\frac{A_e}{A_g} \geq \frac{(1.2)(1.0)(408)}{(0.75)(58.0 \times 25.4)} \geq 0.44$$

$A_e \geq 0.44 A_g = (0.44)(32.0) = 14.2 \text{ in}^2$

$A_{e\,actual} = 25.4 > 14.2$ \qquad OK

Use W14×109 girder.

Gusset Plate Thickness ■ Try ½-in-thick gusset plates. Base analysis upon "Whitmore's Method" and the "Method of Sections."
 (i) First, use "Whitmore's Method" and investigate design tension strength at the base of section 1-1, Fig. E.22. For strength capacity use LRFD Sec. D1.

Effective width = $2(18.0 \times \tan 30°) + 5.5 - 2(1.125)$

Effective width = 23.9 in

$A_e = 2(0.50)(23.9) = 23.9 \text{ in}^2$

$\phi_t P_n = 0.75\, F_u A_e = (0.75)(58.0)(23.9) = 1040 \text{ kips}$

or

$\phi_t P_n = 0.90\, F_y A_g = (0.90)(36.0)(26.4) = 855 \text{ kips}$ \qquad (governs)

$P_u = 555 \text{ kips} < \phi_t P_n = 855$ \qquad OK

 (ii) Second, use the "Method of Sections" and investigate both design shear and flexural strength at section 3-3 of Fig. E.22.

$V_u = (P_{u\,de} + P_{u\,ef}) = 785 \text{ kips}$

Per LRFD (J5-3), design shear strength = ϕR_n

$\phi R_n = (0.80)(0.7\, A_g F_y)$
    $= (0.80)(0.7 \times 2 \times 0.50 \times 60)(36.0) = 1210 \text{ kips}$

$1210 > 785$ \qquad OK

$M_u = (V_u)(11.0 \text{ in}) = (785)(11.0) = 8640 \text{ kip-in}$

Per LRFD (F1-5), (F1-10), and (F1-14), where design flexural strength = $\phi_b M_n$
Using two plates ½-in thick by 60 in wide:
$L_b \approx 20$ in as shown on Fig. E.22.

$$L_p = \frac{3750 r_y}{M_p}\sqrt{JA}$$

$r_y = 0.29t = 0.145$ in; $\quad J = \frac{bt^3}{3} = \frac{(2)(60)(0.50)^3}{3} = 5.0$ in^4

$A = (2)(60)(0.5) = 60$ in^2; $\quad M_p = \frac{(2)(0.50)(60)(36.0)}{4} = 32{,}400$ kip-in

$$L_p = \frac{(3750)(0.145)}{32{,}400}\sqrt{(5.0)(60)} = 0.29 < L_b$$

$$L_r = \frac{57{,}000 r_y}{M_r}\sqrt{JA} = \frac{(57{,}000)(0.145)}{0.67 \times 32{,}400}\sqrt{(5.0)(60)} = 6.7 < L_b$$

Hence LRFD F1-14 is determined:

$$M_{cr} = \frac{57{,}000 C_b \sqrt{JA}}{L_b / r_y} = \frac{(57{,}000)(1.75)\sqrt{(5.0)(60)}}{20/0.145} = 12{,}500 \text{ kip-in}$$

$\phi M_n = \phi M_{cr} = (0.90)(12{,}500) = 11{,}250 > M_u = 8640$

Use ½-in gusset plate.

E.4.10 Connection of Brace to Column and Girder

Utilize the double gusset plate connection as shown in Fig. E.23. Design of connection is similar to Sec. E.4.9.

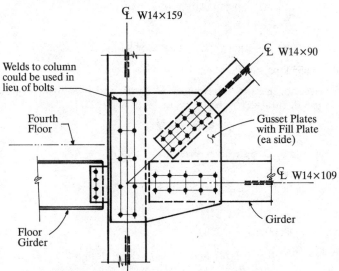

Fig. E.23 Brace to column and girder connection.

E.4.11 Alternate Braced Frame Connection Details

In lieu of the connection details just shown, the details depicted in this section, which have a single gusset plate, could be utilized. These connections differ from those previously shown in the following respects:

1. A single gusset plate is used rather than double gussets.
2. The web of the girder lies in a vertical plane rather than in a horizontal plane, thus providing greater gravity load carrying capacity for the girder.
3. The connection does not possess the torsional rigidity that is provided by the double gusset plate system.
4. These connections are best suited for braced frames resisting "light to moderate" lateral loads, while the double gusset plate connections are best suited for frames resisting "moderate to heavy" loads.

See Figs. E.24, E.25, and E.26 for details.

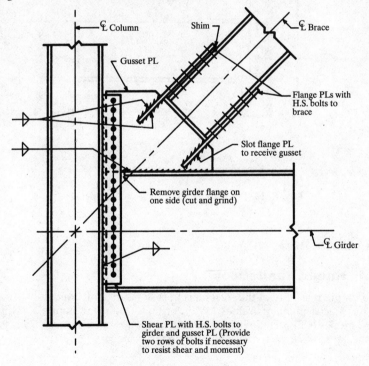

Notes:
1. Shop connections made with welds and field connections with H.S. bolts
2. Shear PL may be replaced by a pair of framing angles.

Fig. E.24 Brace and girder to column flange detail.

E-52 ■ Appendix E

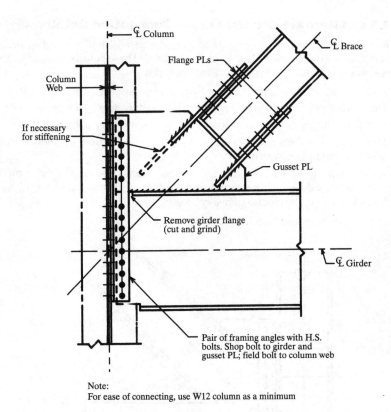

Fig. E.25 Brace and girder to column web detail.

E.5 Conclusion

E.5.1 Height Limitations

Based on the reliability of the steel bracing systems, height, limitations are specified for special moment frames (SMF), ordinary moment frames (OMF), and braced frames (CBF). These height limitations are specified in ASCE Table 9.2.2.2, and they are as follows for Seismic Performance Category D:

SMF	Unlimited
OMF	160 ft
CBF	160 ft

E.5.2 Energy Absorption and Ductility

The ability of a steel bracing system to absorb energy inelastically through ductile behavior is somewhat proportionate to its R value assigned the system by ASCE Table 9.2.2.2. The larger the R value, the better the ductility of the system. R values are as follows:

SMF   8

OMF $4\frac{1}{2}$

CBF   5

The basic characteristics of the CBF and the lack of rigid design requirements of the OMF do not assure that large quantities of energy can be absorbed inelastically by these bracing systems in the event of a major earthquake. However, the reverse is true of the SMF.

Because of both the basic characteristics and the many design requirements of the SMF, this system is indeed capable of absorbing large quantities of energy inelastically through the formation of plastic hinges. These hinges can undergo large inelastic rotations without the premature failure of the frame's connections and/or premature buckling of its members.

This does not necessarily mean that the SMF will perform better than the OMF or CBF in the event of a major earthquake, as was surprisingly evident in the Northridge Earthquake of 1994. The performance of steel bracing systems will depend largely on both the expertise of the structural engineer and careful attention to details, especially the design of connections.

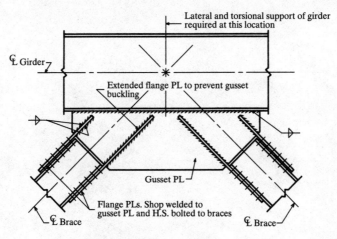

Fig. E.26 Brace to girder detail.

APPENDIX F

NOMENCLATURE

A	Cross-sectional area, in^2
A_B	Loaded area of concrete, in^2
A_b	Nominal body area of a fastener, in^2
A_c	Area of concrete, in^2
A_c	Area of concrete slab within effective width, in^2
A_D	Area of an upset rod based on the major diameter of its threads, in^2
A_e	Effective net area, in^2
A_f	Area of flange, in^2
A_{fe}	Effective tension flange area, in^2
A_{fg}	Gross area of flange, in^2
A_{fn}	Net area of flange, in^2
A_g	Gross area, in^2
A_{gt}	Gross area subject to tension, in^2
A_{gv}	Gross area subject to shear, in^2
A_n	Net area, in^2
A_{nt}	Net area subject to tension, in^2
A_{nv}	Net area subject to shear, in^2
A_{pb}	Projected bearing area, in^2
A_r	Area of reinforcing bars, in^2
A_s	Area of steel cross section, in^2
A_{sc}	Cross-sectional area of stud shear connector, in^2
A_{sf}	Shear area on the failure path, in^2
A_w	Web area, in^2
A_1	Area of steel bearing concentrically on a concrete support, in^2
A_2	Total cross-sectional area of a concrete support, in^2

B	Factor for bending stress in tees and double angles
B	Factor for bending stress in web-tapered members
B_1, B_2	Factors used in determining M_u for combined bending and axial forces when first-order analysis is employed
C_{PG}	Plate-girder coefficient
C_b	Bending coefficient dependent on moment gradient
C_m	Coefficient applied to bending term in interaction formula for prismatic members and dependent on column curvature caused by applied moment
C_m'	Coefficient applied to bending term in interaction formula for tapered members and dependent on axial stress at the small end of the member
C_p	Ponding flexibility coefficient for primary member in a flat roof
C_s	Ponding flexibility coefficient for secondary member in a flat roof
C_v	Ratio of "critical" web stress, according to linear buckling theory, to the shear yield stress of web material
C_w	Warping constant, in²
D	Outside diameter of circular hollow section, in
D	Dead load due to the weight of the structural elements and permanent features on the structure
D	Factor dependent on the type of transverse stiffeners used in a plate girder
E	Modulus of elasticity of steel ($E = 29{,}000$ ksi)
E	Earthquake load
E_c	Modulus of elasticity of concrete, ksi
E_m	Modified modulus of elasticity, ksi
F_{BM}	Nominal strength of the base material to be welded, ksi
F_{EXX}	Classification number of weld metal (minimum specified strength), ksi
F_L	Smaller of $(F_{yf} - F_r)$ or F_{yw}, ksi
F_{by}	Flexural stress for tapered members
F_{cr}	Critical stress, ksi
$F_{crft}, F_{cry}, F_{crz}$	Flexural-torsional buckling stresses for double-angle and tee-shaped compression members, ksi
F_e	Elastic buckling stress, ksi
F_{ex}	Elastic flexural buckling stress about the major axis, ksi
F_{ey}	Elastic flexural buckling stress about the minor axis, ksi
F_{ez}	Elastic torsional buckling stress, ksi
F_{my}	Modified yield stress for composite columns, ksi
F_n	Nominal shear rupture strength, ksi
F_r	Compressive residual stress in flange (10 ksi for rolled; 16.5 ksi for welded), ksi
F_{sy}	Stress for tapered members, ksi
F_u	Specified minimum tensile strength of the type of steel being used, ksi
F_w	Nominal strength of the weld electrode material, ksi
$F_{w\gamma}$	Stress for tapered members, ksi

Nomenclature ■ F-3

F_y	Specified minimum yield stress of the type of steel being used, ksi. "Yield stress" denotes either the specified minimum yield point (for those steels that have a yield point) or specified yield strength (for those steels that do not have a yield point)
F_{yf}	Specified minimum yield stress of the flange, ksi
F_{yr}	Specified minimum yield stress of reinforcing bars, ksi
F_{yst}	Specified minimum yield stress of the stiffener materials, ksi
F_{yw}	Specified minimum yield stress of the web, ksi
G	Shear modulus of elasticity of steel, ksi (G = 11,200)
H	Horizontal force, kips
H	Flexural constant
H_s	Length of stud connector after welding, in
I	Moment of inertia, in^4
I_d	Moment of inertia of the steel deck supported on secondary members, in^4
I_p	Moment of inertia of primary members, in^4
I_s	Moment of inertia of secondary members, in^4
I_{st}	Moment of inertia of a transverse stiffener, in^4
I_{yc}	Moment of inertia about y axis referred to compression flange, or if reverse curvature bending referred to smaller flange, in^4
J	Torsional constant for a section, in^4
K	Effective length factor for prismatic member
K_z	Effective length factor for torsional buckling
K_γ	Effective length factor for a tapered member
L	Story height, in
L	Length of connection in the direction of loading, in
L	Live load due to occupancy and moveable equipment
L_b	Laterally unbraced length; length between points which are either braced against lateral displacement of compression flange or braced against twist of the cross section, in
L_c	Length of channel shear connector, in
L_e	Edge distance, in
L_p	Limiting laterally unbraced length for full plastic bending capacity, uniform moment case (C_b = 1.0), in
L_p	Column spacing in direction of girder, ft
L_{pd}	Limiting laterally unbraced length for plastic analysis, in
L_r	Limiting laterally unbraced length for inelastic lateral-torsional buckling, in
L_r	Roof live load
L_s	Column spacing perpendicular to direction of girder, ft
M_A	Absolute value of moment at quarter point of the unbraced beam segment, kip-in
M_B	Absolute value of moment at centerline of the unbraced beam segment, kip-in
M_C	Absolute value of moment at three-quarter point of the unbraced beam segment, kip-in

M_{cr}	Elastic buckling moment, kip-in
M_{lt}	Required flexural strength in member due to lateral frame translation only, kip-in
M_{max}	Absolute value of maximum moment in the unbraced beam segment, kip-in
M_n	Nominal flexural strength, kip-in
M'_{nx}, M'_{ny}	Flexural strength for use in alternate interaction equations for combined bending and axial force, kip-in or kip-ft as indicated
M_{nt}	Required flexural strength in member assuming there is no lateral translation of the frame, kip-in
M_p	Plastic bending moment, kip-in
M'_p	Moment for use in alternate interaction equations for combined bending and axial force, kip-in
M_r	Limiting buckling moment, M_{cr}, when $\lambda = \lambda_r$ and $C_b = 1.0$, kip-in
M_u	Required flexural strength, kip-in
M_y	Moment corresponding to onset of yielding at the extreme fiber from an elastic stress distribution ($=F_y S$ for homogeneous sections), kip-in
M_1	Smaller moment at end of unbraced length of beam or beam-column, kip-in
M_2	Larger moment at end of unbraced length of beam or beam-column, kip-in
N	Length of bearing, in
N_r	Number of stud connectors in one rib at a beam intersection
P_{e1}, P_{e2}	Elastic Euler buckling load for braced and unbraced frame, respectively, kips
P_n	Nominal axial strength (tension or compression), kips
P_p	Bearing load on concrete, kips
P_u	Required axial strength (tension or compression), kips
P_y	Yield strength, kips
Q	Full reduction factor for slender compression elements
Q_a	Reduction factor for slender stiffened compression elements
Q_n	Nominal strength of one stud shear connector, kips
Q_s	Reduction factor for slender unstiffened compression elements
R	Load due to initial rainwater or ice exclusive of the ponding contribution
R_{PG}	Plate girder bending strength reduction factor
R_e	Hybrid girder factor
R_n	Nominal strength
R_v	Web shear strength, kips
S	Elastic section modulus, in^3
S	Spacing of secondary members, ft
S	Snow load
S'_x	Elastic section modulus of larger end of tapered member about its major axis, in^3
S_{eff}	Effective section modulus about major axis, in^3
S_{xt}, S_{xc}	Elastic section modulus referred to tension and compression flanges, respectively, in^3

T	Tension force due to service loads, kips
T_b	Specified pretension load in high-strength bolt, kips
T_u	Required tensile strength due to factored loads, kips
U	Reduction coefficient, used in calculating effective net area
V_n	Nominal shear strength, kips
V_u	Required shear strength, kips
W	Wind load
X_1	Beam buckling factor
X_2	Beam buckling factor
Z	Plastic section modulus, in^3
a	Clear distance between transverse stiffeners, in
a	Distance between connectors in a built-up member, in
a	Shortest distance from edge of pin hole to edge of member measured parallel to direction of force, in
a_r	Ratio of web area to compression flange area
a'	Weld length, in
b	Compression element width, in
b_e	Reduced effective width for slender compression elements, in
b_{eff}	Effective edge distance, in
b_f	Flange width, in
c_1, c_2, c_3	Numerical coefficients
d	Nominal fastener diameter, in
d	Overall depth of member, in
d	Pin diameter, in
d	Roller diameter, in
d_L	Depth at larger end of unbraced tapered segment, in
d_b	Beam depth, in
d_c	Column depth, in
d_o	Depth at smaller end of unbraced tapered segment, in
e	Base of natural logarithm = 2.71828...
f	Computed compressive stress in the stiffened element, ksi
f_{b1}	Smallest computed bending stress at one end of a tapered segment, ksi
f_{b2}	Largest computed bending stress at one end of a tapered segment, ksi
f'_c	Specified compressive strength of concrete, ksi
f_o	Stress due to $1.2D + 1.2R$, ksi
f_{un}	Required normal stress, ksi
f_{uv}	Required shear stress, ksi
f_v	Required shear stress due to factored loads in bolts or rivets, ksi
g	Transverse center-to-center spacing (gauge) between fastener gauge line, in
h	Clear distance between flanges less the fillet or corner radius for rolled shapes; and for built-up sections, the distance between adjacent lines of fasteners or the clear distance between flanges when welds are used, in
h	Distance between centroids of individual components perpendicular to the member axis of buckling, in

F-6 ■ Appendix F

h_c	Twice the distance from the centroid to the following: the inside face of the compression flange less the fillet or corner radius, for rolled shapes; the nearest line of fasteners at the compression flange or the inside faces of the compression flange when welds are used, for built-up sections, in
h_r	Nominal rib height, in
h_s	Factor used for web-tapered members
h_w	Factor used for web-tapered members
j	Factor for minimum moment of inertia for a transverse stiffener
k	Distance from outer face of flange to web toe of fillet, in
k_v	Web plate buckling coefficient
l	Laterally unbraced length of member at the point of load, in
l	Length of bearing, in
l	Length of connection in the direction of loading, in
l	Length of weld, in
m	Ratio of web to flange yield stress or critical stress in hybrid beams
r	Governing radius of gyration, in
$r_{\overline{T}o}$	For the smaller end of a tapered member, the radius of gyration, considering only the compression flange plus one-third of the compression web area, taken about an axis in the plane of the web, in
r_i	Minimum radius of gyration of individual component in a built-up member, in
r_{ib}	Radius of gyration of individual component relative to centroidal axis parallel to member axis of buckling, in
r_m	Radius of gyration of the steel shape, pipe, or tubing in composite columns. For steel shapes it may not be less than 0.3 times the overall thickness of the composite section, in
$\overline{r}_o$	Polar radius of gyration about the shear center, in
r_{ox}, r_{oy}	Radius of gyration about x and y axes at the smaller end of a tapered member, respectively, in
r_x, r_y	Radius of gyration about x and y axes, respectively, in
r_{yc}	Radius of gyration about y axis referred to compression flange, or if reverse curvature bending, referred to smaller flange, in
s	Longitudinal center-to-center spacing (pitch) of any two consecutive holes
t	Thickness of connected part, in
t_f	Flange thickness, in
t_f	Flange thickness of channel shear connector, in
t_w	Web thickness of channel shear connector, in
t_w	Web thickness, in
w	Plate width; distance between welds, in
w	Unit weight of concrete, lb/ft^3
w_r	Average width of concrete rib or haunch, in
x	Subscript relating symbol to strong-axis bending
x_o, y_o	Coordinates of the shear center with respect to the centroid, in
$\overline{x}$	Connection eccentricity, in

Nomenclature ■ F-7

y	Subscript relating symbol to weak-axis bending
z	Distance from the smaller end of tapered member used, for the variation in depth, in
α	Separation ratio for built-up compression members $= h/2r_{ib}$
Δ_{oh}	Translation deflection of the story under consideration, in
γ	Depth tapering ratio. Subscript for tapered members
ζ	Exponent for alternate beam-column interaction equation
η	Exponent for alternate beam-column interaction equation
λ_c	Column slenderness parameters
λ_e	Equivalent slenderness parameter
λ_{eff}	Effective slenderness ratio
λ_p	Limiting slenderness parameter for compact element
λ_r	Limiting slenderness parameter for noncompact element
ϕ	Resistance factor
ϕ_b	Resistance factor for flexure
ϕ_c	Resistance factor for compression
ϕ_c	Resistance factor for axially loaded composite columns
ϕ_{sf}	Resistance factor for shear on the failure path
ϕ_t	Resistance factor for tension
ϕ_v	Resistance factor for shear

Index

AASHTO LRFD code, **11**-6 to **11**-7, **11**-13 to **11**-48
  bridges, **8**-6, **11**-13 to **11**-48
    composite slab-on-girder structures, **11**-23 to **11**-30
    design example, **11**-30 to **11**-45
    limit states design, **11**-6 to **11**-7
    load and resistance factors, **11**-13 to **11**-17
    load classification, **11**-17 to **11**-21
    other structure loads, **11**-21 to **11**-23
  design example, **11**-30 to **11**-47
    deck, **11**-42
    load modifiers, **11**-42 to **11**-43
    problem definition, **11**-30 to **11**-31
    resistance factors, **11**-42
    section and design basis, **11**-32 to **11**-42
  development of, **11**-6 to **11**-7
  force effects:
    from live load, **11**-43 to **11**-44
    effects from other loads, **11**-44
Abrupt impact loading, **B**-3
Absolute maximum permissible load, **11**-48
Acceleration limits, vibrations due to rhythmic activities, **9**-30
Acceptance criteria, weld, **12**-26 to **12**-27
Access holes:
  section properties at, **B**-9
  (*See also* Bolt holes; Weld access holes)
ACI Building Code, **11**-6, **11**-7
Acoustic deck, **A**-10 to **A**-11, **A**-13
Acoustic emission testing of welds, **12**-38
Advance bill, **12**-1 to **12**-2
Advanced inelastic analysis, **6**-16
AISC limit states design, **11**-6
AISC Quality Certification Program, **12**-40

Alignment, field tolerances, **12**-30
Alignment chart method, **6**-10 to **6**-11
Allen and Murray criterion for walking vibrations, **9**-23 to **9**-29, **9**-35 to **9**-38
Allowable stress design, deck material, **A**-1
Alloy elements, **8**-31
Alternate design fastener (twist-off bolt) installation method, **12**-12 to **12**-13
Aluminum, **1**-31
American Institute of Steel Construction Quality Certification Program, **12**-40 to **12**-41
Amperage, and weld quality, **B**-27
Analysis:
  basis of, **1**-17
  framing, single-story multibay braced, **6**-32 to **6**-33
  graphical, **1**-8 to **1**-9
Anchor bolts:
  in concrete, fatigue strength, **8**-17 to **8**-18
  erection plans, **12**-3
  field tolerances, **12**-29
Angle legs, simple beams under shear and axial loads, **7**-112, **7**-113, **7**-116
Angle:
  connections:
    hangers, **7**-71 to **7**-73
    mechanical shearing, **12**-6
  section, materials tolerances, **12**-27
Arc-air gouging for welds, **7**-16
Arch bridges, **11**-12, **11**-13
Arc puddle welds, form decks, **A**-22
Arc voltage, and weld quality, **B**-27
Assembly, shop fabrication, **12**-6 to **12**-7
A307 bolts, **7**-3, **7**-11
Attachments, form deck, **A**-16

I-2 ■ Index

Automated welding, **12**-7
Autostress method, **11**-13
Average daily truck traffic (ADTT), bridge design, **11**-20
Axial compressive strength, **6**-6
Axial force connections, **7**-30 to **7**-86
  bracing, **7**-20 to **7**-30
    beam-to-column, **7**-37 to **7**-39, **7**-52 to **7**-55
    brace-to-gusset, **7**-32 to **7**-35, **7**-46 to **7**-50
    frame action, **7**-55 to **7**-60
    gusset plate design, **7**-39 to **7**-45
    gusset-to-beam, **7**-36 to **7**-37, **7**-50 to **7**-52
    gusset-to-column, **7**-36, **7**-50
    for seismic loads, **7**-45 to **7**-55
  column base plates, **7**-77 to **7**-79
  hanger connections, **7**-71 to **7**-77
  reinforcement of, **7**-113 to **7**-116
  simple beams under shear and axial loads, **7**-107 to **7**-113
  splices, **7**-79 to **7**-86
    columns, **7**-79 to **7**-82
    truss chords, **7**-82 to **7**-86
  truss, **7**-60 to **7**-71
    gusset-to-top chord, **7**-66 to **7**-67
    gusset-to-truss vertical, **7**-68 to **7**-70
    splices, **7**-82 to **7**-86
    truss vertical-to-top chord, **7**-70 to **7**-71
Axial loads, **6**-8
  beam-column members, **6**-15
  column design, **3**-45 to **3**-50
  composite columns, **10**-42
  compression members (*see* Compression members)
  gusset plate design, **7**-40, **7**-42
  simple beam connections under, **7**-107 to **7**-113
Axial strength, **6**-6, **6**-7

Backing, weld, **B**-17
Backing bar notch, **8**-6
Backing bar splices, fractures in, **8**-4
Bar stock, **12**-2
Base metals, welds:
  detail categories for, **8**-13 to **8**-15

Base metals, welds (*Cont.*):
  filler material requirements, **7**-23 to **7**-24
  inspection, **12**-22
  for seismic loading, **B**-21 to **B**-23
  size of welds, **7**-16
Base plates:
  axial force connections, **7**-77 to **7**-79
  cutting, **12**-6
Bay, **3**-29
Beam analysis, composite girders, **10**-26
Beam-and-column flange, restrained region, **B**-15
Beam behavior:
  bending moments, **2**-1 to **2**-3
  continuous beam design, **2**-20 to **2**-23
  design charts, **2**-17 to **2**-20
  design of beams, zone 1, **2**-5 to **2**-10
    beam weight estimates, **2**-5 to **2**-7
    holes in beams, **2**-7 to **2**-10
  elastic buckling, zone 3, **2**-3, **2**-15 to **2**-17
  inelastic buckling, zone 2, **2**-2 to **2**-3, **2**-10 to **2**-12
  moment capacities, zone 2, **2**-13 to **2**-15
  plastic buckling, zone 1, **2**-2, **2**-3 to **2**-4
  ponding, **2**-28 to **2**-32
  shear, **2**-23 to **2**-25, **2**-26
  unsymmetrical bending, **2**-26 to **2**-28
Beam-column interaction equations, **6**-6 to **6**-7, **6**-13
Beam columns:
  in braced frames, **3**-75 to **3**-79
  design of, **3**-81 to **3**-88, **6**-14
  equivalent axial load form, **6**-13
  in unbraced frames, **3**-80 to **3**-81
Beam frequency, **9**-8 to **9**-9
Beam resistance factor, **6**-6
Beams:
  bridge, **11**-10, **11**-11, **11**-44
  composite:
    deflection, **10**-10, **10**-17
    reinforcing, **10**-26 to **10**-36
    sample problem, **10**-12 to **10**-19
    (*See also* Composite construction)
  compact, plastic buckling, **2**-2
  connections:
    axial force (*see* Axial force connections)
    complete joint penetration (CJP) welds, **12**-37, **B**-9, **B**-11, **B**-17

Index ■ I-3

Beams, connections (*Cont.*):
    moment, **7**-89, **7**-92 to **7**-93, **7**-94 to **7**-95
    shear splices, **7**-103 to **7**-106
    simple beams under shear and axial
        loads, **7**-107 to **7**-113, **7**-114, **7**-115,
        7-116
    welded, **7**-12, **B**-11
  continuous, design of, **2**-20 to **2**-23
  design charts, **2**-17 to **2**-20
  doublers, **2**-25
  floor system, **4**-2
  frame design (*see* Braced frames; Frame
    design)
  gravity frames, **6**-73 to **6**-74, **6**-78 to **6**-79,
    6-44 to **6**-45
  holes in beams, **2**-7 to **2**-10
    fabrication, **12**-8 to **12**-9
    (*See also* Bolt holes; Weld access holes)
  mill cambering limits, **12**-7
  single-story multibay braced, **6**-31
  tolerances, **12**-28, **12**-30
  transformed section properties, **9**-10
  vibration (*see* Serviceability limit states
    design)
Beam weight estimates, **2**-5 to **2**-7
Beam yield-to-tensile ratio, seismic design,
  B-22
Bearing joints, cutting, **12**-6
Bearing-type connections, bolted, **7**-6 to **7**-7
Beedle, Lynn, **1**-17
Belt truss, **6**-62
Bending:
  deck load tables, **A**-6, **A**-3
  gusset plate design, **7**-40, **7**-42
  shear due to, **4**-28
Bending analogy, **4**-23 to **4**-25
Bending coefficients, **2**-10 to **2**-12
Bending moments:
  framing:
    single-story multibay braced, **6**-32
    single-story multibay unbraced, **6**-37
  full plastic moment, **1**-24 to **1**-26
Bendixen, Axel, **1**-14
Bernoulii, **1**-3
Bett-Rayleigh reciprocity theorem, **1**-13
Bill, advance, **12**-1 to **12**-2
Block shear, tension member, **3**-24 to **3**-28
Block shear failure, **2**-25

Block shear fracture, gusset plate design,
  7-40
Block shear rupture (*see* Axial force connections; Tearout)
Bolted connections:
  axial force connections:
    hangers, **7**-71 to **7**-73
    reinforcement of simple beams, **7**-114,
      7-115
    (*See also* Axial force connections)
  bearing type and slip resistant, **7**-6 to **7**-7
  clearance for tightening, **7**-9 to **7**-10
  detail categories for, **8**-15 to **8**-18
  edge distance and spacing, **7**-8 to **7**-10
  fabrication of, **12**-8 to **12**-15
    hole punching and drilling, **12**-8 to **12**-9
    inspection, **12**-14 to **12**-15
    installation methods, **12**-9 to **12**-14
    materials control, **12**-5
  fatigue strength in, **8**-16
  field tolerances, **12**-29
  fully tensioned and snug tight, **7**-3 to **7**-6
  holes for (*see* Bolt holes)
  seismic design:
    braced frame systems, **E**-46 to **E**-49
    special moment frame, **E**-25, **E**-28
  seismic resistance, **B**-18
  shear, **7**-107 to **7**-108
  shear splices, **7**-104
  tension load, **3**-9 to **3**-11
  types of bolts, **7**-2 to **7**-3
  welds with, **7**-7, **B**-17 to **B**-18
Bolt holes, **7**-7 to **7**-8
  and beam behavior, **2**-7 to **2**-10
  direct-tension indicator tightening, **7**-5
  edge distance and spacing, **7**-8 to **7**-10
  layout and assembly, **12**-6
  tension members, **3**-2 to **3**-4, **3**-1
    net area, **3**-2 to **3**-4
    staggered holes, **3**-4 to **3**-7
Bolt tension calibrator, **12**-11, **12**-14
Boron, **1**-31
Box column welds, distribution of stress,
  B-3
Box girder bridges, **11**-10, **11**-11
Box girder fractures, **8**-4, **8**-5
Braced frames:
  alignment chart factors, **6**-10

Braced frames (*Cont.*):
  beam columns in, **3**-75 to **3**-79
  effective lengths for columns in, **3**-30, **3**-31
  interaction with moment frame, **6**-62
  lateral force resistance, **6**-15
  multistory, **6**-41, **6**-42, **6**-49 to **6**-53, **6**-71 to **6**-82
    bracing frame, **6**-74 to **6**-76
    serviceability deflections, **6**-81, **6**-82
    sizing of beams, **6**-78 to **6**-79
    sizing of braces, **6**-77
    sizing of columns, **6**-76 to **6**-77
    structural analysis, **6**-79 to **6**-81
  multistory gravity frames, **6**-74 to **6**-76, **6**-77
    beams in, **6**-73 to **6**-74
    columns in, **6**-71 to **6**-73
  portal frames, **6**-20 to **6**-21
  seismic design, **E**-36 to **E**-37, **E**-34 to **E**-51
    b/t ratio, **B**-8
    energy absorption and ductility, **E**-53
    height limitations, **E**-52
    special moment frames (SMF), **E**-22, **E**-23
    (*See also* Seismic design)
  seismic resistance, **B**-8 to **B**-11
  single story:
    lattice frame, **6**-24 to **6**-27, **6**-28, **6**-29
    multibay braced frames, **6**-32
  stability, **6**-2
Braces:
  axial force connections, **7**-20 to **7**-30
    beam-to-column, **7**-37 to **7**-39, **7**-52 to **7**-55
    brace-to-gusset, **7**-32 to **7**-35, **7**-46 to **7**-50
    frame action, **7**-55 to **7**-60
    gusset-to-beam, **7**-36 to **7**-37, **7**-50 to **7**-52
    gusset-to-column, **7**-36, **7**-50
    for seismic loads, **7**-45 to **7**-55
  compression members:
    columns, **3**-43
    sidesway prevention, **3**-32 to **3**-38
  connections:
    bolts, **7**-2

Braces, connections (*Cont.*):
    fatigue cracks in gusset connections, **8**-40, **8**-41
    (*See also* Braces, axial force connections)
    forms of, **6**-1, **6**-2 to **6**-3
    framing members (*see* Braced frames)
    roof deck as, **A**-1
    shop detail drawings, **12**-3
    weld quality, **7**-28
Bridges:
  AASHTO codes, **11**-13 to **11**-48
    composite slab-on-girder structures, **11**-23 to **11**-30
    design example, **11**-30 to **11**-45
    load and resistance factors, **11**-13 to **11**-17
    load classification, **11**-17 to **11**-21
    other structure loads, **11**-21 to **11**-23
  AISC quality certification, **12**-41
  bolt inspection, torque testing, **12**-15
  constituent components, **11**-9 to **11**-10
  construction-related benefits, **11**-13
  fracture in, **8**-2
  limit states design overview, **11**-1 to **11**-8
    AASHTO code development, **11**-6 to **11**-7
    Canadian design codes (OHBDC), **11**-7 to **11**-8
    working stress, **11**-2 to **11**-6
  load rating, **11**-48 to **11**-53
  types and functionality, **11**-10 to **11**-13
  working stress approach, **11**-54
Brittle failure, **1**-32, **B**-12
Brittle fracture, **8**-2, **8**-27, **8**-33, **8**-44, **B**-16
  equations to predict, **8**-43
  seismic loading and, **B**-11
  temperature and, **8**-25 to **8**-26
Brittle fracture surface, **8**-4
Buckling:
  web plate, **7**-86
  (*See also* Beam behavior)
Buckling analysis:
  column end restraints, **3**-30
  gravity columns, **6**-45
  single-story portal frames, **6**-23, **6**-24
  story, **6**-11 to **6**-13
  system, **6**-13

Buckling curves, **1-3**, **C-4**
Buckling load, **6-7**, **6-8**
  residual stresses and, **1-30**
  single-story portal frames, **6-23**
Buckling strength, **3-50**
Building code special inspections, **12-39**
Building design
  loads, **5-10** to **5-29**
    dead, **5-10** to **5-12**
    live, **5-12**, **5-13**, **5-14**
    snow, **5-12**, **5-14** to **5-18**
    wind, **5-18** to **5-23**
  openings and voids, **5-29**
  planning structures, **5-1** to **5-9**
  thermal and seismic movement, **5-29** to **5-32**, **5-33**
Building period check, **E-33**
Built-up sections:
  AISC requirements, **7-2**
  columns:
    with components not in contact with each other, **3-59** to **3-61**
    compression members, **3-53** to **3-54**
    tension members, **3-17** to **3-19**
    tolerances, **12-28**
Bundled tubes, **6-62**
Buried penny-shaped crack, **8-44** to **8-45**
Burning, weld access holes, **B-20**
Button punches, **A-16**

Cable-stayed bridges, **11-13**
CAD—CAM interface, **12-3** to **12-4**
Calibrated-wrench method, **7-3** to **7-4**, **12-11** to **12-12**
Camber, **6-4**
  shop fabrication, **12-7** to **12-8**
  tolerances, **12-28**, **12-30**
Canadian standards:
  bridge design specifications, **11-7** to **11-8**
  for walking vibrations, **9-17** to **9-20**
Cantilever members:
  field tolerances, **12-30**
  roof decks, **A-4** to **A-5**
Cantilever roof construction, shear splices in, **7-104**, **7-105**
Carbon, **1-31** to **1-32**
Carbon equivalent (CE) method, **1-33**

Carbon steel bolts, **7-2**
Carbon steels, **6-18**
Castigliano, Alberto, **1-12** to **1-13**
Cast-in-place concrete (*see* Composite construction)
CBF (*see* Braced frames)
Cellular decks, **A-10**, **A-11**
Center crack, fracture mechanics analysis, **8-36** to **8-38**
Certification, **12-37** to **12-41**
  of compliance with ASTM specifications, **12-5**
  of materials, **12-4**, **12-5**
Channel load, bridge, **11-21**
Channel-type diaphragms, bridge, **11-10**
Charpy V-notch (CVN) impact test, **7-2**, **8-2**, **8-4**, **8-25**
  fracture mechanics analysis correlation, **8-34**, **8-35**, **8-36**
  at P frequency, **8-31**
  of steel and filler metal, **8-28** to **8-32**
Chemical composition of steel, **1-30** to **1-32**
Chimneys, **5-23**
Chromium, **1-32**
Circle of stress, **1-8** to **1-9**
Circles, principal, **1-11** to **1-12**
Circular crack, **8-44** to **8-45**
Circular tubular sections, fatigue design of, **8-19** to **8-20**
CJP (*see* Complete joint penetration beam)
Clapeyron's theorem, **1-9**
Cleavage cracks, **8-40**
Clip angles, **12-3**
  layout and assembly, **12-6**
  thickness of, **7-112** to **7-113**, **7-114**
Coating, **12-30** to **12-36**
  inspection, **12-35** to **12-36**
  materials control, **12-4**
  surface preparation, **12-30** to **12-33**
  systems, **12-33** to **12-35**
  welded surfaces, **7-27**
Cold forming, deck material, **A-1**
Cold-saw equipment, shop fabrication, **12-5**
Cold work, and tensile properties, **1-27**
Columbium, **1-32**
Column base plates, axial force connections, **7-77** to **7-79**
Column effective length, **6-9** to **6-13**, **6-23**

Column flange:
  seismic design, braced frame systems, E-51
  welds, **B**-11
Column formulas, **3**-40 to **3**-41
Column frequency, **9**-9
Column K factors, **3**-37
Column resistance factor, **6**-6
Columns:
  bay establishment, **3**-29
  connections:
    beam-to-column bracing, **7**-37 to **7**-39, **7**-52 to **7**-55
    bolted, **7**-2
    moment, **7**-92, **7**-95
    spliced, **7**-79 to **7**-82
    welded, **7**-12, **B**-11
  composite, **10**-38 to **10**-45
  compression:
    effective length, **3**-29 to **3**-38
    (*See also* Compression members)
  elevator shaft, **12**-29 to **12**-30
  field tolerances, **12**-29 to **12**-30
  framing:
    gravity frame, **6**-76 to **6**-77
    multistory unbraced rigid frames, **6**-53, **6**-54
    single-story lattice frames, **6**-29
    single-story multibay braced, **6**-31 to **6**-32, **6**-34 to **6**-36
    (*See also* Braced frames; Frame design)
  seismic design, braced frame systems, E-51
  special moment frames:
    doublers, E-18 to E-19, E-21
    local yielding, E-32
    strength check, E-14 to E-17
  tolerances, **12**-28
Column splices:
  bolted, **7**-2
  cutting, **12**-6
  welded, space requirements, **7**-24
Columns stiffeners:
  compression columns, **3**-38 to **3**-40
  moment connections, **7**-90
  special moment frames, E-30 to E-32
Column stiffness, **3**-34 to **3**-37
Column story shear, moments inducing, **7**-92

Column to girder ratio, E-19 to E-21
Column to panel zone ratio, E-19 to E-21
Column web:
  braced frame systems, seismic design, E-51, E-52
  special moment frame, E-18 to E-19, E-21, E-32
Common bolts, **7**-3, **7**-4
Compact beam, plastic buckling, **2**-2
Compact sections, **2**-1
  bridges, **11**-30
  plastic stress distribution, **3**-38 to **3**-39
Compact-tension (CT) specimen, **8**-45 to **8**-46
Complete joint penetration (CJP) beam welds, **12**-37, **B**-9, **B**-11, **B**-17
Complete penetration weld, defined, **7**-14
Composite construction, **10**-1 to **10**-45
  beams, **10**-3 to **10**-26
    columns, **10**-38 to **10**-45
    computer program (HCOMPL), **10**-37 to **10**-38
    deflection of, **10**-10
    design procedure, **10**-4 to **10**-7
    floor vibrations, **10**-10 to **10**-11
    partial composites, **10**-8 to **10**-9
    reinforcing existing composite beams, **10**-26 to **10**-36
    sample problem, composite beam, **10**-12 to **10**-19
    sample problem, composite girder, **10**-19 to **10**-26
    stud shear connectors, **10**-7 to **10**-8
  bridges, **11**-45, **11**-46
    composite section module, **11**-38 to **11**-39
    loading of, **11**-26
    slab-on-girder, **11**-23 to **11**-30
  decks, **A**-16 to **A**-24
    connections, **A**-22 to **A**-23
    diaphragms, **A**-23, **A**-24
    floor, **A**-1
    loads, **A**-16 to **A**-21
Compression:
  braced frame system girders, seismic design, **10**-5, E-45
  composite beams, **10**-14
  concrete stress-strain diagram, **11**-4

Compression (*Cont.*):
  depth of web in, **11**-29
  reversal loading, fatigue testing, **8**-8
Compression flange:
  bridge design, **11**-44, **11**-45
  moment connections, strong axis beam stiffeners, **7**-89
Compression members, **6**-8
  buckling, types of, **3**-28 to **3**-29
  built-up columns, **3**-53 to **3**-54
  built-up columns with components in contact with each other, **3**-54 to **3**-59
    connection requirements for, **3**-56 to **3**-59
    deformation modes, **3**-54 to **3**-55
  built-up columns with components not in contact with each other, **3**-59 to **3**-61
  column formulas, **3**-40 to **3**-41
  design of axially loaded columns, **3**-45 to **3**-50
  end restraint and effective lengths of columns, **3**-29 to **3**-38
  example problems, **3**-41 to **3**-45
  flexural-torsional buckling of, **3**-61 to **3**-64
  with flexure, **3**-64 to **3**-88
  maximum slenderness ratios, **3**-41
  single-angle, **3**-53
  stiffened and unstiffened columns, **3**-38 to **3**-40
  stiffness reduction factors, **3**-50 to **3**-53
  torsional rotation, **4**-6
Computer-aided manufacturing (CAM), **12**-4
Computer numerically controlled equipment, **12**-4
Computer program, **10**-37 to **10**-38, **10**-42
Concentrically braced frames, seismic resistance, **B**-8
Concrete:
  anchor bolts, **8**-17
  bridge members, **11**-8, **11**-10
  composite construction (*see* Composite construction)
  creep, **11**-28
  stress-strain diagram, **11**-4
Concrete decks:
  composite, **A**-16 to **A**-24
  form, **A**-11 to **A**-16

Concrete decks (*Cont.*):
  simple beams under shear and axial loads, **7**-109, **7**-112
Connections, **6**-56
  bolted, **7**-2 to **7**-12
    bearing type and slip resistant, **7**-6 to **7**-7
    edge distance and spacing, **7**-8 to **7**-10
    fully tensioned and snug tight, **7**-3 to **7**-6
    holes for, **7**-7 to **7**-8
    installation, **7**-10 to **7**-12
    materials control, **12**-4, **12**-5
    types of bolts, **7**-2 to **7**-3
    with welds, **7**-7
    (*See also* Bolted connections)
  bridges, **11**-10
  composite construction, **10**-1
  compression members, built-up columns, **3**-56 to **3**-61
  decks:
    composite, **A**-22 to **A**-23
    roof, **A**-10
  defined, **7**-1
  design of, **7**-29 to **7**-117
    axial force connections (*see* Axial force connections)
    clearance for tightening, **7**-9 to **7**-10
    miscellaneous, **7**-107 to **7**-117
    moment connections, **7**-86 to **7**-97
    philosophy, **7**-29 to **7**-30
    procedure, **7**-30
    shear force connections, **7**-97 to **7**-106
    types of, **7**-30
  detail categories for, **8**-15 to **8**-18
  drawings, **12**-3 to **12**-4
  erection plans, **12**-3
  field tolerances, **12**-29
  layout and assembly, **12**-6
  misaligned, **8**-20
  seismic design:
    braced frames, **E**-43, **E**-44, **E**-46 to **E**-50, **E**-51, **E**-52, **E**-53
    pre- and post-Northridge details, **B**-31, **B**-32, **B**-33 to **B**-34
    special moment frames, **E**-23 to **E**-30
  seismic resistance, **B**-25 to **B**-26
    ductile hinges, **B**-11 to **B**-17

I-8 ■ Index

Connections, seismic resistance (*Cont.*)
    weld acess holes, **B**-18 to **B**-20
    shop detail drawings, **12**-3
    tension members, **3**-13 to **3**-14
        bolted, **3**-9 to **3**-11
        built-up members, **3**-17 to **3**-19
        and choice of member type, **3**-14
        pin-connected, **3**-21 to **3**-23
        rods and bars, **3**-20 to **3**-21
        welded, **3**-11 to **3**-13
    unbraced rigid frames, **6**-55 to **6**-59
    welded steel moment frame, **8**-32
    welds, **7**-12 to **7**-29
        fabrication, **12**-15 to **12**-27
        material, **7**-23 to **7**-24
        materials control, **12**-4, **12**-5
        positions, **7**-24
        procedures, **7**-25 to **7**-28
        quality of, **7**-28 to **7**-29
        symbols on drawings, **7**-18 to **7**-23
        types of, **7**-13 to **7**-18
        (*See also* Welds)
Connection slip, **6**-4
Constant-amplitude fatigue limit (CAFL), **8**-12, **8**-10
Constant-amplitude fatigue test, **8**-23
Continuity plates, load transfer, **B**-3
Continuous beams, design of, **2**-20 to **2**-23
Continuous welds, drawing specification, **7**-19, **7**-20
Cooling rates, residual stress, **1**-30
Cope holes, **8**-4
    crack sources, **8**-13 to **8**-15
    for weld access, **8**-18
    (*See also* Bolt holes; Weld access holes)
Copper, **1**-32
Core braced systems, multistory frames, **6**-51
Corners:
    fillet welds, **7**-17
    groove welds, **7**-14
Corrosion, **1**-30, **1**-32, **6**-4
Costs, weld types, **7**-15 to **7**-18
Coupon tensile strength, **3**-7
Cover plates:
    compressive members, **3**-59, **3**-60
    seismic resistance, **B**-16 to **B**-17, **B**-33
Cracklike notches, **8**-33

Cracks, **8**-4, **8**-13, **8**-18
    composite slabs, **10**-4
    fatigue crack formation and propagation, **8**-1, **8**-2, **8**-22, **8**-40, **8**-41, **8**-43
    fracture mechanics analysis, **8**-32 to **8**-45
        buried penny-shaped crack, **8**-44 to **8**-45
        center crack, **8**-36 to **8**-38
        edge crack, **8**-39 to **8**-44
    microcracks, **B**-20
    sources of, **8**-13 to **8**-15
    welds, **7**-25, **7**-26, **7**-27, **B**-30
        center cracks in joints, **8**-37
        fatigue crack formation and propagation, **8**-2
Crack-Tip Opening Displacement (CTOD) Fracture Toughness Measurement, **8**-47
Crane support connection, bolts, **7**-2
Creep:
    composite bridges, **11**-28
    heat and, **1**-30
Creep strength, **1**-33
Critical damping, **9**-5
Critical K, **8**-34
Cross bracing, multistory system, **6**-49
Cross-section terminology, decks, **A**-2
Culmann, Karl, **1**-8 to **1**-9
Cumulative damage model, **8**-22
Current density, and weld quality, **B**-27
Curtain wall, seismic design, **E**-4, **E**-45
Curving, shop fabrication, **12**-7 to **12**-8
Cutting:
    drawings, **12**-4
    shop fabrication, **12**-5 to **12**-6
Cutting tolerances, **12**-1 to **12**-2
Cyclic loading, **8**-22
    effective stress range for variable amplitude loading, **8**-23 to **8**-25
    fatigue crack formation, **8**-1, **8**-2
    redistribution of stresses under seismic loading, **B**-7
    weld testing, **12**-26
Cyclic plastic deformation, **8**-6
Cyclic slip, **8**-23 to **8**-24

d'Alembert's principle, **1**-3

Damping, **9**-5, **9**-31
    composite beams, **10**-19
    floor system, **9**-21
da Vinci, Leonardo, **1**-2
Dead load, **1**-20, **5**-10 to **5**-12, **11**-18, **11**-19
    bridge design, **11**-34 to **11**-36, **11**-45
        positive flexure, **11**-33
        slab-on-girder, composite, **11**-25 to **11**-26
    composite girders, **10**-19, **10**-20 to **10**-23
    deck, **A**-1
    framing, single-story frames, **6**-17, **6**-37
Dead load moment, **11**-36
Deck, bridge, **11**-8, **11**-9
Deck design, **A**-1 to **A**-24
    composite decks, **10**-4, **A**-16 to 24
        connections, **A**-22 to 23
        diaphragms, **A**-23, **A**-24
        loads, **A**-16 to **A**-21
    form deck, **A**-11 to **A**-16
        attachments, **A**-16
        nonstructural fills, **A**-16
        structural concrete form, **A**-11 to 15
    roof deck, **A**-1 to **11**, A-12
        fastening, **A**-10
        generic profiles, **A**-2, 3
        horizontal load, **A**-7 to 10
        special decks, **A**-10 to **11**, **A**-12, **A**-13
        vertical load, **A**-2 to 7
    rules for, **A**-1
    types of, **A**-1 to 2
Deflection:
    composite beams, **10**-10, **10**-17
    composite girders, **10**-25 to **10**-26
    in design, **5**-3 to **5**-4
Deflection limits, **6**-4, **6**-5
Deformation, in design, **5**-3 to **5**-4
Densities of building materials, **5**-10
Depth of web in compression, bridge design, **11**-29, **11**-41 to **11**-42
Depth-thickness ratio, welded girder flatness, **12**-29
Derived SI units, **D**-2
Design charts, beams, **2**-17 to **2**-20
Design concepts, **1**-17 to **1**-22
Design criterion for rhythmic excitations, **9**-31
Design factored load, single-story portal frames, **6**-23

Design lanes, bridge design, **11**-20
Design strength of axially loaded column, **3**-43 to **3**-44
Design truck, bridge design, **11**-19, **11**-20
Design truck live load plus impact moment, **11**-37 to **11**-38
Detail categories:
    fatigue strength, **8**-11 to **8**-20
        base metal and thermal cut edges, **8**-13 to **8**-15
        mechanically fastened joint, **8**-15 to **8**-18
        welded joints, **8**-18 to **8**-20
    S-N curves, **8**-10
Detailing:
    CAD packages, **12**-4
    connections, clearance for tightening, **7**-9 to **7**-10
    shop, **12**-1 to **12**-4
Detailing drawings, **12**-2 to **12**-3
Detailing rules, **8**-6, **8**-21
Deterioration, **5**-7, **5**-8
Diaphragm:
    bridge, **11**-10
    deck:
        composite, **A**-23, **A**-24
        concrete form, **A**-13, **C**-16
        form, **A**-23, **A**-24
        roof, **A**-7 to **A**-10
Diaphragm-braced system, **6**-1 to **6**-2
Direct stress due to warping, **4**-26, **4**-28, **4**-29
Direct-tension indicator installation method, **7**-5 to **7**-6, **12**-12 to **12**-13
Distortion, fatigue and, **8**-21 to **8**-23
Distortion energy yield criterion, **7**-111
Distortion-induced stresses, **8**-9
Distribution factor for moment, bridge design, **11**-44
Double-angle shear connections, **7**-107
Double-bevel welds, **7**-14, **7**-15, **7**-16, **7**-21
Double gusset plate connections, braced frame systems, **E**-50, **E**-51
Double J groove weld, **7**-14, **7**-15, **7**-16
Doublers:
    increasing shear strength of beam by, **2**-25
    moment connections, **7**-91
Double U groove weld, **7**-14, **7**-15, **7**-16

I-10 ■ Index

Double V groove weld, **7**-14, **7**-15, **7**-16, **7**-21 to **7**-22
Drafting packages, **12**-3 to **12**-4
Draped mesh concrete slabs, deck, **A**-13
Drawings, **12**-2 to **12**-3
   CAD—CAM interface, **12**-3 to **12**-4
   connections, clearance for tightening, **7**-9 to **7**-10
   weld symbols, **7**-18 to **7**-23
Drift, **3**-70, **6**-4, **6**-5
   multistory frames, **6**-50 to **6**-51, **6**-54
   seismic design:
      braced frames, **E**-35 to **E**-36
      special moment frames, **E**-6 to **E**-9
Drill lines, drawings, **12**-4
Ductile fracture, **8**-25, **8**-26 to **8**-27
Ductile hinge formation in connection, seismic loading and, **B**-11 to **B**-17
Ductile tearing, **8**-27
Ductility, **1**-32
   bridge design, **11**-42
   carbon and, **1**-31
   double angle shear connections, **7**-107
   seismic design, **B**-7, **E**-53
   simple beams under shear and axial loads, **7**-113, **7**-116 to **7**-117
   welding principles, **B**-6
Dye penetrant tests:
   weld access holes, **B**-20
   welds, **7**-29, **12**-24 to **12**-25, **B**-29, **B**-31
Dynamic crack propagation path, **8**-43
Dynamic lateral-force procedure, **5**-27
Dynamic load factor, **9**-21
Dynamic load, **B**-3
   bridge design, **11**-43, **11**-45, **11**-37
   detailing rules for, **8**-6
   weld filler material requirements, **8**-31
   weld tab removal, **7**-27

Earth loads, **11**-18
Earthquakes (*see* Seismic design; Seismic load)
Eccentric bracing:
   multistory system, **6**-49
   seismic resistance, **B**-8 to **B**-9, **B**-11
Edge crack, stress-intensity factor for, **8**-39 to **8**-44
Edge panel mode, joist and beam, **9**-26
Edge preparation for welds, **7**-16
Effective axial load method, **3**-81
Effective flange width, bridge design, **11**-33, **11**-46, **11**-26 to **11**-27
Effective length factors, **6**-14
   column, **6**-9 to **6**-13
   compression columns, **3**-29 to **3**-38
   multistory frames, **6**-46
Effective moment of inertia, composite beams, **10**-18
Effective slenderness ratio, **3**-45 to **3**-46, **3**-51
Effective stress range, variable-amplitude loading, **8**-23 to **8**-25
Effective width of slab, bridge design, **11**-33
EGW (*see* Electrogas welding)
Elastic analysis, **1**-19
   bridge I girder, **11**-32
   framing:
      multistory, **6**-59
      single-story multibay braced, **6**-32, **6**-37
Elastic buckling, **2**-3, **2**-15 to **2**-17, **3**-40, **3**-41
   single-story portal frames, **6**-23
   of web, **2**-24
Elastic curves, **1**-3
Elastic design methods, **1**-19
Elastic properties
   bridge design, **11**-45
      depth of web in compression, **11**-29, **11**-41
      fatigue and fracture limit state, **11**-43
   composite beams, **10**-18
   composite girders, **10**-25
   working stress approach, **11**-3
Elastic surface waves, **1**-14
Electrogas welding (EGW), **12**-15
Electromagnetic testing of welds, **12**-38
Electroslag welding (ESW), **7**-24, **12**-15
Element forces, bridge design, **11**-39, **11**-34
Elevator shaft columns, **12**-29 to **12**-30
Elliptical interaction equation, **7**-111
End connection rigidity, tier buildings, **12**-30
Endplates:
   axial force connections (*see* Axial force connections)
   shear, **7**-98 to **7**-100

Index ■ I-11

End restraints, compression members, **3-29** to **3-38**
End returns, fillet welds, **7-17**
End torsion, free end, **4-4** to **5**
Envelope, field tolerances, **12-30**
Environmental-control systems, **5-7**, **5-9**
Environmental zones and conditions, **12-31**
Equivalent axial load, **3-81**, **6-13**
Erection:
    column base plate axial force connections, **7-79**
    column stability, **7-81** to **7-82**
    plans, **12-2** to **12-3**
    tolerances, **12-29** to **12-30**
Erection procedures, **12-36** to **12-37**
Erection tolerances, **12-27** to **12-30**
ESW (*see* Electroslag welding)
Euler, **1-3**
Euler buckling, **3-28**, **3-41**, **3-61** to **3-64**
Euler equation, **3-40**
Evolution of structural design, **1-1** to **1-17**
Expansion joints, seismic movement and, **5-29**, **5-30**, **5-31**
Explosions, **8-6**
Exposure factors, **5-14**
Exterior columns, field tolerances, **12-30**
Exterior frame tubes, **6-62**
Exterior stringer, **11-26**
Exterior wall, seismic design, **E-4**, **E-45**

Fabricated sections, portal frames with, **6-21**
Fabrication:
    built-up members, **7-2**
    certification, qualification, and inspection program, **12-37** to **12-41**
    coating, **12-30** to **12-36**
        inspection, **12-35** to **12-36**
        surface preparation, **12-30** to **12-33**
        systems, **12-33** to **12-35**
    connections:
        bolted, **12-8** to **12-15**
        welded, **12-15** to **12-27**
    erection procedures, **12-36** to **12-37**
    shop, **12-4** to **12-8**
        cambering and curving, **12-7** to **12-8**
        cutting, **12-5** to **12-6**
        layout and assembly, **12-6** to **12-7**

Fabrication, shop (*Cont.*):
    materials control, **12-4** to **12-5**
    shop detailing process, **12-1** to **12-4**
    tolerances, **12-27** to **12-30**
    welding procedures, **7-26**
Fabrication cracks, **8-4**, **8-13**
Fabrication defects, **8-4**
Fabrication plants, AISC quality certification, **12-41**
Factored design moment, **2-5**
Factored load, shop detail drawings, **12-3**
Factor of safety, **11-48**, **11-4**
Fasteners:
    deck:
        form, **A-16**
        roof, **A-7** to **A-8**
    lubication, **12-11** to **12-12**
Fatigue, **8-1** to **8-8**
    bridge design, **11-22** to **11-23**, **11-43**, **11-44**
    categories of fatigue strength, **8-11** to **8-20**
        base metal and thermal cut edges, **8-13** to **8-15**
        mechanically fastened joints, **8-15** to **8-18**
        welded joints, **8-18** to **8-20**
    classification of structural details, **8-11** to **8-20**
    distortion and multiaxial-loading effect, **8-21** to **8-23**
    evaluation of structural details for, **8-25** to **8-48**
    scale effects, **8-20** to **8-21**
    seismic loading, **B-7**
    variable amplitude loading, effective stress range for, **8-23** to **8-25**
Faying, **7-104**
FCAW (*see* Flux-cored arc welding)
Fedderson formula, **8-36**
Field bolting, multistory frames, **6-41**
Field bolt lists, **12-3**
Field legs of connection angles, **7-112**
Field measurements, **11-49**
Field tolerances, **12-29** to **12-30**
Field welds:
    access holes, **B-19**
    drawing specification, **7-20**
    erection plans, **12-3**
    moment connection, **7-87**

Field welds (*Cont.*):
  versus shop welds, **B**-28
Fifty grade columns, seismic design, **E**-19, **E**-32
Filler, bolt spacing, **7**-11 to **7**-12
Filler material for welds, **7**-11 to **7**-12, **7**-23 to **7**-24
  ductile behavior, **8**-27
  fracture toughness specification, **8**-28
  materials control, **12**-5
  and seismic resistance, **B**-23 to **B**-25
Fillet welds, **7**-7
  costs, **7**-15, **7**-16, **7**-17 to **7**-18
  defined, **7**-13 to **7**-14
  drawing symbols, **7**-19
  intermittent, **7**-17
  moment connections, **7**-90
  position, **7**-24
  quality criteria, **7**-28 to **7**-29
  seismic design, special moment frame, **E**-27
  on shear tabs, **B**-18
  space for, **7**-24, **7**-25
Fillet weld throat dimensions, root openings and, **B**-8, **B**-9
Fills, form decks, **A**-16
Finish machining, welds, **7**-27
Fireproofing, **12**-31
Fire rating, decks, **A**-17, **A**-22
Fire resistance, **5**-4 to **5**-7
First-order elastic analysis, **6**-79
  deflection from service wind load, **6**-36
  multistory frames, **6**-59, **6**-65, **6**-68
First-order elastic moments, **6**-7
First-order end moments, **6**-8
First-order moments, **3**-68 to **3**-69
Fit-up of connection:
  turn-of-the-nut method, **7**-4
  and weld quality, **B**-28
Flame cutting:
  shop fabrication, **12**-5
  surface roughness, **12**-6
  weld costs, **7**-16
Flange:
  bridge design:
    bracing, **11**-46
    field measurements, **11**-49
    slenderness of, **11**-46

Flange, bridge design (*Cont.*):
  stress reduction factor, **11**-43, **11**-44, **11**-30
  design strength, special moment frame columns, **E**-30
  fractures, **8**-4, **8**-5
  material, **12**-2
  moment connections
    shear forces, **7**-91 to **7**-93
    strong axis beam, **7**-88 to **7**-89
    welds of stiffeners to column flange and web, **7**-90
    welds to, **7**-93 to **7**-94
  warpage, **12**-29
  truss chord connection, **7**-82 to **7**-85
Flat hardened washers, **7**-5
Flatness tolerance, **12**-28
Flat position welding, **7**-24, **B**-6
Flat roof:
  ponding, **2**-28 to **2**-32
  snow loads, **5**-12
Flat-roof portals, **6**-19, **6**-20, **6**-18
Flexural buckling, **3**-28
Flexural members (*see* Beam behavior)
Flexural rotation, **4**-6
Flexural strengths, **6**-6, **6**-7
Flexural-torsional buckling of compression members, **3**-61 to **3**-64
Flexure:
  bridge design, **11**-46
  built-up members, AISC requirements, **7**-2
  frame drift, **6**-50 to **6**-51
  tension and compresion members with, **3**-64 to **3**-88
Flexure ductility, bridge design, **11**-46
Floor load, seismic design, **E**-4
Floor systems:
  beam torsion effects, **4**-2
  cambering, **12**-7
  dynamic behavior of, **9**-3 to **9**-12
    natural frequency evaluation, **9**-8 to **9**-9
    nature of vibrations, **9**-3 to **9**-4
    structural dynamics, **9**-4 to **9**-8
    transformed section properties, **9**-9 to **9**-12
  minimum live loads for, **5**-13
  multistory frames, **6**-48

Index ■ I-13

Floor vibrations:
  composite beams, **10**-19, **10**-10 to **10**-12
  (*See also* Serviceability limit states design)
Flux-cored arc welding (FCAW), **8**-6, **12**-15,
    12-16 to **12**-18
  access holes for, **B**-19
  field welding, **B**-28
  filler material requirements, **7**-23
  fracture behavior, **8**-30
  preheat and interpass temperatures, **7**-26
  self-shielded (*see* Self-shielded flux-cored
    arc welding)
Force measurement conversion table, **D**-1
Form deck, **A**-11 to **A**-16, **A**-1, 2
  attachments, **A**-16
  nonstructural fills, **A**-16
  structural concrete form, **A**-11 to **A**-15
Foundation overturning, **E**-39 to **E**-41
Fracture, **8**-1 to **8**-8, **8**-25 to **8**-48
  fracture mechanics analysis, **8**-32 to **8**-45
    buried penny-shaped crack, **8**-44 to
      8-45
    center crack, **8**-36 to **8**-38
    edge crack, **8**-39 to **8**-44
  fracture mechanics test methods, **8**-45 to
    8-48
  specification of steel and filler material,
    8-28 to **8**-32
  tension members, bolt holes and, **3**-1
  types of, **8**-25 to **8**-27
Fracture-critical members (FCM), **8**-2
Fracture limit state, bridge design, **11**-43
Fracture mechanics analysis, **8**-32 to **8**-45
Fracture-mechanics calculations, **8**-43
Fracture-mechanics test methods, **8**-45
Fracture toughness, **8**-2, **8**-18, **8**-30
Framed connection:
  shear connections, **7**-97 to **7**-100
  simple beams under shear and axial
    loads, **7**-109
Frame design:
  general principles, **6**-1 to **6**-16
    advanced inelastic analysis, **6**-16
    beam-column interaction equations, **6**-6
      to **6**-7
    column effective length, **6**-9 to **6**-13
    design limit states, **6**-3 to **6**-6
    joints, **6**-3

Frame design, general principles (*Cont.*):
    live-load reduction, **6**-7
    member size selection, **6**-13
    notional load approach for column sta-
      bility assessment, **6**-14 to **6**-16
    second-order effects, **6**-7 to **6**-9
    stability, **6**-1 to **6**-3
  multistory, **6**-41 to **6**-82
    braced frame examples, **6**-71 to **6**-82
    bracing systems, **6**-49 to **6**-53
    combined systems, **6**-61 to **6**-62, **6**-63
    gravity bracing, **6**-43 to **6**-48
    unbraced frame examples, **6**-63 to **6**-71
    unbraced rigid frames, **6**-53 to **6**-61
  single-story, **6**-16 to **6**-41
    lattice frames, **6**-24 to **6**-29
    loading, **6**-17, **6**-18
    multibay braced frame, **6**-29 to **6**-36
    multibay unbraced frame with leaning
      columns, **6**-36 to **6**-41
    portal frames, **6**-19 to **6**-24
Frame drift, multistory frames, **6**-81, **6**-52
Frames:
  axial force connections, **7**-55 to **7**-60, **7**-52
  cambering, **12**-7
  compression column bracing, **3**-43
  connections (*see* Connections)
  height measurement, **7**-2
  seismic resistance, **B**-8 to **B**-11
    (*See also* Seismic design)
  single-story structures, **6**-18 to **6**-19
  stability, erection procedures, **12**-36
  torsion effects, **4**-2
Free end torsion, **4**-4 to **5**
Full plastic moment, **1**-24
Fully tensioned bolts, **7**-3 to **7**-6

Galambos, T.V., **1**-17
Gaps, welding considerations, **B**-8, **B**-9,
    B-28
Gas-metal arc welding (GMAW), **12**-15,
    12-18 to **12**-19
  filler material requirements, **7**-23
  position, **7**-24
  preheat and interpass temperatures, **7**-26
  shielded (*see* Shielded metal arc welding)
Gas shielded, flux cored arc welding, **12**-17

Gas tungsten arc welding (GTIW), **12**-15
Gauge, fastener spacing, **7**-9
Gauge of angle (GOL), **7**-107 to **7**-108
G factors, **3**-37
Girder frequency, **9**-9
Girders:
  bridge design, **11**-11, **11**-10, **11**-47
  composite, **10**-19 to **10**-26
  fractures in, **8**-4
  lattice, **6**-27 to **6**-29
  multistory unbraced rigid frames, **6**-54
  plate, layout and assembly, **12**-6 to **12**-7
  seismic design
    braced frames, **E**-43 to **E**-50
    special moment frames, **E**-13 to **E**-14, **E**-15, **E**-23 to **E**-30
    (*See also* Seismic design)
  tolerances, **12**-28, **12**-30
  transformed section properties, **9**-12
  vibration (*see* Serviceability limit states design)
Girder stiffnesses, **3**-34 to **3**-37
Girder-to-column joint restraint, **E**-22
Girder webs, mill order, **12**-2
Girts, **3**-21
GMAW (*see* Gas metal arc welding)
Gouges, weld access hole cutting, **B**-20
Grade, mill order, **12**-2
Graphic methods, **1**-8
Gravity columns, **6**-45
Gravity frames, **6**-41, **6**-43 to **6**-48
Gravity loads
  multistory frames, **6**-68, **6**-60
  notional load analysis, **6**-14
  seismic design, **E**-3 to **E**-4, **E**-41
Groove weld, **8**-19
  access to, **B**-5
  combination of types of welds, **7**-7
  costs, **7**-16
  distribution of load stress, **B**-3
  drawing specification, **7**-20 to **7**-21
  end treatment, **7**-26
  position, **7**-24
  preparation for, **7**-22
  quality criteria, **7**-29
  residual stresses, **B**-5
  seismic loading, material zones, **B**-21
  types of, **7**-14 to **7**-15

Gruning cross-sectional dimensions, **1**-16
GTAW (*see* Gas tungsten arc welding)
Gusset plates, **7**-39 to **7**-45
  layout and assembly, **12**-6
  seismic design, **E**-49 to **E**-50, **E**-51
  welding, **8**-21 to **8**-22
Gussets:
  axial force connections (*see* Axial force connections)
  bolt spacing, **7**-11 to **7**-12

Hacksaws, **12**-5
Hanger connections, **7**-71 to **7**-77
  angles, **7**-71 to **7**-73
  bolts, **7**-73 to **7**-77
Haris Engineering Software Company, **10**-42, **10**-37
Harmonic motion:
  composite beams, **10**-10
  vibrations due to rhythmic activities, **9**-30
Haunch connections, **B**-33
Haunched plate girder, **11**-11
HCOMPL, **10**-37
HCONCOL, **10**-42
Heat-affected zones (HAZ) of weld areas, **B**-23, **B**-25, **8**-48
Heat input, and weld quality, **B**-27
Heat treated low-alloy steel, **6**-18
Heat treatment, welded assemblies, **7**-27
Hex nut, **7**-7
High-cycle fatigue loading, **B**-3
High-strength low-alloy steel, **6**-18
High-strength structural steel bolts, **7**-2 to **7**-3
Hinge formation, ductile, **B**-11 to **B**-17
History, **1**-1 to **1**-17
  early empirical period, **1**-1 to **1**-2
  18th century, **1**-3 to **1**-6
  evolution of structural design, **1**-1
  1900 to **1950**, **1**-14 to **1**-17
  19th century, **1**-6 to **1**-14
  17th century, **1**-2 to **1**-3
Holes in beams, **2**-7 to **2**-10
  fabrication, **12**-8 to **12**-9
  (*See also* Bolt holes; Weld access holes)
Hooke, Robert, **1**-2
Hooke's law, **1**-2, **1**-5, **1**-6, **1**-15, **4**-9

Index ■ I-15

Horizontal shear, unbraced rigid frames, 6-53
Horizontal welding, **7**-24
Hot spot stress range, **8**-11
Hydrogen, **1**-32

I column welds, **B**-3
Ideal stress-strain diagram, **1**-23
I girder, bridge, **11**-32 to **11**-33
Impact moment, bridge design, **11**-37 to **11**-38
Impact wrench, **12**-12
Imperfection effects, **6**-15
Incomplete-penetration defect, **8**-37
Inelastic analysis, bridge I girder, **11**-32
Inelastic buckling, **2**-2 to **2**-3, **2**-11 to **2**-12, **3**-40
   of web, **2**-24
Inertial moments:
   composite beams, **10**-18
   composite girders, **10**-25 to **10**-26
Inertia of transformed slab, bridge design, **11**-41
Inspection, **11**-49, **12**-37 to **12**-41
   fabrication:
      coating, **12**-35 to **12**-36
      connections, bolted, **12**-14 to **12**-15
      qualifications, **12**-38 to **12**-39
      weld access holes, **B**-20
      welded connections, **7**-29, **12**-21 to **12**-26
         base metal quality, **12**-22
         equipment and consumables, **12**-23
         joint preparation and fit-up, **12**-22 to **12**-23
         nondestructive testing, **12**-24 to **12**-25
         operating conditions, **12**-23
         preheat, **12**-24
         procedures, **12**-22
         qualifications of inspector, **12**-38 to **12**-39
         qualifications of welder, **12**-37 to **12**-38
         for seismic loading, **B**-28 to **B**-31
         during and after welding, **12**-24
         before welding, **12**-22
Institutional buildings. AISC quality certification, **12**-40 to **12**-41
Insulation, deck, **A**-16, **A**-11, **A**-13

Insulation board, deck loads, **A**-1 to **A**-2, **A**-6
Interface forces, axial force connections, **7**-57
Interior beams, bridge design, **11**-44
Interior stringers, **11**-26
Intermediate columns, **3**-40
Intermittent welds, **7**-17, **7**-20
International Organization for Standardization vibration scale, **9**-15 to **9**-17
Interpass temperatures, welding, **7**-26, **B**-27
Interstory drift, multistory frames, **6**-52
Inventory, bridge performance, **11**-48 to **11**-49
Inventory control, **12**-5
ISO scale, vibration perception, **9**-15 to **9**-17
I values, deck material, **A**-1

Joints:
   cutting, **12**-6
   detail categories for, **8**-15 to **8**-18
      mechanically fastened, **8**-15 to **8**-18
      welded, **8**-18 to **8**-20
   frame design, **6**-3, **6**-55 to **6**-56
   weld inspection, **12**-22 to **12**-23
Joist frequency, **9**-8 to **9**-9
Joists:
   transformed section properties, **9**-11 to **9**-12
   vibration (*see* Serviceability limit states design)
Jumbo sections
   Charpy-impact-energy data, **8**-31, **8**-35
   fabrication cracks at cope holes, **8**-13
   tension chord fractures, **8**-3 to **8**-4

K bracing, multistory system, **6**-49
Kerf, **B**-20
KISS method, **7**-60, **7**-63, **7**-64

Lack-of-fusion defect, **8**-4, **8**-37, **8**-41, **8**-42
Lamellar tear, **8**-37
Lanes, bridge design, **11**-38, **11**-43
Lap joints, welded, **7**-14, **7**-15

Lateral bracing, 3-44
Lateral buckling, 2-1
Lateral displacement:
  framing:
    single-story multibay unbraced, 6-37, 6-38
    multistory braced frames, 6-81
    moment diagram, B-9, B-10
Lateral force procedures, 5-23 to 5-27
Lateral load, 5-3
  frames:
    multistory, 6-41, 6-51 to 6-52
    unbraced rigid, 6-53
    moment connections and, 7-86 to 7-87
Lateral torsional buckling, 2-15 to 2-16
Lateral-translation (LT) analysis, 6-37, 6-38
Lattice frames, 6-18, 6-19, 6-24 to 6-29
  bracing, 6-24 to 6-27, 6-28, 6-29
  structural forms, 6-24
  truss design, 6-27 to 6-29
Lattice work, compressive members, 3-59, 3-60
Layout and assembly, shop fabrication, 12-6 to 12-7
Leak testing of welds, 12-38
Leaning columns, single-story multibay frames:
  braced, 6-34 to 6-35
  unbraced, 6-36 to 6-41
Length of weld, drawing specification, 7-19
Lenzen's Modified Reiher-Meister scale, 9-13 to 9-14, 9-35
Limit states, 6-3 to 6-6
Limit states analysis, 1-17, 7-29 to 7-30
Limit states design, 11-5
  (*See also* Bridges; Frame design)
Limit states equation, 11-5
Linear-elastic fracture mechanics (LEFM), 8-27, 8-2, 8-32, 8-33
Liquid penetrant testing of welds, 12-24 to 12-25, 12-26, 12-38
Live load, 1-20, 5-12, 5-13, 5-14
  bridge design, 11-36 to 11-38, 11-19 to 11-21
    fatigue, 11-22
    force effects, 11-43 to 11-44
    I girder, 11-32 to 11-33
  deck, A-1

Live load (*Cont.*):
  serviceability requirements, 9-2
  single-story frames, 6-17
Live-load reduction, 6-6
Load:
  bridge design, 11-16 to 11-17, 11-34 to 11-39, 11-44
    classification of, 11-17 to 11-21
    combinations, 11-42
    I girder, 11-32
    ratings (*see* Load ratings of bridges)
  building design, 5-10 to 5-29
    combinations, 5-28 to 5-29
    dead, 5-10 to 5-12
    live, 5-12, 5-13, 5-14
    snow, 5-12, 5-14 to 5-18
    wind, 5-18 to 5-23
  connections, 7-1
    and welded joint fininshing, 7-27
    weld quality, 7-28 to 7-29
  decks, A-1
    composite, A-18 to A-21, A-22, A-23, A-16 to A-21
    form deck as load carrier, 3-6
    roof, A-2 to A-10
  fatigue testing, 8-8
  frames:
    lattice truss forces, 6-28, 6-29
    single-story frames, 6-17, 6-18
    single-story multibay braced, 6-30
    (*See also* Frame design)
  measurement, conversion table, D-1
  seismic design:
    gravity load, E-3 to E-4
    (*SSee also* Seismic load)
  serviceability requirements, 9-2
  shop detail drawings, 12-3
Load factor design, 1-17
Load factor design selection table, C-1 to C-6
Load factors, 1-20, 5-28 to 5-29
  bridge design, 11-8, 11-13 to 11-17, 11-42
  composite beams, 10-4 to 10-5
  composite girders, 10-19, 10-20 to 10-23
  limit states, 6-3, 6-4, 11-5
  LRFD specification, 1-20 to 1-21, 8-1
Load paths:
  connection design, 7-29, 7-30, 7-31

Load paths (*Cont.*):
  frame design, **6**-3
  moment connections, **7**-87 to **7**-88
Load ratings of bridges, **11**-48 to **11**-53
  field measurements and inspection, **11**-49
  inventory and operating ratings, **11**-48 to **11**-49
  loading structure, **11**-49 to **11**-50
  strength evaluation, **11**-51 to **11**-53
  working stress method, **11**-50 to **11**-51
Load and resistance factor design (LRFD), **1**-19, **1**-20 to **1**-21, **8**-1
  HCOMPL program, **10**-37 to **10**-38
  idealized stress-strain characteristics, **1**-23
  material behavior, **1**-23 to **1**-33
  seismic design (*see* Seismic design)
  torsion, **4**-25 to 26
  (*See also* Bridges; Frame design; Seismic design)
Load resistance factor design specification, **1**-20 to **1**-21, **8**-11 to **8**-12
  deck material, **A**-1
  high-cycle fatigue, **8**-3, **8**-6, **8**-7 to **8**-8, **8**-9
Local buckling, **3**-28
Locknuts, **7**-3
Lock pin and collar installation method, **12**-14
Lock washers, **7**-3
Long columns, **3**-40
Long grips, bolt spacing, **7**-11 to **7**-12
Longitudinal fillet welds, **7**-18
Long slotted holes, **7**-8 to **7**-9
Lot numbers—lot identification, **12**-5
Low-cycle fatigue, **8**-6
Lower limit bound theorem of limit analysis, **7**-29 to **7**-30
Low-rise buildings, moment connections, **7**-91
Low slope roof, **5**-14 to **5**-15
LRFD (*see* Load and resistance factor design)
Lubrication, fastener, **12**-11 to **12**-12

Machine bolts, **7**-3, **7**-4
Machining, welds, **7**-27
Magnetic flux gauge, **12**-35

Magnetic particle testing:
  weld access holes, **B**-20
  welds, **7**-29, **12**-25, 38, **B**-29 to **B**-30, **B**-31
Magnification factors, **3**-69 to **3**-71
Maier Leibnitz test, **1**-1
Maintenance, bridges, **11**-8
Manganese, **1**-32
Masonry, as bracing system, **3**-44, **6**-26
Match-marking for bolt tightening, **12**-11
Material behavior, **1**-23 to **1**-33
  cold work effects on tensile properties, **1**-26 to **1**-27
  elemental composition of steel, **1**-31 to **1**-33
  idealized stress-strain characteristics, **1**-23
  moment curvature graphs, **1**-26
  plastic modulus of wide-flange and compound sections, **1**-24 to **1**-26
  properties, **1**-33
  residual stresses, **1**-30, **1**-31
  temperature elevation and, **1**-27 to **1**-30
Material control:
  shop fabrication, **12**-4 to **12**-5
Material list, **12**-1 to **12**-2
Material properties:
  densities of, **5**-10
  pre- and post-Northridge details, **B**-31, **B**-32
  seismic resistance, **B**-25 to **B**-26
Material takeoff, **12**-1 to **12**-2
Material test reports, **12**-4, **12**-5
Material tolerances, **12**-27 to **12**-30
Maximum compresive elastic flexural stress, **11**-44
Maximum slenderness ratios, **3**-41
Maximum yield strength, **8**-43
Maxwell, James Clerk, **1**-9 to **1**-10
Measuring units, metric conversion table, **D**-1 to **D**-2
Mechanically fastened joints, detail categories for, **8**-15 to **8**-18
Membrane effect, bridges, **11**-8
Metal arc welding, **7**-24, **12**-15, **12**-18 to **12**-19
  field welding, **B**-28
  filler material requirements, **7**-23
  preheat and interpass temperatures, **7**-26

Metal arc welding (*Cont.*):
  shielded (*see* Shielded metal arc welding)
  submerged (*see* Submerged-arc welding)
Method of sections, **E**-49
Metric conversion table, **D**-1 to **D**-2
Microcracks, **B**-20
Midspan deflection, multistory braced frames, **6**-81
Mill cambering limits, **12**-7
Mill cutting tolerances, **12**-1 to **12**-2
Milling, base plates, **12**-6
Mill material tolerances, **12**-27 to **12**-28
Mill order, **12**-1 to **12**-2
Mill rolling tolerances, **12**-29
Mill scale:
  fireproofing and, **12**-31
  welding considerations, **7**-25
Mill test reports, **12**-4
Miner's rule, **8**-23
Minimum angle gauges, **7**-107 to **7**-108
Minimum specified yield strength (MSYS), **8**-37
Minimum toughness specifications, **B**-31, **B**-32
Misalignment, **8**-20
Modification factor, **3**-71 to **3**-75
Modular ratio values, **11**-35, **11**-32, **11**-27
Modulus of elasticity:
  composite beams, **10**-17
  composite sections of bridge, **11**-27
  temperature and, **1**-29
Modulus of rigidity, **4**-9
Mohr, Otto, **1**-10 to **1**-12
Molybdenum, **1**-32
Moment amplification, **3**-68
Moment amplification factor, single-story portal frames, **6**-23, **6**-24
Moment capacity:
  composite beams, **10**-5, **10**-17 to **10**-18
  special moment-resisting frames, **B**-15
  welded webs and, **B**-18
  zone 2, **2**-13 to **2**-15
Moment connections, **4**-6, **7**-86 to **7**-97, **B**-18
  shear connections, **7**-90 to **7**-97
  stiffener size determination, **7**-89
  stiffener weld to column flange and web, **7**-89 to **7**-90
  strong axis beam, **7**-88 to **7**-89

Moment frames, seismic design:
  ordinary, **E**-33 to **E**-34
    energy absorption and ductility, **E**-53
    height limitations, **E**-52
  special (*see* Special moment frames)
Moment curvature graphs, **1**-26
Moment diagram, lateral displacement and, **B**-9, **B**-10
Moment frame, interaction with braced frame, **6**-62
Moment joints, multistory frames, **6**-41
Moment measurement, conversion table, **D**-1
Moment modification, **3**-71 to **3**-75
Moment of inertia:
  bridge design, **11**-41, **11**-45
  composite beams, **10**-18
  composite girders, **10**-25 to **10**-26
Multiaxial loading, **8**-9, **8**-21 to **8**-23
Multibay frame, **6**-29 to **6**-36
  portal frames, **6**-22, **6**-23
  single-story:
    braced, **6**-29 to **6**-36
    unbraced, **6**-36 to **6**-41
Multiple presence factor, bridge design, **11**-43, **11**-20 to **11**-21
Multistory frames, **6**-41 to **6**-82
  braced frame examples, **6**-71 to **6**-82
  bracing systems, **6**-49 to **6**-53
  combined systems, **6**-61 to **6**-62, **6**-63
  connections, bolts, **7**-2
  drift, **3**-70
  elevator shaft columns, **12**-29 to **12**-30
  gravity, **6**-43 to **6**-48
  notional load analysis, **6**-14
  stability, **6**-2
  unbraced frames, examples of, **6**-63 to **6**-71
    multistory, **6**-68 to **6**-71
    two-story, **6**-63 to **6**-68
  unbraced rigid frames, **6**-53 to **6**-61
Murray criterion for walking vibrations, **9**-20 to **9**-23
  composite beams, **10**-19
  Lenzen's Modified Reiher-Meister scale modification, **9**-14

Nails, form deck, **A**-16

Natural frequency, **9**-5, **9**-7, **9**-8 to **9**-9
Necking down, **B**-12
Negative flexure, bridge design, **11**-33
Net areas, tension members, **3**-2 to **3**-4, **3**-5, **3**-7 to **3**-13
Neutron radiographic testing of welds, **12**-38
Nickel, **1**-32, **8**-31
No lateral translation (NT) analysis, **6**-7
Nomenclature, **F**-1 to **F**-7
Nominal fatigue resistance, bridges, **11**-22 to **11**-23
Nominal strength, limit states design, **11**-5
Nominal stress range times, **8**-14
Noncompact sections, bridges, **11**-30
Nondestructive tests, welds, **7**-29, **8**-18, **12**-24 to **12**-25, **B**-30 to **B**-31
Nonstructural fills, form decks, **A**-16
Nontranslation (NT) analysis, **6**-37, **6**-38
Nontubular load, weld testing, **12**-26
Northridge earthquake, **E**-53
  experience based on, **B**-31 to **B**-34
  moment connection performance, **7**-86
  weld access hole cracking, **B**-20
Notches, weld access hole cutting, **B**-20
Notch impact resistance, temperature and, **1**-30
Notchlike conditions, **B**-11
Notch toughness, **1**-32, **8**-2, **8**-4
  aluminum and, **1**-31
  jumbo shapes, **8**-35
  seismic design
    base metal, **B**-23
    pre- and post-Northridge details, **B**-31, **B**-32
    redistribution of stresses under seismic loading, **B**-7
    weld metals, **B**-24 to **B**-25
  of steel and filler metal, **8**-28 to **8**-32
  weld deposits, **8**-31
Notional load analysis, **6**-14 to **6**-16

Oiled fasteners, **12**-11 to **12**-12
One bay braced system, **E**-37, **E**-39 to **E**-40
Ontario Highway Bridge Design Code (OHBDC), **11**-7 to **11**-8
Openings, **5**-29

Operating conditions, weld inspection, **12**-23
Operating ratings, bridge performance, **11**-48 to **11**-49
Operational importance, bridge design, **11**-42
Operations sequencing, erection procedures, **12**-36 to **12**-37
Opposite hand details, **12**-3
Ordinary bolts, **7**-3, **7**-4
Ordinary moment frames, seismic design, **E**-33 to **E**-34
  energy absorption and ductility, **E**-53
  height limitations, **E**-52
Outrigger truss, **6**-62
Overhead welds, **7**-24, **B**-6
Overturning moment, **5**-23
  braced frame systems, **E**-37 to **E**-40
  multistory frames, **6**-51, **6**-52

Paint:
  bearing-type connections, **7**-6
  welded surfaces, **7**-27
  (*See also* Coating)
Panel joints, multistory frames, **6**-57 to **6**-59
Panel mode properties, **9**-26 to **9**-28, **9**-35 to **9**-37
Panel zones:
  moment connections, **7**-91
  special moment frames (SMF), **E**-17 to **E**-22
Partially restrained connections, seismic loads, **B**-34
Partial penetration welds, defined, **7**-14 to **7**-15
Partition load, seismic design, **E**-4
P-delta effects, **6**-5, **6**-8, **6**-15, **E**-12
Peening, welds, **7**-27
Pencil-type pull-off gauge, **12**-35
Penetrant testing, welds, **7**-29, **12**-24 to **12**-25, 26, 38, **B**-29, **B**-31
Penny-shaped crack, **8**-44 to **8**-45
Perforated cover plates, compressive members, **3**-59, **3**-60
Period check, **E**-33
Period of building:
  east-west stress analysis, **E**-4, **E**-5

Period of building (*Cont.*):
    ordinary moment frames, **E**-33 to **E**-34
    special moment frames, **E**-6 to **E**-7
Permanent loads, bridge design, **11**-18, **11**-44, **11**-16, **11**-17
P frequency, CVN testing at, **8**-31
Phosphorus, **1**-32
Piece marks, erection plans, **12**-3
Pin connections:
    composite construction, **10**-1
    compression members, **3**-30
    end, torsion, **4**-5
    tension members, **3**-21 to **3**-23
Pipe column, **3**-47
Pipe sections, manufacturing tolerances, **12**-28
Pitched-roof portals, **6**-18, **6**-20, **6**-21
Plane-strain fracture toughness, **8**-34
Plans, **12**-2 to **12**-3, **7**-9 to **7**-10
    CAD—CAM interface, **12**-3 to **12**-4
    weld symbols, **7**-18 to **7**-23
Plastic analysis, **1**-17, **2**-5
Plastic deformation, **11**-4
    buckling zone, **2**-2, **2**-3 to **2**-4
    seismic loading, **B**-7
Plastic hinge, **1**-24, **E**-53
Plastic modulus, **1**-24 to **1**-26, **2**-5
Plastic moment, bridge design, **11**-33, **11**-40
    composite bridges, **11**-28 to **11**-29
    depth of web in compression calculation for, **11**-29, **11**-42
Plastic neutral axis, bridge design, **11**-33, **11**-39, **11**-40
Plastic range, stress-strain diagrams, **11**-3
Plastic shear strain, ductility conditions, **B**-12
Plastic stress distribution
    compact and noncompact sections, **3**-38 to **3**-39
    compact section, **2**-1
Plate girders:
    bridge superstructure, **11**-11
    layout and assembly, **12**-6 to **12**-7
Plate gross shear, moment connections, **7**-91
Plates:
    composite beam reinforcement, **10**-26 to **10**-36
    materials tolerances, **12**-27 to **12**-28

Plates (*Cont.*):
    mill order, **12**-2
    tension member net area, **3**-3
    thickness of and fatigue, **8**-20
Plate washers, **7**-6
Plug and slot welds, **7**-15, **7**-13
Plug welds, **7**-18, **7**-7
Polarity, and weld quality, **B**-27
Polystyrene form deck fills, **A**-16
Ponding, **2**-28 to **2**-32
Porosity of weld, **B**-30
Portal frames
    analysis and design, **6**-23 to **6**-24
    bracing, **6**-20 to **6**-21
    with fabricated sections, **6**-21
    multibay, **6**-22, **6**-23
    single-story structures, **6**-18, **6**-19 to **6**-24
    structural form, **6**-19 to **6**-20
Positive flexure, bridge design, **11**-32
Powder actuated pins, **10**-1
Precast concrete (*see* Composite construction)
Predictive modeling, fatigue cracking, **8**-22
Preheat, weld, **7**-26, **12**-24, **B**-27
Premature inelastic action, residual stresses and, **1**-30
Prequalified joints, **7**-13
Pressure coefficients, **5**-21
Prestressed steel bridges, **11**-11 to **11**-12
Principal circles, **1**-11 to **1**-12
Probability density function for load effect and strength, **1**-21
Probability of failure, **1**-22
Proportional limit, **11**-3
Prying action:
    hangers, **7**-74 to **7**-77
    simple beams under shear and axial loads, **7**-112 to **7**-113, **7**-114, **7**-115, **7**-116, **7**-117
Prying action theory, **7**-69 to **7**-70
Pull-off gauge, **12**-35
Pulsating loading, fatigue testing, **8**-8
Punching:
    bolted connections, **12**-8 to **12**-9
    drawings, **12**-4
Pure torsion, **4**-12
Pure torsion shear stress, **4**-29, **4**-30, **4**-17, **4**-19, **4**-26, **4**-27, **4**-27

Index ■ I-21

Purlins, **3**-21, **6**-19, **6**-24

Quality control:
  inspection and certification program, **12**-37 to **12**-41
  materials control, **12**-4 to **12**-5
  welds:
    inspection, **7**-28 to **7**-29, **12**-21 to **12**-26
    written specifications, **12**-20 to **12**-21 to **12**-22
  welds for seismic loading:
    inspection, **B**-28 to **B**-31
    workmanship, **B**-25 to **B**-28
Quasi-plate girders, roof deck, **A**-8
Quenching and self-tempering (QST) process, **8**-31

Radiographic testing, welds, **7**-29, **12**-26, **12**-38, **B**-30, **B**-31
Rainwater, **1**-20
Rat holes, **B**-18 to **B**-20
  crack sources, **8**-13 to **8**-15
  (*See also* Bolt holes; Weld access holes)
Rational analysis, **5**-23, **6**-8
Rayleigh equation, E-7, E-33
Rayleigh Ritz method, **1**-13
Rayleigh waves, **1**-14
Reciprocity theorem, **1**-10
Recrystallization, **1**-28
Reduced beam section connections, post-Northridge designs, **B**-33 to **B**-34
Reduction factor, composite girders, **10**-25
Redundancy, bridge design, **11**-42
Rehabilitation of bridges, **11**-8
Reiher-Meister scale:
  for steady-state vibrations, **9**-12 to **9**-13
  for transient vibrations, **9**-13 to **9**-14, **9**-35
Reinforced-concrete deck slabs, bridges, **11**-8
Reinforcing plates, composite beams, **10**-26 to **10**-36
Reliability index, **1**-22
Reserve capacity, **11**-48
Residual stresses, **1**-30, **1**-31, **B**-5 to **B**-6
Resistance factors, **1**-20, **1**-21
  bridge design, **11**-13 to **11**-17, **11**-42
  limit states, **6**-3, **6**-4

Resistance factors (*Cont.*):
  LRFD specification, **1**-20 to **1**-21
Resistance moment capacity, composite beams, **10**-5
Reversal loading fatigue testing, **8**-8
Rhythmic excitations, **9**-29 to **9**-31, **9**-38 to **9**-39
Rigid frame system, **6**-1
  braced (*see* Braced frames)
  unbraced, **6**-53 to **6**-61
    joints, **6**-55 to **6**-56
    multistory, **6**-53, **6**-54
    stability of, **6**-2
Ritz, Walter, **1**-13
Riveted members
  detail categories for, **8**-15 to **8**-18
  fatigue crack formation, **8**-1
Rods and bars, as tension members, **3**-20 to **3**-21
Rolled beam:
  bridge superstructures, **11**-10
  with cover plate, bridge superstructure, **11**-11
  mill cambering limits, **12**-7
Rolled section stringers, bridge, **11**-10
Rolled shapes:
  composite construction (*see* Composite construction)
  notch toughness, **B**-23
Rolling schedule, **12**-2
Roof:
  bolted truss splices, **7**-2
  cambering, **12**-7
  framing:
    single-story multibay braced, **6**-33 to **6**-34
    single-story structures, **6**-19, **6**-20
  lattice structures, **6**-24
  loads, **1**-20
    minimum live, **5**-14
    seismic design, E-3
    snow loads, **5**-12, **5**-14 to **5**-18
  ponding, **2**-28 to **2**-32
  sag rods, **3**-21
  serviceability considerations for, **9**-3
Roof deck, **A**-1 to **A**-11, **A**-12, **A**-1
  fastening, **A**-10
  generic profiles, **A**-2, 3

Roof deck (*Cont.*):
  horizontal load, **A-7** to **A-10**
  special decks, **A-10** to **A-11**, **A-12**, **A-13**
  vertical load, **A-2** to **A-7**
Root openings (gaps), welding considerations, **B-8**, **B-9**, **B-28**
Rotation, welded beam end-column connections, **7-12**
Rotational stiffness, **3-34**
Runoff plates, **7-27**
Rust, bolt tightening, **12-12**

Safe evaluation concept, **11-53**
Safety factors, working stress approach, **11-4**
Sag rods, **3-21**
SAW (*see* Submerged arc welding)
Scale effects in fatigue, **8-20** to **8-21**
Screws:
  form deck, **A-16**, **A-22**
  roof deck, **A-10**
Sea air, and fatigue life, **8-9**
Seated shear connections, **7-100** to **7-103**
Seawater, and fatigue life, **8-9**
Secant formula, **8-36**
Secondary loading, **8-22**
Secondary members, welding principles, **B-5**
Second order distortion-induced stresses, **8-9**
Second order elastic analysis, **6-7**
Second-order moments, **3-68** to **3-69**
Second-order P-delta effect, **6-5**
Second-order plastic hinge analysis, multi-story frames, **6-59**
Sections, mill order, **12-2**
Section selection, tension members, **3-14** to **3-17**
Seismic design, **E-1** to **E-53**
  braced frames, **E-34** to **E-51**, **E-52**, **E-53**
    alternate connection details, **E-51**, **E-52**, **E-53**
    brace design (fourth floor), **E-41** to **E-42**
    bracing system configuration, **E-36** to **E-37**
    column design (third to fifth floors), **E-42** to **E-43**

Seismic design, braced frames (*Cont.*):
    connection of brace to column and girder, **E-50**
    connection of brace to girder, **E-46** to **E-50**
    foundation overturning, **E-37** to **E-39**
    frame analysis, **E-39** to **E-41**
    girder design (fourth floor), **E-43** to **E-46**
    horizontal distribution of seismic forces, **E-36**
    north-south seismic forces (stress and drift), **E-35** to **E-36**
  energy absorption and ductility, **E-53**
  general information, **E-1** to **E-4**
  gravity loads, **E-3** to **E-4**
  seismic design parameters, **E-3**
  specifications and manual, **E-1**
  height limitations, **E-52**
  ordinary moment frames (OMF), **E-33** to **E-34**
  special moment frames (SMF), **E-4** to **E-33**
    building period check, **E-33**
    column to girder and column to panel zone ratio, **E-19** to **E-22**
    east-west forces for drift analysis, **E-6** to **E-9**
    east-west forces for stress analysis, **E-4** to **E-6**
    girder bracing, **E-23**
    girder moment connection, **E-23**, **E-24**
    girder shear connection with moment, **E-28** to **E-30**
    girder shear connection without moment, **E-24** to **E-27**
    panel zone, **E-17** to **E-19**
    P-delta effects, **E-12**
    preliminary frame design with drift control, **E-9** to **E-12**
    stiffener plates, **E-30** to **E-32**
    strength check for frame members, **E-13** to **E-17**
  welding, **B-1** to **B-34**
  (*See also* Welds, for seismic resistance)
Seismic load, **1-20**, **5-23** to **5-28**, **5-29** to **5-32**
  bolts, **7-2**
  bridge, **11-21**
  fracture behavior, **8-6**, **8-30**

Index ■ I-23

Seismic load (*Cont.*):
   moment connections, **7**-86 to **7**-97
   roof deck as bracing, **A**-1
   single-story frames, **6**-17
Self-shielded flux-cored arc welding (FCAW-S):
   Charpy V-notch tests, **8**-34 to **8**-35
   edge cracks, **8**-39
   field welding, **B**-28
   toughness of welds, **8**-32
Sequence of erection operations, **12**-36 to **12**-37
Sequence of loading, bridge design, **11**-32
Serviceability, **4**-25
Serviceability deflections, multistory braced frames, **6**-81, **6**-82
Serviceability limit states design, **6**-4
   Allen and Murray criterion, **9**-23 to **9**-239
   bridges, **11**-14 to **11**-17, **11**-43
   CSA criterion for walking vibrations, **9**-17 to **9**-20
   design examples, **9**-31 to **9**-39
   dynamic behavior of floor systems, **9**-3 to **9**-12
   ISO scale, **9**-15 to **9**-17
   Murray criterion for walking vibrations, **9**-20 to **9**-23
   Reiher-Meister scale, **9**-12 to **9**-13
   Reiher-Meister scale, modified, **9**-13 to **9**-14
   requirements for static deflections, **9**-2 to **9**-3
   rhythmic excitations, **9**-29 to **9**-31
   for sensitive equipment and facilities, **9**-17, **9**-18
   Wiss and Parmelee rating factor, **9**-14 to **9**-15
Service load:
   composites for decks, **A**-17
   framing, single-story multibay braced, **6**-36
   limit states design, **11**-5
   shop detail drawings, **12**-3
   single-story frames, **6**-17
Settlement, and fractures, **8**-6
Shear center, **4**-6 to **4**-7, **4**-14
Shear force connections, **7**-97 to **7**-106
   beam shear splices, **7**-103 to **7**-106

Shear force connections (*Cont.*):
   composite construction, **10**-1
   framed, **7**-97 to **7**-100
   moment connections, **7**-90 to **7**-97
   seated, **7**-100 to **7**-103
   simple beams under shear and axial load, **7**-107 to **7**-113
Sheared edge plates, residual stresses, **1**-30
Shear fracture, **2**-25, **3**-25
   axial force connections (*see* Axial force connections)
   gusset plate design, **7**-40
Shear lag, **3**-8
Shear modulus, **4**-9
Shear plate, special moment frame, **E**-25 to **E**-27, **E**-28 to **E**-30
Shear racking, unbraced rigid multistory frame, **6**-54, **6**-53
Shear rupture (*see* Tearout)
Shear strength
   deck material, **A**-1
   tension members, **3**-28
Shear stress, **4**-9
   beams, **2**-23 to **2**-25, **2**-26
   bridge design, **11**-47
   connections:
      gusset plate design, **7**-40 to **7**-41, **7**-52
      plug and slot welds, **7**-15
      shear force connections, **7**-97 to **7**-106
      simple beam connections, **7**-107 to **7**-113
      (*See also* Axial force connections)
   decks, **A**-1
      deck load basis, **A**-3, **A**-4
      roof deck deflection, **A**-8 to **A**-9
   drawings, **12**-4
   ductility conditions, **B**-12
   in flange and web, **4**-24
   frame drift component, **6**-50
   pure torsion, **4**-17, **4**-19, **4**-26, **4**-27
   seismic movement and, **5**-29
   and tensile stress, **B**-13 to **B**-14, **B**-15
   warping torsion, **4**-17, **4**-19, **4**-26, **4**-27
Shear studs:
   composite beams, **10**-7 to **10**-8, **10**-17
   composite decks, **A**-17, **A**-18, **A**-19, **A**-20, **A**-21
   composite girders, **10**-23 to **10**-25

I-24 ■ Index

Shear studs:
  form decks, **A**-22
Shear tabs, fillet welds on, **B**-18
Shear yielding, **3**-25
Sheet steel, deck material, **A**-1
Shielded metal arc welding (SMAW), **12**-15, **12**-16, **12**-17
  field welding, **B**-28
  filler material requirements, **7**-23
  position, **7**-24
  preheat and interpass temperatures, **7**-26
  weld access holes for, **B**-19
Shielded metal arc welding electrodes, **12**-5
Shielding gases, material control, **12**-5
Shop bolting, simple beams under shear and axial loads, **7**-109
Shop detail drawings, **12**-2 to **12**-3
Shop detailing process, **12**-1 to **12**-4
Shop fabrication, **12**-4 to **12**-8
Shop tolerances, **12**-28 to **12**-29
Shop welding, **7**-26
  versus field welding, **B**-28
  shear splices, **7**-104
Shored construction, composite slab-on-girder bridges, **11**-25
Short columns, **3**-40
Short slotted holes, **7**-8 to **7**-9
Shrinkage stress, welds, **7**-25, **7**-26
Side lap fastenings, form decks, **A**-22 to **A**-23
Sidesway:
  compression members, column end restraints, **3**-30 to **3**-31
  multistory frames, **6**-51
  unbraced rigid frame, **6**-53
Signs, wind loads, **5**-23
Silicon, **1**-32
Silicon-killed steel, **1**-31
SI measuring units, metric conversion table, **D**-1 to **D**-2
Single-angle compression members, **3**-53
Single bevel welds, **7**-14, **7**-15, **7**-16
Single-degree-of-freedom system, **9**-4
Single edge-notched bend (SENB) bars, **8**-45 to **8**-46
Single J groove weld, **7**-14, **7**-15, **7**-16
Single plate shear connections, **7**-107

Single-story frames, **6**-16 to **6**-41
  drift assessment, **6**-50
  lattice frames, **6**-24 to **6**-29
  loading, **6**-17, **6**-18
  multibay braced frames, **6**-29 to **6**-36
  multibay unbraced frames with leaning column, **6**-36 to **6**-41
  portal frames, **6**-19 to **6**-24
  stability to, **6**-1 to **6**-2
Single U groove weld, **7**-14, **7**-15, **7**-16
Single V groove weld, **7**-14, **7**-15, **7**-16, **7**-21
Sinusoidal loading, vibrations due to rhythmic activities, **9**-30
Skew weld, shear endplate, **7**-99
Slab design, form deck, **A**-13, **A**-15
Slab-on-girder bridges, composite, **11**-23 to **11**-30
Slab-on-stringer bridge, **11**-25, **11**-9
Slabs:
  bridge design, **11**-8, **11**-28, **11**-35 to **11**-36
  composites, **10**-4
Slag, welding considerations, **7**-27, **B**-30
Slender beam theory, **7**-41, **7**-43
Slender compression elements, **3**-39
Slender member model, gusset plate design, **7**-40
Slenderness ratios, **3**-41
Slip-critical connections, **7**-6
  bolt holes for, **7**-8
  shear endplate, **7**-98
Slip-critical joints:
  bolt installation, **12**-10
  holes for, **12**-8
Slip planes, **B**-12
Slip-resistant bolts, **7**-6
Slope, S-N curves, **8**-10
Slope deflection method, **1**-14
Sloped roof snow, **5**-16 to **5**-17
Slotted holes, **7**-8 to **7**-9, **12**-8
Slot welds, **7**-18, **7**-7
SMAW (*see* Shielded metal arc welding)
S-N curve, **8**-9 to **8**-10, **8**-11 to **8**-12
  bolts, **8**-17
  effective stress range for variable amplitude loading, **8**-25
Snow load, **1**-20, **5**-12, **5**-14 to **5**-18
  deck, **A**-1
  single-story frames, **6**-17, **6**-37

Index ■ I-25

Snug-tightened bolts, **7**-2, **7**-3 to **7**-6
  installation of, **12**-9, 10
  turn-of-the-nut method, **7**-4
Solar heating, form deck membrane, **A**-16
Spacing of bolts, **7**-8 to **7**-10
Spacing of welds, **7**-18
Spandrel, torsion effects, **4**-2
Span ratios, serviceability requirements, **9**-2
Spans, roof decks, **A**-4
Special moment frames
  seismic design, **E**-4 to **E**-33
    comparison with ordinary moment frames, **E**-34
    energy absorption and ductility, **E**-53
    height limitations, **E**-52
    (*See also* Seismic design)
  seismic resistance, **B**-18, **B**-8 to **B**-9, **B**-10, **B**-11, **B**-31 to **B**-32
Splices:
  axial force connections, **7**-79 to **7**-86
    columns, **7**-79 to **7**-82
    truss chords, **7**-82 to **7**-86
  beam shear, **7**-103 to **7**-106
  bolted, **7**-2, **7**-11 to **7**-12
  cutting, **12**-6
  fractures in, **8**-4
  welded, **7**-24, **7**-26
Spot measurement, coatings, **12**-35
Square groove weld, **7**-14, **7**-15, **7**-16, **7**-20 to **7**-21
Stability analysis, **6**-10
Stability index (theta), **E**-12
Standard holes for bolts, **7**-8 to **7**-9
Static lateral-force procedure, **5**-23 to **5**-26
Static loading, weld testing, **12**-26
Steel arch, **11**-12
Steel composition, **1**-30 to **1**-32
Steel Deck Institute (SDI), **A**-2
Steel frame tolerances, **12**-29
Steel properties:
  fracture toughness specification, **8**-28
  welding principles, **B**-4 to **B**-5
Steel rigid strut frame bridge, **11**-13
Stiffened elements:
  compression columns, **3**-38 to **3**-40
  weld seat connections, **7**-100
Stiffeners:
  layout and assembly, **12**-6

Stiffeners (*Cont.*):
  moment connections, **7**-86, **7**-88 to **7**-90
  merging of, **7**-93
  stresses in, **7**-94
  plates:
    seismic design, **E**-30 to **E**-32
    special moment frames, **E**-30 to **E**-32
  tolerances of built-up sections, **12**-29
  transfer of loads, **B**-3, **B**-4
Stiffness, and torsion, **4**-5
Stiffness factors, **3**-37
Stiffness reduction factors, **3**-50 to **3**-53
Stitch fasteners, **7**-9
Stitch welds, drawing specification, **7**-20
Storage of welding materials, **12**-5
Story buckling analysis, **6**-11 to **6**-13
Story drift:
  multistory frames, **6**-54
  single-story braced frame, **6**-50
  special moment frames, **E**-10
Straightness tolerance, cambering, **12**-7
Strain-aging effects, **1**-28
Strain hardening, **1**-19, **1**-27 to **1**-28
  stress in transition region and, **3**-7 to **3**-8
  tension member, **3**-1
Strain stress analysis, composite columns, **10**-42 to **10**-45
Strength, LRFD specification, **1**-20 to **1**-21
Strength check, special moment frames, **E**-13 to **E**-17
Strength evaluation, bridge rating, **11**-51 to **11**-53
Strength limit state, **9**-1
  bridge design, **11**-33, **11**-42, **11**-43
    AASHTO code, **11**-14 to **11**-17
    yield moment, **11**-28 to **11**-29
  frame design, **6**-3
Strength reduction factor, limit states design, **11**-5
Stress:
  measurement unit conversion table, **D**-1
  seismic (*see* Seismic load)
Stress, circle of, **1**-8 to **1**-9
Stress-concentration factor (SCF), **8**-11, **8**-32
  copes, blocks, or cuts, **8**-14
  at weld, **8**-18
Stress-corrosion cracking, anchor bolts, **8**-18
Stress indexes, **2**-2 to **2**-29

Stress-intensity factor, **8**-33 to **8**-34, **8**-39
Stress range:
  bolts, **8**-16
  copes, blocks, or cuts, **8**-14
  fatigue tests, **8**-9
Stress reversal, bolted connections, **7**-2
Stress-strain characteristics:
  for A36 steel in tension, **11**-3
  idealized, **1**-23
Stress-strain curve, **B**-14
Stress-strain diagram, **11**-4
Stringers, bridge, **11**-10
  dead load, **11**-34, **11**-35, **11**-36
  transformed composite section, **11**-28
Structural concrete form, form decks, **A**-11 to **A**-15
Structural dynamics, **9**-4 to **9**-8
Structural materials, **5**-4
Structural steel, **6**-18
Structural steel bolt and nut, **7**-2 to **7**-3
Structural strength, LRFD specification, **1**-20 to **1**-21
Studs, composite construction, **10**-1
  beams, **10**-4, **10**-17
  decks, **A**-17, **A**-18, **A**-19, **A**-20, **A**-21
  girders, **10**-23 to **10**-25
Stud shear connectors, **10**-7 to **10**-8
St. Venant torsion, **2**-15, **4**-12, **4**-13, **4**-19
Subframes, multistory frames, **6**-60
Submerged-arc welding (SAW), **8**-6, **12**-15, 19 to **12**-20
  edge cracks, **8**-39
  filler material requirements, **7**-23
  fluxes, storage of, **12**-5
  position, **7**-24
  preheat and interpass temperatures, **7**-26
Sulfur, **1**-32 to **1**-33
Superframe, **6**-62
Superimposed dead loads, bridge design, **11**-35, **11**-33, **11**-36
Superstructure of bridge, **11**-9 to **11**-10
Surface preparation
  coating, **12**-30 to **12**-33
  cut edges, **12**-6
  for welding, **7**-25
Surface treatment, welded surfaces, **7**-27
Suspension bridges, **11**-13
S values, deck material, **A**-1

Sway:
  multistory frames, combined characteristics, **6**-61 to **6**-62, **6**-63
  single-story multibay braced frames, **6**-36
Symbols, welds, **7**-18 to **7**-23
Systematic creep test, **1**-30
System buckling analysis, **6**-13

Tabs, shear, **7**-107
Tabs, weld, **B**-17
Tack welds, **8**-6
  defined, **7**-13
  edge cracks, **8**-39
  pretreatment and finishing, **7**-27
Tall building frame systems, **6**-62, **6**-63
Tanks, wind loads, **5**-23
Teal's formula, **E**-33 to **E**-34, **E**-7
Tearout, **3**-28
  axial force connections (*see* Axial force connections)
  brace-to-gusset connection, **7**-48
  gusset plate design, **7**-40, **7**-44 to **7**-45
  simple beams under shear and axial loads, **7**-110 to **7**-113
Tee joints:
  groove welds, **7**-14
  shear connections, **7**-107
Tee section materials tolerances, **12**-27
Temperature, **1**-30, **5**-14, **5**-29
  Charpy impact test, **8**-29 to **8**-30
  expansion and contraction, **6**-4
  form deck membrane, **A**-16
  and fracture behavior, **8**-44, **8**-25, **8**-36
  and tensile properties, **1**-28
  welding procedures, **7**-26, **12**-24, **B**-27
Tensile ductility, cracklike defects and, **8**-1
Tensile failure stress, **3**-7
Tensile fracture, **3**-25
  (*See also* Axial force connections)
Tensile properties:
  cold work and, **1**-27 to **1**-28
  temperature elevation and, **1**-29 to **1**-30
Tensile strength, **1**-28, **1**-29, **1**-30
  increase of axial capacity in presence of prying action, **7**-117
  residual stresses and, **1**-30
  seismic design, **B**-22

Tensile strength (*Cont.*):
  temperature and, **1**-28, **1**-29
Tensile stress:
  AISC requirements, **7**-2
  brittle fracture, **B**-16
  and shear stress, **B**-13 to **B**-14, **B**-15
Tension flange:
  fractures in, **8**-4
  moment connections, strong axis beam stiffeners, **7**-89
Tension members, **3**-1 to **3**-28
  connections:
    bolt installation, **12**-9, **12**-10
    connecting elements for, **3**-13 to **3**-14
    staggered holes, **3**-4 to **3**-7
  design specifications, **3**-1 to **3**-2
  design topics, **3**-14 to **3**-18
    block shear, **3**-23 to **3**-28
    built-up members, **3**-17 to **3**-19
    pin-connected members, **3**-21 to **3**-23
    rods and bars, **3**-20 to **3**-21
    selection of sections, **3**-14 to **3**-17
  effective net areas, **3**-7 to **3**-13
  with flexure, **3**-64 to **3**-88
    beam column design, **3**-81 to **3**-88
    beam columns in braced frames, **3**-75 to **3**-79
    beam columns in unbraced frames, **3**-80 to **3**-81
    first- and second-order moments with axial compression and bending, **3**-68 to **3**-69
    magnification factors, **3**-69 to **3**-71
    members subject to axial tension and flexure, **3**-65 to **3**-67
    moment modification, **3**-71 to **3**-75
  net areas, **3**-2 to **3**-4
Tension (pulsating loading), fatigue testing, **8**-8
Terminology, **F**-1 to **F**-7
Testing, welds, **12**-24 to **12**-25m 38 to **12**-39, **B**-20, **B**-30 to **B**-31
Theory of structures, **1**-14
Thermal cutting:
  detail categories for, **8**-13 to **8**-15
  weld access holes, **B**-20
Thermal factors (*see* Temperature)
Thermal load, bridge, **11**-21

Thickness, bar stock, **12**-2
Thickness effect, in fatigue, **8**-20
Threaded rod, as tension members, **3**-20 to **3**-21
Thread length for high-strength bolts, **7**-3
Threads in shear planes, bearing-type connections, **7**-6 to **7**-7
Three bays braced system, foundation overturning, **E**-39
Tier buildings, field tolerances, **12**-30
TIGW (*see* Tungsten inert gas welding)
Timoshenko, S., **1**-16
Tolerances, fabrication, **12**-27 to **12**-30
Top chord connections, **7**-66 to **7**-70
Torque control methods, **12**-5
Torque testing, bolt inspection, **12**-15
Torsion:
  avoidance or limitation of, **4**-5 to **4**-6
  bending analogy, **4**-23 to **4**-25
  combined stresses, **4**-17 to **4**-23
  equations, **4**-26 to **31**
  in flange and web, **4**-24
  LRFD design, **4**-25 to **4**-26
  pure, **4**-7 to 10, **4**-11
  response to, **4**-3 to **4**-4
  serviceability, **4**-25
  shapes and, **4**-6
  shear center, **4**-6 to **4**-7
  support definitions, **4**-4 to **4**-5
  warping, **4**-11 to **4**-17
Torsional buckling, **3**-28, **3**-61 to **3**-64
Torsional resistance, **2**-15
Toughness specifications, pre- and post-Northridge details, **B**-31, **B**-32
Towers, wind loads, **5**-23
Traceability of materials, **12**-4 to **12**-5
Transfer forces, shop detail drawings, **12**-3
Transfer of loads, **B**-3, **B**-4
Transformed section:
  bridge design, **11**-27
    I girder, **11**-32
    width of slab, **11**-35 to **11**-36
  properties of, **9**-9 to **9**-12
Transient loads, bridges, **11**-16 to **11**-17, **11**-18 to **11**-21
Transition-range fracture, **8**-26
Translation analysis, two-story unbraced frames, **6**-65, **6**-66

Transverse loading:
  braced frame system girders, **E-45**
  multistory frames, **6-51** to **6-52**
Transverse stiffeners, welding, **8-21** to **8-22**
Triaxial stresses, welding principles, **B-6**
Truss bridges, **11-13**
Trusses:
  axial force connections, **7-60** to **7-71**
    gusset-to-top chord, **7-66** to **7-67**
    gusset-to-truss vertical, **7-68** to **7-70**
    splices, **7-82** to **7-86**
    truss vertical-to-top chord, **7-70** to **7-71**
  camber tolerances, **12-28**
  connections:
    bolted, **7-2**
    welded, **7-12**, **7-28**
  lattice, **6-27** to **6-29**
  layout and assembly, **12-7**
Tubular load, weld testing, **12-26**
Tungsten inert gas welding (TIGW), **12-15**
Turn-of-the-nut bolt installation, **7-4**, **12-10** to **12-11**
Twist angle, **4-16**
Twist-off bolts, **7-5**, **7-6**, **12-12** to **12-13**
Two bays braced system, **E-37**, **E-39**
Two-story unbraced frame, **6-63** to **6-68**

Ultrasonic tests:
  welds, **7-29**, **12-25**, 26, 38, **B-30**, **B-31**
  WSMF, **8-37**
Unbraced frames, **6-63** to **6-71**
  alignment chart factors, **6-10** to **6-11**
  beam-columns in, **3-80** to **3-81**
  multistory, **6-68**
  rigid, stability of, **6-2**
  two-story, **6-63** to **6-68**
Undercutting of welds, **7-28**, **7-29**
Unfinished bolts, **7-3**, **7-4**
Uniform force method, **7-58**
  axial force connections (*see* Axial force connections)
  bracing connections:
    gusset plate design, **7-44**
    special case 2, **7-47**, **7-49** to **7-50**
    truss connections, **7-60**, **7-61**, **7-62**, **7-65**

Uniformly distributed load, single-story multibay framing:
  braced, **6-30**
  unbraced, **6-37**
Uniqueness theorem, **1-18** to **1-19**
Unshored construction, slab-on-girder bridges, **11-25**
Unstiffened elements, compression columns, **3-38** to **3-40**
Unstiffened weld seat connections, **7-100**
Unsymmetrical bending, **2-26** to **2-28**
Uplift, **5-23**
Upper bound theorem, **1-17** to **1-18**

Vanadium, **1-33**
Vapor pressure, form deck membrane, **A-16**
Variable-amplitude loading, effective stress range for, **8-23** to **8-25**
V bracing, seismic design, **E-41**
Vehicle live load, bridges, **11-19**
Vented form deck, **A-16**
Vermiculite, form deck fills, **A-16**
Vertical braced bays, **6-1**
Vertical welding, **7-24**
Vibrations, **6-4**
  composite beams, **10-19**, **10-10** to **10-12**
  floor:
    Allen and Murray criterion, **9-23** to **9-29**, **9-35** to **9-38**
    Murray criterion, **9-20** to **9-23**, **9-31** to **9-35**
    (*See also* Serviceability limit states design)
Visual acceptance criteria, weld, **12-26**
Visual inspection, welds, **12-24**, **12-26**, **12-38**, **B-29**
Vitruvius, **1-2**
V notches, **12-6**
Voids, **5-29**
von Mises yield criterion, **7-111**

Walking vibrations:
  Allen and Murray criterion for, **9-23** to **9-29**, **9-35** to **9-38**
  Murray criterion, **9-20** to **9-23**, **9-31** to **9-35**
Warpage, flange, **12-29**

Warping constant, **4-14**
Warping torsion, **4-11** to **17**
  direct stress, **4-26**, **4-28**
  normal stress, **4-19**
  shear stress, **4-17**, **4-24**, **4-26**, **4-27**, **4-29**, **4-30**
Washer plates, **7-113** to **7-114**, **7-115**, **7-116**
Washers, **7-3**
  direct-tension indicator tigtening, **7-5** to **7-6**
  materials control, **12-5**
Watch-the-wrench-chuck method, **12-11**
Water:
  and fatigue life, **8-9**
  roof ponding, **2-28** to **2-32**
  welding materials, **12-5**
Waviness tolerance, **12-28**
Wearing surface, **11-9**
Web crippling, **7-50**, **7-52**
  hangers, **7-77**
  weld seat connections, **7-103**
Web crippling design strength, **7-37**
Web doubler, special moment frames, **E-18** to **E-19**, **E-21**
Web-flange junction, fractures, **8-5**
Web gap in girders, **8-21**
Web off-center tolerance, **12-29**
Webs:
  girder, **12-2**
  roof deck, **A-8**
  truss chord connections, **7-85** to **7-86**
  welds:
    seismic resistance, **B-18**
    shear plate, E-29 to **E-30**
Web slenderness, bridge design, **11-30**, **11-46**, **11-49**
Web tear out, **3-28**
Web yielding, **2-24**, **E-32**
Weight of beam, estimation of, **2-5** to **2-7**
Weldability of steel, **1-33**
Weld access holes:
  crack sources, **8-13** to **8-15**
  design for seismic resistance, **B-18** to **B-20**
  fractures, **8-3**, **8-4**
  plug welds, **7-18**
  weld joint, **B-5**
Weld backing, **B-17**

Weld defects, **8-33**
Welded joints:
  center cracks in, **8-37**
  unbraced rigid frames, **6-55** to **6-56**
Welded members:
  cracks:
    center cracks in joints, **8-37**
    fatigue crack formation and propagation, **8-2**
    girders, welding stiffeners and gusset plates, **8-21**
    rigid joints, unbraced rigid frames, **6-55** to **6-56**
    splices, AISC requirements, **7-2**
Welded steel moment frames (WSMF), **8-6**, **8-7**
  connections, **8-32**
  fracture in, **8-7**, **8-41**, **8-42**, **8-43**
Welding position, **7-16**
Welding procedure specifications, **12-20** to **12-21**, **12-22**, **B-26**
Weld metals, edge cracks, **8-39**
Weld metal toughness, **8-30**, **8-30** to **8-32**, **8-44**
Welds:
  axial force connections (*see* Axial force connections)
  with bolts, **B-17** to **B-18**
  built-up section tolerances, **12-29**
  composite construction, **10-1**
  connections:
    with bolts, **7-7**
    material, **7-23** to **7-24**
    positions, **7-24**
    procedures, **7-25** to **7-28**
    quality of, **7-28** to **7-29**
    symbols on drawings, **7-18** to **7-23**
    types of, **7-13** to **7-18**
  erection plans, **12-3**
  fabrication, **12-15** to **12-27**
    acceptance criteria, **12-26** to **12-27**
    inspection, **12-21** to **12-26**
    procedures, **12-20** to **12-21**
    types of welding, **12-15** to **12-20**
  factors affecting weld quality, **B-26**
  and fatigue life, **8-10** to **8-11**
  filler, **7-11** to **7-12**, **7-23** to **7-24**
    ductile behavior, **8-27**

Welds, filler (*Cont.*):
    fracture toughness specification, **8**-28
    materials control, **12**-5
    and seismic resistance, **B**-23 to **B**-25
    form decks, **A**-16, **A**-22 to **A**-23
    fracture mechanics analysis, **8**-32 to **8**-45
    fractures in, **8**-4, **8**-5, **8**-6, **8**-7
    layout and assembly, **12**-7
    moment connections, **7**-86 to **7**-97
    multiaxial fatigue experiments, **8**-22 to **8**-23
    qualifications of welder and welding inspector, **12**-38 to **12**-39
    residual stress from, **8**-21
    seismically loaded structures, **B**-26
    seismic design:
        braced frame systems, **E**-50
        details of, **B**-17 to **B**-20
        special moment frame, **E**-27, **E**-29 to **E**-30, **E**-31
        undersized, effects of, **B**-8
    for seismic resistance, **B**-1 to **B**-34
        basic engineering principles, **B**-3 to **B**-6
        design, **B**-8 to **B**-20
        ductile hinges in connections, **B**-11 to **B**-17
        inspection, **B**-28 to **B**-31
        materials, **B**-21 to **B**-26
        metal properties, **B**-23 to **B**-25
        post-Northbridge details, **B**-31 to **B**-34
        system options, **B**-8 to **B**-11
        unique features of seismically loaded structures, **B**-7 to **B**-8
        workmanship, **B**-26 to **B**-28
    shear splices, **7**-104
    tension loads, **3**-11
Weld tabs, **7**-26, **7**-27, **B**-17
Weld volume minimization, **B**-3
Wheel load distribution factor, **11**-44
Whitmore section:
    brace-to-gusset connection, **7**-48 to **7**-50, **7**-33
    gusset plate design, **7**-40, **7**-45
    hangers, **7**-73
Whitmore's method, **E**-49

Wide-flange section, materials tolerances, **12**-27
Wide-flange stringers, bridge, **11**-10
Wind, roof exposure factors, **5**-14
Wind load, **1**-20, **5**-18 to **5**-23
    bridge, **11**-21
    connections:
        bolted, **7**-2
        moment, **7**-86 to **7**-87
        multistory braced frames, **6**-81
    and fatigue cracking, **8**-2 to **8**-3
    framing, **6**-81
        multistory frames, **6**-60
        multistory unbraced frames, **6**-68
        single-story frames, **6**-17
        single-story multibay braced, **6**-36
        single-story multibay unbraced, **6**-37, **6**-38
    roof deck as bracing, **A**-1
    seismic design, **E**-6
Wiss and Parmelee rating factor, **9**-14 to **9**-15
Workers:
    clearance for tightening bolts, **7**-9 to **7**-10
    raising gang, **12**-37
    welder qualifications, **12**-37 to **12**-38
    weld orientation, **B**-6
Working stress design (WSD), bridges, **11**-2 to **11**-4, **11**-50 to **11**-51, **11**-54
Workmanship, welds for seismic loading, **B**-26 to **B**-27
Wrench, calibrated, **12**-11 to **12**-12
Written specifications, welding, **12**-20 to **12**-21, **12**-22, **B**-26
WSMF (*see* Welded steel moment frame)

X-ray tests, welds, **7**-29 , **12**-26, **12**-38, **B**-30, **B**-31

Yield deformations, **8**-44
Yield moment, bridge design, **11**-28 to **11**-29, **11**-33
Yield point, **1**-30, **11**-4
Yield strength, **1**-28, **1**-29, **1**-30
    bridge design, **11**-44
    carbon and, **1**-31

Yield strength (*Cont.*):
    maximum, **8**-43
    seismic design, **B**-7, **B**-21 to **B**-22
    special moment frame column, E-32
    temperature and, **1**-28, **1**-29, **1**-30

Yield stress, **1**-23, **B**-7
Young, Thomas, **1**-5 to **1**-6

Zone zero, **12**-31

About the Editor

Akbar R. Tamboli is a senior project engineer with CUH2A in Princeton, New Jersey, a well-recognized company specializing in architecture, engineering, and planning. He is a former vice president and project manager for Irwin G. Cantor, PC, Consulting Engineers, in New York City, and was the principle consuling engineer on a number of noteworthy projects, including Morgan Guarantee Bank Headquarters at 60 Wall Street and Salomon Brothers World Headquarters, Seven World Trade Center.

How to Use the Accompanying STAAD-III CD-ROM Disk

SYSTEM REQUIREMENTS

Pentium.
Graphics Card of 800 × 600 or higher resolution.
26 MB free on hard disk:
    10 MB to load the program.
    16 MB to operate the program.
16 MB free RAM.
Dos 6.2 or higher.
Windows 3.1/Windows 3.11/Windows 95/Windows NT.
Recommended Virtual Memory:
    Windows 3.1/Windows 3.11:30 MB.
    Windows NT/Windows 95: 60 MB.

INSTALLATION

1. Place the STAAD-III CD-ROM in your CD-ROM drive.
2. Windows 3.1/Windows 3.11:   In Program Manger, choose File/Run.
   Windows 95/Windows NT: In Start menu, choose Run.
3. Type d:\setup. (If your CD-ROM drive is not drive D, type the appropriate letter instead.)
4. Choose OK.
5. Follow the instructions on the screen.

VIEWING THE ONLINE MANUALS

The following manuals are included on the STAAD-III CD-ROM:
    STAAD-III Getting Started Manual.
    STAAD-III Reference Manual.
To view the online manuals, put the STAAD-III CD-ROM into the CD-ROM drive, and then double-click the icon of the manual your want.

TECHNICAL SUPPORT

For questions about STAAD-III, contact Research Engineers, Inc. at:
    Phone: (714) 974-2500
    Fax: (714) 921-2543